Have you ever wondered how an FM radio or a television actually works? Even in electrical engineering course, ubiquitous products like these are not given detailed treatment. *Consumer Electronics for Engineers* is the first book of its kind to explain clearly the operating principles of "real world" electronic devices, including video recorders, compact disc players, and mobile phones.

Each chapter begins with a brief historical overview of the device concerned. The key principles of the device's operation are then described, and a block circuit diagram is presented. Next, these "real world" circuits are analyzed in detail and, finally, the present state-of-the-art is discussed. This approach will help to integrate the many different aspects of an electrical engineer's coursework, from physical optics to digital signal processing, as never before.

The book is very accessible and contains over 350 illustrations and many exercises. It will be an ideal textbook for undergraduate students of electrical engineering and will also appeal to practicing engineers.

Consumer electronics for engineers

Consumer electronics for engineers

PHILIP HOFF
California State University, Chico

PUBLISHED BY THE PRESS SYNDICATE OF THE UNIVERSITY OF CAMBRIDGE
The Pitt Building, Trumpington Street, Cambridge, CB2 1RP, United Kingdom

CAMBRIDGE UNIVERSITY PRESS
The Edinburgh Building, Cambridge CB2 2RU, United Kingdom
40 West 20th Street, New York, NY 10011-4211, USA
10 Stamford Road, Oakleigh, Melbourne 3166, Australia

First published 1998

Printed in the United States of America

Typeset in Times Roman 10/12 pt. in LaTeX 2_ε [TB]

Library of Congress Cataloging-in-Publication Data

Hoff. Philip Herbert, 1941–
Consumer electronics for engineers / Philip Hoff.

p. cm.

ISBN 0-521-58207-5 (hb)

1. Household electronics. 2. Electronic circuits – Design and construction. I. Title.

TK7880.H64 1998
621.382 – dc21 97-24468
CIP

A catalog record for this book is available from the British Library

ISBN 0 521 58207 5 hardback
ISBN 0 521 58817 0 paperback

Contents

Preface

As far as this author is aware, this book blazes new ground. It covers a subject that has been much neglected in traditional electrical engineering education. As such, it may have value as a professional book, as well as serving its original purpose as a text. For such a book, it is mandatory that both the purpose of writing the book and the philosophy governing its structure be articulated.

When I was in high school, I had a home-based business of repairing stereos, radios, and televisions. The knowledge to do this was obtained by reading various technician-level books and magazines on the subject and by experience obtained in the field.

Upon beginning college as an electrical engineering (EE) major, I had eager expectations that the curriculum would teach the principles upon which the equipment worked. Alas, those expectations remained unfulfilled. The "real world" circuits, whose schematics a service technician is used to seeing, are generally far more complex than those in the electronics textbooks. I had a certain sense of having been cheated. Not until later did I realize that presenting simplified circuits in the electronics textbooks was thought of as an act of mercy by the textbook authors. It is obviously easier to understand a simplified circuit than a "real-world" one.

This led to Phase II of this book's origins. After graduation I realized that an EE education at least provided the tools needed to understand consumer electronic circuits at a more advanced level. Commencing a career in college teaching gave additional impetus. A survey of seniors showed a great deal of interest in a course in consumer electronics. This led to a search for books that would give the information needed at the level of understanding desired. This search engendered disappointment of a different sort. The available materials, without exception, fell into three categories:

(1) The first type comprised technician-level books. These are generally the first information to be published on new consumer devices, but they are often fraught with errors and misconceptions.
(2) The second type of publication was written by engineers but was more like a reference handbook. Although some valuable information could be obtained from a few such books, they were not at all suitable as textbooks.
(3) The third type of useful publication consisted of technical journals – particularly the *IEEE Transactions on Consumer Electronics*. Although such journals might serve well as reference material, they clearly would not suffice as texts either.

No text was ever found explaining consumer electronics at an engineering level. This led to Phase III. There was apparently no alternative but to start putting together notes on the subject. This was done and the course was offered as a senior-level elective. The initial notes used "real-world" circuits that were not too difficult and simply skipped over those sections that were not readily comprehensible. The response from the students was very enthusiastic. This was for many years the most popular elective offered in the EE department (until DSP began to be offered).

Phase IV began when it became apparent that there needed to be an updating of the circuits analyzed and an inclusion of some of the plethora of new consumer items that were appearing. Furthermore, since the notes were to be expanded into a book, it would no longer be possible to slide over the tougher sections of the circuitry. Thus began (in 1982) the task of writing this book. From the beginning it has been a labor of love. At that time, the California State University at Chico (CSU-Chico) did not have personal computers or word processors or good circuit-analysis programs, so a great deal of hand analysis was done. It is not to be understood that this is considered bad. Much more insight to a circuit can be obtained by hand calculations than by computer analysis in most cases. This is particularly true if the analyst is forced to make physically reasonable simplifying assumptions to render the circuit analyzable. Later, as these tools became available, those few analyses that had not yielded to hand calculations were finished.

The structure of the book can be subdivided into considerations of what topics receive a chapter's worth of coverage, what particular unit should be analyzed for each selected type of equipment, and how the chapter itself is structured.

Two criteria have been used in deciding chapter topics. After all, someone may wonder why pagers are not covered. Another may think that satellite TV should be covered. Yet another will think that the most glaring omission of all is that personal computers are not covered. It should be obvious that the size of the book, even without including all of these things, weighs against including everything else that might otherwise qualify.

Beyond that, however, the first criterion used in deciding what should be covered is this: If the potential subject of a chapter is found in at least half the homes in America, it qualifies. This in itself rules out the pager, the satellite TV system, and the personal computer. Someone might object that the cellular phone is not found in half the homes in America, even if you count the auto as an extension of the home. So why is it covered? It is covered because the telephone, the subject of Chapter 10, is found in more than half the homes and the cellular phone is just one variety of telephone.

Someone else might ask why monochrome TV has a chapter of its own, since it is no longer found in half the homes. This brings us to a second criterion. If the material in one chapter is foundational for that in another, it would have to be covered anyway, so it makes sense to break up coverage of the more recent equipment into its historical antecedent and then just detail the differences. The reader of Chapters 6 and 7 will note time and again in Chapter 7 that the circuitry that accomplishes a certain function is virtually the same as that used for the same purpose in Chapter 6. Much the same thing is true as regards coverage of AM-only and FM-only radios. Although neither is manufactured now (with the few exceptions mentioned in the text) the information in these chapters provides a foundation for what is covered in Chapter 3.

There is another reason for the exclusion of personal computers. While digital electronics is growing more important almost daily, it is not amenable to the type of coverage given in this book. The fundamental hardware structure of a digital system is the same whether it is a personal computer or the system control

David Layer of the National Association of Broadcasters sent voluminous information on the RBDS system, as well as answering a few other questions about broadcasting.

David Koo of Philips Consumer Electronics, and one of the inventors of their ghost-canceling technology, was kind enough to review the section of the manuscript dealing with that subject.

Bill Mirhahem of Marantz provided access to the technical information on the Marantz tape deck analyzed in Chapter 5.

Special thanks are due to Sueyuki Hirooka and Vernon P. Brisson for providing the technical information needed for the analysis of the Sharp VCR in Chapter 8. After getting the "run-around" at the lower levels of the Sharp organization, as a desperation measure, I addressed a letter to Mr. Hirooka, the President of Sharp Electronics in America, asking for his intervention. He referred the request to Mr. Brisson, who not only sent to Japan for the required information, but who arranged to have key notations in the IC data sheets translated by K. Ideda.

Numerous electronic technicians have provided access to their schematic files and have given valuable insights into consumer electronic performance. These include Jerry Bennett and Steve Loney of Payless TV and Electronics, Karl Delmatier of The Electron Factory, Neil Ragsdale of Shastronics, Roger Schreur of Roger's Home Electronics, and Frank Dato of Cellular Plus.

Sean Kane, a student here at CSU-Chico, labored long and hard producing most of the line drawings used throughout the text.

Mimi Huie of CalTrans and the author's wife and daughter, Regina and Elaine Hoff, provided valuable help with their proofreading efforts.

Mrs. Andrea Bowman and Mrs. Ona Goodwin spent long hours typing the original manuscript, and Mrs. Jane Greene has made corrections to the typed manuscript as they were necessary. Since much of this work was done before word processing was available on this campus, it all had to be scanned into a word processor to ready the manuscript for submission to the publisher. The work of these ladies made that task as easy as possible.

Fred Zerega proposed what we have come to call the "road atlas" method of fitting large schematic diagrams onto much smaller book pages. For this we are much indebted to him.

Thanks are also due to CSU-Chico, who granted the author a Research Award. This allowed full-time work on the book in the Fall 1996 semester.

Finally, the author owes a huge debt of thanks to his family, who have been deprived of time and attention from their husband and father and son over a period of years that has stretched far beyond what it should have. During all of this time, they have remained supportive.

"And further, by these, my son, be admonished: of making many books there is no end; and much study is a weariness of the flesh."

Ecclesiastes 12:12

Philip Hoff
March 1997

(2) To the reader who is reading for comprehension, but whose mind might be wandering, the exercise can serve as a "wake-up" stimulus. These exercises have served well as home problems for the course. They range from the very simple to the very difficult.

The entirety of the book is too much to cover in a single three-unit semester course. It is helpful for the students to have had at least an introduction to AM and FM communications systems. This could allow Chapters 1 and 2 to be covered more lightly than is otherwise possible. Where this is not the case, it will probably be necessary to skip chapters. If this is done, the teacher should be aware that Chapter 3 builds on Chapters 1 and 2; Chapter 7 builds on Chapter 6; and both build on many of the concepts in Chapters 1 and 2. Chapter 8 builds on Chapter 5. Probably Chapters 9 and 10 would be most suitable for "stand-alone" study.

As the title suggests, this is a book for electrical engineers. It should be understood that the text assumes that the reader has a proficiency in calculus and transform circuit analysis, as well as a thorough background in electronic circuits and some familiarity with electromagnetics and control systems. Readers without this background may still profit from parts of the book, but they can also expect to have some large gaps in their understanding of the material.

The first major occurrence of a term in the text is generally italicized. This will hopefully be an aid to learning the vocabulary of consumer electronics.

Where standards are used, the dominant American system has been adopted for analysis.

The author welcomes comments and suggestions (and even corrections where needed!) from readers.

Acknowledgments

Many people have made contributions to this book besides the author. In the acknowledgments that follow, the affiliations given will be those in effect at the time the help was given. People change jobs a lot, and there is no implication that they are still where they were then.

One of the first individuals to whom I owe thanks is Oliver Richards, of Sprague Semiconductor (now Allegro Microsystems). He not only provided information about some of their chips but also about receiver design in general. Furthermore, he was excited about the prospect of there being a textbook such as this one.

My colleague Prof. Richard Bednar gave help of several kinds with some of the messier calculations in the book. Professors Ralph Gagnon and Jeff Leake reviewed portions of Chapter 9 that dealt with their specialties of optics and error-correcting codes respectively. Professor Harold Peterson provided stimulating discussions and suggestions about several points in the book. In addition, the students in his course on Advanced Instrumentation performed measurements that determined the dynamical characteristics of the optical read head in Chapter 9 and the ARL and limit circuits of Chapter 5.

Emil Torick of Broadcast Technology Partners (BTP), who was the coinventor of FMX™, provided the information needed to understand that system, as did Murray Bod, who was Director of Product Development on the project. More recently, John Browne of BTP provided an updated status report on this remarkable system.

analysis of a representative "real world" circuit with as much of its circuitry being discrete as possible, and
modern updating as to circuitry and systems.

There is an inevitable tension between covering the latest in integrated circuit realizations of the equipment and covering earlier models where there were no ICs used at all or where those used were smaller-scale devices that allow for an analysis that is much more "up close and personal." The chapter structure just set forth attempts to present both types of analyses where possible. There can be little doubt, however, that students learn the principles better by studying discrete circuitry. The difficulties of analysis for equipment with large-scale ICs are well illustrated by Chapter 8.

One of the most important features of the book lies in the breadth of the material that had to be plumbed to make it an integrated whole. Every circuit analysis technique an electrical engineer ever learns was pressed into use in explaining the circuits. The early chapters require insight into antennas. The reader will see dozens of examples of feedback amplifiers, which one must learn to recognize and characterize as to their gain. The reader will also encounter many oscillators of types never seen in textbooks and have to discover what components in them actually determine the frequency of oscillation. Control system concepts are required to assess system stability. Hall-effect devices had to be explained for the brushless motors used in the VCR and the CD. Magnetic principles had to be drawn in to explain the audio and video tape recorders. The physics of color had to be explained to understand the color TV. Many optical principles had to be reviewed to be able to explain the optical pickup head in a CD player, as did information theory and error-correcting techniques. As various chapters in the book have been completed and taught through, this eclecticism has enabled the book to achieve an effect far beyond what was originally conceived for it. That is the effect of drawing together the disciplines within electrical engineering (that many undergraduates previously perceived as isolated entities) into an integrated whole. To maximize this effect, all of the required electronics courses have been established as prerequisites for this one.

As part of their educational mission, academic programs at CSU-Chico are required to do an outcomes assessment. A large part of this consists of contacting the graduates themselves a while after they have taken a job and having them fill out a questionnaire. Several respondents have said that they felt that the course in consumer electronics was their most valuable course while in college. Some graduates have been told they were hired in preference to others primarily because they had taken this course.

As the material was being written, each time an intriguing question arose, it would be framed as an exercise embedded in the body of the text. Some of these exercises deal with design issues, and more deal with analysis problems. This book has no end-of-chapter problems, which have always seemed less than ideal. Instead, the problems are presented in the context in which they arise. This serves a dual purpose:

(1) The issues that bear on the problem are all present in the reader's mind. If the problem is at the end of the chapter, the reader must find the context again and then make a "running start" at it to grapple with it in the same way.

Chapter 6: Monochrome TV

This chapter begins with a presentation of the principles unique to TV, and proceeds to complete analysis of a discrete TV set and then to an integrated one where most of the circuitry is on a single chip, the Motorola "Monomax."

Chapter 7: Color TV

Building, of course, on the previous chapter, Chapter 7 begins with a rather detailed discussion of the physics of color as they relate to color TV. Unlike other chapters, this one does not analyze an entire discrete set, but only those stages of it that deal directly with color signals. This is followed by an analysis of an entire integrated color TV set.

Chapter 8: Video cassette recorders

The VCR remains, by a large margin, the most complex electromechanical system ever to be mass-produced. This is the first device discussed in the book that was never produced with most of the circuitry in discrete transistors. Its complexity is such that its creation had to await the advent of the large-scale integrated circuit to make it a viable consumer item. Even without covering a discrete circuit and with many of the magnetic principles having been covered in Chapter 5, this is easily the longest chapter of the book.

Chapter 9: The digital compact disc

This chapter begins with a brief discussion of the video laser disc, which actually preceded the CD as we know it today. The laser disc, though still with us, has never become a large marketing success. Prior to discussing the circuitry itself, extensive explanations of the error-correcting codes used in the system and the tracking mechanisms are given. Like the VCR, there has never been a discrete version of the CD player, but an integrated one is analyzed in detail.

Chapter 10: Telephones

As the introduction to the chapter explains, the telephone was never considered a consumer device until AT&T was broken up and people could buy their own phones. Virtually anyone outside the Bell System who wanted to understand the telephone better has been frustrated by the difficulty of finding technical information on telephones. Even worse, when such information is found, it is rarely presented in standard schematic notation. These faults are corrected in this chapter. The major phones from the Bell System are analyzed, and then a single-chip telephone from Motorola is studied. The cellular phone system is explained toward the end of the chapter, and an analysis of the RF portion of a cellular phone is presented.

Most chapters are structured as follows:

historical background,
new principles used in this type of equipment,
block diagram overview,

module of a VCR. The differences lie in the input source, the output destination, and (most importantly) in the code. This is not a book about analyzing software code. On these grounds, in spite of the fact that the personal computer is a marvelous instrument that will probably impact all of our lives more than any of the items examined in the book, an exposition of it at the level of this book would make for a pretty short chapter.

The chapters chosen and comments on them follow.

Chapter 1: AM receivers

This chapter begins with the humble crystal set and proceeds through examples of the regenerative receiver, the reflex receiver, the TRF, and both discrete and integrated superhets. It concludes with an explanation of AM stereo.

Chapter 2: FM receivers

In the 1950s, FM was the "weak sister" of AM, and relatively few receivers had FM capability. Of course, all commercially available receivers then were vacuum tube models, because the transistor was in its infancy. By the time transistor radios became the norm, FM and AM were much more nearly on a par as to demand, and most solid-state receivers offered both bands. It was therefore difficult to find an FM-only receiver, but it was considered necessary in order to simplify the initial presentation of the topic as much as possible. The analysis of the circuit of this receiver is followed by an exposition of FM Multiplex stereo. This, in turn, is followed by a presentation of the FMX® noise reduction system and the RDS system.

Chapter 3: Receivers that work over both the AM and FM bands

A "Walkman" is analyzed in this chapter. Various topics related to modern receivers are also covered, the most important of which is quartz-synthesized tuning.

Chapter 4: Equalization

This is the only chapter that does not deal with a piece of equipment, but with a specific topic. The reasons for and means of achieving it are explained. Also shown is the distinction between a true tone control capable of both boost and cut at both the upper and lower ends of the audio spectrum versus a "losser"-type of tone control, which can do nothing more than cut treble.

Chapter 5: Audio tape recorders

A background on magnetic fundamentals is presented before the audio tape recorder is analyzed. This chapter also covers the following noise-reduction systems: Dolby® B and C systems, the dbx system, the DNR® system from National Semiconductor and the HUSH® system from Analog Devices.

Index of abbreviations

A/D	Analog/Digital
ABL	Automatic Brightness Limiter
ABLC	Automatic Bias Level Control
AC	Alternating Current
ACC	Automatic Color Control
AF	Audio Frequency
AFC	Automatic Frequency Control
AFT	Automatic Fine Tuning
AGC	Automatic Gain Control
ALC	Automatic Level Control
ALPC	Automatic Laser Power Control
ALU	Arithmetic Logic Unit
AM	Amplitude Modulation
AMPS®	Advanced Mobile Phone Service
ANL	Automatic Noise Limiter
APC	Automatic Phase Control
ARL	Automatic Record Level
ASCII	American Standard Code for Information interchange
ATC	Automatic Tint Control
AVC	Automatic Volume Control
BCD	Binary-Coded Decimal
BGPG	Burst Gate Pulse Generator
CATV	Community Antenna Television (cable TV)
CB	Common Base or Citizen's Band
CD	Compact Disc
CD-I	Compact Disc-Interactive
CE	Common Emitter
CIRC	Cross-Interleaved Reed-Solomon Coding
CLV	Constant Linear Velocity
CMOS	Complementary Metal-Oxide-Semiconductor (logic)
CQAM	Compatible Quadrature Amplitude Modulation
CRT	Cathode-Ray Tube
CW	Continuous Wave
D/A	Digital/Analog
DC	Direct Current
DOC	DropOut Compensation
DSB	Double SideBand
DSB-SC	Double SideBand-Suppressed Carrier
DSP	Digital Signal Processing
DTMF	Dual-Tone Multi-Frequency
EAROM	Electrically-Alterable Read-Only Memory
ECL	Emitter-Coupled Logic
EE	Electric-Electric (transfer)
EFM	Eight-to-Fourteen Modulation
EIA	Electronic Industries Association
EP	Extended Play (6 hours)
ETC	Electronic Telephone Circuit
FCC	Federal Communications Commission
FET	Field-Effect Transistor
FG	Field Generator
FIR	Finite Impulse Response (digital filter)
FM	Frequency Modulation
FV	False Vertical (sync)

GCR	Ghost Cancellation Reference
HDTV	High-Definition TeleVision
HF	High Frequency
HPF	High-Pass Filter
HSWP	Head-Switch Pulse
HV	High Voltage
I	In phase
I^2L	Integrated Injection Logic
ID	IDentification (pulse)
IDTV	Intermediate-Definition TeleVision
IF	Intermediate Frequency
IGMF	Infinite-Gain Multiple-Feedback (filter)
IHVT	Integrated High-Voltage Transformer
IIR	Infinite Impulse Response (digital filter)
IR	Infra-Red
LC	inductor-Capacitor
LED	Light-Emitting Diode
LP	Long-Playing or Long Play (4 hour)
LPF	Low-Pass Filter
LSI	Large-Scale Integration
MOSFET	Metal-Oxide-Semiconductor Field-Effect Transistor
MSI	Medium-Scale Integration
MTSO	Mobile Telephone Switching Office
MX	MultipleX
NAB	National Association of Broadcasters
NBFM	Narrow-Band FM
NPO	Negative Positive zerO (temperature coefficient)
NRZ	Non-Return to Zero (pulse train format)
NTSC	National Television System Committee
OEM	Original Equipment Manufacturer
OSD	On-Screen Display
OTA	Operational Transconductance Amplifier
PAL	Phase Alternation Line (TV system)
PB	PlayBack
PCM	Pulse-Code Modulation
PG	Pulse Generator
PLL	Phase-Locked Loop
PWM	Pulse-Width Modulation
Q	Quiescent point or Quadrature phase
QAM	Quadrature Amplitude Modulation
RAM	Random Access Memory
RC	Resistor-Capacitor
RF	Radio Frequency
RFC	Radio-Frequency Choke (inductance)
RIAA	Recording Industries of America Association
RLC	Resistor-inductor-Capacitor
ROM	Read-Only Memory
RSSI	Received Signal-Strength Indicator
S&H	Sample and Hold
S-VHS	Super-Video Home System
S/N	Signal/Noise ratio
SAP	Second Audio Program
SAT	Supervisory Audio Tones
SAW	Surface Acoustic Wave
SCA	Subsidiary Communications Authorization
SCR	Silicon Controlled Rectifier
SECAM	SEquential Color And Memory (TV system)
SP	Standard Play (2 hour) or Signal Processing
SPDT	Single Pole Double Throw (switch)
SPST	Single Pole Single Throw (switch)
ST	Signaling Tones

TRF	Tuned Radio Frequency (receiver)	VHS	Video Home System
TTL	Transistor-Transistor Logic	VIR	Vertical Interval Reference
TV	TeleVision or Tuning Voltage	VLSI	Very-Large-Scale Integration
UHF	Ultra-High Frequency	VV	Video-Video (transfer)
VCO	Voltage-Controlled Oscillator	VVC	Voltage-Variable Capacitor
VCR	Video Cassette Recorder	VVR	Voltage-Variable Resistor
VHF	Very-High Frequency	XOR	eXclusive OR (logic gate)

Publisher's note

Electronic versions of several of the figures from this book are available at the book's website, http://www.cup.org/titles/58/0521582075.html.

Figure Credits

Figures 1.17, 6.12a, and 6.17 courtesy of Howard W. Sams & Company. Figure 1.32 courtesy of Motorola Inc. Figure 2.6 courtesy of Signetics Corporation. Figures 2.20 and 2.30 courtesy of Emerson Radio Corporation. Figures 3.1 and 7.12 courtesy of Sony Electronics Inc. Figures 5.8 and 5.20 courtesy of Superscope Technologies Inc. Figure 6.2 courtesy of the Coyne American Institute. Figures 6.19 and 6.51 courtesy of Zenith Electronics Corporation. Figures 7.19, 7.21, 7.22, 7.23, 7.29, 7.31, 7.35, 7.37, 7.38, 7.40, 7.46, and 7.55 courtesy of RCA Corporation. Figure 7.56 courtesy of S. Barton and B. Sadler in IEEE Transactions on Consumer Electronics, Vol. CE 24, No. 3, 1978. Figures 8.13, 8.14, 8.20, 8.22, 8.23, 8.24, 8.31, 8.49, 8.57, 8.59, and 8.61 reproduced with the permission of Sharp Electronics Corporation. Figures 9.26, 9.27, 9.28, 9.29, 9.32, 9.36, and 9.38 reproduced with the permission of Yamaha Corporation. Figure 10.4 reprinted with permission of AT&T.

1 AM receivers

1.1 The crystal set

From both the conceptual and historical points of view, our study of AM receivers must start with the "crystal set." This is the simplest AM receiver, consisting of only three essentials: a tuned circuit, a detector, and a transducer (usually headphones or an earphone) to provide sound output. A simple crystal set schematic is shown in Fig. 1.1. To get usable volume, it is generally necessary to use a long (50′ or more) antenna and an earth ground. The capacitor is variable so that the parallel resonant tuned circuit composed of it and the inductance L can tune over the AM broadcast band (535–1,705 kHz). Two types of variable capacitors used for tuning are shown in Fig. 1.2.

Exercise 1.1 A once-common variable capacitor has a maximum capacitance of 365 pF.

(a) What must the inductance be if the receiver is to tune the entire AM band?

(b) What must the minimum value of the variable capacitance be to tune the entire AM band?

The sound section of any consumer electronic equipment is referred to as *audio*, and the amplifier that amplifies electrical signals in the range of human hearing is called an *audio amplifier*.

The diode in Fig. 1.1 demodulates the audio signal from the AM carrier and feeds it into the headphones.

The crystal set has two serious shortcomings: poor selectivity and poor sensitivity. *Selectivity* is the ability to separate two stations whose carrier frequencies are close together – especially if one of them delivers significantly more signal to your antenna than the other. If you thought upon reading this definition that selectivity must depend on the Q of the tuned circuit, you are absolutely correct.

Sensitivity is a measure of how strong a signal must be delivered to the antenna for a radio to receive it properly. It is usually expressed in μvolts/meter needed to drive the radio receiver to full output for AM. For FM, it is usually given as μvolts needed for full quieting. Note that for AM the units of sensitivity are those of electric field. In other words, the sensitivity specification is a measure of the E field strength of the station's transmitted electromagnetic wave needed by the receiver's antenna.

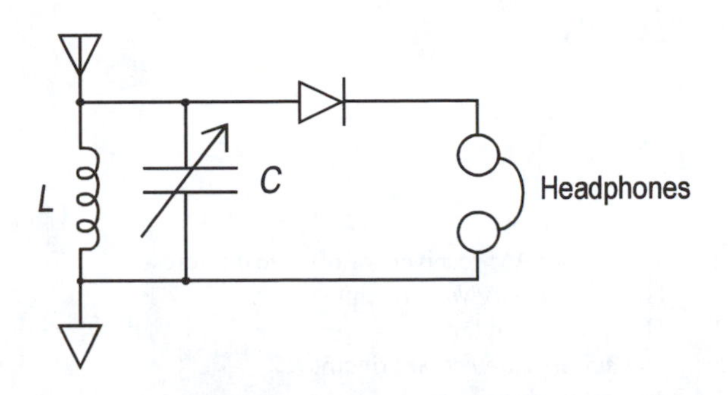

Fig. 1.1: A simple crystal set.

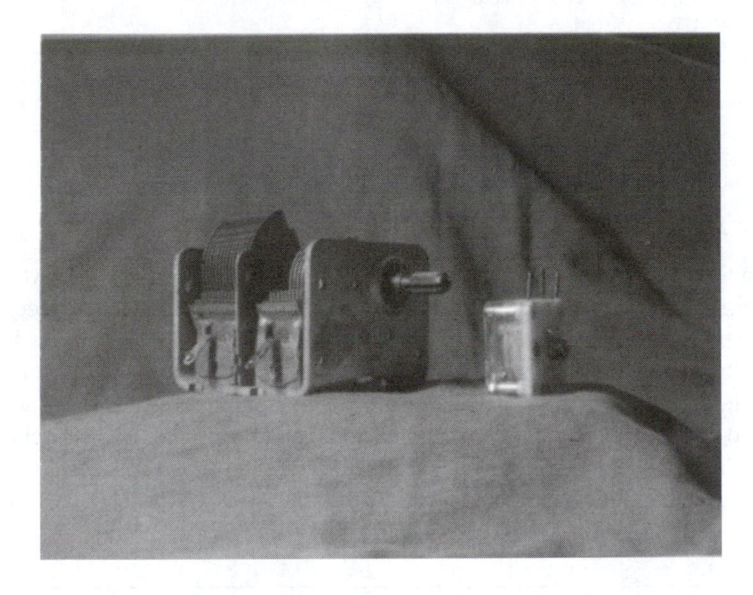

Fig. 1.2: Two older types of variable capacitors.

We now wish to look at the detection process prior to examining why the crystal set has both poor selectivity and sensitivity.

The wave transmitted by a broadcast AM station would look like Fig. 1.3a in the time domain if a single sinusoidal tone were being transmitted. This, therefore, is also the waveform that will appear across the tuned circuit. The signal is then applied to the series combination of the diode and the headphones. If we assume for simplicity that the diode is ideal, it will conduct only on the positive part of the waveform of Fig. 1.3a.

Let us assume 100% modulation and unity amplitude. We will also assume that the carrier frequency is 1,000 times the modulating frequency. The mathematical form of the AM wave is then

$$v(t) = \sin 1000\omega_{\mathrm{m}}t \cdot (1 + \sin \omega_{\mathrm{m}}t).$$

Passing through an ideal diode clips the negative excursions and causes the voltage of Fig. 1.3b to be applied to the load. A Fourier analysis of this waveform yields the following amplitudes:

DC component: .130,
component at ω_{m}: .102,

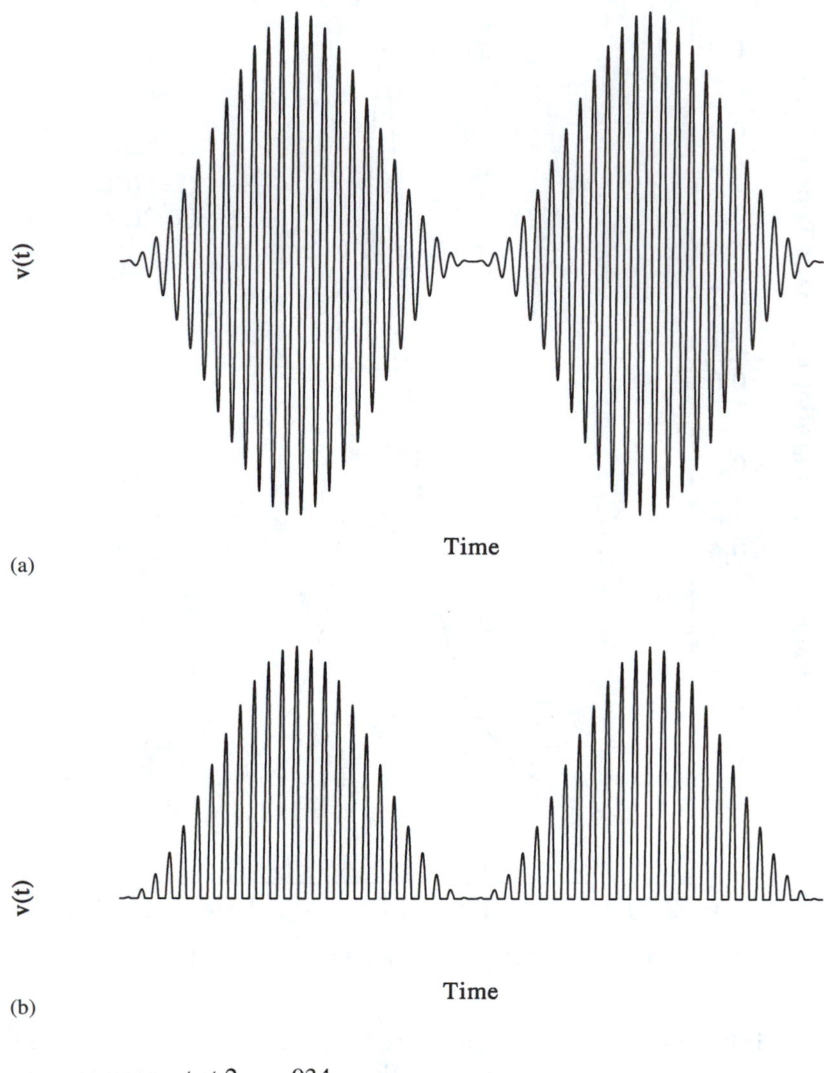

Fig. 1.3: (a) An AM waveform. (b) The same waveform after passing through an ideal diode.

component at $2\omega_\mathrm{m}$: .034,
component at $3\omega_\mathrm{m}$: .034.

Higher-order components get progressively smaller. Since the only low-pass filtering is that due to the mechanical response of the headphones, the higher-order terms are not removed but remain as distortion. Excluding the DC component, about 90% of the signal power is at the fundamental frequency, ω_m. The other 10% is distortion.

There are two reasons for the poor sensitivity of the crystal set. The first is that there is relatively little signal to work with, there being no amplification. The second reason is that the series combination of the diode and headphones constitutes a nonlinear resistance in parallel with the tuned circuit. The effect of this nonlinear resistance is beyond the scope of this book to model quantitatively, but we know qualitatively that the effect of resistive loading is to spread out the response curve (degrading selectivity) and lower its peak (degrading sensitivity). This is illustrated in Fig. 1.4. The upper curve corresponds to a

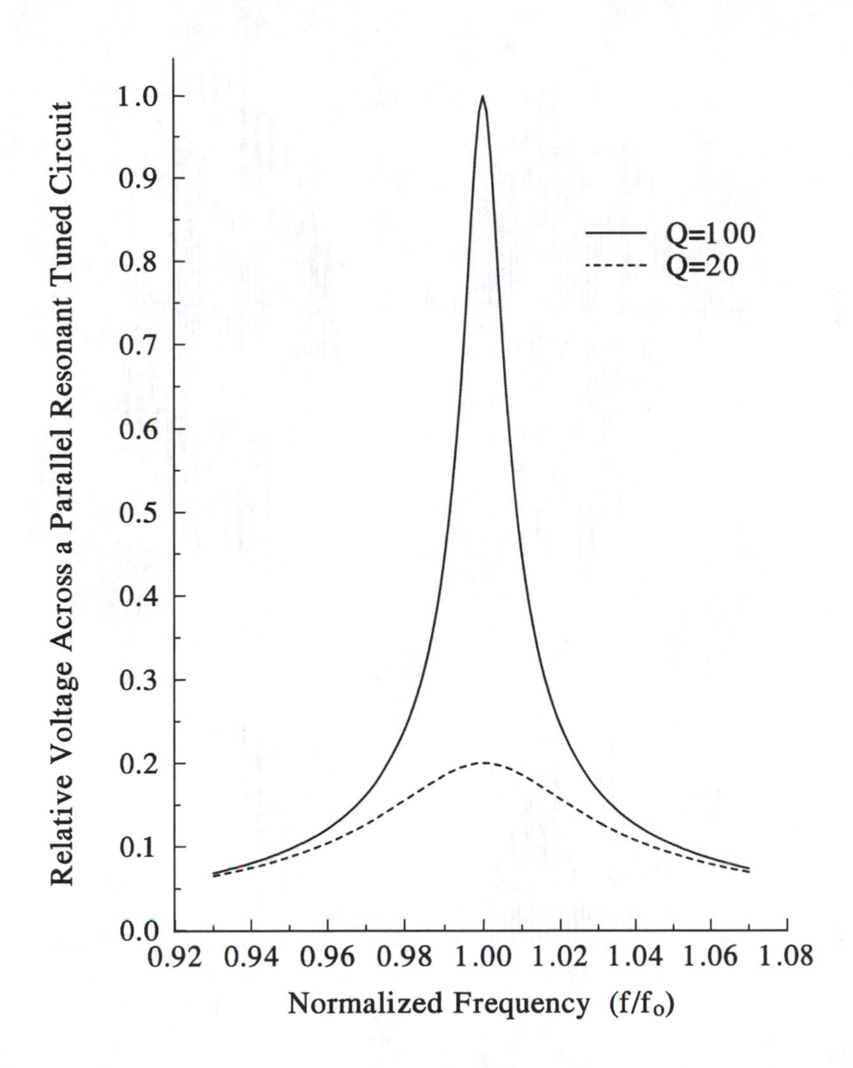

Fig. 1.4: The effect of the Q of a tuning circuit on the sensitivity and selectivity of the receiver.

Q that is five times greater than that of the lower curve. Since broadcast AM stations are assigned carrier frequencies 10 kHz apart, and our reference carrier frequency is 1 MHz, "nearest neighbor" stations would have carrier frequencies of .99 and 1.01 MHz. Therefore, for the smaller Q, a station at .99 MHz would need to present only 16% more signal power to the antenna than one at 1.00 MHz to develop the same voltage across the tuned circuit. For the larger Q, a signal at .99 MHz would need to be five times as large as one at 1.00 MHz to develop the same signal voltage. This bears directly on the selectivity. It is also difficult to get good selectivity with a simple crystal set, because for AC purposes, the relatively low impedance of the antenna is effectively in parallel with the tuned circuit.

The effect of the lower Q on the sensitivity is even more obvious, because the peak voltage developed at resonance is directly proportional to the Q, assuming the inductance is constant.

Figure 1.5 shows a modified crystal set that gives significant improvement in selectivity as well as some improvement in sensitivity. In this circuit, L_1 is driven by current going from antenna to ground. L_2 is a voltage (and thus impedance) step-up winding which is tuned by C_1. Mutual inductance is used

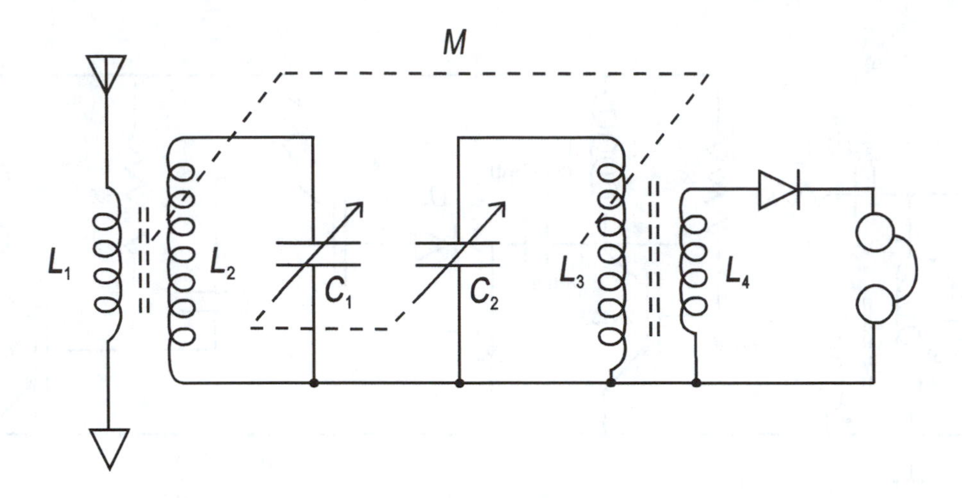

Fig. 1.5: An improved version of the crystal set.

to couple the signal from L_2 to L_3, the latter being tuned by C_2. Thus, the two tuned circuits improve selectivity by:

(1) operating with high shunt impedance, which keeps the Q high;
(2) the use of ferrite core coils, which also increases the coil Q; and
(3) the presence of two tuned circuits, which cause "bandwidth shrinkage" just like cascaded amplifiers.

The L_3–L_4 combination steps down the impedance to more nearly match the diode–headphone load. The voltage is, of course, stepped down also.

It should be noted in conclusion that the headphones (and earphones) in common use today would be useless in an unamplified crystal set, since almost all have an impedance of 8–10 Ω. This is entirely too heavy a load for a crystal set. An impedance of 2,000 Ω for the headphones or earphone should be considered minimal.

1.2 Some other AM receivers of historical interest

1.2.1 The regenerative receiver

In the early days of electronics, tubes were relatively expensive, and it was important to extract as much gain from each stage as possible. This suggested the use of positive feedback. Radio receivers that used positive feedback were called *regenerative receivers*. The technique worked very well in some respects, but positive feedback also had distinct disadvantages, including poor fidelity, high noise, and gain instability. The latter problem could be particularly annoying. Regenerative receivers invariably had a regeneration control to vary the amount of positive feedback. To achieve maximum gain, one usually tried to set the control as high as possible without oscillation. However, since the gain also becomes increasingly unstable for large positive feedback, it was common for the receiver to drift over into oscillation, "treating" the headphone-listener to a sudden burst of high-amplitude sound. This could be avoided by using less feedback. The results were still well above what could be expected from one or two tubes without feedback. With the advent of transistors, the scenario was

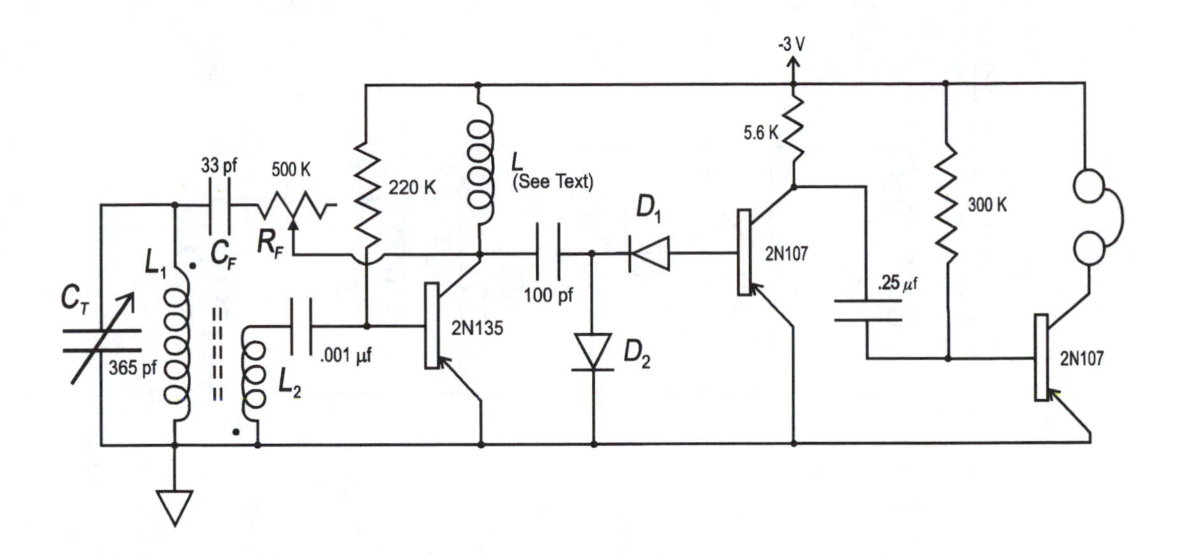

Fig. 1.6: A regenerative receiver.

repeated, although it was primarily the experimenters who were interested at this point. An effective regenerative receiver is shown in Fig. 1.6. L_1 and L_2 are actually the two sides of a tapped winding on a *loopstick antenna*. This is a ferrite rod with wire wound around it.

This circuit has several points of interest besides the regeneration. Let us begin by finding the impedance of the 33 pF, 100 pF, and .001 μF capacitors at 1 MHz, which is near the center of the AM band:

$$|\mathbf{Z}_{33\,\text{pF}}| = 4{,}823\,\Omega,$$

$$|\mathbf{Z}_{100\,\text{pF}}| = 1{,}592\,\Omega,$$

$$|\mathbf{Z}_{.001\,\mu\text{F}}| = 159\,\Omega.$$

The 33 pF capacitor is part of the feedback loop. The impedance of the .001 μF capacitor is less than a tenth of the expected input impedance of the 2N135. Hence, it may be considered an AC short.

The 100 pF capacitor coupling the AM signal to the detector is, of course, also blocking the DC collector current of the 2N135 from flowing to ground through D_2. The magnitude of this capacitor's impedance leads us to suspect that it also limits the loading of the first stage. Note that D_1, when it conducts, will forward bias the base of the first audio amplifier, and that there is no other source of base bias for this stage. This is our first example of what is called *signal-developed bias*. The other diode, D_2, allows the 100 pF capacitor to discharge on positive half cycles. If this were not done, no signal could be coupled to the audio amplifier after the 100 pF capacitor charged to the peak signal amplitude on negative swings. D_1, of course, is also used as the detector. The two diodes, the 100 pF capacitor, and the input capacitance of the 2N135 form a voltage doubler.

With some crude assumptions, we can use feedback theory to arrive at a "quick and dirty" analysis of the first stage.

We consider the antenna as a current source. The tuned circuit is considered lossless so at resonance its impedance is infinite and no signal current flows

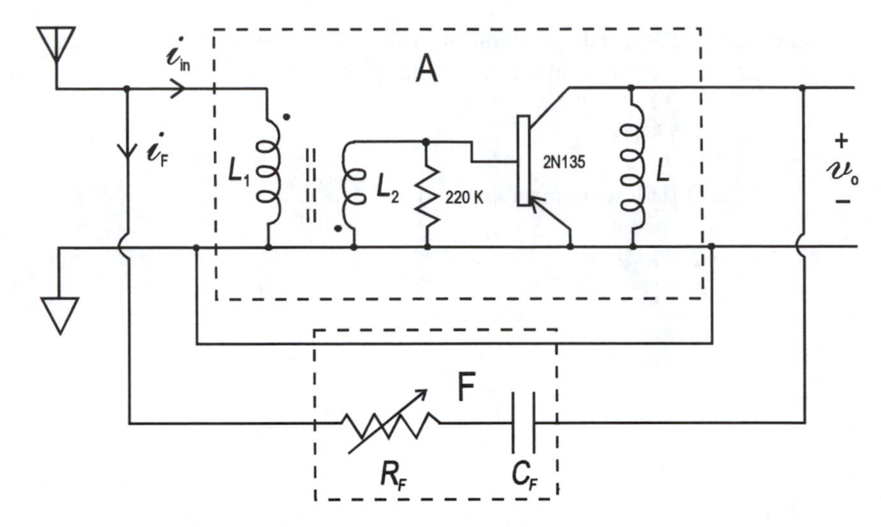

Fig. 1.7: The regenerative receiver of Fig. 1.6 put into standard feedback form.

into it. The transformer comprised of L_1 and L_2 is assumed to be ideal. This effectively means that both of its windings have infinite inductance. Strictly speaking, this would require $C_T = 0$ to resonate the tuned circuit. If C_T is small and L_1 is large, neither of these assumptions is terribly unreasonable. Subject to these assumptions and our discussion of the fixed capacitors, we can draw the first stage in feedback form. This is shown in Fig. 1.7.

The generalized form for a feedback amplifier's gain is

$$A_f = \frac{A}{1 + AF}, \tag{1.1}$$

where A is the gain of the base amplifier and F is the transfer function of the feedback network. We should recognize this circuit as a transimpedance amplifier. Ignoring the loadings of A on F and F on A we can find the separate transfer functions

$$\mathbf{A} = \frac{\mathbf{V}_o}{\mathbf{I}_{in}} = h_{fe} \cdot \frac{N_1}{N_2} \cdot \mathbf{Z}_L. \tag{1.2}$$

In this equation, N_1 and N_2 are the turns on L_1 and L_2, respectively. The dot conventions shown will introduce a 180° phase shift that cancels that which we would otherwise expect from the CE (common-emitter) configuration. The inductive load is very noncritical; a TV peaking coil can be used for this inductor. *Peakers*, as they are called, are small coils wound around resistors of a few kΩ and connected in shunt across the resistors. Their purpose is generally to serve as loads in high-frequency circuits. Their impedance, which rises with frequency, will tend to keep the output amplitude constant even though the output from the active device is dropping off at higher frequencies.

A typical peaker has an inductance on the order of 100 μH and a series resistance so small that the shunt resistance determines the Q. The circuit of Fig. 1.6 was built as shown, except that C_F was 47 pF. The peaker used as the load for the first stage was 100 μH in parallel with 5.6 kΩ. A station at 1,060 kHz was tuned in, and R_F was adjusted to give maximum volume without oscillation. The measured value of R_F for these conditions was 30 kΩ. The following parameters and specifications are also relevant: For the transistor

$$f_\alpha = 4.5\,\text{MHz},$$

$$h_{fe} = 20 \text{ at low frequencies, and}$$

$$f_\beta = 225\,\text{kHz}.$$

Thus,

$$h_{fe}(f) = \frac{20}{1 + jf/f_\beta}. \tag{1.3}$$

Since $f = 1.06$ MHz

$$h_{fe}(1.06\,\text{MHz}) = \frac{20}{1 + j4.71}.$$

For the transformer

$$N_1/N_2 = 10. \tag{1.4}$$

For the peaker

$$\mathbf{Z}_L = \frac{j3.73 \times 10^6}{j666 + 5600}. \tag{1.5}$$

For the feedback network (ignoring loading), Fig. 1.7 shows that

$$\frac{\mathbf{I}_F}{\mathbf{V}_o} = \mathbf{F} = \frac{-1}{R_F + 1/jC_F\omega} = \frac{-jC_F\omega}{1 + jR_FC_F\omega}$$
$$= \frac{-j3.13 \times 10^{-4}}{1 + j9.39}. \tag{1.6}$$

Equations (1.3)–(1.6) may be combined with (1.2) to give the loop gain,

$$\mathbf{AF} = \frac{20}{1 + j4.71} \cdot 10 \cdot \frac{j3.73 \times 10^6}{5600 + j666} \cdot \frac{-j3.13 \times 10^{-4}}{1 + j9.39}$$
$$= \frac{233{,}500}{(1 + j4.71)(5600 + j666)(1 + j9.39)} = \frac{233{,}500}{-251{,}461 + j50170}$$
$$= .91\angle 169^\circ = -.91\angle -11^\circ. \tag{1.7}$$

The significance of this result can be appreciated by looking back to Eq. (1.1). That equation shows that when $\mathbf{AF} = -1\angle 0^\circ$, oscillation will occur. If $|\mathbf{AF}| < 1$, we have positive feedback. It has already been observed that this is the case for the regenerative receiver. Furthermore, because the receiver was adjusted fairly close to the verge of oscillation, it should be expected that **AF** would be close to -1, and its phase close to zero. The agreement is, in fact, so good that it must be regarded as largely fortuitous, since the parts tolerance is only 10% itself, and since our non-loading assumption is quite questionable for an amplifier with heavy positive feedback. This is because for a transimpedance amplifier such as this,

$$\mathbf{Z}_{\text{inF}} = \frac{\mathbf{Z}_{\text{in}}}{1+\mathbf{AF}} \gg \mathbf{Z}_{\text{in}},$$

$$\mathbf{Z}_{\text{outF}} = \frac{\mathbf{Z}_{\text{out}}}{1+\mathbf{AF}} \gg \mathbf{Z}_{\text{out}}.$$

For large positive feedback, the denominator of both expressions approaches zero. A high input impedance will tend to load **F**, which is sourcing current to the input node, whereas a high output impedance is susceptible to loading by **F**.

1.2.2 The reflex receiver

The *reflex receiver* is another design capable of achieving high performance with low parts count. The basic principle is that a transistor initially amplifies the signal as transmitted. This is called RF for Radio Frequency amplification. After detection, the Audio Frequency (AF) signal is fed back through the same transistor again. Frequency-selective components are used to channel both RF and AF signals to the proper points from the output.

At first our intuition might suggest that having two such completely different signals in one transistor might cause them to "get all messed up together." We must recall, however, that if the amplifier is linear, the principle of superposition applies, and thus the two signals can be amplified together and still maintain their integrity. This is true only to the extent that the amplifier is actually linear. There will be a second-order error in that the larger AF signal will cause parameter variation in the equivalent circuit. Since this occurs slowly compared to the RF, the effect will be to modulate the (already modulated) RF signal at the audio rate. But the audio will not have changed significantly in the time it takes the signal to be reflexed back. Therefore it is unlikely that the detected audio will be affected audibly.

Figure 1.8 shows the schematic of an efficient two-transistor reflex receiver. The greater part of the analysis of this receiver will be left as an exercise. However, let's note three things about the circuit first:

(1) R_7 is set up as a voltage divider on the audio signal, controlling how much of it is sent on. It is a *volume control*, and its circuit placement is typical of almost all volume control circuitry.

(2) The 2N168A transistor performs the reflex function. C_3 in its emitter circuit functions as a bypass for the audio frequencies. At radio frequencies, electrolytic capacitors – especially aluminum electrolytics – may be far from ideal capacitors. Their impedance can be orders of magnitude above what an ideal capacitor of the same value would exhibit. This can be seen clearly in Fig. 1.9. These curves are adapted from Fink and Carroll (1968). Among other things, the curves show that over the entire broadcast band, an ideal .02 μF capacitor has a lower impedance and is thus a better bypass than a 200 μF etched-foil aluminum electrolytic! If we scale these curves to the 25 μF value in

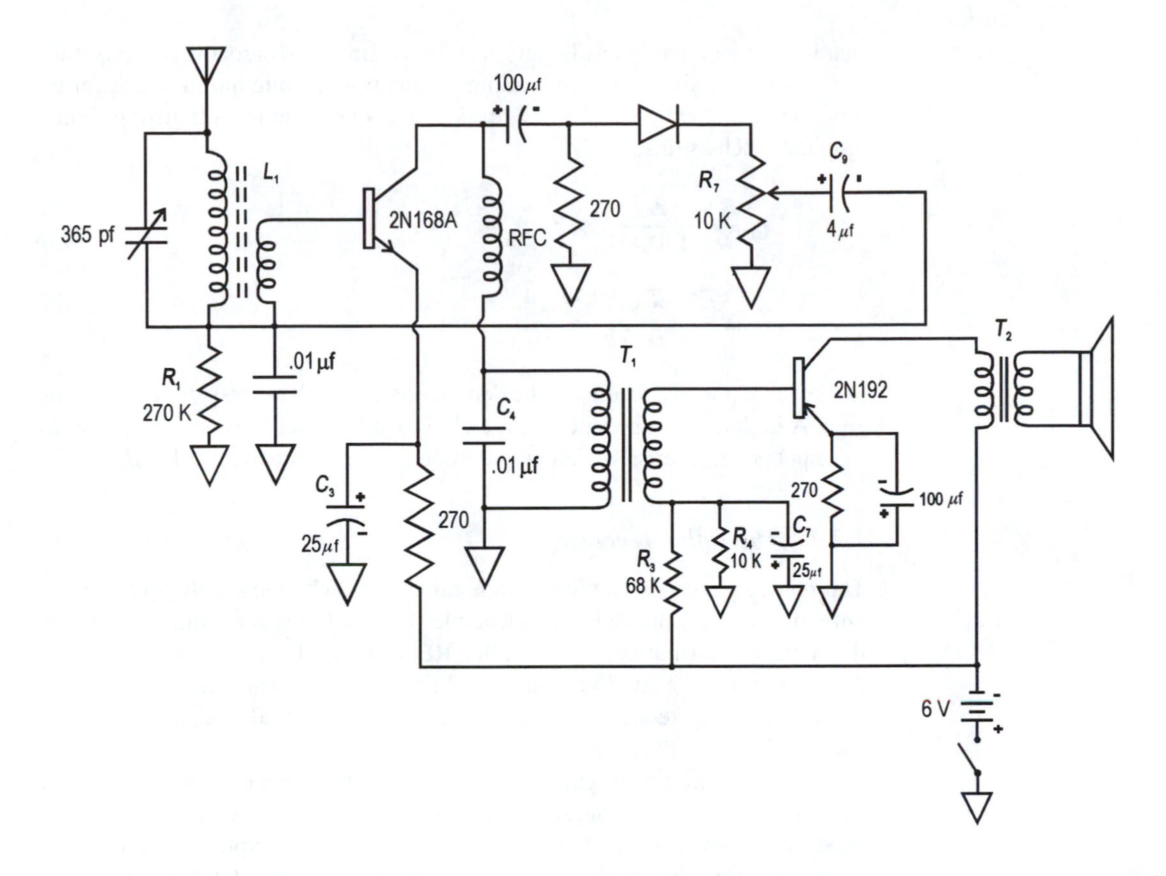

Fig. 1.8: A reflex receiver.

the emitter circuit, we see that a plain foil capacitor would have an impedance of about .7 Ω at 1 MHz, which is an adequate bypass. However, an etched-foil unit would have about 160 Ω impedance, which would make it a very poor bypass for a 270 Ω resistor. It is for this reason that one frequently sees an aluminum electrolytic in parallel with a ceramic disc in IF and RF circuits.

(3) The inductance in the collector circuit of the 2N168A is called an *RFC* or *RF choke*. It presents a high impedance to the RF signal and is thus effectively the collector load. It presents a low impedance to the audio so that little audio voltage appears across it. The audio frequency current flows through the *RFC* into the primary of T_1.

Exercise 1.2

(a) What is the function of C_4?
(b) The main signal flow path is from the volume control through C_9 to where?
(c) What is the purpose of R_1?
(d) Why isn't the lower end of T_1's secondary directly grounded?
(e) Estimate the low-end audio frequency breakpoint of the 2N168A stage due to C_3.

1.2.3 The tuned radio frequency (TRF) receiver

When the first commercial AM broadcast was made in 1920, the only listeners were those who had built their own radios. This situation dramatically

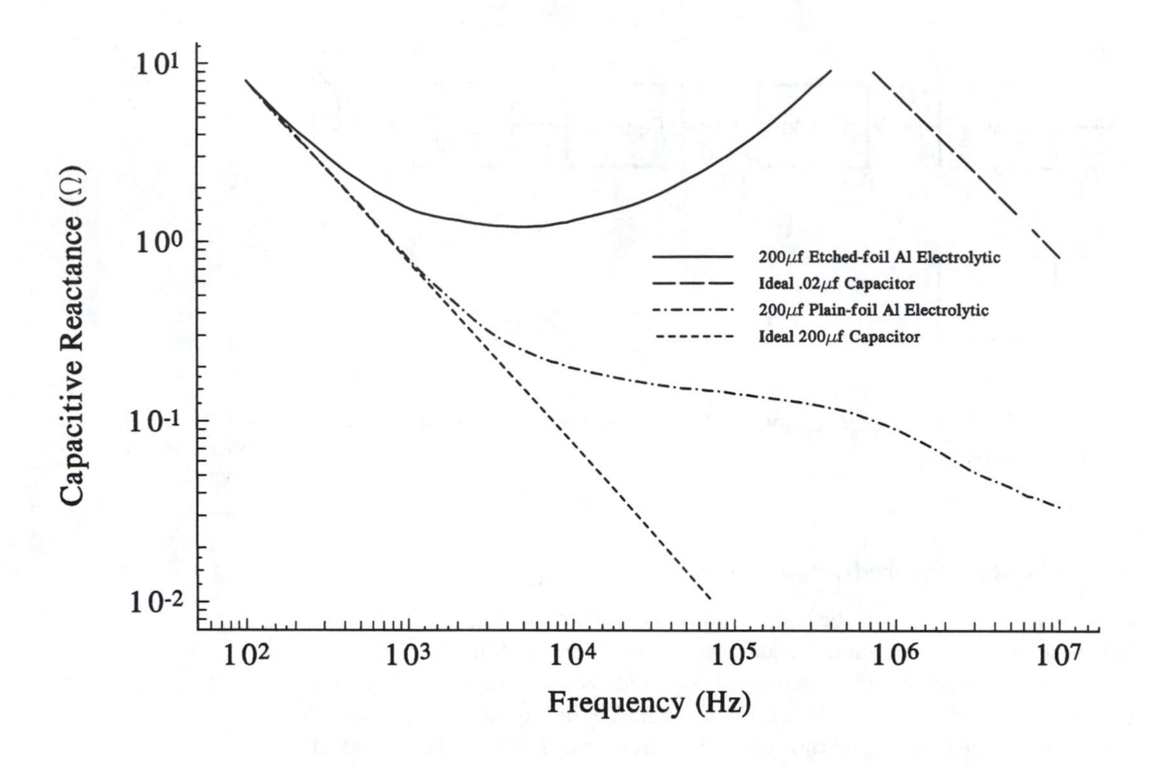

Fig. 1.9: The reactance of various capacitors versus frequency.

changed in a very short time as radio production facilities sprang up everywhere.

Although simple receivers with high performance/cost ratio such as the regenerative and reflex receivers had been dominant with those who built their own, the general public wanted something a little less critical to use and with better sound quality. Therefore, most of the earliest commercially produced receivers were of the *TRF* (*Tuned Radio Frequency*) variety. Another form of receiver called the superheterodyne, invented in 1918 by Major Edwin Armstrong, was destined to achieve an almost total dominance in the receiver market – a dominance that it still retains today. This circuit will be covered starting in Section 1.3.

The TRF featured one, two, or even three stages of RF amplification with each stage being tuned. The multiple tuning gave a high degree of selectivity. The earliest models used separate controls to tune each stage, which was a painstaking procedure. Multi-gang variable capacitors became available and were incorporated into the TRF sets, making tuning appreciably easier. The output of the RF stage(s) was fed to a detector and hence to an audio amplifier, which drove a speaker through an impedance step-down transformer.

A simple TRF receiver will be shown in Section 1.8.1.

The TRF did not have a long period of dominance. The advantages of the superheterodyne (or *superhet* as it came to be called) were too pronounced to ignore. The TRF tended to oscillate and to require careful neutralization of the tuned amplifier stages for stability. Furthermore, the simplicity that single-knob tuning provided had its negative side. The user was not able to "tweak" the tuning of the stages individually to optimize performance. The superhet provided greater sensitivity, a selectivity at least equal to that of the singly

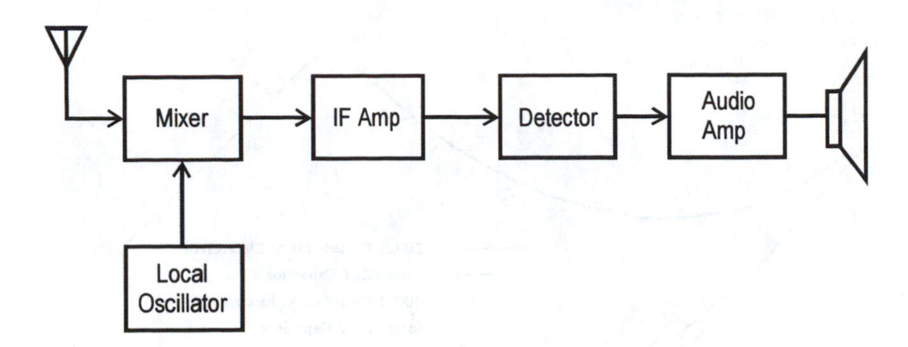

Fig. 1.10: Block diagram of a superheterodyne receiver.

tuned TRF, and a lower parts count. By 1930 the superhet became the standard for radio receivers.

1.3 The *superheterodyne* principle

Heterodyning may perhaps be better known to the electrical engineer as mixing. When two signals of different frequencies are mixed in the nonlinear sense, the output of the mixer will have frequency components at the sum and difference of the two inputs. Typicaly the difference signal is in the audio range. In the superheterodyne receiver, it is not, but lies at a frequency between the RF and AF ranges. Thus, it is called *intermediate frequency* or *IF*. Because the difference signal is at a superaudio frequency, it is called a superheterodyne. A block diagram of a superheterodyne receiver is shown in Fig. 1.10.

The local oscillator provides a more-or-less sinusoidal output that is mixed with the RF signal from the antenna. (Once in a while you may see a superhet with a stage of RF amplification between the antenna and the mixer.) The two-gang variable capacitor used to tune the input of the mixer also tunes the local oscillator. The purpose of this is to keep its frequency at exactly the IF frequency above the station frequency. For example, the IF frequency for most American-made AM radios is 455 kHz. Thus, if the radio is tuned to 1,000 kHz, the local oscillator should run at 1,455 kHz. If a station at 1,550 kHz is tuned in, the local oscillator should run at 2,005 kHz. This principle can be demonstrated readily. Most radios emit a certain amount of radiation at the local oscillator frequency. Place two radios as close together as possible. Tune one to a station below 1,100 kHz. Its local oscillator will then run 455 kHz higher. Calculate and tune the second radio to that frequency. As the second receiver is tuned past the local oscillator frequency of the first, you should hear a pronounced whistle. Verify its source by turning off the first radio. It may help if at least the first radio is a portable, so you can vary the orientation between the radios as you seek for the local oscillator signal. What happens to the sum frequency? We want the difference frequency, which stays constant at 455 kHz, no matter where we tune the radio, but the sum frequency will vary and is not wanted. The answer is that there is a parallel tuned circuit between the mixer and the IF amplifier. This tuned circuit is adjusted for maximum transmission at 455 kHz in a process called *alignment*. The IF amplifier generally has two and occasionally three stages in AM receivers. The input and output of each stage are also tuned to the same frequency range during alignment.

It has been noted that the local oscillator is designed to operate at 455 kHz above the RF signal selected. Why not below it? Because it is much easier for the local oscillator to track the RF if its frequency is above the RF. *Tracking* refers to staying exactly 455 kHz from any selected RF signal. Consider why this is true. In 1996 the span of carrier frequencies in the broadcast AM band was extended from its original range (540 kHz to 1,600 kHz) to a new one (540–1,700 kHz). If the local oscillator operated below the RF signal it would need to be able to tune from 85 kHz to 1,245 kHz. Operating above the RF, it must tune between 995 kHz and 2,155 kHz. These figures should suggest two tuning advantages to the latter scheme. The first advantage is that it requires a frequency variation by a factor just over 2. The first scheme requires a variation by just over a factor of 13. Recalling that the resonant frequency of a parallel LC circuit varies as $C^{-1/2}$, a factor of 13 variation in frequency would require a factor of 169 variation in capacitance. This would be difficult if not impossible to achieve with a single control.

The second advantage of operating the local oscillator above the RF signal is seen by considering the reception of a single station near midband – say 1,000 kHz. The local oscillator would need to operate at either 545 kHz or 1,455 kHz. The ratio between these two frequencies is 2.67. Therefore, the capacitor required to tune the local oscillator to 545 kHz would be over 7 times as large as that required to tune to 1,455 kHz. Thus, the nominal capacitor size required to tune the lower range would be prohibitively large and expensive, if indeed it were available at all.

Let us consider another issue. Suppose we tuned the receiver to 640 kHz. The local oscillator would then run at 1,095 kHz. Also suppose there was a very strong station at 1,550 kHz. Presumably if the receiver input were tuned to 640 kHz, very little signal at 1,550 kHz would be passed by it. But any signal that did get into the mixer and mixed with the 1,095 kHz signal from the local oscillator would also generate a 455 kHz difference signal that would be processed normally along with the desired signal at 640 kHz. The undesired signal (at 1,550 kHz in this case) is called an *image signal*. The simplest measure that can be taken against images is to improve the selectivity of the RF circuit, which practically means raising the Q of the tuned circuit and/or adding a tuned RF stage.

Before leaving the subject of the superheterodyne principle, let us note that information carried on the original AM wave is all present and in the same form in the IF signal. In effect, all we have done is to place the program material onto a different frequency carrier – a 455 kHz carrier. This can be shown mathematically but is perhaps more easily seen from a spectral energy diagram for sinusoidal modulation such as that of Fig. 1.11. Each component of the original spectrum forms sum and difference frequencies with the 1,455 kHz local oscillator output. The lower triad is selected out by the tuned IF amplifier and is thus simply a downshifted replica of the original signal.

1.4 The *AVC* principle

AVC stands for *automatic volume control*. The purpose of AVC is to somewhat equalize the volume received from stations that may vary considerably in the signal they present to the receiver. This circuit function is particularly important in automobile receivers, where the additional factors of antenna orientation and

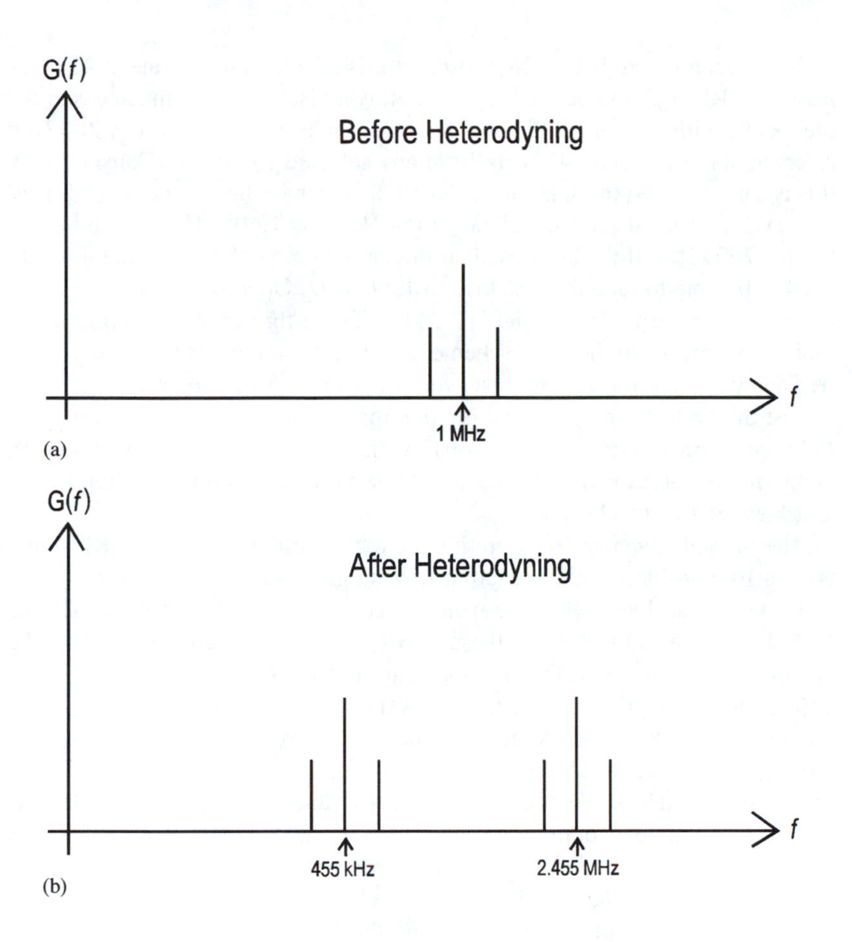

Fig. 1.11: Spectral energy diagram of a broadcast AM signal (a) before heterodyning and (b) after heterodyning.

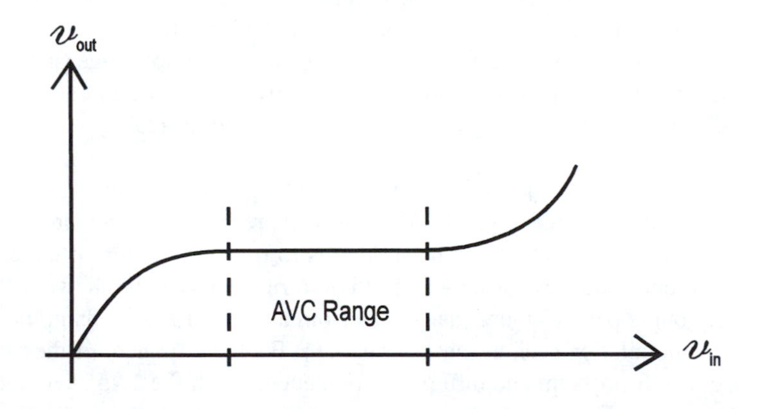

Fig. 1.12: Transfer function of an amplifier with AVC.

environmental obstructions are continually changing, causing even more extensive variation in signal strength. Even a relatively cheap home receiver should have an AVC range of 20 dB, whereas a 40 dB range would be a relatively low value for a car radio. Figure 1.12 shows the transfer function of an amplifier with AVC.

AVC is achieved by sampling the AC output, converting it to DC, and using the DC to control the gain of an amplifier – usually by varying its transconductance. Normal AVC involves using the DC to reduce the quiescent collector

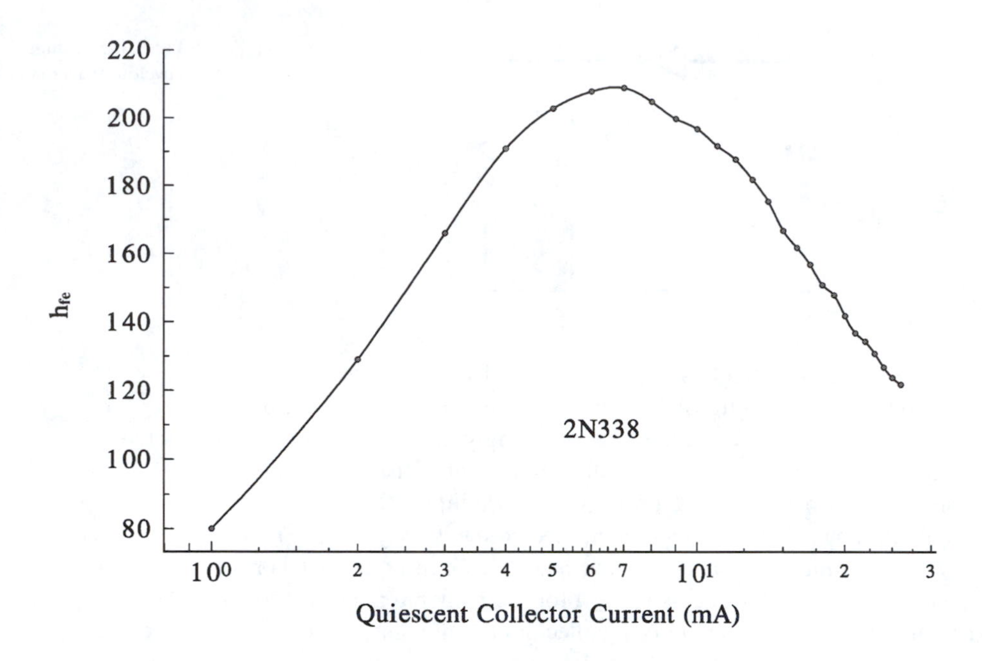

Fig. 1.13: Variation of the h_{fe} of a transistor versus the quiescent collector current.

current of the transistor. The gain is approximately

$$A_v = -g_m R_L,$$

where

$$g_m = I_{CQ}/V_T.$$

Here I_{CQ} is the quiescent collector current and V_T is the thermal voltage, kT/q. Combining these two equations gives

$$A_v = -I_{CQ} R_L / V_T.$$

Thus, reducing the quiescent collector current reduces the gain. The collector current is reduced by reducing the base-emitter bias. Although this approach works very well, it is suited only to stages that handle relatively small signals, since large signals can drive the weakly biased stage into cutoff. When this is the case, the options are to forego AVC on that stage or to use forward AVC. In this scheme, the gain is reduced by increasing the collector current. Different mechanisms are at work here. In forward AVC, the increasing collector current reduces r_π in the transistor equivalent circuit. The pertinent equation is

$$A_v = \frac{-h_{fe} R_L}{R_S + r_\pi + r_x}, \tag{1.8}$$

where R_S is any source resistance on the amplifier input. The CE short-circuit current gain, h_{fe}, will exhibit a peak at a given value of collector current and will fall off on either side of that peak. The location and breadth of the peak will vary widely between transistor types. Those devices intended for use in IF amplifiers generally have a fairly rapid decline in h_{fe} at higher currents. Figure 1.13 shows the I_{CQ} dependence of h_{fe} for a 2N338 transistor. The effect,

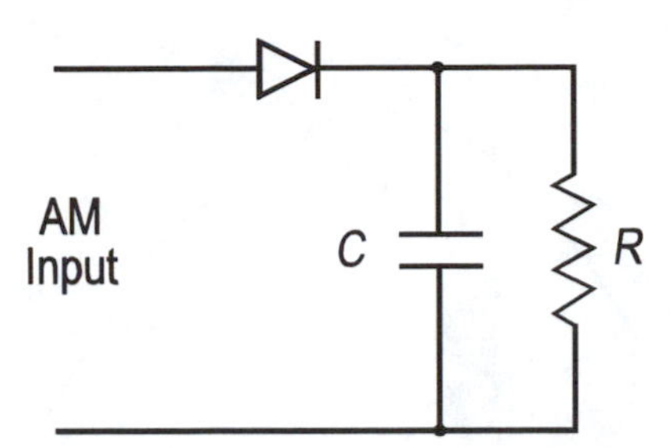

Fig. 1.14: Simple envelope detector.

while clearly evident, is actually somewhat masked by the fact that these data were taken after allowing the transistor to thermally stabilize at each data point. At elevated temperatures h_{fe} increases fairly strongly, and this effect would mask the falloff of h_{fe} due to I_{CQ}. An excellent treatment of the semiconductor mechanisms responsible for this falloff is given by Phillips (1962).

Equation 1.8 shows that if both r_π and h_{fe} decrease with I_{CQ}, R_S and r_x will establish a lower limit on the denominator, but no such limit exists (short of burning out the transistor by excessive I_{CQ}) for the numerator. Thus the gain will eventually decrease as the quiescent collector current is increased.

1.5 The *envelope detection* principle

In discussing the crystal set, it was shown that the primary signal component in the rectified AM wave is at the modulating signal frequency, but that there is also a relatively high level of distortion. It was discovered very early in the history of radio that the signal component could be greatly increased and distortion greatly reduced by adding a suitably chosen capacitor in parallel with the load as shown in Fig. 1.14.

Now the load voltage, rather than consisting of a series of pulses as in Fig. 1.3b, cannot drop rapidly with each pulse, since the diode becomes reverse-biased and thus opens the discharge circuit. As long as the diode is reverse-biased, the capacitor discharges through the resistor. Proper design requires that the discharge rate be rapid compared to the modulating signal but slow compared to the carrier. If this is the case, the load voltage will look like Fig. 1.15. The ratio ω_c/ω_m has been made unrealistically low (=50) to better show the nature of the inter-cycle decay.

Clearly the load voltage (Fig. 1.15b) is essentially the modulating voltage with some much-higher-frequency ripple on it. The ripple is at the IF frequency and is thus totally insignificant in a first-order analysis. In practice, second-order effects such as active device parameter modulation make it desirable to keep the ripple down to as small a value as is practical.

Exercise 1.3 If the RC decay constant is too large, the load voltage cannot drop as rapidly as the envelope does. This leads to serious distortion. Assuming that the decay starts at one of the peaks of the carrier and that $m = 1$, find a constraint on RC to prevent this occurrence.

Exercise 1.4 If $RC\omega_c = 200$, find the worst case ripple voltage as a fraction of the initial voltage from which the decay begins. State your assumptions!

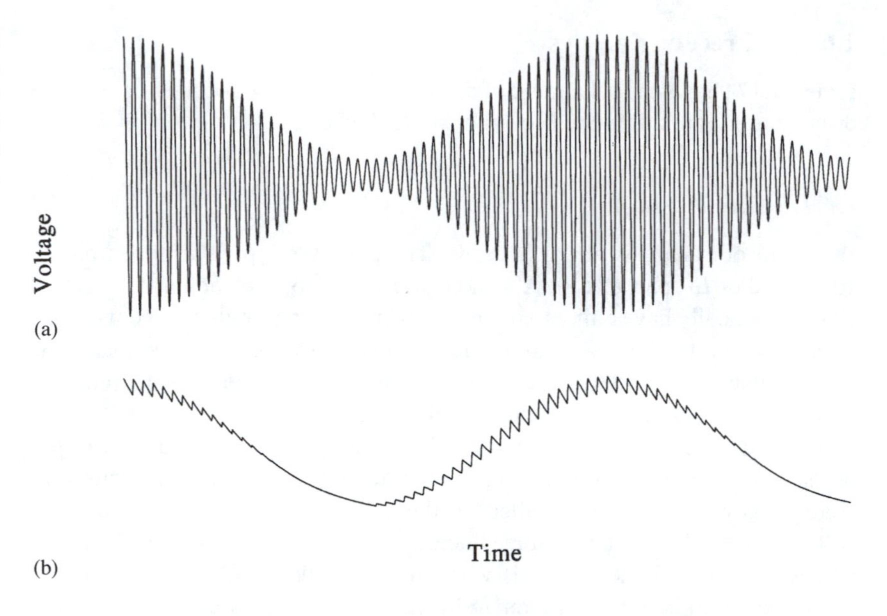

Fig. 1.15: (a) Input waveform to the envelope detector. (b) Output of the envelope detector showing high-frequency ripple riding on the demodulated signal.

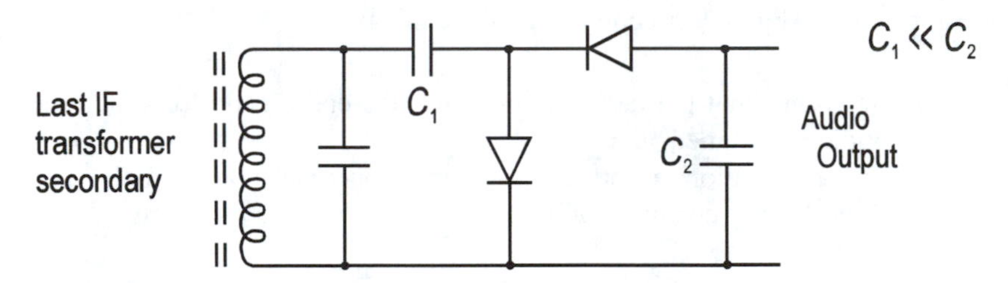

Fig. 1.16: Improved envelope detector.

1.5.1 Improved envelope detector

A drawback of the envelope detector is that it causes a DC component of current to flow through the secondary of the IF transformer that is driving it. These are interstage coupling transformers for the IF amplifier. Because of the ferrite cores in these transformers a DC current will move the quiescent flux toward saturation, thus tending to cause a little distortion. This problem can be circumvented and a larger output voltage obtained by using the circuit shown in Fig. 1.16.

This circuit will be recognized as a voltage doubler. Aside from the obvious advantage of doubling the output without amplification, this circuit has two additional advantages:

(1) It loads the IF transformer equally on both half cycles and allows no DC to flow in the winding. These both reduce distortion.

(2) The loading is now primarily capacitive. This means the tuned circuit has a higher Q, and thus it provides an even larger output voltage. Due to the capacitive loading, it is necessary to retune the IF transformer secondary if a radio is modified to incorporate this improved envelope detector.

1.6 AM receiver analysis

Figure 1.17 shows the schematic of an excellent older AM receiver using eight discrete transistors. It is the Motorola Model XP7CE.

1.6.1 The RF stage

We begin our analysis at the left side. There we see a parallel tuned circuit comprised of L_1, part of C_{27}, A8, and C_{10}. L_1 is a loopstick antenna. This will characteristically have either two separate windings or a single tapped winding. In either case, the purpose is to provide an impedance step-down to match the high-Q tuned circuit to the transistor's low input impedance. The dashed lines between the windings are indicative of the ferrite core.

We might wonder why three parallel capacitors are used to tune it. C_{27} is one section of a three-section variable main-tuning capacitor. Each section (or *gang* as they are commonly called) of this capacitor will have, as an integral part, a small *trimmer* capacitor, which is adjusted as part of the alignment process. The one for this gang is what we have called A8. Strictly speaking, this designation is not a component identifier but rather is an indication that this is an adjustment point during the alignment. That is why it is enclosed in a square box. Nevertheless, since this component has no other designator, we treat it as one. Alignment accomplishes three things:

(1) It assures that the radio can tune over the entire band for which the radio is designed, and no further.
(2) It makes station reception occur at the proper place on the dial.
(3) It peaks the sensitivity of the entire radio.

The value of A8 is not even given on the schematic, since it is only a few pF; its setting is not important – only its effect. C_{10} is relatively distinctive. The key as to its function is the notation N750. This indicates a controlled *temperature coefficient* (*tempco*). It compensates for thermal drift of the other tuned circuit components.

It is somewhat unusual to see the low end of C_{27} and associated capacitors returned to 3.9 V, rather than directly to ground as usual. The 3.9 V supply will, of course, be the same as ground for AC purposes. The reason for not returning the capacitor to DC ground will soon be seen. The signal from the secondary of L_1 is coupled into the base of X_1. (Note: The use of X as a designator for transistors was an early and not unreasonable choice. In time, however, the designator became standardized as Q. However, we shall refer to the transistors and diodes by the X designation shown on the schematic to avoid confusion.)

The low end of this winding is held at AC ground by C_{11}. The base bias for X_1 can be coming only from R_5. Tracing back from the low end of R_5, we see that the bias originates with X_9. It is thus AVC voltage. We see further that the AVC voltage filter affords us another example of the paralleling of an electrolytic capacitor with a ceramic or some other small-value high-Q unit. C_{17} bypasses any high frequencies on the AVC line while C_3 bypasses any audio. R_5 and C_{11} provide another stage of high-frequency bypassing.

X_1 has its Q point stabilized by R_6, which is bypassed by C_{12}. Since this stage is biased by AVC, it is a formidable task to find the quiescent collector current. We can see, however, from the orientation of the detector diode, that a

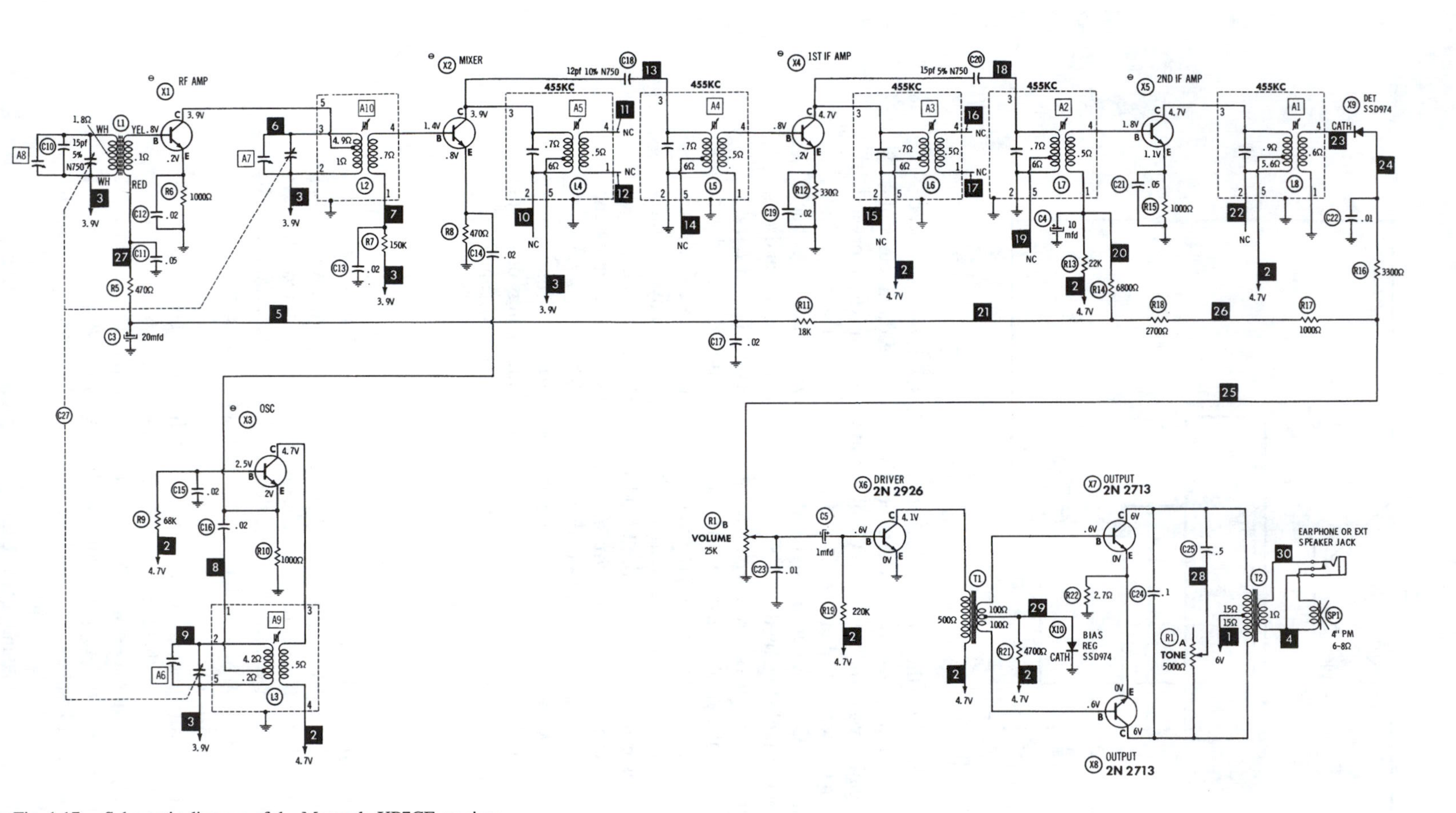

Fig. 1.17: Schematic diagram of the Motorola XP7CE receiver.

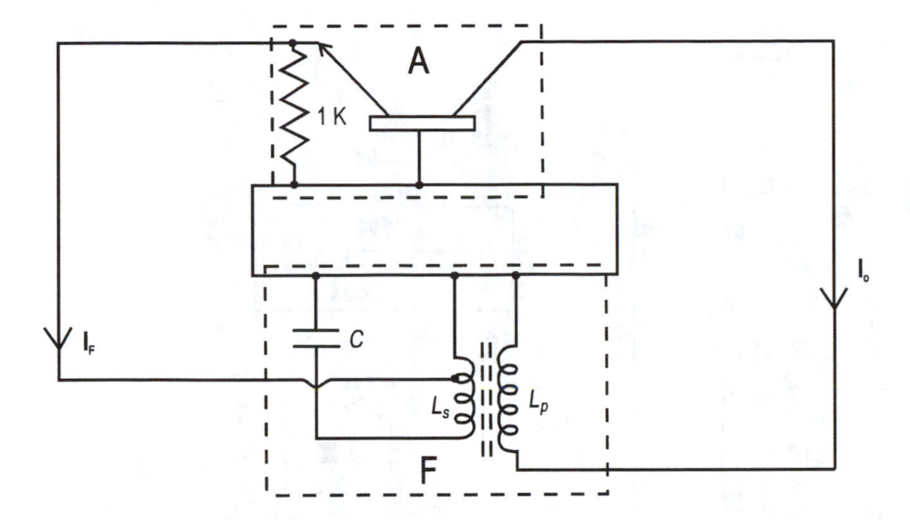

Fig. 1.18: Local oscillator of the XP7CE receiver drawn in standard feedback form.

stronger signal will make the AVC line less positive. Thus, this is normal AVC. We might also inquire how this bias gets positive, since the diode can only generate a negative voltage. Yet the AVC voltage must be positive to forward bias X_1 (among others). The answer is that R_{13} and R_{14} sum some positive voltage into the AVC line. Thus, the signal-developed AVC voltage must buck the positive bias established by the 4.7 V supply applied through R_{13} and R_{14}.

The output of X_1 appears across a parallel tuned load consisting of L_2, A7, and another gang of C_{27}. Again, A7 is the trimmer capacitor associated with the second gang of C_{27}. This stage provides an output at the same frequency as its input since it is simply a linear amplifier. Thus L_2, as tuned by C_{27}, must be capable of resonating over the entire broadcast band. It is operating at RF frequencies and is thus called an *RF transformer*. Because both the input and output of the X_1 stage are parallel tuned, this is a *tuned RF amplifier stage*.

The collector of X_1 receives its bias through the primary of L_2. This requires that the low end of C_{27} be connected to +3.9 V. But the three gangs of C_{27} all have their low ends common. This is why the low end of the first gang of C_{27} was returned to 3.9 V rather than ground.

1.6.2 The local oscillator

The signal input to X_2 is exactly analogous to that of X_1. Bias is different. Rather than receiving AVC, X_2 is biased from the 3.9 V supply through R_7. The emitter of X_2 is not bypassed. This is because there is a signal fed into this point from X_3. Based on what we know of the superheterodyne principle and the fact that the X_3 circuitry has no input – only an output – we deduce that this stage is the *local oscillator*.

Although this hardly appears to be a textbook oscillator, it can be analyzed fairly readily. We first note that the base of X_3 is AC-grounded by C_{15}. Bias is provided by R_9 in conjunction with R_{10}, the emitter resistor. The knowledge that the configuration of the *A* block is common-base (CB) allows us to draw the oscillator in standard form as shown in Fig. 1.18. This is a form of the tuned-input oscillator.

The very low input impedance of the CB means that any signal current going through the 1 kΩ emitter resistor will be negligible. The resistive load on the

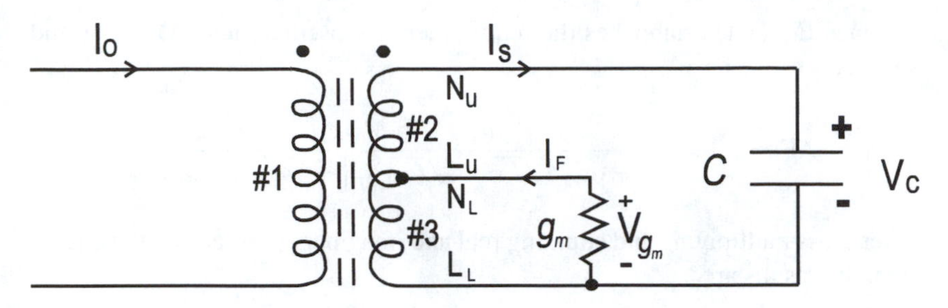

Fig. 1.19: The feedback network of the oscillator shown in Fig. 1.18.

secondary of the oscillator coil will then be approximately $1/g_{\mathrm{m}} = R_{\mathrm{in}}$ (CB). Note that $A = \alpha$.

The sign convention shown for $\mathbf{I}_{\mathrm{F}}$ is consistent with Eq. (1.1), as may be seen by referring back to Fig. 1.7. We need to find the transfer function of F. To do so we will assume perfect coupling between the two parts of L_{s}, which we shall call L_{U} and L_{L}. We will also denote the respective number of turns as N_{U} and N_{L}. The feedback network may then be drawn as shown in Fig. 1.19.

We want to find $\mathbf{F} = \mathbf{I}_{\mathrm{F}}/\mathbf{I}_{\mathrm{o}}$. By the assumed perfect flux coupling within the secondary we have

$$\frac{\mathbf{I}_{\mathrm{s}}}{sC(N_{\mathrm{U}} + N_{\mathrm{L}})} = \frac{\mathbf{V}_{\mathrm{c}}}{N_{\mathrm{U}} + N_{\mathrm{L}}} = \frac{d\Phi}{dt} = \frac{\mathbf{V}_{g_{\mathrm{m}}}}{N_{\mathrm{L}}} = \frac{-\mathbf{I}_{\mathrm{F}}}{g_{\mathrm{m}} N_{\mathrm{L}}}$$

so that

$$\mathbf{I}_{\mathrm{s}} = -\mathbf{I}_{\mathrm{F}} \cdot \left(\frac{sC}{g_{\mathrm{m}}}\right) \cdot \left(\frac{N_{\mathrm{U}} + N_{\mathrm{L}}}{N_{\mathrm{L}}}\right) = -\mathbf{I}_{\mathrm{F}} \left(\frac{sC}{g_{\mathrm{m}}}\right) \cdot \left(\frac{\sqrt{L_{\mathrm{U}}} + \sqrt{L_{\mathrm{L}}}}{\sqrt{L_{\mathrm{L}}}}\right). \tag{1.9}$$

Because the primary is current driven, mutual coupling from the secondary into the primary will not affect the current transfer ratio. We will write the loop equation that includes N_{L} and g_{m}. Voltage is induced in this loop from both the primary and the upper part of the secondary. The equation is

$$sM_{13}\mathbf{I}_{\mathrm{o}} - s\sqrt{L_{\mathrm{U}} L_{\mathrm{L}}}\mathbf{I}_{\mathrm{s}} = -\mathbf{I}_{\mathrm{F}}/g_{\mathrm{m}} + sL_{\mathrm{L}}(\mathbf{I}_{\mathrm{s}} - \mathbf{I}_{\mathrm{F}}), \tag{1.10}$$

where we have used the perfect secondary flux coupling to say $M_{23} = \sqrt{L_{\mathrm{U}} L_{\mathrm{L}}}$.

If (1.9) is inserted into (1.10), we get

$$\begin{aligned} sM_{13}\mathbf{I}_{\mathrm{o}} &= -\mathbf{I}_{\mathrm{F}}/g_{\mathrm{m}} - sL_{\mathrm{L}}\mathbf{I}_{\mathrm{F}} - s(L_{\mathrm{L}} + \sqrt{L_{\mathrm{U}} L_{\mathrm{L}}})\mathbf{I}_{\mathrm{F}} \\ &\quad \cdot \left(\frac{sC}{g_{\mathrm{m}}}\right) \cdot \left(\frac{\sqrt{L_{\mathrm{U}}} + \sqrt{L_{\mathrm{L}}}}{\sqrt{L_{\mathrm{L}}}}\right) \\ &= -\mathbf{I}_{\mathrm{F}} \left[\frac{1}{g_{\mathrm{m}}} + sL_{\mathrm{L}} + s(\sqrt{L_{\mathrm{L}}} + \sqrt{L_{\mathrm{U}}})^2 \cdot \left(\frac{sC}{g_{\mathrm{m}}}\right)\right], \end{aligned} \tag{1.11}$$

$$\begin{aligned} \frac{\mathbf{I}_{\mathrm{F}}}{\mathbf{I}_{\mathrm{o}}} &= \frac{-sM_{13}g_{\mathrm{m}}}{[1 + s^2(\sqrt{L_{\mathrm{U}}} + \sqrt{L_{\mathrm{L}}})^2 C] + sL_{\mathrm{L}} g_{\mathrm{m}}} \\ &= \mathbf{F} = \frac{-j\omega M_{13} g_{\mathrm{m}}}{[1 - \omega^2(\sqrt{L_{\mathrm{U}}} + \sqrt{L_{\mathrm{L}}})^2 C] + jL_{\mathrm{L}}\omega g_{\mathrm{m}}}. \end{aligned}$$

Since Eq. (1.1) establishes the requirement for oscillation as $\mathbf{AF} = -1$ and since $A = \alpha$, we have

$$\mathbf{AF} = -1 = \frac{-j\omega_{\text{osc}} \cdot \alpha \cdot M_{13} g_{\text{m}}}{\left[1 - \omega_{\text{osc}}^2 (\sqrt{L_{\text{U}}} + \sqrt{L_{\text{L}}})^2 C\right] + j\omega_{\text{osc}} L_{\text{L}} g_{\text{m}}}.$$

After cross multiplying and equating real and imaginary parts across the equals sign, the results are

$$\frac{1}{g_{\text{m}}}\left[1 - \omega_{\text{osc}}^2 (\sqrt{L_{\text{U}}} + \sqrt{L_{\text{L}}})^2 C\right] = 0 \Rightarrow \omega_{\text{osc}}^2 = \frac{1}{(\sqrt{L_{\text{U}}} + \sqrt{L_{\text{L}}})^2 C}$$

and

$$L_{\text{L}} \omega_{\text{osc}} = \omega_{\text{osc}} \cdot \alpha \cdot M_{13} \Rightarrow L_{\text{L}} = \alpha \cdot M_{13}.$$

The oscillator is tuned by the third gang of C_{27} and trimmed by A6.

The relationship

$$L_{\text{L}} = \alpha \cdot M_{13}$$

represents a lower limit on the value of α that will allow oscillation. The circuit should work for

$$\alpha \geq \frac{L_{\text{L}}}{M_{13}}.$$

Since $\alpha \approx 1$, this provides a design equation for the oscillator coil.

It is interesting to note that neither the condition for oscillation nor the frequency of oscillation exhibits any dependence on g_{m}. This is as it should be if A does not load F, which in our case means that $Z_{\text{in}}(A)$ must be near zero. This assumption is implicit in our analysis. Thus no resistance remains in the circuit to cause losses, and consequently no g_{m} is needed to overcome such losses.

1.6.3 The mixer

The output from the oscillator is taken from the emitter of X_3 to ensure low output impedance. It is input through C_{14} to the emitter of X_2. This stage has the RF signal applied to its base and the local oscillator connected to its emitter. It is therefore the *mixer*.

Let us examine the principle of operation of this mixer.

The Ebers–Moll equation for the collector current when the B–C junction is reverse biased is

$$i_{\text{C}} = \alpha_{\text{F}} I_{\text{ES}} \left(e^{\frac{v_{\text{BE}}}{V_{\text{T}}}} - 1\right) + I_{\text{CS}}.$$

In this equation, I_{CS}, I_{ES}, and α_{F} are the Ebers–Moll parameters. The variables i_{C} and v_{BE} denote the total variables – DC and AC components combined.

If $v_{BE} \gg V_T$ then

$$i_C = \alpha_F I_{ES} \cdot e^{\frac{v_{BE}}{V_T}} = \alpha_F I_{ES} \cdot e^{\frac{V_{BE}+v_{be}}{V_T}}.$$

In this equation we have separated the total variable v_{BE} into AC and DC components, v_{be} and V_{BE}, respectively. However, $I_{CQ} = \alpha_F I_{ES} \cdot e^{v_{BE}/V_T}$, where I_{CQ} is the quiescent collector current. Thus we can write the total collector current as

$$i_C = I_{CQ} e^{v_{BE}/V_T}.$$

For linear amplifier analysis, one assumes $v_{be} \ll V_T$ and so we can approximate

$$i_C = I_{CQ}(1 + v_{be}/V_T).$$

When the amplitude of v_{be} does not satisfy this inequality, it is necessary (at least in theory) to include higher-order terms in the expansion of the exponential. These higher-order terms represent nonlinearities. The mixer is designed so that the output of the local oscillator will be large enough to assure nonlinear operation. The nature of the nonlinearity is critical only insofar as it governs the amplitude of the output. For simplicity we will assume that we can express i_C as the first three terms of a Maclaurin series:

$$i_C = I_{CQ}\left(1 + \frac{v_{be}}{V_T} + \frac{v_{be}^2}{2V_T^2}\right). \tag{1.12}$$

Let us call the signal input to X_2's base $v_s(t)$. Assume

$$v_s(t) = V_s \sin \omega_c t \cdot (1 + m \sin \omega_s t)$$

and

$$v_o(t) = V_o \sin \omega_o t.$$

Since $v_{be} = v_s(t) - v_o(t)$, we have

$$v_{be} = V_s \sin \omega_c t \cdot (1 + m \sin \omega_s t) - V_o \sin \omega_o t.$$

Using this in (1.12) yields

$$i_C = I_{CQ}\left[1 + \frac{V_s \sin \omega_c t \cdot (1 + m \sin \omega_s t)}{V_T} - \frac{V_o \sin \omega_o t}{V_T}\right.$$

$$+ \frac{V_s^2 \sin^2 \omega_c t \cdot (1 + m \sin \omega_s t)^2}{2V_T^2}$$

$$\left. - \frac{V_o V_s \sin \omega_c t \cdot \sin \omega_o t \cdot (1 + m \sin \omega_s t)}{V_T^2} + \frac{V_o^2 \sin^2 \omega_o t}{V_T^2}\right].$$

Table 1.1. *Spectral amplitude versus frequency for I_C in a square-law mixer*

Frequency	Relative amplitude
DC	$V_o^2/2V_T^2$
1 kHz	$V_s^2/4V_T^2$
2 kHz	$V_s^2/8V_T^2$
454 kHz	$V_o V_s/4V_T^2$
455 kHz	$V_o V_s/2V_T^2$
456 kHz	$V_o V_s/4V_T^2$
.999 kHz	$V_s/2V_T$
1.000 MHz	V_s/V_T
1.001 MHz	$V_s/2V_T$
1.455 MHz	V_o/V_T
1.998 MHz	$V_s^2/16V_T^2$
1.999 MHz	$V_s^2/4V_T^2$
2.000 MHz	V_s^2/V_T^2
2.001 MHz	$V_s^2/4V_T^2$
2.002 MHz	$V_s^2/16V_T^2$
2.454 MHz	$V_o V_s/4V_T^2$
2.455 MHz	$V_o V_s/2V_T^2$
2.456 MHz	$V_o V_s/4V_T^2$
2.910 MHz	$V_o^2/2V_T^2$

By a long laborious trigonometric expansion, this expression may be broken down into its frequency components. If we assume $f_s = 1$ kHz and $f_c = 1$ MHz, then $f_o = 1.455$ MHz. Further assume $m = 1$. Finally, assume $V_o \gg V_s$ and $V_o \gg V_T$. Rather than crowd the results onto a spectral energy plot, the amplitudes of the various frequency components are shown in Table 1.1. The three terms between the horizontal lines are the ones we want. They will be selected out as the collector current composed of these components passes through a 455 kHz parallel resonant circuit.

1.6.4 Mismatching

In our receiver, the nature of this resonant circuit is a little unusual. This is because there are actually two resonant circuits in the collector load of the mixer.

Collector current for the mixer flows through L_4. This is the primary of an IF transformer, but since it is tapped, it is serving as an autotransformer.

It may be recalled that tuned amplifiers have a tendency to oscillate. This may be explained by observing that at a frequency just below resonance, a parallel *RLC* circuit has an impedance that is large and slightly inductive. In general,

$$\mathbf{Z}_{\text{LOAD}} = \frac{jL\omega R}{R(1 - \omega^2 LC) + jL\omega}. \tag{1.13}$$

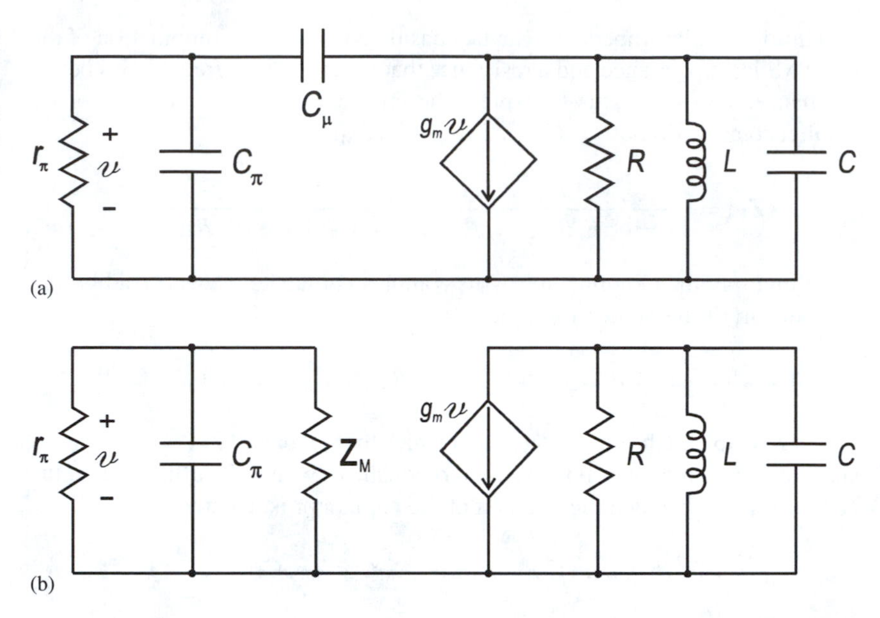

Fig. 1.20: (a) The equivalent circuit of a tuned amplifier with a parallel-tuned *RLC* load. (b) The same equivalent circuit after applying Miller's Theorem.

We shall show shortly that the Miller effect will cause this inductance to appear to the input port as a negative resistance. If the magnitude of this component is greater than the positive resistances in the input circuit, there will be oscillation, but not the stable, well-defined oscillation produced by a well-designed positive feedback amplifier. The negative resistance is a very nonlinear component, which makes analysis of the nature of the oscillation very difficult. We can, however, readily determine the conditions for oscillation.

Figure 1.20a shows the equivalent circuit for a transistor tuned amplifier with a parallel *RLC* load, and Fig. 1.20b shows the equivalent circuit after applying Miller's Theorem.

The Miller impedance, $\mathbf{Z}_\mathrm{M}$, is given by

$$\mathbf{Z}_\mathrm{M} = \frac{1}{(1 + g_\mathrm{m}\mathbf{Z}_\mathrm{LOAD})j\omega C_\mu}.$$

Since it will always be true that $g_\mathrm{m}\mathbf{Z}_\mathrm{LOAD} \gg 1$ over the range of operating frequencies, we can simplify this to

$$\mathbf{Z}_\mathrm{M} = \frac{1}{jg_\mathrm{m}C_\mu\omega\mathbf{Z}_\mathrm{LOAD}},$$

where $\mathbf{Z}_\mathrm{LOAD}$ is given by (1.13). Thus,

$$\mathbf{Z}_\mathrm{M} = \frac{1}{jg_\mathrm{m}C_\mu\omega \cdot \dfrac{jL\omega R}{R(1-\omega^2 LC) + jL\omega}} = \frac{R(1-\omega^2 LC) + jL\omega}{-\omega^2 LC_\mu \cdot g_\mathrm{m}R}$$

$$= -\left(\frac{1}{g_\mathrm{m}}\right)\cdot\left(\frac{1-\omega^2 LC}{\omega^2 LC_\mu}\right) + \frac{1}{jC_\mu\omega g_\mathrm{m}R} \equiv R_\mathrm{M} + \frac{1}{jC_\mathrm{M}\omega}. \tag{1.14}$$

Thus the Miller impedance can be considered the series combination of the usual Miller capacitance and a resistance that is negative for frequencies below resonance. This allows us to express the impedance of the input port as the parallel combination of r_π, C_π, and $R_{\mathrm{M}} + l/jC_{\mathrm{M}}\omega$ or

$$\mathbf{Z}_{\mathrm{IN}} = \frac{r_\pi(1 + jR_{\mathrm{M}}C_{\mathrm{M}}\omega)}{[1 - \omega^2 R_{\mathrm{M}}C_{\mathrm{M}}r_\pi C_\pi] + j\omega[r_\pi(C_\pi + C_{\mathrm{M}}) + R_{\mathrm{M}}C_{\mathrm{M}}]}.$$

If we multiply the denominator by its complex conjugate, we can readily find the real part of the numerator to be

$$r_\pi\{(1 - \omega^2 C_{\mathrm{M}}R_{\mathrm{M}}C_\pi r_\pi) + \omega^2 C_{\mathrm{M}}R_{\mathrm{M}}[r_\pi(C_{\mathrm{M}} + C_\pi) + R_{\mathrm{M}}C_{\mathrm{M}}]\}.$$

We want to find the conditions under which the real part of $\mathbf{Z}_{\mathrm{IN}} < 0$. However, since the denominator is positive after rationalization, it will suffice to find the conditions under which the real part of the numerator is negative:

$$1 + \omega^2 R_{\mathrm{M}}C_{\mathrm{M}}[r_\pi C_{\mathrm{M}} + \cancel{r_\pi C_\pi} + R_{\mathrm{M}}C_{\mathrm{M}} - \cancel{r_\pi C_\pi}] < 0,$$

$$1 + \omega^2 R_{\mathrm{M}}C_{\mathrm{M}}^2(r_\pi + R_{\mathrm{M}}) < 0.$$

Although this looks like a simple enough condition, it is enormously complicated by the pronounced frequency dependence of R_{M}. Inserting the explicit expressions for R_{M} and C_{M} from (1.14) gives

$$1 - \frac{g_{\mathrm{m}}R^2 C_\mu \cdot (1 - \omega^2 LC)}{L} \cdot \left(r_\pi - \frac{(1 - \omega^2 LC)}{g_{\mathrm{m}}\omega^2 LC_\mu}\right) < 0.$$

Let us replace the inequality with an equality to find the limiting condition. We will also make the substitution

$$x \equiv \omega^2 LC,$$

giving

$$1 - \frac{R^2 C_\mu}{L} \cdot (1 - x)\left[\beta - \frac{(1 - x)C}{xC_\mu}\right] = 0.$$

Algebraic rearrangement gives

$$x^2\left(\beta\frac{C_\mu}{C} + 1\right) - x\left(\beta\frac{C_\mu}{C} + 2 - \frac{L}{R^2 C}\right) + 1 = 0. \tag{1.15}$$

The term L/R^2C may be expressed as

$$\frac{L}{R^2 C} = \left(\frac{L\omega_{\mathrm{o}}}{R}\right) \cdot \left(\frac{1}{RC\omega_{\mathrm{o}}}\right) = \frac{1}{Q^2},$$

since both terms in parentheses are alternate expressions for the reciprocal Q of a parallel resonant circuit. We will also define

$$a \equiv \beta C_\mu / C.$$

We can thus express (1.15) as

$$x^2(a+1) - x(a+2-1/Q^2) + 1 = 0.$$

We may use the quadratic equation to solve for the limiting values of x from this equation:

$$x = \frac{(2+a-1/Q^2) \pm \sqrt{a^2 - 2(2+a)/Q^2 + 1/Q^4}}{2(1+a)}. \tag{1.16}$$

Since for AM we might usually expect a to be a few tenths and Q to be 50, it seems reasonable to assume that the Q terms are negligible, giving

$$x = \frac{(2+a) \pm a}{2(1+a)} \Rightarrow x = 1, \qquad 1/(1+a).$$

It will be recalled that these are the limiting values of x. The reader should convince himself that $\mathbf{Z}_{\mathrm{IN}}$ will have a negative real part for any frequency between those corresponding to these values of $\mathbf{x}$, that is,

$$\mathbf{Z}_{\mathrm{IN}} < 0 \quad \text{for} \quad \frac{1}{LC\left[1 + \left(\frac{\beta C_\mu}{C}\right)\right]} = \frac{1}{L(C + \beta C_\mu)} < \omega^2 < \frac{1}{LC}.$$

This inequality suggests that oscillation could occur at any frequency in this range and illustrates the ill-determined character of negative-resistance oscillation.

This result could be pretty discouraging, however. It says there *will* be oscillation for any $a > 0$. Obviously, we should not believe this, since nonoscillating tuned amplifiers do exist. Maybe we should take another look at the Q terms in (1.16). If the argument of the square root were less than zero, we would have to conclude that there is no physical value of x (and thus of ω) for which $\mathrm{Re}(\mathbf{Z}_{\mathrm{IN}}) < 0$. Thus there would be no oscillation. This would occur if

$$a^2 - 2(2+a)/Q^2 + 1/Q^4 < 0.$$

We will again replace the inequality by an equality to find the limiting values. The solutions of the quadratic equation for Q^2 are then

$$Q^2 = \frac{2 + a \pm \sqrt{1+a}}{a^2} \tag{1.17}$$

from which we see that

$$\mathrm{Re}(\mathbf{Z}_{\mathrm{IN}}) > 0 \quad \text{for} \quad \frac{2+a-\sqrt{1+a}}{a^2} < Q^2 < \frac{2+a+\sqrt{1+a}}{a^2}.$$

For a typical value of $a = .5$, this condition gives

$$.45 < Q < 4.45 \tag{1.18}$$

as the range of Q values for which there will be no oscillation.

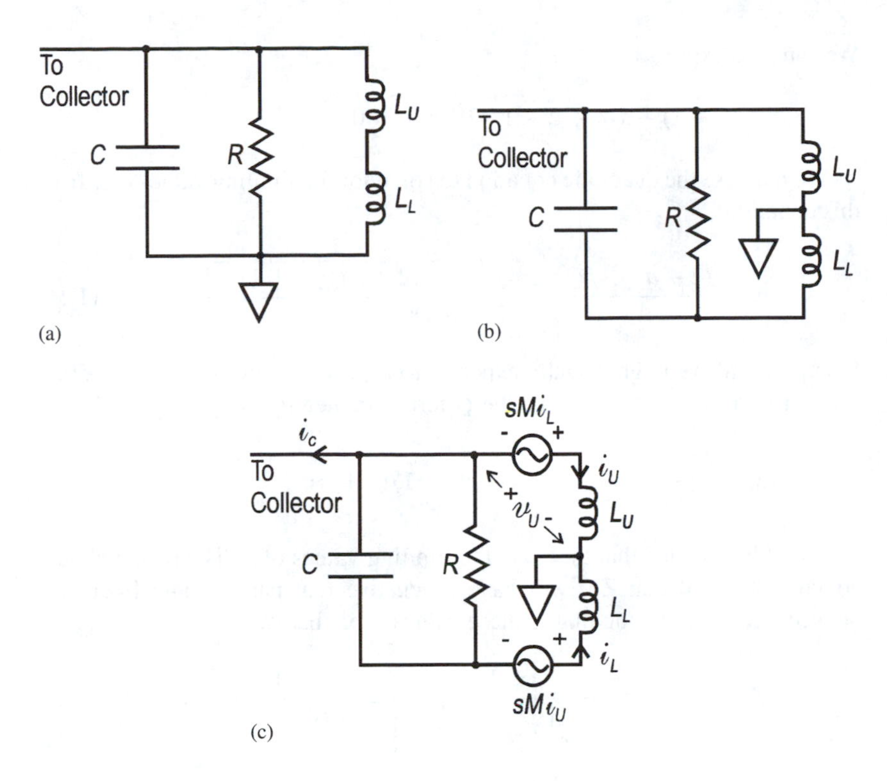

Fig. 1.21: The application of mismatching. (a) A tuned R-L-C output load. The inductor has been separated into two parts to accommodate the next step. (b) The junction of the two inductors is grounded. This effectively configures the inductor as an autotransformer. (c) The equivalent circuit showing the mutual inductance effects explicitly.

Unfortunately these values of Q are so low as to be useless for performing adequate frequency discrimination in the tuned circuits. The solution to this dilemma lies in making the effective resistance of the tuned circuit *seem* small to the transistor. We can accomplish this by adding a tap to the primary of the IF transformer. The technique is called *mismatching*. To understand exactly how mismatching works, refer to Figs. 1.21a and b, which represent two different but similar AC configurations for an IF transformer primary as seen by the collector of the transistor driving it. Figure 1.21c is an equivalent circuit for Fig. 1.21b, with the mutual-inductance effects explicitly shown. In all of these circuits, the resistance is often implicit rather than explicit. It models the losses in the tuned circuit. In Fig. 1.21a, we have somewhat artificially represented the total inductance as the sum of two smaller ones, L_{U} and L_{L}. This is to show the two configurations in Figs. 1.21a and 1.21b in as parallel a form as possible. In fact, the only physical difference in these circuits is the addition of the ground connection. The parameters of the circuit of Fig. 1.21a are well known. It has a resonant frequency of $\omega_{\mathrm{o}} = 1/\sqrt{(L_{\mathrm{U}} + L_{\mathrm{L}})C}$, an impedance at resonance of $\mathbf{Z}(\omega_{\mathrm{o}}) = R$, and a Q at resonance of $R/L\omega_{\mathrm{o}}$.

Most of the parameters of Fig. 1.21b are a little less obvious. It should be clear that the actual Q of both circuits will be the same, since Q simply relates the energy storage and dissipation of a resonant circuit. We do need to calculate the other circuit parameters, however. We shall assume perfect flux coupling between L_{U} and L_{L} and analyze Fig. 1.21c. The equations for this are

$$\mathbf{V}_{\mathrm{U}}(s) = -sM\mathbf{I}_{\mathrm{L}}(s) + sL_{\mathrm{U}}\mathbf{I}_{\mathrm{U}}(s),$$

$$\mathbf{V}_{\mathrm{U}}(s) = \mathbf{I}_{\mathrm{L}}(s)R/(1 + sCR) - sM\mathbf{I}_{\mathrm{U}}(s) + sL_{\mathrm{L}}\mathbf{I}_{\mathrm{L}}(s),$$

$$\mathbf{Z} = -\mathbf{V}_{\mathrm{U}}(s)/\mathbf{I}_{\mathrm{C}}(s) = \mathbf{V}_{\mathrm{U}}(s)/[\mathbf{I}_{\mathrm{U}}(s) + \mathbf{I}_{\mathrm{L}}(s)].$$

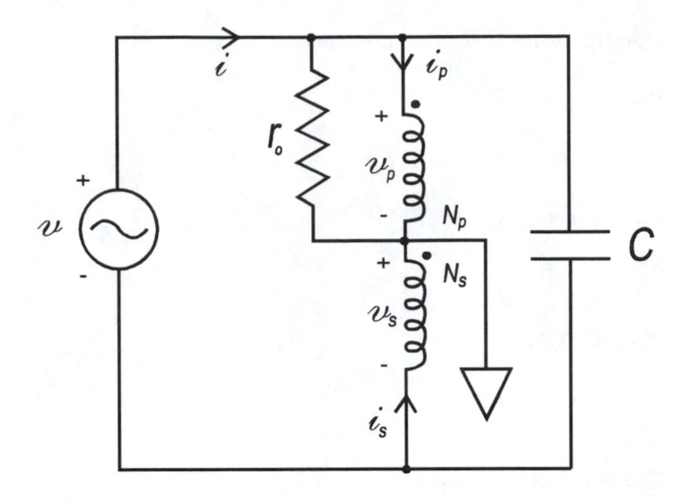

Fig. 1.22: Circuit used for computing the transformation of the output impedance of a transistor in a mismatched tuned amplifier.

Some algebraic work yields the expression

$$\mathbf{Z} = \frac{s^2(L_U L_L - M^2) + sL_U \cdot R/(1+sCR)}{s(L_U + L_L) + R/(1+sCR)}.$$

Because we assumed perfect flux coupling, the first term in the numerator is zero. Multiplying through by $1 + sCR$, and converting to complex notation, we get

$$\mathbf{Z} = \frac{j\omega L_U R}{R[1-\omega^2(L_U+L_L)C] + j(L_U+L_L)\omega}.$$

At resonance, the first term in the denominator is zero, yielding the two results

$$\omega_o = \sqrt{\frac{1}{(L_U+L_L)C}} \quad \text{and} \quad \mathbf{Z}(\omega_o) = \frac{L_U}{L_U+L_L} \cdot R \equiv R_{\text{eff}}.$$

We note that the resonant frequency is also unchanged, but the impedance at resonance seen by the transistor is dropped by a factor of $L_U/(L_U + L_L)$ and

$$Q_{\text{eff}} = \frac{R_{\text{eff}}}{L\omega_o} = \frac{L_U}{L_U+L_L} \cdot \frac{R}{L\omega_o} = Q \cdot \frac{L_U}{L_U+L_L} = Q\frac{N_U^2}{N_U^2+N_L^2},$$

where N_U and N_L are the number of turns on the respective windings.

Let us imagine that $Q = 50$. By (1.18), there will be oscillation (in the absence of mismatching). If we simply apply mismatching with $N_L = 7 \cdot N_U$, Q_{eff} gets reduced down to 1, thus guaranteeing freedom from oscillation for $a < 3$. The mismatching technique has a further advantage. It actually transforms up the output resistance of the transistor as seen by the tuned circuit, thereby raising its Q. This may be verified by again redrawing the circuitry of T_2 showing the output resistance of the transistor as shown in Fig. 1.22.

We want to find the ratio $v/i = -v/i_s$, which represents the impedance of the tuned circuit. This time the transformer is assumed ideal, and the circuit equations are (ignoring C for the moment)

$$v = v_p + v_s,$$

$$i_p + v_p/r_o = -i_s,$$

$$v_p/N_p = v_s/N_s,$$

and

$$N_p i_p = N_s i_s.$$

Solving for $-v/i_s$ yields

$$\mathbf{Z} = \frac{-v}{i_s} = \left(\frac{N_s + N_p}{N_p}\right)^2 r_o.$$

This equation represents the impedance seen by v exclusive of C. Since C is right across v, its impedance can be combined in parallel with the one just calculated. However, what we wanted to see was how $\mathbf{Z}$ depends on r_o. Not surprisingly, the transistor output resistance as seen by the tuned circuit is increased by the same ratio that the load of the tuned circuit seen by the transistor is decreased. This measure typically makes r_o appear to the tuned circuit as something on the order of a megohm and means that the main factor governing Q will be losses in the windings themselves.

In most receivers, the mismatching is done by the primary winding, and a secondary winding couples signal onto the next stage. This receiver capacitively couples signal from the mixer to X_4 via L_5, which is used in a straightforward input–output manner. The effect of developing the signal across one tuned circuit (L_4) and coupling it through another (L_5) is to sharpen the bandpass characteristics and thereby improve the selectivity. Alignment of L_5 will cause a small change in the load seen by X_2, thus detuning it slightly. Such an effect is of no great significance, because good alignment practice calls for repeating the adjustments until no further improvement is obtained anyway.

1.6.5 The IF amplifiers

The IF signal is coupled by L_5 to the base of X_4, which is thus the *first IF amplifier*. This stage gets base bias from the AVC line through the secondary of L_5. The emitter of X_4 is bypassed to ground by C_{19}.

The circuitry coupling the output of the first IF amplifier (X_4) to the input of X_5 is exactly the same as that just analyzed coming from the mixer to the first IF. X_5 is the *second IF amplifier*. It is also biased by both the AVC line and the 4.7 V supply through R_{13}, making its base bias higher and thereby allowing the use of a larger emitter resistor (R_{15}), which is also bypassed by C_{21}. The output of the second IF is coupled more conventionally to the subsequent circuitry by L_8, which provides mismatching by its primary.

1.6.6 The detector

From the secondary of L_8, the signal is applied to a conventional diode envelope detector comprised of X_9 and C_{22}. R_{16} limits the loading on the envelope detector when the volume control is set near maximum. The signal present at its low end will have both AC and DC components as may be seen by referring to Fig. 1.15b. This signal is low-pass filtered by R_{17}, R_{18}, R_{11}, C_{17}, and C_3 to extract the DC component, which constitutes the AVC for the first and second IF amplifiers and (after further filtering by R_5 and C_{11}) for the RF amplifier. Having three stages with AVC should afford superlative AVC action.

1.6.7 The audio amplifiers

The AC component at the output of the envelope detector receives additional high-pass filtering by means of the volume control and C_{23}, and is then coupled through the blocking capacitor C_5 to the base of the *audio amplifier*, X_6. This stage features fixed bias via R_{19}. The use of fixed bias on a stage that handles a large voltage swing is the only feature of this entire design that is not first rate. The collector load for X_6 is an audio transformer primary (T_1).

The secondary of T_1 is center-tapped to provide drive to the class B push-pull output transistors, X_7 and X_8. These transistors are very slightly biased into conduction by a voltage derived from R_{21} and X_{10} and applied through the secondary of T_1. A diode (X_{10}) is used in the divider, rather than another resistor, because the diode drop will depend on temperature in the same way as the V_{BE} drops of X_7 and X_8. This keeps their quiescent collector currents at a low and more-or-less constant value even though temperature changes.

The push-pull stage is configured as CE, providing both voltage and current gain. R_{22} provides a small amount of emitter degeneration, improving both gain and bias stability somewhat, as well as improving linearity. C_{24} shunts out some of the high frequencies above the audio range. The combination of C_{25} and R_1, called a *losser tone control*, is analyzed further in Section 4.1.

1.7 Variations

Any number of variations upon this design may be found in discrete transistor radio designs. Most have the reduction of cost as their purpose. Here is a brief discussion of a few of them.

(1) The RF amplifier stage is usually absent, yet the number of stations that can be received when one is used is markedly greater. Most decent automobile radios do have an RF stage. If it is absent, the antenna couples to the mixer in exactly the same way as it couples to the RF amplifier in our circuit.

(2) In most receivers, the mixer and oscillator functions are performed by a single transistor, which is then called the *converter stage*. Figure 1.23 shows a circuit of this type. The components C_{1A}, C_{1B}, C_2, L_1, C_4, C_5, R_7 and T_1 all have the same function as the corresponding parts in the Motorola receiver. Q_1 receives base bias from the R_5–R_6 divider through the secondary of L_1.

Oscillation is output from the #2 winding of L_2, and input to the base of Q_1 through C_3 and the secondary of L_1. C_3 is a .01 μF capacitor, so its impedance over the broadcast band is 29 Ω or less. It is effectively an AC short and serves the purpose of preventing the shorting out of the base bias by winding #2 of L_2.

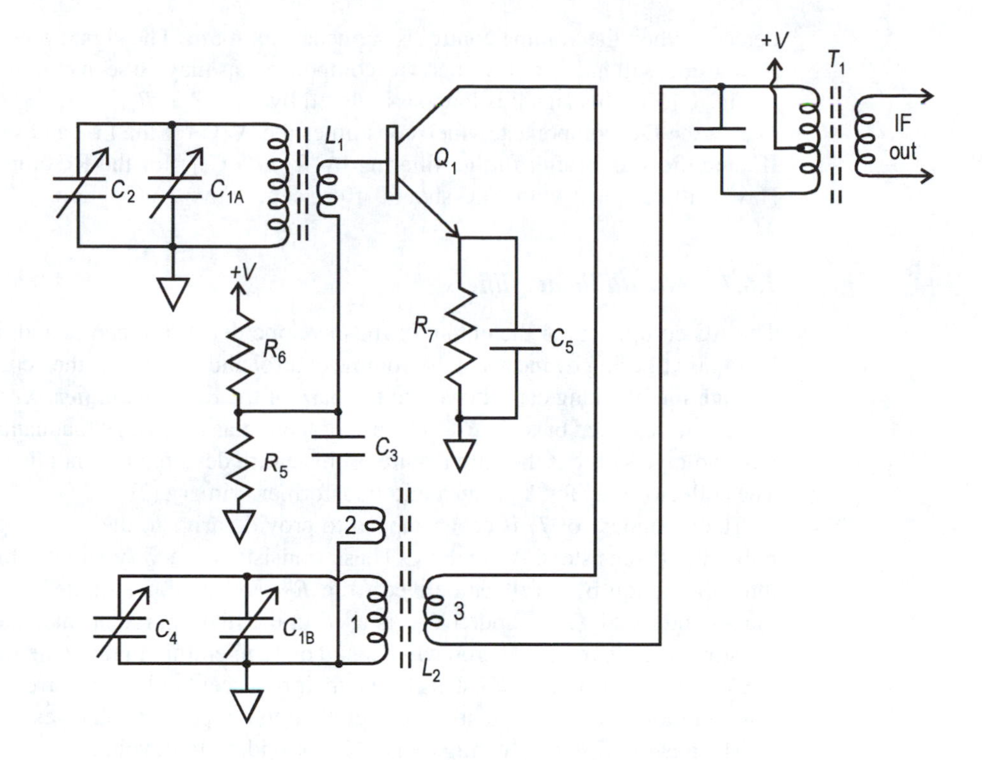

Fig. 1.23: A converter stage, which incorporates the functions of both the local oscillator and mixer in a superhet receiver.

Note that this arrangement places the local oscillator output from winding #2 of L_2 in series with the RF signal, which is found on the secondary of L_1. These signals thus add, and an analysis such as that performed for the Motorola receiver mixer is applicable. The collector current is sampled by winding #3 of L_2 and is of the proper phase to sustain oscillation. Although this circuit is satisfactory, it does introduce a little more distortion than when the oscillator and mixing are done separately. This should be readily understandable. Amplitude stabilization in an oscillator is accomplished when the peaks of oscillation drive the transistor slightly into saturation. If the same transistor is to act as a linear amplifier, it can be seen that there may be significant parameter variation.

Exercise 1.5 The oscillator function of this stage can be analyzed from this feedback circuit:

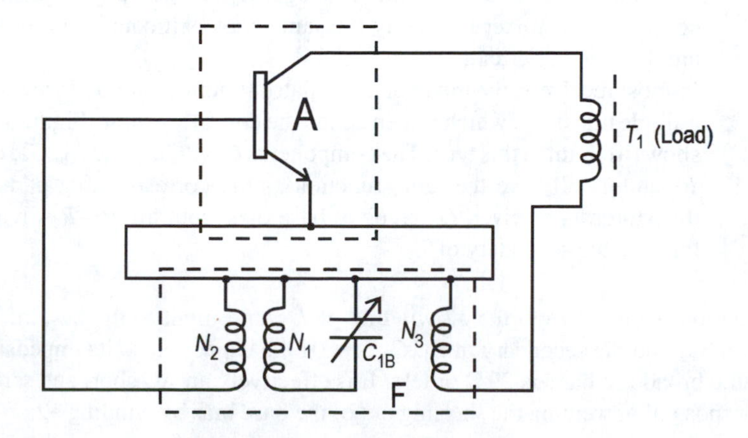

The transistor should be analyzed as a transadmittance amplifier ($A = i_{out}/v_{in}$) and with the loading of A on F and F on A neglected. Analysis can be further simplified by assuming the following:

(1) The same flux couples all three coils.
(2) Coil #3 has pure current drive.
(3) Coil #2 has no current through it.

The latter two assumptions are the consequence of the no-loading assumption.

In addition, it will be necessary to introduce a resistance in series with coil #l to avoid having a lossless circuit. This assumed series R will govern the Q of the tuned circuit.

Find:

(a) the minimum gain for oscillation to occur,
(b) the frequency of oscillation.

(3) The more conventional use of IF transformers to couple IF amplifiers has already been discussed and illustrated by the connection between X_5 and L_8 in the Motorola receiver.
(4) Most inexpensive receivers provide AVC only to the first IF amplifier, resulting in a much smaller AVC range. Automobile receivers will generally feed it to both IF stages because of the greater RF signal variation they see.
(5) It is not uncommon to have two stages of audio amplification prior to the audio output stage. If this is the case, they are usually called the audio preamplifier and the audio driver stage. This may give more audio output, but it does not improve the sensitivity, which can only be done at the "front end." Tuning to a weak station with extra audio gain just gets you louder noise along with your louder signal.
(6) There may still be a few receivers in which the audio output (power amplifier) stage consists of a single transistor in class A. Although class A amplifiers are markedly inferior to class B types in efficiency and distortion, using them reduces the parts count. In particular, it allows us to eliminate the driver transformer and, in some cases, even the output transformer. Push-pull output stages featuring paralleled transistors or Darlingtons may also be encountered in some older designs where high audio output power was a design priority.

1.8 Integrated circuits for AM

1.8.1 A TRF IC

The first AM radio IC we shall look at is an older and somewhat obscure TRF chip, the Ferranti ZN414. The simplicity of this chip is demonstrated by the fact that it only has three leads. Of necessity, when the receiver is this simple, not only must the pins be multipurpose, but certain performance concessions must be made.

Figure 1.24 shows the circuit of a simple TRF receiver based on this chip. Unlike commercial TRF receivers made in the time of its dominance, this circuit does not have separate tuning for each stage. However, since the entire signal processed through the IC is at RF frequencies until it reaches the detector, and since there is an RF tuning operation at the input, it is properly classified as a TRF receiver.

The parallel LC combination tunes the receiver. L is a loopstick and thus serves as the antenna as well. At the middle of the AM band, the .01 μF

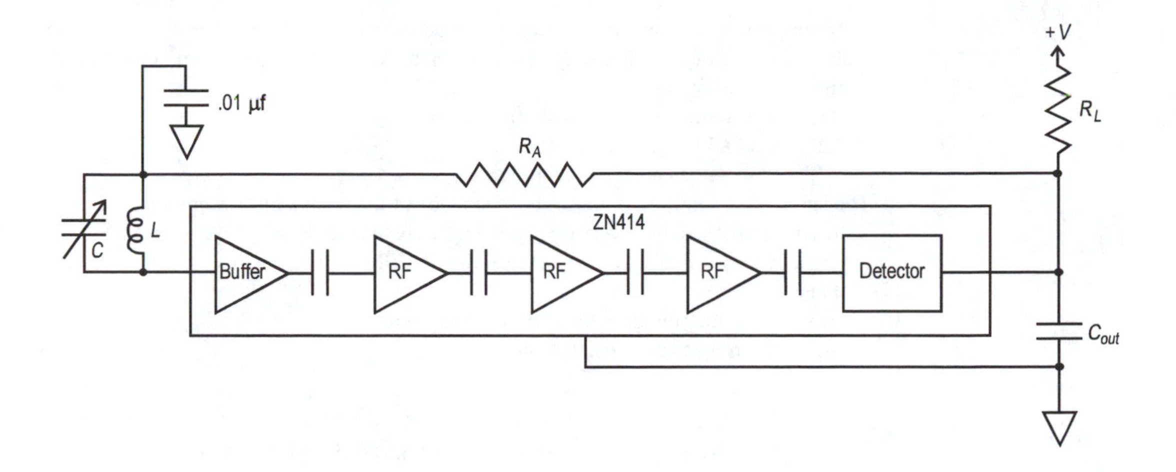

Fig. 1.24: A simple TRF receiver designed around the Ferranti ZN414 IC.

capacitor has an impedance of about 16 Ω. Since this is much less than the resonant impedance of the tuned circuit, it may be considered an AC short, and thus it AC grounds one end of the tuned circuit. The other end feeds the input of the buffer amplifier. This stage has an input impedance of >2 MΩ to allow the resonant circuit to have a high Q. The signal then goes through three stages of RF amplification. These stages are not narrow-band tuned, which costs some gain and degrades the S/N ratio. The amplified RF is then detected and applied to the load R_L, which could be an earphone of a few kilohms impedance.

The resistor R_A and the .01 μF capacitor form a low-pass filter (LPF) which extracts the DC component of the output from the audio and applies it to the input through L. This is the AVC voltage. It is found to decrease monotonically with increasing input signal amplitude. As this DC voltage component of the output voltage decreases, the AC amplitude first rises sharply and then falls more slowly. Another consequence of the circuit's simplicity is that the bandwidth increases (and thus selectivity decreases) with increasing input signal level.

The output capacitance is selected to give a rolloff at the AM station bandwidth in conjunction with R_L, thereby reducing noise and oscillatory tendencies.

Used by itself, the chip is designed to operate on a single 1.34 V mercury cell. When used with a larger supply voltage, a larger value of R_L can be used to provide a greater AVC range. In fact, since the usable frequency range of the chip extends between 150 kHz and 3 MHz it is also usable as an AM IF amplifier. A receiver circuit published by the manufacturer shows a superhet with a combined mixer–oscillator stage such as that shown in Fig. 1.23 followed by the ZN414 and an audio amplifier. The output of this circuit is adequate to drive a speaker. Coupling between the mixer output and the ZN414 is by means of a ceramic resonator, rather than an IF transformer. Ceramic resonators are small inexpensive devices that operate on the piezoelectric effect in much the same way as crystals but with Q values more like those of IF transformers. They may be used in place of or in conjunction with IF transformers. Their use is growing, because of their smaller size and good performance.

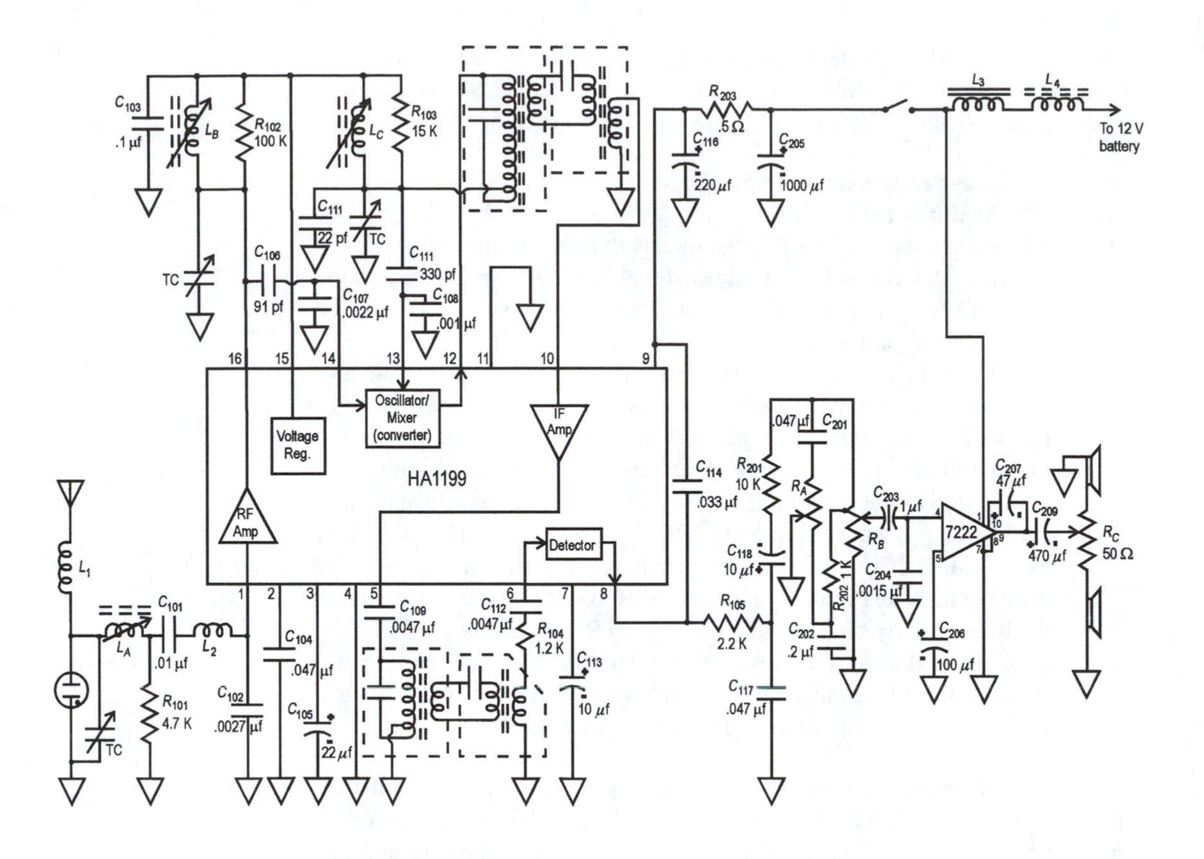

Fig. 1.25: Clarion AM automobile receiver designed around the HA1199 IC.

1.8.2 *Superhet ICs*

As far as portable radios are concerned, about the only niches left for "AM only" receivers are novelty radios and very inexpensive "throwaway" radios. These radios are so called because they are not designed to be repaired when they fail, but replaced. Since very low cost is a prime criterion, it is mandatory that these receivers be as integrated as possible. There are a number of AM radio ICs available. They are quite similar in many respects. The chief differences are that some have an RF amplifier and others do not and some include a detector whereas others do not. All include the mixer, local oscillator, and IF stages. All require the use of IF interstage transformers or ceramic resonators, and all have AVC. Since we have learned that AVC is derived from the detector, and since several chips have no detector, one might wonder how they generate AVC. The answer is that they include one or more dedicated diodes on the chip whose only function is to serve as AVC rectifiers. Audio detection is performed by a separate diode off chip in these cases.

One chip actually has a stage of audio preamplification following the detector (RCA CA 3088), but it has no RF stage. One of the better chips is the HA1199. The chip contains both an RF stage and a detector, as well as all the other circuit blocks common to all AM superhet chips. This chip, and others using an RF amplifier stage, are used primarily for automobile radios. Figure 1.25 shows the schematic of a Clarion RE-158A automobile radio using this chip.

The simplicity of the circuit relative to a discrete design such as the Motorola portable (Fig. 1.17) should be immediately obvious. Although "top of the line" auto radios are all AM–FM stereo now, AM receivers of reasonable quality may still be found as original equipment in some less expensive cars. This receiver was designed for that application. Here are a few of its specifications:

(1) The image rejection ratio >50 dB.
(2) The AVC range is >45 dB.
(3) The selectivity is >20 dB. This means that the received signal for a channel adjacent to the one tuned for will be attenuated by at least a factor of 100 in power over what would be received if the receiver were tuned to that station.
(4) The IF frequency is 262.5 kHz – a value quite different from the 455 kHz value we have discussed previously. Let us consider why this value is common in car radios. There are several good reasons for making the IF frequency as high as possible, but none of them is as compelling as the fact that it improves image rejection, since the image station is always twice the IF frequency above the station tuned to. Yet, recall from our original discussion of image signals that the best measure that can be taken against them is to improve the selectivity of the RF circuit by such means as adding an RF stage. The HA1199 has an RF stage that does improve image rejection. This makes it possible to lower the IF frequency to achieve whatever benefits that measure will yield and still retain a respectable image rejection of >50 dB.

To see the advantage of the lower IF frequency, let us consider a situation where we are trying to receive a signal at 1,550 kHz, and there is an interfering signal at 1,530 kHz. For a 455 kHz IF, the undesired signal will come through the heterodyning process at 475 kHz. The 20 kHz deviation relative to the 455 kHz center frequency represents a 4.4% change. If the IF were at 262.5 kHz, the same 20 kHz deviation represents a 7.6% change. If the Q of the IF transformers were the same in both cases, the unwanted station would be further down from the peak of the response curve for the lower IF frequency. This effect would provide about 5 dB of additional selectivity if the Q of both tuned circuits were 50 or more.

Let us now refer again to Fig. 1.25 and analyze the circuitry. To begin, we note that only pins 2, 3, and 7 of the IC are not identified as to function. All three pins are connections for bypass capacitors for various parts of the circuit. As might be suspected from the small value of C_{104}, this pin is the bypass connection for the RF amplifier. C_{113} is the first filter capacitor for AVC voltage, and C_{105} is a second filter for the AVC path that controls the RF amplifier. We shall soon have more to say about the unconventional AVC circuitry on this chip.

It may also be noted that, although the designations are not called out on the schematic, on the chip data sheets, pins 4 and 11 are labeled "Low-level ground" and "High-level ground" respectively. The former is the ground connection for the RF and converter stages, whereas the latter is for the IF and detector stages. The grounds are brought out separately, because in some cases it may be necessary to provide additional *decoupling* between these two circuit groups. Suppose the connection between pin 11 and supply ground has a resistance of 1 mΩ and that a signal current of 10 mA flows through it. The voltage drop between pin 11 and supply ground is then 10 μV. But this is comparable to the input signal level from the antenna, and if pin 4 (RF ground) were tied to pin 11, the RF amplifier would be responding to the difference between supply ground and the antenna signal. Since the difference voltage is derived from the IF amplifier and detector, what we thus have is an unwelcome feedback connection from the output stages back to the input ones. In this receiver, the pins are tied together on the schematic. In the actual receiver, they are returned to a common ground point, so the high-level ground currents do not flow out of the low-level ground pin. The heavy ground is achieved by the use of a printed circuit board and a heavy connection from it to the radio chassis. While on the subject of ground, it should be noted that even though the schematic shows many discrete connections to ground, this is done to somewhat unclutter the schematic. In the circuit itself, great care is taken to design a ground path on the circuit board that will hold all of these points at as nearly the same potential as possible.

1.8.2.1 RF circuitry

The signal from the antenna comes through L_1 (whose value is not given) and passes through an *RLC* network to the input of the RF amplifier. The inductance L_A in this network is shown as being variable. This radio (as was common in car radios before the advent of quartz-synthesized tuning) is tuned by ganged variable inductors. This facilitated push-button tuning. In addition to the variable inductor in the input circuit, there is also one in the RF amplifier output circuit (L_B) and one in the oscillator circuit (L_C). The input circuit is certainly not a conventional parallel *LC*, but analysis shows that it has two resonant frequencies. One is that given by L_2 with C_{101} and C_{102} in series. Since the value of L_2 is not given either, we assume it is non-critical and is hence not a tuning element. Thus this resonant frequency is assumed to be outside the AM broadcast band. The other resonant frequency of the input network is that given by the capacitor labeled *TC* and the parallel combination of L_1 and L_A. This frequency is certainly the one we want because it involves both L_A and *TC*, which is a trimmer capacitor to be adjusted during alignment.

The bulb shown in the input circuit is a gas-discharge bulb, which is normally an open circuit, but which will ionize and clamp the voltage across it if the antenna should pick up a static voltage large enough to damage the IC. In this respect it acts like a Zener diode, but it is usually less expensive and more tolerant of high surge currents.

The chip's RF amplifier, stripped of its bias circuitry, is shown in Fig. 1.26 as it appears for AC. Here $\mathbf{Z}_L$ represents the load impedance comprised of L_B, R_{102}, and a trimmer C plus the effective capacitance of the voltage divider $C_{106} - C_{107}$. We recognize this circuit as a differential amplifier but of an unusual sort. The inputs to the differential pair are DC only. One DC value is derived from the internal voltage regulator and is fixed. The other value varies with the AVC voltage. The signal is applied to the current source base!

Exercise 1.6 Show that if $g_m \gg 1/r_\pi$ for all three transistors then $\mathbf{A}_v = -g_1\mathbf{Z}_L$, where g_1 is the transconductance of Q_1 and $A_v = v_{out}/v_{in}$.

1.8.2.2 The AVC circuitry

As the AVC voltage drops, I_{C1} drops, g_1 drops, and A_v drops. We shall now show that a larger signal will cause the AVC voltage to drop. The essential AVC circuit internal to the chip is shown in Fig. 1.27.

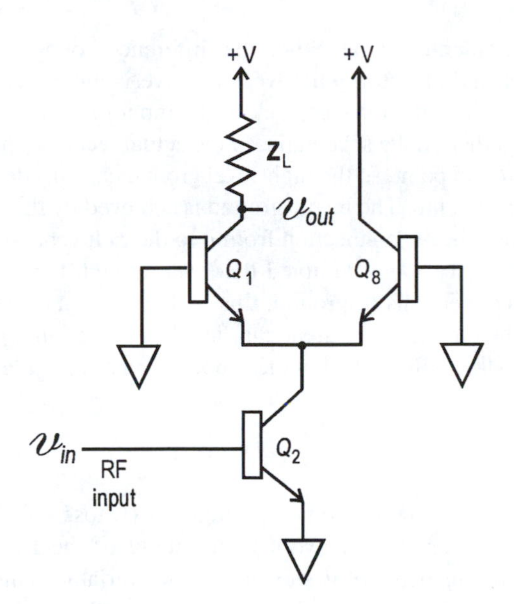

Fig. 1.26: RF amplifier section of the HA1199 IC.

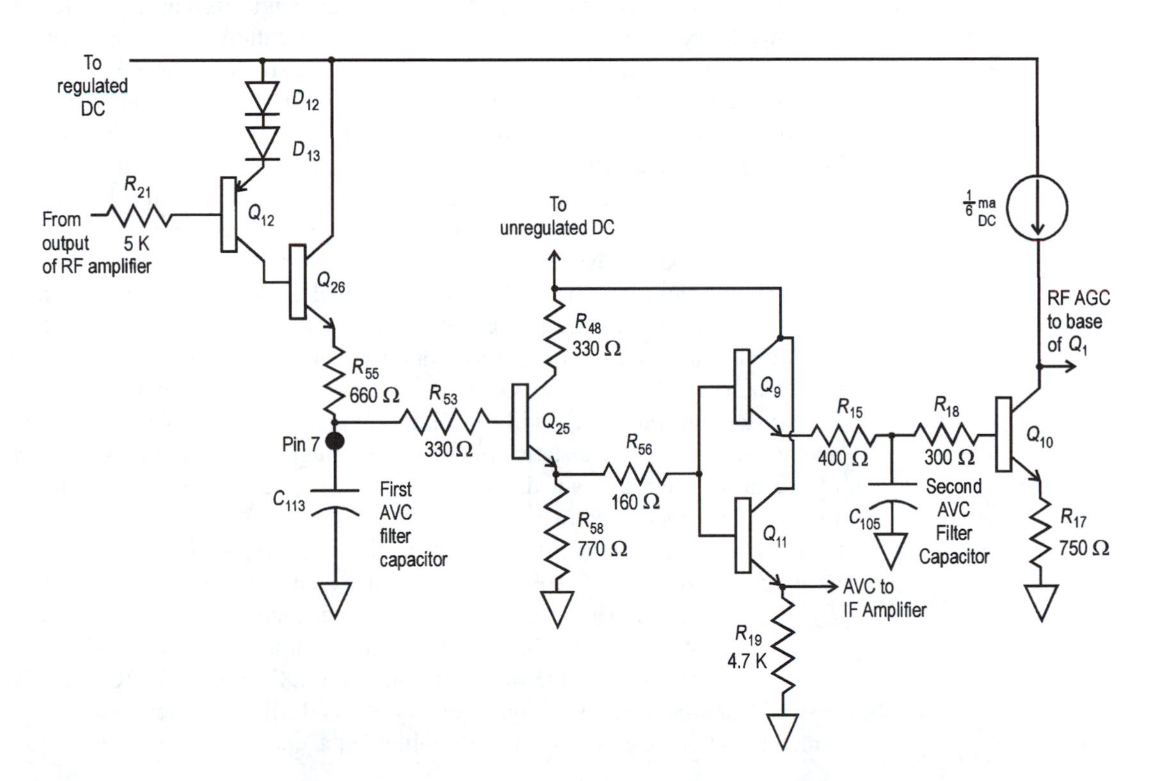

Fig. 1.27: AVC circuitry of the HA1199 IC.

This circuit is fairly self-explanatory. It is not conventional inasmuch as the AVC voltage is not derived from the detector. As the output of the RF amplifier swings to its negative half cycle, current flows from the regulated DC supply through D_{12} and D_{13} into the emitter of Q_{12}. This current will flow in pulses, whose amplitude will increase with larger signal levels. These pulses are current-amplified by Q_{26} and filtered by R_{55} and C_{113} to extract the DC component. This DC is buffered by another emitter follower, Q_{25},

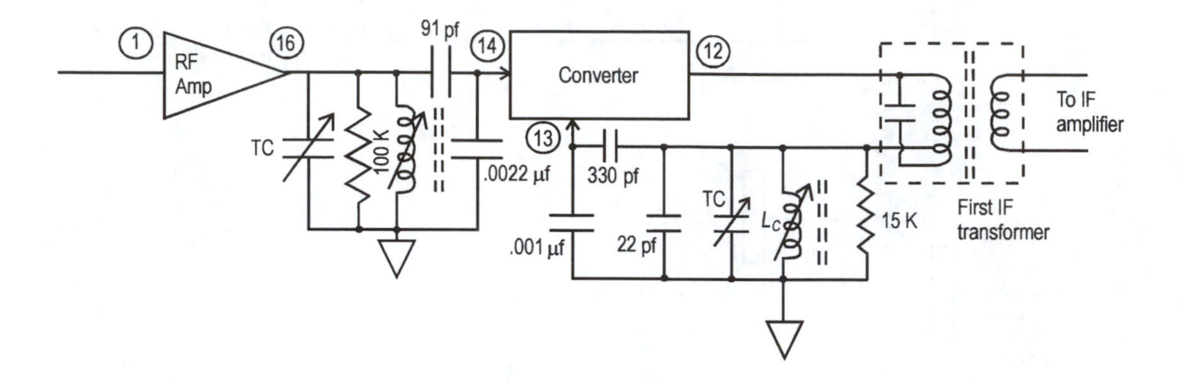

Fig. 1.28: Converter circuitry of the HA1199.

and then applied to Q_9 and Q_{11}, which provide the RF and IF AVC voltages, respectively. The RF AVC output from Q_9 goes through one more section of low-pass filtering (C_{105} and R_{15}) and is then applied to the base of Q_{10}. As this voltage increases due to the increasing signal at the RF amplifier output, Q_{10} conducts more heavily, presenting a higher-conductance path to the current source and thus lowering the base bias to the RF amplifier, Q_1. This will lower the gain of the RF stage as we have already seen.

The output of the RF amplifier is tuned by $\mathbf{Z}_L$ as it was previously defined. The tops of L_B and R_{102} as drawn are returned to the regulator output. C_{103} assures that this will be a good AC ground.

1.8.2.3 The converter circuitry

From the RF amplifier, the signal goes to the converter via the capacitive voltage divider, C_{106}–C_{107}. The AC circuit of the converter has nearly the same topology as that of the RF amplifier (Fig. 1.25) except that:

(1) The input to the current–source transistor is the amplified RF from pin 16 of the chip.
(2) The local oscillator input is applied where the AVC was in the previous circuit.
(3) The output is taken from the opposite transistor of the differential pair.

It is a fairly straightforward extension of the previous exercise to show that the output of this stage will, to the first approximation, be the sum of these two inputs. If the amplitudes are large enough, then nonlinear mixing occurs and the sum and difference frequencies (among others) are generated.

The chip does not have a separate oscillator. It is possible to use an external oscillator and feed its output to pin 13, but in the Clarion receiver, a different technique is used. This is shown in Fig. 1.28, which is an AC equivalent schematic.

The current at the converter output contains a component at the IF frequency, which is selected as usual by the IF transformer primary and its internal resonating capacitor. But the converter output current also contains a component at the local-oscillator frequency. This current, having a frequency well removed from that of the IF, passes through the IF transformer with very little voltage drop but then passes into the tuned circuit that feeds pin 13 and develops a

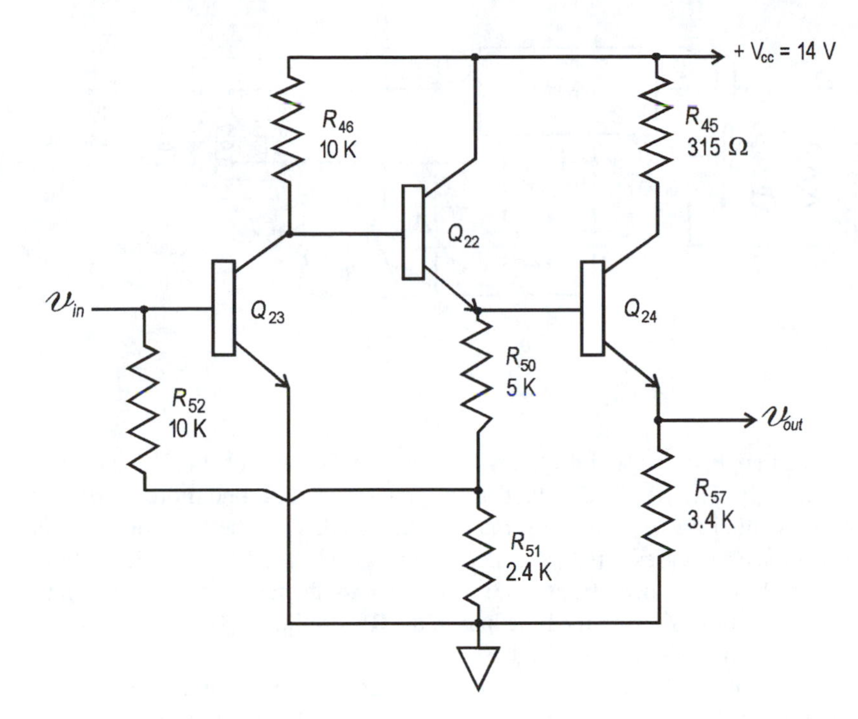

Fig. 1.29: Detector circuitry of the HA1199.

significant voltage there at the oscillation frequency. This is the feedback that sustains the oscillation. The secondary of the IF transformer shown in Fig. 1.28 is capacitively coupled to the input of another one, whose output goes to the input of the chip's IF amplifier. This configuration is somewhat unusual in an AM radio. It is probably used so that the transformers can be *stagger-tuned*. This means that one transformer is tuned slightly below the IF frequency, and the other is tuned an equal amount above it. This has the desirable effect of flattening the top and "steepening" the sides of the IF response if the offset is properly chosen.

Exercise 1.7 On the same set of axes, plot the frequency response of two cascaded tuned circuits each with a Q of 50

(a) when they are both tuned to the IF frequency and

(b) when they are stagger-tuned to IF frequency $\times$ $(1 \pm .01)$.

1.8.2.4 The IF amplifier and detector

The IF amplifier is also a differential amplifier. Its output passes through an emitter follower and then proceeds to pin 5. The IF AVC is used to control the base bias on the current–source transistor by means of another transistor that functions exactly analogously to Q_{10} in Fig. 1.27. The output of the IF amplifier is capacitively coupled to another pair of linked IF transformers, whose output is capacitively coupled to the detector.

The detector is a direct-coupled array of three transistors with the latter two being in the Darlington configuration. The circuit is shown in Fig. 1.29. As can be seen, there is negative DC feedback from the emitter circuit of Q_{22} back to the base of Q_{23} to stabilize the bias. The quiescent value of base voltage on

Q_{23} as determined by SPICE® is .7753 volts. If an external voltage of .7801 volts is applied to V_{IN}, Q_{24} completely cuts off. Thus less than a 5 mV swing on the base of Q_{23} will cut off Q_{24}. For all practical purposes, the stage may be thought of as cut off at the quiescent point. Positive-going swings will keep it in cutoff whereas negative-going input swings will result in positive output swings that are an amplified version of the input. This is the behavior required of an AM detector, but with gain added. C_{114} is the detector's filter capacitor and C_{117}–R_{105} constitutes another section of low-pass filtering to remove ripple from the detector's audio output. The DC component of the detector output is removed by C_{118}, and the combination of R_{201}, C_{201}, R_{202}, C_{202}, and R_A form a tone-control circuit. Note that the volume control, R_B, has a tap on it in this type of circuit. C_{203} is a non-polarized electrolytic, which serves as a blocking capacitor so as not to give pin 4 of the 7222 a DC path to ground. This chip is, of course, the audio amplifier. Its output goes through C_{209} to keep DC off of the output. The potentiometer, R_C, is called a *fader* control and allows the user to continuously vary the portion of the output signal applied to each of the two speakers (front and back of the car).

1.8.2.5 Power supply

It might be appropriate in concluding our analysis of this receiver to say a few words about its power supply. Ultimately the supply is, of course, the car battery/alternator system, but the electrical environment in a car is extremely noisy. Thus the incoming DC to the radio passes first through an *LC* filter (L_4, L_3, and C_{205}). L_4 has a ferrite core and L_3 has a laminated iron core. L_4 will be of most use for short-duration variations. The DC out of the *LC* filter also goes through an *RC* filter (R_{203}–C_{116}) to further clean it up before applying it to the V_{cc} pin of the radio chip. Even though the resistor is only .5 Ω, this section will give additional noise suppression for frequency components above 1.5 kHz.

Then, as already noted, the chip has an internal voltage regulator to supply the more critical RF and converter stages with bias, providing further protection from noise.

1.9 AM stereo

In the face of a continuously declining share of the listener market, AM stations have looked at AM stereo as a possible means of increasing listener appeal. Several competing systems were presented to the FCC by manufacturers who wanted theirs to be accepted as the standard. The FCC, faced with a dearth of technically qualified people at the time, chose the coward's way out and opted for what they called a "Market Decision."

This decision (or non-decision!) seriously damaged the future of AM stereo. Receiver and chip manufacturers were reluctant to commit themselves to designs that might be useless if another system were ultimately to win out. Some receivers were introduced that claimed to decode all three of the front-running systems, but which actually decoded none of them properly.

The three leading contenders were systems from Harris, Motorola, and Khan-Hazeltine. The Motorola system seemed from the beginning to be in the best position to prevail for the following reasons:

(1) It is fully compatible with existing AM receivers with envelope detectors. The envelope detector will output $L + R$. (The two stereo signals are called $L(t)$ and $R(t)$ or L and R to denote the left and right channels, respectively.)

(2) Motorola is a vertically integrated company. Not only do they manufacture the broadcast equipment, but they produce the MC13020P chip, which properly decodes the AM stereo signal produced by their broadcast system. It accepts the IF signal as its input and outputs the L and R audio signals. In addition to the chip itself, a typical application will require 11 resistors, 15 capacitors, and a 3.6 MHz ceramic resonator. The chip also includes an LED stereo indicator driver. After discussing the system itself, we'll take a quick look at the chip's block diagram.

(3) Early in the out-working of the "market decision," the Motorola system was chosen by GM-Delco for incorporation into their automobile receivers. This is an enormous market and it made the Motorola system more attractive to AM stations that were looking at converting to stereo.

(4) The Motorola system is probably the most tolerant of all of the systems to nonideal reception conditions if all things are considered. The Harris system had superior channel separation and gave the most distortion-free stereo reception but gave an average of 2.5% distortion for mono signals. The Harris system could give in excess of 6 dB more noise if the L and R channels were not nearly equal.

This combination of advantages proved to be adequate to give Motorola the victory, but it was a Pyrrhic victory. The battle between the Motorola and Khan-Hazeltine systems went on so long that although Motorola won the battle, they may have lost the war. Stations more or less lost interest while the battle was going on. Today, the few stations that broadcast in AM stereo use the Motorola system. The system is still supported by the National Association of Broadcasters, however, and the FCC has finally brought out regulations dealing with AM stereo transmissions in the format Motorola proposed.

The basic modulation scheme used for the Motorola system is a clever variation of what is known as Quadrature Amplitude Modulation (QAM). QAM "enables two DSB-SC modulated waves (resulting from the application of two independent message signals) to occupy the same transmission bandwidth, and yet it allows for the separation of the two message signals at the receiver output" (Haykin, 1983). This scheme sounds like it was made to order for AM stereo, except that one of the waves is DSB-SC and the other is DSB with a carrier. The phasor diagram of a QAM system is shown in Fig. 1.30.

The diagram corresponds to the time-domain signal

$$\begin{aligned} v(t) &= (1 + S)\cos\omega_c t + D\sin\omega_c t \\ &= \sqrt{(1+S)^2 + D^2}\cdot\cos\left(\omega_c t - \arctan\frac{D}{1+S}\right). \end{aligned}$$

When this signal is applied to an envelope detector, the output is $\sqrt{(1+S)^2+D^2}$, whereas we would like to recover $1 + S$. The D term represents distortion in

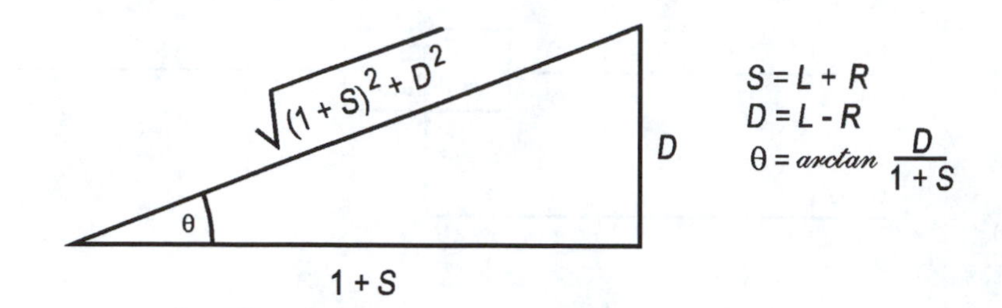

Fig. 1.30: Phasor diagram of CQUAM® AM stereo system.

the recovered monophonic signal. Conceptually, we could multiply both D and $1 + S$ by $\cos\theta$. It would still be true that $\theta = \arctan D/(1 + S)$, but now we would have

$$v(t) = \sqrt{(1+S)^2 + D^2} \cdot \cos\theta \cdot \cos\left(\omega_c t - \arctan\frac{D}{1+S}\right).$$

However, from the phasor diagram,

$$\cos\theta = \frac{1+S}{\sqrt{(1+S)^2 + D^2}},$$

which makes

$$v(t) = (1+S) \cdot \cos\left(\omega_c t - \arctan\frac{D}{1+S}\right).$$

This signal will be properly recovered by an envelope detector giving the desired output of $1 + S$. Motorola calls this system CQUAM® (Compatible Quadrature Amplitude Modulation).

Figure 1.31a shows a monophonic AM transmitter setup, and Fig. 1.31b shows the corresponding CQUAM® apparatus. The QAM signal is generated conventionally. After limiting, the amplitude variation is removed, but the $L - R$ information is still present in the form of phase modulation. The final modulation step (which is completely analogous to monophonic AM) results in the generation of CQUAM®. Two additional refinements are found in a real system:

(1) To maintain reasonable channel separation over the audio range, phase equalizers must be inserted in series with the inputs to the two balanced modulators of Fig. 1.31b. These will compensate for unequal propagation times through the audio (upper) and RF (lower) paths to the final modulator.

(2) A 25 Hz stereo pilot tone is also summed with D before it enters the balanced modulator. If D were zero, the 25 Hz signal would provide an output with 4% modulation of the carrier.

The block diagram of the MC13020P decoder chip is shown in Fig. 1.32. As shown in Fig. 1.31b, the CQUAM® output has the form $(1 + S) \cdot \cos(\omega_c t - \theta)$. This, then, is the mathematical form of the input applied to pin 3 of the decoder chip, where it is understood that the "carrier" frequency is 455 kHz by virtue of the IF translation. The heart of the circuitry is in the upper left-hand corner

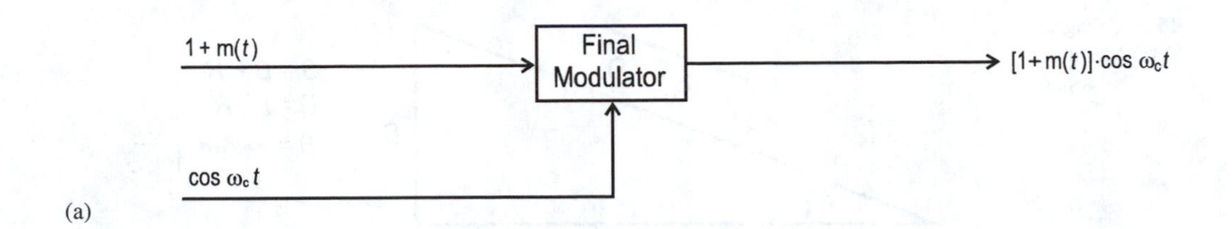

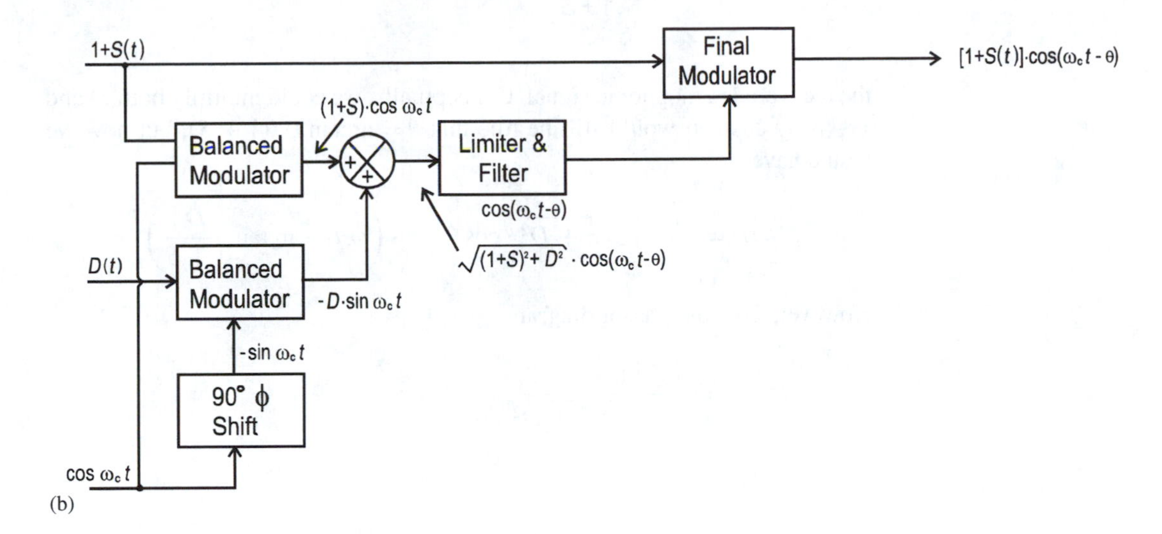

Fig. 1.31: Block diagrams of AM transmitters. (a) Monophonic, (b) CQUAM® stereo.

of the figure. The input is envelope detected to recover $1 + S$. The R and C connected to pin 2 comprise an LPF for the envelope detector. The input also goes to a variable-gain amplifier whose gain can be written as

$$A_v = 2V_c,$$

where V_c is the gain-controlling voltage, which is shown as $1/\cos\theta$. Thus

$$A_v = 2/\cos\theta.$$

Since $\cos\theta$ varies at an audio rate, it changes slowly compared to the IF frequency being amplified; therefore linear amplification is readily achieved.

The output of the variable-gain stage is thus equal to the CQUAM® input multiplied by this gain:

$$v_{\text{out}} = \frac{2(1+S)\cdot\cos(\omega_c t - \theta)}{\cos\theta}.$$

This voltage is input to the I (in-phase) detector along with a regenerated carrier from the VCO and ÷8 counter. The VCO is run at a frequency so high that it is above the broadcast band for any of the IF frequencies found in AM receivers. If, for example, the ÷8 circuitry were not included and the VCO operated at 455 kHz, its square wave output would have a large harmonic at 3 × 455 kHz = 1,365 kHz. But this is right in the heart of the broadcast band, and radiation

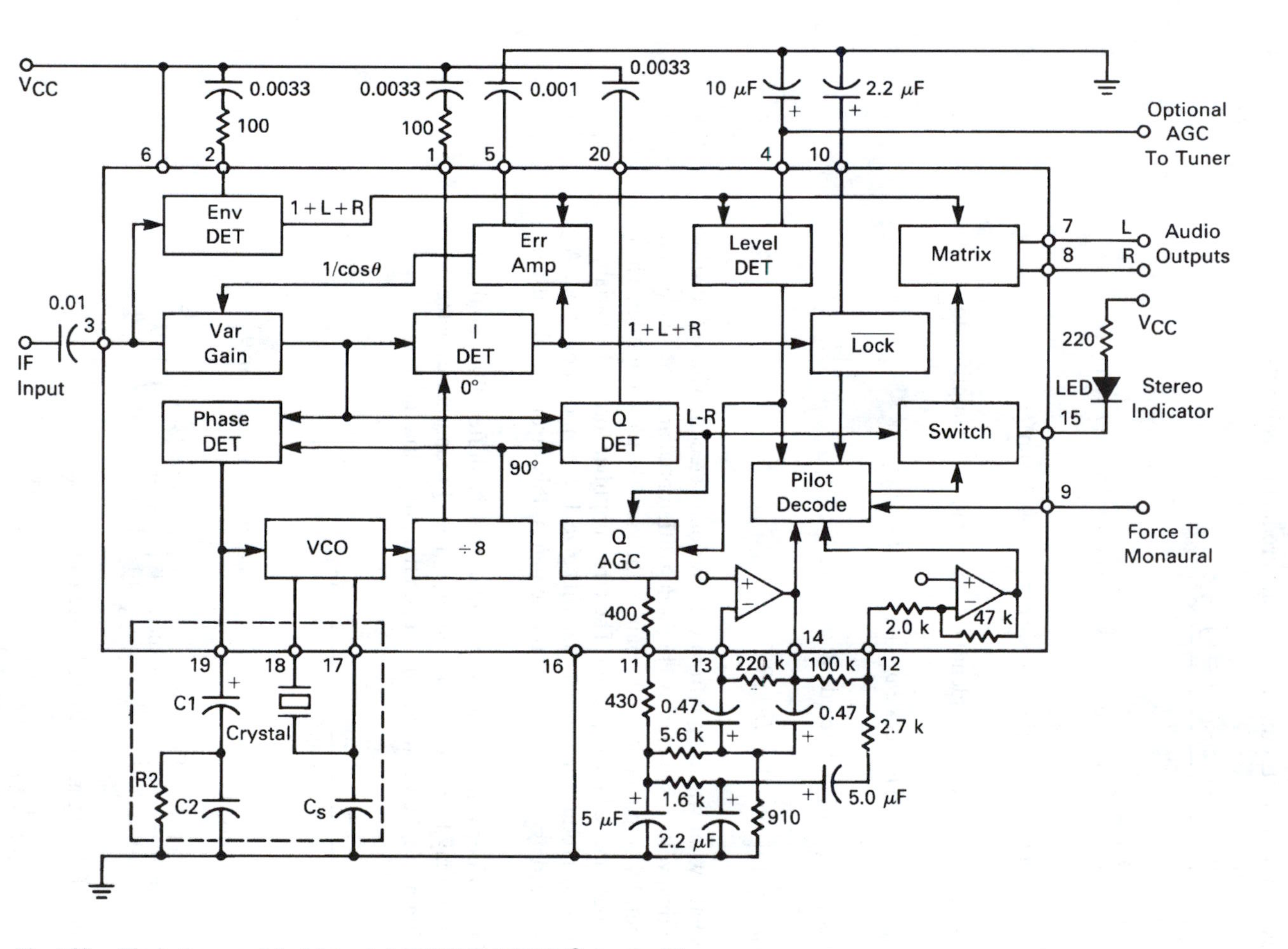

Fig. 1.32: Block diagram of the Motorola MC13020P CQUAM® decoder IC.

from the VCO might not be tolerable. Although the output from the VCO fed to the I detector is a square wave, its fundamental is a cosine. Higher-order terms will give high-frequency output components, which will be removed by the LPF connected to pin 1. Thus we consider the wave a cosine and model the I detector as a multiplier. Its output is then the product of its inputs:

$$\begin{aligned} v_{\text{out}_\text{I}} &= \frac{2(1+S)\cdot\cos(\omega_c t-\theta)}{\cos\theta}\times\cos\omega_c t \\ &= \frac{2(1+S)}{\cos\theta}[\cos^2\omega_c t\cdot\cos\theta+\sin\omega_c t\cdot\cos\omega_c t\cdot\sin\theta] \\ &= 2(1+S)\big[\cos^2\omega_c t+\sin\omega_c t\cdot\cos\omega_c t\cdot\tan\theta\big]. \end{aligned}$$

Using well known trigonometric identities, this expression simplifies to

$$v_{\text{out}_\text{I}} = (1+S)[1+\cos 2\omega_c t+\sin 2\omega_c t\cdot\tan\theta].$$

The terms at $2\omega_c$ will be filtered out, leaving

$$v_{\text{out}_\text{I}} = 1+S = 1+L+R,$$

which is exactly what was output from the envelope detector. These two supposedly identical voltages are fed to a high-gain differential amplifier called the error amplifier. Its output controls the gain of the variable-gain amplifier. Thus the loop has been closed conceptually. The error amplifier forces the variable-gain amplifier to produce an output that will give $1+S$ upon detection. As indicated on the diagram and by the mathematics, this error signal must be $1/\cos\theta$.

The VCO divider also produces an output that is shifted 90° relative to the carrier. This may be represented as $\sin\omega_c t$. The Q (quadrature) detector is exactly analogous to the I detector in its operation. Its output is

$$\begin{aligned} v_{\text{out}_\text{Q}} &= \frac{2(1+S)\cdot\cos(\omega_c t-\theta)}{\cos\theta}\cdot\sin\omega_c t \\ &= \frac{2(1+S)}{\cos\theta}\big[\cos\omega_c t\cdot\sin\omega_c t\cdot\cos\theta+\sin^2\omega_c t\cdot\sin\theta\big] \\ &= \frac{2(1+S)}{\cos\theta}\left[\frac{\sin 2\omega_c t}{2}\cdot\cos\theta+\left(\frac{1}{2}-\frac{\cos 2\omega_c t}{2}\right)\sin\theta\right]. \end{aligned}$$

Again low-pass filtering at pin 20 will leave us with

$$v_{\text{out}_\text{Q}} = \frac{2(1+S)}{\cos\theta}\cdot\left(\frac{1}{2}\sin\theta\right) = (1+S)\tan\theta$$

but, as shown in deriving the CQUAM® signal,

$$\tan\theta = D/(1+S)$$

so

$$v_{\text{out}_Q} = D = (L - R).$$

It should also be remembered that the recovered signal from the Q detector will actually be $D + 25$ Hz stereo pilot signal. It is also appropriate to point out here that the name "pilot signal" originated with FM. Although AM came before FM, FM stereo came before AM stereo. The FM pilot signal has the primary purpose of regenerating the carrier. It was also used for giving a visible indication that the receiver was tuned to a stereo signal. In AM stereo, the pilot signal is not used for regenerating the carrier and is thus not necessary in this sense. It is used to give a visible stereo indication as in FM, and it is also used to switch from stereo back to mono reception when the signal strength is inadequate for good stereo reception. (This same kind of automated switching can be done in FM receivers also.) The *RC* network connected to pin 19 performs the low-pass filtering for the loop phase detector and sets loop damping. These subjects are covered more quantitatively in Chapter 2.

As the amplitude of the input signal varies, the amplitude of the Pilot $+ (L - R)$ signal varies also. This variation is compensated for by the Pilot $+ (L - R)$ AGC (Automatic Gain Control) block, which is like an AVC circuit for this signal. The Pilot $+ (L - R)$ information is low-pass filtered (to increase the pilot relative to the $L - R$) by the 400 Ω internal resistance that feeds pin 11, by the external resistance connected to it, and by the 5 μF capacitance to ground. The AGC action is only effective over a finite range of input amplitude. When the input falls below this range, the pilot component across the 5 μF capacitor drops. The signal here is filtered much more effectively by the active bandpass filter enclosed in dashed lines and is fed to the pilot decoder. This block makes the logic decisions as to whether the output will be mono or stereo. In making this decision, the pilot decoder block also utilizes inputs that tell it whether or not the Phase Locked Loop (PLL) is locked, what the input signal level is, and whether the output is being manually forced to the mono mode. If all conditions are correct, the pilot decoder activates a two pole switch that:

(1) turns on the stereo indicator light and
(2) allows $L - R$ to pass into the matrix, where it combines with $L + R$ to generate the L and R signals.

When conditions are not correct for good stereo reception, only $L + R$ from the envelope detector is input to the matrix, and this signal appears at both outputs.

The chip has one more clever capability: If the receiver picks up signals from two stations transmitting at the same nominal carrier frequency, the unwanted station is said to constitute *co-channel interference*. Since the carrier frequencies will not be at exactly the same frequency, the (Q) detection operation will result in frequency components at the carriers' difference frequency. This would result in a frequency so low that it would pass through the pilot filter. From there it goes through another section of low-pass filtering (1.6 K and 2.2 μF) and is amplified by the co-channel amplifier. Any 25 Hz pilot present will also be amplified here, but the inverted output of the pilot bandpass filter is summed in here at the proper level to null out the 25 Hz component, making the output of the co-channel amplifier a valid indicator of the level of co-channel interference. This output is fed to the pilot decoder along with the other inputs previously enumerated and has a role in determining whether the output will be mono or stereo.

Exercise 1.8 The active bandpass pilot filter is of the Infinite-Gain Multiple Feedback (IGMF) form. For this realization,

$$A_0 = \text{passband gain} = -R_C/2R_A,$$

$$Q = \text{quality factor} = \frac{1}{2}\sqrt{\frac{R_C}{R_A \parallel R_B}},$$

and

$$f_0 = \text{center frequency} = \frac{1}{2\pi\sqrt{(R_A \parallel R_B)R_C} \cdot C}.$$

For the parts values shown in Fig. 1.32, evaluate these three parameters.

1.10 AMAX™ certification

AMAX™ is a certification mark awarded by the Electronic Industries Association and the National Association of Broadcasters (EIA/NAB) to an AM receiver design that meets the following criteria:

(1) It must comply with the standards for bandwidth and distortion set forth by the National Radio Systems Committee as summarized in the document EIA/IS-80 dated March 1991.

(2) The receiver must have some provision for control of the receiver bandwidth. This control can be manual or automatic. In its simplest form, it can be a manual switch between "narrow band" and "wide band."

(3) If the AM receiver has stereo capability and otherwise meets AMAX™ requirements, it receives certification for AMAX™ Stereo.

(4) AM receivers must employ noise blanking unless they are battery-only receivers. An overview of noise-blanking circuitry can be found in Section 3.8.

(5) If the receiver has provision for attaching an external FM antenna, it must also have provision for attaching an external AM antenna. Receivers whose FM antenna is a telescoping whip antenna or a headphone cord are excluded from this requirement. A Walkman receiver using the latter antenna option is covered in Section 3.1.

(6) An AMAX™ receiver must also have the capability to tune the recent extension of the AM band, which incorporates carrier frequencies between 1,610 and 1,700 kHz.

References

Fink, D.G. and Carroll, J.M. 1968. *Standard Handbook for Electrical Engineers*, 10th ed., p. 5–36. New York: McGraw-Hill.

Haykin, S. 1983. *Communication Systems*, 2nd ed., p. 135. New York: Wiley.

Phillips, A.B. 1962. *Transistor Engineering*, p. 226ff. New York: McGraw-Hill.

2

FM receivers

2.1 Superheterodyning revisited

Just as the superheterodyne receiver dominates AM receiver design, so it dominates FM receiver design and has done so since its infancy. It is interesting to note that Major Edwin Armstrong, who invented the superheterodyne principle, also invented FM. This invention arose from his investigation of the causes and elimination of noise in AM receivers.

Because the frequency spectrum of FM is so much wider than that for AM, the FM band is placed much higher in the frequency spectrum. Broadcast FM stations are assigned frequency slots 0.2 MHz wide – 20 times that for commercial AM. These stations have carrier frequencies between 87.9 and 107.9 MHz. Thus both the AM and FM broadcast bands will accommodate about 100 discrete channels. The IF frequency used for broadcast FM receivers is 10.7 MHz.

As was the case for AM, it will be instructive to look at the spectral energy diagram for sinusoidal modulation (Fig. 2.1). As was also the case for AM, the mixing process results in two complete FM signals. The difference frequencies are clustered about 10.7 MHz and pass on into the IF amplifier.

Exercise 2.1 Assuming the station tuned to has a carrier at 99.3 MHz, find the frequency about which the sum frequencies are clustered.

As might be expected, the sum frequencies are eliminated by the tuned amplifier. All of the original FM information is thus in the IF signal.

2.2 The limiting principle

Because all of the information in an FM signal is encoded as frequency variation, it is a foregone conclusion that any variation of amplitude represents noise. This source of noise can be virtually eliminated by passing the FM signal through a *limiter* prior to detection. A simple and effective limiter consists of an amplifier stage driven to saturation on one polarity of signal swing and cutoff on the other polarity. There will be, of course, a finite slope to the rise and fall of the squared-off output, but the greater the overdrive the FM signal presents to the limiter, the more nearly square will be its output. Limiters are also cascaded to give more perfect (or *harder*) limiting.

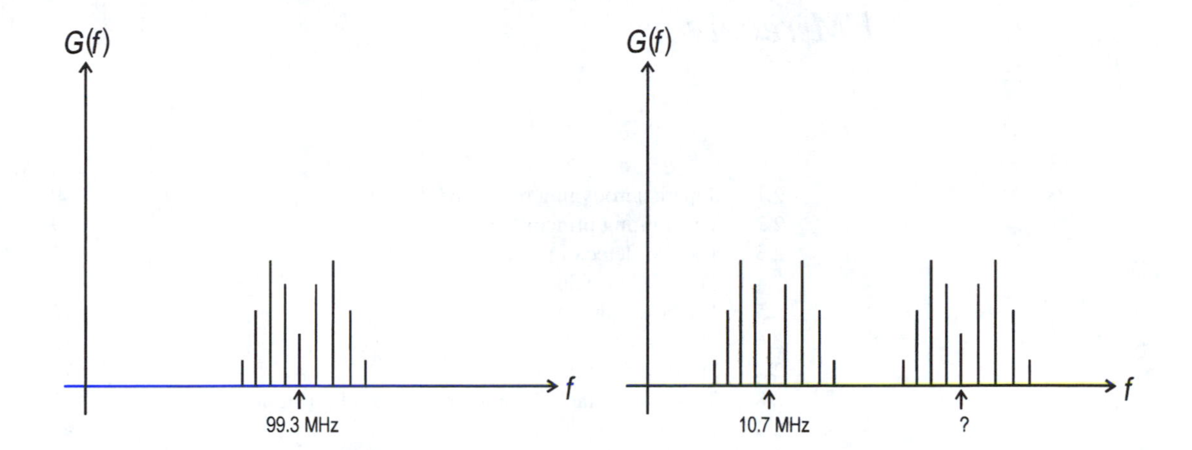

Fig. 2.1: Spectral energy diagram for frequency modulation of a carrier by a single-frequency sinusoid.

2.3 The FM detector principle

FM detectors are usually called *discriminators* or *demodulators*. Unlike AM, where the form of the detector is almost universal, there is a wide variety of workable FM detectors.

2.3.1 Ratio detector and Foster–Seeley discriminator

Early receivers used the Foster–Seeley discriminator or the ratio detector. These two circuits are similar in principle. The Foster–Seeley gives twice the output of the ratio detector in theory, but the ratio detector also has built-in AM rejection that the Foster–Seeley does not have. Thus the ratio detector saw much wider usage than the Foster–Seeley. However, neither of these circuits is amenable to integration, and consequently they are only found in older equipment. For this reason, we will not analyze either the ratio detector or the Foster–Seeley Discriminator in great detail in this text. The FM receiver analyzed later in this chapter uses one of the less-common among the many forms of the ratio detector, and we will outline in that context an approach to analysis of this circuit. The reader interested in actually seeing some of the calculations for a more-conventional form of the ratio detector is referred to Chirlian (1981) or other texts pre-dating his.

2.3.2 The phase-locked loop

The use of an integrated phase-locked loop (PLL) as an FM detector is well known. A typical circuit is shown in Fig. 2.2.

2.3.2.1 v–f relationship

The phase detector is characterized by a constant, K_d, which relates its low-frequency output voltage to the difference in phase between the inputs from the VCO and the "outside world":

$$\mathbf{V}_d(s) = K_d \cdot [\boldsymbol{\theta}_i(s) - \boldsymbol{\theta}_o(s)]. \tag{2.1}$$

This equation holds exactly for square wave inputs and is a good approximation for sinusoidal ones if $\sin(\boldsymbol{\theta}_i(s) - \boldsymbol{\theta}_o(s)) \approx \boldsymbol{\theta}_i - \boldsymbol{\theta}_o$. Typically, K_d will increase

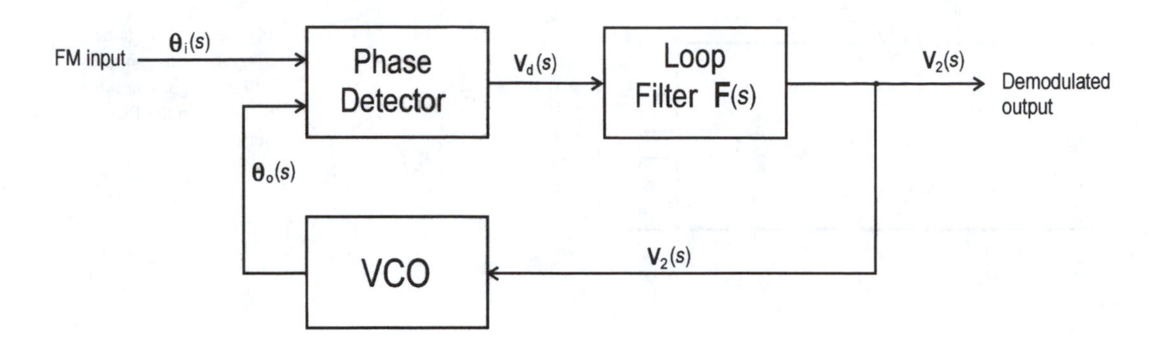

Fig. 2.2: Block diagram of a phase-locked loop (PLL) FM demodulator.

linearly with the input signal amplitude until saturation is reached, and K_d becomes independent of input amplitude.

The VCO is characterized by a constant, K_o, that relates the output frequency to the applied voltage:

$$\boldsymbol{\theta}_o(s) = K_o \mathbf{V}_2(s)/s$$

$$K_o \mathbf{V}_2(s) = s\boldsymbol{\theta}_o(s) = \omega_o(s).$$

The product $K_d \cdot K_o$ is the loop gain and is called K_v. It has dimensions of frequency and is typically on the order of the free running frequency.

Gardner (Gardner, 1966, p. 8) gives the following relationship between the variables in Fig. 2.2:

$$\frac{\boldsymbol{\theta}_o(s)}{\boldsymbol{\theta}_i(s)} = \frac{K_v \cdot \mathbf{F}(s)}{s + K_v \cdot \mathbf{F}(s)}.$$

This may be combined with the VCO equation to give

$$K_o \mathbf{V}_2(s) = \frac{s\boldsymbol{\theta}_i(s) \cdot K_v \cdot \mathbf{F}(s)}{s + K_v \cdot \mathbf{F}(s)},$$

$$\mathbf{V}_2(s) = \frac{\omega_i(s)}{K_o} \cdot \left[\frac{K_v \cdot \mathbf{F}(s)}{s + K_v \cdot \mathbf{F}(s)}\right] \equiv \frac{\omega_i(s)}{K_o} \cdot \mathbf{H}(s). \tag{2.2}$$

If $\mathbf{H}(s)$ were unity, this equation would represent ideal detection. This is generally a good approximation in a well-designed circuit. Let's consider the linearity problem quantitatively.

2.3.2.2 *Linearity considerations*

It is difficult to generalize about the linearity error of a second-order loop, since so much depends on the exact form of the loop filter used. However, some important principles can be learned by examining a specific case. The simplest loop filter giving a second-order response is shown in Fig. 2.3. For this filter (Gardner, 1966, p. 12)

$$\omega_n = \sqrt{K_v \omega_p}, \qquad \zeta = \frac{1}{2}\sqrt{\frac{\omega_p}{K_v}},$$

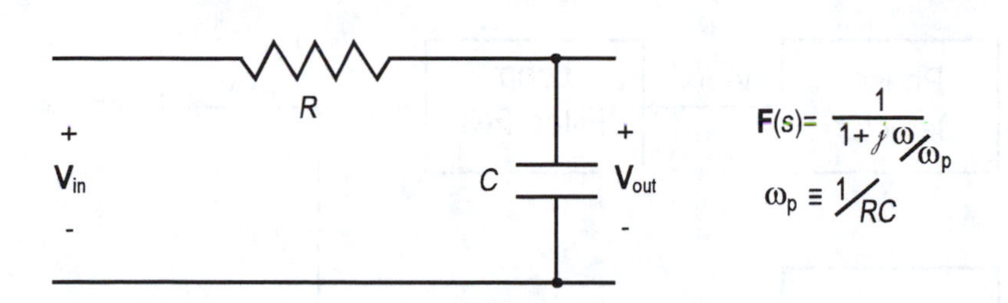

Fig. 2.3: Schematic diagram of the simplest LPF giving a second-order response in the PLL demodulator.

and

$$\mathbf{H}(s) = \frac{K_{\mathrm{v}}\omega_{\mathrm{p}}}{s^2 + s\omega_{\mathrm{p}} + K_{\mathrm{v}}\omega_{\mathrm{p}}}. \tag{2.3}$$

The linearity error is $|1 - |\mathbf{H}(s)||$, where

$$|\mathbf{H}(s)| = \left|\frac{K_{\mathrm{v}}\omega_{\mathrm{p}}}{(K_{\mathrm{v}}\omega_{\mathrm{p}} - \omega^2) + j\omega\omega_{\mathrm{p}}}\right| = \frac{K_{\mathrm{v}}\omega_{\mathrm{p}}}{\sqrt{(K_{\mathrm{v}}\omega_{\mathrm{p}} - \omega^2)^2 + \omega^2\omega_{\mathrm{p}}^2}}$$

$$= \frac{1}{\sqrt{\left(1 - \dfrac{\omega^2}{K_{\mathrm{v}}\omega_{\mathrm{p}}}\right)^2 + \dfrac{\omega^2}{K_{\mathrm{v}}^2}}}.$$

Since $\zeta = \frac{1}{2}\sqrt{\frac{\omega_{\mathrm{p}}}{K_{\mathrm{v}}}}$ it follows that $4\zeta^2 K_{\mathrm{v}} = \omega_{\mathrm{p}}$ and

$$|\mathbf{H}(s)| = \frac{1}{\sqrt{\left(1 - \dfrac{\omega^2}{4\zeta^2 K_{\mathrm{v}}^2}\right)^2 + \dfrac{\omega^2}{K_{\mathrm{v}}^2}}},$$

and the linearity error is thus

$$\varepsilon_{\mathrm{L}} = \left|1 - \frac{1}{\sqrt{\left(1 - \dfrac{\omega^2}{4\zeta^2 K_{\mathrm{v}}^2}\right)^2 + \dfrac{\omega^2}{K_{\mathrm{v}}^2}}}\right|. \tag{2.4}$$

The absolute value is necessary, because $|\mathbf{H}(s)|$ can be greater than unity if the response is underdamped.

This error is a function of the modulation frequency and the loop damping. For a fixed upper limit of ω/K_{v}, we can find the value of ζ that will give the smallest RMS error over that range of frequencies:

$$\varepsilon_{\mathrm{R}}(\zeta, x_{\mathrm{MAX}}) = \sqrt{\frac{1}{x_{\mathrm{MAX}}}\int_0^{x_{\mathrm{MAX}}} \left[1 - 1\Big/\sqrt{\left(1 - \frac{x^2}{4\zeta^2}\right)^2 + x^2}\,\right]^2 \cdot dx},$$

where $x \equiv \omega/K_{\mathrm{v}}$ and $(0 < x_{\mathrm{MAX}} \leq 1)$.

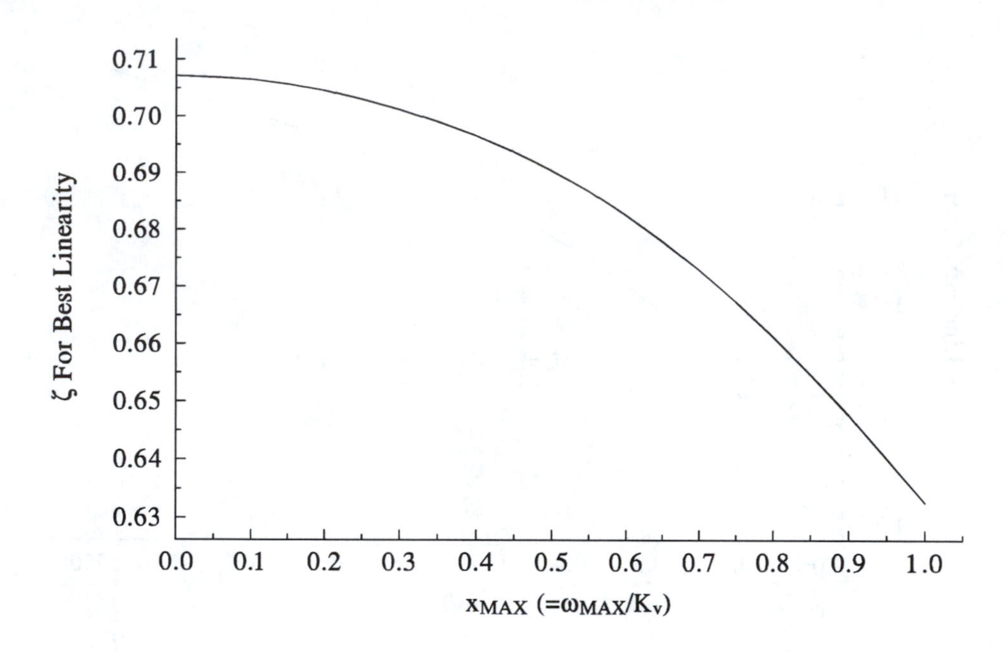

Fig. 2.4: Value of damping factor giving the smallest RMS linearity error for a second-order PLL demodulator as a function of maximum frequency of operation.

Using integration by Simpson's Rule and $x_{\mathrm{MAX}} = 1$, the following data were generated:

ζ	$\varepsilon_{\mathrm{R}}(\zeta, 1)$ (in percent)
0.61	2.25
0.62	2.01
0.63	1.90
0.64	1.93
0.65	2.08

Interpolation gives the minimum value of ε_{R} for $\zeta = 0.633$.

Similarly, the value of ζ corresponding to minimum ε_{R} can be found for each value of x_{MAX}. These data are presented graphically in Fig. 2.4. From this graph we see that for $x_{\mathrm{MAX}} \ll 1 (\omega_{\mathrm{MAX}} \ll K_{\mathrm{v}})$, the optimum damping is .707, and that even for $x_{\mathrm{MAX}} = 1$ $(\omega_{\mathrm{MAX}} = K_{\mathrm{v}})$, the optimum damping differs from this value by no more than 10%. Thus, for linear detection of FM with a PLL, we would like to have $\zeta = 0.707$ and to make K_{v} large compared to ω.

This value of ζ reduces (2.4) to

$$\varepsilon_{\mathrm{L}} = \frac{1}{\sqrt{1 - \dfrac{\omega^4}{4K_{\mathrm{v}}^4}}} - 1. \tag{2.5}$$

This expression is plotted versus ω/K_{v} in Fig. 2.5, which shows that if $\omega \ll K_{\mathrm{v}}$, the linearity is excellent. Demodulation is much less linear if the signal input to the phase detector is sinusoidal.

Let us now find ω/K_{v} for a typical FM detection application. For the Signetics 560 PLL, $K_{\mathrm{v}} = 0.96\omega_{\mathrm{o}}$ Hz (Signetics, 1972, p. 40), where ω_{o} is the free-running frequency in radians/sec. This value presupposes that the phase detector is limiting, which requires an input of at least 50 mV RMS. For broadcast FM

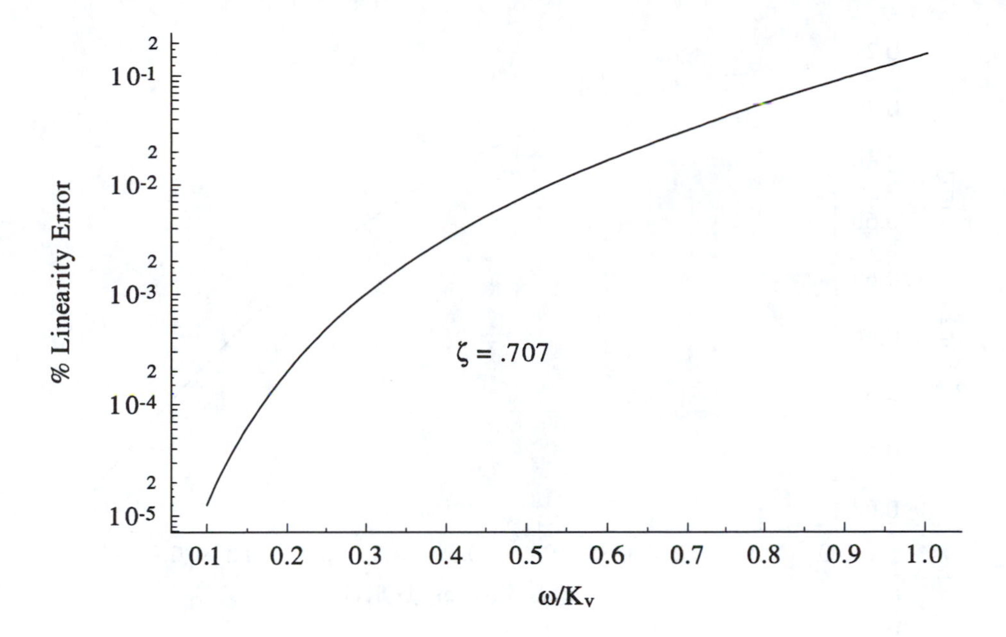

Fig. 2.5: Linearity error of a second-order PLL demodulator with the optimum value of ζ as a function of frequency of operation.

detection, the loop would be set to run at 10.7 MHz, giving

$$K_v = 0.96 \times 2\pi \times 10.7 \times 10^6 = 64.54 \times 10^6.$$

The maximum value of the modulating frequency is set by the FCC as 15 kHz, so $\overline{\omega} = 2\pi \times 15 \times 10^3 = 94.25 \times 10^3$.
Thus

$$\omega / K_v = 0.00146.$$

From Eq. (2.5), this yields

$$\varepsilon_L = 5.67 \times 10^{-13}.$$

This nonlinearity is so small as to be meaningless. In fact, it is so small that it would be advantageous to increase it. This is because from Eqs. (2.3), $\omega_n = \sqrt{2} K_v$, and for $\zeta = 0.707$, the loop noise bandwidth (B_L) is (Gardner, 1966, p. 20):

$$B_L = 0.53\omega_n = 0.53 \cdot \sqrt{2} \cdot K_v = 48.37 \times 10^6 \text{ Hz}.$$

This enormous bandwidth will mean noisy detection. The 560 PLL can have its bandwidth reduced by placing a resistor between pins 14 and 15 (Signetics, 1972, p. 52). The loop gain will then be reduced by the factor

$$\frac{R_s}{12\,\text{k} + R_s},$$

where R_s is the value of the added shunt resistor, and 12 k is the resistance internal to the IC. The addition of this shunt resistor will also affect ω_p strongly.

If the pole is formed by placing a capacitor between pins 14 and 15 (the simplest approach), then

$$\omega_{\mathrm{p}} = \frac{1}{2 \cdot 12\,\mathrm{k} \cdot C}$$

with no shunt resistance and

$$\omega_{\mathrm{p}} = \frac{1}{2 \cdot (12\,\mathrm{k} \parallel R_{\mathrm{s}}) \cdot C}$$

with shunt resistance. If the gain is reduced by a large factor, as in our example, the capacitance must be increased substantially to retain the same ω_{p}.

What emerges from all of this is that there is a trade-off between linearity and noise. A maximum nonlinearity of 0.01% should be more than adequate for even the most discriminating listener. From Fig. 2.5, this corresponds to $\omega/K_{\mathrm{v}} < 0.17$. Since $\omega \leq 94.25 \times 10^3$, K_{v} need be no greater than 589 kHz. This provides an 82-fold reduction in the noise bandwidth. Finally, from Eq. (2.3), for $\zeta = 0.707$, we have

$$\omega_{\mathrm{p}} = 2K_{\mathrm{v}} = 1.078 \times 10^6 \text{ rad/sec}$$

and since

$$\mathbf{H}(s) = \frac{K_{\mathrm{v}}\omega_{\mathrm{p}}}{s^2 + s\omega_{\mathrm{p}} + K_{\mathrm{v}}\omega_{\mathrm{p}}} = |\mathbf{H}(s)|\angle\theta,$$

where

$$\theta = -\arctan\frac{\omega\omega_{\mathrm{p}}}{(K_{\mathrm{v}}\omega_{\mathrm{p}} - \omega^2)}$$

we have

$$\theta = -\arctan\frac{2K_{\mathrm{v}}\omega}{2K_{\mathrm{v}}^2 - \omega^2} = -\arctan\frac{2 \cdot \dfrac{\omega}{K_{\mathrm{v}}}}{2 - \left(\dfrac{\omega}{K_{\mathrm{v}}}\right)^2},$$

which we will write as

$$\theta = -\arctan\frac{2 \cdot \dfrac{\omega}{K_{\mathrm{v}}}}{2 - \left(\dfrac{\omega}{K_{\mathrm{v}}}\right)^2} = \frac{-\omega}{K_{\mathrm{v}}} - \Delta\theta.$$

The first term on the right represents a small time delay. The second is a frequency-dependent phase error. For $\omega/K_{\mathrm{v}} = 0.16$, the phase error is 6.77×10^{-4} radians decreasing down to zero for $\omega = 0$. This magnitude of phase error is well below that needed to cause perceptible distortion.

2.3.2.3 *Limiting and detection threshold*

A final consideration with PLL FM demodulation is that of threshold. If the amplitude of the input signal is not adequate to cause the IF stages to limit, the gain of the phase detector declines. This reduces K_{v}, which will cause degradation of the linearity. But two other factors are even more serious:

(1) The failure of the IF stages to limit allows a large amount of AM noise to accompany the detected FM signal. This obviously degrades the received signal. FM receivers often have their sensitivity stated in terms of microvolts required for full quieting. This is the antenna signal level required to cause the IF stages to limit.

(2) The most serious nonideality is the catastrophic failure of the detector when the PLL goes out of lock. For a second-order loop, the phase error is (Gardner, 1966, p. 29)

$$\theta = \Delta\omega / K_{\mathrm{v}}.$$

For square-wave inputs, this must have a magnitude of less than $\pi/2$ so

$$\frac{\pi}{2} \geq \left| \frac{\Delta\omega}{K_{\mathrm{v}}} \right|$$

or

$$|4\Delta f| \leq K_{\mathrm{v}},$$

where Δf is the difference between the instantaneous frequency input and the loop's free-running frequency. It was just mentioned that K_{v} is reduced when the IF stages are not limiting. Assuming that the loop free-running frequency is squarely set at 10.7 MHz, Δf will be 75 kHz, so the lock requirement is $K_{\mathrm{v}} > 300$ kHz.

In our example $K_{\mathrm{v}} = 589$ kHz, but this assumed full limiting, which required an input signal amplitude to the phase detector of at least 50 mV RMS. This need drop to only 25 mV RMS to cause the loop gain to drop low enough to lose lock.

This problem is even more critical if free-running frequency is not properly set. On this account, adequate limiting is even more important for FM receivers with PLL demodulators than for other types.

2.3.3 The quadrature detector

The quadrature detector is another integrated FM discriminator that is very popular. It is used chiefly in the low to middle price range receivers. Although it is not the most linear form of detection, its low price and ease of setup (there being only one adjustment made to peak the output) contribute to its popularity. The integrated quadrature detector is really nothing more than a phase detector to which an outboard tuned circuit is added. It is most often integrated with a high-gain IF amplifier to insure limiting. Figure 2.6 shows a schematic of one such device, the ULN2111.

2.3.3.1 IC analysis

Examination of the ULN2111 schematic shows that the IF input is applied to pin 4, which is one input of a differential amplifier. The other input (pin 5) is AC grounded through two 0.1 μF capacitors. The gain of this differential

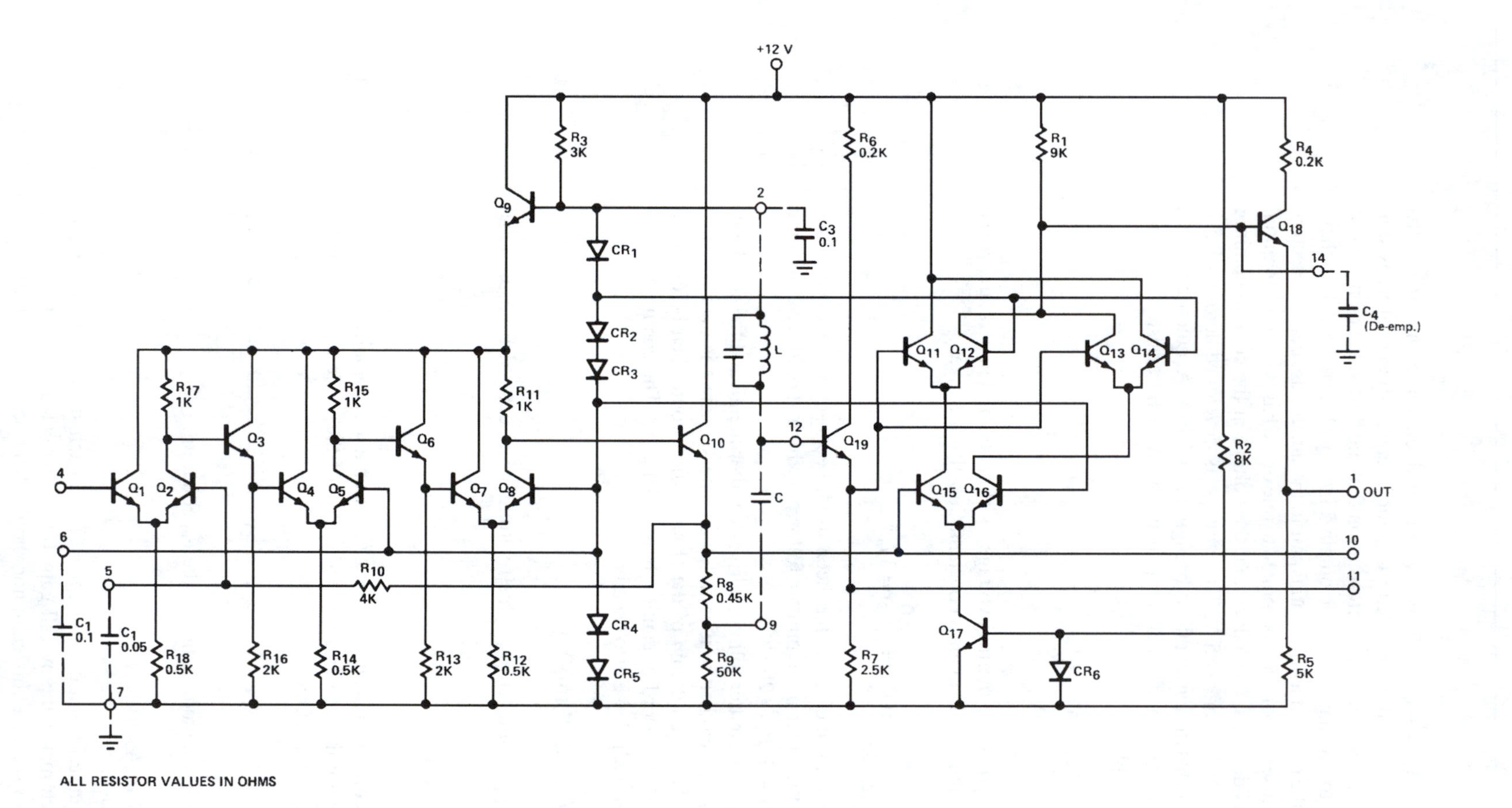

Fig. 2.6: Schematic diagram of the ULN2111 FM quadrature detector IC.

amplifier stage is $A_v = -g_m R_L/2$. We wish to evaluate this gain numerically. To do so, we need to find the collector current in each differential transistor. The Signetics data sheet gives the information that the bias voltage on pin 6 is 1.45 V. It is safe to say that the voltage on pin 5 is very close to this, since these pins connect to the bases of a differential pair in the quadrature detector section (Q_{15}, Q_{16}). Because $V_5 \approx 1.45$ V is applied to one of the bases in the differential stage under consideration, we deduce that the voltage at the common emitter terminal is ~0.6 V less or ~.85 V. This voltage is developed across a 0.5 k resistor, which means the current flow through it is 1.7 mA. At quiescence, this current should be the sum of equal components from each transistor, meaning $I_{CQ} = 0.85$ mA. Thus

$$|A_v| = \frac{I_{CQ}(\text{mA})}{2 \cdot V_T(\text{mV})} \cdot R_L(\Omega),$$

where $V_T \equiv kT/q$ is the thermal voltage, with T being temperature, q the electron charge, and k Boltzmann's constant. Substituting gives

$$|A_v| = \frac{0.85}{2 \cdot 25} \cdot 1000 = \frac{850}{50} = 17,$$

where a load resistance of 1 kΩ has been used, because the only other load on this stage is the input of an emitter follower, which should have an input impedance of $(\beta + 1)R_e \approx 200$ kΩ.

The output of this emitter follower drives another identical differential amplifier, which drives another emitter follower. The sequence is then repeated one more time. The overall gain of the IF amplifier would then be expected to be $17^3 = 4{,}913$. A word of caution is in order here. The commonly used expression for g_m (which we also used here),

$$g_m = I_c(\text{mA})/V_T(\text{mV}),$$

is actually an upper limit on g_m. This relationship is derived from the ideal diode equation

$$I = I_s\left(e^{V/V_T} - 1\right).$$

Practical silicon junctions have some generation and recombination in their space charge region that requires a modification of the diode equation by the "η factor" (Millman and Grabel, 1987):

$$I = I_s\left(e^{V/\eta V_T} - 1\right) \qquad (1 \le \eta \le 2).$$

This in turn requires a modification of the g_m equation:

$$g_m = I_c(\text{mA})/\eta V_T.$$

In the worst case for our example, $\eta = 2$ and each stage has half the gain calculated, which makes the overall gain 1/8 of that calculated or 614. There is no feedback to stabilize this gain, nor should there be. If the amplifier will limit in the worst case ($A = 614$), it will limit even better for the larger gain. The worst case (temperature-wise) will be at 85°C (specified as the highest operating temperature) when $V_T = 30$ mV. If this condition prevails along with $\eta = 2$, the individual stage gain is 7.08 and the overall gain is 355.

The data sheet indicates that limiting will be achieved for 500 μV RMS input. In the worst case this would produce an output voltage at pin 10 of 178 mV RMS = 0.25 V peak. This deviation is coupled to the base of Q_{15}, which varies about 1.45 V. The base of Q_{16}, in the meantime, is held at a constant 1.45 V. Thus the Q_{15}, Q_{16} pair is subject to a differential input voltage of ±0.25 V.

Exercise 2.2 Q_{17} and its associated diode form a 1.5 mA current source.

(a) Estimate the base bias on $Q_{11} - Q_{14}$ (all about the same).

(b) When the base voltage on Q_{15} is 1.70 V, estimate the base–emitter voltages of Q_{15} and Q_{16}.

(c) Do the same for a signal swing of +0.10 V.

If you have done this exercise, you should be led to the conclusion that the circuit will work for a $0 \pm .25$ V swing but not for any swing much less because neither Q_{15} nor Q_{16} would ever turn completely off.

2.3.3.2 v–f relationship

Now let us develop the v–f relationship for the quadrature detector portion of the circuit.

As already noted, the output of the IF amplifier is fed directly to the base of Q_{15}, whereas the base of Q_{16} is held at AC ground. Signetics, who second-source this chip, call the basic quadrature detector circuit (Q_{11}–Q_{14}, Q_{15}, Q_{16}, and Q_{17} with associated components) a doubly balanced multiplier and give the equation (Signetics, 1972, pp. 15, 16)

$$v_{\text{out}} = |v_{14}| \approx |v_1| = \frac{2A_{\text{d}}}{\pi} \cdot |v_{12}| \cos\theta. \tag{2.6}$$

The voltage subscripts refer to pin numbers. A_{d} is the differential gain of the phase detector, and θ is the phase angle between v_{12} and v_9, which is $1/10$ of the IF amplifier output. This equation assumes that the differential voltage swing applied to the bases of Q_{16} and Q_{15} from the IF amplifier is adequate to drive one of the bases fully into saturation and the other fully into cutoff. The previous reference gives an excellent quantitative discussion of the operation of the phase detector itself, with particular reference to its use as a PLL phase detector.

We may schematically represent the relationship between the relevant voltages and components as shown in Fig. 2.7. The resistor may, for the moment, be considered to model the losses in the tuned circuit. We shall see later that there are good reasons for adding it explicitly. The impedance of the parallel circuit is

$$\frac{sLR}{s^2LCR + sL + R}$$

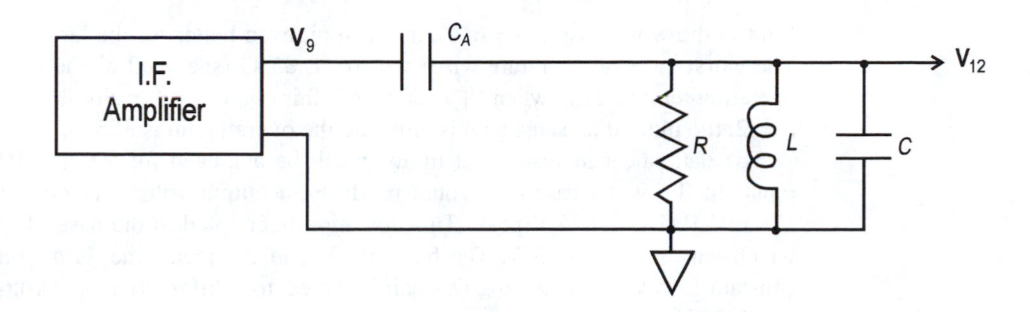

Fig. 2.7: Simplified schematic diagram of a quadrature detector.

so that

$$\frac{\mathbf{V}_{12}}{\mathbf{V}_9}(s) = \frac{\frac{sLR}{s^2LCR+sL+R}}{\frac{1}{sC_\mathrm{A}} + \frac{SLR}{s^2LCR+sL+R}} = \frac{s^2LC_\mathrm{A}R}{s^2LCR + sL + R + s^2LC_\mathrm{A}R}.$$

Define $C_\mathrm{p} \equiv C + C_\mathrm{A}$; then

$$\frac{\mathbf{V}_{12}}{\mathbf{V}_9}(s) = \frac{s^2LC_\mathrm{A}R}{s^2LC_\mathrm{p}R + sL + R}$$

and

$$\frac{\mathbf{V}_{12}}{\mathbf{V}_9} = \frac{-\omega^2LC_\mathrm{A}R}{R(1-\omega^2LC_\mathrm{p}) + jL\omega} = \mathbf{M}\,\angle\theta. \tag{2.7}$$

Now from Eq. (2.6) we can write

$$|\mathbf{V}_1| = \frac{2A_\mathrm{d}}{\pi} \cdot M \cdot |\mathbf{V}_9| \cos\theta. \tag{2.8}$$

Rewriting (2.7) in polar form, we have

$$\frac{\mathbf{V}_{12}}{\mathbf{V}_9} = \frac{\omega^2LC_\mathrm{A}R}{\sqrt{R^2(1-\omega^2LC_\mathrm{p})^2 + L^2\omega^2}}\angle 180^\circ - \arctan\frac{L\omega}{R(1-\omega^2LC_\mathrm{p})},$$

from which we obtain

$$\mathbf{M} = \frac{\omega^2LC_\mathrm{A}R}{\sqrt{R^2(1-\omega^2LC_\mathrm{p})^2 + L^2\omega^2}} \tag{2.9a}$$

and

$$\theta = 180^\circ - \arctan\frac{L\omega}{R(1-\omega^2LC_\mathrm{p})}. \tag{2.9b}$$

We wish to get a zero output voltage from the detector when the frequency input is 10.7 MHz. This requires that $\theta = 90^\circ$ or upon substituting into (2.9b) we have

$$90^\circ = \arctan \frac{L\omega_o}{R\left(1 - \omega_o^2 LC_p\right)},$$

which can only be true if $\omega_o^2 LC_p = 1$ or

$$\omega_o^2 = \frac{1}{LC_p}, \tag{2.10}$$

where $\omega_o = 2\pi \times 10.7 \times 10^6$.

The circuit will receive input frequencies that represent small variations about ω_o. Thus we can effect a simplification by setting

$$\arctan \frac{L\omega}{R(1 - \omega^2 LC_p)}$$

$$= \arctan \frac{L\omega_o}{R\left(1 - \omega_o^2 LC_p\right)} + \arctan\delta = 90^\circ + \arctan\delta, \tag{2.11}$$

where $\arctan\delta$ represents a small-angle variation around 90°. Trigonometric manipulation will show that

$$\arctan\delta = -\arctan \frac{R(1 - \omega^2 LC_p)}{L\omega}.$$

This equation may be inserted into (2.11) to give

$$\arctan \frac{L\omega}{R(1 - \omega^2 LC_p)} = 90^\circ - \arctan \frac{R(1 - \omega^2 LC_p)}{L\omega}.$$

This equation may in turn be used with (2.9b) to yield

$$\theta = 90^\circ + \arctan \frac{R(1 - \omega^2 LC_p)}{L\omega}.$$

Finally, if this equation and (2.9a) are put into (2.8), the result is

$$|\mathbf{V}_1| = \frac{2A_d}{\pi} \cdot \frac{\omega^2 LC_A R|\mathbf{V}_9|}{\sqrt{R^2(1 - \omega^2 LC_p)^2 + L^2\omega^2}}$$

$$\cdot \cos\left(90^\circ + \arctan \frac{R(1 - \omega^2 LC_p)}{L\omega}\right).$$

By trigonometric expansion of the cosine

$$|\mathbf{V}_1| = \frac{-2A_d}{\pi} \cdot \frac{\omega^2 LC_A R|\mathbf{V}_9|}{\sqrt{R^2(1 - \omega^2 LC_p)^2 + L^2\omega^2}}$$

$$\cdot \sin\left(\arctan \frac{R(1 - \omega^2 LC_p)}{L\omega}\right).$$

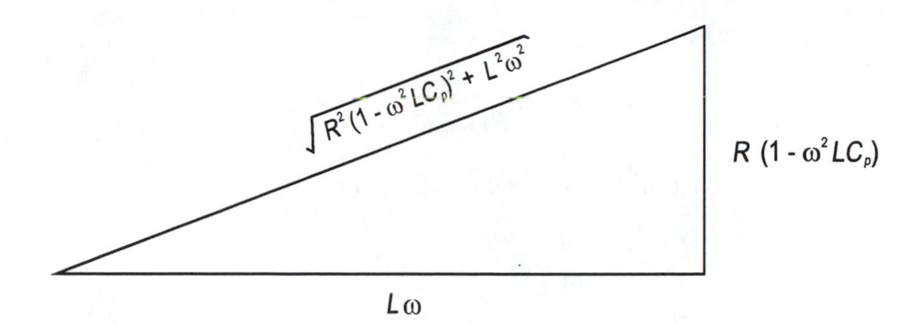

Fig. 2.8: Phase diagram showing how to find the sine of an arc tangent.

We can resolve the sine of the arc tangent by reference to Fig. 2.8, obtaining

$$|\mathbf{V}_1| = \frac{-2A_\mathrm{d}}{\pi} \cdot \frac{\omega^2 L C_\mathrm{A} R |\mathbf{V}_9|}{\sqrt{R^2(1-\omega^2 L C_\mathrm{p})^2 + L^2\omega^2}}$$

$$\cdot \frac{R(1-\omega^2 L C_\mathrm{p})}{\sqrt{R^2(1-\omega^2 L C_\mathrm{p})^2 + L^2\omega^2}}.$$

$$= \frac{-2A_\mathrm{d}}{\pi} \cdot \frac{\omega^2 L C_\mathrm{A} R^2 (1-\omega^2 L C_\mathrm{p})}{R^2(1-\omega^2 L C_\mathrm{p})^2 + L^2\omega^2} \cdot |\mathbf{V}_9|.$$

Introducing ω_o from (2.10), we get

$$|\mathbf{V}_1| = \frac{-2A_\mathrm{d}}{\pi} \cdot \frac{\frac{\omega^2}{\omega_\mathrm{o}^2}\left(\frac{C_\mathrm{A}}{C_\mathrm{p}}\right) R^2 \left(1 - \frac{\omega^2}{\omega_\mathrm{o}^2}\right)}{R^2\left(1 - \frac{\omega^2}{\omega_\mathrm{o}^2}\right)^2 + L^2\omega^2} \cdot |\mathbf{V}_9|$$

$$= \frac{-2A_\mathrm{d}}{\pi} \cdot \frac{\left(\frac{R^2}{L^2\omega_\mathrm{o}^2}\right)\left(\frac{C_\mathrm{A}}{C_\mathrm{p}}\right)\left(\frac{\omega_\mathrm{o}^2 - \omega^2}{\omega_\mathrm{o}^2}\right)\left(\frac{\omega^2}{\omega_\mathrm{o}^2}\right)}{\left(\frac{R^2}{L^2\omega_\mathrm{o}^2}\right)\left(1 - \frac{\omega^2}{\omega_\mathrm{o}^2}\right)^2 + \frac{\omega^2}{\omega_\mathrm{o}^2}} \cdot |\mathbf{V}_9|.$$

But $R/L\omega_\mathrm{o} = Q$ so

$$|\mathbf{V}_1| = \frac{-2A_\mathrm{d}}{\pi} \cdot \frac{\left(\frac{C_\mathrm{A}}{C_\mathrm{p}}\right) Q^2 \left(1 - \frac{\omega^2}{\omega_\mathrm{o}^2}\right)\left(\frac{\omega^2}{\omega_\mathrm{o}^2}\right)}{\left[Q\left(1 - \frac{\omega^2}{\omega_\mathrm{o}^2}\right)\right]^2 + \frac{\omega^2}{\omega_\mathrm{o}^2}} \cdot |\mathbf{V}_9|. \tag{2.12}$$

2.3.3.3 Linearity considerations

Equation (2.12) is as exact as our model. If we recognize that $|\omega - \omega_\mathrm{o}| \ll \omega_\mathrm{o}$ (that is, the deviation from the center frequency will be much less than the center frequency itself), we can consider that $\omega = \omega_\mathrm{o}$ in (2.12) except for the terms $1 - \omega^2/\omega_o^2$, which may be expressed as

$$1 - \frac{\omega^2}{\omega_o^2} = \frac{(\omega_o + \omega)(\omega_o - \omega)}{\omega_o^2} \cong \frac{2\omega_o(\omega_o - \omega)}{\omega_o^2}$$
$$= \frac{2(\omega_o - \omega)}{\omega_o} \equiv \frac{2\Delta\omega}{\omega_o}.$$

Using this and approximating ω as ω_o in (2.12) yields the result

$$|\mathbf{V}_1| = -\frac{2A_d}{\pi} \cdot \left(\frac{C_A}{C_p}\right) Q^2 \frac{2\omega_o \Delta\omega}{4Q^2(\Delta\omega)^2 + \omega_o^2} |\mathbf{V}_9|.$$

From this equation, it is clear that if $2Q \cdot \Delta\omega \ll \omega_o$, the result reduces to one yielding linear demodulation of the FM signal, that is, one where the recovered voltage is directly proportional to the deviation from the center frequency,

$$|\mathbf{V}_1| = -\frac{4A_d}{\pi} \cdot \left(\frac{C_A}{C_p}\right) Q^2 \frac{\Delta\omega}{\omega_o} |\mathbf{V}_9|. \tag{2.13}$$

We want to find the linearity error of the quadrature detector. To do this, we will take the exact expression for output voltage in (2.12) and subtract the linear approximation thereto from (2.13). The difference will then be divided by the exact expression for output voltage from (2.12) to yield a fractional error. To this end we define $r \equiv \omega/\omega_o = f/f_o$. Then $\Delta\omega/\omega_o = 1 - r$. Since the fraction could be positive or negative, depending on the sign of $\Delta\omega$, we will take the absolute value of the result, yielding

$$\varepsilon_L = \left| (r - 1) \left[\frac{r}{r+1} + 2Q^2 \frac{(1 - r^2)}{r^2} \right] \right|. \tag{2.14}$$

As can be seen from this result, low Q is conducive to good linearity. Q can be lowered by parallel resistive loading of the tuned circuit. Accompanying the improved linearity, however, is a decrease in the detected signal amplitude since, as Eq. (2.13) shows, this amplitude is proportional to Q^2. It might be suggested that this could be compensated for by increasing the ratio $C_A/C_p \cdot [= C_A/(C + C_A)]$. Let us first note that this ratio is ordinarily quite small (between 0.01 and 0.1) and with good reason. If the ratio is made much larger, the detector begins to load the IF amplifier heavily. Another problem with too large a ratio is that it will violate one of the key assumptions made in the derivation.

Exercise 2.3 Which assumption is violated if C_A is not $\ll C$? What will be the effect on the recovered signal?

The ULN2111 is specified as having a linearity error of 1.5% for FM broadcast band detection.

2.3.4 Monostable FM Detection

It is also possible to make an FM detector using a monostable multivibrator triggered by the limited FM signal. The output is then averaged to recover a voltage

Table 2.1. *A comparison of FM demodulators*

Name	Ideal v-f characteristic	Fractional linearity error	Comments
Ratio detector	$\left\|\frac{dv_o}{d\omega}\right\| = 2\|V_{in}\| \cdot \left(\frac{M}{L_1}\right) \cdot \frac{Q^2}{\omega_o}$	$\frac{(1+r-2r^2) - 2Q^2(1-r^2)^2}{1+r}$ where $r \equiv \omega/\omega_o$	Assumes ideal diodes. Assumes $\|V_2\| \ll \|V_t\|$
Phase-lock ed loop–2nd order	$\left\|\frac{dv_o}{d\omega}\right\| = \frac{1}{K_o}$	$\frac{1}{\sqrt{1-\omega^4/K_v^4}} - 1$	K_o = VCO Gain
Quadrature detector	$\left\|\frac{dv_o}{d\omega}\right\| = \frac{4}{\pi} \cdot A_d\|V_{in}\| \cdot \left(\frac{C_A}{C_A+C}\right) \cdot \frac{Q^2}{\omega_o}$	$\frac{r(r-1) - 2Q^2(1-r^2)^2}{r^2(1+r)}$ where $r \equiv \omega/\omega_o$	A_d = Differential amplifier gain
Monostable detector	$\left\|\frac{dv_o}{d\omega}\right\| = \frac{V_p t_p}{2\pi}$	Limited only by timing stability and recovery time of monostable	V_p = Pulse amplitude; t_p = Pulse width

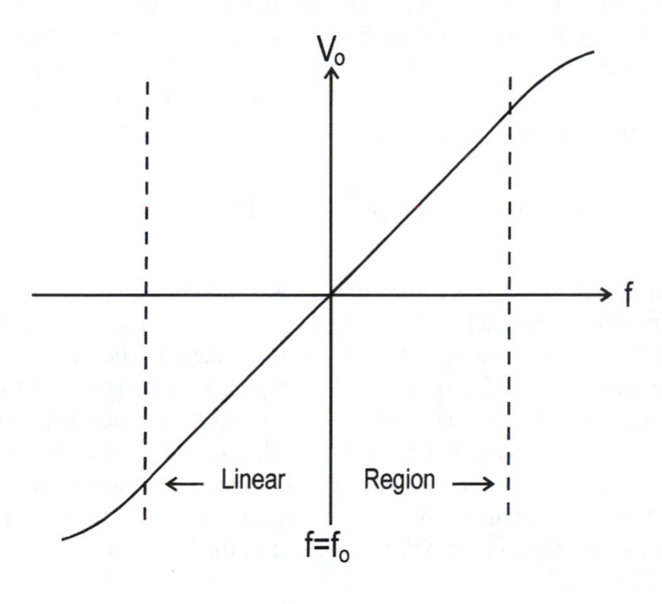

Fig. 2.9: Typical v–f curve for an FM demodulator.

directly proportional to frequency. This circuit has the advantage of being inherently linear in its v–f characteristic. Only the recovery time of the monostable multivibrator and the choice of output-pulse width can limit the usable range. Its disadvantages are low output and the need for a very well-regulated power supply so that the output reflects only variations in frequency and not in amplitude.

The main requirement of any broadcast FM detector circuit is that it provide an output voltage that varies linearly with input frequency over a frequency range of ±75 kHz about a center frequency of 10.7 MHz. Table 2.1 summarizes some of the salient characteristics of the detector circuits just discussed. A typical v_o versus f curve is shown in Fig. 2.9.

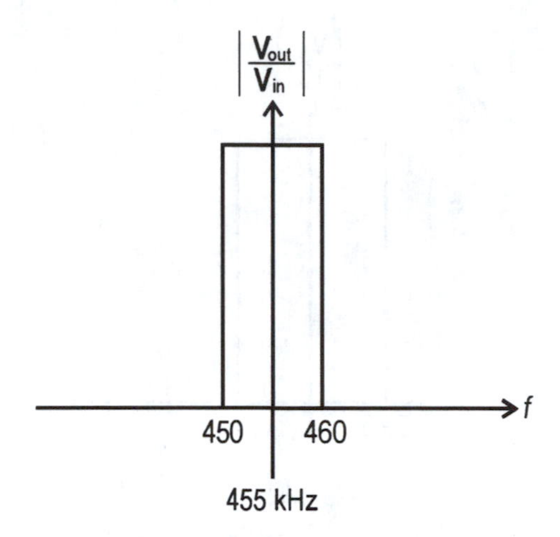

Fig. 2.10: Ideal response of the IF amplifier in an AM receiver.

2.4 The AFC principle

AFC stands for *automatic frequency control*. It is a means of keeping the local oscillator running at the proper frequency. Let's look first at why it is necessary and then at how it works.

2.4.1 The need justified

In the AM receiver, we considered a midband station at 1 MHz corresponding to a local oscillator frequency of 1,455 kHz. The ideal IF amplifier response is as shown in Fig. 2.10. This response would give full output for all frequencies in the station's spectrum, no output for any other station, and minimum noise. (Of course, the actual response more closely resembles that of Fig. 1.4.) Let us now consider what would happen if the local oscillator were to run at 1,460 kHz instead of 1,455 kHz. The difference frequency applied to the tuned IF amplifier would then be centered at 460 kHz. From Fig. 2.10 we can see clearly that if a signal centered about 460 kHz were applied, half the signal power would be lost. This would result, we recall, if the frequency of the local oscillator were off by 5 kHz out of 1,455 kHz or 1 part in 291.

A similar analysis for a midband FM broadcast station at 99.9 MHz and an FM IF response like that of Fig. 2.11 shows that an error of only 1 part in 1,106 would cause half the signal power to fall outside the amplifier passband. This in itself would suggest that FM is about four times as susceptible to local oscillator drift as AM.

In fact, the situation is even worse. To see why, let's consider the detection process for both AM and FM if the local oscillator frequency drifts by an amount equal to half the ideal IF amplifier passband. Consider first an AM signal with unity carrier amplitude and sinusoidal modulation with index m $(0 \leq m \leq 1)$:

$$v_{\text{in}}(t) = \cos \omega_{\text{c}} t \cdot (1 + m \cdot \cos \omega_{\text{m}} t). \tag{2.15}$$

Trigonometric expansion gives

$$v_{\text{in}}(t) = \cos \omega_{\text{c}} t + \frac{m}{2} \cos(\omega_{\text{c}} - \omega_{\text{m}})t + \frac{m}{2} \cos(\omega_{\text{c}} + \omega_{\text{m}})t.$$

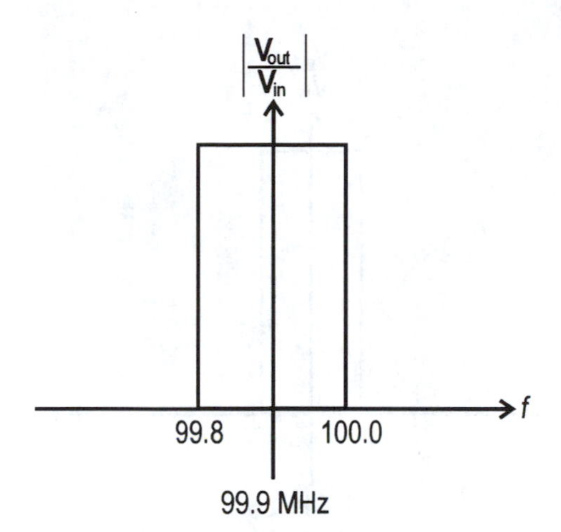

Fig. 2.11: Ideal response of the IF amplifier in an FM receiver.

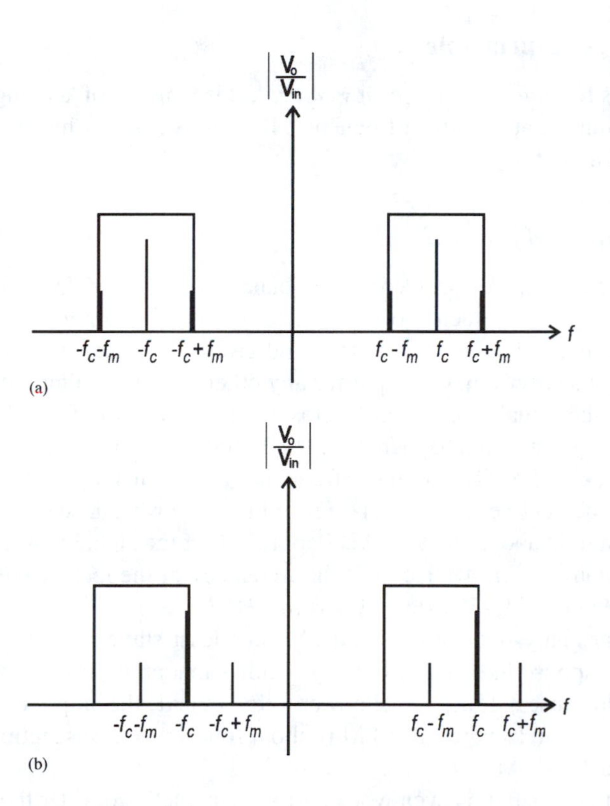

Fig. 2.12: Spectral energy diagram of an AM IF signal superimposed upon the ideal response curve of an AM IF amplifier (a) when the IF signal is properly centered at 455 kHz and (b) when the IF signal is offset by just 5 kHz to 460 kHz.

Figure 2.12a shows the power density spectrum superimposed on the ideal IF passband when the carrier $\omega_c = 2\pi \times 455$ kHz as it should. Figure 2.12b shows the corresponding plot when ω_c is at $2\pi \times 460$ kHz. It is clear from an examination of these figures that the local oscillator drift will cause the upper sideband of the AM IF frequency to fall outside the passband of the IF amplifier. Thus the output of the IF amplifier would be

$$v_{\text{out}}(t) = A\left[\cos\omega_c t + \frac{m}{2}\cos(\omega_c - \omega_m)t\right]. \tag{2.16}$$

We next need to know what kind of output we will have from an envelope detector that receives this signal. The procedure for calculating this is as follows (Haykin, 1983): Equation (2.16) may be written

$$\begin{aligned} v_{\text{out}}(t) &= A\left[\cos\omega_{\text{c}}t + \frac{m}{2}\cos\omega_{\text{c}}t\cdot\cos\omega_{\text{m}}t + \frac{m}{2}\sin\omega_{\text{c}}t\cdot\sin\omega_{\text{m}}t\right] \\ &= \text{Re}\left[\tilde{v}_{\text{out}}(t)\cdot e^{j\omega_{\text{c}}t}\right], \end{aligned} \tag{2.17}$$

where $\tilde{v}_{\text{out}}(t)$ is called the complex envelope of $v_{\text{out}}(t)$. From the latter equation we may write

$$\begin{aligned} v_{\text{out}}(t) &= \text{Re}[\{\text{Re}[\tilde{v}_{\text{out}}(t)] + j\text{Im}[\tilde{v}_{\text{out}}(t)]\}\cdot(\cos\omega_{\text{c}}t + j\sin\omega_{\text{c}}t)] \\ &= \text{Re}[\text{Re}[\tilde{v}_{\text{out}}(t)]\cdot\cos\omega_{\text{c}}t - \text{Im}[\tilde{v}_{\text{out}}(t)]\cdot\sin\omega_{\text{c}}t] \\ &\quad + j[\text{Re}[\tilde{v}_{\text{out}}(t)]\cdot\sin\omega_{\text{c}}t + \text{Im}[\tilde{v}_{\text{out}}(t)]\cdot\cos\omega_{\text{c}}t] \\ &= \text{Re}[\tilde{v}_{\text{out}}(t)]\cdot\cos\omega_{\text{c}}t - \text{Im}[\tilde{v}_{\text{out}}(t)]\cdot\sin\omega_{\text{c}}t. \end{aligned}$$

This expression for $v_{\text{out}}(t)$ can be equated to that in (2.17):

$$\begin{aligned} &A\left[\cos\omega_{\text{c}}t + \frac{m}{2}\cos\omega_{\text{c}}t\cdot\cos\omega_{\text{m}}t + \frac{m}{2}\sin\omega_{\text{c}}t\cdot\sin\omega_{\text{m}}t\right] \\ &\quad = \text{Re}[\tilde{v}_{\text{out}}(t)]\cdot\cos\omega_{\text{c}}t - \text{Im}[\tilde{v}_{\text{out}}(t)]\cdot\sin\omega_{\text{c}}t. \end{aligned}$$

Equating coefficients of $\sin\omega_{\text{c}}t$ and $\cos\omega_{\text{c}}t$ gives the equations

$$A\left[1+\frac{m}{2}\cos\omega_{\text{m}}t\right] = \text{Re}[\tilde{v}_{\text{out}}(\text{t})], \qquad A\frac{m}{2}\sin\omega_{\text{m}}t = -\text{Im}[\tilde{v}_{\text{out}}(\text{t})],$$

from which we obtain

$$\tilde{v}_{\text{out}}(t) = A\left[1 + \frac{m}{2}\cos\omega_{\text{m}}t\right] - j\frac{Am}{2}\sin\omega_{\text{m}}t.$$

The time-dependent envelope, $a(t)$, is then given by the magnitude of the complex envelope:

$$\begin{aligned} a(t) = |\tilde{v}_{\text{out}}(t)| &= \left[A^2\left(1 + \frac{m}{2}\cos\omega_{\text{m}}t\right)^2 + \frac{A^2m^2}{4}\sin^2\omega_{\text{m}}t\right]^{1/2} \\ &= A\left[1 + m\cos\omega_{\text{m}}t + \frac{m^2}{4}\cos^2\omega_{\text{m}}t + \frac{m^2}{4}\sin^2\omega_{\text{m}}t\right] \\ &= A\sqrt{1 + \frac{m^2}{4} + m\cos\omega_{\text{m}}t}. \end{aligned} \tag{2.18}$$

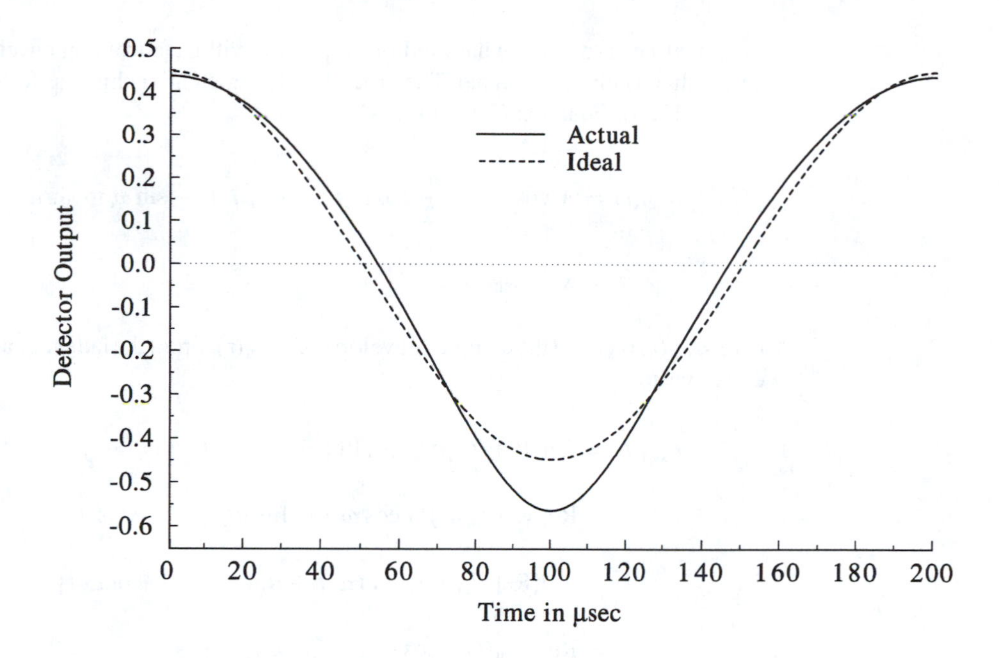

Fig. 2.13: Demodulated output of an AM receiver in the time domain for a 5 kHz modulating signal. Solid line: Spectrum offset from the IF passband by 5 kHz. Dashed line: Spectrum centered on the IF passband.

This envelope function is obviously not proportional to the modulating signal, $\cos \omega_{\mathrm{m}} t$. Yet, for small m, we may approximate (2.18) using the first two terms of its binomial expansion to yield

$$a(t) = A\sqrt{1 + (m^2/4)} + \frac{Am \cos \omega_{\mathrm{m}} t}{2\sqrt{1 + (m^2/4)}}, \tag{2.19}$$

which consists of a DC term plus the desired signal. Figure 2.13 shows the envelope from Eq. (2.18) plotted along with the desired output from Eq. (2.19) for $m = 1$. This is, of course, the worst case. Agreement improves for smaller m.

Now let us follow an essentially analogous procedure for FM. For FM with unity carrier amplitude and sinusoidal modulation

$$v_{\mathrm{in}}(t) = \cos[\omega_{\mathrm{c}} t + \beta \sin \omega_{\mathrm{m}} t]. \tag{2.20}$$

This can be expressed in terms of discrete frequency components as

$$v_{\mathrm{in}}(t) = \sum_{\mathrm{n}=-\infty}^{\infty} \mathrm{J_n}(\beta) \cdot \cos(\omega_{\mathrm{c}} + n\omega_{\mathrm{m}})t. \tag{2.21}$$

Thus in theory an infinite bandwidth would be required for exact transmission of this signal. In practice, an acceptably small distortion can be achieved with a limited bandwidth. The minimum bandwidth required is given by Carson's Rule as $f_{\mathrm{m}}(\beta + 1)$ sidebands on either side of the carrier. Thus we can approximate our input as

$$v_{\mathrm{in}}(t) \cong \sum_{\mathrm{n}=-(\beta+1)}^{\beta+1} \mathrm{J_n}(\beta) \cdot \cos(\omega_{\mathrm{c}} + n\omega_{\mathrm{m}})t. \tag{2.22}$$

Figures 2.14a and b show the power density spectra imposed upon the ideal FM IF passband when the local oscillator is on frequency and when it is one

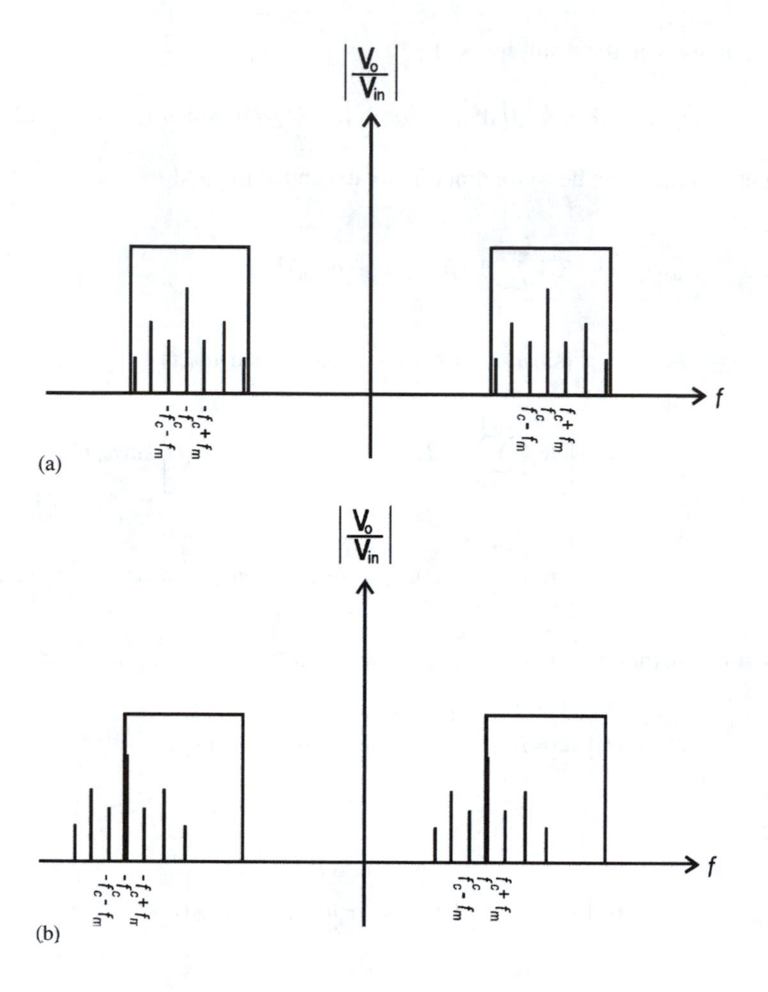

Fig. 2.14: Spectral energy diagram of an FM IF signal superimposed upon the ideal response curve of an FM IF amplifier (a) when the IF signal is properly centered at 10.7 MHz and (b) when the IF signal is offset by just .1 MHz to 10.6 MHz.

half the IF passband (100 kHz) below its proper frequency. Thus the signal out of the IF amplifier is

$$v_{\text{out}}(t) = A \sum_{\text{n}=0}^{\beta+1} \text{J}_\text{n}(\beta) \cdot \cos(\omega_\text{c} + \omega_\text{m})t. \tag{2.23}$$

We will model the FM detector as an ideal differentiator followed by an envelope detector. The differentiator will be assumed to have a transfer function given by

$$v_{\text{od}} = K \frac{dv_{\text{id}}}{dt}.$$

However, $v_{\text{id}} = v_{\text{out}}$ from the tuned amp. Accordingly,

$$v_{\text{od}} = -KA \sum_{\text{n}=0}^{\beta+1} \text{J}_\text{n}(\beta) \cdot (\omega_\text{c} + n\omega_\text{m}) \cdot \sin(\omega_\text{c} + n\omega_\text{m})t \tag{2.24}$$

This is the signal whose envelope we need to find. Set

$$v_{\text{od}}(t) = \text{Re}\big[\tilde{v}_{\text{od}}(t) \cdot e^{j\omega_\text{c}t}\big].$$

By analogy with the result for AM

$$v_{od}(t) = \text{Re}[\tilde{v}_{od}(t)] \cdot \cos\omega_c t - \text{Im}[\tilde{v}_{od}(t)] \cdot \sin\omega_c t. \tag{2.25}$$

Equation (2.24) can be trigonometrically expanded to yield

$$\begin{aligned} v_{od}(t) &= -KA \sum_{n=0}^{\beta+1} \text{J}_n(\beta) \cdot (\omega_c + n\omega_m) \\ &\quad \times [\sin\omega_c t \cdot \cos n\omega_m t + \cos\omega_c t \sin n\omega_m t] \\ &= -\left[KA \sum_{n=0}^{\beta+1} \text{J}_n(\beta)(\omega_c + n\omega_m) \cdot \cos n\omega_m t \right] \sin\omega_c t \\ &\quad - \left[KA \sum_{n=0}^{\beta+1} \text{J}_n(\beta)(\omega_c + n\omega_m) \sin n\omega_m t \right] \cos\omega_c t. \end{aligned} \tag{2.26}$$

Equating coefficients of $\sin\omega_c t$ and $\cos\omega_c t$ yields

$$\text{Re}[\tilde{v}_{od}(t)] = -KA \sum_{n=0}^{\beta+1} \text{J}_n(\beta)(\omega_c + n\omega_m) \sin n\omega_m t$$

and

$$\text{Im}[\tilde{v}_{od}(t)] = KA \sum_{n=0}^{\beta+1} \text{J}_n(\beta)(\omega_c + n\omega_m) \cos n\omega_m t,$$

so that

$$\tilde{v}_{od}(t) = -KA \sum_{n=0}^{\beta+1} \text{J}_n(\beta)(\omega_c + n\omega_m)[\sin n\omega_m t - j\cos n\omega_m t]$$

and

$$\begin{aligned} a(t) = |\tilde{v}_{od}| &= \left\{ K^2A^2 \left[\sum_{n=0}^{\beta+1} \text{J}_n(\beta)(\omega_c + n\omega_m) \sin n\omega_m t \right]^2 \right. \\ &\quad \left. + K^2A^2 \left[\sum_{n=0}^{\beta+1} \text{J}_n(\beta)(\omega_c + n\omega_m) \cos n\omega_m t \right]^2 \right\}^{1/2} \\ &= KA \left\{ \left[\sum_{n=0}^{\beta+1} \text{J}_n(\beta)(\omega_c + n\omega_m) \sin n\omega_m t \right]^2 \right. \\ &\quad \left. + \left[\sum_{n=0}^{\beta+1} \text{J}_n(\beta)(\omega_c + n\omega_m) \cos n\omega_m t \right]^2 \right\}^{1/2}. \end{aligned} \tag{2.27}$$

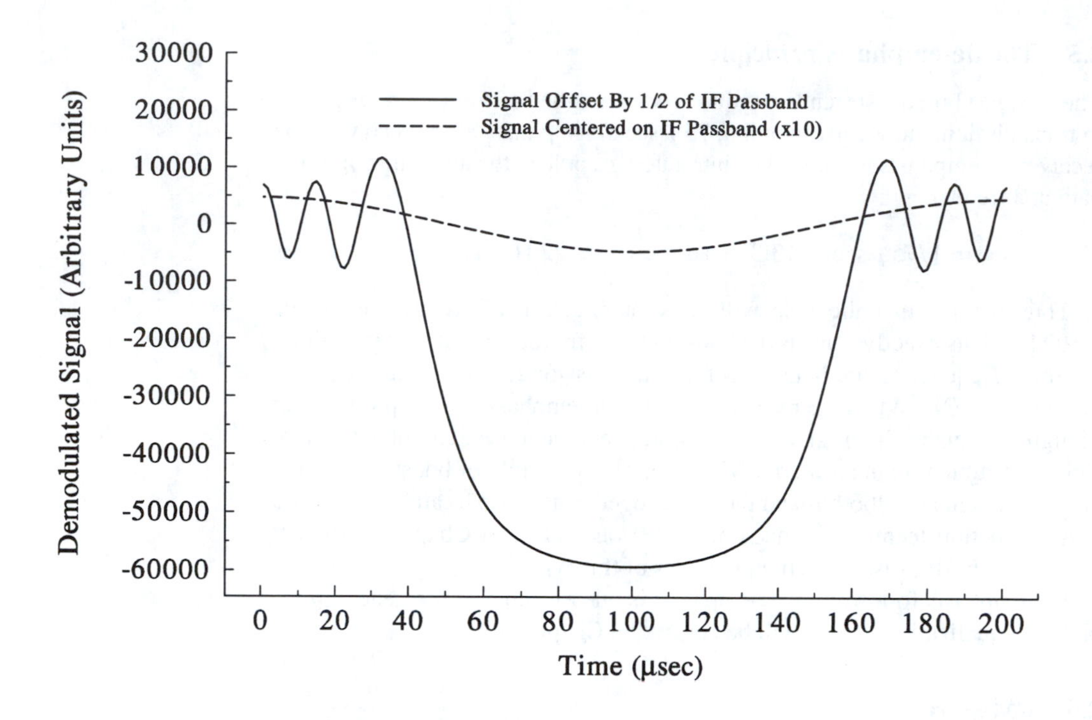

Fig. 2.15: Demodulated output of an FM receiver in the time domain for a 5 kHz modulating signal. Solid line: Spectrum offset from the IF passband by .1 MHz. Dashed line: Spectrum centered on the IF passband. Note the difference in scale between the two curves.

For FM broadcast of a 5 kHz signal, a β of 15 gives a frequency deviation of 75 kHz. In Fig. 2.15 we plot Eq. (2.27) versus time for $\beta = 15$. The signal obviously bears little resemblance to the cosine wave that we would ideally like to recover.

We conclude from all this that equal local oscillator drifts (as a fraction of the IF passband) in AM and FM are not equal in their deleterious effects on the recovered signal because FM is significantly more vulnerable to sideband attenuation. A plot similar to Fig. 2.15 was generated for $\beta = 1$, and the result was qualitatively the same.

2.4.2 *Principle of implementation*

Having established the need for AFC, we turn to a consideration of how it is implemented. As shown in Fig. 2.9, any workable FM detector displays a linear $\mathcal{V}$–f characteristic. The circuit is adjusted so that the IF frequency, f_o, lies in the center of the linear region. When the local oscillator is at the correct frequency, the IF signal will be at f_o. When the local oscillator frequency is too high, the IF signal will be above f_o, and the average output will be higher. Likewise, a lower local oscillator frequency will result in a lower average output. The average output is simply the DC component of the output. Thus this DC component can be used to adjust the local oscillator frequency by means of a voltage-variable capacitor. These devices are reverse-biased diodes that have been designed to optimize the variation of capacitance with voltage. Although this capacitance is often in the pF range, since the FM local oscillator operates at upwards of 100 MHz for the most part, the required tuning capacitance is small, and the varicap, which need do nothing more than "tweak," can do its job very well with a few pF of capacitance.

2.5 The de-emphasis principle

The U.S. FM broadcast standard requires what is called a 75 μsec *pre-emphasis* in transmission and a corresponding 75 μsec *de-emphasis* at the receiver. The receiver de-emphasis is realized by introducing a pole in the audio transmission path at a frequency

$$\omega_p = 1/75\ \mu\text{sec} = 13{,}333\ \text{rad/sec} \Rightarrow 2{,}122\ \text{Hz}.$$

Thus we see that the pole will cause attenuation of frequencies above 2,100 Hz. This exactly cancels the boost in those frequencies at the transmitter. The high frequencies are boosted before transmission to boost their signal-to-noise ratio (S/N). When the received signal is de-emphasized, the proper level of high frequencies in the signal gets restored, but the noise gets cut with them. This helps to maintain a superior S/N ratio. This principle of boosting a signal and then attenuating both it and the noise together is basic to almost all audio noise-reduction techniques and some video ones also. In Chapter 5 we will discuss Dolby-B noise reduction, which is of this type.

Pre-emphasis followed by de-emphasis is one example of the general principle of equalization, which will be covered in Chapter 4.

2.6 FM stereo

The great majority of FM receivers now come with stereo capability. Let us consider briefly how the stereo signal is generated at the transmitting station and then how it is recovered at the receiver.

2.6.1 The compatible stereo signal

Before FM stereo transmission was authorized by the FCC they established the constraint that people without FM stereo capability were not to "lose anything" in their reception. This translated into the requirement that the unmodified receiver must receive a signal that is the sum of the two stereo channel signals. These signals have come to be known as $L(t)$ and $R(t)$, which stand for the left and right channel signals respectively. Most often they are shortened to L and R, with the time dependence being implicit.

Another signal had to be transmitted that could be combined with the $L + R$ after detection but prior to being fed to the audio amplifiers. This signal was settled on being the $L - R$ signal. If you add $(L + R)$ to $(L - R)$ you get $2L$. If you subtract $(L + R) - (L - R)$ you get $2R$. Thus the channel signals are symmetrically decoded. The one remaining decision that had to be made was how to transmit the $L - R$ signal on the same carrier as the $L + R$ without increasing the bandwidth requirement. As already noted, the FCC specifies an allowed frequency deviation of ± 75 kHz about the assigned carrier frequency.

In the system adopted, a balanced AM modulator encodes the $L - R$ signal in the form of a DSB-SC signal centered about 38 kHz. This is a form of frequency-domain multiplexing and thus is called FM Multiplex (MX) stereo. A spectral power density plot showing both signals is given in Fig. 2.16.

This signal, all of which will be FM modulated onto the assigned carrier frequency, is called the *baseband* signal. Figure 2.16 also shows a discrete frequency component at 19 kHz. The purpose of this *pilot signal* will be discussed

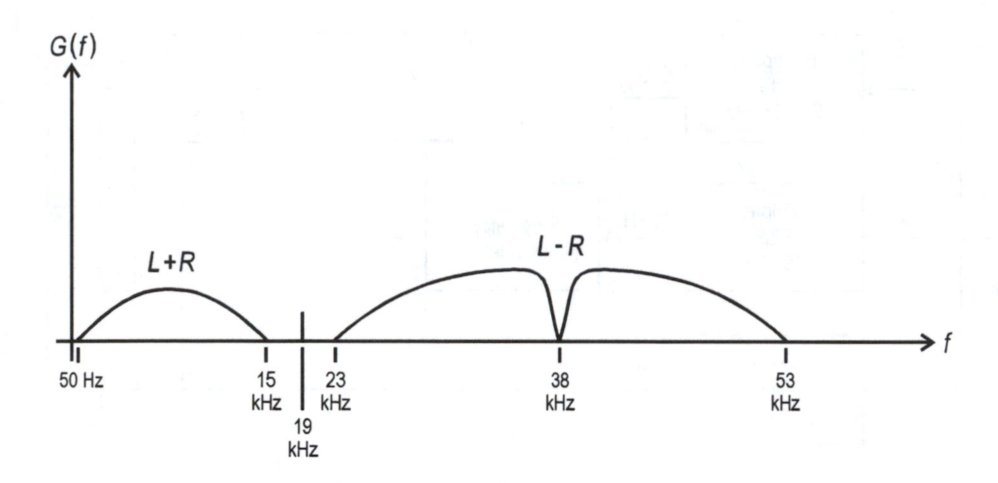

Fig. 2.16: Spectral energy plot of the baseband signal of a station broadcasting FM Multiplex stereo.

shortly. The "humps" in the spectral density plot are representative of the spectral characteristics of the program material. If the L and R signals were both simple sinusoids, the baseband spectrum would be composed of just six discrete frequency components plus the pilot.

Exercise 2.4 If $R = \sin \omega_1 t$ and $L = \sin \omega_2 t$ ($\omega_2 > \omega_1$), plot the spectral density of the baseband signal. Label all frequencies.

The frequency deviation is a function of the amplitude of the modulating signal for FM. When the baseband signal includes the DSB-SC modulated $L - R$, it might seem necessary to reduce the amplitude of $L + R$ to allow some "swing room" for $L - R$. It can be shown (Taub and Schilling, 1971) that, practically speaking, this is not the case. The pilot component *will* require some reduction of the signal components, but in practice, its amplitude is only about 10% of the signal amplitudes, requiring a reduction of only 1 dB for stereo transmission.

2.6.2 The stereo decoding circuitry

In the FM receiver, the FM stereo signal is processed from the antenna through the detector exactly like a mono signal. However, unlike a mono receiver, the stereo receiver does not have a de-emphasis network on the detector output, because the entire baseband signal is present, and the higher-frequency components that carry the $L - R$ signal would be shunted to ground. De-emphasis networks for each channel are placed between the L and R decoders and their respective audio amplifiers.

Once the baseband signal is recovered from the FM detector, we can begin to process it to extract the L and R components. Figure 2.17 shows the method used. The operation of the circuit is basically explained by the block diagram itself. The multiplier serves as a synchronous demodulator for the DSB-SC modulated $L - R$ signal. A synchronous demodulator requires that the carrier supplied to it have exactly the same phase and frequency as the carrier used to generate the modulated signal in the first place. This is why the 19 kHz pilot signal is transmitted. Frequency doubling is used both at the studio and in the receiver to generate the 38 kHz carrier from the 19 kHz pilot signal. This may

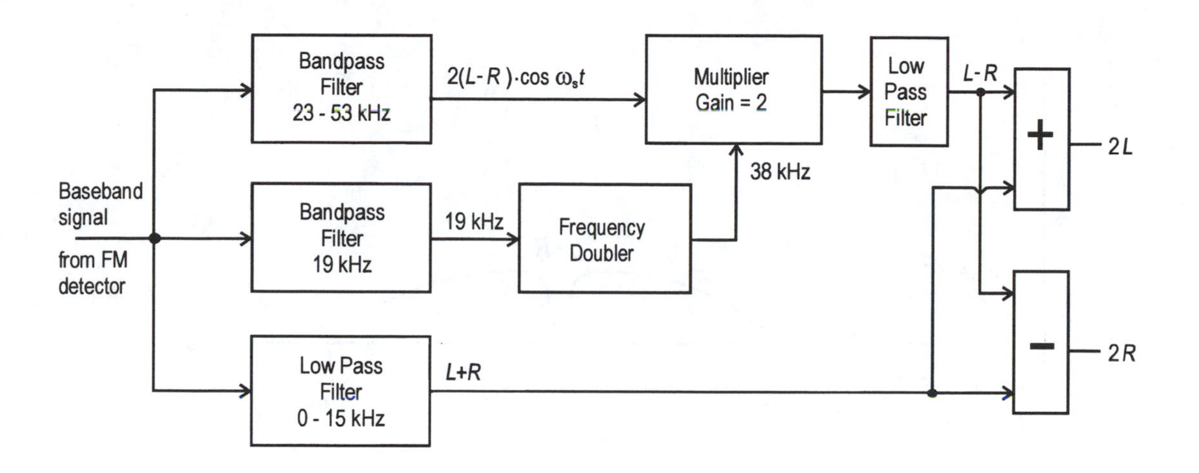

Fig. 2.17: Block diagram of an early FM Multiplex stereo decoder.

lead some to question why a pilot signal at 38 kHz is not used, thus eliminating the frequency doubler. The answer is that it would be extremely hard to extract from the $L - R$ encoded signal, which may lie as little as 50 Hz on either side from a carrier that might be transmitted at 38 kHz.

Exercise 2.5 If the multiplier is ideal, theoretically the 23–53 kHz bandpass filter is unnecessary. To show this, refer to Fig. 2.17, and make a spectral power plot showing the output of the multiplier if its two inputs are the 38 kHz carrier and the baseband signal. Label all frequencies.

2.6.3 PLL stereo decoding

Early FM MX decoders were expensive and complex, because all of the filtering was done with critically tuned passive filters. With the growth of linear ICs came the integrated phase-locked loop, which greatly simplified the filtering requirements by eliminating the bandpass. The PLL was able to lock on the 19 kHz pilot and generate the 38 kHz carrier required with its VCO. Figure 2.18 illustrates this technique.

The consumer market is, of course, a huge one, and general-purpose PLLs gave way to those especially designed for this purpose and which contain almost everything needed to implement FM stereo on a single chip. A good example of such a chip was the LM1800. Figure 2.19 shows a block diagram for this chip along with outboard components needed in conjunction with it. As the level of integration has continued to increase, chips of this sort have been supplanted by even larger ones in which the circuitry shown here is only a subset.

Several of the principles used in the AM stereo decoder chip studied in Chapter 1 are also used here, even though the method of stereo encoding used is completely different.

The composite (baseband) signal is AC-coupled to an integrated audio amplifier with a typical input impedance of 45 kΩ. This eliminates loading of the FM detector to a large extent. This amplifier's output feeds pin 2 and sees a capacitance to ground (0.0027 μF) whose function is to filter out anything above the range of the stereo frequencies. This filtered signal also goes to one

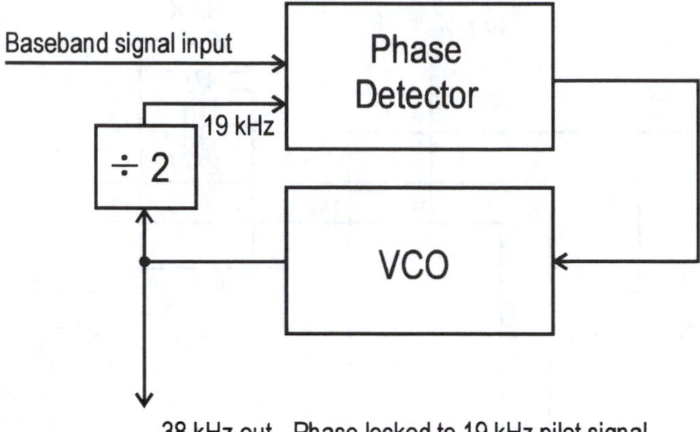

Fig. 2.18: Block diagram of a PLL subcircuit, which can regenerate the 38 kHz subcarrier required in an FM Multiplex decoder.

input of the quadrature phase detector, which is part of the phase-locked loop. The other input to this phase detector is the VCO output divided by four. The only component of the baseband signal that is always present in a stereo broadcast is the 19 kHz pilot signal. This is what the PLL locks to. This means that $f_{\mathrm{VCO}}/4 = 19$ kHz, which means $f_{\mathrm{VCO}} = 76$ kHz.

The free-running frequency of the VCO is set to 76 kHz by the components connected to pin 15. This insures the widest possible range of lock conditions for an FM broadcast signal. The VCO is designed to run at 76 kHz rather than 38 kHz because this way the 38 kHz generated by dividing down will have a duty cycle of exactly 50%. This is important because cycle asymmetry will result in impaired separation between the stereo channels.

The normal phase-locked loop establishes lock when the signals from the VCO and the "outside world" are 90° apart. Thus the loop phase detector is called a quadrature phase detector. The output of this phase detector goes through the loop filter connected to pins 13 and 14.

If the stereo signal is too weak or is absent, it should not be processed as a stereo signal, since the $L - R$ "information" consists mainly of noise. The decoding process must be self-disabled under these circumstances. Since the pilot-signal amplitude will decline as the signal strength declines, this amplitude can be used as an indicator of signal strength. However, the pilot signal comprises only a small part of the baseband signal. How can we examine just the pilot's amplitude? Perhaps the technique most readily integrated is to use another phase detector whose inputs are the baseband signal and the 19 kHz signal derived from the VCO. The PLL will insure that these inputs are at exactly the same frequency. Thus the difference frequency component in the output of this phase detector will be DC, which is readily extracted by low-pass filtering (the 0.33 μF capacitor between pins 9 and 10). Why could we not use the loop phase detector instead of adding a separate one? Because the loop phase detector is designed to operate in the limiting mode, in which case its output (the VCO correction voltage) depends only on phase and not on amplitude of the pilot signal.

The additional phase detector (called the "in-phase" detector on the circuit diagram) does not operate in the limiting mode. Therefore its DC output will depend linearly on the input pilot amplitude and will accordingly vary linearly

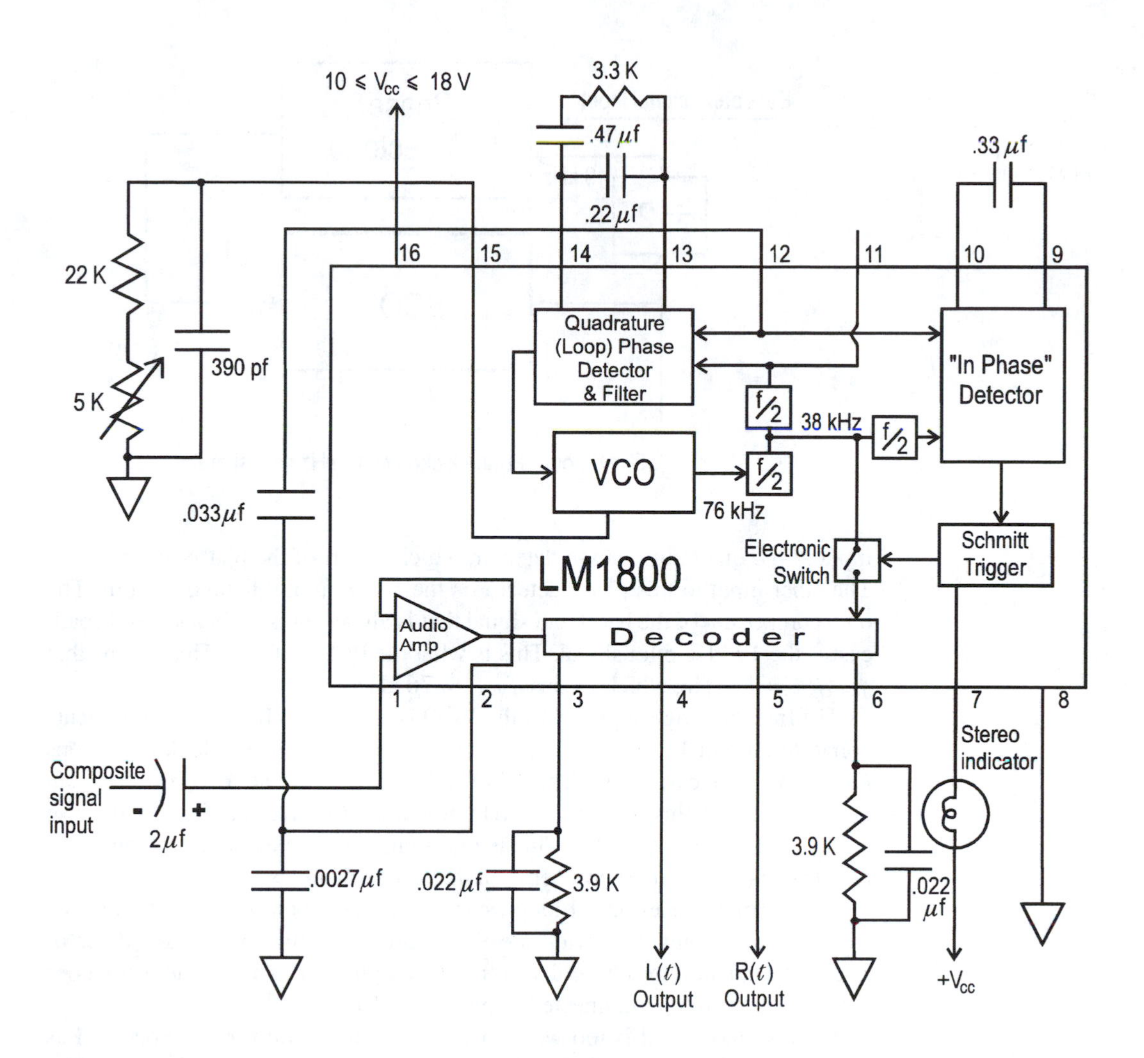

Fig. 2.19: Block diagram of the circuitry in an LM1800, an FM Multiplex stereo decoder IC.

with signal strength. The in-phase detector gets its divided-down VCO signal from a different flip-flop than the loop phase detector does. These two flip-flops provide outputs that are 90° out of phase. The 90° phase difference in these flip-flop outputs means that the inputs of the "in-phase" detector will be just that – in phase – whenever the loop is locked.

The DC output from the "in-phase" detector goes to a Schmitt Trigger whose threshold is set so that it changes state whenever the pilot signal amplitude is inadequate. When it is adequate, the Schmitt enables not only the synchronous detector but a stereo-indicator lamp output that can sink up to 100 mA as well.

When enabled, the synchronous detector recovers the $L - R$ signal by multiplying the composite signal and the regenerated 38 kHz subcarrier. This $L - R$ and the $L + R$ from the composite signal are then combined in what is called the matrix to generate the sum and difference frequencies, which are L and R. Pins 3 and 6 are for the 75 μsec de-emphasis networks for the two channels.

No method of FM stereo demultiplexing can compare to the PLL for simplicity, economy, stability, and performance. Every major analog IC manufacturer has produced one or more chips of this type.

2.7 FM receiver analysis

For purposes of analysis, it would be desirable to have an "FM only" receiver. However, this happens to be a fairly rare piece of equipment. Until the advent of FM stereo, FM was always the "weak sister" to AM and was sold almost exclusively in the form of AM–FM receivers. By the time FM might have had the stature to stand on its own, integrated circuits had advanced to the point where it cost very little more to add FM capability, and now, except for the very cheap "throwaway" and novelty radios, AM only is almost as rare as FM only. An integrated AM–FM radio is covered in Chapter 3.

The one area where FM has always been able to stand on its own has been in high-fidelity audio systems, where it has always been preferred because of its lower noise susceptibility and its higher frequency response. In high-fidelity systems, FM reception is provided by a tuner. A tuner is a receiver without any audio section or perhaps with one stage of audio preamplification. The low-level audio output is fed into the main system power amplifier. This gives far better sound quality than that which can be obtained with the small amplifiers and speakers used in radios. Both radios and tuners may properly be called receivers.

The FM receiver we will be analyzing is an older stereo tuner by H. H. Scott. This receiver features modular design. Its four modules are:

(1) front end, which includes RF amplification and IF conversion;
(2) IF amplifier and detector;
(3) multiplex stereo decoder, which we will not analyze because it uses what is now the markedly inferior technique of passive filtering to process the baseband signal components; and
(4) Squelch circuit, which will not be analyzed either because its operation is dependent in part on some of the circuitry on the MX board.

It is, however, well worth our while to look at the purpose and principle of squelch. When the FM receiver is between stations, a relatively high level of pure noise is heard. The squelch circuit is designed to cut off the FM stereo outputs when this condition exists. It does so by amplifying the noise and using the DC level of the amplifier's output to enable or disable the squelch circuitry. Squelch is also almost universally used in receivers for CB, police, fire, and taxis.

The schematic of the front end is shown in Fig. 2.20. The antenna generally used with FM and TV is a dipole, which has two elements that are balanced above ground. To apply the signal from such an antenna to a single-input amplifier, some measure must be taken to unbalance it without shorting or distorting its output. The approach used in this receiver is to connect it between ground and one of the taps of a doubly tapped tuning coil, L_{201}. The autotransformer action of this coil matches the low characteristic impedance of the antenna and its twin lead and steps it up by a large turns ratio squared so that it appears as a large shunt impedance to this input tuned circuit. The other tap on L_{201} feeds Q_{201} and Q_{202}, which together comprise the RF amplifier. The tuning circuit is exactly analogous to the input tuned circuit of the AM receiver analyzed in the last chapter. The three capacitors are one section of the main tuning capacitor, a trimmer, and a temperature compensating capacitor (NPO).

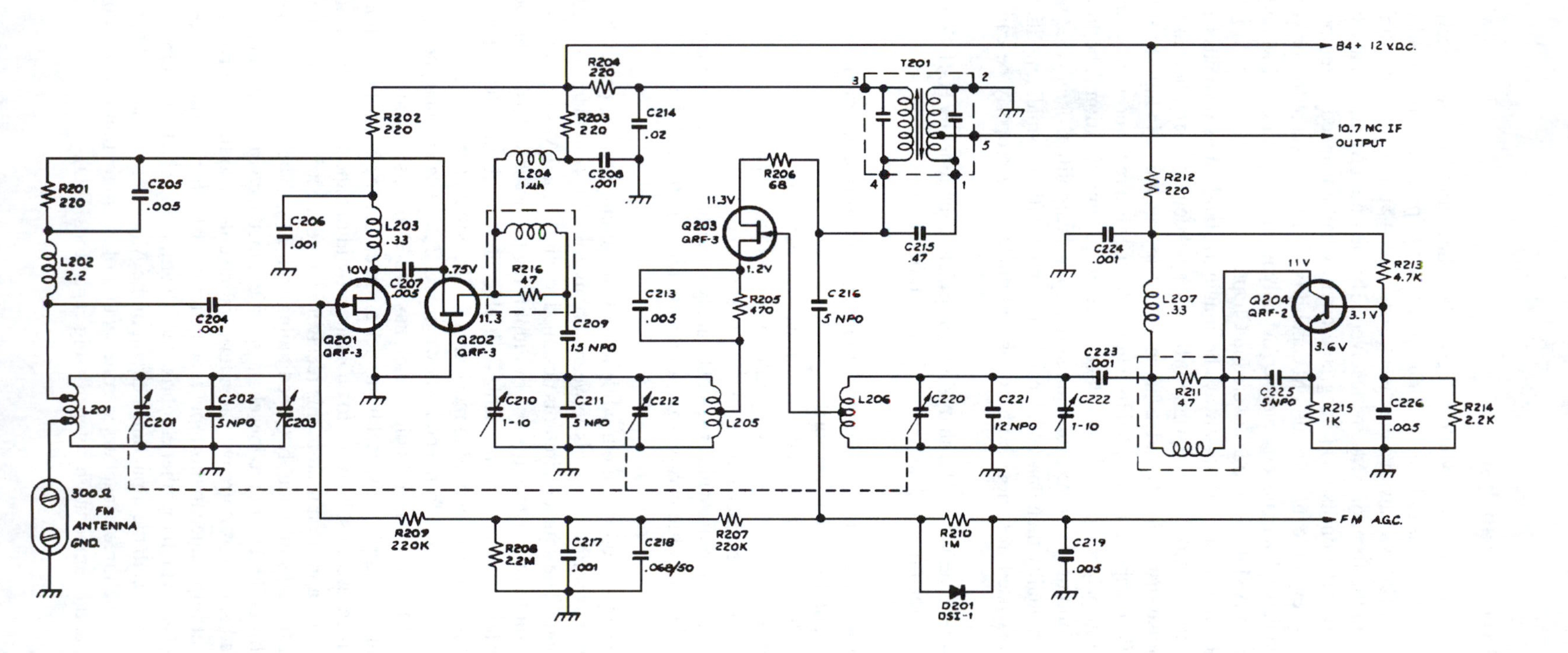

Fig. 2.20: Schematic diagram of the "front end" circuitry in the H. H. Scott FM tuner.

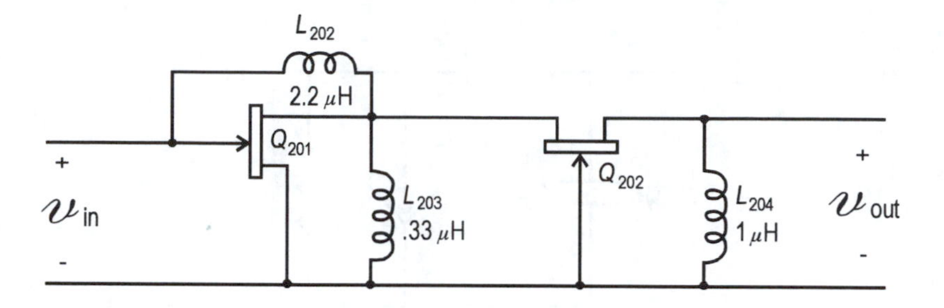

Fig. 2.21: Schematic diagram of the RF amplifier portion of Fig. 2.20 after "AC simplification," which consists of treating all capacitors of 1 nf or larger as shorts for AC.

2.7.1 *The RF amplifier*

As we contemplate AC analysis of this stage, it seems much more complicated than it is. This is because most of the capacitors are effectively AC shorts. In particular, C_{205} is part of the feedback loop. It is there to short out R_{201}, which is the source resistor of Q_{202} and through which its bias current flows. C_{204} is a blocking capacitor to keep the AGC (automatic gain control) voltage from being grounded out through L_{201}. The term AGC is a generalization of AVC and is probably to be preferred. The volume is controlled only because the gain of the IF stage(s) is. C_{206} and R_{202} form a decoupling network. Decoupling is necessary because the DC supply is not a perfect AC short. Typically there might be a millivolt or less of signal AC on the supply line. If this were allowed to enter a stage amplifying 2–60 microvolts, its effects could be very deleterious, ranging from oscillation to distortion to modulation. Thus decoupling networks are universally used in low-level stages. They are actually just another section of low-pass filtering, with the capacitor forming a "local" AC short for each stage. C_{208} is also an AC short, working with R_{203} to decouple the drain supply of Q_{202}. C_{207} is a coupling capacitor that acts as an AC short to couple Q_{201} and Q_{202}.

Most of the inductances on the schematic have no units after their numerical values, but L_{204} is specified as 1 μH. It may, therefore, be reasonable to assume that the other inductance values are in microhenries also. This conclusion, it will be seen, is supported by the circuit analysis.

When all of the external capacitors are considered shorts, the AC equivalent circuit simplifies to Fig. 2.21. It is obvious from this figure that the second stage is in the common-gate configuration, and the first is in common-source. But a common-source feeding a common-gate is just an FET cascode pair, which is certainly a reasonable design choice for an FM RF amplifier. This, however, presents two problems – one minor and one major. The minor problem is that the loads for both stages appear to be inductive, which would imply that the gain over the FM band will not be constant. However, because this band is only 0.09 decades wide, the gain variation need not be large. We shall see that other factors further reduce the significance of this effect. The second and more serious problem concerns the effects of L_{202}. In the first place, a cascode has no transimpedance feedback such as this inductance tends to provide. Furthermore, the use of negative transimpedance feedback on an FET will lower the input impedance of the stage dramatically. It seems a little strange to use a high input impedance device like an FET for a voltage amplifier and then apply a form of feedback that will lower the input impedance and thus make it a *less* nearly ideal voltage amplifier. Finally, negative feedback, such as L_{202} seems to provide,

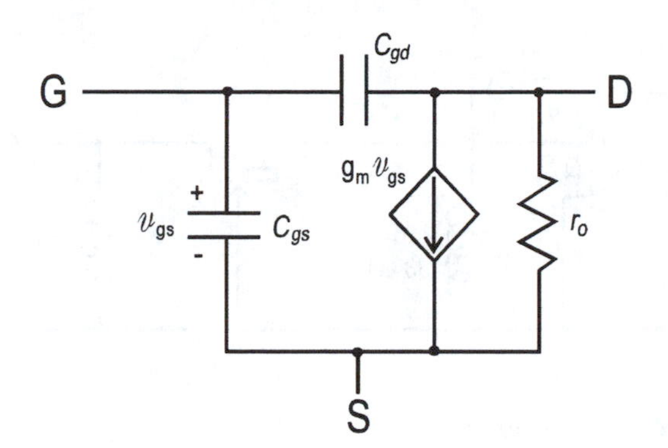

Fig. 2.22: Simplified equivalent circuit for an FET.

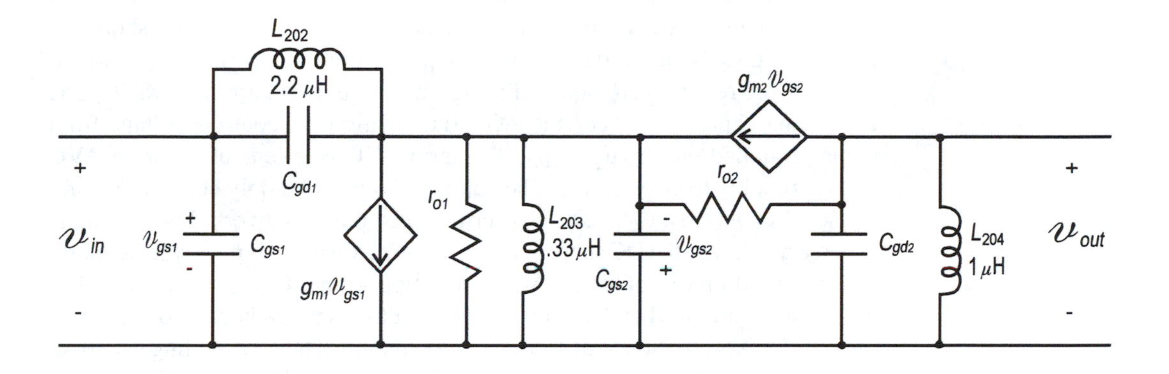

Fig. 2.23: Schematic diagram of the same circuit shown in Fig. 2.21 after replacing the FET with the equivalent circuit of Fig. 2.22.

would broaden the frequency response of the stage. This is not desirable in this application, because the FM band is quite narrow, and any extra bandwidth will only allow a broader spectrum of noise into the amplifier. What we want to do is restrict the bandwidth. This will require capacitors. But of course! Figure 2.21 does not incorporate the FET AC equivalent circuit, which has several capacitors in it. We shall use the equivalent circuit for the FET (Sedra and Smith, 1991) shown in Fig. 2.22. When Fig. 2.22 is inserted into Fig. 2.21 in place of the FET, we have Fig. 2.23. The most significant observation we can make about this circuit is that every inductance is now seen to be in shunt with one of the inter-electrode capacitances.

It would be desirable to be able to simplify this circuit further to facilitate analysis. We begin with the most obvious simplifications. The input capacitor, C_{gs1}, can be omitted since v_{in} will not depend on it. The output shunt resistance, r_{o1}, should be negligible since it is in parallel with the input impedance of a common-gate stage, which will be

$$\mathbf{Z}_{\mathrm{in}} = \frac{r_{\mathrm{o2}} + \dfrac{jL_{204}\omega}{1 - \omega^2 L_{204} C_{\mathrm{gd2}}}}{g_{\mathrm{m2}} r_{\mathrm{o2}}}.$$

Because r_{o2} might typically be 100 kΩ and $L_{204}\omega$ would be about 600 Ω at the middle of the FM band, the complex term in the numerator should be negligible unless the frequency is very close to the resonant frequency of the output L C. We shall return and consider this in more detail shortly. For now we say

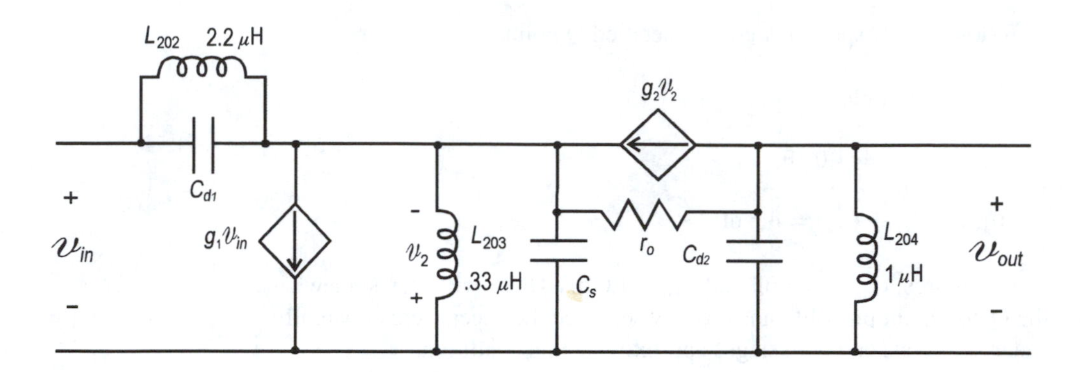

Fig. 2.24: Further simplified version of Fig. 2.23 obtained by ignoring r_{o1} relative to the input impedance of the common-gate stage.

that if $L_{204}\omega \ll r_{o2}$, then $\mathbf{Z}_{\rm in} = 1/g_{\rm m}$ for the common-gate stage. Since this is on the order of a few hundred ohms, it should be $\ll r_{o1}$, rendering the latter negligible.

From Fig. 2.23, we therefore derive the somewhat simplified circuit of Fig. 2.24 with somewhat simplified notation.

We observe that from the standpoint of gain, it would be desirable to have all three of the parallel *LC* circuits at or near resonance. Let us consider each.

(1) L_{202}–$C_{\rm d1}$. If this impedance were infinite, it would totally eliminate the Miller effect and maximize gain. The cascode configuration is used in the first place to minimize the Miller effect, and the use of L_{202} will further minimize it if these two elements are near resonance. This in itself could explain why transimpedance feedback is provided via L_{202}. It already exists via $C_{\rm d1}$, and L_{202} tends to cancel it.

(2) L_{203}–$C_{\rm s}$. If this pair of elements is chosen to be near resonance in the FM band, it will maximize the load impedance seen by the first stage and thus its gain. The parallel impedance only need be much greater than the common-gate input impedance (with which it is in parallel) to accomplish this end.

(3) L_{204}–$C_{\rm d2}$. This pair of elements is the explicit load for the common-gate stage. If the *LC* is near resonance, its impedance will be large, maximizing gain.

To assess the frequencies at which these *LC* circuits are resonant, we will have to know something about the values of the FET capacitors. Unfortunately, the QRF-3 is an obsolete device, and the values of its capacitances are hard to determine. However, a number of manufacturers of replacement semiconductors list replacement devices for the QRF-3, and we can use the specs of a replacement device as an approximation to those of the QRF-3. The most complete such set of specifications is for the Sylvania ECG312, which gives us the two fundamental DC specs,

$$5\,\text{mA} \leq I_{\rm DSS} \leq 15\,\text{mA}$$

and

$$|V_{\rm p}| \leq 6\,\text{V},$$

and the following AC specs at some unspecified Q point

$$g_m = 5.5\,\text{mmhos},$$

$$C_{rss}(= C_{gd}) = 1.0\,\text{pF},$$

$$C_{iss}(= C_{gd} + C_{gs}) = 4.5\,\text{pF}.$$

From these figures, $C_{gs} = 3.5$ pF and $C_{gd} = 1.0$ pF. These will vary somewhat with the Q point, but probably not strongly so, since the specs were apparently taken near zero bias, and both stages approximate this condition.

Exercise 2.6 Using the relationships

$$g_m = \frac{2\sqrt{I_D I_{DSS}}}{|V_p|}, \qquad \frac{I_D}{I_{DSS}} = \left(1 - \frac{V_{GS}}{V_p}\right)^2,$$

both of which apply in the pinchoff region, and the specs just given, justify the assertion that the specs were probably taken near zero gate bias.

If we use the specified values of the ECG312 as the actual values for the QRF-3, we have $C_{d1} = 1.0$ pF, $C_s = 3.5$ pF, and $C_{d2} = 1.0$ pF. The three L–C pairs then have resonant frequencies given by:

$$(1/2\pi) \cdot (1/L_{202}C_{d1})^{1/2} = 107\,\text{MHz},$$

$$(1/2\pi) \cdot (1/L_{203}C_s)^{1/2} = 148\,\text{MHz},$$

$$(1/2\pi) \cdot (1/L_{204}C_{d2})^{1/2} = 159\,\text{MHz}.$$

The resonant frequency of the L_{202}–C_{d1} combination is just at the top of the FM band. Even at the low end of the band its impedance is almost 4 kΩ, which allows us to approximate this parallel combination as an open circuit over the FM band. In other words, it allows us to ignore the Miller effect. This is not surprising since the cascode configuration is used for just that reason.

Since the other two resonant frequencies are well above the FM band, we can conclude that the impedances of their tuned circuits will be rising throughout the FM band, and thus the amplifier gain will do so as well. Since these other two resonant circuits are operating over frequencies well removed from their resonant frequencies, their impedances over the FM band would be expected to be low enough to allow us to neglect r_o. A numerical calculation confirms this. Using the values of capacitance assumed previously, neither LC ever has an impedance greater than 1.5 kΩ in the FM band. Using these further simplifications, the circuit in Fig. 2.24 can be reduced to that of Fig. 2.25.

The circuit equations are

$$g_1\mathbf{V}_{in} = \frac{\mathbf{V}_2}{sL_{203}} + sC_s\mathbf{V}_2 + g_2\mathbf{V}_2, \qquad g_2\mathbf{V}_2 + \frac{\mathbf{V}_o}{sL_{204}} + sC_d\mathbf{V}_o = 0,$$

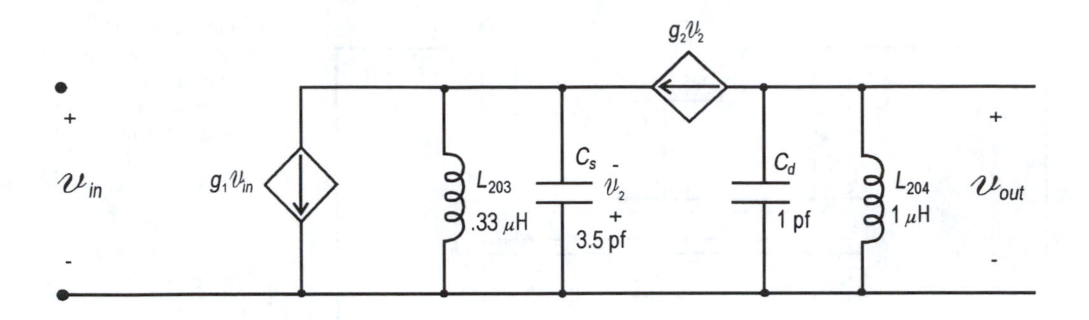

Fig. 2.25: Further simplified version of Fig. 2.24 obtained by considering numerical values of the impedances of the resonant circuits.

from which we obtain

$$\frac{\mathbf{V}_o}{\mathbf{V}_{in}}(s) = \frac{-s^2 L_{203} L_{204} g_1 g_2}{(1 + s^2 L_{203} C_2 + s L_{203} g_2)(1 + s^2 L_{204} C_d)}.$$

From this expression it would be easy to find poles and zeros and make a Bode amplitude plot, but it would be of little use because the Bode plot is a poor approximation to the actual response of high-Q circuits near resonance. Accordingly, we find the magnitude of the gain as a function of ω. The result is

$$|A_v| = \frac{\omega^2 L_{203} L_{204} g_1 g_2}{(1 - \omega^2 L_{204} C_d)\sqrt{(1 - \omega^2 L_{203} C_s)^2 + (L_{203} g_2 \omega)^2}}.$$

Using the numerical values, we can write this expression as

$$|A_v| = \frac{9.98 \times 10^{-18}\, \omega^2}{(1 - 10^{-18}\, \omega^2)\sqrt{(1 - 1.155 \times 10^{-18}\, \omega^2)^2 + 3.29 \times 10^{-18}\, \omega^2}}.$$

A numerical calculation based on this expression shows $|A_v|$ increasing monotonically from 11.3 dB at 88 MHz to 16.2 dB at 108 MHz. This gain variation of 4.9 dB would be considered excessive for good design in a modern receiver. There are, however, a couple of effects that might at least reduce this large variation in gain:

(1) The internal capacitances of the ECG312 may be significantly different than those of the QRF-3. The latter device, being the older, may have been physically larger and had larger capacitances. This is not probable, however, because to shift from near 150 MHz down to somewhere around 100 MHz would require both C_s and C_{d2} to be more than double the values we have used. This seems unlikely. However, it should be noted that, regardless of what the specs of the QRF-3 were, if it were to be replaced in circuit by the ECG312, the receiver's performance would be expected to be something close to what we have calculated here.

(2) Since Fig. 2.25 shows no effective resistances in parallel with the two LC circuits to make Q finite for each, it might be supposed that adding these would lower the resonant peaks and thus reduce the gain variation over the FM band. Although this effect is a real one, and

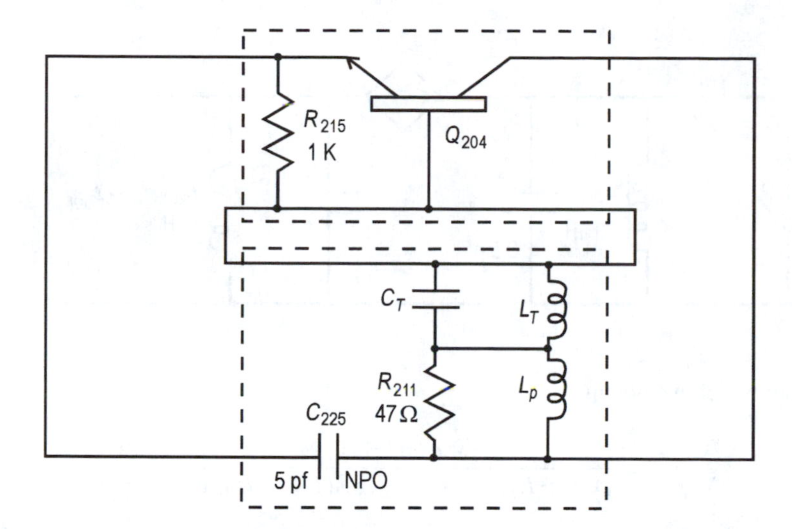

Fig. 2.26: AC equivalent circuit of the local oscillator in the H. H. Scott FM receiver drawn in standard feedback form.

it proceeds in the right direction, its magnitude is inadequate. The Q values must drop from infinity down to about ten to reduce the gain variation by 1 dB over the FM band. Further lowering of Q will cause much greater reduction in gain variation, but Q values of 10 or less at FM frequencies are not reasonable.

We, therefore, conclude that it is probable that the amplifier designer was willing to use an RF amplifier with a rising gain characteristic over the FM band.

The output of Q_{202}, which is in the CG configuration, is developed across L_{204}, which will have an impedance of 600–700 Ω over the range of RF frequencies. The low-potential end of L_{204} is AC-grounded by C_{208}. The output from this stage is coupled through a parallel RL and C_{209} to the output tuned circuit, C_{210}, C_{211}, C_{212}, and L_{205}. C_{212} is part of the three-gang tuning capacitor (C_{201}, C_{212}, C_{220}) as indicated. It is difficult to determine the function of the parallel RL since the value of the inductance is not given. C_{209}, which is also temperature compensated, has a nominal impedance in the FM band of only about 100 Ω. This is much less than that of the output tuned circuit, so almost all of the output voltage of Q_{202} is developed across the tuned circuit. C_{209} will, however, have an effect not only on the resonant frequency, but more importantly, on the temperature coefficient of the resonant frequency.

2.7.2 The local oscillator

This circuit is made up of Q_{204} and its associated components. Capacitors C_{224}, C_{223}, and C_{226} are all AC shorts at the frequency of oscillation. This means that the stage is operated in the common-base configuration and bias resistors R_{213} and R_{214} are bypassed. The stage can be drawn in standard feedback configuration as shown in Fig. 2.26.

The tuning inductance, L_{T}, is the parallel combination of L_{207} (0.33 μH) and L_{206}, which is not identified as to value. The tuning capacitance, C_{T}, is composed of C_{220}, C_{221}, and C_{222} in parallel. L_{p} and R_{211} are combined in parallel. As was the case in the RF amplifier, the value of L_{p} is not given, and the effect of the combination is second order – that is, its impedance is expected to be small

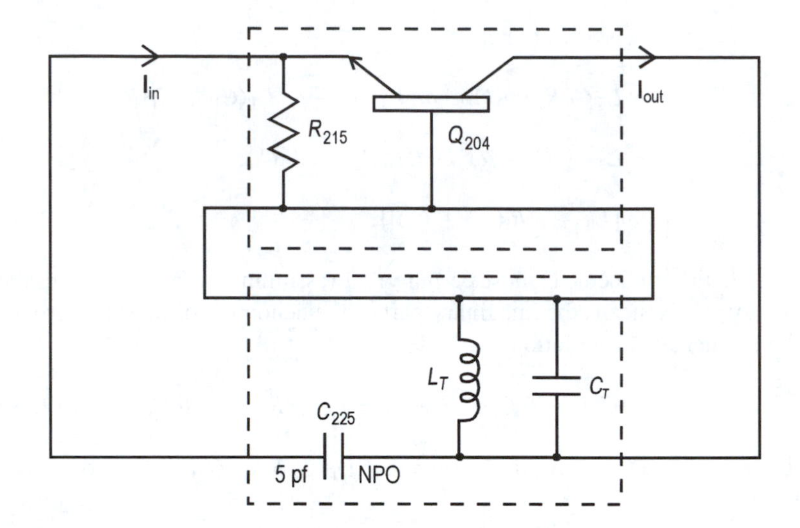

Fig. 2.27: Simplified version of Fig. 2.26 obtained by ignoring the impedance in series with the parallel resonant circuit $L_T - C_T$.

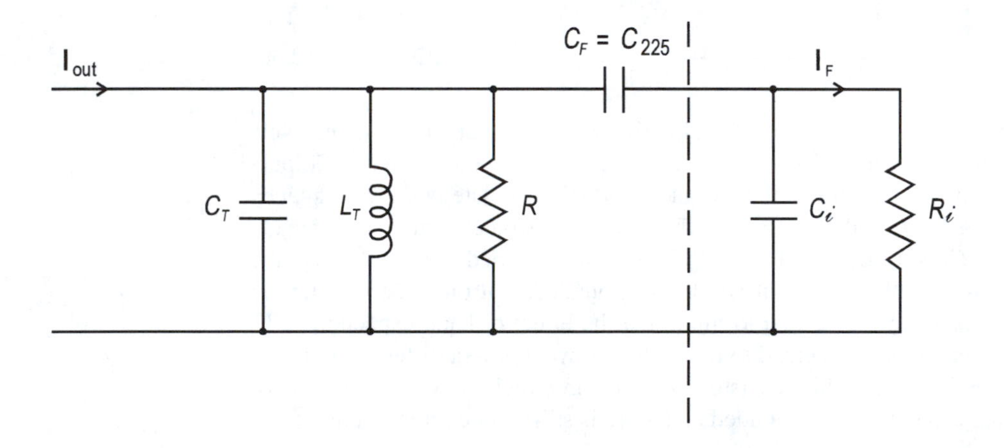

Fig. 2.28: Schematic diagram of the feedback network of Fig. 2.27 including the loading effect of the input of the amplifier block and with the implicit losses of the L-C being modeled by R.

compared to that of the L_T–C_T combination at the frequency of oscillation. If this is the case, the equivalent circuit becomes that of Fig. 2.27.

A first-order analysis of this circuit yields no condition under which oscillation will occur. It is necessary to introduce an output resistance into the formulation, even though it drops out of the final result if it is the only nonideality considered. We note that if there were a capacitance in parallel with R_{215}, Fig. 2.27 would represent a deviant form of the common-base Colpitts. But there actually is such a capacitance. It is C_i, the input capacitance of Q_{204}. Ignoring r_x, this is simply C_π, which is generally quite large compared to C_{225}. As long as we are considering the capacitive loading of the amplifier input on the feedback network's output, we ought to include the resistive component, which we will call R_i. The emitter resistance, R_{215}, is rightly a part of the loading on F. It can be lumped with R_i, but practically, it is so much larger than R_i that it is negligible for AC.

The deviant character of this "Colpitts-like" oscillator is evidenced by the presence of C_T in the circuit. The reason for this will soon become evident.

The feedback network that we have for analysis is thus represented by Fig. 2.28, where R models the losses in L_T (and C_T). The transfer function

$\mathbf{F} = \mathbf{I}_F/\mathbf{I}_{out}$ may be shown to be

$$\mathbf{F} = -\omega^2 L_T C_F R/\{[R(1-\omega^2 L_T C_F) - \omega^2 L_T(C_F + C_i)R_i - \omega^2 L_T C_T R] - j\omega^3 L_T R R_i C_T(C_F + C_i) + j\omega[L_T + RR_i(C_F + C_i)]\}. \tag{2.28}$$

For oscillation to occur, **F** must be pure real (assuming no phase shift in the transistor). This means the imaginary part of the denominator must be zero, or at the frequency of oscillation

$$\omega_o^2 L_T R R_i [C_T(C_F + C_i) + C_F C_i] = L_T + RR_i(C_F + C_i). \tag{2.29}$$

In the exercise to follow, it is shown that $L_T \ll RR_i \cdot (C_F + C_i)$, so

$$\omega_o^2 = \frac{RR_i(C_F + C_i)}{L_T R R_i [C_T(C_F + C_i) + C_F C_i]} = \frac{1}{L_T(C_T + C_F C_i/(C_F + C_i))} = \frac{1}{L_T(C_T + C_S)}, \tag{2.30}$$

where $C_s \equiv C_F C_i/(C_F + C_i)$ is the effective value of C_F and C_i in series. Now we are in a position to see why C_T was added to the common-base Colpitts circuit. If it were not there, the circuit would still oscillate but at a frequency determined by L and C_s. To tune this oscillator would require varying C_s. However, $C_s = C_F C_i/(C_F + C_i)$. If C_F were to be varied, we would have the problem that neither end of it would be grounded. But one side of a tuning capacitor is always grounded to minimize the effect of hand capacitance. If, on the other hand, we decided to tune with C_i, we'd have a different problem, since C_i is internal to the transistor! With C_T in circuit, however, one side of the tuning capacitance is grounded and there is still a wide tuning range.

Exercise 2.7 Assume that the local oscillator operates 10.7 MHz above the FM station carrier to which the receiver is tuned. Also assume that $C_i \gg C_F$ and that $C_{222} = 5$ pF.

(a) If $C_{220} = 0$ when the receiver is tuned to 108 MHz, what is its value when the receiver is tuned to 88 MHz?
(b) What is the value of C_T at midband?
(c) What is the value of L_T?
(d) What is the value of L_{206}?
(e) If $R = 10$ kΩ, $R_i = 20$ Ω, and $C_i = 100$ pF, compare the magnitudes of the two terms on the right-hand side of (2.29).

If the imaginary part of (2.28) is zero then

$$\mathbf{F} = \frac{-\omega_o^2 L_T C_F R}{R\left(1 - \omega_o^2 L_T(C_F + C_T)\right) - \omega_o^2 L_T(C_F + C_i)R_i}.$$

Inserting (2.30) into this equation yields

$$\mathbf{F} = \frac{-C_F R}{R(C_s - C_F) - R_i(C_F + C_i)}.$$

Reinserting the defining relationship for C_s gives

$$\mathbf{F} = \frac{-C_F R}{R\left(\frac{C_F C_i}{C_F + C_i} - C_F\right) - R_i(C_F + C_i)}$$

$$= \frac{-C_F R(C_F + C_i)}{R\left(C_F C_i - C_F^2 - C_F C_i\right) - R_i(C_F + C_i)^2}$$

$$= \frac{C_F(C_F + C_i)R}{C_F^2 R + (C_F + C_i)^2 R_i}.$$

The loop gain is $\mathbf{AF} = \alpha\mathbf{F}$. For the sign convention shown in Fig. 2.27, this gain must equal or exceed unity in order to have oscillation:

$$1 \leq \frac{\alpha C_F(C_F + C_i)R}{C_F^2 R + (C_F + C_i)^2 R_i}$$

and so

$$\alpha \geq \frac{C_F^2 R + (C_F + C_i)^2 R_i}{C_F(C_F + C_i)R} = \frac{C_F}{(C_F + C_i)} + \frac{(C_F + C_i)R_i}{C_F R}.$$

If oscillation is to be obtained with $\mathbf{A} = \alpha < 1$ for a common-base configuration, the terms on the right must sum to less than unity. Even though α, the open loop current gain, is less than unity, oscillation is still achievable since $\mathbf{F} > 1$, and we only require $\mathbf{AF} > 1$ for oscillation.

2.7.3 The mixer

The outputs of the RF amplifier and the local oscillator are both coupled to the mixer stage, Q_{203}, via taps on their output coils, L_{205} and L_{206} respectively. The local oscillator is fed to the gate and the RF is fed to the source through C_{213}, which is an AC short. This capacitor bypasses R_{205}, which is a bias resistor for the mixer. The combination of Q point and signal levels applied to this stage is designed to assure nonlinear operation so that mixing is achieved.

The use of R_{206} in series with the drain lead might seem pointless, since the impedance of the tuned circuit at resonance is not only resistive but much higher than that of R_{206}, with which it is in series. We showed in Chapter 1 that the coupling capacitor between input and output (C_μ in a bipolar or C_{gd} in an FET) gives rise to the Miller effect, which dramatically cuts an amplifier's high-frequency response and can easily lead to oscillation if the amplifier load is a tuned circuit. In an amplifier that has both the input and output tuned, this same capacitor can cause mischief of yet another type. Consider the schematic of a JFET amplifier with both input and output tuned as shown in Fig. 2.29.

Whether the active device is a bipolar or a JFET, C_F is the capacitance of what is nominally a reverse-biased junction. However, like any junction capacitance, it varies with applied voltage and can increase by orders of magnitude if it becomes forward biased. If $R = 0$ and C_F becomes comparable to C_i and C_o, it couples the two tuned circuits, and instead of having two single-tuned circuits with a narrow passband yielding good selectivity, you have a double-tuned

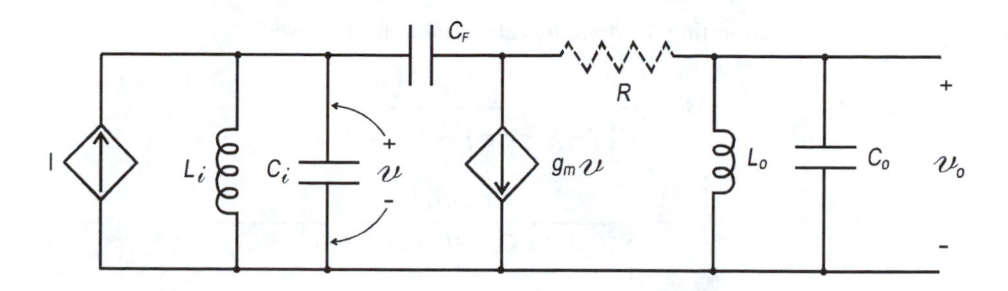

Fig. 2.29: Schematic diagram of an FET amplifier with both input and output tuned.

circuit with a widened passband. This unwanted coupling can often lead to oscillation as well. The capacitance on the input of the mixer is on the order of 10 pF, so C_F need not be very large for this to be a problem. The addition of R simply puts a limit on the amount of coupling possible between the input and output tuned circuits. Because R is in series with the output tuned circuit, even a fairly small resistor will develop a significant voltage difference between the tuned circuits, thus decoupling them.

2.7.4 *The IF amplifier*

Referring to Fig. 2.30, we see that the signal out of the front end is FM-modulated around 10.7 MHz. It is coupled to Q_{301}, the first IF amplifier through C_{302}, whose impedance at 10.7 MHz is 15 Ω. This capacitor may thus be considered an AC short. The base of Q_{301} is biased in part by the AGC through R_{304} and in part by the +12 V supply through R_{305}. The AGC will be analyzed in more detail later. The emitter is conventionally biased and bypassed. R_{306}, which also connects to the emitter, feeds the signal strength meter. The collector current flows through the primary of T_{301}. This transformer mismatches Q_{301} in exactly the same way and for the same reasons as the IF transformers in the AM receiver we analyzed. C_{305} is an AC short to ground the primary tap of T_{301}. It also decouples the collector supply to Q_{301} in conjunction with R_{307}.

On the secondary of T_{301} we see a new configuration. Base bias is derived exclusively from the 12 V supply via the R_{310}–R_{312} divider. This stage receives no AGC. C_{307} grounds the low end of the secondary. We thus have another example of the circuit technique of placing a resistor (R_{311}) across the tuned secondary to *lower* the Q. In some instances this is necessary because too high a Q will shrink the bandwidth to the exclusion of some of the sideband frequencies. The secondary tap better matches the low input impedance of Q_{302}. The emitter bypass of this stage is identical to that of the first stage. Again a resistor connected to this emitter goes to the signal strength meter. In the absence of a signal, there is no AGC voltage, and stages 1 and 2 should have the same Q point. Thus there is no voltage difference between the emitters and the meter does not deflect. When a signal is tuned in, the bias on Q_{301} drops as does its emitter potential. This places a net voltage drop across the meter, and it deflects upscale. The stronger the signal, the larger the deflection. The main tuning capacitor should be adjusted for maximum meter reading. Both ends of the meter have a capacitor to ground (C_{304}, C_{308}), which provides (in conjunction with the series resistors) additional low-pass filtering. The meter should see DC only.

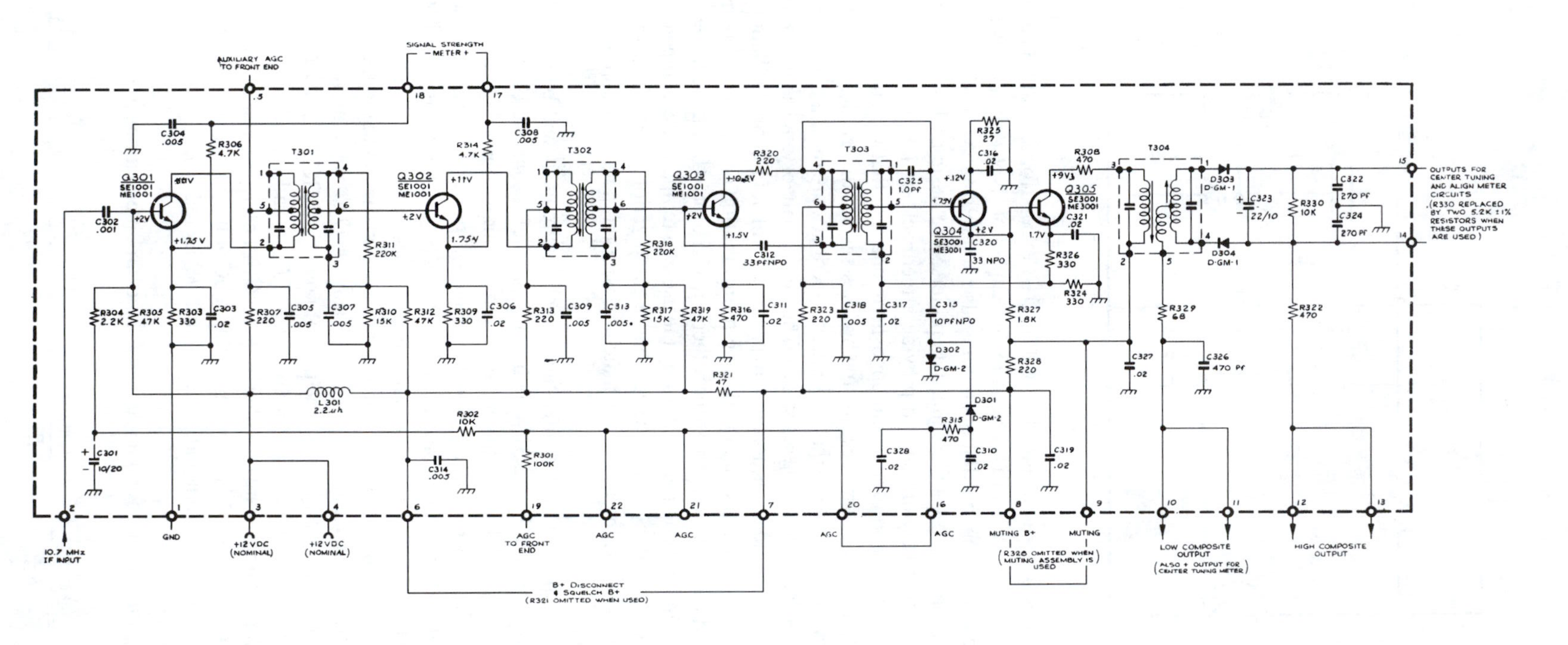

Fig. 2.30: Schematic diagram of the IF amplifier and detector circuitry used in the H. H. Scott FM tuner.

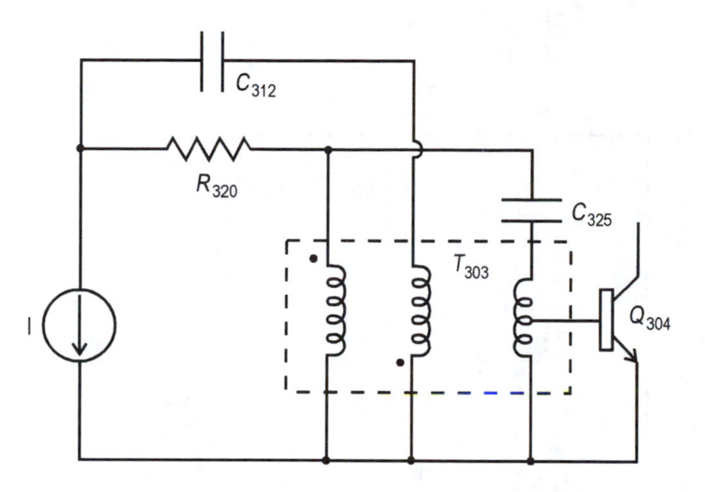

Fig. 2.31: Schematic diagram isolating the coupling circuitry between Q_{303} and Q_{304} in Fig. 2.30.

The input to Q_{303} is identical to that of Q_{302} except for an additional connection from C_{312}, which is a neutralizing capacitor. Not only is this stage mismatched to prevent oscillation, but it is also neutralized. *Neutralization* is a less-used technique to stabilize tuned amplifiers. The basic philosophy is that the neutralization network should provide a feedback signal to the transistor base that is equal in amplitude and opposite in phase to that which is fed back through C_{μ} and r_{μ}. In practice, the method has several difficulties, stemming in part from the fact that internal and external feedback are returned to opposite ends of r_{x}, the base-spreading resistance. Thus, even in theory, the neutralization cannot be perfect, and if it is not close to perfect, it will not inhibit oscillation. In practice there is also a problem with long-term variation of critical component values by aging as well as short-term thermal variation. (In this receiver, the latter is presumably compensated for by the use of a temperature-compensated neutralizing capacitor.) For these reasons, neutralization by itself is seldom used. In this circuit, the use of mismatching with the neutralization makes it less critical.

To analyze the rest of this interstage coupling circuit, it will be helpful to redraw it as shown in Fig. 2.31. The internal parallel resonating capacitors of T_{303} are not shown. The current source is from the output port of Q_{303}. The left and center windings of T_{303} are actually its tapped primary. They, along with C_{312}, provide the neutralization. The function of R_{320} is the same as that of R_{206} in the mixer stage, which we have already discussed.

Signal from Q_{303} is not only coupled into Q_{304} by the mutual inductance of T_{303}, but also by capacitance C_{325}. Before leaving Q_{303} we want to note that the right-hand side of R_{320} is also the takeoff point for AGC. The AGC rectifier is a voltage doubler, consisting of C_{315}, D_{302}, D_{301}, and C_{310}. Two stages of low-pass ripple filtering are provided by R_{315}–C_{328} and R_{302}–C_{301} before being applied to Q_{301}. Figure 2.32 shows how this AGC voltage fits into the total bias circuitry for the FET front end. In this figure R_{s} includes R_{315} plus the AGC source resistance.

Ignoring the current drain through R_{301}, the circuit equations are

$$\frac{V_{B}-V_{C}}{R_{304}}=\frac{V_{C}-V}{R_{302}}=\frac{V+V_{AGC}}{R_{s}} \quad \text{and} \quad \frac{V_{CC}-V_{B}}{R_{305}}=I_{B}+\frac{V_{B}-V_{C}}{R_{304}}.$$

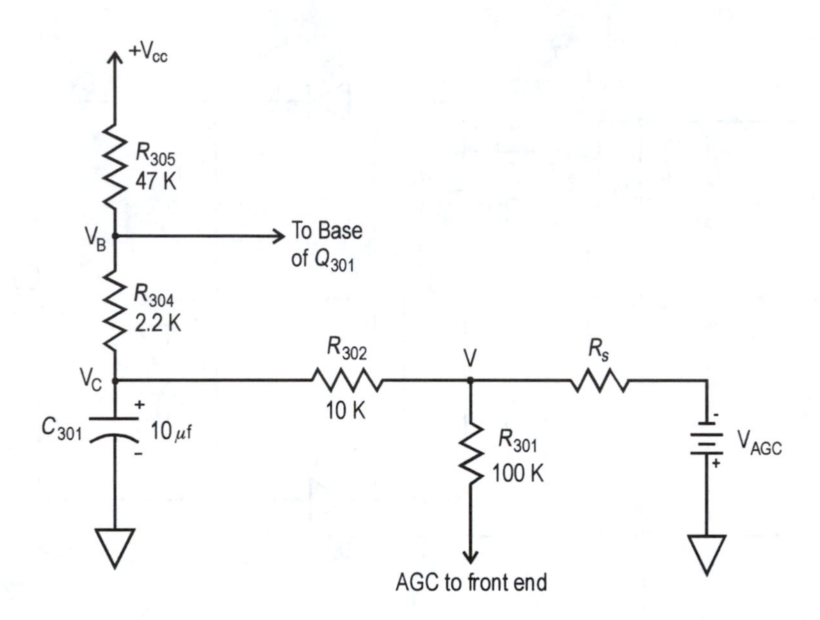

Fig. 2.32: Schematic diagram isolating the AGC circuitry in Fig. 2.30.

Solving for V and inserting known values gives

$$V = \frac{[V_{cc} - 47\,\text{k}\Omega \cdot I_B(\text{mA})]R_s(\text{k}\Omega)}{59.2 + R_s(\text{k}\Omega)} - \frac{59.2 \cdot V_{AGC}}{59.2 + R_s(\text{k}\Omega)}.$$

In the absence of a signal, $V_{AGC} = 0$. To determine I_B, we first find I_C from the emitter voltage in Fig. 2.30: 1.75 V/330 Ω = 5.30 mA. The beta of currently available transistor substitutes for the SE1001 is 30. This suggests $I_B = 5.30/30 = .177$ mA. The source impedance, R_s, would be the output impedance of Q_{303} in parallel with a (presumably matched) reflected impedance from Q_{304}. We shall use $R_S = 15\,\text{k}\,\Omega$. Using these values for the voltage, V, gives .747 volts.

Now referring back to Fig. 2.20, we see the RF AGC voltage, V, entering at the lower right corner. For $V > 0$, D_{201} is reverse biased and R_{210}, R_{207}, and R_{208} form a voltage divider which will place a positive voltage of $.747 \cdot 2.2/(2.2 + .22 + 1) = .48$ volts on the gate. This, then, is the order of magnitude of the no-signal bias applied to the FET RF amplifier, Q_{201}. This small forward bias will maximize g_m for the Q_{201} stage in the absence of a signal. As soon as a signal is received, the AGC swings negative, reducing the stage gain according to the signal strength.

The last IF amplifier stage comprises Q_{304} and Q_{305}, which are direct-coupled together. The collector load of Q_{304} is R_{327} (1.8 kΩ) in parallel with the input impedance of Q_{305}, which will be less than 1 kΩ. The divided emitter resistance on Q_{305} is bypassed by C_{317} and C_{321}. Thus only DC is fed back to bias Q_{304} through the secondary of T_{303}. This DC feedback stabilizes the Q points of this amplifier's two transistors.

Exercise 2.8 Confirm that if a temperature rise causes the collector current of Q_{304} to increase, the feedback will cause it to decrease – in other words, show that the DC feedback is negative.

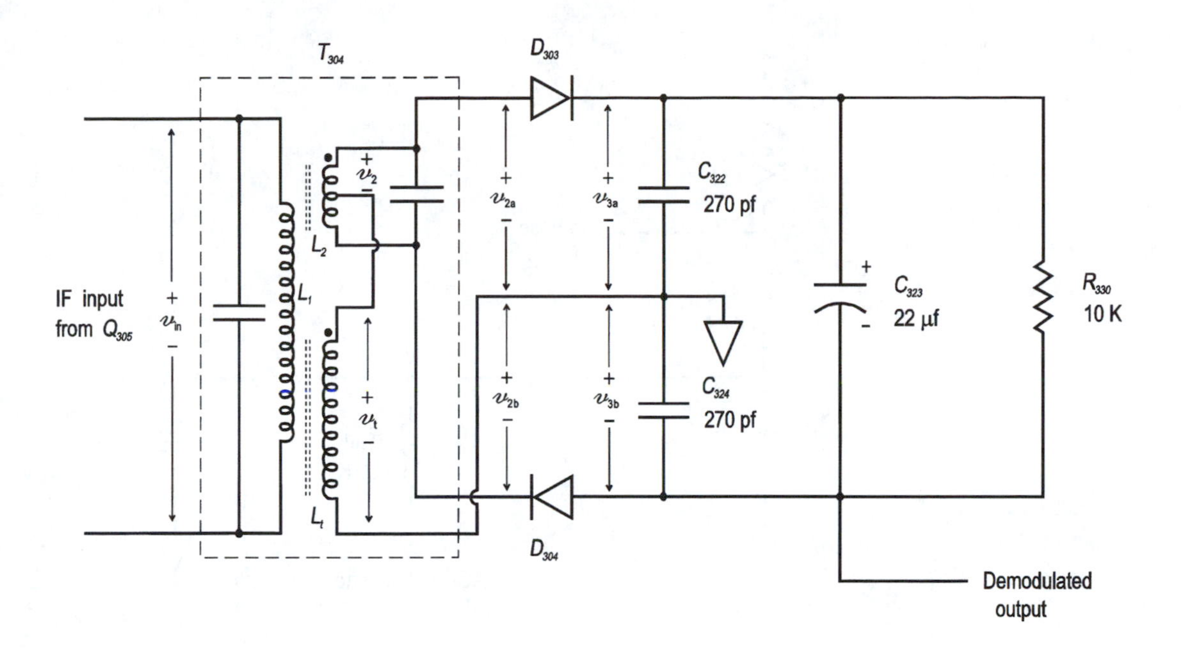

Fig. 2.33: Schematic of a conventional ratio detector circuit obtained by treating R_{329} and C_{326} in Fig. 2.30 as short circuits.

The collector supply for both transistors comes from the R_{328}–C_{327} decoupling network. C_{319} is an on-board bypass for the positive supply. Lead inductance between the supply and the IF circuit board can appreciably raise the effective supply impedance seen from the board. This necessitates a local bypass to restore a low supply impedance to ground.

2.7.5 The detector

The output of Q_{305} is fed to T_{304}, which is part of a ratio detector. This now-obsolete type of detector was mentioned in Section 2.3.1. There are numerous variants of it. The one used in this receiver is one of the most obscure of those forms. If we were to replace both R_{329} and C_{326} with shorts, we would have a much more conventional ratio detector, as depicted in Fig. 2.33. We shall perform an analysis of this circuit and then look at what effect the inclusion of R_{329} and C_{326} has on our findings.

The ratio detector basically consists of three parts:

(1) the phase discriminator circuitry,
(2) the rectifier circuitry, and
(3) the ratio circuitry.

The phase discriminator consists of the tuned and tapped secondary of the detector transformer and the tertiary winding of this same transformer. Subject to the following reasonable assumptions, the voltages between the center tap of the secondary and either end of the winding are 180° out of phase with each other and 90° out of phase with the primary voltage at the resonant frequency. The assumptions are:

(1) The primary self-inductance of T_{304} is much larger than that of the secondary. This is necessarily so, since it is a step-down transformer

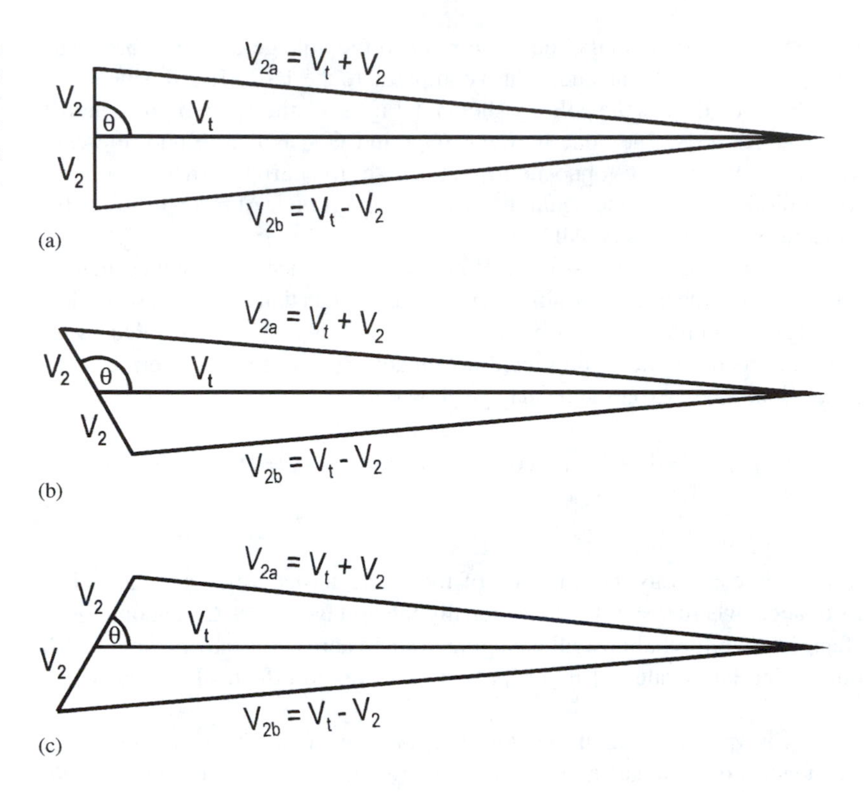

Fig. 2.34: Phasor diagram showing the phases of the voltages in a ratio detector (a) at resonance, (b) below resonance, and (c) above resonance.

to match the high output impedance of the IF amplifier to the low impedance of the diode detectors. This means that the effect of M on the primary will be much less important than its effect on the secondary. Thus we will ignore not only the M term in the primary but also the primary series resistance in writing circuit equations.

(2) We will assume that the voltage on the tertiary winding is much larger than that on the tapped secondary winding and that it is so closely coupled to the primary that its voltage will always be in phase with the primary voltage.

(3) We will assume that the forward diode voltage drop in conduction is zero. This is tantamount to assuming that it is much less than the phasor sum of the voltages on the secondary and tertiary windings.

The voltage on the tertiary winding, v_t, is added and subtracted "phasorially" with the voltages on the two halves of the secondary. Figure 2.34a shows the phase relationship at resonance. At frequencies off resonance, the phase of the secondary voltages changes with frequency as shown in Figs. 2.34b and c. Analytically, one can use the law of cosines to find the amplitudes of the phasor sum voltages, $\mathbf{V}_{2a}$ and $\mathbf{V}_{2b}$.

For the rectifier circuitry, we again refer to Fig. 2.33. The voltages $\mathbf{V}_{2a}$ and $\mathbf{V}_{2b}$ are applied to a rectifier circuit consisting of D_{303}, D_{304}, C_{322}, and C_{324}. Subject to assumption (3) above, this circuit will produce voltages across C_{322} and C_{324} equal to the amplitudes of $\mathbf{V}_{2a}$ and $\mathbf{V}_{2b}$. We may therefore speak of the voltages v_{3a} and v_{3b} interchangeably with $\mathbf{V}_{2a}$ and $\mathbf{V}_{2b}$ even though the former are AC voltages at a frequency of 10.7 MHz and the latter are "DC" voltages of the same magnitude. We refer to the voltage as "DC," because it varies very slowly (at an audio rate) compared to the 10.7 MHz IF signal.

Whenever $\mathbf{V}_{2a}$ has the polarity shown in Fig. 2.33, the upper diode conducts and charges C_{322}. The node between C_{322} and C_{324} is the ground reference. When $\mathbf{V}_{2b}$ has polarity opposite to that shown, the current flows through the lower diode and C_{324} and again returns through ground. At resonance, the two capacitors are charged equally.

The ratio circuitry consists of R_{330} and C_{323}. Since we assumed that the voltage across the tertiary winding is much larger than that across the secondary, the phasor amplitudes given by the law of cosines may be expanded by the binomial theorem and represented adequately by the first two terms of the expansion. These expressions may be shown to be

$$v_{3a} = |\mathbf{V}_t| + |\mathbf{V}_2| \cdot \cos\theta \qquad \text{and} \qquad v_{3b} = |\mathbf{V}_t| - |\mathbf{V}_2| \cdot \cos\theta \tag{2.31}$$

It should be obvious that the sum of these two terms is the same as it is at resonance, where $\theta = 90°$. Hence ideally the voltage across C_{323} should never change. Therefore, almost all changing voltage across it is attributable to AM noise. The large value of this capacitance serves to effectively suppress this noise.

C_{323} is variously called the *storage capacitor* or *stabilizing capacitor*. At frequencies off resonance, either v_{3a} or v_{3b} will increase, but the other will decrease, as shown by Eq. (2.31) and the phasor diagrams (Fig. 2.34). Thus although the sum of v_{3a} and v_{3b} remains constant, their ratio continuously changes depending on the phase angle between the secondary and tertiary voltages. This is the genesis of the name "ratio detector." Resistor R_{330}, which appears across C_{323}, serves to slowly discharge it so that it can respond to changing signal conditions such as a change of station. It is only required that the R_{330}–C_{323} time constant be much longer than the period of the 10.7 MHz IF signal.

To derive the v–f relationship of the ratio detector, it is necessary to find the explicit form of the relationship between $\mathbf{V}_2$ and f and to insert this into (2.31). Analysis of the phase discriminator portion of the circuitry leads to the equation

$$\mathbf{V}_2 = \frac{M\mathbf{V}_{in}}{L_1[(1-\omega^2 L_2 C) + jRC\omega]},$$

where R is the resistance of the secondary winding. Conversion to polar form yields

$$|\mathbf{V}_2| = \frac{M\mathbf{V}_{in}}{L_1\sqrt{[(1-\omega^2 L_2 C)^2 + R^2C^2\omega^2]}} \qquad \text{and} \qquad \theta = -\arctan\frac{RC\omega}{1-\omega^2 L_2 C}$$

and so

$$\cos\theta = \cos\left\{-\arctan\frac{RC\omega}{1-\omega^2 L_2 C}\right\} = \frac{1-\omega^2 L_2 C}{\sqrt{(1-\omega^2 L_2 C)^2 + R^2C^2\omega^2}}.$$

Inserting these results into (2.31) yields

$$\mathcal{V}_{\text{out}} = -|\mathcal{V}_{3b}| = -|\mathbf{V}_t| + |\mathbf{V}_{\text{in}}| \frac{(M/L_1)(1-\omega^2/\omega_o^2)Q^2}{\{[Q(1-\omega^2/\omega_o^2)]^2 + \omega^2/\omega_o^2\}}. \tag{2.32}$$

In this equation, $\mathcal{V}_{\text{out}}$ is taken from the bottom of C_{324} in agreement with Fig. 2.30, and we have defined $\omega_o^2 \equiv 1/L_2C$ and used $RC\omega = RC\omega_o(\omega/\omega_o) = (1/Q)\cdot(\omega/\omega_o)$.

Of course, we want to see a linear relationship between $\mathcal{V}$ and f, but Eq. (2.32) does not look at all linear. Yet satisfying two reasonable inequalities will suffice to linearize it:

(1) $\Delta\omega \equiv \omega_o - \omega \ll \omega_o$,
(2) $\omega/\omega_o \gg Q(1-\omega^2/\omega_o^2)$.

The first condition says that the frequency deviation around the resonant frequency will be small compared to the resonant frequency itself. The second condition is even more stringent. By factoring the right-hand side of the inequality, we can restate the requirement as

$$\omega \gg Q \cdot \Delta\omega \cdot (1+\omega/\omega_o) \cong 2Q \cdot \Delta\omega.$$

This same condition had to be imposed on the quadrature detector (Section 2.3.3.3) to linearize it and, as was the case for that detector, we conclude that low Q is conducive to linearity in the ratio detector. For this reason, it is not unusual to see ratio detector circuits that have a resistance added in series with the leads of the transformer secondary winding to lower the Q.

Subject to these assumptions, (2.32) can finally be written as

$$\mathcal{V}_{\text{out}} = -|\mathbf{V}_t| - 2|\mathbf{V}_{\text{in}}|(M/L_1)Q^2\Delta\omega/\omega_o. \tag{2.33}$$

Equation (2.33) shows the desired linear relationship between the output voltage and the frequency deviation from ω_o. It also shows that even when the input frequency is at ω_o, the output voltage has a large DC component equal to $-|\mathbf{V}_t|$. The linearity error in Table 2.1 for the ratio detector is found by subtracting (2.33) from (2.32), converting to percent error, and taking the absolute value.

It is finally time to go back and look at the role of the components omitted from this analysis: R_{329} and C_{326}. The contribution of C_{323} to AM suppression has already been mentioned. R_{329} simply improves AM suppression (Krauss, Bostian, and Raab, 1980). As for C_{326}, we should first note that a larger-scale schematic of the same tuner shows this capacitor as .001 μF, rather than 470 pF. If this value is correct, the capacitive reactance at 10.7 MHz is less than 15 Ω. Hence its perturbation of the circuit operation we have described will be minimal. It can be verified readily by referring to Fig. 2.30 that conduction through D_{303} will tend to charge C_{326} negatively, and conduction by D_{304} will charge it positively. Thus, at resonance, when the diodes conduct equally, there will be no net charging of C_{326}. As we move away from resonance, the capacitor will acquire a net charge of one polarity or the other, depending on the input

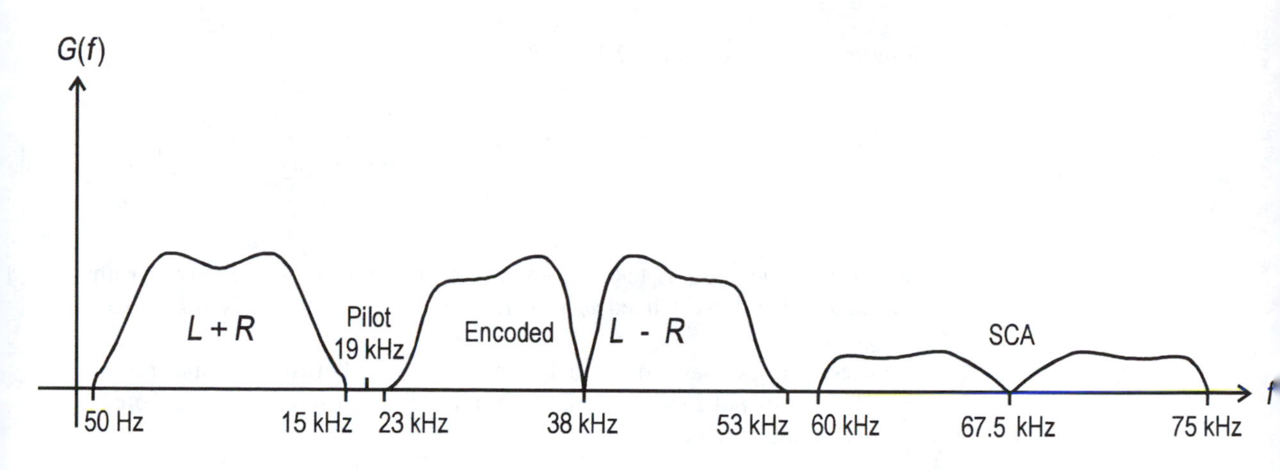

Fig. 2.35: Spectral energy plot of the baseband signal of a station broadcasting FM Multiplex stereo and an SCA channel.

frequency and thus the polarity of the phase difference. This suggests that the output taken from the top of C_{326} is also an FM demodulated signal, but one that is centered about zero, unlike that of Eq. (2.33). As indicated on the schematic, however, this is achieved at the expense of a lower amplitude of the demodulated output. The particular form of the MX decoder used in this tuner requires both the high-amplitude output not referenced to zero and the low-amplitude one, which is referenced to zero.

2.8 SCA (Subsidiary Communications Authorization)

This communications channel is best known as the "background music service." It provides commercial-free music to subscribers and has also been used for other programming.

The SCA signal is yet another component added to the baseband signal before it is modulated onto the station carrier. The SCA signal is a narrow-band FM signal centered about a subcarrier frequency of 67.5 kHz and with an audio bandwidth of 7 kHz. Thus, as Fig. 2.35 shows, it lies above the FM stereo component of the baseband signal.

As was the case with the FM pilot signal, the SCA component does contribute to the frequency deviation of the station carrier. Therefore, like the pilot, it is transmitted with just 10% of the amplitude of the stereo signal, which is reduced another 10% to accommodate it without exceeding the maximum frequency deviation specification imposed by the FCC.

You can legally build an SCA decoder for use in your own home only. This used to be a formidable task, but PLLs have made it quite simple. The baseband signal right out of the FM demodulator must be used as input (before any de-emphasis, which would eliminate the SCA signal). A suitable circuit published by Signetics can be seen in Fig. 2.36 (Signetics 1985). The output simply goes to an audio amplifier. The 5 k pot is adjusted to set the VCO free-running frequency to 67 kHz. Pins 2 and 3 are differential inputs to the phase detector.

Exercise 2.9 Consider the AC characteristics of the input filter.

(a) Given that pins 2 and 3 each have a nominal impedance of 10 kΩ to ground, find the transfer function from the input to pin 2.

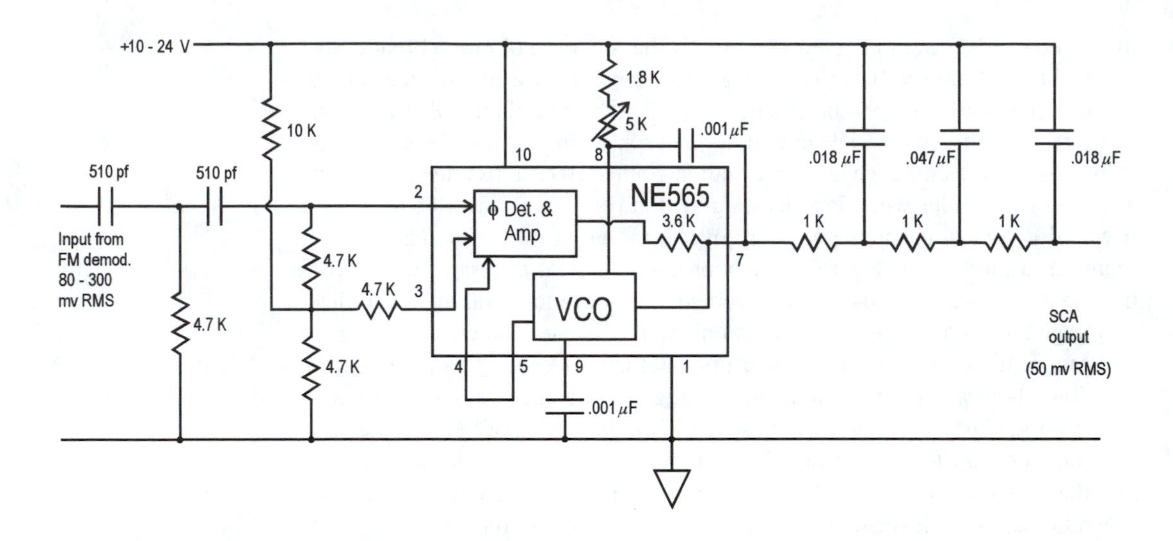

Fig. 2.36: Schematic diagram of a simple PLL circuit that can be used to extract an SCA signal from a demodulated FM baseband signal.

(b) Classify the filter as to type (Butterworth, Chebyshev, inverse Chebyshev, elliptic).

2.9 The FMX® system

The FMX® system represents a significant advance in FM technology, but one that is currently mired. At its peak at the end of the eighties, over a hundred FM stations in the United States were broadcasting in FMX®. Yet this was not good enough for the Japanese firms that make the majority of the radio receivers used in the United States. Only a couple of high-end receivers reached the market with FMX® capability. This made the listening market so small that as the broadcast equipment required service, most stations opted to discontinue FMX® broadcasting. There are now only a handful of stations in the this country that continue to do so. The commercial rights to the technology now reside with Broadcast Technology Partners, who are negotiating with a major automotive manufacturer to incorporate FMX® into their auto receivers.

It has been noted that transmission of FM stereo requires an attenuation of the monophonic signal to accommodate the frequency deviation caused by the pilot signal (and the SCA signal if one is transmitted). However, this is just the "tip of the iceberg" as far as degradation of the signal-to-noise ratio in FM stereo is concerned. The most important effect is that the wider bandwidth of the stereo signal must necessarily admit much more noise than the narrower bandwidth of monophonic FM. The stereo signal requires a bandwidth 3.5 times as large as the monophonic transmission. To make matters worse, the noise spectrum is not flat over the bandwidth of the FM signal but increases at 20 dB/decade as you move away from the carrier signal (Torick and Keller, 1985). The net effect is that when all things are considered, the total degradation of signal-to-noise ratio when converting from mono to stereo reception can be as much as 20 dB (Torick and Keller, 1985). This translates to a deterioration in the quality of a received signal at a given location relative to the transmitter. This is why virtually all FM stereo receivers have manual and/or automatic switching to mono when the received signal is too weak. But even operating the stereo receiver in mono does not recover the S/N ratio of mono broadcasts. Stereo broadcasting will typically degrade reception in a mono receiver by 1–7 dB relative to monophonic broadcasting (Torick and Keller, 1985). Another measure of the effect of S/N degradation when converting to stereo broadcasting is to consider the coverage area. If a station could deliver a usable signal over a radius of 128 miles while broadcasting in mono and then converted to stereo, it would deliver a usable signal only over a radius of 60 miles. This would represent a more than four-fold reduction in theoretical coverage area. Even if a mono receiver were used, the coverage radius would shrink to 100 miles. Coverage area is of prime concern not only to the listener, but even more so to the station. If stations can show sponsors that their coverage area is increased four-fold, they can get much more advertising revenue because they will reach a much larger audience with their message. Until the advent of FMX®, the best that could be done to counter the S/N penalty of stereo was to incorporate *blend* into the receiver. This allows a gradual transition from stereo to mono as the S/N ratio degrades, instead of the binary action of mono/stereo switching. The blend concept can be incorporated in the FMX® system also, though it will not be activated nearly as often.

Block diagrams of the transmitter and receiver circuitry needed to implement FMX® are shown in Figs. 2.37a and b respectively. As shown in Fig. 2.37a, the transmitter circuitry is modified by an "add-on" that compresses the $L - R$ signal and then modulates it onto a 90°-phase-shifted 38 kHz carrier. The three signals going to the summer in normal FM stereo are thus joined by a fourth (S') in quadrature with the normal $L - R$ signal (S). There is also an FMX® ID tone, which is not shown in Fig. 2.37a. The fourth signal component and the ID tone will be rejected by the stereo demodulator of a receiver not equipped for FMX® stereo reception, thus assuring compatibility with existing equipment.

Circuitry is included in the FMX® system to decode this additional signal. In a properly equipped receiver, the S' component undergoes an expansion process that is the exact complement of the compression process at the transmitter, and the expanded quadrature component is the one used to generate the L and R signals from the matrix. The compression/expansion process is what contributes so significantly to the improvement of the S/N ratio. The mechanics of this are explained in Chapter 5.

Since the S' component is used to reconstruct the L and R signals, we might ask if the old $L - R$ signal (S) is simply ignored in an FMX® system. It is not. It is demodulated as usual and its output gets used as the "reference" in the control system that gets used to accomplish the complementary expansion of the compressed $L - R$ signal. This adaptive companding makes it theoretically possible to use any compression process and still properly recover the $L - R$ signal. The recommended "re-entrant compression characteristic" is specified because it gives the maximum improvement in S/N ratio consonant with eliminating the possibility of the modulator being overdriven by the addition of the S' signal.

The first company to produce an FMX® decoder IC was Sanyo. Their IC is still being marketed as of this writing and carries the number LA3440. Sometime later, Samsung brought out a "next-generation" chip, the KA1260, but this chip is no longer being marketed. Figure 2.38 shows a block diagram of the Sanyo chip. We shall briefly describe its operation. The baseband signal is input in the upper right-hand corner. This signal is found to consist of five frequency components:

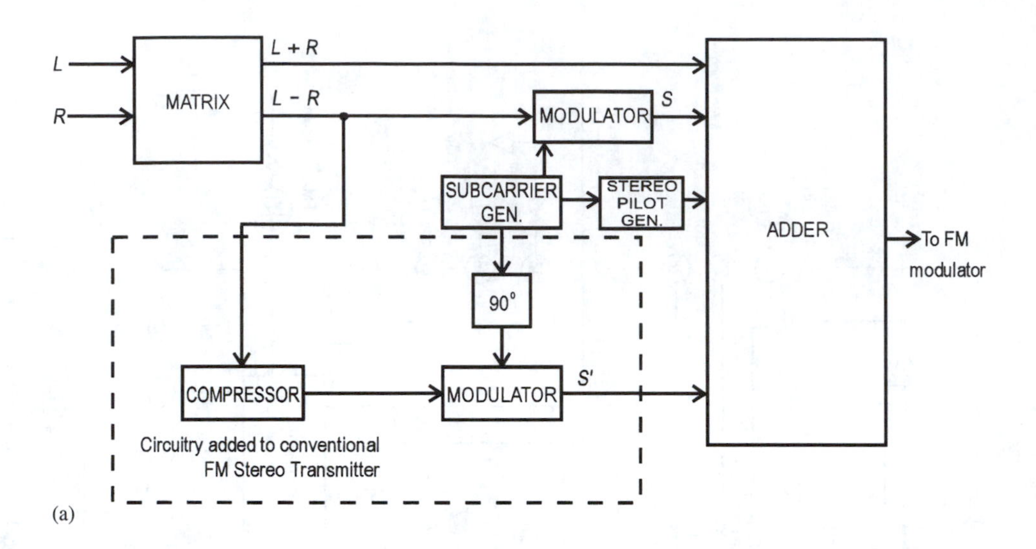

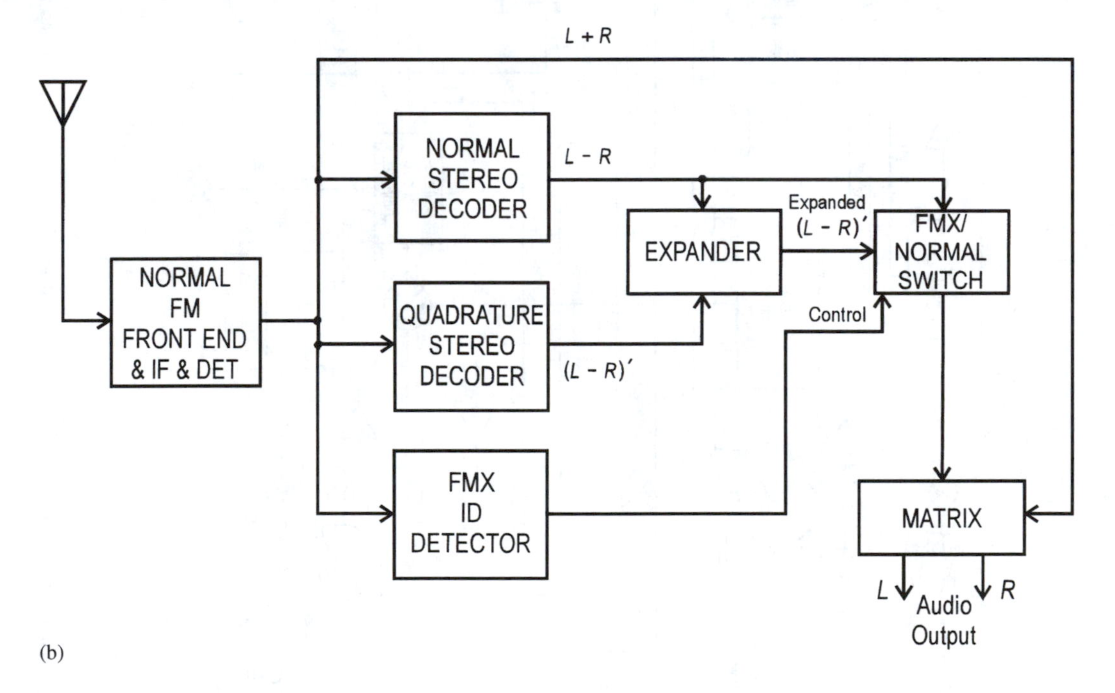

Fig. 2.37: Block diagrams of the circuitry required for FMX®: (a) the transmitter; (b) the receiver.

(1) $L + R$,
(2) 19 kHz pilot,
(3) $L - R$ DSB-SC modulated onto a 38 kHz carrier (S),
(4) $L - R$ compressed and then DSB-SC modulated onto a 38 kHz quadrature carrier (S'), and
(5) 9.9 Hz FMX® ID tone.

The series LC leading to pin 3 is made series resonant at 19 kHz to select out the pilot signal. In the vicinity of 19 kHz, the phase shift going through this network will vary enormously, so the variable inductance also provides phase

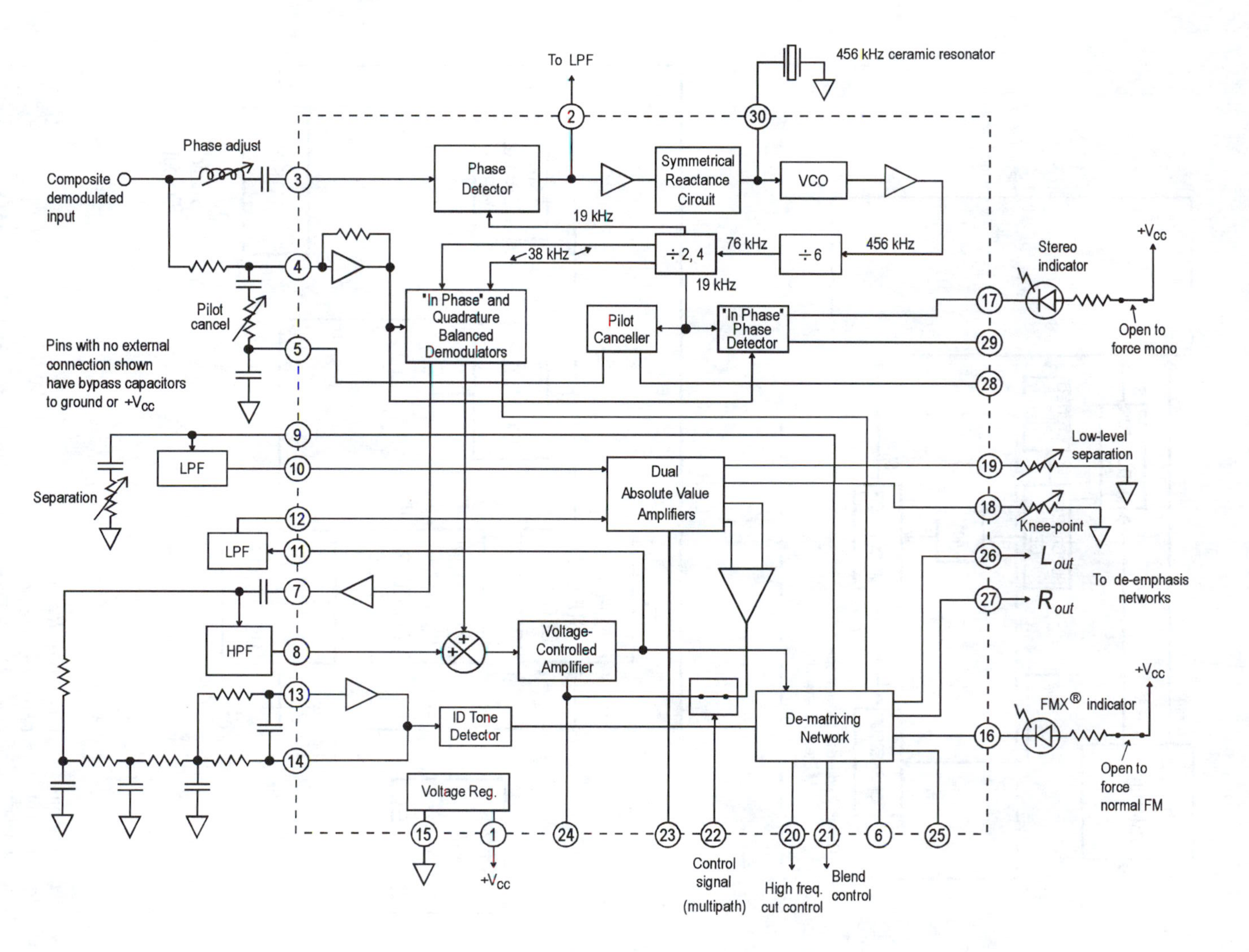

Fig. 2.38: Block diagram of the LA3440, which demodulates the FM signal and also recovers FMX® where it is being transmitted.

control. The other input to the phase comparator is, as we shall see, a 19 kHz signal derived from the VCO. At lock, these two 19 kHz signals will be in quadrature.

The output of the phase comparator is low-pass filtered by a network attached to pin 2; it then passes through a buffer to a symmetrical reactance circuit. Reactance circuits will be discussed in some detail in Chapter 6. For now, let it suffice to say that they are circuits whose reactance can be made to vary with a DC voltage. A varicap is the simplest reactance circuit. The reactance circuit works with the 456 kHz ceramic resonator attached to pin 30 and the VCO to give oscillation that is not only in the vicinity of the intended frequency, but one that can be electrically "pulled" to be exactly what is needed. A ceramic resonator works on the same principle as a quartz crystal, but it has a much lower Q, and so its frequency is more readily modified by external circuitry. This is, of course, exactly what we want for this application.

The output of the VCO goes to a $\div 6$ divider, which outputs 76 kHz. This in turn goes into another divider that produces two outputs at 38 kHz and two at 19 kHz. The duplication is necessary, because signals are needed with differing phases. Recall that in Section 2.6.3 an integrated PLL stereo decoder was analyzed. In that circuit also there was a need for 19 kHz signals in quadrature. As was the case there, the in-phase detector in the Sanyo chip drives a stereo indicator. In the LM1800, however, there was no need for two 38 kHz signals – only one. In the FMX® system, however, the S' signal is in quadrature with the S signal and needs a separate phase to demodulate it. A portion of the divided-down VCO output at 19 kHz is applied at pin 5 to the network that feeds the quadrature demodulator input (pin 4). A trimmer resistance allows the phase to be adjusted so as to cancel the pilot component of this input. Recall that the amplitude of this component is only about 10% that of the FM signal itself.

The S signal leaves the in-phase demodulator and goes three places:

(1) straight to the final de-matrixing block,
(2) to an internal summing point, and
(3) through some low-pass filtering and then to an absolute-value amplifier.

The S' signal also leaves the demodulator and, after buffering and high-pass filtering to remove the ID tone, is applied to the same summing point. The summing point outputs $S + S'$. The purpose of the next section of circuitry is to servo this sum signal ($S + S'$) to force its amplitude to be exactly the same as that of S itself. This is the step that undoes the compression applied to the S' signal at the transmitter. Thus the sum signal goes to a voltage-controlled amplifier whose output becomes the new S' signal. This signal goes through the LPF connected between pins 11 and 12 and thence to its own absolute-value amplifier.

The outputs of the two absolute-value amplifiers are applied to the inputs of a comparator. The output of the comparator goes through a normally closed switch to the gain-control pin of the VCA. A high-quality capacitor is also connected from this pin to ground. This capacitor has a dual function: (1) It provides a time constant for the gain control of the VCA and (2) if multipath is detected and triggers the gate (pin 22), the normally closed switch opens and the capacitor connected to pin 24 holds the previous control voltage on the VCA. Multipath can be detected as amplitude modulation of the FM stereo pilot signal.

The VCA output is, as mentioned, the new S' signal, and it is applied to the de-matrixing circuitry along with the S already transmitted there and the mono $(L + R)$ signals.

The only other large block of circuitry is that used to detect the ID tone. The original S' signal exits pin 7 and enters a filter comprised of two sections of passive RC low-pass filtering followed by a second-order active IGMF LPF. Thus we have a fourth-order LPF used for extracting the ID tone. The output of the ID detector is also fed to the de-matrixing circuitry to automatically disable FMX® decoding if the ID tone is not present.

Although far from exhaustive, this description of the LA3440 hopefully suffices to give an overview of the design philosophy of the chip and the steps required to decode an FMX® signal.

2.10 The Radio Broadcast Data System (RBDS)

The RBDS system has been developed by the National Radio Systems Committee of the Electronic Industries Association and the National Association of Broadcasters. The specifications of the system form a very lengthy publication (NRSC, 1993). Its purpose is to allow the transmission of text or other information on the carrier of an FM signal.

A similar system (called RDS) had been defined in Europe almost ten years earlier. There, auto radios are equipped with RDS. Some stations have repeaters throughout a country, and as a car travels out of the range of one of the repeaters, the RDS system is capable of automatically changing the tuning of the car radio to a nearer repeater to keep reception optimum. This feature gives some idea of the sophistication that RBDS is capable of bringing to a receiver.

In the United States, the uses of the system are not yet standardized. It could obviously be used for such things as displaying the call letters of the station to which you are listening. It could also display the name and other information about the selection being played. Other uses within the capability of the system as it is now structured include radio paging and transmission of location and navigation information. The government wants to use it as a means of transmitting "emergency information." Since the type of programming carried by the station is routinely included as part of the RDS code, one could readily conceive of a system that would allow your radio to search only through stations that are programming in the format in which you are interested (classical, country/western, etc.).

The reader who will refer to Fig. 2.35 will see that there remains a small unused portion of the baseband signal spectrum between 53 and 60 kHz. RBDS uses it. The FM pilot signal is tripled up to 57 kHz, and this becomes the RBDS subcarrier. If a station is not transmitting in stereo, the frequency specification for the subcarrier remains the same. In Section 2.10.3 we shall look at the nature of the data stream. The digital data stream is DSB-SC modulated onto the 57 kHz carrier just as the SCA signal (if present) is frequency modulated onto a 67.5 kHz carrier.

Figure 2.39a shows the circuitry used at the transmitter to incorporate the RBDS data into the transmitted FM signal. Figure 2.39b shows the receiver circuitry necessary to extract this information. Note that we are not concerned

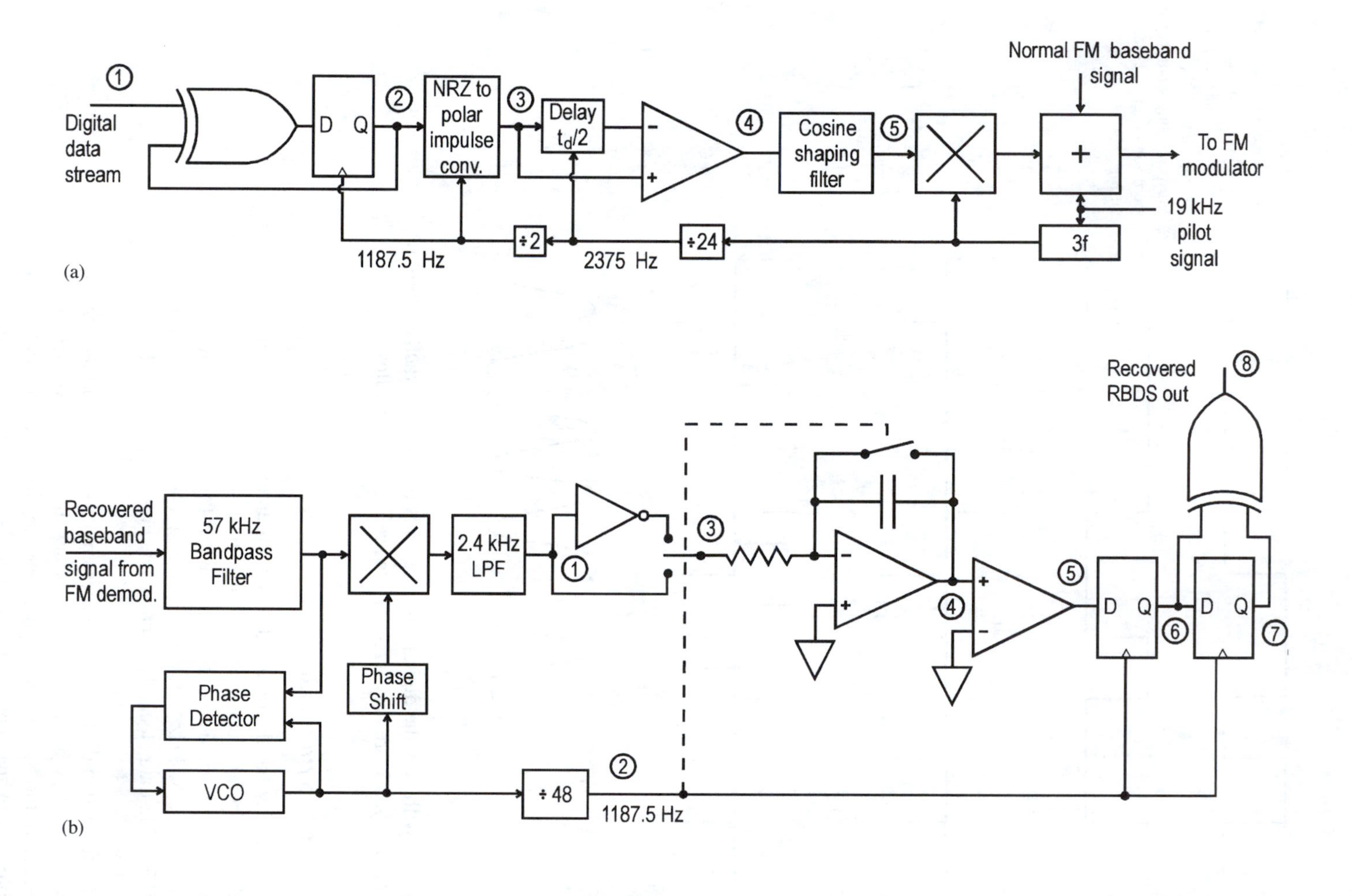

Fig. 2.39: Block diagrams of the circuitry required for RBDS: (a) the transmitter; (b) the receiver.

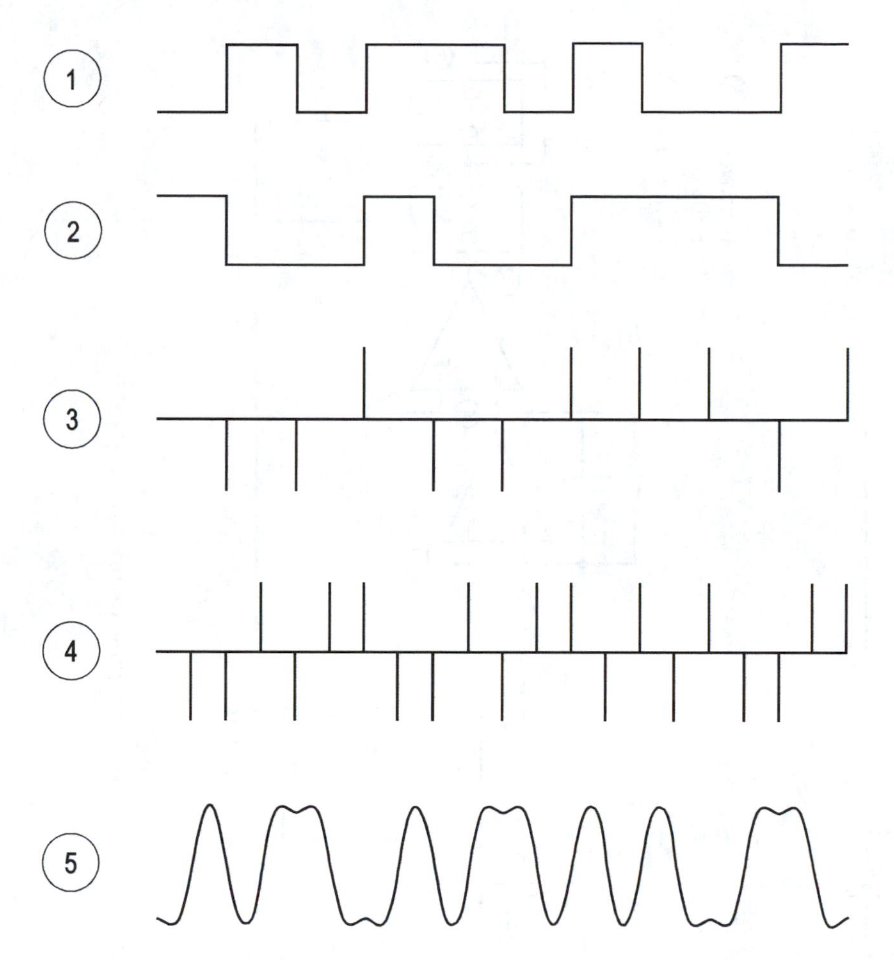

Fig. 2.40: Timing diagram for the RBDS transmitter circuitry. The circled numbers on the waveforms are measured at the correspondingly numbered points in the circuit diagram of Fig. 2.39a.

in these figures with how the digital data stream is created or exactly where it goes after recovery. We are only interested in seeing the signal flows.

2.10.1 The RBDS transmitter

In Fig 2.39a we see that the digital data stream enters an exclusive OR gate, which feeds a D flip-flop. The feedback added to this combination converts the data to NRZ (non-return to zero) format. The flip-flop is clocked by a divided-down version of the 57 kHz subcarrier. In total, this carrier is divided by 48 to yield a 1.1875 kHz data clock. The actual synchronizing of the clock and the data is not shown.

The NRZ-formatted data then undergoes conversion to a format consisting of "impulses" that may be either positive or negative going. At each positive-going transition of the NRZ signal, a positive-going impulse is output from this converter block. The analogous behavior is true for negative-going inputs. The timing diagram in Fig. 2.40 illustrates this. Consider a moment when the input to the delay block is positive. The output of the delay block is then zero, since it could obviously have had no input a half clock-time before. The positive impulse causes the output of the comparator to drive positive, but as soon as the input "impulse" is gone, the output drops back down also. Note, however, that at a time $t_d/2$ later, this positive value appears at the inverting input of the comparator. At that same time, the noninverting input is receiving the then-current value of the impulse, which is, of course, zero, since this is between the clock pulses and thus between the impulses. This will make the comparator rail negative. If we start with a negative impulse, the corresponding process will cause the comparator to pulse negative initially and then positive a time $t_d/2$ later. In both cases, however, the railing only lasts for the duration of the input "impulse." This whole process is called *bi-phase symbol conversion* because each switching event is converted to a pair of impulses of opposite polarity. This admittedly cumbersome technique is used because it moves the spectral energy of the encoded signal away from the 57 kHz subcarrier frequency, enabling one to avoid data-modulated cross talk in PLL stereo decoders.

The bi-phase impulse train is fed to a filter with a cosine response. Not only does this accomplish bandlimiting, but the peculiar form of the filter is capable of minimizing intersymbol interference. For a much more detailed account of the issues involved here, the reader may wish to refer to a good text on communications (Couch, 1987).

The output of the cosine-shaping filter is fed to a balanced modulator along with the 57 kHz subcarrier. The output of this block is summed in with all of the usual components of the baseband signal shown in Fig. 2.30.

2.10.2 The RBDS receiver

The FM signal received will go through normal heterodyning, IF amplification, and FM demodulation, and these processes will recover the baseband signal, which now includes RBDS data. This baseband signal will pass through a 57 kHz bandpass filter to extract the RBDS sidebands. A PLL is locked to this signal to force the VCO to run at the 57 kHz subcarrier frequency. This oscillation goes through a $\div 48$ stage to recreate the 1.1875 kHz data clock. The subcarrier is also phase shifted by 90° and sent to a balanced demodulator along with the sidebands. The output of the balanced modulator contains the encoded RBDS data, which is removed by a 2.4 kHz LPF.

The output of this filter goes to a circuit that behaves something like an analog XOR gate. When the data clock is positive, the SPDT electronic switch is in the lower position and passes the analog signal without a sign change. When the data clock is negative, the switch is in the upper position, and the analog data has its phase inverted. As long as the data clock and the data itself have the same sign, the input to the integrator stage will be positive. When they have opposite signs, the input to the integrator becomes negative. This can be verified by examining the timing chart in Fig. 2.41.

Since the integrator is of the inverting variety, it will change the sign of its input and cause a ramping action of one polarity or the other for the duration of each cycle of the data clock. At each negative-going transition of the data clock, the SPST electronic switch in the integrator circuit shorts out the integrating capacitor, driving the integrator output back toward zero. The switch is closed for a short enough time that the voltage actually never goes all the way to zero unless it is passing through zero due to a sign change in the input.

The integrator output is input to a zero-referenced comparator and then sent to a pair of D flip-flops whose outputs are the inputs of an XOR gate. The gate will provide an output only when the flip-flops have the opposite outputs, and

Fig. 2.41: Timing diagram for the RBDS receiver circuitry. The circled numbers on the waveforms are measured at the correspondingly numbered points in the circuit diagram of Fig. 2.39b.

this will happen only when the data stream out of the comparator has stayed in one state for at least two data-clock times.

2.10.3 Data encoding format

The data are encoded into *groups* of 104 bits. Each group is broken up into four *blocks* of twenty-six bits each. The words are further divided up into sixteen bits of information and a ten-bit *checkword*. In general, different information is encoded into each block.

Block 1

The entire sixteen bits of information comprise what is called PI (Program Identification) code. This might better be called station identification code, since it contains the station's call letters plus other information such as whether two stations carry the same programming.

Block 2

The first four bits are called *group code*. This is a very important block of code. It codes which of sixteen possible functions or collections of functions are being transmitted at the time. For instance, a group code of 0100_2 says that time and date are being transmitted. A code of 1001_2 is used for emergency warnings. It would obviously be far beyond the scope of this book to exhaustively detail the function of all 16 bits of all four blocks for all 16 group codes. We shall content ourselves with an examination of the format of group 0000_2, which is titled "Basic Tuning and Switching Information."

The fifth bit in this word is said to be the version type. Most of the sixteen group codes have two versions, called the A and B versions. For group 0, version B can only be used when no alternative frequencies exist for a given program. Version A is usually used when there are alternative frequencies.

Bits 6 and 12 are called the traffic-program and traffic-announcement code bits. These indicate whether a station provides traffic announcements and whether such announcements are currently being transmitted.

Bits 7 through 11 carry PTY (program type) code. These five bits can, of course, code thirty-two possible station formats or current programming. At present, twenty-five of them are allocated.

Bit 13 is a music/speech switch code.

Bit 14 is actually part of a four-bit code that is distributed with one bit in this position in each of four successive groups. This is called a DI (decoder identification) control code.

Bits 15 and 16 code which of the four bits in the DI code are included in the block currently being received.

Block 3

For version B coding, block 3 duplicates the PI information sent in block 1. In version A coding, the third block codes as two 8-bit numbers the alternate frequencies on which the currently received programming can be found. Version A coding of group code 0 is called Group Type 0A.

Block 4

These sixteen bits are allocated to two 8-bit characters that form part of the identification of the radio network of which the transmitting station is a part. Every group of 104 bits will contain two characters' worth of this code in block 4. After the four groups are received, the cycle is complete and a total of eight characters can be displayed. Then the same transmission is repeated for the next four groups. We saw that bits 15 and 16 of block 1 are used as a pointer to which of four groups containing a distributed DI code is being transmitted. These same two bits code which of the Program Service characters are being transmitted in block 4 of each of four successive groups. Figure 2.42 focuses on the details of how this information is distributed.

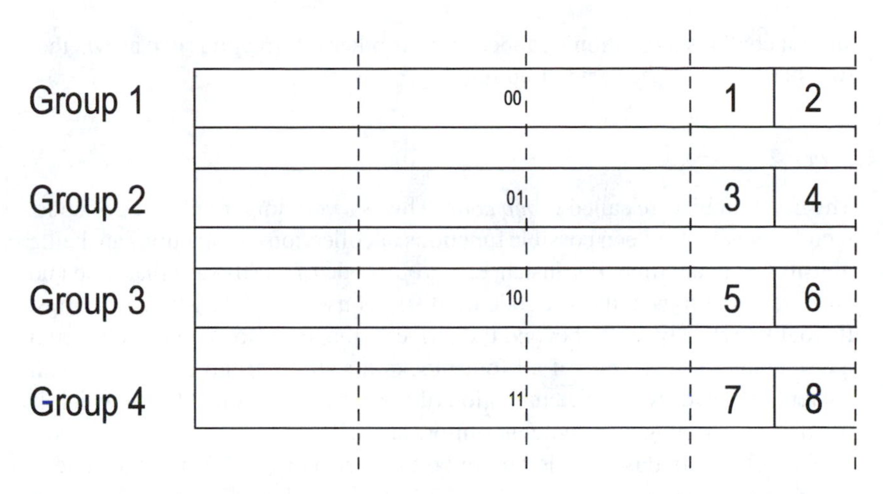

Fig. 2.42: Structure of the distributed RBDS block that contains the Program Service codes. For simplicity, this diagram omits the 10-bit checkwords that follow every sixteen bits of data. The checkwords go where the dashed lines are shown. The digital codes shown in bits 15 and 16 of the second block identify which two of the eight words of Program Service characters will be found in block 4 of the group being read.

References

Chirlian, P. 1981. *Analysis and Design of Integrated Electronic Circuits*, pp. 1007–1011. New York: Harper and Row.

Couch II, L.W. 1987. *Digital and Analog Communication Systems*, 2nd ed., pp. 157–161. New York: Macmillan Publ. Co.

Gardner, F. 1966. *Phaselock Techniques*, New York: Wiley.

Haykin, S. 1983. *Communications Systems*, 2nd ed., p. 89. New York: Wiley.

Krauss, H.L., Bostian, C.W., and Raab, F.H. 1980. *Solid State Radio Engineering*, p. 309. New York: Wiley.

Millman, J. and Grabel, A. 1987. *Microelectronics*, 2nd ed., p. 46. New York: McGraw-Hill.

National Radio Systems Committee. 1993. *United States RBDS Standard*, pp. 1–104.

Sedra, A.S. and Smith, K.C. 1991. *Microelectronic Circuits*, 3rd ed., p. 344. Philadelphia: Saunders College Publishing.

Signetics Corporation 1972. *Linear Phase Locked Loops Application Book*, Sunnyvale, CA: Signetics Corporation.

Signetics Corporation 1985. *Linear LSI Data and Applications Manual*, p. 9–127. Sunnyvale, CA: Signetics Corporation.

Taub H. and Schilling, D.L. 1971. *Principles of Communications Systems*, pp. 149–50. New York: McGraw-Hill.

Torick, E. and Keller, T. 1985. Improving the Signal-to-Noise Ratio and Coverage of FM Stereophonic Broadcasts. *J. Audio Eng. Society* 33(12): 939.

3

Modern receiver circuitry

As indicated in the introduction, new types of receivers and receiver circuitry are being introduced continually. In this chapter we will look at some of them.

3.1 An integrated AM/FM receiver

The receiver we consider here is one of the Sony Walkman series that is designed to drive lightweight stereo headphones rather than a speaker. The headphones are so intrinsic to the design that their cord also serves as the FM antenna! The schematic is shown in Fig. 3.1.

We begin by directing our attention to the lower right-hand corner of the schematic. Rather than returning one headphone lead from each channel directly to ground, the ground return for both channels is through L_7. This has a reactance of about 1.4 kΩ in the FM band but less than an ohm for audio frequencies. Thus one side of each headphone is AC grounded for audio, but a significant FM signal is developed across L_7 and coupled via C_{39} to the RF input of the receiver. This capacitor's reactance is less than 50 Ω in the FM band. Diodes D_1 and D_2 serve to prevent overdriving IC_1. For reception of distant FM stations, *S1-1* couples the FM antenna circuitry directly to a bandpass filter. For local reception, the coupling is via R_{16}, which attenuates the larger local signals. In the AM position, C_1 AC grounds any FM signal received by the FM antenna. The DC potential on this line is .85 V for FM, but it gets pulled down to zero in the AM position of *S1-1*. This line is tied to the internal voltage regulator of IC_1 and may be a shutdown line. The 3 V input to the regulator (pin 26 of IC_1) provides a regulated 1.3 V output on pin 8 when the voltage on pin 15 is at the normal value of .85 V (for FM). When pin 15 is grounded in the AM position, the regulator output shuts down, removing supply voltage from the FM front-end portion of IC_1. This supply line is bypassed to ground by C_2 and C_{40}. The grounded end of each of the parallel tuned circuits is returned to this 1.3 V regulated supply line. Therefore, they are AC grounded but not DC grounded. This is analogous to the ZN414 TRF IC receiver of Chapter 1.

The AM front-end part of IC_1 is driven by a signal from the loopstick antenna, which also serves as the inductance for the input AM resonant circuit.

It may be assumed with safety that both the AM and FM front ends include the local oscillator, mixer, and, at least for the FM, an RF amplifier stage. At the time this receiver was designed, an analog integrated FM front end was still

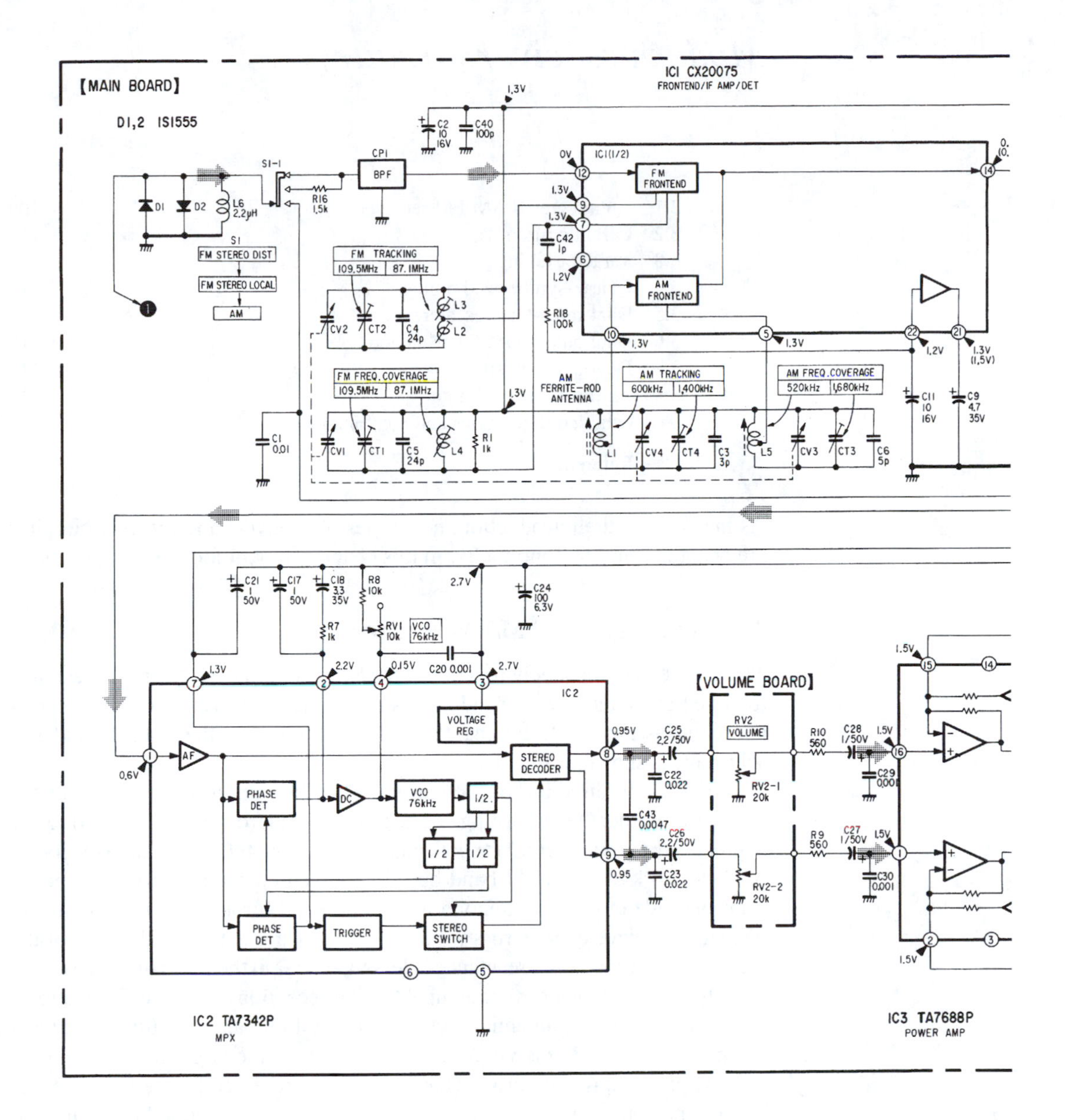

Note:

- All capacitors are in μF unless otherwise noted. pF : μμF 50WV or less are not indicated except for electrolytics and tantalums.
- All resistors are in ohms, 1/6W unless otherwise noted. kΩ : 1000Ω, MΩ : 1000kΩ
- △ : internal component.
- ▭ : panel designation.
- ▭ : adjustment for repair.
- Voltages are dc with respect to ground unless otherwise noted.
- Readings are taken under no-signal (detuned) conditions with a VOM.
  no mark : FM
  ( ) : AM
- Voltage variations may be noted due to normal production tolerances.

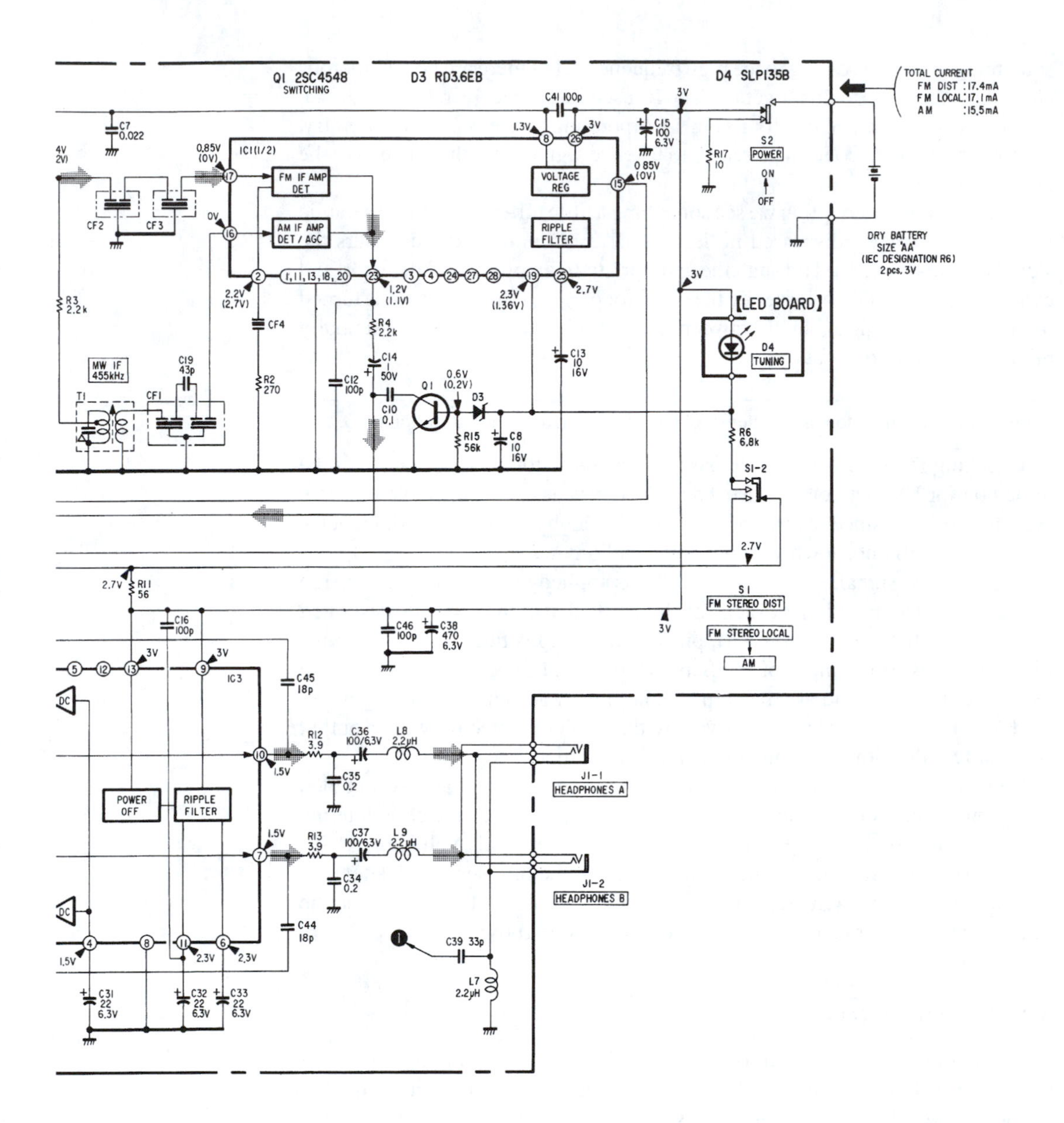

• Switch

Ref. No.	Switch	Position
S1	FM DIST/FM LOCAL/AM	FM DIST
S2	POWER	OFF

• ⇨ : Signal path

Fig. 3.1: Complete schematic diagram of a Sony SRF-33 W Walkman receiver.

something of a rarity, owing to the high frequencies, but there are now numerous chips that can perform this function. The IC used in this receiver, the CX20075, was a very sophisticated one for its era, incorporating all the necessary circuitry, except for the tuned coupling circuits, to take the signal from the antenna to the detector outputs.

In this receiver's coupling we see something a little different, namely ceramic filters. These will be discussed in Section 3.1.2. From the ceramic filters the signal reenters IC_1, goes through the appropriate IF amplifier and detector, and exits from pin 23. It then passes by the collector of Q_1. If this transistor is turned on, it will place C_{10} in shunt between the audio line and ground, essentially putting an LPF into the audio line.

Exercise 3.1 Find the break frequency of the LPF formed by the saturation of Q_1.

Checking the potentials on the base of this transistor, we see that it will be turned on for FM but not for AM. Thus Q_1 serves as a sort of squelch switch. We say "sort of" since it just knocks down the higher frequency components of the interstation noise where most of the audio power is found.

The detected signal, whether AM or FM, is applied to IC_2, which is a stereo decoder chip. Its operation is exactly analogous to that of the LM1800 described in Chapter 2. If *S1-2* is set for AM, pin 7 of this chip is tied to the 2.7 V line. Pin 7 connects to the input of the Schmitt trigger in IC_2, which will disable the stereo decoder and pass the chip's audio input through to both outputs. In the FM position, the Schmitt is driven by the in-phase detector, whose output apparently falls with increasing pilot signal amplitude.

The outputs of this chip pass through volume controls and are AC coupled into an integrated stereo amplifier chip, IC_3. As is most often the case, internal circuitry biases both the inputs and outputs to half the supply voltage.

The RC LPF section on the output of each channel serves to attenuate any large-amplitude high-frequency oscillation that might otherwise reach the headphones. The series inductance serves the same purpose.

3.2 Ceramic filters

These filters serve the same purpose as IF transformers but are finding increasing use because they are small and inexpensive. In addition, they have quality factors comparable to IF transformers and, because they are nonadjustable, they eliminate a tedious alignment step. One disadvantage is that they have an insertion loss of 5–6 dB, which is inferior to that of transformers. However, with integrated circuits, gain is cheap, and this is a relatively minor problem.

3.3 Varactor tuning

Varactors are otherwise known as varicaps or VVCs (Voltage Variable Capacitors). They are actually diodes that are operated in the reverse-bias region. The depletion capacitance in an abrupt junction diode varies with the reverse bias as

$$C_d = \frac{A}{2} \cdot \sqrt{\frac{2q\varepsilon}{(\Psi_0 + |V_R|)(1/N_A + 1/N_D)}}$$

where A is the cross-sectional area of the junction, q is the electronic charge, ε is the dielectric constant of the semiconductor, Ψ_0 is the zero bias junction barrier height in volts, N_A is the acceptor density on the p side, N_D is the donor density on the n side, and V_R is the reverse bias voltage.

The varicaps that were originally available had capacitances that were both small and limited in range. Although capacitance can be increased by doping both sides of the junction heavily, this will also raise the maximum E field at the junction for a given reverse bias, thus limiting the range of V_R (and thus C_d) that can be applied without causing breakdown of the junction. On these grounds, it appears that the best route to increasing C, while still allowing a good variation of it with V_R, is to increase A.

Varactors that have maximum capacitances of well over 100 pF are now available, making them suitable for tuning AM receivers.

Although there have been circuit applications for varicaps ever since they became available, none of them can compare in importance to circuitry in which the varicap serves as an element in a closed loop. The advent of quartz-synthesized tuning using varicaps set new standards for tuning reliability and precision.

3.4 Quartz-synthesized tuning

One of the very earliest pieces of consumer equipment to use quartz-synthesized tuning was what Heath Co. called their Computer FM Tuner. Although the state of the art has now advanced beyond this circuitry, it was a remarkable innovation and still illustrates clearly the basic principles involved.

Figure 3.2 shows the circuitry in block diagram form.

The circuit is essentially a distributed phase-locked loop, which will attempt to equalize the frequencies at the two inputs of the phase detector. The programmable divider can divide its input frequency by any integer between 494 and 593. Mathematically, the circuit is described by

$$\frac{f_{LO}}{8n} = \frac{f_Q}{4}, \qquad f_{LO} = 2nf_Q \qquad (494 \le n \le 593).$$

Since $f_0 = .1$ MHz, $f_{LO} = .2n$ MHz. When $n = 494$, $f_{LO} = 98.8$ MHz, and when $n = 593$, $f_{LO} = 118.6$ MHz. Thus the user's choice of n determines the local oscillator frequency. To the first order, this will not produce any output from the mixer unless an RF signal 10.7 MHz removed from the local oscillator output is also present.

Of course, the tuner was more "user friendly" than to require programming of the value of n. One of three programming modes allowed the station frequency to be entered from a calculator-type keyboard. There were also three preset station frequencies that could each be selected by pushing a single button and a search tuning mode wherein the dial would be scanned and the search would stop at each station producing a clean signal.

One might ask why the upper signal path in Fig. 3.3 did not utilize a fixed $\div 2$ block rather than the $\div 8$. This would have allowed the elimination of the $\div 4$ in the lower path. The equilibrium condition would then be

$$\frac{f_{LO}}{2n} = f_Q \quad \text{or} \quad f_{LO} = 2nf_Q.$$

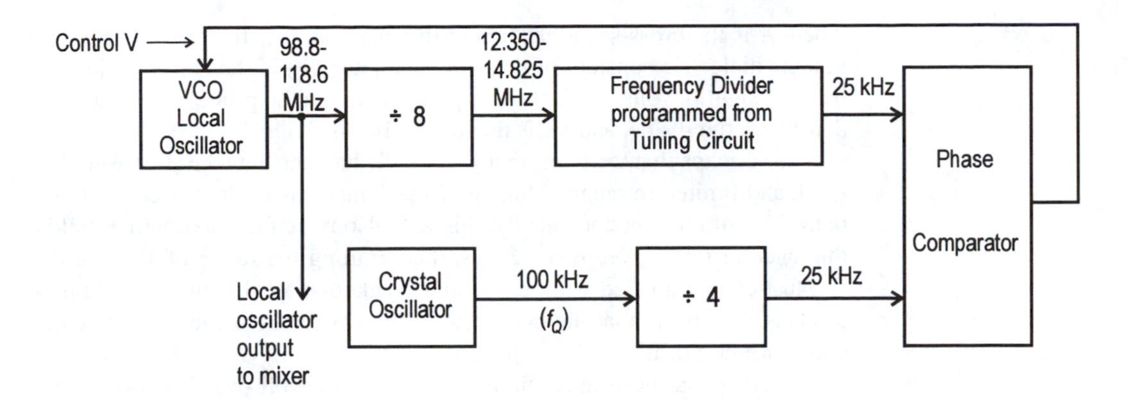

Fig. 3.2: Frequency-synthesis loop for Heath "Computer Tuner."

The same result is obtained with the circuit actually used. This option was not used because the frequency into the programmable divider would be raised by a factor of four. It would range between 49.4 MHz and 59.3 MHz. These frequencies are out of the reach of standard TTL, and the newer high-speed versions of TTL were not available when the circuit was designed.

Modern circuitry can accomplish the same thing in a single chip that operates at frequencies up to 120 MHz. The two technologies that make such performance possible are the incorporation of emitter-coupled logic (ECL) in the front end and the dual-modulus prescaler.

3.5 Dual-modulus prescaler

The dual-modulus prescalar is the key module in the high-speed programmable counters now used in the front ends of quartz-tuned receivers. The prescaler itself is fabricated of ECL so it can function at frequencies well over 100 MHz, allowing it to be used for FM as well as AM. The circuit itself consists of three main functional blocks. In addition, the circuitry includes some gate logic, ECL-TTL level translation, and in most cases a microprocessor, which, among other things, converts from the tuning information input by the user to the programming needed by the dividers. The interconnection between the blocks is shown in Fig. 3.3.

The swallow counter is so called because it can apparently "swallow" one pulse of f_{in} for each time the prescaler circles through its count sequence. We will give a brief description of the circuit operation and then show an example of how the circuit counts. In practice, f_{in} might be the output of the local oscillator with signal conditioning to render it acceptable to the ECL prescaler. The output should be at a fixed reference frequency, which would be compared to a divided-down quartz oscillator as shown in Fig. 3.4, which is simply a more generic version of Fig. 3.2. When the loop is locked, $f_{\text{OUT}} = f_{\text{ref}}$.

Referring back to Fig. 3.3, when the cycle starts, the following conditions apply:

(1) Dual-modulus prescaler is set in high modulus mode by control line.
(2) Swallow counter starts counting up from its preset output state.
(3) Main counter also starts counting up from its preset output state.

This state of affairs continues until the swallow counter outputs are all at logic 1. At this point, the carry output goes high. This causes two things to happen:

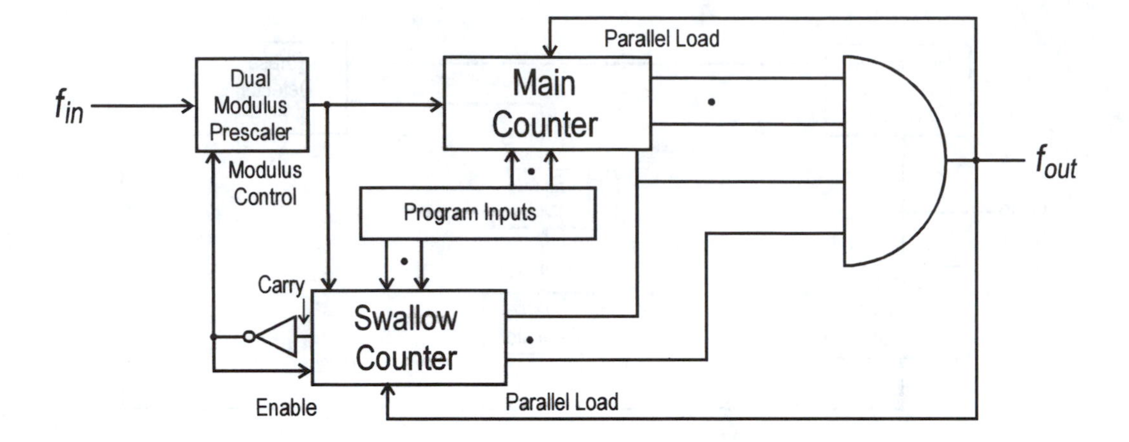

Fig. 3.3: High-speed programmable counter using a dual-modulus prescaler.

(1) The prescaler is dropped from its higher to its lower modulus.
(2) The swallow counter is disabled. Its outputs will all remain high until the counters are again preset.

The prescaler, now operating at its lower modulus, continues to increment the main counter until its outputs are all logic 1 also. At this point, all inputs to the AND gate are unity, and its output causes the swallow and main counters to be preset again, starting the cycle over again.

This sequence may all be very clear, but does it do us any good? To see how it works, we shall consider an example that, although too simple to be of practical use, well illustrates the principles of the technique.

The three main blocks of Fig. 3.3 may be labeled to indicate their count ranges or logic states as shown in Fig. 3.5. The ÷3,4 block is the dual-modulus prescaler. It is assumed that when its modulus control line is high, it will ÷4, and when it is low, it will ÷3. The main counter is to be a modulo-3 counter, so we parallel load in the ones complement of this, or set it to 00_2. The swallow counter is to be a modulo-2 counter, so it is preset to 01_2. Thus at the start of a cycle we have the prescaler in the high-modulus mode and the overall situation is as shown in Fig. 3.6a. After four input pulses, the prescaler outputs a pulse that increments both the main and swallow counters (see Fig. 3.6b). After four more pulses (eight total), we have the situation depicted in Fig. 3.6c. However, since the swallow counter outputs are all at logic 1, the prescaler modulus drops to 3, and the swallow counter is disabled (Fig. 3.6d). After three more input pulses (eleven total) we have the situation shown in Fig. 3.6e. But this is the condition under which the AND gate outputs a pulse and reloads the main and swallow counters and thus restores the prescaler modulus to 4. Hence we return to the situation of Fig. 3.6a, and one reference period has been completed. We thereby conclude that this is a ÷11 counter.

Let us make the following observations about this result:

(1) The circuit could be replaced by a single ECL ÷11 counter, although it might not be so readily programmable and would certainly consume more power.
(2) Making just the prescaler out of ECL allows the rest of the circuitry to be well handled by Advanced Low-Power Schottky (ALS) TTL.

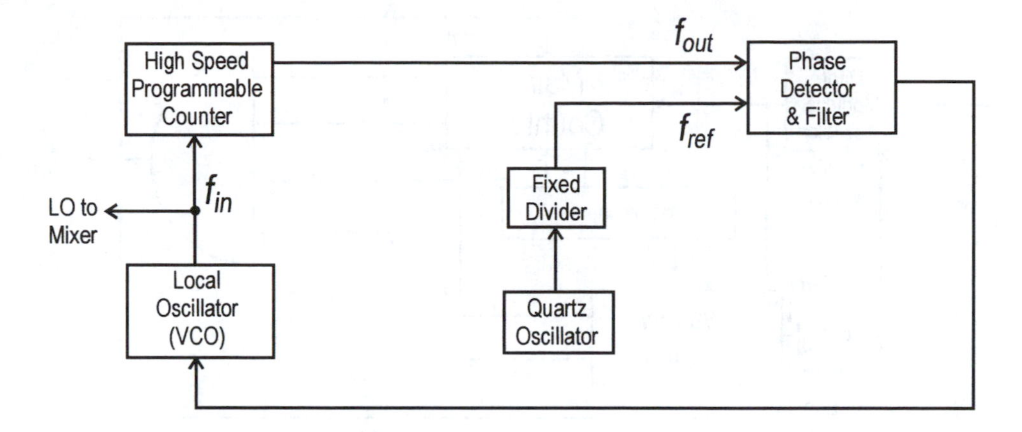

Fig. 3.4: Block diagram showing the use of a high-speed programmable counter in a frequency-synthesis loop.

(3) Although we have used five counting stages to achieve a ÷11 function, it is still a fairly effective approach from a hardware point of view, since four stages is the minimum no matter what binary technique is used.

(4) A general expression for N (where $f_{OUT} = f_{in}/N$) is

$$N = N_{ps}N_M + N_S,$$

where N_{ps} is the lower modulus of the prescaler, N_M is the desired modulus of the main counter (which is 1's complemented and parallel loaded into the main counter), and N_S is the desired modulus of the swallow counter. (Which is also 1's complemented and parallel loaded into the swallow counter.)

In our example

$$11 = 3 \cdot 3 + 2.$$

(5) The swallow counter "swallows" one pulse out of every four when in the high-modulus mode. It is essentially programmed with the number of pulses to be swallowed during each reference cycle. When they are completed, the modulus drops back to three and no more swallowing is done until the reference cycle is completed. The need for a swallowing capability may be seen as follows: Suppose we had a fixed modulo-10 prescaler followed by a programmable divider. Then we might have $N = 190, 200, 210\ldots$ as the programmable divider is set for 19, 20, 21 .... Frequency division factors between these are not available in the system. If the dual-modulus prescaler approach is used instead, and it has a swallow counter with $N_S \leq 9$, it can swallow from 0 to 9 counts, adding all integer division ratios between two adjacent values of N to the repertoire of the unenhanced counter. It is true that we could accomplish the same thing by adding another decade divider in series with the prescaler and programmable divider, but this would lower f_{OUT} by a factor of 10. This would require that the reference oscillator also undergo an additional ÷10 step. The net effect of this lower reference frequency is an increased lock time, which is highly undesirable.

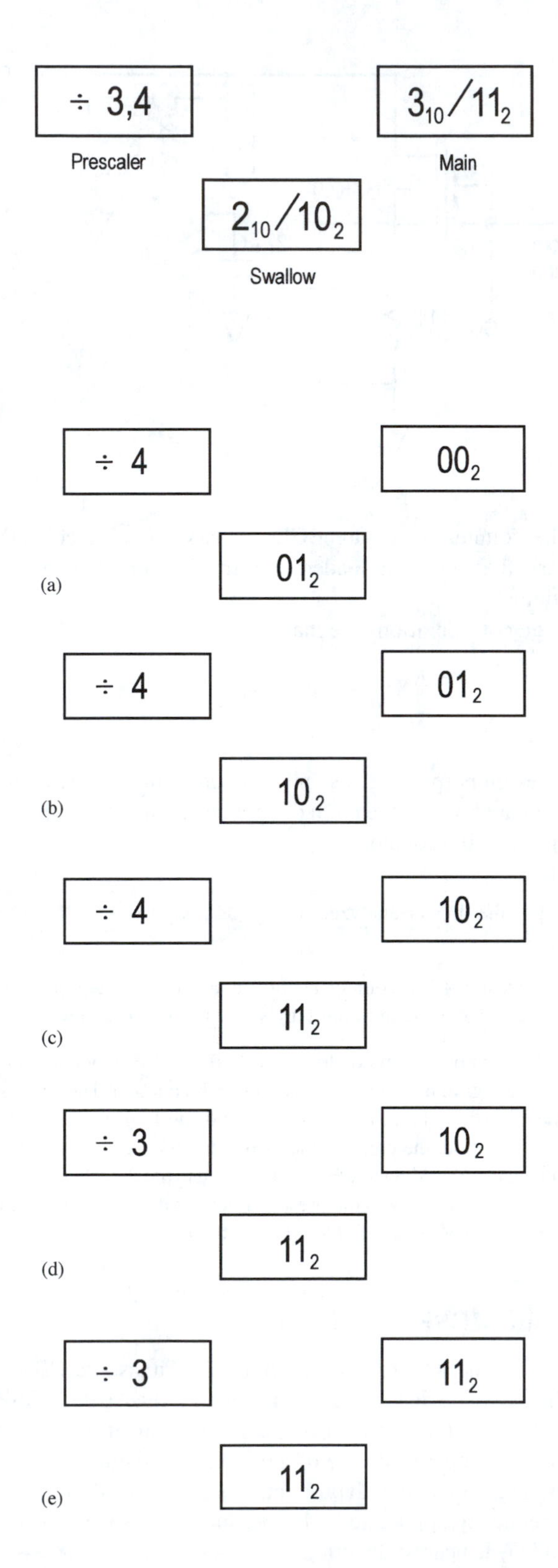

Fig. 3.5: Block representation of a high-speed programmable counter using a dual-modulus prescaler. The block on the left shows the two possible values for the modulus of the prescaler; the other two show the preset count values set into the swallow (center) and main (right) counters.

Fig. 3.6: (a) The situation in the configuration of Fig. 3.5 before any pulses are input. (b) After four pulses are received and both the main and swallow counters are incremented once. (c) After four more pulses are received (eight total). (d) Since the swallow counter outputs are all at logic 1, the prescaler modulus drops to 3 and the swallow counter is disabled. (e) After three more input pulses (eleven total), both the main and swallow counters all have outputs at logic 1. This generates a pulse which reloads both the main and swallow counters to their preset values and the cycle repeats.

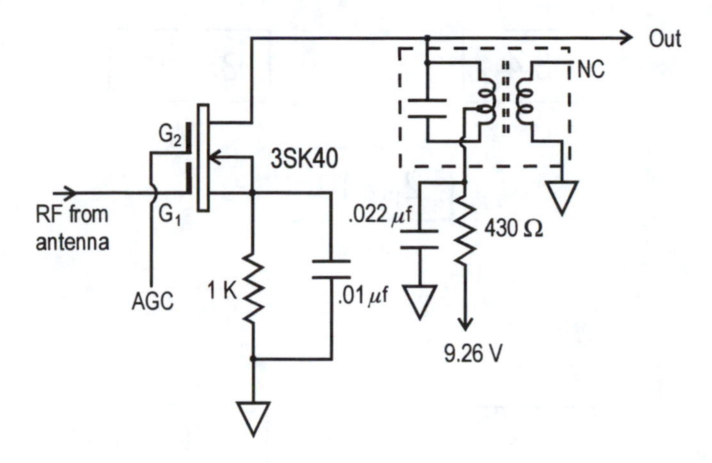

Fig. 3.7: RF amplifier using a dual-gate MOSFET.

(6) If the counter stages are BCD (Binary Coded Decimal) instead of binary, they are to be loaded with the 9's complement of the desired modulus.

(7) Design considerations are that

$$N_S < N_M,$$

$$N_{ps} \leq N_M + 1.$$

(8) For design purposes, the swallow counter may be treated as the least-significant bits and the main counter as the most-significant bits of the programmable counter.

Exercise 3.2 For the divider analyzed in Fig. 3.7, what is N if the main counter is loaded with 01_2?

Exercise 3.3 Devise a ÷200 counter using a dual-modulus prescaler with the least hardware possible. What is the minimum division ratio it can achieve?

Exercise 3.4 The National Semiconductor DS8907 AM/FM synthesizer has a 13-bit latch to present the programming information to a 13-bit counter. The six least-significant bits may be thought of as the swallow counter and the higher-order bits as the main counter. For FM operation, the prescale factor is ÷63, 64. Suppose we wish to use this chip in a system such as that shown in Fig. 3.4. If we use the 25 kHz reference frequency for which the chip is designed, find the preset values for the main and swallow counters for the extremes of the FM band, 87.9 MHz and 107.9 MHz.

3.6 Dual-gate MOSFET front end

Figure 3.7 depicts the RF amplifier section of a Panasonic CB radio (Model RJ-3050). The active device in Fig. 3.7 is called a dual-gate MOSFET. In this amplifier the MOSFET is used for a specific reason. It will be recalled that the Miller effect can operate on C_μ of a bipolar transistor or C_{gd} of an FET to severely limit the gain-bandwidth product. It is obviously desirable to minimize C_{gd} in high-frequency applications. In the dual-gate MOSFET, G_1 is nearest the source, and G_2 is nearest the drain. G_2 is held at a more or less "fixed bias" (the AGC voltage) and so shields G_1 from the drain, making C_{gd} very small.

The 430 Ω resistor and .022 μF capacitor decouple the supply from the drain circuit. Cascaded tuned circuits are used to improve the frequency selectivity of the receiver.

3.7 Double conversion

The use of two mixing stages is called double conversion. This technique is commonly used in better CB and communications receivers because it improves image rejection. If we were to try to translate from 27 MHz down to 455 kHz in a single conversion step, the local oscillator would have to run at 27.455 MHz. But a non-CB signal at 27.910 MHz would also give a 455 kHz difference signal, which would pass right through the IF amplifier. Furthermore, the signals at 27 MHz and 27.91 MHz are close enough together that the tuned circuit resonant at 27 MHz would be expected to have significant transmission at 27.91 MHz also.

In double conversion, the first mixing operation brings the 27 MHz signals down to about 10.7 MHz, and the second mixing operation brings them down to 455 kHz. This system uses a 16.3 MHz local-oscillator signal to give an output at 10.7 MHz from the first mixer when the receiver is tuned to 27 MHz. An image signal at 5.6 MHz would also give a 10.7 MHz output, but in this case 27 MHz and 5.6 MHz are far enough apart that the tuned circuits could make a significant separation between them. This is particularly true when, as is the case for this receiver, there is an RF amplifier prior to the first mixer that has both input and output tuned to 27 MHz.

Frequencies have been chosen to allow the use of standard AM and FM IF transformers or resonators.

3.8 Automatic noise limiter

The operation of an automatic noise limiter (ANL) is based on the fact that noise in the signal generally has the form of narrow spikes whose amplitude is greater than that of the signal. This is particularly true in the automotive environment in which most CBs are operated. Thus if the output amplitude is clipped at the proper level, the signal will get through essentially unclipped, but the noise will be clipped. Since the noise power increases as the square of the spike amplitude, an ANL can make a significant difference in the audible noise level of the received signal. This type of ANL circuit is shown in Fig. 3.8. In this circuit, C_{32} is chosen large enough to serve as an AC short for the audio. Thus any audio that passes through the ANL must also pass through the diode. This will require it to have a forward bias greater than the expected signal amplitude so that both positive and negative excursions of the signal may pass. This bias can be calculated by incorporating R_1, R_2, R_3, R_4, and V_{DC} into a Thevenin equivalent circuit.

The orientation of the ANL diode in the signal path will always be opposite that of the detector diode. This will insure that the polarity of detected noise spikes will be such as to overcome the forward bias on the ANL diode and reverse bias it. Of course, that portion of the noise spike with amplitude greater than the forward bias level will thus be cut off, since the diode is reverse biased during that time.

A more elegant and more expensive noise reduction scheme is called *noise blanking*. It is based on the fact that the noise is fairly broadband in nature

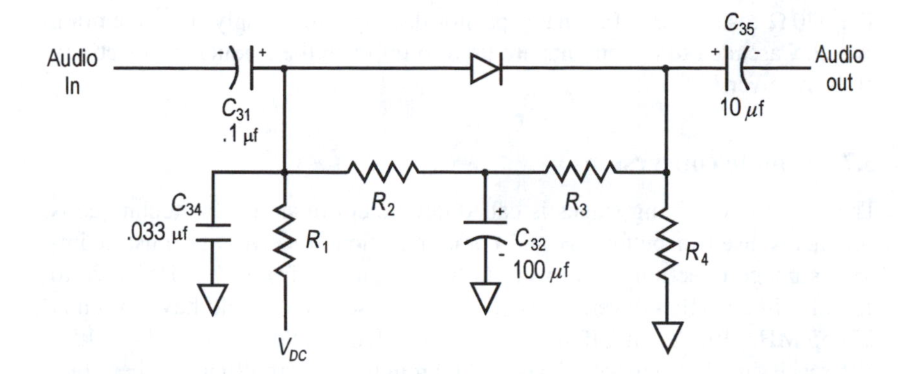

Fig. 3.8: Automatic noise limiter circuitry.

because of the narrowness of the noise pulses in the time domain. A second receiver front end is added to the receiver but is tuned to a vacant frequency near the band for which the receiver is designed. It should pick up no signal but nearly the same noise as the desired channel. A coincidence circuit then detects whenever the two channels have the same output and blanks the audio during that time. This should occur only during the noise pulses in a properly designed and functioning system.

3.9 Other configurations and combinations

Cassette–radio combinations generally don't use any circuitry in common except part of the power supply and the audio amplifier. Audio tape recorders will be covered in Chapter 5.

Clock–radio combinations generally connect the two functions only in that an output from the clock chip is used to enable the power supply to the radio in the alarm mode. The radio can, of course, also be switched on manually.

Top-of-the-line auto radios now feature AM/FM stereo with quartz tuning and search plus cassette and/or CD capability. Search is an operational mode wherein a slowly varying ramp voltage is applied to the varactor, causing it to tune across the band and stop whenever the received signal is strong enough to afford good reception. The user who wants to continue the search, presses a button and the ramp resumes.

All of this capability and more is now available in a two-chip set from Philips Semiconductors. The chips are the TEA6811 and the TEA6821.

The TEA6811 contains RF amplification, mixers for both AM and FM, and a VCO, which serves as the local oscillator for both bands. This chip also contains the frequency synthesizer and an I^2C bus interface for communication with its companion chip and a microprocessor. Double conversion is used in the chip set. To enable as much as possible of the circuitry to be used for both bands, the first mixing operation (which is performed in the TEA6811) generates an AM output at 10.7 MHz (yes, you read that correctly!) and an FM output at 72.2 MHz. These two outputs pass to the second chip, a TEA6821. A crystal oscillator connected to this chip provides one of the inputs to the second mixer for both bands. It drives the FM second mixer directly and also drives the AM second mixer through an integer divider.

Exercise 3.5 Given that the desired output of the FM second mixer is 10.7 MHz and that of the AM second mixer is 450 kHz, find

(a) the required crystal frequency,
(b) the modulus of the integer divider.

Among the high-level capabilities of this chip set are:

(1) ignition noise blanking;
(2) multipath detection and correction;
(3) search capability;
(4) RBDS and RDS capability;
(5) interface for a separate precision adjacent-channel suppression IC;
(6) built-in diagnostics available on power-up;
(7) automatic stereo to mono blend for FM in weak signal areas;
(8) field-strength-dependent bandwidth reduction (as the signal weakens, the audio bandwidth is cut to reduce the noise level);
(9) 42 preset frequencies;
(10) Interface for a cassette mechanism (the chip provides selectable Dolby, metal tape equalization, and automatic music search), and
(11) CD interface.

All of these capabilities are available for about $10 per set in production quantities. It is not to be inferred that every radio made with these chips will have all of the capabilities listed, only that the chips have all of these capabilities if the radio makers should choose to avail themselves of them.

Reference

Egan, W.F. 1981. *Frequency Synthesis by Phase Lock*, p. 156. New York: Wiley.

4

Equalization

The "jargon gap" is exemplified by the word "equalization," a common term in the high-fidelity audio industry. We also saw an example of it in the FM receiver we analyzed. The electrical engineer in the field of communications would recognize the term, but engineers outside these two areas might need to have the *equalizer* identified for them as a filter.

Perhaps electrical engineers as a whole have too limited a conception of a filter. For example, we tend to think of active filters as classified by an associated function (high pass, low pass, band pass, etc.), realization (VCVS, IGMF, biquad, etc.), ripple characteristics (Butterworth, Chebyshev, elliptic, etc.), and order. This is in fact the classical approach to a study of filters and can be justified because some kind of orderly classification system is necessary to grasp the plethora of types of filters.

Yet other than the simple parallel LC tuning circuit and the low-pass ripple filter on power supplies, many of the filtering circuits found in consumer electronic equipment do not fit neatly into any of the pigeon-holes established by filter theory. We need then to broaden our conception of what a filter is. In the simplest definition, it is a circuit designed to alter the frequency spectrum of a signal input. This definition certainly encompasses the equalizing circuits commonly used in consumer electronics.

We will take a look here at three important types of equalizing circuits. Others will be covered in less detail as we encounter them. The three we will examine are the tone control, the RIAA equalizer, and the NAB equalizer.

4.1 The tone control

Practically every piece of sound equipment costing more than \$10 has a knob labeled "Tone." To be sure, these controls affect the tone, but in most equipment less than \$100 (and even some over \$100) the tone change is accomplished with what is called a "losser" type of control. Such a control is depicted in Fig. 4.1.

To see quantitatively what this stage does, we need to include the output impedance of the nth stage and the input impedance of the $(n + 1)$st stage. For simplicity we'll assume that the latter is infinite. The relevant schematic is shown in Fig. 4.2. This circuit has the transfer function:

$$\frac{\mathbf{V}_{n+1}}{\mathbf{V}_o} = \frac{R + \dfrac{1}{sC}}{R + R_o + \dfrac{1}{sC}} = \frac{1 + sCR}{1 + sC(R + R_o)}.$$

A Bode amplitude plot is shown in Fig. 4.3.

From this figure it can be seen that the control has no effect at low frequencies. However, at higher frequencies, there is a drop in the amplitude response. As R

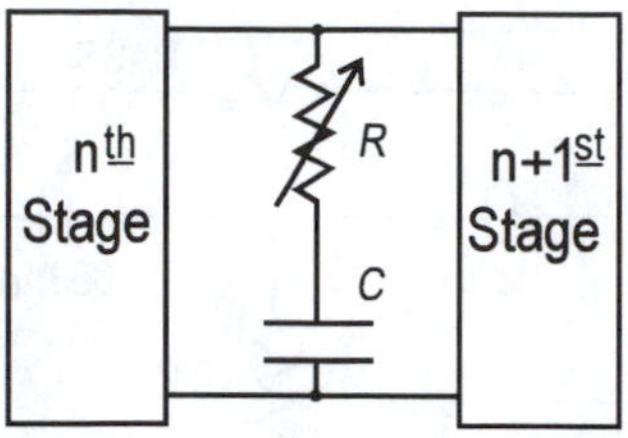

Fig. 4.1: Circuit configuration for a "losser"-type tone control.

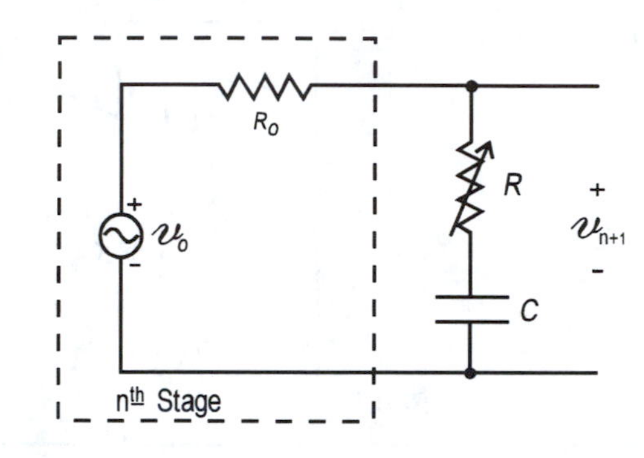

Fig. 4.2: Equivalent circuit used to analyze the "losser" control.

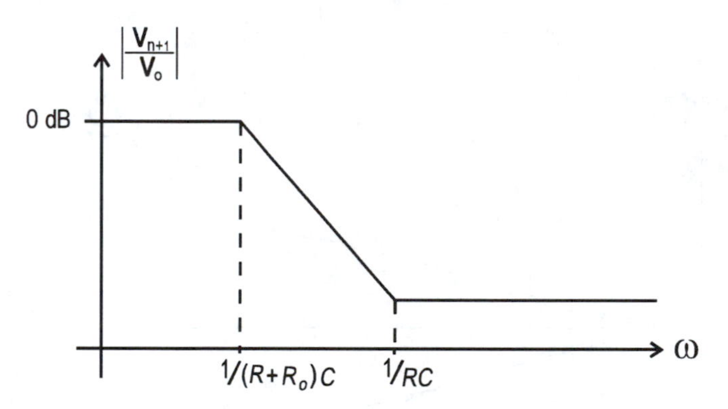

Fig. 4.3: Bode amplitude plot of a "losser"-type control.

is reduced, both the pole and zero move to the right, but the zero moves without limit as $R \to 0$, whereas the pole approaches $1/R_oC$. One might typically choose this pole frequency to be somewhere in the range between 1 and 3 kHz. As R increases, the pole frequency actually moves lower, but the zero moving in from infinity causes a limitation in the high-frequency drop-off.

The net effect is that the ear perceives far less high frequency when $R = 0$, even though the pole frequency is at a maximum. This control is a treble-cut control. Having less treble gives the illusion of having more bass. It also makes the sound "muddy" in most cases when set for minimum treble.

This circuit, though popular, has little to recommend it except its low cost.

A true tone-control circuit will have separate adjustments for treble and bass and will, in fact, be able to provide boost as well as cut for treble and bass separately. This is accomplished by introducing extra gain in the circuit that is not utilized unless the treble and/or bass controls are set to provide boost.

It is common practice now to implement not only tone controls but most equalization circuits in the feedback loop of an operational amplifier (op amp). Figure 4.4a shows one such tone-control circuit. Figure 4.4b shows the response curves for various settings of the treble control, and Fig. 4.4c shows the response

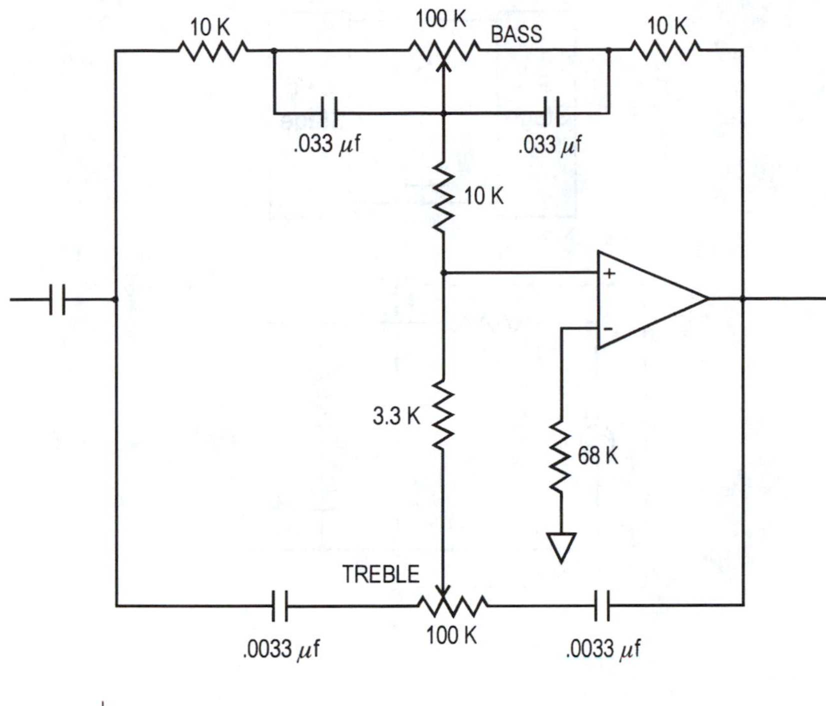

Fig. 4.4(a): Schematic diagram of a full-range tone control.

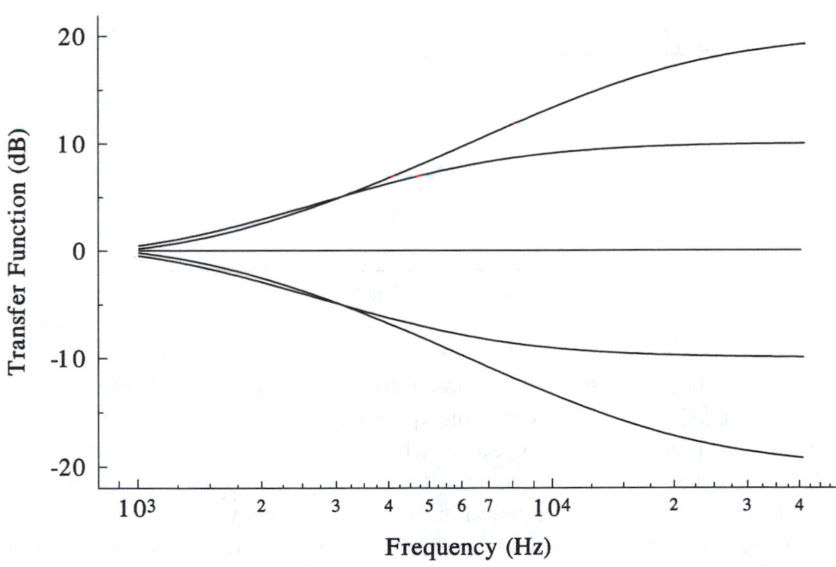

Fig. 4.4(b): Response curves showing the effect of the treble control.

for various settings of the bass control. Note from Fig. 4.4a that when both controls have their sliders to the extreme right, the impedance in the feedback loop is minimum while the impedance in series with the input is maximum. Both of these will minimize gain. Thus for both controls, the right end of rotation produces a cut, and the left end produces a boost.

Suppose we did not have Fig. 4.4b, and we wished to know something quantitative about the performance of this circuit. Since it is fairly complex for straightforward circuit analysis, let's see what information we can get by looking at the limits of operation.

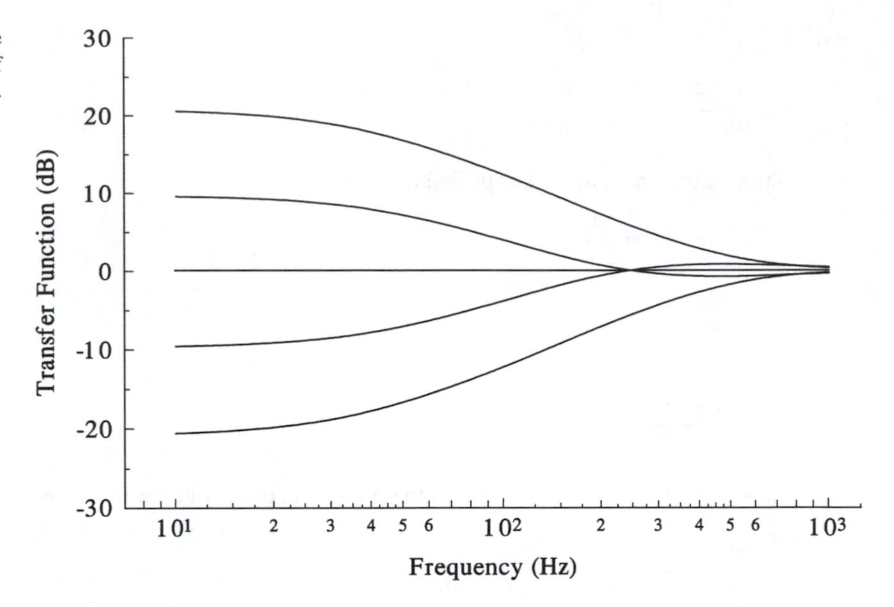

Fig. 4.4(c): Response curves showing the effect of the bass control.

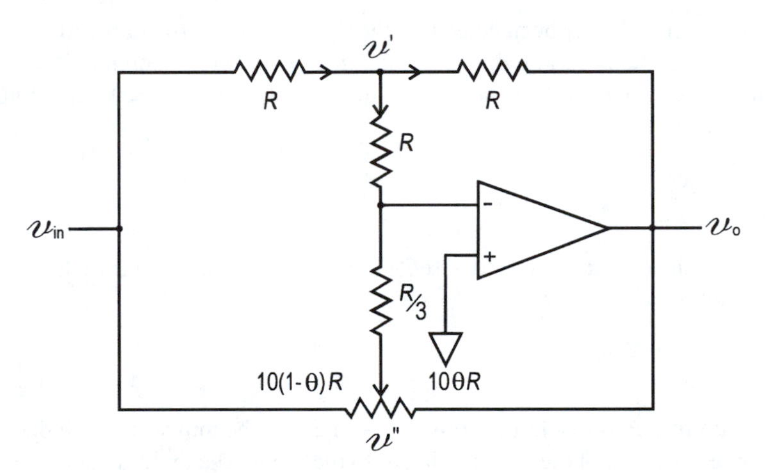

Fig. 4.5: Simplified circuit for finding the gain limits of the circuit for frequencies so high that the circuit capacitors may all be considered to be short circuits.

Let us first consider the operation for input frequencies high enough that all capacitors can be considered as AC shorts. Figure 4.5 shows the effective circuit in this case. Note that the bass control does not even appear. All resistances have been expressed in terms of one to which we assign a value of $R = 10\ \text{k}\Omega$. Note also that the 68 kΩ resistor in series with the noninverting input has been eliminated, since it serves only to balance DC output offset due to input bias current. The factor θ represents the fraction of rotation of the pot ($0 \le \theta \le 1$).

This circuit is fairly easily analyzed. The relevant equations are:

$$\frac{v_{\text{in}} - v'}{R} = \frac{v' - v_{\text{o}}}{R} + \frac{v' - 0}{R} \quad \text{(current at the } v' \text{ node)},$$

$$\frac{v'}{R} = \frac{-v''}{R/3} \quad \text{(current at the inverting input)},$$

and

$$\frac{v_{\text{in}} - v''}{10(1-\theta)R} + \frac{0 - v''}{R/3} = \frac{v'' - v_{\text{o}}}{10\theta R} \quad \text{(current at the } v'' \text{ node).}$$

These three equations may be simplified to give

$$v_{\text{in}} + v_{\text{o}} = 3v',$$

$$v' = -3v'',$$

and

$$\frac{v_{\text{in}} - v''}{10(1-\theta)} = 3v'' + \frac{v'' - v_{\text{o}}}{10\theta}.$$

Straightforward algebraic solution gives the high-frequency transfer function:

$$\frac{v_{\text{o}}}{v_{\text{in}}} = -\frac{-30\theta^2 + 39\theta + 1}{-30\theta^2 + 21\theta + 10}. \tag{4.1}$$

Since it has already been noted that the right end of pot rotation will produce a treble cut, and since Fig. 4.5 shows that this would correspond to $\theta = 0$, we would expect that Eq. (4.1) would predict a treble cut for $\theta = 0$. Substituting $\theta = 0$ gives

$$\frac{v_{\text{o}}}{v_{\text{in}}} = -\frac{1}{10}.$$

This, of course, corresponds to a treble cut. If, on the other hand, we let $\theta = 1$, the result is

$$\frac{v_{\text{o}}}{v_{\text{in}}} = -10,$$

which corresponds to a treble boost. In both cases, the minus sign is indicative of phase inversion of the output relative to the input. The tenfold cut and boost can be expressed as a ± 20 dB dynamic range for the treble control. Somewhere in between we would expect to find the "flat" position of the treble control, which corresponds to neither cut nor boost, but to a gain of unity, or 0 dB.

Exercise 4.1

(a) Find the value of θ corresponding to the "flat" position.

(b) Does your answer to part (a) predict a linear relationship between high frequency gain and θ? Discuss.

Of course, the same kind of limiting analysis can be performed for frequencies low enough that all of the capacitors are effectively open circuits.

Exercise 4.2 Find the transfer function versus θ of the bass control for low-frequency inputs. Compare the dynamic range of the bass control to that of the treble control.

The limiting expressions we have derived here represent the extreme right-hand and left-hand edges of the graphs in Figs. 4.4b and 4.4c.

4.2 Record (RIAA) equalization

RIAA (Recording Industry Association of America) was the standard equalization curve for LP records produced in America. Although production of LP records has ceased for all practical purposes, large collections of records remain in the hands of collectors, and hi-fi equipment will have to be able to provide RIAA equalization for years to come. Furthermore, a study of the need for and the nature of RIAA will give us the best possible insight into equalization.

Before examining the technical aspects, it would be informative to see why equalization is necessary for cutting and reproducing phonograph records. This, in turn, requires an elementary understanding of how records were recorded and reproduced.

The recording system was somewhat similar in principle to a phonograph as far as its mechanical aspects were concerned. The disc placed on the turntable was not grooved like a phonograph record, but was completely smooth. The stylus that rode on this disc was designed for cutting it. The assembly that carried the stylus and the apparatus that made it vibrate comprised the cutting head.

Being a mechanical system, the cutting head is governed by Newton's laws of motion. This assembly is assumed to be driven by a sinusoidal driving function, which, for generality, may be expressed as

$$F = F_o(\omega) \cdot \sin \omega t,$$

where the driving amplitude may in general be a function of ω. The mechanical system will have mass, damping, and restoring force. If it is assumed to be critically damped, the displacement amplitude will be given by

$$|X| = \frac{F_o(\omega)/k}{1 + (\omega/\omega_o)^2}, \tag{4.2}$$

where k is the "spring constant" and $\omega_o = \sqrt{k/m}$ where m is the mass of the vibrating assembly. Equation (4.2) tells us that if we were to make a Bode plot of displacement versus frequency, it would have the same form as that for an LPF. Above the corner frequency, the plot would drop off at -40 dB/decade.

This system is exactly analogous to an analog meter movement with zero at center scale that is driven by a variable-frequency sinusoid. At very low frequencies, the movement can follow the sinusoid exactly, but as the frequency increases, the meter will not reach one limit of its swing before the sinusoidal excitation reverses sign. The meter then reverses also without having deflected as far as it did for lower frequencies. As the frequency continues to increase, the amplitude of the deflection continues to decrease. In the limit, of course, no deflection is visible, and we have the familiar result that AC applied to a DC meter movement produces no reading since the meter mechanically averages the torque due to the sinusoid.

The important point for our purposes here is that for both the critically damped cutting head and meter movement, frequencies well above the pole frequency cause an amplitude response that drops off as $1/\omega^2$ for constant-amplitude sinusoidal excitation.

This could have disastrous consequences for the record. Suppose we set up our cutting-head drive to give a reasonable displacement at 500 Hz. The displacement will govern how many grooves we can put on the record. A typical groove density might be 100/inch of record radius. This means displacements must be limited to about $\pm.005''$ for all conditions of amplitude and frequency. Therefore, suppose we set up for $\pm.005''$ displacement at 500 Hz for the highest amplitude to be encountered. What will happen if we get a 20 Hz signal of the same amplitude? This frequency is $1/25$ of 500 Hz. Because the displacement response varies as $1/\omega^2$, this means that, all other things being equal, the 20 Hz signal will cause a displacement 625 times as large ($\pm3.125''$!) as the 500 Hz signal.

Nor is the high-end situation more encouraging. Consider a recorded signal of equal amplitude at 20 kHz. This is 40 times larger than our reference frequency of 500 Hz. Thus the amplitude of the stylus vibrations would be only $1/1{,}600$ of the amplitude (±781 Å!) of those due to 500 Hz drive. A displacement this size would be only about $1/7$ of a wavelength of light. Such a small signal would be obtained from it on playback that it would be totally lost in the surface noise due to microscopic scratches and other imperfections on the record surface.

Clearly this is an intolerable situation. The way out of it has already been hinted at by allowing F_o to be a function of ω. We can make F_o a monotonically increasing function of ω, so that the drive is minimized at low frequencies, where the mechanical response is maximized, and maximized at higher frequencies, where the response is minimized. Beyond this requirement there is no one intrinsically "right" function relating F_o to ω. "Right" has now been defined by the RIAA as having $F_o \propto \omega^2$ at the low-frequency end of the audio spectrum, $F_o \propto \omega$ in the middle, and $F_o \propto \omega^{3/2}$ on the high-frequency end. This means that on the low end of the audio spectrum, Eq. (4.2) will give a stylus displacement independent of frequency, as long as ω_o is at a subaudio frequency. The reason an unphysical response like $\omega^{3/2}$ is used on the upper end instead of ω^2 is that the power requirements for the cutter would become enormous if $F_o \propto \omega^2$ over the entire audio range.

Because the mechanical force applied to the cutter head is directly proportional to the applied signal, the frequency response of F_o is tailored by equalizing the signal to be recorded before it is applied to the cutter head. The Bode plot of the recording equalizer is given in Fig. 4.6. This figure also shows a plot versus frequency of the corresponding displacement of the stylus for a constant input amplitude to the equalizer. The plot exhibits a 20 dB drop over the audio range, or about a factor of 10 in groove spacing, for a constant-amplitude input signal to the recording apparatus.

Now let us direct our attention to the playback process. This begins with the stylus that follows the grooves in the record and vibrates according to the sound pattern impressed in those grooves. In the vast majority of quality sound systems still capable of playing LP records, that stylus is part of a magnetic phono cartridge. There are several varieties of this cartridge, but the underlying principle is the same in all of them. The stylus vibration is coupled to a magnetic circuit involving a coil and a magnet. A magnetic flux that varies with time in response to the music is induced in the coil. Leads from this coil are brought out of the cartridge so that the small voltage induced in accordance with Faraday's Law (usually units of mV) can be amplified.

It is important to realize that not only has the frequency response of the recorded material been systematically distorted, but there will also be magnetic

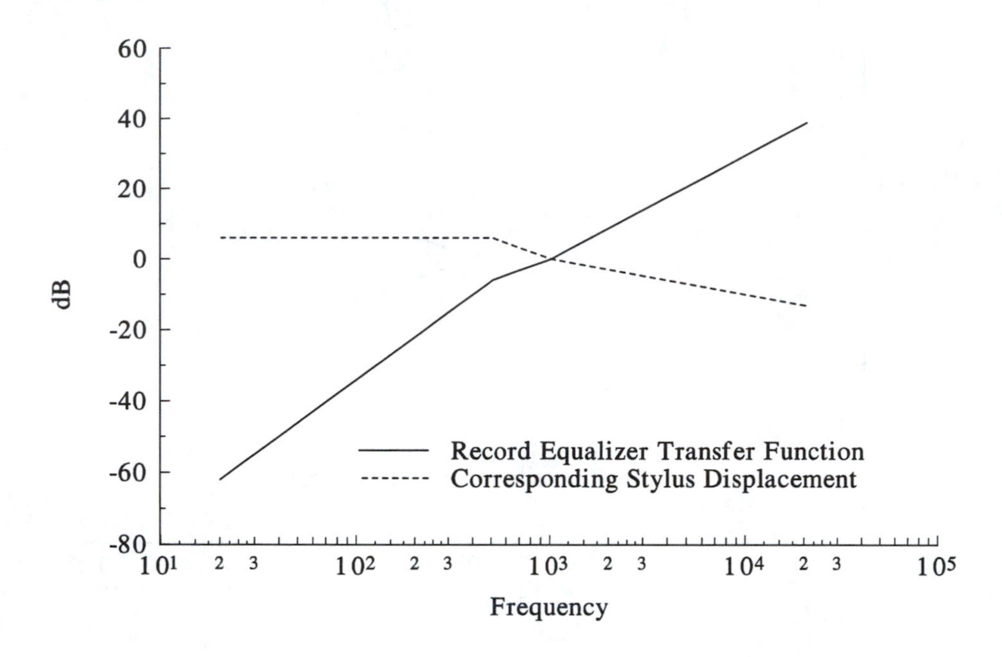

Fig. 4.6: Record equalizer transfer function (solid line) and corresponding stylus displacement (dashed line).

frequency-response considerations. Faraday's Law states

$$\frac{v_o}{v_{in}} = -N\frac{d\phi}{dt} \Rightarrow \mathbf{V} = -j\omega N\mathbf{\Phi}$$

for sinusoidal signals. The flux (ϕ) will be proportional to the displacement such that

$$\begin{aligned} \mathbf{\Phi} &= \alpha\mathbf{X}, \\ \mathbf{V} &= -j\alpha N\omega\mathbf{X}. \end{aligned} \tag{4.3}$$

Referring back to Fig. 4.6, we see that for $20 \le f < 500$ Hz, X is a constant for a constant-amplitude input. Thus, in this range, the recovered voltage from the cartridge will, by Eq. (4.3), be linear in ω. Likewise, in the frequency range $500 < f < 1{,}000$ Hz, $X \propto 1/\omega$, so v is flat. In the range $1{,}000$ Hz $< f \le 20{,}000$ Hz, $X \propto 1/\sqrt{\omega}$ (and v is proportional to $\sqrt{\omega}$). This frequency dependence is plotted in Fig. 4.7.

It should be clear from this plot that a playback equalization step is necessary to flatten out the frequency response characteristic and restore it to that of the original program material. It is this function that is performed within the consumer electronic equipment.

It might be questioned how the desired $1/\sqrt{\omega}$ response for the high end can be achieved. It can't. This same problem was glossed over in discussing the record equalizer also. There we would like to have had a response of $\omega^{3/2}$. In both cases the best we can do is to approximate the desired response. The truth of the matter is that even if the desired slope were an integer multiple of 20 dB/decade, we could still do no better than approximate it. This is because we have been dealing with Bode plots, which are only asymptotic approximations in the first place.

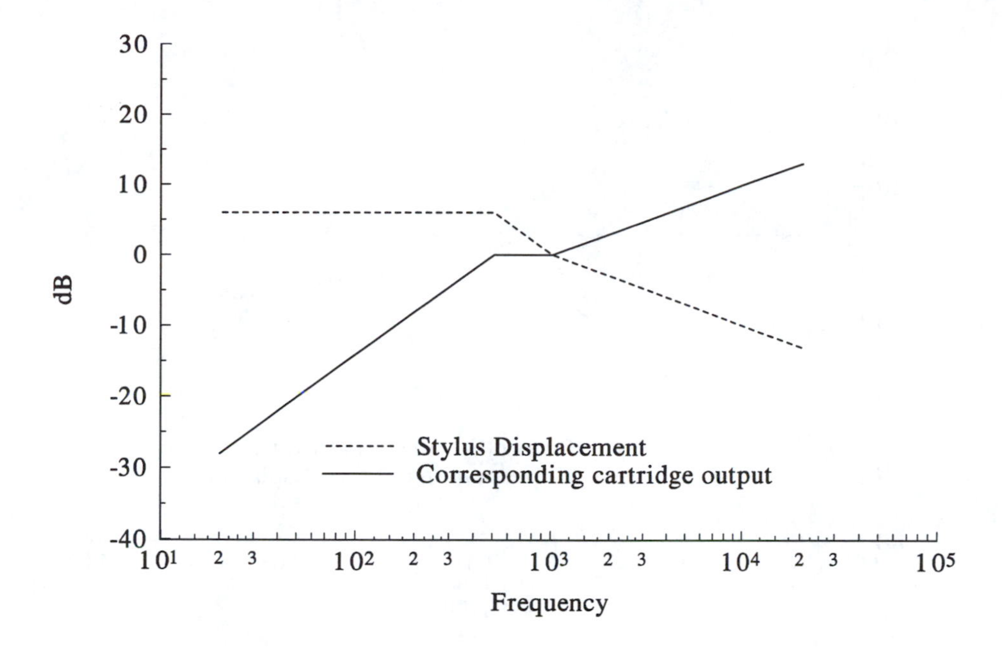

Fig. 4.7: Stylus displacement (dashed line) and corresponding phono cartridge output (solid line).

The desired response is approximated by using an equalizer with poles at 50 Hz and 2,100 Hz and a zero at 500 Hz. Figure 4.8 shows the actual response along with the Bode asymptotic response and the ideal response needed to equalize the output of Fig. 4.7. The asymptotic curves coincide between 50 Hz and 1 kHz. There are two differences between these curves. The first difference is that the actual circuit has a pole at 50 Hz that the ideal circuit doesn't. This pole keeps the op amp from the open loop condition at DC. The second difference is that the 1 kHz break point and −10 dB/decade slope of the ideal are replaced with a 2.1 kHz break point and −20 dB/decade slope in the actual realizable circuit. Hence the actual circuit provides more output than the ideal circuit between frequencies of 1 and 4.5 kHz and less output above 4.5 kHz.

The failure of the actual circuit to exactly equalize the output shown in Fig. 4.7 does not mean we are doomed not to have flat frequency response. The curve of Fig. 4.7 is no more physically realizable than the one needed to equalize it. All that matters is that the playback equalizer have a pole everywhere the record equalizer has a zero and vice versa; thus their effects will cancel perfectly.

An actual circuit capable of realizing the evenly dashed Bode plot of Fig. 4.8 is shown in Fig. 4.9. The transfer function is

$$\frac{\mathbf{V}_\mathrm{o}}{\mathbf{V}_\mathrm{i}}(s) = \left\{ s^2 R_\mathrm{L} R_\mathrm{R} C_\mathrm{L} C_\mathrm{R} + s \left[R_\mathrm{L} C_\mathrm{L} + R_\mathrm{R} C_\mathrm{R} + (C_\mathrm{L} + C_\mathrm{R}) \cdot \frac{R_\mathrm{L} R_\mathrm{R}}{R} \right] \right.$$

$$\left. + \left(1 + \frac{R_\mathrm{L} + R_\mathrm{R}}{R}\right) \right\} \Big/ \{(1 + sR_\mathrm{L}C_\mathrm{L}) \cdot (1 + sR_\mathrm{R}C_\mathrm{R})\}.$$

The numerator, being second order, will have two zeros. Since only one is desired in the range of audio frequencies, the other will need to be well outside that range. The values $C_L = 1.5$ nF, $R_L = 51$ kΩ, $R_R = 750$ kΩ, $C_R = 5.6$ nF, and $R = 1$ kΩ give poles at 38 Hz and 2,080 Hz and zeros at 463 Hz and

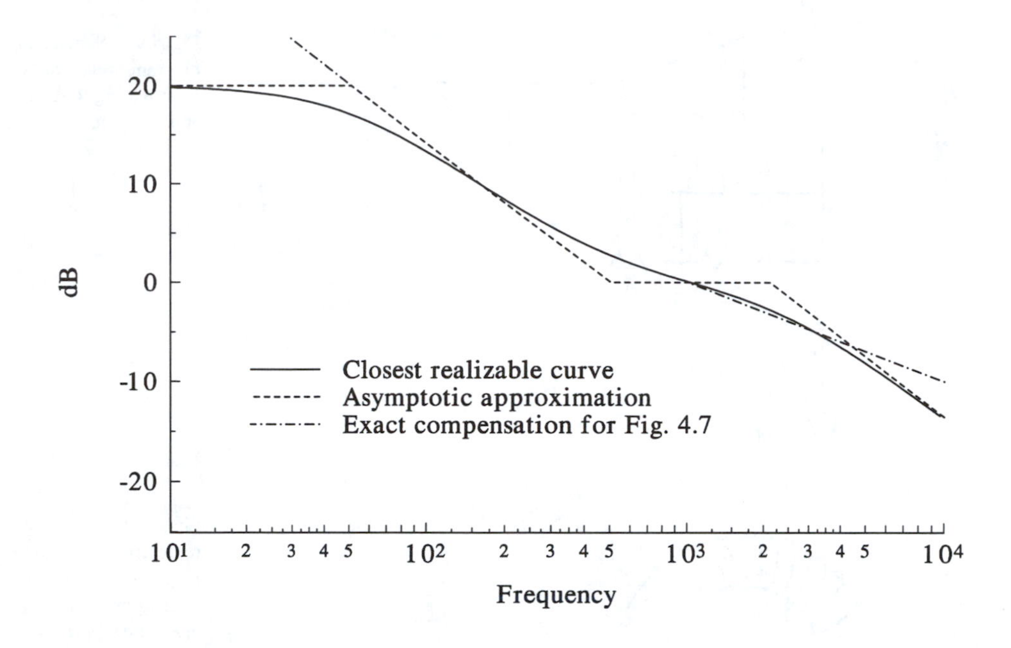

Fig. 4.8: Playback amplifier equalization. Shown are the exact compensation for the solid curve in Fig. 4.7 (unevenly dashed line), the asymptotic Bode amplitude plot of the closest approximation readily realized (evenly dashed line), and the actual response plot corresponding to the Bode plot (solid line). Note that between 50 Hz and 1 kHz, the two dashed curves coincide, though only the evenly dashed one shows.

137 kHz. The latter, as desired, is well out of the audio range. This is about as close as we can expect to come with standard components.

On a stereo record, there is different information on each of the two walls of the record groove. These two information streams correspond to the right (R) and left (L) stereo channels. One stylus with degrees of freedom normal to both groove walls picks up both signals and couples them to two different magnetic circuits. These produce the stereo output voltages, which are separately equalized and amplified.

4.3 Tape (NAB) equalization

NAB stands for National Association of Broadcasters, which developed the equalization standards for tape.

The principles of magnetic tape recording and playback will be covered more thoroughly in the next chapter. For now, we need to understand just enough of the principles to understand how equalization enters the picture.

Magnetic tape is composed of a strong plastic strip of material called the tape base, which has a powdered magnetic material coated on one face of the strip. The record process consists of impressing a magnetic pattern that corresponds to the desired program material onto the tape. The playback process consists of converting that magnetic pattern on the tape back into an electrical signal that, hopefully, is a faithful replica of the original recorded signal.

Both of these conversions are realized using what are called *tape heads*. Better-quality machines have separate heads for the record and playback operations. Less-expensive recorders have a single head that does both. Figure 4.10 shows the basic geometry of a tape head.

In recording, a signal applied to the coil produces a time-varying magnetic field around the magnetic loop, which is closed by the magnetic coating on the tape. This field will magnetize the grains of the coating to a degree that varies from one instant to the next as the signal amplitude and polarity vary. The tape

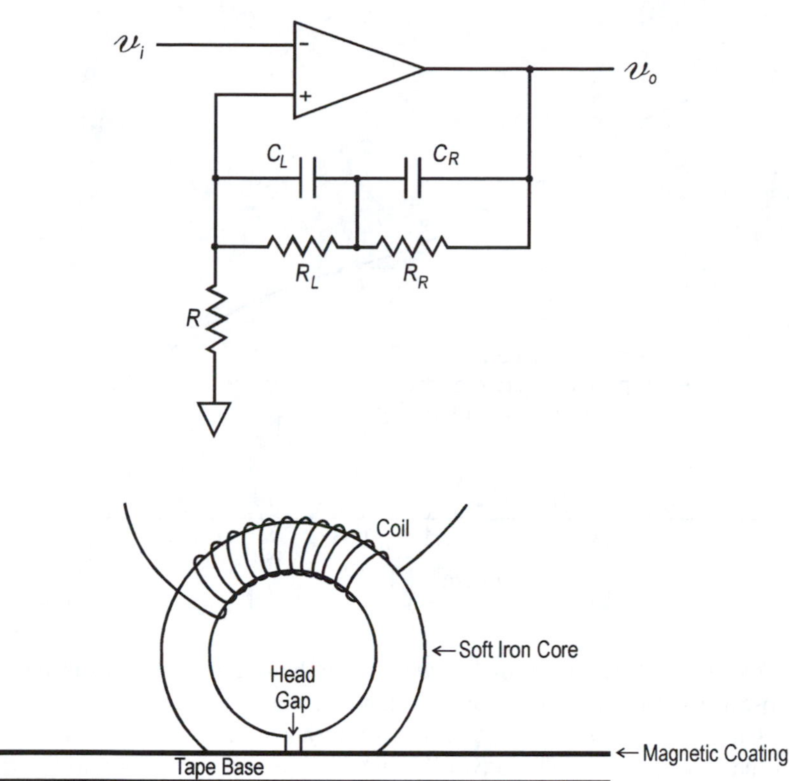

Fig. 4.9: Schematic diagram of a circuit capable of providing RIAA equalization.

Fig. 4.10: Schematic representation of the magnetic circuit involved in reading or writing magnetic tape.

is moving by the head at a constant linear velocity, so this varying magnetic field is always being impressed on a new segment of the tape. On playback, the magnetized tape moving by the head couples a time-varying magnetic field into the soft iron and thus through the coil. This field will then induce a voltage across the coil by Faraday's Law.

Having covered the physical aspects of how the storage and retrieval system works, let's go back and look at the equalization aspects of the system. Equalization is, in fact, used on both record and playback, but unlike phonograph recording, the transfer functions are not ideally inverses.

4.3.1 Record cycle

Tape heads are relatively low impedance devices. Therefore, they are readily driven from a current source. Current source drive has two additional advantages: (1) The flux generated by the head and that impressed on the tape are directly proportional to the current through the head. (2) Use of current drive eliminates any frequency-related phase shifts between the signal voltage and the head flux that would be present with voltage drive.

What is ideally needed for the record amplifier is a transconductance amplifier whose high input impedance will not load the source being recorded from and whose high output impedance will provide the current drive the record head needs. It is not uncommon to use an amplifier with low output impedance and to drive the head through a large series resistance (20–200 kΩ) to simulate current drive. This is particularly true of less-expensive tape recorders where one power amplifier serves to drive both the record head in the record function and the speaker in the playback function. The speaker requires a low impedance output from the driving amplifier, so the use of a series resistance to drive the record head is an inexpensive, if imprecise, way to satisfy both requirements. The record amplifier must be able to magnetically saturate the tape without going into saturation itself. It is not desirable to saturate the tape, of course, as this causes distortion, but in a well-designed system, the tape is always the limiting factor.

On the basis of what has been said so far, it would appear that there is no call for record equalization. There are two reasons why this is not true. The first is that at high frequencies, hysteresis and eddy-current losses in the head cause the flux amplitude to decrease. Therefore, it is a universal practice to use a record equalization that boosts the high frequencies prior to applying them to the record head. This high-frequency boost usually starts in the low-kilohertz range. The second reason for record equalization is that, as we shall soon see, playback equalization is needed to achieve even a first-order approximation to the correct frequency response. Unfortunately, however, boosting some frequencies on playback will also boost the noise that has accrued to the signal to that point. This problem can be circumvented by using record equalization not only to achieve a flat response, but to deliberately boost both low and high frequencies above the levels that would give a flat response. This technique, while serving its purpose well, is not without its drawbacks. The first is that this practice is not standardized. Thus, tapes recorded on a recorder so equipped will be properly equalized only for playback on the same recorder. This may possibly be compensated for by tone controls and the individual's ear, but the second objection is not so easily disposed of. Magnetic tape has, as has been alluded to, a saturation level above which distortion is introduced into the recorded program material. One wants to have the peak tape flux as close to saturation as possible without being in it, as this will give the strongest recorded signal and thus the best signal-to-noise ratio. Now if the low and high frequencies are pre-emphasized by the record equalizer, it is they, rather than the middle frequencies, that will limit the maximum signal that can be recorded. This may actually cause a degradation of the signal-to-noise ratio in the mid-frequency range. The optimum amount to boost the low and high frequencies can actually be determined exactly only if we know the exact spectral distribution of the program material.

4.3.2 Playback equalization

We shall imagine for simplicity that record equalization was used to achieve a constant flux amplitude on the tape, rather than having any additional record pre-emphasis of high and low frequencies. By the nature of the playback head, Faraday's Law is the governing principle in playback, just as it was for the magnetic phono cartridge:

$$v = -N\frac{d\phi}{dt} \Rightarrow \mathbf{V} = -j\omega N\mathbf{\Phi}.$$

The assumed constancy of flux amplitude on the tape versus frequency enables us to say that v will be directly proportional to ω. This proportionality is experimentally observed – up to a certain frequency. Once that frequency is

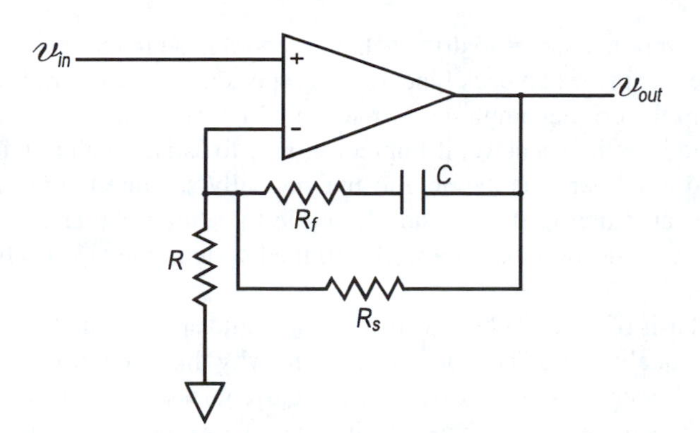

Fig. 4.11: Schematic diagram of a circuit capable of providing NAB equalization.

reached, the response begins to fall off rapidly. A great deal of research has gone into finding ways to raise that frequency. It is found that two factors strongly affect where the response of the head drops off, and they do so by means of their functional relationship to the same phenomenon. They are the tape speed and the head gap width. The drop-off begins when the half-wavelength of the recorded signal reaches the order of magnitude of the head gap. For example, suppose that the tape is moving at 3.75″/second past a 2 μm gap. The tape speed is 95,250 μm/sec. The relevant equation is

$$v = f\lambda$$

or

$$f = v/\lambda.$$

When $\lambda/2 = g$,

$$f = \frac{v}{2g} = \frac{95{,}250\ \mu\text{m/sec}}{4\ \mu\text{m}} = 23{,}800\text{ Hz}.$$

We would expect, therefore, to observe a departure from the ideal head response about an order of magnitude below this or at about 2.5 kHz. It is the purpose of playback equalization to compensate for this drop-off and maintain an essentially flat response out to a significantly higher frequency. Unlike record equalization, playback equalization is well standardized in quality equipment. Because the response of the head features a zero at zero frequency, we might expect to use an equalizing network with a pole at zero frequency. As was the case for RIAA compensation, this is not generally done, to prevent the op amp from going open loop at DC. Not only would this be undesirable for stability reasons, but the open loop gain of the op amp itself exhibits a pole – usually between 3 and 30 Hz for internally compensated op amps. Thus, we sacrifice nothing by introducing a pole below the audio frequency range. A circuit suitable for doing this is shown in Fig. 4.11.

The transfer function is

$$\frac{\mathbf{V}_{\text{out}}}{\mathbf{V}_{\text{in}}}(s) = \frac{(1 + R_{\text{s}}/R) + sC(R_f + R_{\text{s}} + R_f R_{\text{s}}/R)}{1 + sC(R_{\text{s}} + R_f)}.$$

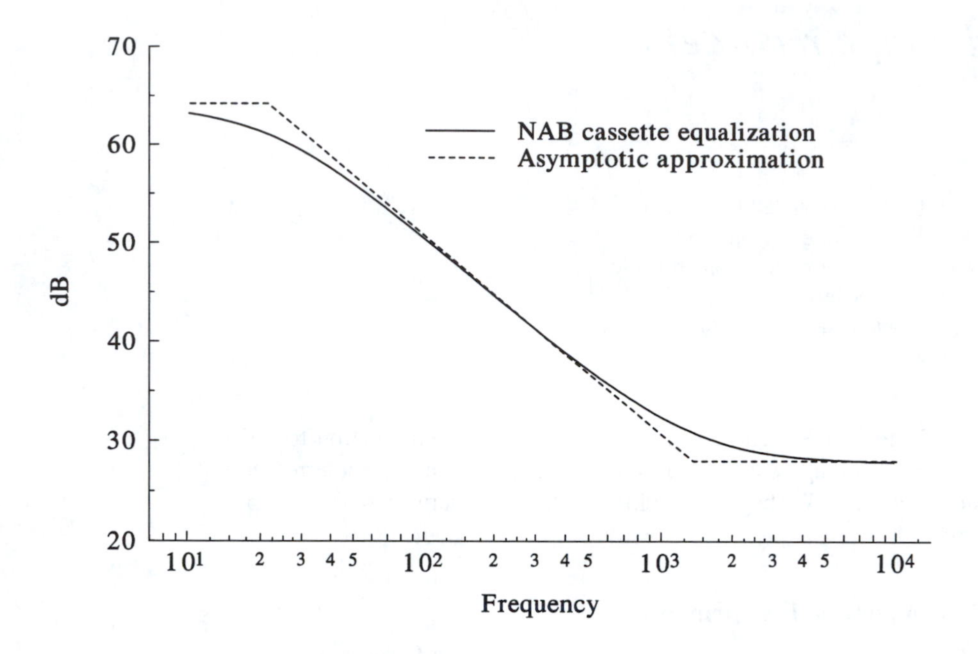

Fig. 4.12: Frequency response of the NAB preamp of Fig. 4.11.

The values $R = 1\ \text{k}\Omega$, $R_s = 1.6\ \text{M}\Omega$, $R_f = 24\ \text{k}\Omega$, and $C = 4.7$ nF provide a pole frequency of 21 Hz and a zero at 1,355 Hz. This is the standard equalization breakpoint for cassette tape recorders.

It is customary to specify the zero location by a time constant rather than by a frequency. This time constant will be the value of RC for which $RC\omega_z = 1$ or

$$RC = \frac{1}{\omega_z} = \frac{1}{2\pi f_z} = 118\ \mu\text{sec}$$

for the design above. Thus 120 μsec is standard for ordinary ferric-oxide magnetic tape. Chrome and metal tape use 70 μsec in cassette recorders.

The zero is introduced to limit the high-frequency attenuation of the preamp, thus compensating for the high-frequency rolloff of the playback head.

Figure 4.12 shows the actual frequency response of the circuit of Fig. 4.11 with the asymptotic approximation thereto.

5

Audio tape recorders

A study of audio tape recorders will also serve as an introduction to video tape recorders, although coverage of that subject will have to be deferred until television is covered. We begin by looking at three distinctive circuit principles in tape recorders.

5.1 The magnetic-bias principle

The principle of biasing an active device to place it at a point about which variation is linear is one of the most fundamental in electronics. An analogous problem occurs in magnetic tape recording. The problem stems from the fact that the B_r–H curve of magnetic tape exhibits not only a saturation nonlinearity as we might expect, but also a "crossover distortion" type of nonlinearity. This is shown in Fig. 5.1. B_r is the remanent magnetic flux that remains on the tape after it leaves a record head gap with field strength H in it.

Recall that in the record process, the tape head is driven by a current proportional to the signal we wish to record:

$$i_{\text{record head}} \propto v_{\text{signal}}.$$

The field strength in the head gap will be proportional to the record head current, so it will also be true that

$$H_{\text{record gap}} \propto v_{\text{signal}}. \tag{5.1}$$

The value of H in the record gap impresses a remanent magnetic field on the tape, which we shall call B_r and which varies from point to point on the tape.

During playback if we assume that the tape remanent field is longitudinal, it is easy to see that the remanent flux is also longitudinal and that

$$\phi_r = B_r \cdot w \cdot c,$$

where w is the track width of the recording, and c is the magnetic coating thickness (see Fig. 5.2). For low recorded frequencies, all of this flux may be approximated as being coupled through the playback head and inducing a voltage that varies linearly with the flux magnitude:

$$v_{\text{play}} \propto \frac{d\phi_r}{dt} \propto \omega \cdot \phi_r \propto \omega \cdot B_r. \tag{5.2}$$

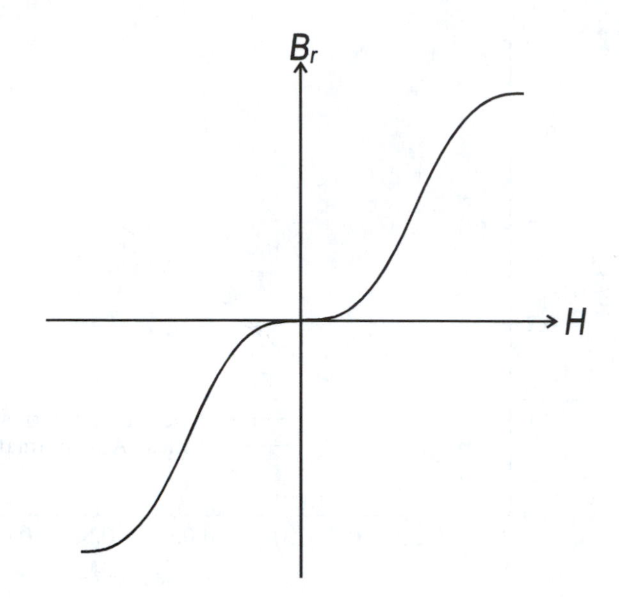

Fig. 5.1: A typical B–H curve for magnetic tape.

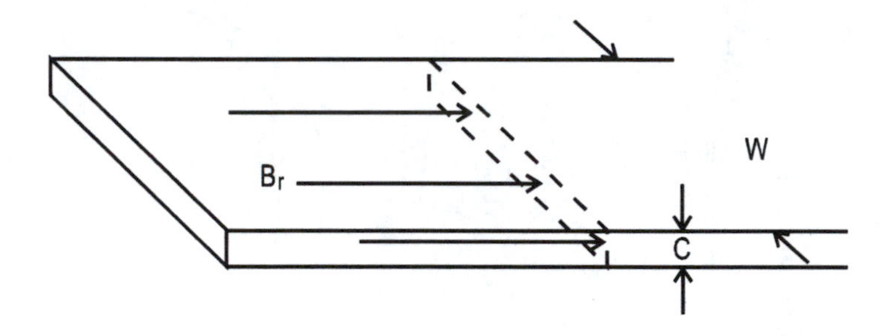

Fig. 5.2: Longitudinal B field in a magnetic tape coating.

To summarize: If

$$v_{\text{signal}} \propto H_{\text{record gap}} \quad \text{and} \quad \omega \cdot B_{\text{r}} \propto v_{\text{play}}$$

and if we could establish a linear relationship between $H_{\text{record gap}}$ and B_{r}, it would follow that

$$v_{\text{play}} \propto \omega \cdot v_{\text{signal}}, \tag{5.3}$$

which is the ideal linear relationship. The multiplied factor of ω is readily compensated for by equalization, as discussed in Chapter 4.

There are two main assumptions used in arriving at this conclusion. The first is the assumption that the tape flux is totally longitudinal; the second is that the recorded frequencies are low. This means that their wavelength on the tape, as defined in Chapter 4, is much greater than c, the tape coating thickness. When this is not the case, three or more loss mechanisms reduce the amount of tape flux that is coupled into the tape head. Consequently, the playback signal is also reduced. The resulting high-frequency loss may also be compensated for by equalization. But even if this were not done, there would still be a linear relationship between the magnitudes of the flux and the induced voltage.

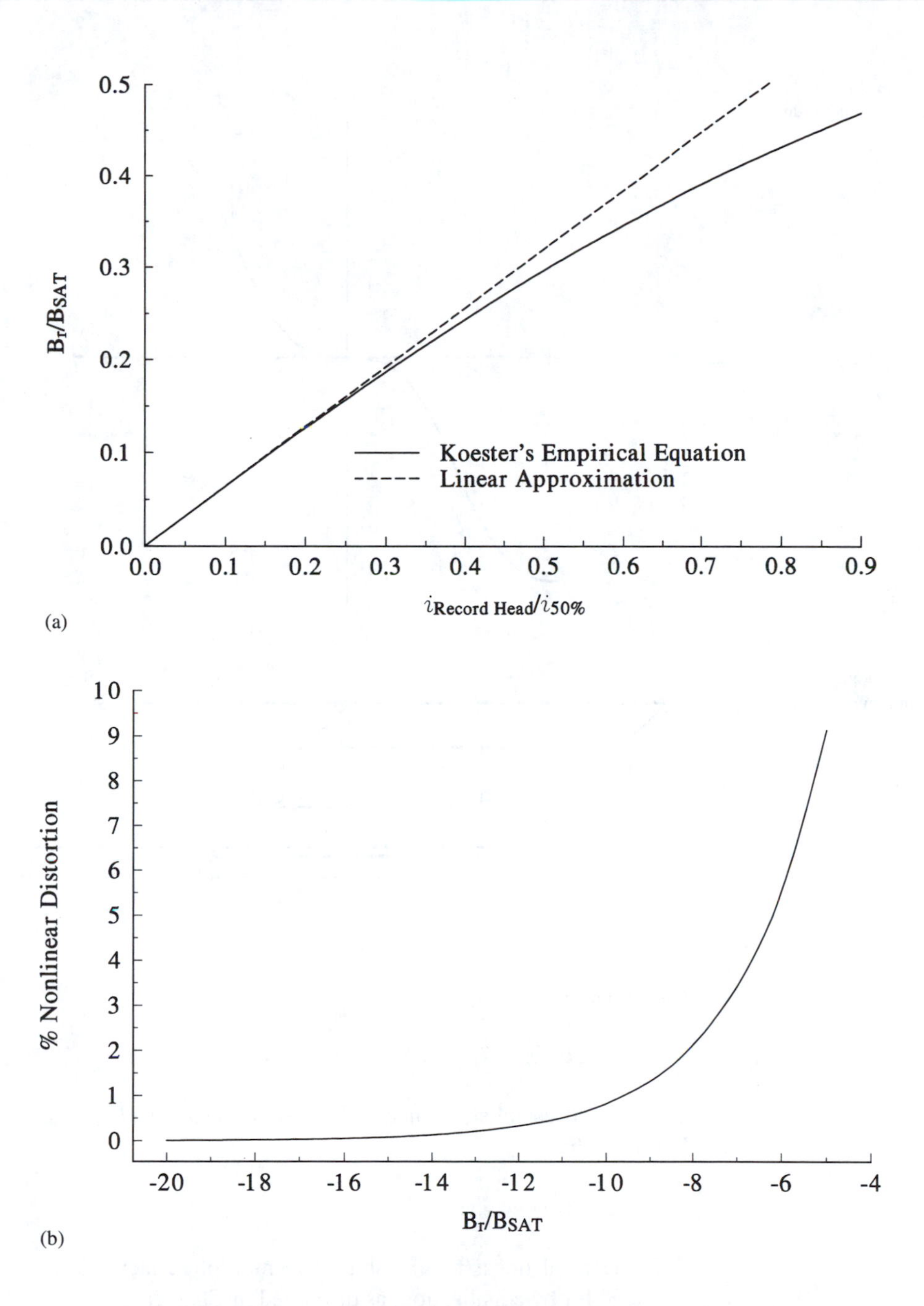

Fig. 5.3: (a) The linearizing effect of AC bias on the relationship between record-head current and remanent tape magnetism. (b) The percent nonlinear distortion of a signal recorded on magnetic tape versus the strength of the magnetic field written to the tape.

Discussion of the implications of assuming longitudinal flux will be deferred until after the discussion of AC bias.

As we have just indicated, if there is a linear relationship between $H_{\text{record gap}}$ and B_r, then there will also be a linear relationship between the record and playback record gap signals. Yet Fig. 5.1 shows that this is far from being the case. Closer inspection shows, however, that there is a segment of the curve in the first and third quadrants that is fairly linear. This might suggest adding a DC current to the signal current in order to place the operating point in the middle of the linear region. This method of linearization actually works, but it is only used in very cheap recorders. It has two serious drawbacks for high-quality sound reproduction. The first drawback is that the allowable signal swing is so restricted that the signal-to-noise ratio is poor. The second drawback is that since the tape coating is not perfectly homogeneous, the flux will not be uniform everywhere even if DC bias is used and no signal is applied. The resulting flux variation is sensed by the playback head and appears as noise along with the signal.

Both of these problems are circumvented by the use of AC bias, or more specifically, ultrasonic bias. As this name implies, a sinusoid above the audible range is fed to the tape head along with the signal to be recorded. The ultrasonic component must be over twice the highest audio frequency the recorder can process in order to prevent the generation of any difference frequency component lying in the audio range when mixed with the signal. Typically, the bias component has an amplitude ten times that of the signal.

There is no general closed-form solution for the linearizing effect of the AC bias. Neither is there a really good physical explanation of how it works, although some excellent qualitative results have been obtained by the use of the Preisach diagram, which represents a systematic ordering of the magnetic properties of all the magnetic domains in an assembly (Jorgensen, 1980a).

It is important to realize that the ultrasonic bias not only linearizes the relationship between the record head's H field and the remanent tape magnetism, but it actually greatly increases the remanent field for a given value of H.

For wavelengths long compared to the coating thickness, Koester has arrived at an empirically-based universal curve relating the remanent magnetism to the record gap field strength with proper bias (Jorgensen, 1980b):

$$\frac{B_{\mathrm{r}}}{B_{\mathrm{SAT}}} = \frac{2}{\pi} \cdot \arctan\left[\frac{i_{\text{record head}}}{i_{50\%}}\right], \tag{5.4}$$

where $i_{50\%}$ is the record head current needed to make the remanent tape magnetism one half of the saturated magnetic remanence. Equation (5.4) is plotted in Fig. 5.3a. The linear region may seem to be rather limited in its extent, but Fig. 5.3b shows that if the record current is such that the tape magnetization is kept 10 dB below saturation, the total departure from linearity is only 0.8%. This means that there is about a 60 dB dynamic range below this level on the tape before the noise level is reached. Note also that if the 0.8% distortion is that obtained on the highest-amplitude peaks, the average distortion will be much smaller.

Since Fig. 5.3a shows a linear relationship between B and i at low current levels, it is clear that the crossover distortion is totally absent. This, it may be recalled, was said to be true for proper bias level. We now wish to inquire as to what constitutes proper bias level.

Figure 5.4 represents the distribution of the magnetic field in the vicinity of the head gap with magnetic tape in contact with it. The portion of the tape shown is just the magnetic coating. The tape base would lie above this and be many times thicker. The shaded areas are called *recording zones* and comprise the portion of the tape that records the magnetization corresponding to the value of H in the gap at the instant shown. These zones extend into the paper for the width of the recording track and thus have the form of thin cylindrical shell sections. The recording zones are centered upon a constant field contour $H = H_{\mathrm{c}}$, where H_{c} is the critical value of field for changing the magnetic polarization of a tape

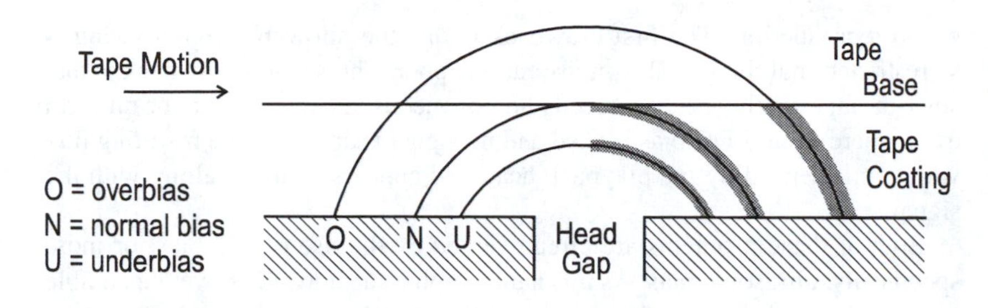

Fig. 5.4: Distribution of remanent magnetic field in tape coating as a function of the AC bias used in recording.

particle. The width of the record zones is a function of the particles' B–H curves and the radius of the field lines.

The following observations can be made from Fig. 5.4:

(1) For inadequate AC bias, the entire thickness of the tape coating is not magnetized. This will naturally reduce the total recorded flux and thus the voltage that can be recovered from the tape on playback. In addition, both noise and distortion are intolerably high.
(2) The assumption that the recorded field is purely longitudinal is more valid at the back part of the magnetic layer than in the part closest to the head. This assumption is strictly valid if the tape is saturated, but of course this condition is to be avoided for the type of recording done in consumer equipment.
(3) The central field line for normal bias is tangent to the back of the magnetic layer.
(4) With overbias the field is less longitudinal and more perpendicular. The perpendicular component generates nothing in the tape playback head. Thus, even though the record zone is wider, indicating a higher overall tape flux, its more perpendicular orientation actually leads to a lower output voltage. Overbias is also unfavorable to good high-frequency response. Nevertheless, both noise and distortion decline enough with overbias that it is not unusual to see a tape deliberately overbiased by a small amount.

5.2 The erase principle

If tape is to be reusable after having once been recorded upon, there has to be a way of erasing what was previously on it, prior to rerecording it. In practice, this is accomplished by positioning an erase head so the tape passes over it just prior to reaching the record head.

The erase head can be driven by DC. DC erasing will align the magnetic orientation of the particles on the tape. Since ϕ will be constant, $\dot{\phi}=0$ and no voltage is induced in a tape head by it. The drawback here is that if the erase level is near saturation to insure full erasure, the record level must be as large to fully overcome the high level of erase magnetization. If this is done, we're right back where we started. The necessary high record level leads to nonlinear distortion.

As was the case with bias, however, AC is the hero in the plot. The same high-frequency oscillator that generates the bias also does a very effective job of erasing the tape. The erase mechanism is different, however. With erasure at

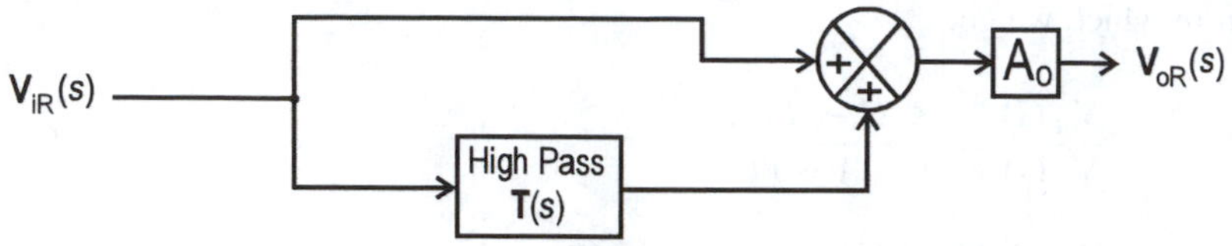

Fig. 5.5: Simplified block diagram of the Record circuitry used in a Dolby® B system.

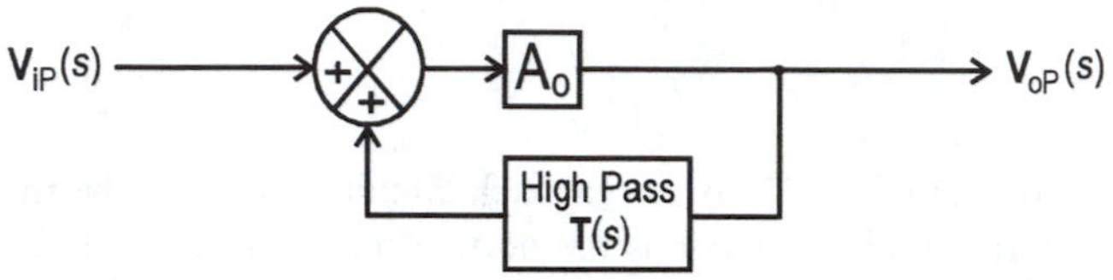

Fig. 5.6: Simplified block diagram of the Playback circuitry used in a Dolby® B system.

these frequencies, the field at the erase head gap will reverse before a given point on the tape traverses the gap. The effect of this is to randomize the orientation of the individual particle magnetizations, making the net magnetic field zero.

5.3 The noise-reduction principle

It was mentioned in connection with FM pre-emphasis that the basic principle of noise reduction is to boost the amplitude of signals in a certain frequency range before passing them through a "noisy channel" and then to cancel out the boost at the receiving end of the channel before converting back to sound. In the case under consideration here, the "noisy channel" is the record–playback process.

5.3.1 The Dolby® system

Although it is by no means the only plausible noise-reduction convention, the Dolby® B process is in such common use as to virtually constitute a standard. In this system, the high frequencies are boosted by passing the signal to be recorded through the circuitry shown in block form in Fig. 5.5. We may thus write

$$\mathbf{V}_{\text{oR}}(s) = \mathbf{V}_{\text{iR}}(s) \cdot [1 + T(s)] \cdot A_{\text{o}}. \tag{5.5}$$

The Dolbyized signal, $\mathbf{V}_{\text{oR}}$, is then recorded. On playback, the recovered $\mathbf{V}_{\text{oR}}$ signal is applied to the Dolby® playback circuit shown in Fig. 5.6. Thus we can write

$$\begin{aligned} [\mathbf{V}_{\text{iP}}(s) + \mathbf{V}_{\text{oP}}(s) \cdot T(s)] \cdot A_{\text{o}} &= \mathbf{V}_{\text{oP}}(s), \\ A_{\text{o}} \cdot \mathbf{V}_{\text{iP}}(s) &= \mathbf{V}_{\text{oP}}(s) \cdot [1 - A_{\text{o}} \cdot T(s)]. \end{aligned} \tag{5.6}$$

Now if $\mathbf{V}_{\text{iP}}(s) = \mathbf{V}_{\text{oR}}(s)$ (which means that a Dolby® recorded signal is applied to this Dolby® playback circuit) then (5.5) and (5.6) can be combined to yield

$$A_{\text{o}}\{\mathbf{V}_{\text{iR}}(s) \cdot [1 + T(s)] \cdot A_{\text{o}}\} = \mathbf{V}_{\text{oP}}(s) \cdot [1 - A_{\text{o}} \cdot T(s)],$$

from which we get

$$\frac{\mathbf{V}_{\mathrm{oP}}(s)}{\mathbf{V}_{\mathrm{iR}}(s)} = \frac{A_{\mathrm{o}}^2(1 + T(s))}{1 - A_{\mathrm{o}} \cdot T(s)}. \tag{5.7}$$

From Eq. (5.7) it can be seen that if $A_{\mathrm{o}} = -1$, then

$$\frac{\mathbf{V}_{\mathrm{oP}}(s)}{\mathbf{V}_{\mathrm{iR}}(s)} = 1. \tag{5.8}$$

That is, the output of the Dolby® playback circuit is equal to the input to the Dolby® record circuit, which was the normal program material. The simple elegance of this concept, however, is only half the story. The genius of the Dolby® system lies in the fact that the pole in the high-pass filter is movable. To benefit from this requires that a dynamic filter be used to provide the high-pass function. Let us see why this is advantageous.

Recall that the white noise passing through a channel is directly proportional to the bandwidth of the channel. For all practical purposes, this means that each octave of frequency response on the upper end of the audio spectrum doubles the noise power passed through the audio channel relative to what it would be without the extra octave. Hence it is the high-frequency end of the signal spectrum that needs to be boosted to preserve the signal-to-noise ratio. It is also obviously true that when the signal has a low amplitude, the S/N ratio will decline, since the noise will remain constant. Therefore, ideally, we would like more high-frequency boost at lower signal levels than at higher ones.

In the Dolby® system, this level-dependent boost is provided by moving the pole frequency to lower frequencies at lower levels of signal amplitude. This is the function of the dynamic filter. The exact mechanism for doing this will be studied later in the chapter. We can state generally here, however, that the audio output is rectified and filtered to produce a DC voltage that varies the operating point of an active device, which is part of the filter. Adding the rectifier–filter combination to Figs. 5.5 and 5.6 gives the diagrams shown in Figs. 5.7a and b.

The important point to note here is that if $v_{\mathrm{oP}}(s) = v_{\mathrm{iR}}(s)$, as in Eq. (5.8), the rectifier–filter will supply equal DC voltages to the dynamic filter on both record and playback, thus insuring that the pole frequency will be the same on playback as it was on record. This in turn will insure the integrity of the decoded signal.

The Dolby® B system provides about 10 dB of noise reduction with relatively simple circuitry. It is now incorporated in almost all quality tape equipment and is widely available in IC form.

5.4 Tape recorder analysis

The unit chosen for analysis is the Marantz-Superscope CD 320 stereo cassette recorder. It is a portable unit with many of the features usually found only in stereo system tape decks, such as switch-selectable bias, equalization, and Dolby®, as well as a good variety of input and output options and a recording level meter.

Figure 5.8 shows the schematic of one of the stereo channels, and Fig. 5.9 shows the equivalent block diagram. We will begin the analysis with the com-

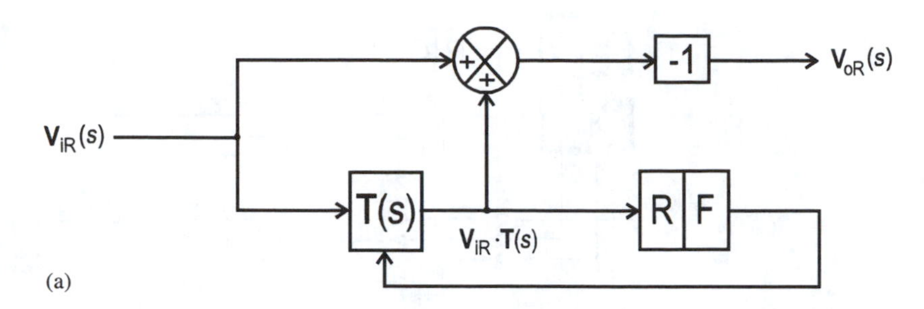

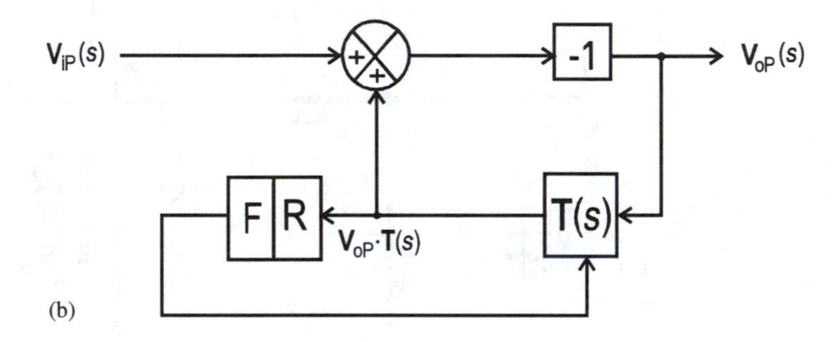

Fig. 5.7: Block diagram of the circuitry used in a workable Dolby® B system: (a) Record circuitry; (b) Playback circuitry.

posite amplifier stage Q_{101}–Q_{102}, which supposedly accomplishes record and playback equalization. This circuitry is shown in Fig. 5.10.

We should note first that the two stages are direct-coupled. This necessitates that measures be taken to stabilize the DC operating point. To this end, both stages have large emitter resistances. The combinations R_{121}–C_{155} and R_{109}–C_{102} are decoupling networks. All of the other electrolytic capacitors are AC shorts as well. The fact that C_{105} is an AC short means that Q_{101} sees an emitter resistance of 10 kΩ for DC bias and a resistance of ~150 Ω for AC. The entire circuit can then be drawn in feedback form as shown in Fig. 5.11.

The capacitance from base to emitter of Q_{101} is omitted. Its value is not given, but it is almost certainly negligible at audio frequencies.

$\mathbf{Z}_f$ is the impedance of the feedback network elements. As can be seen in Fig. 5.10, these will vary depending not only on whether the "Play" or "Record" function is selected, but also on the setting of S_{304} in the "Play" mode. The function of this switch is to select the playback equalization, as we shall soon see more quantitatively.

Figure 5.11 represents a negative-feedback voltage amplifier. Using Figs. 5.10 and 5.11, we get a first-order gain of

$$\mathbf{A}_f = 1/\mathbf{F} = 1 + \mathbf{Z}_f/R_{108}.$$

For playback with S_{304} open, we have

$$\mathbf{Z}_f = R_{115} + R_{114} + \frac{R_{113}}{1 + sC_{104}R_{113}}$$

$$= \frac{R_{113} + R_{114} + R_{115} + sC_{104}R_{113}(R_{114} + R_{115})}{1 + sC_{104}R_{113}}.$$

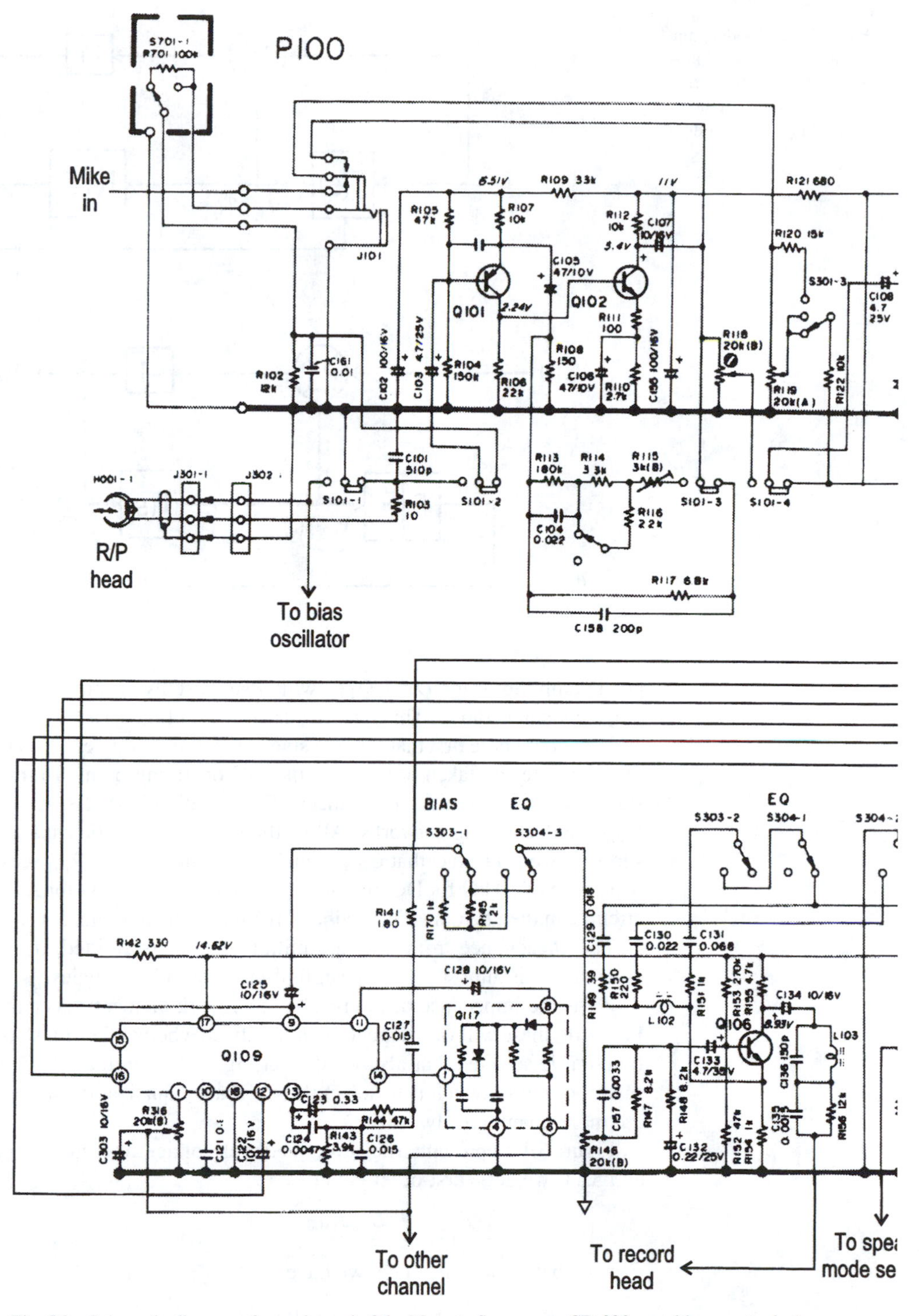

Fig. 5.8 Schematic diagram of one channel of the Marantz-Superscope CD-320 portable cassette deck.

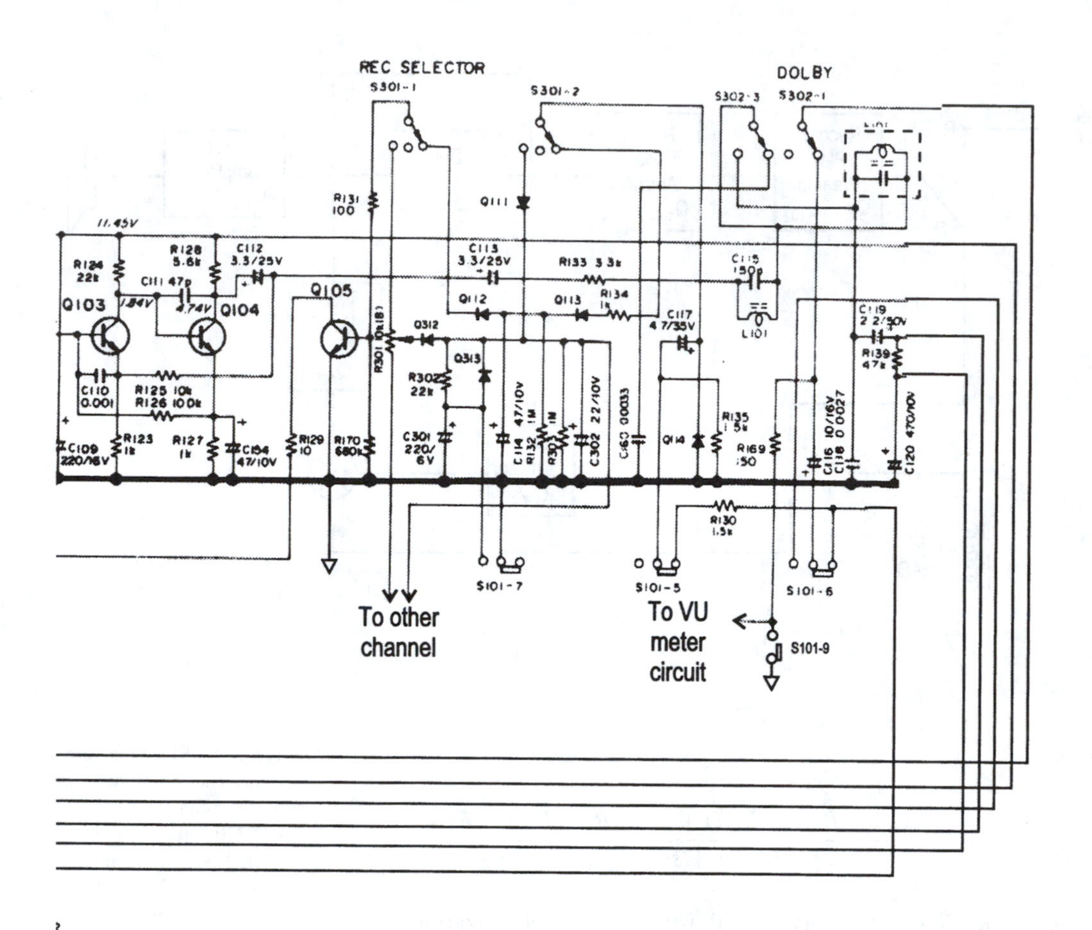
REC SELECTOR
S301-1
S301-2
DOLBY
S302-3
S302-1
Q103
Q104
Q105
To other
channel
S101-7
S101-5
S101-6
To VU
meter
circuit
S101-9

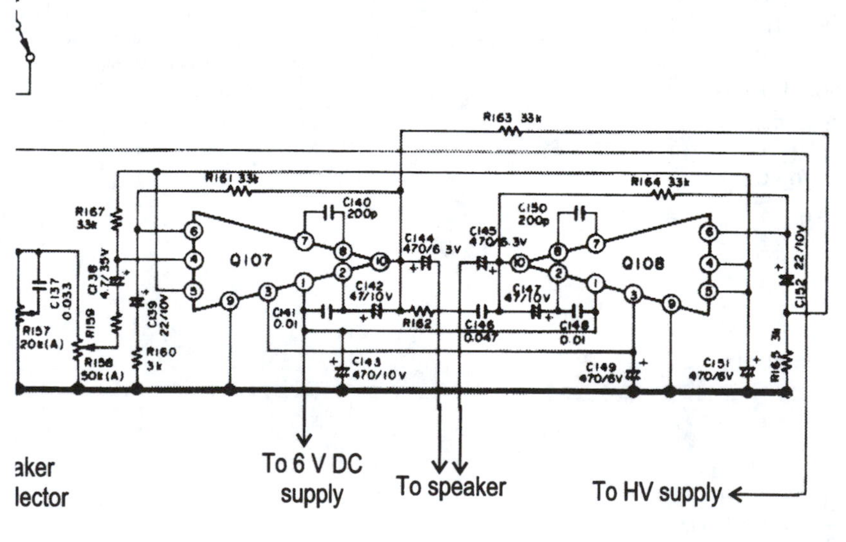
Q107
Q108
aker
lector
To 6 V DC
supply
To speaker
To HV supply

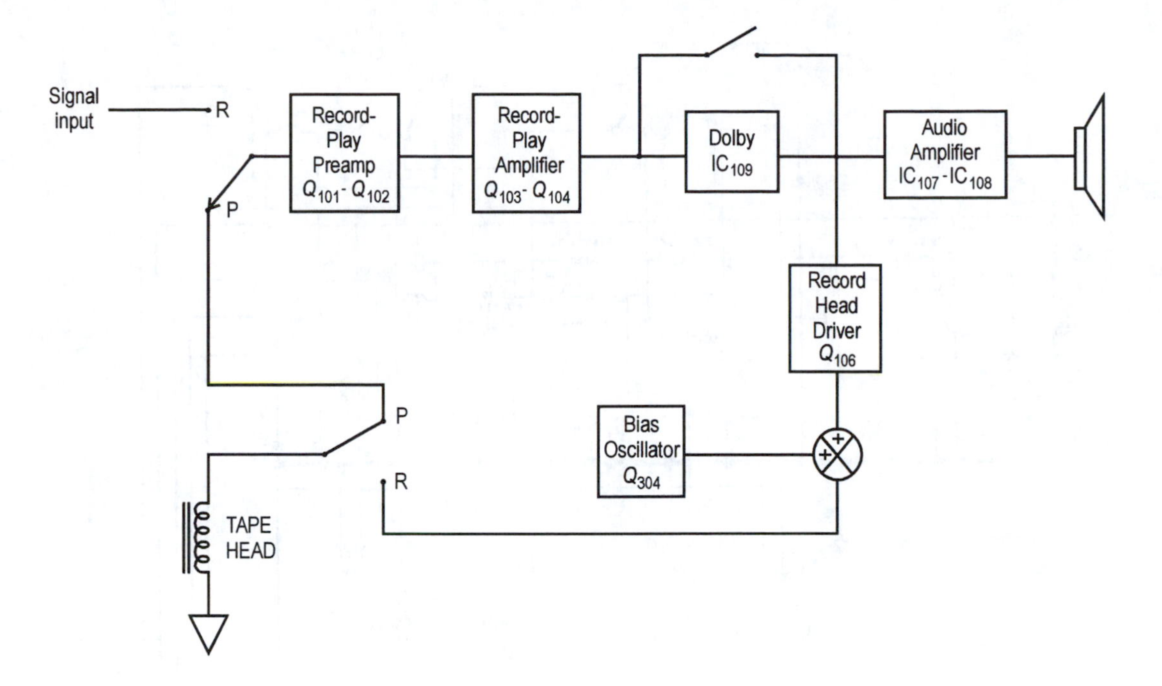

Fig. 5.9: Block diagram of one channel of the Marantz-Superscope CD-320 portable cassette deck. Note that the same bias oscillator serves both channels and that the unit contains only one speaker, which can be switched to output either channel or a sum of the two channels.

Therefore

$$\mathbf{A}_f = 1 + \frac{(R_{113} + R_{114} + R_{115}) + sC_{104}R_{113}(R_{114} + R_{115})}{R_{108}(1 + sC_{104}R_{113})}.$$

The resistance R_{115} is a factory trim adjustment on the equalization. Assume it is set to its midpoint. Then $R_{115} = 1.5\ \mathrm{k}\Omega$ and

$$\mathbf{A}_f = 1 + \frac{184{,}800 + 19.0s}{150 + .594s}$$

$$= \frac{184{,}950 + 19.59s}{150 + .594s},$$

from which we find

$$f_{\text{pole}} = \frac{150}{2\pi \cdot .594} = 40\,\text{Hz}$$

and

$$f_{\text{zero}} = \frac{184{,}950}{2\pi \cdot 19.59} = 1{,}502\,\text{Hz}.$$

These values should be compared to those for the circuit of Fig. 4.11. The zero frequency corresponds to a time constant of 106 μsec, which is about 12% less than the ideal value of 120 μsec. Adjustment of R_{115} could easily correct this.

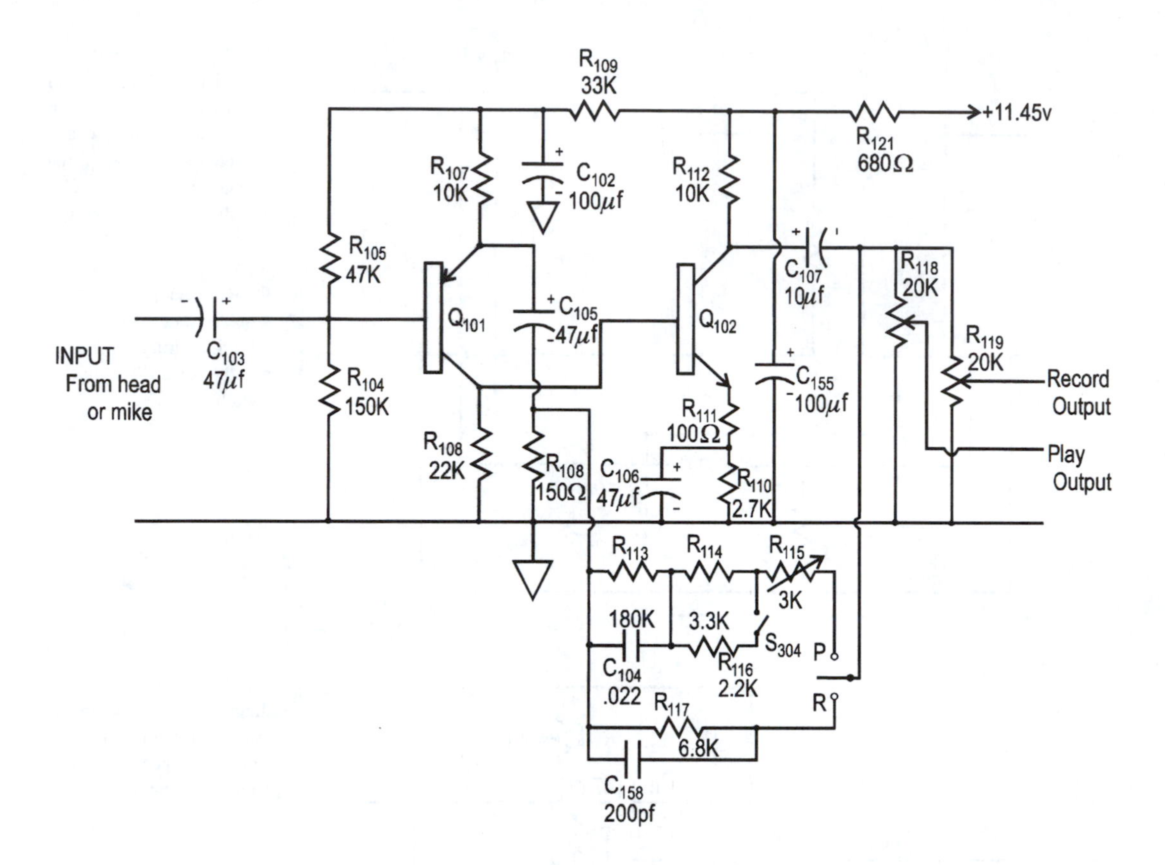

Fig. 5.10: Schematic diagram of the Record/Play preamplifier used in the Marantz-Superscope CD-320 portable cassette deck.

Exercise 5.1 Still assuming that R_{115} is set at its midpoint find the pole and zero frequencies with S_{304} closed. Also find the equalization time constant.

If an analogous analysis is performed for the "Record" mode, the result is

$$\mathbf{A}_f = \frac{(R_{117} + R_{108}) + sC_{158}R_{117}R_{108}}{R_{108}(1 + sC_{158}R_{117})},$$

from which we obtain

$$\omega_{\mathrm{p}} = \frac{1}{C_{158}R_{117}} \Rightarrow f_{\mathrm{p}} = 117\,\mathrm{kHz}$$

and

$$\omega_{\mathrm{z}} = \frac{R_{117} + R_{108}}{C_{158}R_{117}R_{108}} \Rightarrow f_{\mathrm{z}} = 5.41\,\mathrm{MHz}.$$

These values accomplish no useful record equalization. Although it is possible that one of the feedback element values was listed in error or that one or more feedback elements was omitted from the schematic, we shall see later that record equalization is actually provided in another stage altogether.

The next stage is also a negative-feedback voltage amplifier. The schematic is shown in Fig. 5.12. The function of Q_{105} will be discussed last. All of the

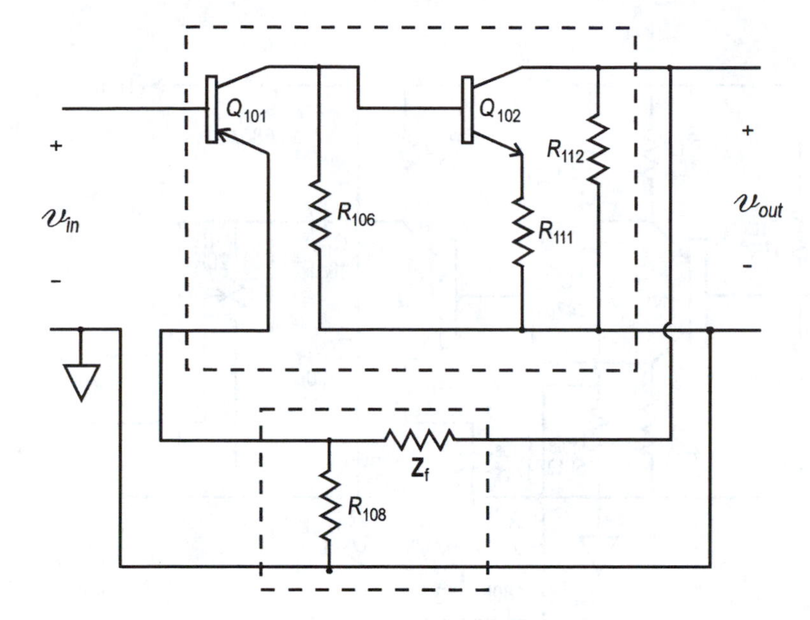

Fig. 5.11: Standard feedback configuration for the circuit of Fig. 5.10. This clearly shows that it is a voltage feedback amplifier. The circuit has been AC simplified prior to being put in this form. Note that Z_f represents the elements in the feedback path of the amplifier and will change with the settings of S_{304} and the Record/Playback switch.

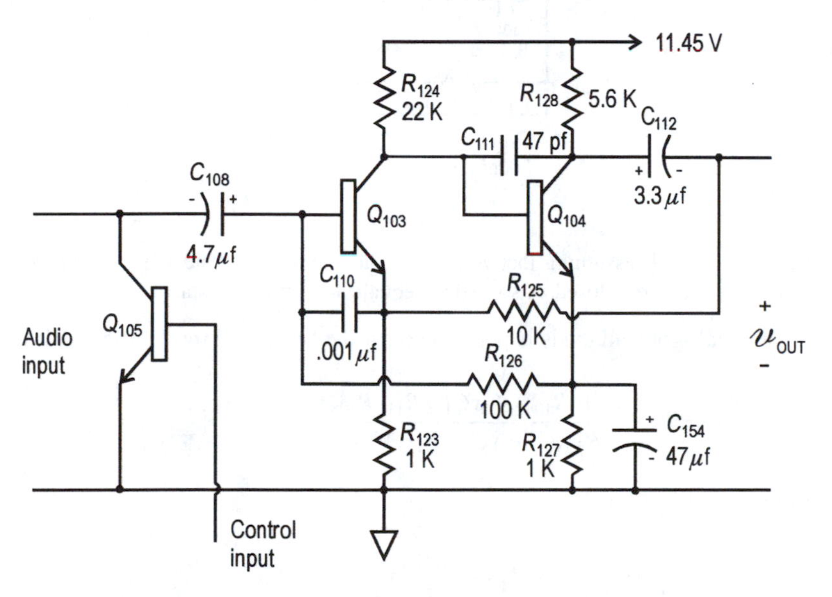

Fig. 5.12: Schematic diagram for the Record/Play amplifier of the Marantz-Superscope CD-320 portable cassette deck.

electrolytics can again be considered AC shorts, although it should be noted that for the open-loop amplifier, C_{154} gives a lower pole frequency of 135 Hz. This in itself would not give high-fidelity response, but the negative feedback will lower this by a factor equal to the loop gain, which will place the closed-loop pole well below the audio range.

Because the emitter of Q_{104} is thus at AC ground, no signal can be fed back through R_{126}, and it is effectively in shunt across the input. The voltage feedback is not frequency-selective for this stage. The pertinent feedback elements are R_{123} and R_{125}, which together give a first-order closed loop gain of 11. The individual-stage feedback capacitors C_{110} and C_{111} each cause the gain of their respective stages to roll off in the low MHz range. Their primary purpose is probably to prevent high-frequency oscillation and/or reduce noise bandwidth.

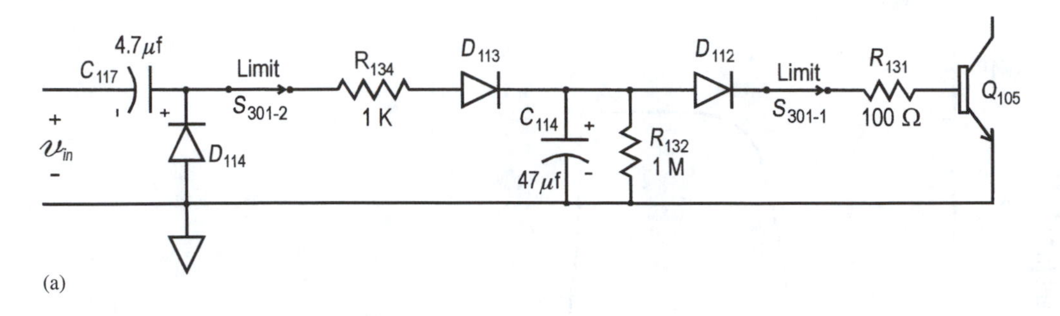

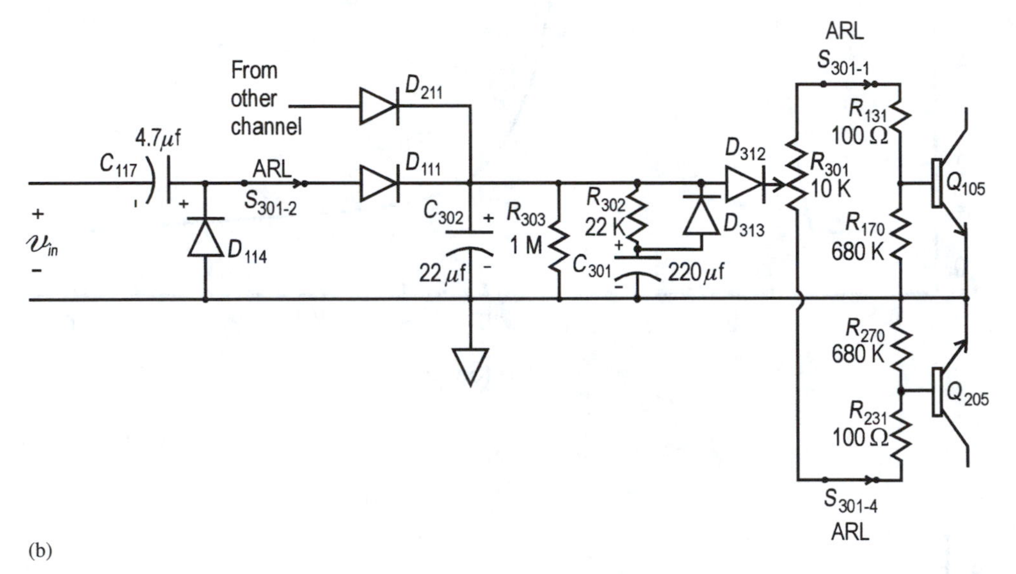

Fig. 5.13: Overdrive-protection circuitry for the Marantz-Superscope CD-320 portable cassette deck. (a) Limit mode circuitry, which clips signal peaks high enough to drive the tape into saturation. (b) ARL (Automatic Record Level) circuitry, which dynamically adjusts gain to keep the signal level below tape saturation.

The input of the stage just analyzed will always come from the previous stage, but the path it takes to arrive will depend on whether the machine is in the "Record" or "Play" mode, and if the former, whether the signal being recorded is from the mike or line input. *Line input* refers to a signal that is at a fairly high level (tenths of volts to volts) and that has a flat frequency response. In any case, the signal arriving at this point will have a source resistance in the kilohm range. The C–E conductance of Q_{105} effectively constitutes the lower leg of a voltage divider. The source resistance is the upper leg. Q_{105} thus serves as a specialized limiter or attenuator. The exact nature of its limiting action will depend on the source of the input applied to its base. In this recorder the user has the option of selecting: (1) automatic record level (ARL), which is analogous to AVC, (2) manual volume control, where the overall audio gain stays constant until manually changed, or (3) limiter mode. In this mode, only the signal peaks, which might otherwise overload subsequent circuitry, are compressed by the limiter circuitry, leaving the rest of the signal unaffected. Each of these three options will present a different input to Q_{105}. In the manual mode (as well as for playback) the base of Q_{105} is held near ground potential by R_{170}. There is no other connection to the base, so Q_{105} always looks like an open circuit. Figure 5.13a shows the circuitry that drives the base in the limit mode; Fig. 5.13b shows the circuitry in the ARL mode.

Aside from the obvious and reasonable difference that ARL action is based on the signal in both channels, whereas the limit action is separate for both

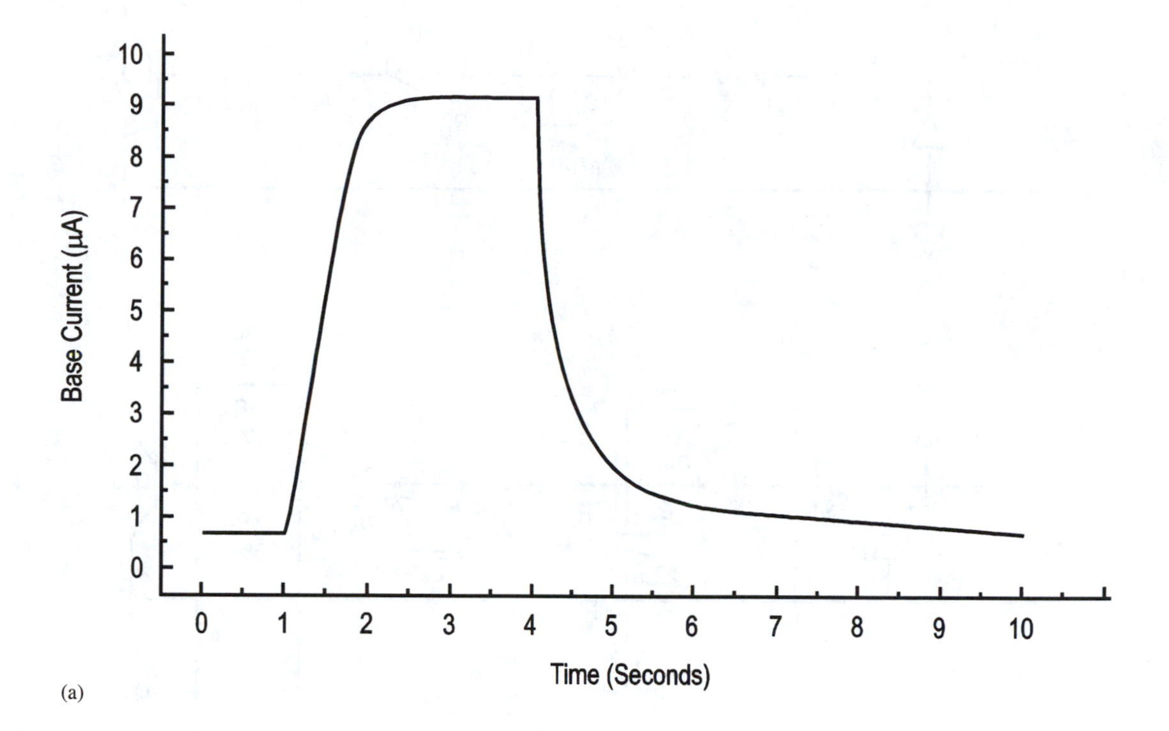
Base Current (μA)
10
9
8
7
6
5
4
3
2
1
0
0
1
2
3
4
5
6
7
8
9
10
Time (Seconds)

(a)

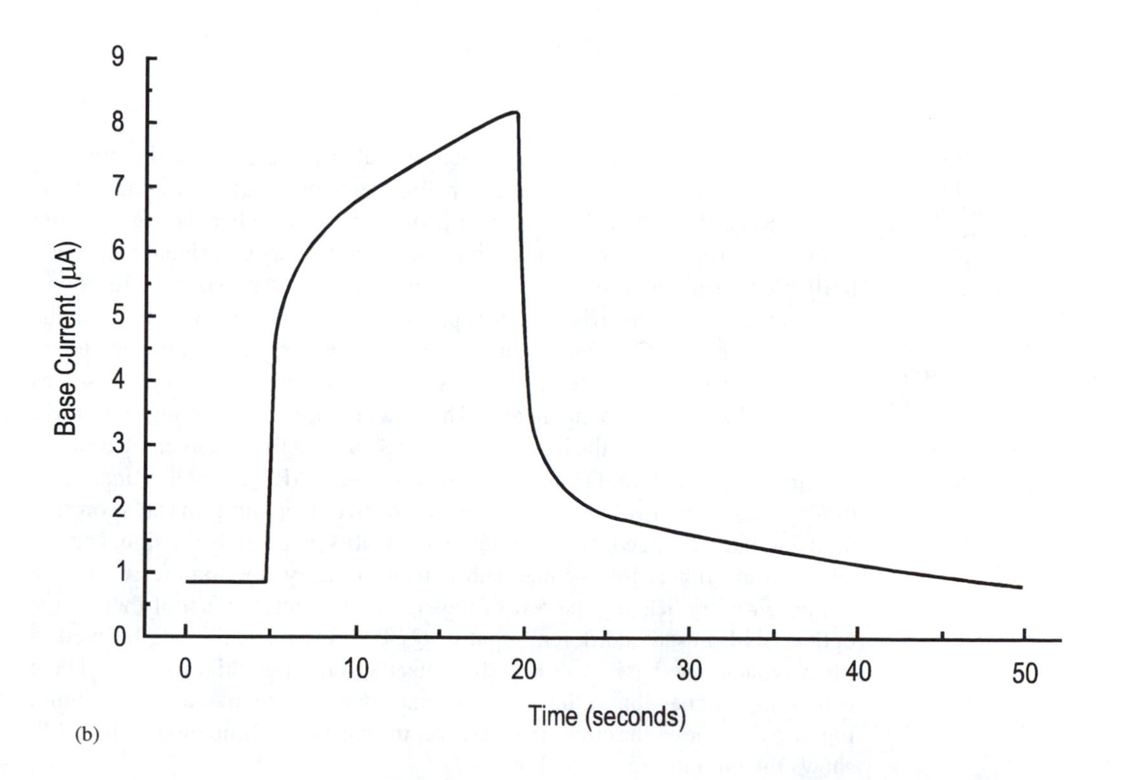
Base Current (μA)
9
8
7
6
5
4
3
2
1
0
0
10
20
30
40
50
Time (seconds)

(b)

Fig. 5.14: Transient response of the overdrive-protection circuits shown in Fig. 5.13: (a) limit mode; (b) ARL mode.
Preceding pages

channels, the main difference is in the time constants for the two modes. This makes it necessary to introduce the concepts of attack and release times. *Attack time* is a measure of how long an overload signal must persist before it actuates the signal-reduction mechanism. *Release time* determines how long it takes for the signal-reduction mechanism to be deactivated after the overload condition has ended.

Because both of these circuits function over the entire range of output currents from zero up to a maximum determined by the signal amplitude, it is evident that an analysis based on the assumption that the diode and V_{BE} voltage drops equal 0.6 V will be invalid, since the output current to the base of Q_{105} can never be zero when the junctions have 0.6 V drops across them. It therefore seems reasonable to do a computer circuit analysis to find the attack and release characteristics. However, simulation results using PSPICE® were not in good agreement with experimental results, therefore these characteristics were determined experimentally.

For both the Limit and ARL modes, measurements were made at an input frequency of 1 kHz. The waveform was initially a .60 VRMS sinusoid. After allowing time for the output to stabilize, this input voltage was stepped to .85 VRMS, and the attack transient was measured. The output variable was taken as the base current of Q_{105}. After this current stabilized at its upper level, the input sinusoid was stepped back down to .60 VRMS and the release transient measured. The measured responses are shown in Fig. 5.14a for the limit mode and in Fig. 5.14b for the ARL mode. Clearly the attack time in the limit mode is very small. The release is also fairly fast although its initial value of di_B/dt is only about half as great as that for the attack. These results are in marked contrast to those for the ARL mode. Here the initial rate of rise of base current for the attack cycle is one quarter that in the limit mode, but after only about 0.2 seconds, the base current seems to level out, although it actually continues to rise at a much slower rate for about 30 seconds. A similar behavior is observed in the release cycle. Half of the decay is completed in .3 seconds, whereas the other half requires about 55 seconds to settle to within 1%. The long time constants are due to the R_{303}–C_{301} time constant.

The potentiometer, R_{301}, is a service adjustment that is set to produce the same ARL effect in both channels. Q_{205} serves the same purpose in the other stereo channel as does Q_{105} in the channel we are analyzing.

The signal is next delivered to Q_{109}, which is half of a dual-channel Dolby® B noise-reduction IC. A block diagram of one channel of this chip (plus the elements common to both channels) is shown in Fig. 5.15 along with the outboard components from the CD-320. The amplifier labeled A_o not only corresponds in function to the one so labeled in Figs. 5.5 and 5.6, but it performs the summing operation as well. The block with a resistor denotes the voltage-variable resistance that forms the heart of the dynamic filter. The DC voltage that controls this filter is applied to pin 14.

The input signal is capacitively coupled to pin 15, which is biased from the internal regulator through R_{139}. The output of this input buffer amplifier is one of the two inputs to the summing amplifier that provides the chip output. The signal path from pin 15 through the input buffer and A_o and out pin 9 is sometimes called the *main chain*. If it were the only input to A_o, the output would just be a non-Dolbyized version of the input. As may be seen from the

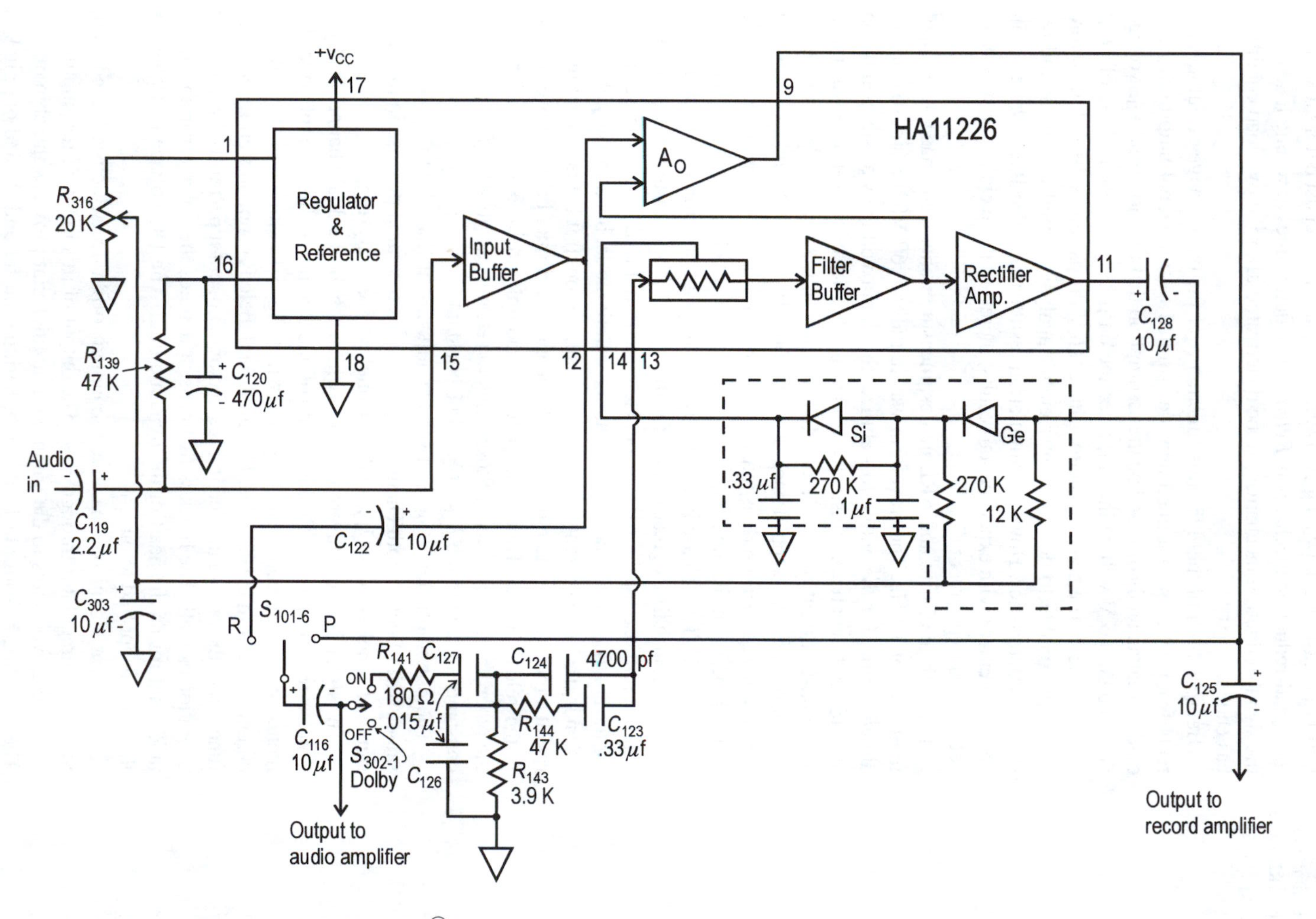

Fig. 5.15: Block diagram of the HA11226 Dolby® integrated circuit and its associated outboard circuitry. Note that this is a dual chip, and only one channel is shown.

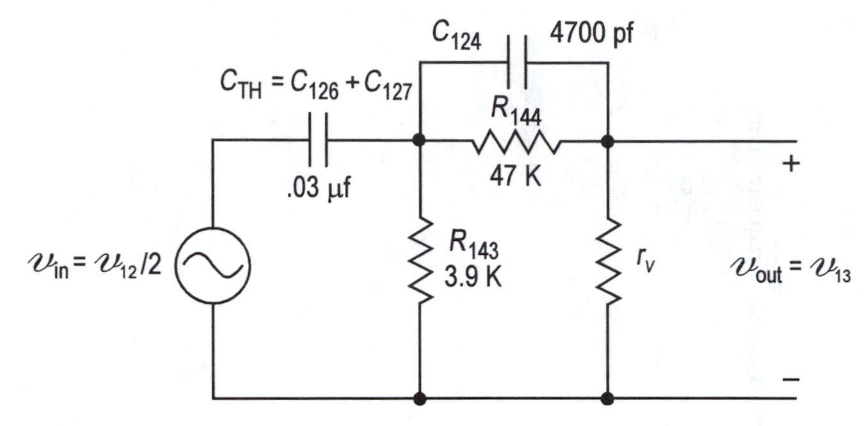

Fig. 5.16: Simplified version of the dynamic filter circuitry shown at the bottom of Fig. 5.15.

bottom of the schematic, if S_{302-1} is in the OFF position, this is exactly what happens, since there is then no input to the dynamic filter or to the circuitry that follows it. This *side chain* component, as it is called, is thus zero.

5.4.1 Record mode

If, however, the Dolby® switch is ON, the buffered signal to be recorded (which is output at pin 12) is applied to the dynamic filter.

It will aid our analysis of this filter if we note the following:

(1) For any frequency below 60 kHz, $R_{141} < 1/(C_{127} \cdot \omega)$. Hence, for a first-order analysis we will assume $R_{141} = 0$.

(2) The above simplification means that C_{126} and C_{127} form a capacitive voltage divider, which can be replaced by a Thevenin equivalent voltage of one half the original and a capacitance of 0.03 μF as the source impedance.

(3) C_{123} will have an impedance over the audio range that is much less than that of R_{144} and can thus be considered an AC short.

(4) The gain-control element is effectively a variable shunt resistance to ground.

These simplifications allow the dynamic filter schematic to be represented as shown in Fig. 5.16. This is clearly a high-pass filter, since at high frequencies $v_{13} \rightarrow v_{12}/2$, but at lower frequencies the transfer function is smaller. The transfer function of this circuit is quadratic in both numerator and denominator:

$$\frac{\mathbf{V}_{\text{out}}}{\mathbf{V}_{\text{in}}} = sC_{\text{TH}}R_{143}r_{\text{v}}(1 + sC_{124}R_{144})/\{s^2C_{\text{TH}}C_{124}R_{143}R_{144}r_{\text{v}}$$
$$+ s[r_{\text{v}}(C_{124}R_{144} + C_{\text{TH}}R_{143}) + R_{143}R_{144}(C_{124} + C_{\text{TH}})]$$
$$+ (r_{\text{v}} + R_{143} + R_{144})\}.$$

From this equation, the 3 dB frequency can be calculated as a function of r_{v}. A plot of $f_{3\,\text{dB}}$ versus r_{v} is given in Fig. 5.17. This plot shows that if a small signal fed to pin 14 causes r_{v} to rise, then the corner frequency will fall, as it should for proper Dolby® action.

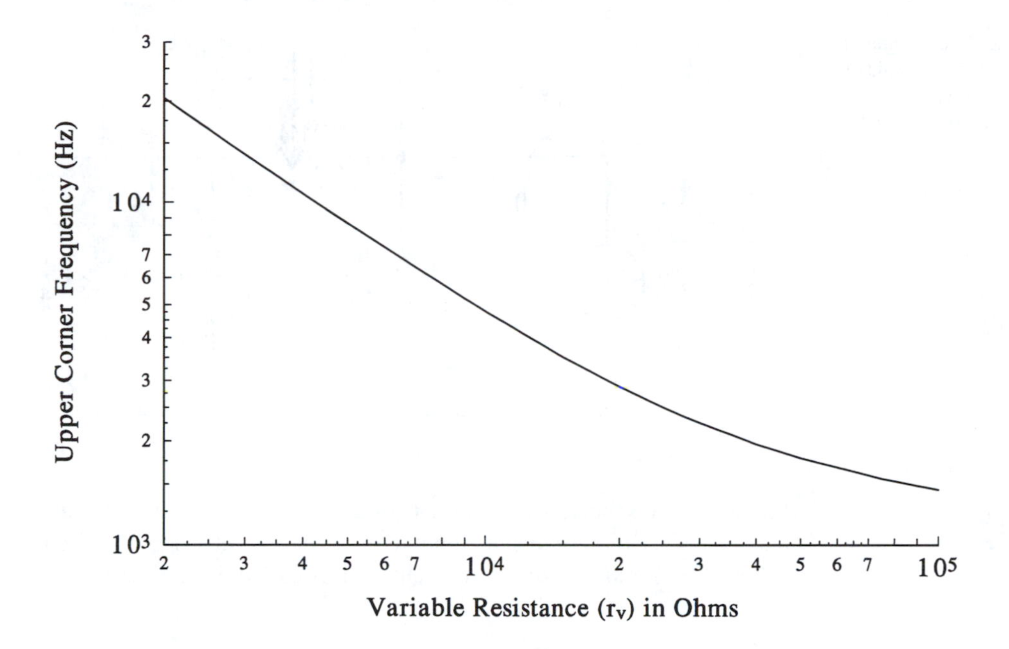

Fig. 5.17: High-frequency breakpoint of the dynamic filter shown in Fig. 5.16 as a function of the variable resistance, r_V.

The output of the dynamic filter is buffered and fed back to A_o, where it is summed with the main chain signal and fed out pin 9 to the Q_{106} stage, which is, as Fig. 5.9 shows, the record amplifier.

The dynamic filter is controlled by a DC voltage derived from rectification and filtering of the side chain. The rectifier–filter is shown enclosed in dashed lines in the lower right corner of Fig. 5.15. The parts values are not supplied by Marantz but are those recommended by Hitachi for this chip. The 0.33 μF capacitor and the diode feeding it are the actual rectifier–filter combination. The Ge diode and the 12 kΩ resistor provide a DC offset at the anode of the Si diode. This offset can be varied by R_{316}, which is called the LAW control. This control effectively sets the lower threshold above which the audio signal will begin to affect the DC voltage on pin 14. Above that threshold the DC output varies linearly with the AC input amplitude, having a constant of proportionality that is fairly independent of the LAW setting. This control is not user-accessible but is a service adjustment. The DC output from the ripple filter is fed back to pin 14 of the IC, where it controls r_V.

5.4.2 Playback Mode

In playback mode the buffered input signal is also fed to A_o as one of its inputs. The inverted output of A_o is now fed to the dynamic filter, and the side chain signal is summed back into A_o. Since the A_o stage is inverting, the side chain signal has the opposite phase on playback from its phase on record, and the recorded boost is subtracted out along with a significant amount of noise, producing a noise-reduced version of the original signal.

On playback, the output of the Dolby® system goes to the audio amplifier ICs, Q_{107} and Q_{108}. This stage will not be analyzed in detail, but it is a configuration that will be strange to some. It is called a *bridge amplifier*. The

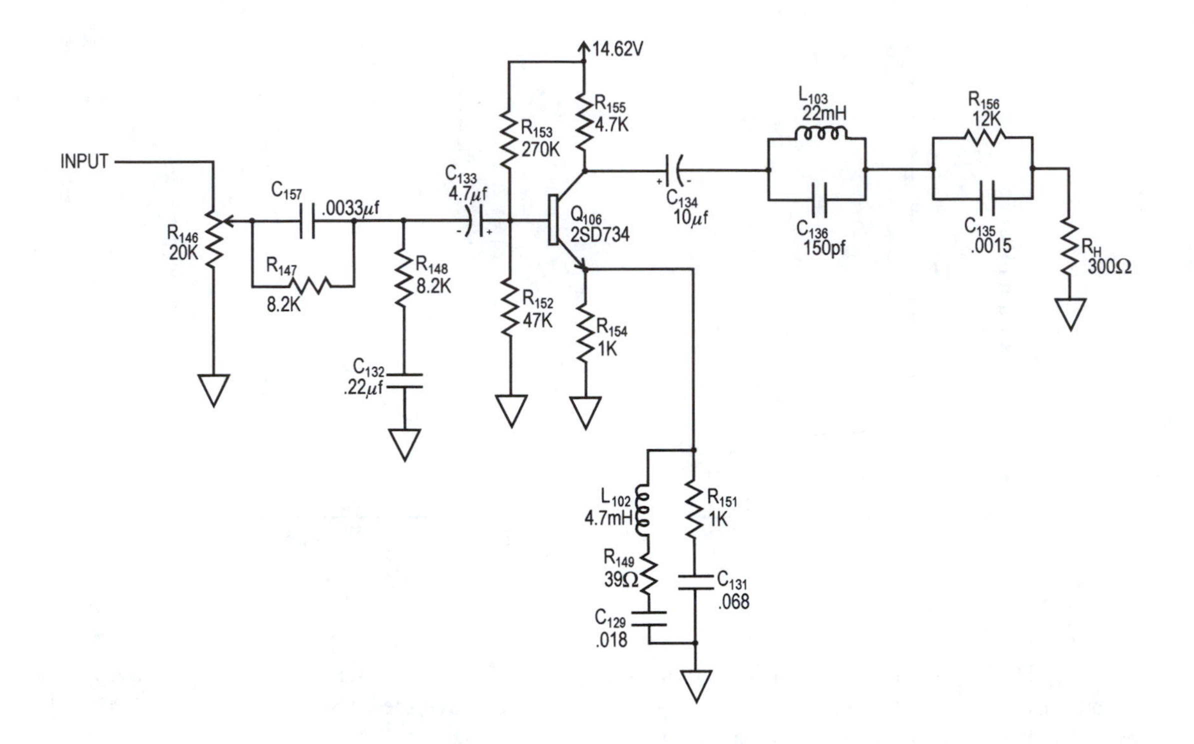

Fig. 5.18: Schematic diagram of the record-head driver for the Marantz-Superscope CD-320 portable cassette deck.

two amplifiers are fed input signals that are equal in amplitude but opposite in phase. Then the outputs are connected to two sides of the load. Thus as one output swings above the quiescent output voltage, the other swings below it by an equal amount. This effectively doubles the voltage swing across the load (and quadruples the power) relative to what could be obtained with a single amplifier of the same type operating from the same supply. Although it is not done in this circuit, it is even possible to eliminate capacitive coupling to the load if the quiescent voltages of the two amplifiers are nearly equal. This is generally true if both amplifiers are on a single chip. This type of amplifier is often used in computer audio equipment, which must operate with supplies of 5 V or less.

Although we found that no meaningful record equalization was accomplished by the Q_{101}–Q_{102} stage, as was pointed out in Chapter 4, there is always record equalization to compensate for the dropoff of the playback head's high-frequency response, and there is often additional boost at high and low frequencies to improve the signal-to-noise ratio.

The Q_{106} stage provides the record equalization and drives the record head. This stage has the form of a transconductance feedback amplifier, but both the collector and emitter impedances are actually networks of several elements. The configuration of the emitter network will depend on the settings of the bias and equalization switches. Figure 5.18 shows the schematic of this stage with the switches set for low bias and 70 μsec equalization (these are the normal settings for recording FeCr and metal tapes).

It should be evident that this is a circuit best analyzed by computer. A PSPICE®-generated plot of the voltage across the head versus frequency for unity input voltage and an assumed record-head impedance of 300 Ω is shown in

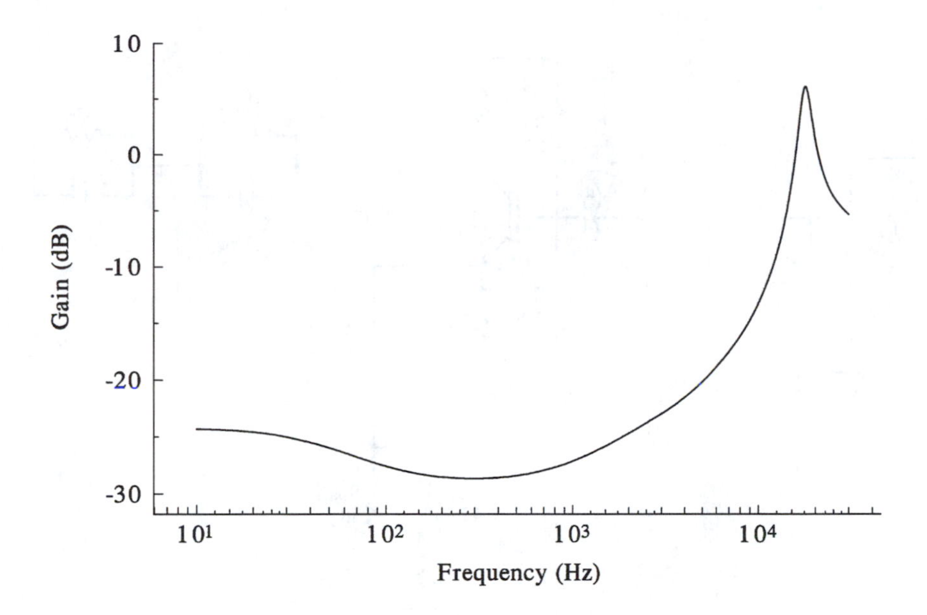

Fig. 5.19: Frequency response of the record-head driver of Fig. 5.18.

Fig. 5.19. The boost at low frequencies and the large boost at high frequencies are both evident. The former raises low-frequency response above the noise level. The latter compensates for the falloff of tape head response.

5.4.3 *Erase Oscillator*

Referring to Fig. 5.20, we now wish to analyze the circuitry surrounding transistors Q_{301}–Q_{305}. Their function is by no means immediately obvious. A good clue is to observe that the erase head is driven from this block of circuitry, so it must contain the erase/bias oscillator. Tracing back from the head, we see it is driven from the secondary of T_{302}, whose primary is driven by Q_{304}. The tap on the secondary of T_{302} also supplies AC bias to the record heads. This stage is therefore most likely the bias oscillator or at least a part of it. Closer inspection reveals that the center tap of the primary of T_{302} is AC grounded by C_{318}. Therefore this winding will couple signal from the collector back into the base by autotransformer action. The overall circuit may thus be drawn as shown in Fig. 5.21a for AC. The secondary winding of T_{302} is not shown.

The feedback network is shown in Fig. 5.21b. The equations for its analysis are:

$$\mathbf{I} = \mathbf{I}_1 + \mathbf{I}_2 + sC(\mathbf{V} + \mathbf{V}_\mathrm{F}),$$

$$\mathbf{V} = sL_2\mathbf{I}_2 - sM\mathbf{I}_1,$$

$$sC_1R \cdot \mathbf{V} = -(2sC_1R + 1) \cdot \mathbf{V}_\mathrm{F},$$

and

$$sL_1\mathbf{I}_1 - sM\mathbf{I}_2 = (\mathbf{I} - \mathbf{I}_1)/sC + \mathbf{V},$$

where $C_1 \equiv C_{315} = C_{316}$, $C \equiv C_{317}$, and R is the combined resistance of R_{324} and the transistor input resistance.

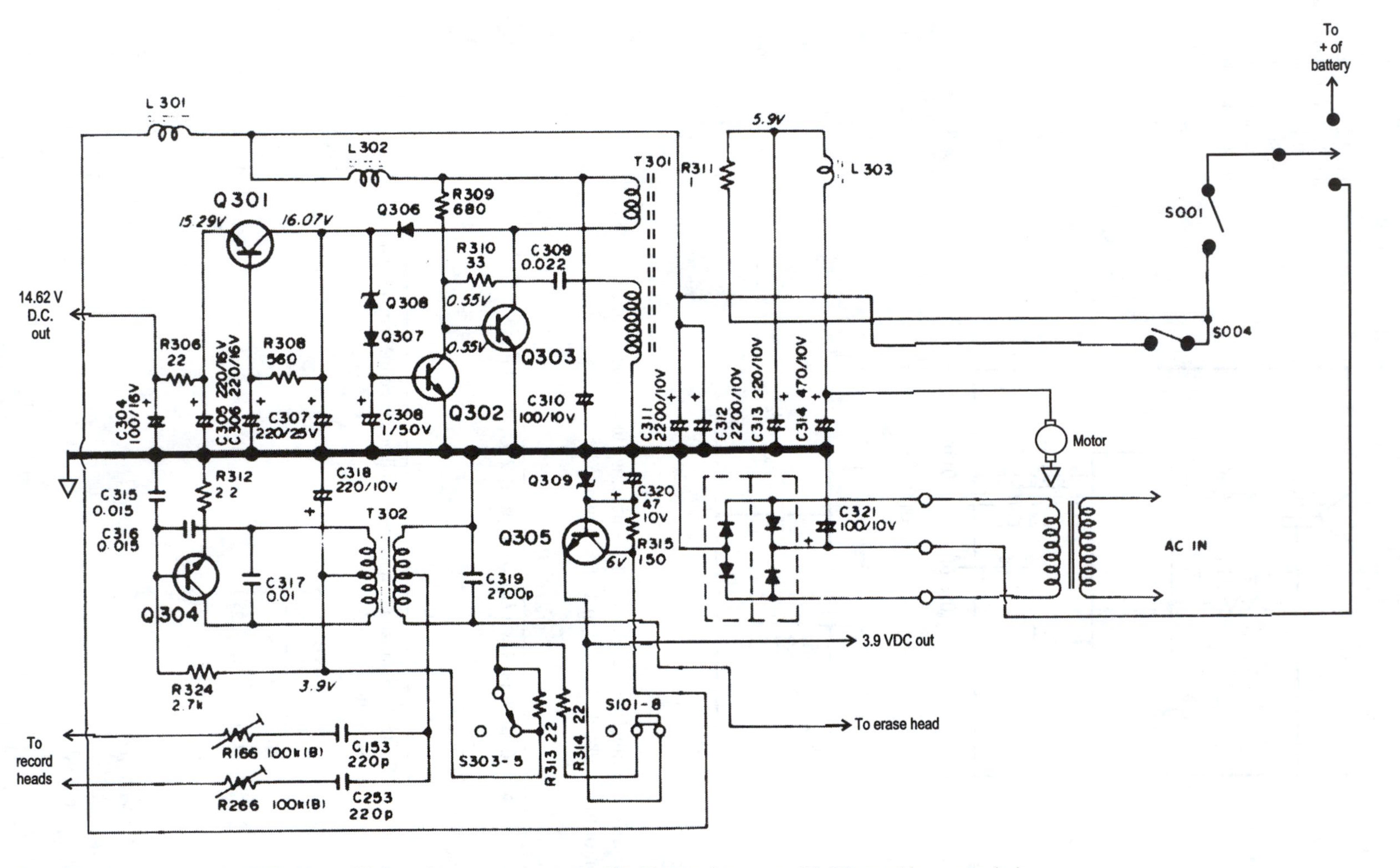

Fig. 5.20: Schematic diagram of the bias-oscillator and power-supply stages of the Marantz-Superscope CD-320 portable cassette deck.

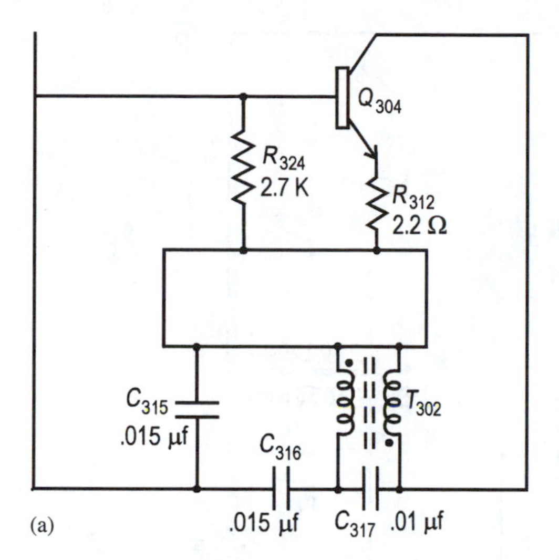

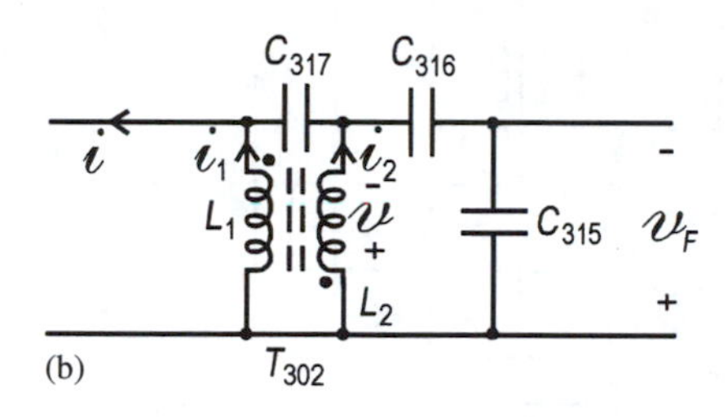

Fig. 5.21: Standard feedback form of the bias oscillator after AC simplification. (a) The entire circuit. (b) The feedback network.

If the flux coupling between L_1 and L_2 is assumed to be perfect, the transfer function of the feedback network may be shown to be

$$\mathbf{F} = \frac{\mathbf{V}_\mathrm{F}}{\mathbf{I}} = \frac{s^2 M C_1 R}{-s^3 L_2 C_1^2 R + [1 + s^2(L_\mathrm{e} C + L_2 C_1)](2sC_1 R + 1)}, \tag{5.9}$$

where $L_\mathrm{e} \equiv (\sqrt{L_1} + \sqrt{L_2})^2$.

Exercise 5.2 Use Eq. (5.9) to find the frequency of oscillation and the minimum transistor gain to ensure oscillation.

The correct answers to this exercise are

$$\omega_\mathrm{osc} = \sqrt{\frac{1}{(L_\mathrm{e} C + L_2 C_1/2)}}$$

and

$$g_\mathrm{m} \geq \frac{N_2}{N_1 R},$$

where we have used

$$\sqrt{\frac{L_1}{L_2}} = \frac{N_1}{N_2}.$$

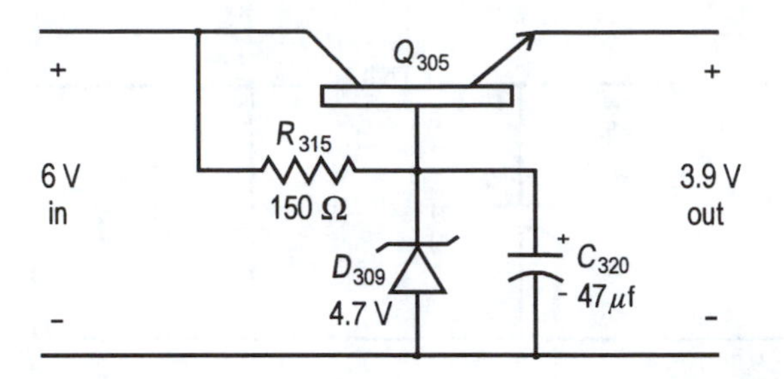

Fig. 5.22: Power-supply stage using the "impedance divider" configuration. This is a subset of the circuitry shown in Fig. 5.20.

If we had not assumed perfect flux coupling, the problem would have become quite unwieldy, but the minimum g_m would have increased significantly. Nevertheless, the actual g_m would typically be over an order of magnitude greater than that needed to sustain oscillation.

5.4.4 Power Supply Circuitry

It can be noted that the positive DC supply for both the collector and base circuits of Q_{304} (3.9 V) is derived from Q_{305}. This transistor is in the classic "impedance divider" circuit as shown in Fig. 5.22.

A DC analysis of this circuit yields the result

$$v_{\text{out}} = \frac{V_z R_{315} + v_{\text{in}} r_z - V_{\text{BE}}(r_z + R_{315})}{(r_z + R_{315}) + r_z R_{315}/(\beta + 1) R_{\text{L}}},$$

where R_{L} is the load resistance connected across the output, r_z is the Zener resistance, and V_z is the Zener voltage. In a properly designed circuit, $R_{315} \gg r_z$ so

$$v_{\text{out}} \approx \frac{V_z + v_{\text{in}} \cdot (r_z/R_{315}) - V_{\text{BE}}}{1 + r_z/(\beta + 1) R_{\text{L}}}.$$

For comparison note that if the transistor were omitted, and R_{L} were connected across the Zener diode, v_{out} would be given by

$$v_{\text{out}} = \frac{V_z R_{315} + v_{\text{in}} r_z}{r_z + R_{315} + \dfrac{r_z R_{315}}{R_{\text{L}}}} \approx \frac{V_z + v_{\text{in}} \cdot [r_z/R_{315}]}{1 + r_z/R_{\text{L}}}.$$

The two significant differences in these two expressions are the V_{BE} drop in the transistor version and the fact that r_z is replaced by $r_z/(\beta + 1)$ in the transistor version. Thus the effective impedance in the base lead is divided by a factor of $(\beta + 1)$. This improves the load regulation significantly. In the same manner, if we were to do an AC analysis, we would find that C_{320} would be effectively multiplied by $(\beta + 1)$, thus reducing its impedance by the same factor.

Note that between the output of Q_{305} and the bias oscillator, Q_{304}, are two series resistances, R_{313} and R_{314}, with a resistance of 22 Ω each. One section of switch S_{303} (bias selector) can be used to short out R_{313}. This is the high AC-bias position used for recording CrO_2 and metal tapes. With both resistors in circuit, the AC bias is lower.

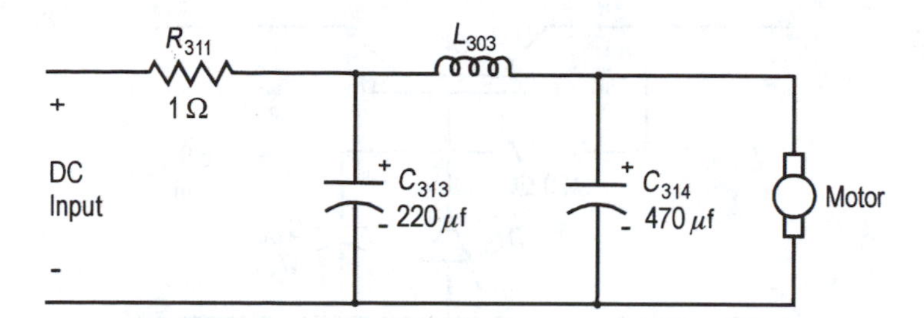

Fig. 5.23: The low-pass filtering used for the motor power supply. This is a subset of the circuitry shown in Fig. 5.20.

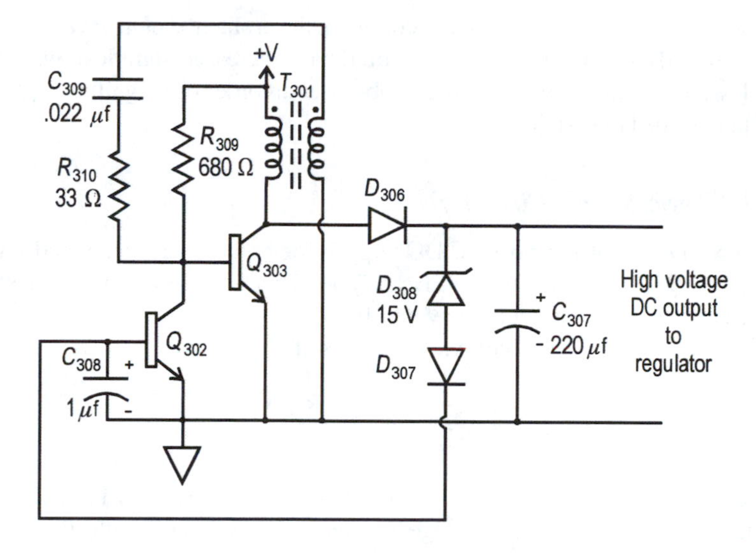

Fig. 5.24: Schematic of the "chopper" circuit used in the Marantz-Superscope CD-320 portable cassette deck to generate high voltage from a lower voltage source.

To the right of Q_{305} are seen a bridge rectifier and filter capacitor. Inserting the line plug into the recorder activates the SPDT switch, which selects battery or AC power for the recorder. When AC is selected, the batteries are charged through a 22 Ω resistor. Whichever source is selected, the DC from it is applied to S_{001}, which is the master off–on switch that is part of the keyboard assembly. When this switch is closed by activating the keyboard, DC is applied to the motor through a two-section LPF shown in Fig. 5.23. Before power is applied to the electronics, however, S_{004} must be closed. This is the "Pause" key. It is normally closed, but pushing it, opens it. This allows the user to begin recording almost instantly upon releasing the Pause key. Otherwise, there would be a couple of seconds delay while the motor comes up to speed and blocking and bypass capacitors charge.

We now turn our attention to the remaining circuitry in this section. Q_{301} looks very similar to Q_{305} in its connections except that there are four additional filtering components – namely C_{307}, C_{305}, R_{306}, and C_{304}.

When we note the voltages on the collector and emitter, we see that they are both well above the 6 V DC supply of the recorder. This may mean that there is a "chopper" oscillator in the circuit whose AC output is transformed up to a higher voltage and rectified back to a higher DC voltage. To investigate this possibility, the circuitry driving the Q_{301} regulator is redrawn in Fig. 5.24. Analysis of this circuitry will in turn be facilitated by reviewing the classical astable blocking oscillator of Fig. 5.25.

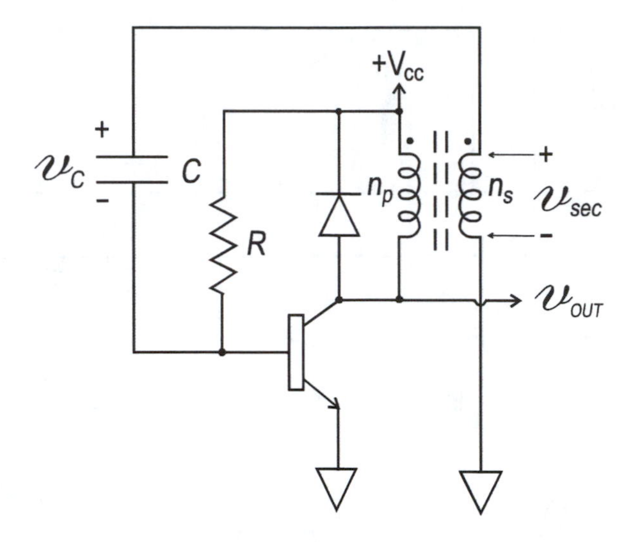

Fig. 5.25: Schematic diagram of a classical blocking oscillator.

We start our analysis of this circuit by assuming that $v_c > 0$ and that the transistor is in cutoff. If the transistor is cut off, there is no voltage across the transformer primary and thus none across the secondary. If there is no voltage across the transformer secondary, Kirchoff's voltage law around the outside loop will insure that the assumed polarity of v_c will actually produce the cutoff situation assumed. But this state of affairs cannot remain for long, because C will begin to charge in such a direction as to reduce and finally change the polarity of v_c. The charging path is through the transformer secondary, C, R, and the DC supply voltage. It might be objected that this charging current violates the assumption that the current through the transformer windings is zero. Strictly speaking, this is true. Practically speaking, however, the effect is insignificant. The transformer used in this application is called a pulse transformer. Its self-inductance is small enough to cause only a very small voltage drop due to the relatively low amplitude and slow rate of change of the base current flowing through it.

Since the target voltage for the charging cycle is $+V$, this voltage rises fairly linearly. The charge cycle terminates when the base-emitter junction becomes forward biased. Once this occurs, it enables the collector current, which flows through the transformer primary and generates a secondary voltage. The dots in Fig. 5.25 indicate that the induced secondary voltage will be of the proper polarity to drive the base harder. This positive feedback saturates the transistor very rapidly, causing the output voltage to drop to nearly zero.

Once the transistor is saturated, the entire supply voltage is dropped across the transformer primary, which commences an exponential rise of current whose parameters are determined by the inductance and resistance of the transformer primary. This primary current pulse is coupled to the secondary, which keeps the base forward biased *even as it charges C with a polarity that will soon cut off the transistor*. As the primary current approaches its maximum value, $di_p/dt \to 0$, causing both the primary and secondary voltages to drop. But the drop of the secondary voltage, coupled to the base through C, drives the transistor further toward cutoff. A regenerative switching again takes place, putting the transistor into cutoff very quickly. As the transformer voltages return to zero, the accumulated negative charge on C just as quickly drives the base potential negative from whence it begins its quasi-linear charge cycle again.

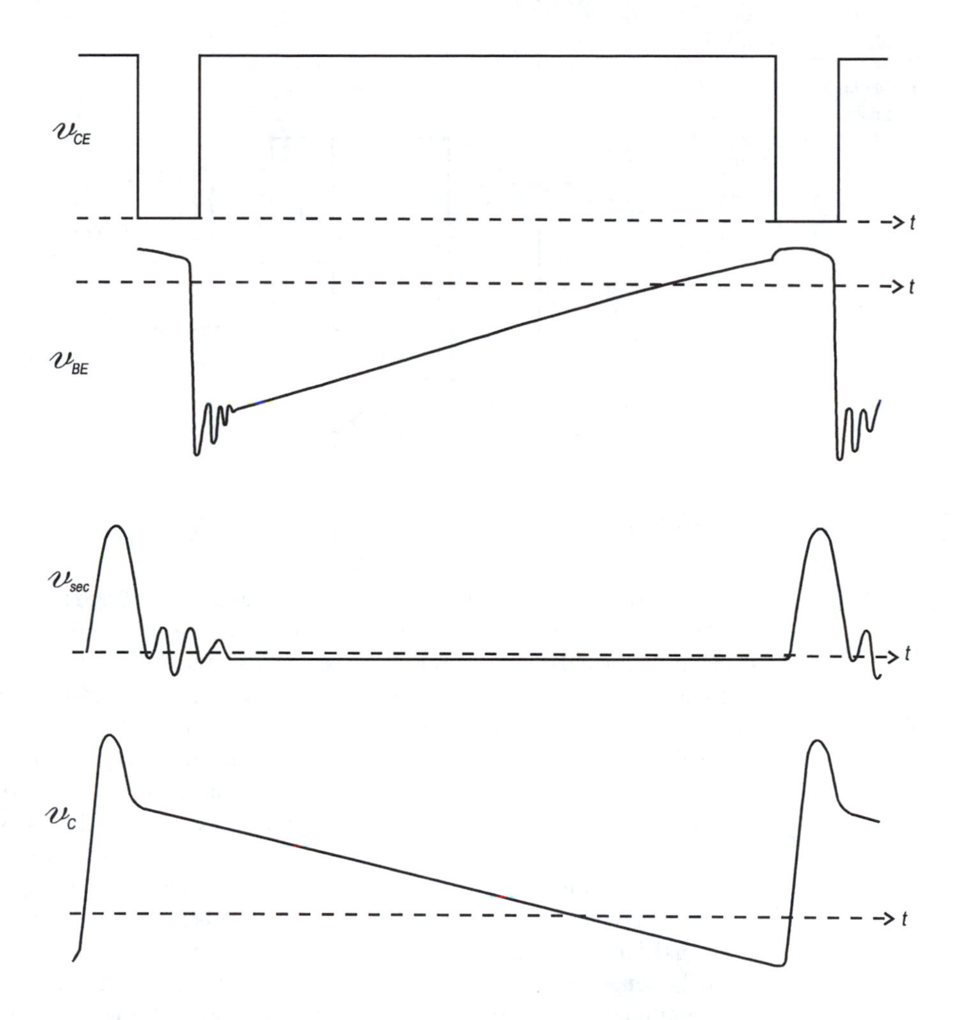

Fig. 5.26: Waveforms at various points in the classical blocking oscillator.

The most important aspect of the oscillatory cycle for our purposes here, however, is that even though both transformer windings feature $di/dt = 0$ and $v = 0$ upon entering cutoff, the primary has just been subject to a large current whose magnetic energy remains in the inductor and must be dissipated. In the circuit of Fig. 5.25 this energy flows back into the DC supply through the diode, giving a fairly clean pulse train at the output as shown in the upper trace of Fig. 5.26.

There is no diode, however, in Fig. 5.24. Instead, a capacitor (C_{307}) absorbs this magnetic energy. D_{306} keeps C_{307} from discharging back into the blocking oscillator. If C_{307} and D_{306} were not in circuit, there would be large-amplitude "ringing" oscillations at the collector of Q_{303} each time it went into cutoff. This diode–capacitor combination effectively serves as a rectifier–filter for this oscillation, resulting in a high DC output.

This output is reduced by the drops of D_{308} and D_{307} (about 15 V and 0.6 V respectively), and the difference between the output voltage and the diode drops (about 0.5 V) is applied to the base of Q_{302}. If the output voltage gets too high, Q_{302} will begin to conduct, reducing the available base drive to Q_{303}. This will cause saturation of the transistor at a lower current level, thus reducing the output voltage. Thus this circuit acts as a preregulator for the output voltage as well as serving to reduce the demands on the oscillator when its full capability is not needed.

Since the pulse transformer generally has a turns ratio between 1:1 and 3:1, a very large base current may result. The purpose of R_{310} is to limit the amplitude of this current pulse. It does not materially affect the principles of operation outlined.

5.5 Other noise reduction technologies

The Dolby® B system discussed earlier in the chapter premiered in 1970. (Dolby® A is a commercial-grade system that is not for consumer electronic use.) It generated a great deal of interest and activity in the area of noise reduction. Subsequent developments have been spurred by the development of IC technology, which makes increasingly complex schemes practical. Like the Dolby® B system, but to an even greater degree, more recent developments have hinged on what is called "psychoacoustic" effects – in other words, how are music and noise perceived by the listener, and what does this allow us to do with the audible spectrum to get rid of noise without perceptibly changing the program material?

5.5.1 The dbx system

The dbx system was introduced in 1975. Its noise reduction capabilities are markedly superior to Dolby® B, yet it has never had anywhere near the universal acceptance of that system. The predominant reason for this is probably that dbx-encoded media are totally unusable without a dbx decoder. This is in contrast to Dolby® B, where encoded tapes can be made to sound quite good to the uncritical ear by simply turning down the treble. In spite of the limited acceptance that dbx has enjoyed, we will still look at its operation, because it is a conceptually simpler system that well illustrates the increasingly-important electronic principle of *companding*. This word is a contraction of *compressing* and *expanding*. What we are trying to do is readily appreciated by studying Fig. 5.27.

Live music sources can easily have a dynamic range of over 100 dB. A good tape recorder might have 60 dB between the levels corresponding to tape saturation at the upper end and the noise floor at the lower end. It is obvious from Fig. 5.27 that the dbx system takes a 100 dB dynamic range, compresses it down to 50 dB for recording, and then expands it back to a 100 dB range on playback. This is called 2:1 compression and 1:2 expansion. Without dbx, the full dynamic range can never be recorded, and the music playback will of necessity always be... well, less dynamic.

But the expansion of the dynamic range is not the most important effect of the system. Noise reduction is the name of the game here. Note that the 50 dB dynamic range of the compressed signal will record comfortably even on a modest tape recorder, and the −80 dB signal, which might otherwise be 30 dB or more below the tape noise floor, is boosted by 40 dB. Now this signal is passed through a "noisy channel" – namely the tape recording process – and acquires a noise component that might be comparable to this signal.

In the expansion process this signal is attenuated by 40 dB, and the noise is also! Thus the system is capable of improving the S/N ratio by up to 40 dB. If

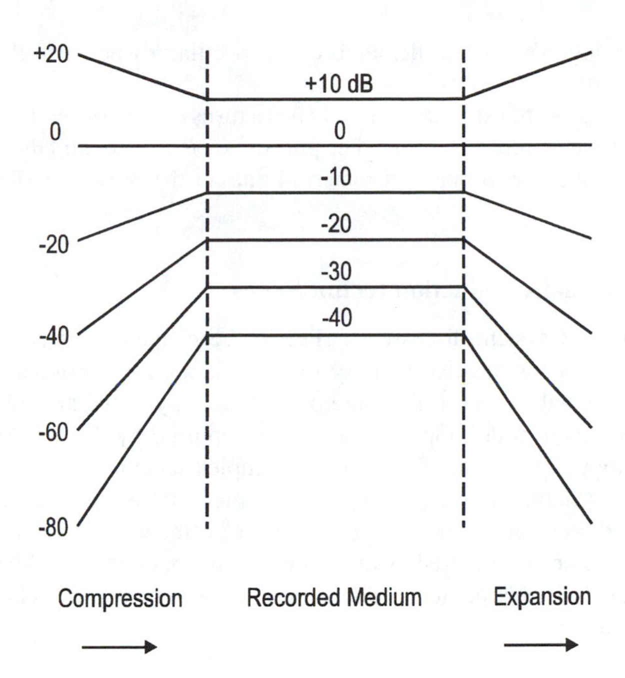

Fig. 5.27: Diagrammatic representation of how a signal with a dynamic range of 100 dB can be compressed by the dbx system to fit on magnetic tape with a dynamic range half that large and then can be recovered again.

the quietest sounds to be recorded are at −60 dB, the S/N improvement will be more like 30 dB. On the upper end, the compression introduces a "buffer zone" that keeps the recorded amplitude away from tape saturation, thus reducing the distortion due to this mechanism, as shown in Fig. 5.3b.

A compressor is specified by a *compression characteristic* showing the transfer function between input and output amplitudes in dB. The compression characteristic for dbx is shown in Fig. 5.28. The Dolby® system is also a compander of sorts, but a frequency-dependent one with the frequency range being determined by the input amplitude.

The entire dbx system is little more than a pre-emphasis stage (which is just an equalizer that boosts high frequencies) and the compressor stage. The pre-emphasis provides additional noise suppression because the high frequencies are subject to complementary de-emphasis after expansion, and this cut in high frequencies cuts the noise also.

This only leaves unanswered the question of how to implement a compressor or an expander. In the first place, we may note that IC companders such as the Signetics NE570 and NE571 are available. However, a compressor and an expander are nothing more than a voltage-controlled amplifier (VCA) and an RMS-to-DC converter, as can be seen in Figs. 5.29a and b. The voltage translator changes the output voltage range of the RMS converter to the proper range for the gain-control input of the VCA. An analog multiplier serves nicely as the VCA but represents "overkill" in this application.

5.5.2 *Dynamic noise reduction*

In 1982, National Semiconductor introduced the LM1894. This chip, although only providing 10 dB improvement in S/N ratio, still represented a remarkable advance in the state of the art because it accomplished this with unencoded signals. Not only can it match Dolby® B in S/N reduction for tape, but unlike

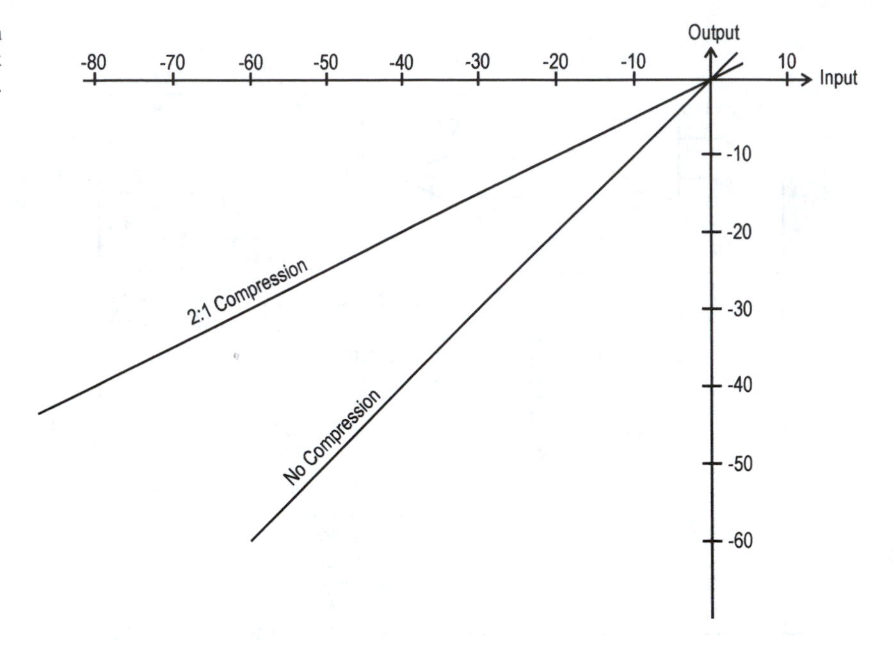

Fig. 5.28: Compression characteristic of the dbx noise-reduction system.

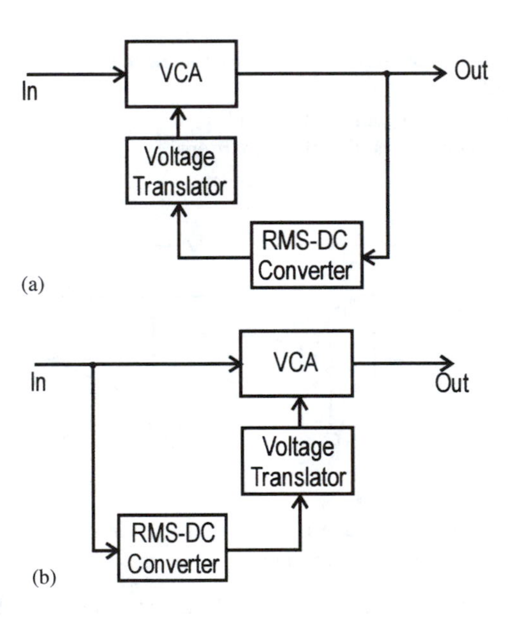

Fig. 5.29: Simplified block diagram of how a compression system such as dbx works.

Dolby®, it requires no licensing fees. Furthermore, it will also work with FM, phonograph, and other audio signal sources. The chip is dual for stereo applications, but in Fig. 5.30 below, only one channel is shown. Only the circuitry above the dashed line is duplicated for the other channel.

The two-stage amplifier has an operational transconductance amplifier (OTA) as its first stage. This device provides an output current that can be characterized as $g_m v$. The transconductance, g_m, will vary linearly with the current supplied to the control pin, where v is the voltage input to the amplifier. The RCA CA3080 is a stand-alone OTA whose data sheet can be studied by those unfamiliar with the genre. The second op amp is used as an integrator.

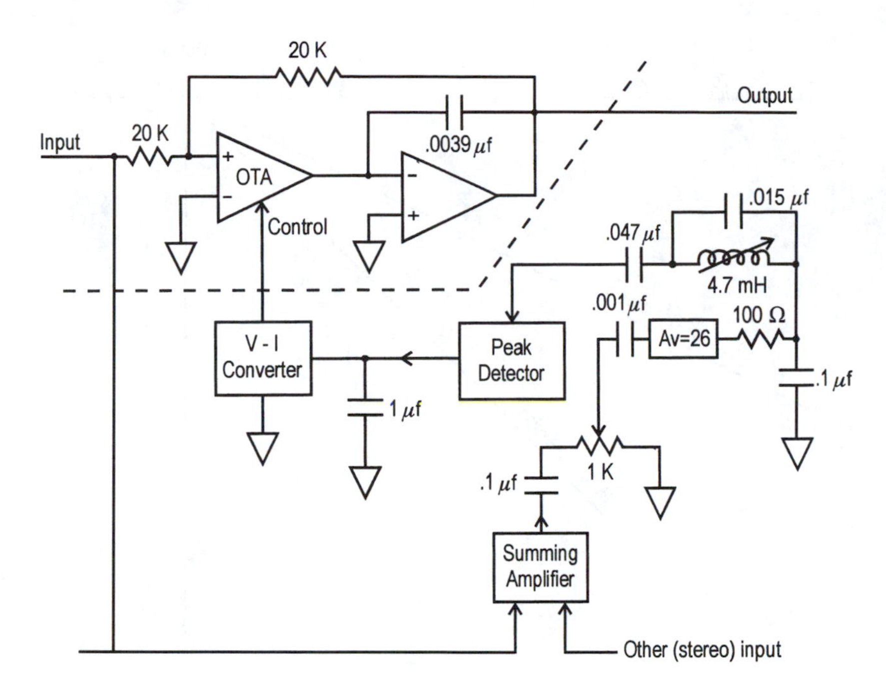

Fig. 5.30: Noise-reduction circuit based upon the National Semiconductor DNR® integrated circuit and its associated outboard circuitry. Only one channel is shown.

Exercise 5.3 Utilizing the circuit shown below, show that

(a) it is an LPF,

(b) it has a maximum gain of unity, and

(c) the pole frequency is directly proportional to i_c.

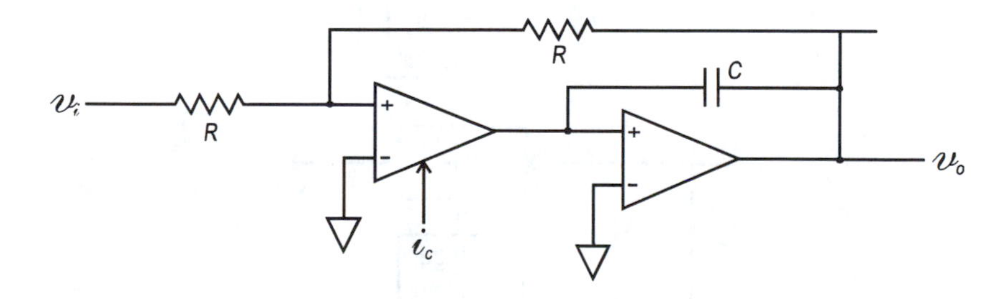

The OTA may be assumed to have infinite input impedance, and the virtual-ground principle does not apply to this device. For this OTA, $g_m = 1.32\ \text{v}^{-1} \cdot i_c$.

The signal path through these two amplifiers is called the main signal path. There is another path through the summing amplifier and associated circuitry that is called the control path. The output of this amplifier is passed through a high-pass filter network consisting of the 0.1 and 0.001 μF capacitors, the 1 kΩ pot, and the 30 kΩ input impedance of the fixed-gain amplifier. Surprisingly, the shape of the Bode amplitude plot for this filter is virtually independent of the pot setting. The output of the fixed-gain amp is shown feeding a parallel-tuned 19 kHz notch filter. This filter removes the pilot signal if the DNR® system is used with FM. In all other applications, the notch filter components are all removed, except for the 0.047 μF capacitor, which not only couples the amplifier output to the peak detector but accomplishes additional high-pass filtering.

Exercise 5.4 The high-pass filtering between the summing amp and the fixed-gain amp produces a Bode amplitude plot as shown. This plot incorporates the input resistance of the ×26 amplifier.

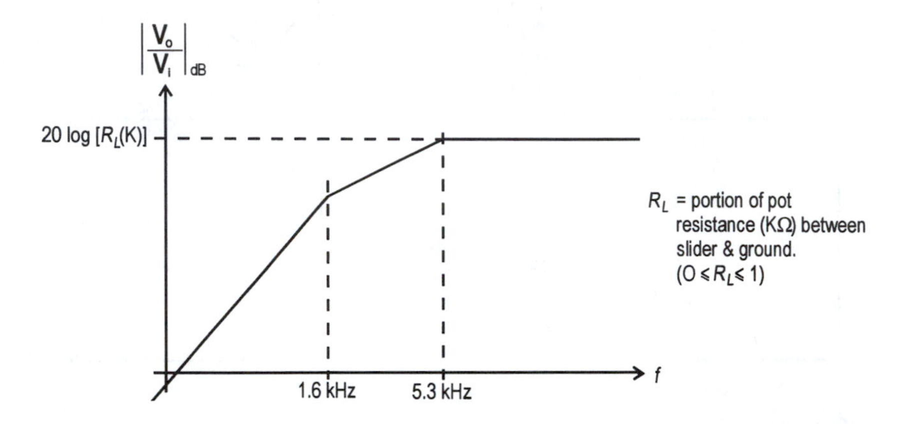

The peak-detector input resistance is nominally 700 Ω. Draw a composite Bode amplitude plot showing the overall frequency response of the control path; assume that all of the active devices in the path have flat frequency response and that the notch filter is removed.

The peak-detector output voltage drives a proportional current into the control pin of the OTA in the main signal path, controlling its gain.

We are now in a position to understand why the signal is run through a high-pass filter. When the program material contains little in the way of high frequencies, the noise becomes more audible. The small high-frequency component traveling through the control path will drive a small control current into the OTA. As shown in Exercise 5.3, this will make for a low corner frequency on the LPF in the main signal path. This in turn will attenuate the high frequencies there. But we just said that this situation is achieved when the high-frequency component is small, so little signal is lost by sliding the low-frequency cutoff downward. Substantial noise may be lost, however, since the main signal bandwidth can be cut from about 34 kHz when there is a strong high frequency component down to about 1 kHz when there is none. Figure 5.31 shows these relationships for the situations of large and small high-frequency content.

5.5.3 *Dolby® C*

In early 1983, Dolby Labs responded to the competition with the announcement of the Dolby® C noise-reduction system. The design goal for this system was a 20 dB improvement in S/N ratio. This goal was achieved by using two stages of Dolby® B type circuitry, but with changed parameters. The first stage operates at levels comparable to the B system (but with about 1/4 of the minimum corner frequency of the B system's 1.5 kHz). The second stage operates at a 20 dB lower threshold.

In addition to improving the S/N performance, Ray Dolby has addressed a few other exotic problems associated with tape recording. These include midband modulation and selective high-frequency tape saturation. Midband modulation can be observed where the dominant spectral content of the music is at high frequencies, and the frequency response of the compander gain control

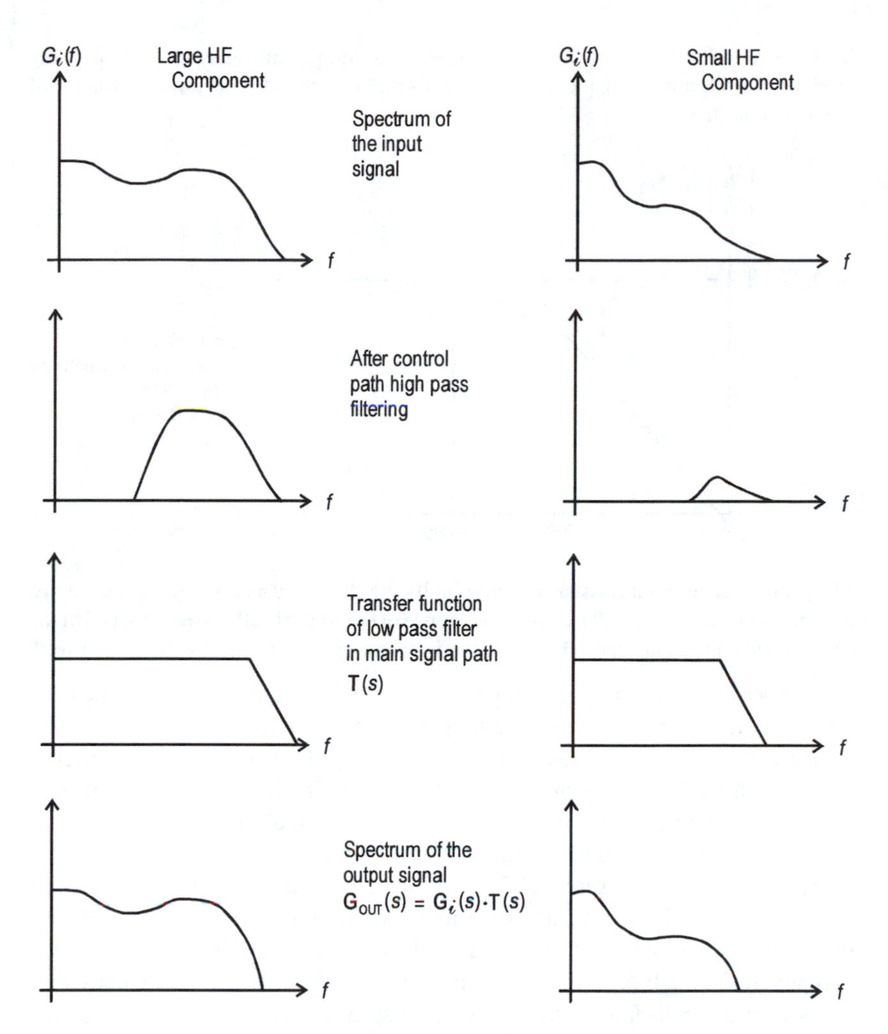

Fig. 5.31: An illustration of the principle upon which the DNR® system works.

circuits is inadequate. This can produce amplitude modulation of any midband components that are present at the same time. High-frequency tape saturation is observed because many cassette recorders have a peak in their high-end frequency response characteristic just before the response drops off sharply. Figure 5.19 illustrates this behavior.

We will not belabor here the techniques Dolby has used to solve these problems, except to say that both are complementary and thus require changes in both the compressor and the expander. The interested reader is referred to Dolby's original paper on the C system (Dolby, 1983).

Whereas listenability without a decoder was a design criterion for the B system, it was not for the C system, which was conceived of as being primarily for audiophiles. Again, the uncritical listener might be able to listen to a C-encoded tape on a recorder with B-encoding capability, but he probably would not buy the C-encoded tape in the first place, and relatively few are available.

The C system has not enjoyed the acceptance by recording companies that the B system did (and still does). Virtually all cassette recorders with C-decoding capability can also decode B with the flick of a switch.

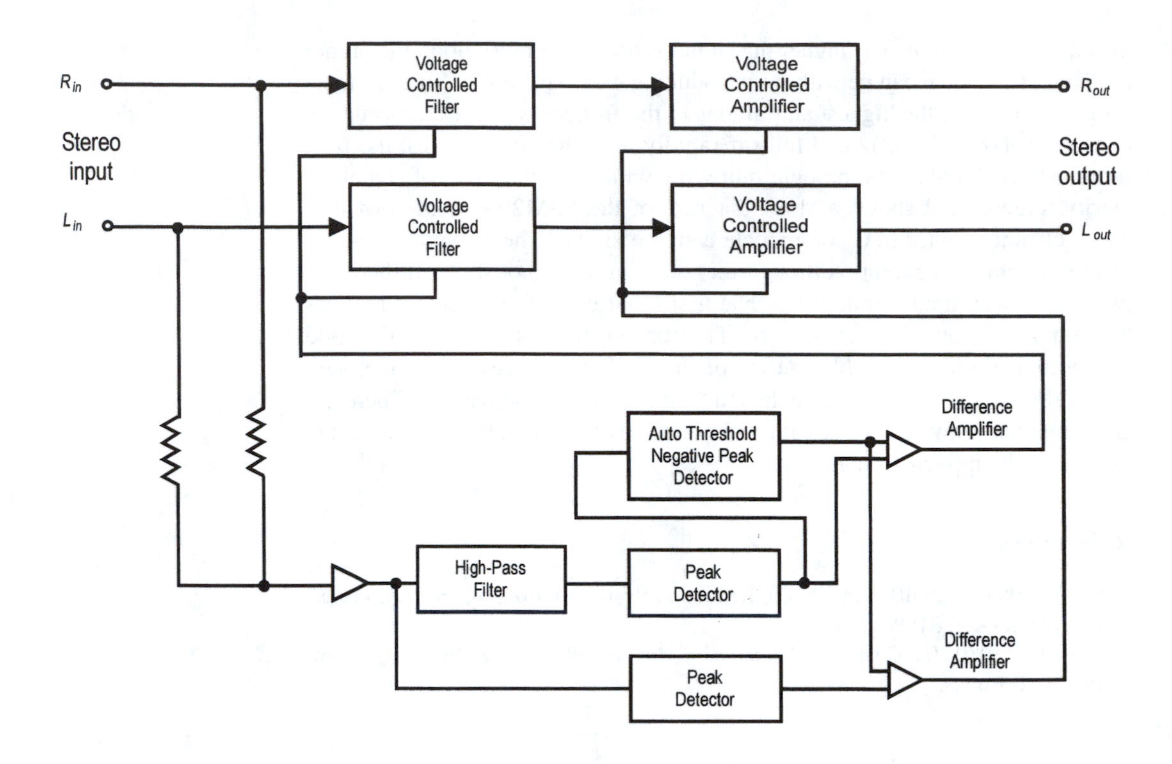

Fig. 5.32: A block diagram of the HUSH® noise-reduction system as implemented in the Analog Devices SSM2000 integrated circuit.

5.5.4 HUSH® noise-reduction system

Talk about leapfrogging technology! Dolby® C is capable of a 20 dB reduction in noise with a preencoded source. The HUSH® system can provide a 25 dB reduction *without* preencoding. This system was announced in 1996 by Analog Devices as implemented in their SSM2000 IC. The system incorporates two noise-reduction principles. The first principle involves the use of a voltage-controlled amplifier. This block can be used to reduce noise by cutting the gain when the input falls below a certain level. It essentially assumes that there is a certain threshold below which anything coming in is noise. The other noise-reduction principle involves the use of a voltage-controlled filter (VCF). It determines an upper frequency limit for the signal and "assumes" that everything above that frequency is noise. The pole frequency of the filter is set so as to eliminate frequencies above this limit.

Both of these principles have been used elsewhere. The genius of the HUSH® system, however, is that it adaptively modifies both the VCA threshold level and the VCF frequency limit. You may wonder about the basis for the continual updating of these two factors. I'm glad you asked. The gain setting is adjusted on the basis that all program material contains brief pauses during which the ONLY component is noise. Most of this noise is in the 3–8 kHz region of the spectrum. A negative peak detector looks at this portion of the spectrum and generates an adaptive threshold voltage that reflects this "noise floor." This noise floor is compared to the total input signal to determine a gain setting for the VCA.

The frequency limit of the VCF is controlled by a difference signal formed by comparing the noise floor to the total signal *in the same band over which the noise is measured.* The higher the signal is above the noise floor, the higher the pole of the VCF can be without introducing excessive noise. Since in most program material, the highest amplitudes in the frequency spectrum occur between 100 Hz and 1 kHz and fall off rapidly at higher frequencies, the pole frequency can actually be brought quite low without a great loss of signal integrity. Figure 5.32 shows a block diagram of the SSM2000. Note that it is a two-channel device to accommodate a stereo signal. The elegant simplicity of the system is apparent. Although this diagram does not show it, there are two logic-level inputs on the chip. The first input is a mute input that reduces the outputs of both channels to zero. The second input is a particularly good psychological ploy: It enables defeat of the HUSH® processing so the user, by toggling the switch, can easily hear a "with/without" comparison. There are also pins that allow the VCA gain and the noise-floor signal to be overridden by externally applied voltages.

References

Dolby, R. 1983. A 20 dB Audio Noise Reduction System. *Journal of the Audio Engineering Society*, 31: 98–113.

Jorgensen, F. 1980. *The Complete Handbook of Magnetic Recording*. Blue Ridge Summit: TAB Books.

6

Monochrome TV

Monochrome or black & white (B/W) TV signals are transmitted by very few stations now. It is still well worth our while, however, to understand monochrome TV. There are several reasons for this:

(1) B/W TV circuitry is a subset of color TV circuitry. If we understand it, we have a 100%-usable foundation for understanding color.

(2) Although B/W transmission is rare, B/W receivers are still manufactured. It is with such receivers that we concern ourselves in this chapter. In Section 6.4 we analyze one of the VLSI chips that makes such receivers so economical. Color signals are compatible with B/W receivers because the receiver circuitry simply ignores the color information.

(3) The mechanisms of picture (video) signal generation and picture formation are much simpler in monochrome sets.

6.1 The nature of monochrome TV signals

6.1.1 The television principle

The picture to be transmitted is focused onto a small photosensitive area. One very small area within the photosensitive device called a pixel (picture element) is probed by a finely focused electron beam, which generates a signal dependent on the light intensity at the point of impact. The exact mechanisms involved in this process are well covered in many books for a variety of imaging devices (e.g., Neuhauser and Cope, 1992). In solid-state imaging arrays, the pixel is addressed digitally rather than by means of an electron beam.

In either case, the end result is the same. The output voltage associated with each pixel is a single-valued function of the light intensity at that pixel. Having determined the light level at that pixel, another pixel – usually an adjacent one – is checked to determine its light level. This process continues over the whole screen until every pixel on it has been evaluated. Then the process starts all over again. The sequence of analog voltage levels corresponding to the point-by-point illumination levels is called the *video* signal. It should be obvious that we must know the light level at every pixel, in order to reproduce the picture. The video signal contains just that information and yet, by itself, it is useless. A little thought will show why this is so. If we have the video signal, we could use its value at a given moment to determine the light level at a pixel on the display device – but *which* pixel? Only if a 1:1 correspondence exists between pixel locations at the camera and at the TV receiver can the latter produce a picture that is a replica of the former.

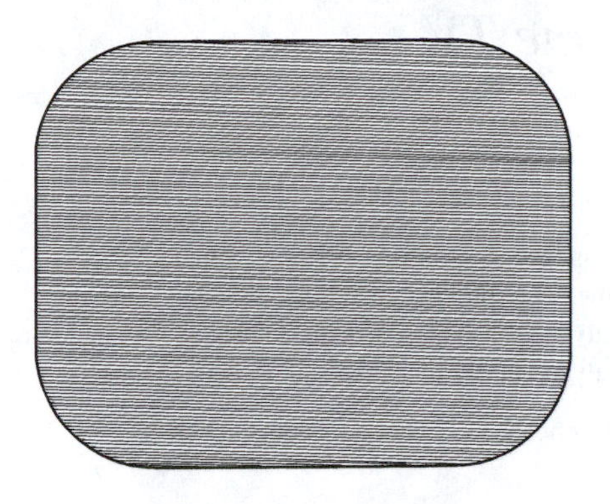

Fig. 6.1: A television raster scan.

Fig. 6.2: A television picture that is not synchronized to the transmitter.

In practice, two measures are necessary to achieve the required correspondence. The first is the practice of *scanning*. Scanning is the establishment of a systematic pattern of pixel reading/writing. The standard scan proceeds from left to right and (much more slowly) from top to bottom at the same time. The resulting scan pattern is shown in Fig. 6.1. On a monochrome set, these lines are easily visible to the naked eye at close range. The standard scan consists of 262.5 lines. After completing a scan, the beam returns to the top of the screen and places another line between each of the previous ones. This practice is called *interlacing*. It gives better resolution than a 262.5 line scan and less picture flicker than a 525 line scan. The 525 lines scanned in two trips over the screen are called a *frame*. The line scan rate is 15,750 Hz. The frame scan rate is 30 Hz. Thus each scan of 262.5 lines takes $1/60$ second.

Yet even the establishment of a standard scan pattern at the transmitter and receiver is not adequate to reproduce the picture. Figure 6.2 shows a received picture when the receiver scan is not *synchronized* to the transmitter scan.

Exercise 6.1 In Fig. 6.2, which is being scanned faster, the receiver or the transmitter? State your reasoning.

The final requirement for picture reproduction is thus scan synchronizing signals (SYNC). These are generated by the transmitter scan circuits and are sent along with the video to the receiver. One of the functions of the receiver is to extract these sync signals and use them to synchronize the receiver scan.

Proper synchronization will cause the beam to hit at the right location on the CRT face at the right time and to produce a brightness proportional to the electron beam current, which is in turn determined by the video signal.

The electron beam produces light when it hits the phosphor on the inside of the CRT faceplate. The light from the phosphor begins to decay exponentially

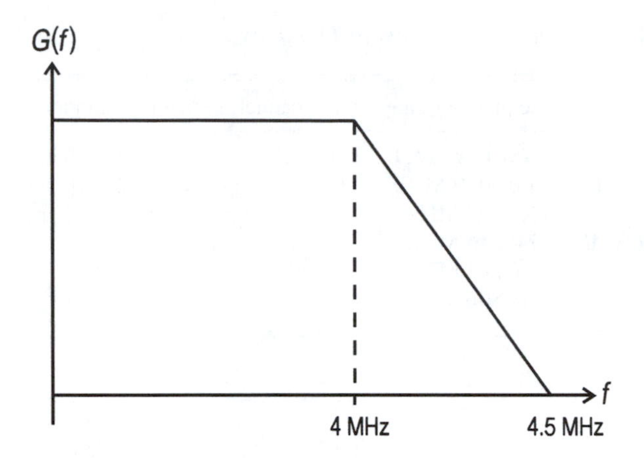

Fig. 6.3: The frequency spectrum of a broadcast-band video signal.

after the beam passes over it. The longer the light-decay time constant, the greater the *persistence* of the phosphor. The phosphor used for monochrome TV has a persistence long enough for each pixel to be rescanned before the light output decays enough to cause visible flicker.

In the *aluminized* CRT, a thin coat of aluminum is evaporated over the back of the phosphor. This reflects light from the phosphor back out the front of the CRT and thus appreciably improves picture brightness.

All that has been discussed so far only suffices to reproduce the picture. The composite TV signal received at the antenna also includes sound. Thus the three signal components at the B/W receiver are: video, sync, and audio.

6.1.2 The monochrome picture baseband signal in the frequency domain

The video is a broad-spectrum signal ranging from DC up to about 4 MHz. Its spectral representation is shown in Fig. 6.3. Generally speaking, the higher frequencies correspond to edges and small objects. The sync actually rides on top of the video as we shall see in the time-domain representation.

6.1.2.1 The video carrier and VSB modulation

The video signal is amplitude modulated (AM) onto the video carrier. The AM signal is then passed through a *vestigial sideband filter*, whose purpose is to pass almost all of the upper sideband and just a little bit, or a vestige, of the lower sideband. This method significantly reduces the bandwidth required to transmit a TV signal. Figure 6.4a shows the output spectrum of the AM modulator. Note that it is the same as that of Fig. 6.3 except that it has been frequency-translated up to a center frequency of f_c, the video carrier frequency. This output is bandlimited to about 4 MHz – not only because this is the bandwidth needed for broadcast-quality video transmission, but also because the video signal must be down to nearly zero at a frequency 4.5 MHz away from the carrier so that it does not interfere with the sound signal.

This spectrum, after being passed through the vestigial sideband filter, emerges as shown in Fig. 6.4b. It should be observed that the entire TV signal thus fits within a bandwidth of 5.75 MHz. In practice, a 6 MHz channel width is

Table 6.1. *Frequency allocations of TV bands*

Name of band	Frequency range	# of channels in band	Channel designations
CATV	5.75–47.75 MHz	7	T7–T13
Low-band VHF	54.0–88.0 MHz	5	2–6
CATV	120–174 MHz	9	A–I
High-band VHF	174–216 MHz	7	7–13
CATV	216–283 MHz	11	J–T
UHF	470–806 MHz	56	14–69

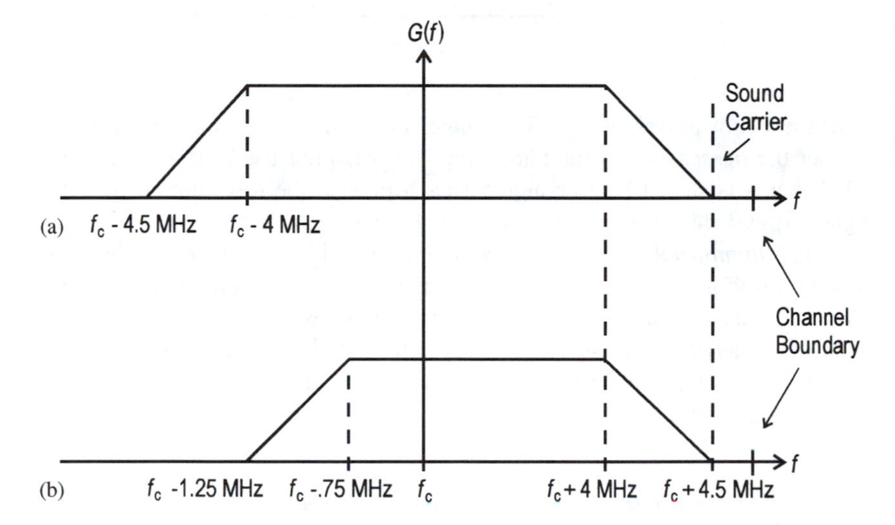

Fig. 6.4: A broadcast-band television signal (a) as it emerges from the AM modulator and (b) as it emerges from the vestigial-sideband filter.

used, with the extra .25 MHz providing more than enough space for the upper sidebands of the sound signal and allowing some space to the next channel.

6.1.2.2 Frequency allocation

The 6 MHz of channel width needed to transmit a TV signal may, in theory, be located anywhere in the frequency spectrum. In practice, certain bands have been allocated for TV by the FCC. These are shown in Table 6.1. CATV means community antenna TV and refers to pay TV stations. It can be verified readily from Table 6.1 that there is at least 6 MHz of channel capacity in every band for each station in that band. Two of the bands have a frequency gap that is not used for TV. Table 6.2 shows a further breakdown of Table 6.1 for VHF.

In every case, as shown in Fig. 6.4b, the video carrier lies 1.25 MHz above the lower edge of the 6 MHz channel.

6.1.2.3 The sound carrier and frequency modulation

The sound transmitter is completely separate from the picture transmitter. In most cases they will share a common antenna. The audio-frequency baseband signal is frequency modulated onto a carrier that is exactly 4.5 MHz above

Table 6.2. *VHF channel allocations*

Channel	Band (MHz)
2	54–60
3	60–66
4	66–72
5	76–82
6	82–88
7	174–180
8	180–186
9	186–192
10	192–198
11	198–204
12	204–210
13	210–216

that of the picture. The allowable bandwidth of the modulating audio and the pre-emphasis characteristic are the same as for broadcast FM, but the allowable carrier-frequency deviation is only 25 kHz, rather than 75 kHz. Furthermore, we shall see that the TV sound IF frequency is 4.5 MHz as compared to the 10.7 MHz value of broadcast FM.

The transmitted sound (*aural*) power must be between 50% and 150% of the transmitted monochrome picture (*visual*) power. Thus the sound spectrum, if it were added to Fig. 6.4b, would have significant amplitude but would appear as a single vertical line, since the allowable frequency deviation of ± 25 kHz would make the sound signal bandwidth less than 1% of the TV channel width.

6.1.3 Time domain representation of the video and sync signals

In the American TV system, the visual signal is said to have a negative polarity. This means that a darker pixel yields a higher video signal and therefore a greater radiated power as it is transmitted.

In the diagram and discussion to follow we will use the notation H for the horizontal scan time $= 1/15{,}750$ Hz $= 63.5$ μsec and V for the vertical scan time $= 1/60$ Hz $= 16.7$ msec. Because of interlace, two successive scans (called the *even* and *odd fields*) will not be identical, even if the video information is. The details of the interlace will not be considered here.

A time segment of the composite video signal is shown in Fig. 6.5, which is drawn to scale on both the voltage and time axes. This figure is adapted from Fink (1957a). We will discuss this time-domain representation starting from the left. There we see horizontal-blanking pulses with horizontal sync on top of them. The horizontal blanking time is the time during which the beam does a rapid retrace from the right of the TV screen back to the left after each horizontal scan. Since these pulses extend up to the black level, the screen will be black for their duration. Ordinarily this black is never seen, but it can be seen in Fig. 6.2 because of the horizontal instability. We note that the width of the blanking pulse is approximately .16H. This correlates very well with the ratio

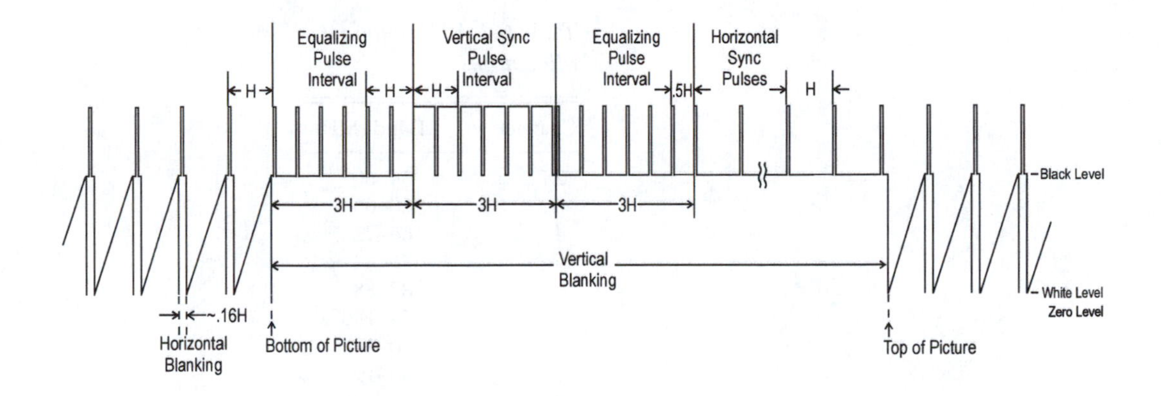

Fig. 6.5: A portion of a broadcast-band television signal in the time domain.

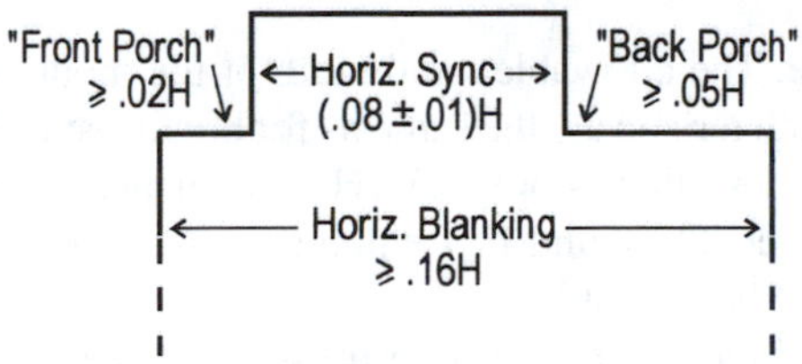

"Front Porch" ⩾ .02H

← Horiz. Sync → (.08 ± .01)H

"Back Porch" ⩾ .05H

Horiz. Blanking ⩾ .16H

Fig. 6.6: The time relationships between the horizontal blanking and sync pulses in the time domain. This is a detail of a portion of Fig. 6.5.

of blanking to picture that can be measured by laying a ruler perpendicular to the blanking bars on Fig. 6.2.

Although the details of the time relationship between the horizontal sync and blanking pulses are not important for our consideration of monochrome, they will be for color, and hence are shown in Fig. 6.6. Since the sync is entirely above the black level, it never shows on the picture. The slanting lines shown between the horizontal-blanking pulses are the video information. They have been drawn to correspond to a particularly trivial screen "picture."

Exercise 6.2 What would the screen pattern look like if the video signal were as shown for the entire scan?

When the beam reaches the bottom of the picture, the vertical-blanking period begins as does an equalizing-pulse interval. The vertical blanking is exactly analogous to the horizontal blanking. The vertical-blanking bar is seen separating frames whenever the TV picture "rolls" vertically. The equalizing pulses have a dual function: to maintain horizontal synchronization even in the absence of video information, so that it does not have to be reacquired later, and to equalize the energy content of the sync pulses during the vertical-blanking periods of the two interlaced scans. Since a single scan contains 262.5 lines, the equalizing pulses must have a repetition rate twice that of the horizontal sync in order to perform their function for both the even and odd fields.

The next waveform encountered during the vertical-blanking period is the vertical-sync pulse, whose width may be seen from Fig. 6.5 to be 3H. The negative-going pulses during this sync interval are called *serrations*. Their

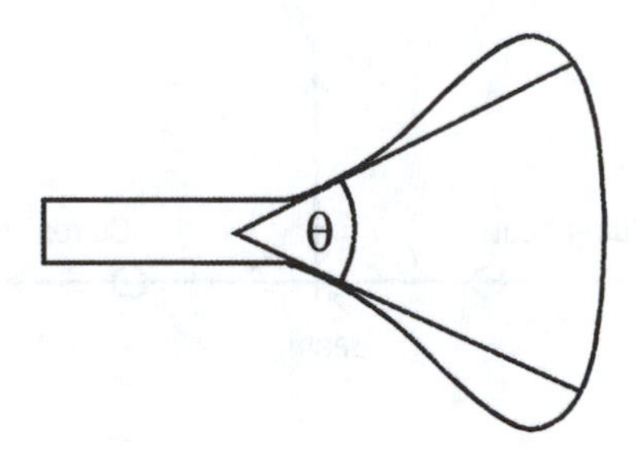

Fig. 6.7: Illustration of how CRT geometry limits the maximum possible deflection angle.

purpose is also to maintain horizontal synchronization during the much-wider vertical-sync pulse. This pulse is followed by more equalizing pulses and then by normally spaced horizontal-sync pulses until the vertical-blanking time (.05 to .08V) is past.

During this time, the beam has scanned through the vertical-blanking bar and is at the top of the picture. From there the visible scan is resumed with the same kind of waveform with which the previous field ended. Of course, in general, the video signal is much more complex and will be slightly different between successive pairs of horizontal-blanking pulses.

6.2 Some principles intrinsic to television

6.2.1 The sound intercarrier principle

As has been noted, the video and audio signals are derived from separate transmitters but are received by the same antenna at the viewer's location. As we shall see, they go through a heterodyning process (in which they retain their 4.5 MHz spacing) and then through an IF amplifier to an envelope detector. Since the envelope detector is a highly nonlinear device, the detection process will also generate sum and difference frequencies between the sound and video carriers. Because the video signal is AM and the sound is FM, the difference signal will be amplitude-modulated FM with a carrier frequency of 4.5 MHz. If this signal is properly limited, it should contain only sound information, because the video information is all in the form of amplitude variation. It is therefore processed through a 4.5 MHz sound–IF amplifier and limited before detection by any of the methods discussed in Chapter 2.

6.2.2 The magnetic deflection principle

Most electrical engineers are familiar with the principles of the cathode ray tube and electrostatic deflection through having studied the oscilloscope. In the cathode ray tubes used for TV, there are two main differences. The phosphor used for monochrome TV must be white and have a longer persistence than that used for an oscilloscope. The longer persistence reduces the flicker during the frame scan. The second major difference is that deflection in a TV is magnetic rather than electrostatic. This is necessitated by the much-larger angles of deflection characteristic of TV picture tubes. Magnetic deflection has smaller aberrations than electrostatic deflection for large deflection angles. The deflection angle concept is made more quantitative in Fig. 6.7

The angle θ is called the deflection angle of the picture tube. Early picture tubes were 70° deflection tubes. For those tubes, the deflection angle was small

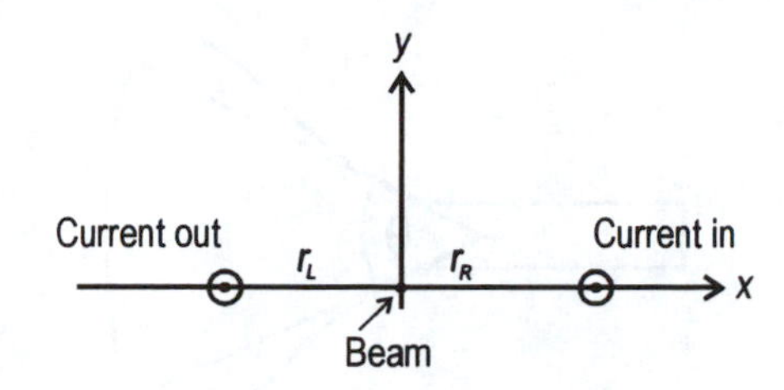

Fig. 6.8: Geometry for determining the magnetic field at the midline of two parallel conductors.

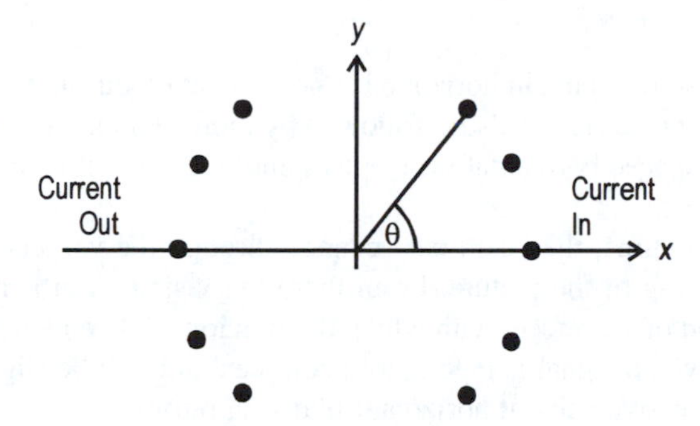

Fig. 6.9: A simplified magnetic-yoke structure composed of numerous parallel pairs of conductors. The turns density can be chosen to make the magnetic field uniform off the axis as well as on it.

enough that a fairly good picture would be obtained by using a magnetic field that was constant across the cross section of the beam from the electron gun. To see how this is achieved, consider first that we have two wires lying parallel to the electron beam and equidistant from it ($r_\mathrm{L} = r_\mathrm{R}$) on opposite sides, as shown in Fig. 6.8. Both of these currents will produce a magnetic field that points in the $+y$ direction. The total H field at any point on the x axis is

$$H = \frac{I_\mathrm{out}}{2\pi r_\mathrm{L}} + \frac{I_\mathrm{in}}{2\pi r_\mathrm{R}}.$$

If $I_\mathrm{out} = I_\mathrm{in} \equiv I$ then

$$H = \frac{I}{2\pi} \cdot \left(\frac{1}{r_\mathrm{L}} + \frac{1}{r_\mathrm{R}}\right).$$

Let $s = r_\mathrm{L} + r_\mathrm{R}$ and $\delta = (r_\mathrm{R} - r_\mathrm{L})/2$. The field on the x axis is then

$$H = (I/2\pi s) \cdot \frac{1}{\left(\frac{1}{4} - \frac{\delta^2}{s^2}\right)},$$

which says that if $|\delta| \ll s$, H is virtually independent of δ. This condition is ordinarily satisfied in practice. The situation is not so favorable off the x axis, however. This suggests placing several conductors on each side of the beam, both above and below the x axis, as shown in Fig. 6.9. It can be shown that if the turns density (at a constant radius from the beam) varies as the cosine of the angle θ, the uniform field is achieved (Schlesinger, 1957b). In this case, the field will be in the $+y$ direction.

An electron going into the plane of the paper and passing through this uniform magnetic field will be deflected in the $+x$ direction. If another pair of coils is

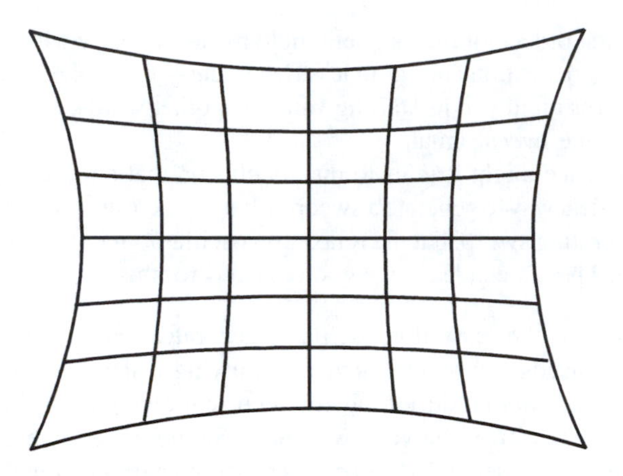

Fig. 6.10: An exaggerated representation of the effect of pincushion distortion. Without distortion, the pattern should be a rectangular grid.

placed at right angles to the first, it can provide $+y$ deflection. In a TV, both sets of coils are enclosed in an assembly called a *deflection yoke* or simply *yoke*. This assembly is so named because it slides on over the neck of the picture tube. The outer surface of the windings is covered with a ferromagnetic shield that serves the triple purpose of (1) providing a low reluctance magnetic return path (thus intensifying the interior field for given excitation), (2) shielding the yoke so that it does not radiate to other points in the TV, and (3) reducing the shock hazard from the yoke windings. The high-frequency components are large enough to give a memorable shock by capacitive coupling from even touching the enamel-insulated winding wires.

There are second-order effects that make a perfectly uniform deflection field not quite ideal. Probably the most serious of these is that portions of the picture tube face that are furthest from the center are also further away from the deflection center near the middle of the yoke. This has the effect of causing a deflection greater than the theoretical one at the points furthest from the center of the picture tube. A rectangular grid imaged by a deflecting system with this characteristic appears in Fig. 6.10 in exaggerated form. This effect is called *pincushion distortion*. It can be eliminated by making the deflecting field produced by the yoke nonuniform in a predetermined manner.

The effort to reduce the size of TV had, as a major component, the shortening of the picture tube from front to back. This could only be accomplished by increasing the deflection angle. After the 70° tube, the 90° tube became prevalent, and 110° tubes are now standard. With each increase in deflection angle, the requirements on the yoke become more stringent, and the field distribution producing the most linear scan pattern becomes less uniform. Whereas yokes for 70° tubes were almost cylindrical, 110° yokes have a larger diameter-to-length ratio and have windings that contour not only to the neck but also to the bell of the picture tube.

6.2.3 The sweep principle

The sweep principle also has its counterpart in the oscilloscope. The deflection of the beam whether it is accomplished electrostatically, as in the oscilloscope, or magnetically, as in the TV, must be done periodically. In the TV, where we want the deflections along both axes to vary linearly with time, we want (at least to the first order) for the magnetic field to vary linearly with time. This in turn requires a deflection current that varies linearly with time during the scan and then returns rapidly to its starting value – in other words, a sawtooth wave. This is called the *sweep* signal.

Our experience might lead us to the conclusion that a relaxation oscillator would be an ideal way to generate a sweep voltage. There are a variety of circuits used for generating sweep, but the relaxation oscillator in its simple form is not one of them. There are at least two good reasons for this:

(1) Whereas the relaxation oscillator generates a sawtooth *voltage*, the yoke needs a sawtooth *current*. Applying a sawtooth voltage to the yoke will *not* produce a sawtooth current owing to the highly inductive character of the yoke windings. Suppose we represent the yoke winding as a series LR and pass a sawtooth current through it. We then have the voltage components shown in Figs. 6.11a–d. Figure 6.11e is an actual oscilloscope waveform of this voltage and shows that our understanding is basically correct.

Of course, the problem of needing a sawtooth current could be solved by passing the output of a relaxation oscillator through a transconductance amplifier if this were the only problem – but it's not.

(2) It is a fundamental requirement of sweep oscillators that they be synchronizable. We have already alluded to the importance of synchronizing the receiver scan with that of the camera to produce a stable picture. Thus the sweep currents that generate the scan must be synchronized or *synced*. This requires an oscillator whose free-running frequency is near the desired value and is variable over a moderate range to achieve synchronization. This sounds a lot like a job for a PLL, and indeed in most modern receivers, PLL-based circuits can be found. Nonetheless we shall see that a number of other good approaches are available to do the job as well.

6.2.4 The flyback principle

The picture tube has an electron-accelerating potential generally between 8 and 20 kV for a monochrome set. This high voltage must be generated somewhere within the set. The universal practice is to feed the horizontal sweep signal not only to the horizontal yoke but also to a step-up transformer called the *flyback transformer* or just *flyback*. Fortunately the current demand is fairly small (50–500 μA), so the wire used in the transformer windings can be of a very light gauge. The turns ratio is typically under 20, which means that the input to the primary may well be in the low kilovolt range itself. Where does this voltage come from? It comes from the collapse of the flyback's magnetic field during the retrace time. The transformer is called a flyback because the high-voltage spike is generated during the time the beam "flies back" to the left-hand edge of the screen. This principle is illustrated by the waveforms in Figs. 6.11c and 6.11e. The decay of magnetic flux induced by the sweep retrace in the primary is also coupled to the secondary and stepped up by the turns ratio. It is then rectified to supply the accelerating potential. On occasion the rectifier is of the voltage-doubling variety. Modern receivers often have the flyback incorporated into a larger module called the *IHVT* or *Integrated High-Voltage Transformer*.

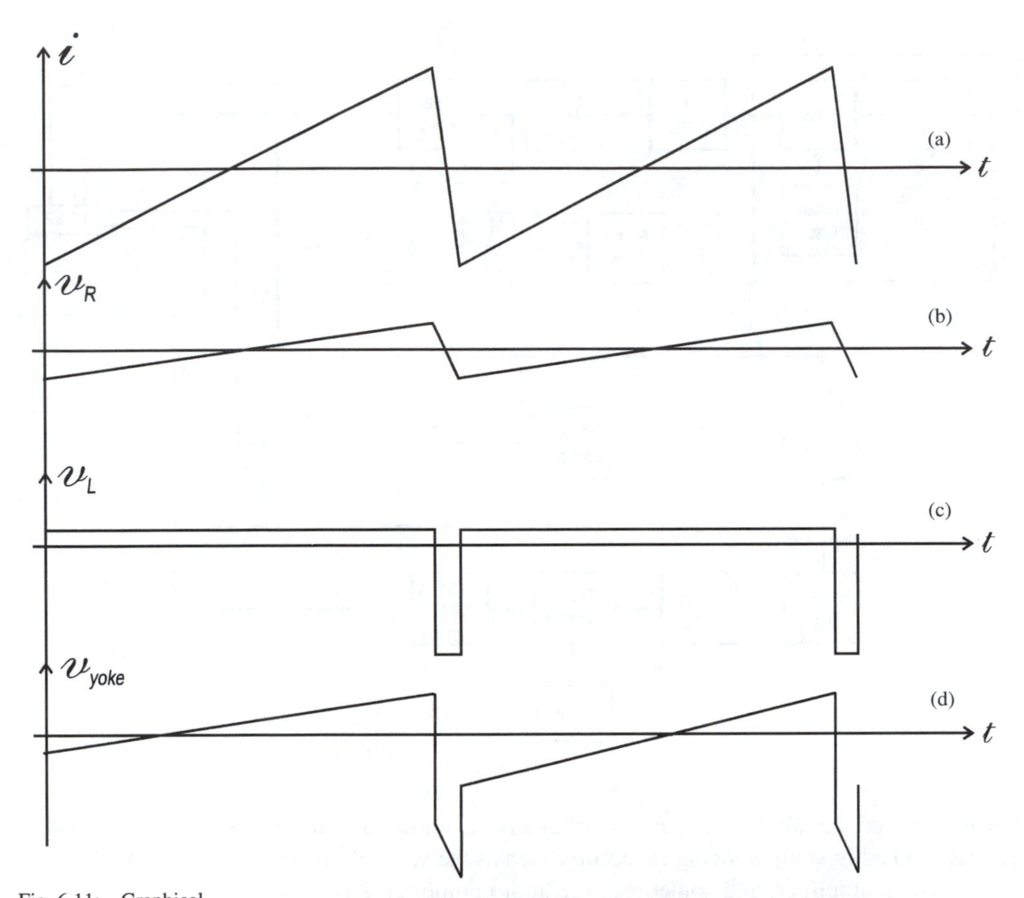

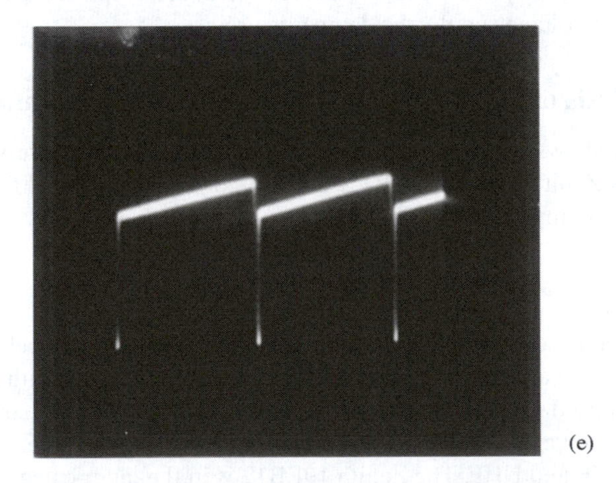

Fig. 6.11: Graphical determination of the voltage waveform across a vertical yoke. (a) The current through the yoke. (b) The component of voltage across the yoke due to its resistance. (c) The component of voltage across the yoke due to its inductance. Inductive voltage is proportional to the slope of the current. (d) The sum of the components of yoke voltage due to its resistive and inductive components. (e) Actual oscilloscope trace of the voltage across a vertical yoke.

6.2.5 *The boost/damper principle*

The damped burst of oscillation generated by the flyback has its highest-amplitude peak rectified to generate the accelerating potential. The next (negative) half cycle can still amount to several hundred volts. Not only can this peak be rectified to provide a "free" high-voltage supply, but this also drains the stored energy out of the flyback and thus helps to damp out the oscillation rapidly. For this reason, the diode used to rectify this *boost* voltage, as it

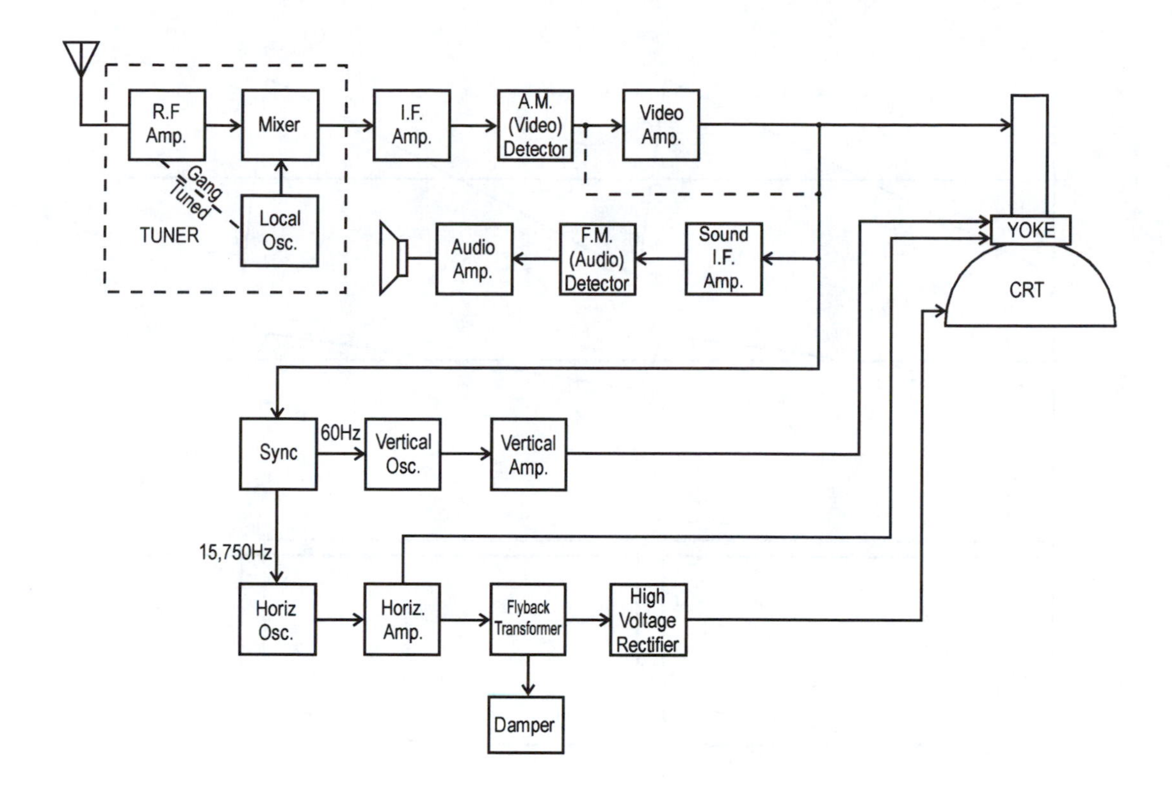

Fig. 6.12: Block diagram of a monochrome TV set.

is known, is called the *damper*. The boost voltage is generally used for those stages that need a high output-swing capability, such as the vertical sweep amplifier, the video amplifier, and sometimes the audio amplifier. Even on sets with no boost voltage supply, the damper is used.

6.3 Analysis of a B/W TV with discrete solid state circuitry

Figure 6.12 shows a block diagram of a typical black-and-white set. As we analyze this Zenith chassis 19EB12, we will correlate its circuit functions to this block diagram.

6.3.1 *The tuners*

Tuners contain the circuitry responsible for selecting one channel out of the many that fill the airwaves. In modern receivers, a single tuner with multiband capability is used to tune all channels the receiver is capable of handling. For many years, however, television sets either had a VHF-only tuner or separate tuners for VHF and UHF. The Zenith 19EB12 is in the latter category.

6.3.1.1 The VHF tuner

As indicated in Table 6.2, this tuner tunes channels 2 through 13. The schematic is shown in Fig. 6.13a. Even a cursory glance will show the familiar configuration of an RF amplifier, local oscillator, and mixer, but it will also reveal some less-familiar circuitry. This is due to antenna impedance-matching circuitry and detent channel selection. *Detent* refers to the fact that the tuner snaps into a distinctive rotational position for each channel.

The circuitry at the left-hand edge of the schematic is the *balun*. It is enclosed in dashed lines, which indicates a shielded metal container. This is a transformer that converts from a balanced TV antenna lead to an unbalanced output. The balanced line has neither conductor grounded, but the unbalanced output has one side grounded. The output is applied to the rest of the tuner circuitry via C_2, which is called a *feedthrough capacitor*. This is shown in cross section in Fig. 6.13b. It provides a means of bringing a lead through a sheet of metal as well as having a controlled capacitance from the lead to ground.

The tuner contains a 6-gang 13-position rotary switch plus two poles of *turret switching*. All switches are shown in the channel 13 position. The odd-shaped strips near each set of terminals are actually rotary contacts and should be envisioned as wrapping around on themselves so that the broken ends join together. As the channel selector is decremented, these strips rotate in the sense that amounts to a downward motion on the schematic. As an example, note that the left-hand strip of S_5 does not make any circuit connections with the channel selector on channel 13 as shown. Nor will it make connection for channel 12, which corresponds to lowering the strip one position. Successive decrementings of the channel selector will eventually result in the situation shown in Fig. 6.14. It can be confirmed that this position corresponds to channel 6. Furthermore, these two switch contacts will be connected for the next four lower channels also (i.e., 2–5). When they are connected, C_1 is switched in parallel with C_3. Recall that channels 2–6 are the low-band VHF channels, and it is precisely for these channels that the extra capacitance is switched in circuit. The entire network switched by S_4 and S_5 is a tuned input circuit for the RF amplifier. The extra capacitance switched into this network by the left-hand strip of S_5 will lower the resonant frequency of this network. The right-hand strip will switch more inductance in shunt to ground as the channel setting is lowered. Note that the five lower inductances are shown schematically with a full turn, whereas the seven upper ones are shown with a half turn each. This is not done to suggest that the seven upper inductances are all equal (or the five lower ones) but rather to show that those coils represented by half turns tune over the high VHF band and those with full turns tune over the low VHF band. This holds true only for those switch gangs that have both types of coils.

We will not trace out all of the switch contacts, as it is too easy to miss the overall picture in the detail. We do need to note, however, that if the right-hand strip of S_4 is moved one position up from that shown, the uppermost and lowermost contacts are connected together by means of the wide part of the strip, which in turn couples to the input of the RF amplifier via C_5. The upper sliding contact is fed (through C_4 and two series inductances) from the UHF–IF input. We shall return to this point shortly. Since the channel selector is shown in the channel 13 position, one position up corresponds to a space between channels 13 and 2 – the UHF position.

The RF amplifier thus receives a signal from the low VHF band, from the high VHF band, or from the UHF tuner. This stage is clearly in the common-base configuration. Its output is resonated by C_{14} in parallel with an inductance selected by S_3. This tuned output signal is coupled to the input tuned circuit of the mixer by the elements between S_3 and S_2. The mixer input is tuned by the inductance selected by S_2 in conjunction with the capacitances C_{10} and C_{19} and the input capacitance of the mixer. Both the RF and local-oscillator signals are sent to the base of the mixer. The difference frequency component of the mixer output is selected by C_{21} and T_3. The little "hook" going from the base

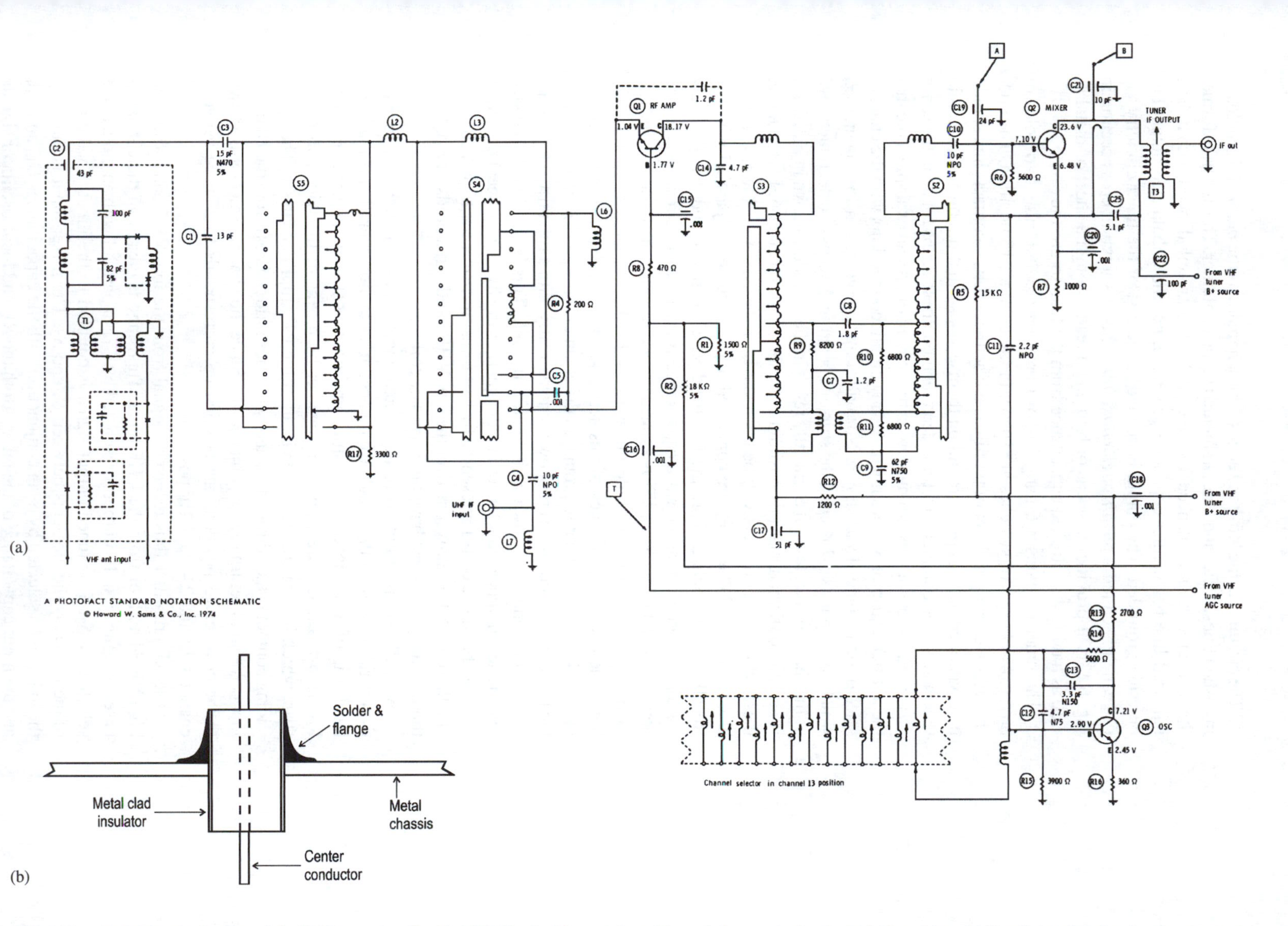

Fig. 6.13: (a) Schematic diagram of the VHF tuner in a Zenith 19EB12 television receiver. The switches are in the channel 13 position. (b) Detail drawing showing the structure of C_2.

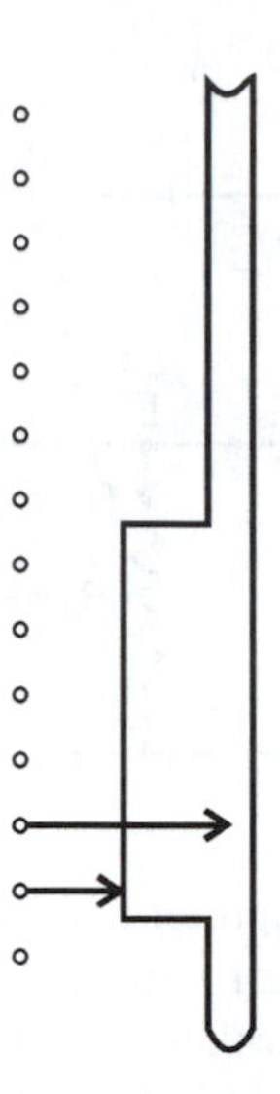

Fig. 6.14: Detail drawing of the left side of S_5 after it has been decremented down to the channel 6 position.

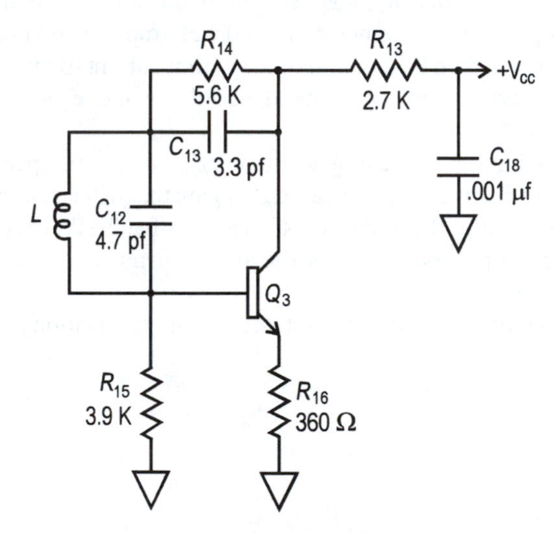

Fig. 6.15: The local oscillator circuitry of Fig. 6.13.

circuit back toward the collector is probably a "*gimmick*" for neutralization. A gimmick is a configuration that exhibits capacitance by proximity rather than by explicit use of a capacitor. Examples are a couple of short pieces of insulated wire twisted together or a short length of cable of some kind.

The analysis of the local oscillator presents some procedural difficulty. The schematic is shown in isolation in Fig. 6.15. The first difficulty is that there are so many components in the equivalent circuit as to preclude an exact analysis. The second difficulty is that the very high frequencies involved mean that reactive effects in the transistor cannot necessarily be ignored. The third and perhaps the worst difficulty is in assessing the loading effects of F on A and A on F when the circuit is viewed from a feedback perspective.

Exercise 6.3 What will be the range of frequencies over which this local oscillator must operate?

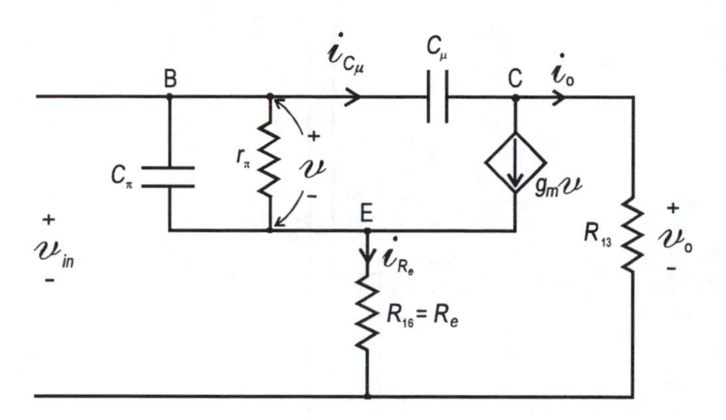

Fig. 6.16: AC equivalent circuit for the local oscillator of Fig. 6.15.

We begin by isolating the amplifier from the feedback network and using the small-signal equivalent circuit for the transistor. The amplifier's equivalent circuit is shown in Fig. 6.16. We will assume that the low frequency $\beta(= \beta_o)$ is infinite. This effectively makes r_π infinite, which drops it out of the formulation. It will also be observed that R_{15} has been omitted from this schematic. This is because the amplifier is of the negative-feedback transconductance type. The transfer function will then be the ratio of the output current to the input voltage. Since R_{15} is simply shunted across the input voltage, it does not enter into the circuit equations.

It might be thought self-defeating to assume $\beta_o \to \infty$, since part of the oscillator analysis is to find the minimum amplifier gain that will allow oscillation. In our case, however, since the amplifier is a negative-feedback transconductance amplifier, its transconductance gain is already independent of β_o to a first-order approximation.

The remainder of the circuit then satisfies the node equation at E,

$$(g_m + sC_\pi)\mathbf{V} = \mathbf{I}_{R_e},$$

the input loop equation,

$$\mathbf{V}_{in} = \mathbf{V} + \mathbf{I}_{R_e} \cdot R_e = \mathbf{V}(1 + g_m R_e + sC_\pi R_e),$$

and the output node equation,

$$g_m\mathbf{V} + \mathbf{I}_o = \mathbf{I}_{C_\mu} = (\mathbf{V}_{in} - \mathbf{I}_o Z_L)sC_\mu.$$

Solving these equations gives the ratio

$$\frac{\mathbf{I}_o}{\mathbf{V}_{in}}(s) = \frac{[sC_\mu - g_m/(1 + g_m R_e + sC_\pi R_e)]}{(1 + sC_\mu Z_L)}.$$

We will also discard the terms $sC_\pi R_e$ and 1 relative to $g_m R_e$. Discarding the latter term is a standard assumption, but the first term may require some elucidation. We are saying in effect that

$$|sC_\pi| \ll g_m = \beta_o / r_\pi$$

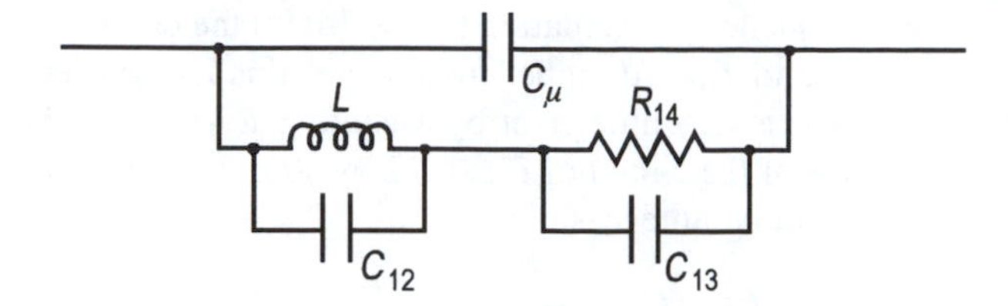

Fig. 6.17: The total impedance found between the base and collector of Q_3 in Fig. 6.13. This assumes that the base-spreading resistance is negligible.

or

$$\left|\frac{sC_\pi r_\pi}{\beta_o}\right| \ll 1.$$

We next recall that the unity-gain frequency of a transistor is

$$\omega_t = \beta_o / r_\pi C_\pi$$

so our inequality becomes

$$\left|\frac{s}{\omega_t}\right| = \frac{\omega}{\omega_t} \ll 1.$$

Thus discarding the sC_π term is equivalent to assuming that the oscillator's operating frequency is well below the unity-gain frequency. The validity of this assumption will, of course, depend on the transistor, but it is probably safe to assume that the oscillator is designed to meet this criterion, since otherwise the frequency of oscillation would depend on C_π. The transconductance gain is thus

$$\mathbf{A}_G = \frac{sC_\mu - 1/R_e}{sC_\mu R_{13} + 1}. \tag{6.1}$$

To find the voltage gain, we need to find the load resistance seen by the amplifier. This is no easy task, since it is not clear whether or not loading of A by F and F by A are negligible. Of course, if the oscillator operated at exactly the resonant frequency of the parallel L–C_{12}, and if this L–C had a high Q, the impedance of the feedback network would be large compared to R_{13} and the loading of F on the output of A could be ignored. However, C_{13} is a temperature-compensating capacitor (N150) and thus plays a role in setting the oscillator frequency. Hence the oscillator frequency is not simply the resonant frequency of the L–C_{12} combination. This in turn means that loading cannot be ignored.

We could take the approach of finding the output impedance of the amplifier in terms of the source resistance seen by the amplifier input, but this will depend on F and the output impedance of the amplifier! There is fortunately a better approach.

It may be verified from Figs. 6.15 and 6.16 that the entire feedback network is in parallel with C_μ. Thus we need only replace sC_μ in (6.1) with the admittance of the entire feedback network, and we will already have found AF. Because this method actually integrates F into A, the only other loading on the amplifier is that due to the resistor R_{13}. Accordingly, we wish to find the admittance of the network of Fig. 6.17.

At the lowest frequency of oscillation (~100 MHz) the conductance of C_{13} is over ten times that of R_{14}. At higher frequencies this factor is even greater. Thus we will introduce very little error by assuming $R_{14} \to \infty$. We can then write the admittance of the circuit of Fig. 6.17 by straightforward parallel and series combinations of admittances.

$$\mathbf{Y}_{\mathrm{F}}(s) = \frac{(1/sL + sC_{12}) \cdot sC_{13}}{1/sL + sC_{12} + sC_{13}} + sC_{\mu}$$
$$= \frac{(1 + s^2 LC_{12}) \cdot sC_{13}}{1 + s^2 L(C_{12} + C_{13})} + sC_{\mu}$$
$$= \frac{s(C_{\mu} + C_{13}) \cdot \left[1 + s^2 L\left(C_{12} + \frac{C_{\mu}C_{13}}{C_{\mu}+C_{13}}\right)\right]}{1 + s^2 L(C_{12} + C_{13})}.$$

This expression, when inserted into (6.1) in place of sC_{μ}, gives

$$\mathbf{A}_{\mathrm{v}}\mathbf{F} = \frac{\frac{s(C_{\mu}+C_{13})\left[1+s^2 L\left(C_{12}+\frac{C_{\mu}C_{13}}{C_{\mu}+C_{13}}\right)\right]}{1+s^2 L(C_{12}+C_{13})} \cdot R_{13} - \frac{R_{13}}{R_{\mathrm{e}}}}{\frac{s(C_{\mu}+C_{13})\left[1+s^2 L\left(C_{12}+\frac{C_{\mu}C_{13}}{C_{\mu}+C_{13}}\right)\right]}{1+s^2 L(C_{12}+C_{13})} \cdot R_{13} + 1},$$

where we have also multiplied $\mathbf{A}_{\mathrm{G}}$ by R_{13}, since this provides the only load. This converts $\mathbf{A}_{\mathrm{G}}$ to a voltage gain, $\mathbf{A}_{\mathrm{v}}$. We then introduce the notation

$$C_{\mathrm{s}} \equiv \frac{C_{\mu}C_{13}}{C_{\mu} + C_{13}} \quad \text{and} \quad A_{\mathrm{vo}} \equiv -R_{13}/R_{\mathrm{e}},$$

where A_{vo} is the gain of the base amplifier. We next convert to complex notation:

$$\mathbf{A}_{\mathrm{v}}\mathbf{F} = \frac{j\omega(C_{\mu} + C_{13})[1 - \omega^2 L(C_{12} + C_{\mathrm{s}})]R_{13} + A_{\mathrm{vo}}[1 - \omega^2 L(C_{12} + C_{13})]}{j\omega(C_{\mu} + C_{13})[1 - \omega^2 L(C_{12} + C_{\mathrm{s}})]R_{13} + [1 - \omega^2 L(C_{12} + C_{13})]}.$$

This equation is of the form

$$\mathbf{A}_{\mathrm{v}}\mathbf{F} = \frac{aA_{\mathrm{vo}} + j\omega b}{a + j\omega b} = \frac{(aA_{\mathrm{vo}} + j\omega b)(a - j\omega b)}{a^2 + \omega^2 b^2}$$
$$= \frac{(a^2 A_{\mathrm{vo}} + \omega^2 b^2) + j\omega ba(1 - A_{\mathrm{vo}})}{a^2 + \omega^2 b^2}. \tag{6.2}$$

$\mathbf{A}_{\mathrm{v}}\mathbf{F}$ will be purely real if

$$\omega ba(1 - A_{\mathrm{vo}}) = 0.$$

Although this condition could be satisfied by $A_{\mathrm{vo}} = 1$, this solution is not physically realizable for the way the amplifier is configured. Therefore the condition for purely real loop gain is $ab = 0$, from which either a or b must be zero:

$$[1 - \omega_{\mathrm{osc}}^2 L(C_{12} + C_{13})](C_{\mu} + C_{13})[1 - \omega_{\mathrm{osc}}^2 L(C_{12} + C_{\mathrm{s}})]R_{13} = 0.$$

This equation clearly has two roots. If $a = 0$, then $\omega_{\text{osc}}^2 = 1/L(C_1 + C_2)$ and from (6.2), $A_v F = 1$. But since the condition for oscillation is that $\mathbf{A}_v\mathbf{F} \leq -1$, the loop phase is incorrect for oscillation even though the transfer function is purely real. If, however $b = 0$, then

$$\omega_{\text{osc}}^2 = \frac{1}{L(C_{12} + C_s)} = \frac{1}{L\left[C_{12} + \frac{C_\mu C_{13}}{C_\mu + C_{13}}\right]},$$

and from (6.2), $\mathbf{A}_v\mathbf{F} = A_{vo} \leq -1$. Since we defined $A_{vo} = -R_{13}/R_e$, it follows that

$$\frac{-R_{13}}{R_e} \leq -1 \quad \text{or} \quad \frac{R_{13}}{R_e} \geq 1.$$

In this particular case, $R_{13} = 7.5\, R_e$, giving a large overdrive. It is probable that only an LC oscillator can tolerate such a large overdrive and still provide a reasonably clean sinusoidal output.

This solution is physically reasonable, because it indicates that the primary factors in determining the local-oscillator frequency are C_{12} and the switch-selected inductance, L. C_μ is expected to be small compared to C_{12}. Furthermore, the relative effect of C_μ is minimized and temperature-compensated by an effective series connection with C_{13}.

The channel selector will switch in a different value of L for each setting, allowing the local oscillator to track the varying RF signal corresponding to the channel selected.

The output of this oscillator is coupled to the mixer through C_{11}, and the output of the mixer passes through a transformer tuned to the video IF frequency and thence to the IF amplifier.

6.3.1.2 The UHF tuner

The UHF tuner circuit is shown in Fig. 6.18. This tuner exhibits some of the specialized circuit construction techniques that are found in ultra-high-frequency circuitry. Perhaps the most notable of these is that the middle compartment (the mixer) is completely shielded from the RF section on the left except for a relatively small hole in the separating wall between them. The hole between the RF and mixer sections simply allows for mutual coupling between the tuning coils in each section. Coupling between the local oscillator and the mixer is obtained by allowing the anode lead of the mixer diode to run parallel to the output coil of the local oscillator for about 0.5″ and then bringing it back into the mixer compartment through the hole between these sections. The mixer output and the local oscillator signal are both found at the anode of the mixer diode, but L_6 serves as an RF choke to effectively attenuate the signal of the local oscillator (which is at a much higher frequency) and to output only the difference signal, which is then fed into the VHF tuner as previously discussed.

The local oscillator is in the common-base configuration. The feedback mechanism to the emitter is not explicitly shown, although it is probably just inductive pickup obtained by dressing R_2 near L_4.

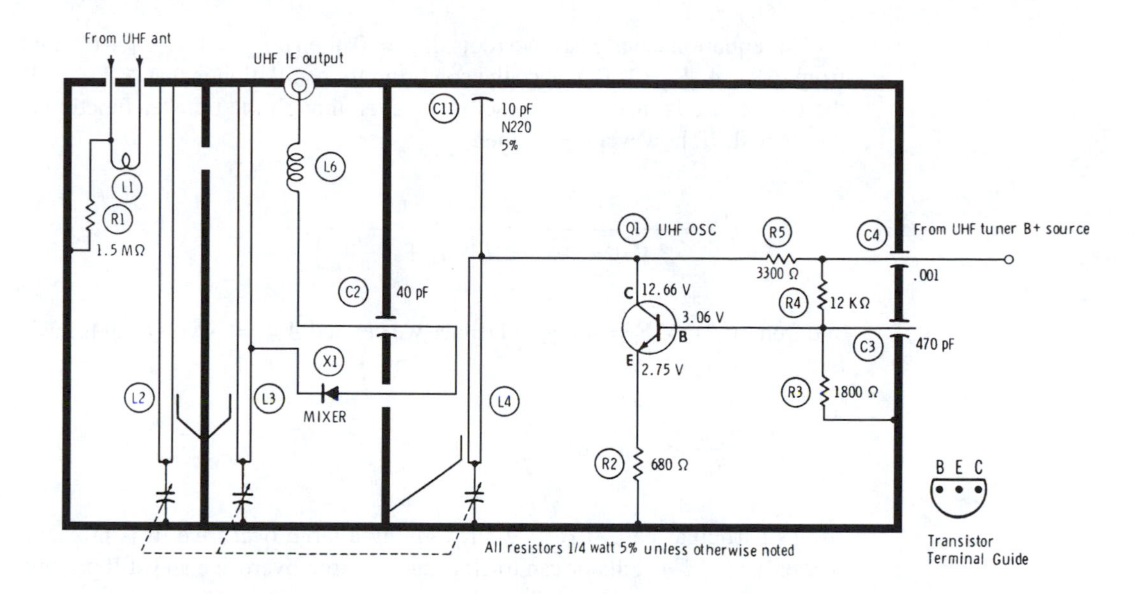

Fig. 6.18: Schematic of the UHF tuner in the Zenith 19EB12 television receiver.

6.3.2 The IF amplifier

Upon leaving the tuner, the signal is routed to the main chassis of the receiver. The schematic is shown in Fig. 6.19. As can be seen, the tuner output first comes into the IF module assembly. This receiver reflects the design philosophy (for the most part) of grouping the components responsible for a certain circuit function into modules, which are usually in the form of printed circuit boards and which can ideally be unplugged and replaced readily. In this receiver, two of the modules (1. Video Output and AGC and 2. Vertical and Horizontal) are pluggable. This can be discerned by the contact terminal identifications around the perimeter of these module schematics.

The IF signal goes through a series-resonant *LC* circuit (C_{101} and L_{101}), which is tuned by means of L_{101} to pass the difference frequency out of the mixer and block the higher-frequency components. AGC is applied to the base of Q_{101} through R_{240}, R_{101}, R_{102}, and R_{103}. R_{240} and C_{103} form a section of low-pass filtering for the AGC voltage.

This set features unusually elaborate circuitry at the input to the IF amplifier. In addition to the C_{101}–L_{101} combination already discussed, signal is also coupled in by means of the mutual inductance of L_{102A} and L_{102B}. The latter inductance is also part of a notch filter whose other components are C_{105}, C_{106}, and R_{102}. This notch filter is tuned to 47.25 MHz. As will be shown, this frequency corresponds to the sound carrier of the station below the one to which the receiver is tuned. Thus this notch filter serves to null out adjacent-channel sound interference. The series resistor, R_{102}, is also part of another notch filter, which includes C_{104} and C_{107}. This filter provides a notch at the video-carrier frequency of the station just above the one tuned to, thus trapping the other adjacent channel's video.

This is a good place to point out that the heterodyning process "frequency inverts" the spectrum of Fig. 6.4b. To see that this is true we note from Table 6.1 that channel 2 starts at 54 MHz. Its 6 MHz bandwidth means it extends up to 60 MHz. The video IF frequency is 44 MHz relative to the center of the passband. Thus if the local oscillator runs at 101 MHz, the center of the RF passband (which is at 57 MHz) is translated down to $101 - 57 = 44$ MHz.

For the same local oscillator frequency, these other frequency translations take place:

lower edge of the passband	$\rightarrow$ 47 MHz,
video carrier frequency	$\rightarrow$ 45.75 MHz,
sound carrier frequency	$\rightarrow$ 41.25 MHz,
upper edge of passband	$\rightarrow$ 41 MHz.

It is readily perceived that the lower edge of the RF passband has been translated to become the upper edge of the IF passband and conversely. Furthermore, for a different channel, the local oscillator will run at a different frequency, but the IF spectrum will be unchanged.

After passing through the first IF amplifier, the signal is applied to another filter network featuring two inductors, L_{105A} and L_{105B}. These are, no doubt, mutually coupled and adjustable. One adjustment (L_{105A}) is used to attenuate the sound carrier as much as possible; the other (L_{105B}) is used to maximize the transmission over the rest of the passband. This information can be determined from the alignment procedure given in the service literature.

The second IF stage has fixed bias set by R_{107} and R_{108}. Although the emitter circuitry is not the more familiar configuration with a split emitter resistor, only part of which is bypassed, it is clear on physical grounds that the low-frequency and midband gains will be essentially the same as if R_{111} were unbypassed and R_{109} were bypassed. A detailed calculation shows that the low-end pole and zero frequencies are also essentially equal in both cases.

The filter network coupling Q_{102} to Q_{103} is almost identical to that coupling the first two IF stages, and it is adjusted to the same criteria.

The third video IF stage is very similar to the second except that the emitter is totally bypassed over the passband, and there is a two-stage decoupler (R_{110}, C_{117}, R_{114}, C_{122}) supplying DC to the stage. The output of this stage is transformer coupled to the envelope detector/filter.

6.3.3 The envelope detector and filter

This circuit is unusually complex compared to the corresponding circuitry in other receivers. It is drawn in simplified form in Fig. 6.20. The Thevenin equivalent involves R_{422}, R_{423}, and the +24 V supply. The secondary of T_{101} and its resonating capacitor, C_{123}, have been replaced by a sinusoidal source. The detected output is fed through L_{109}, another 27 μH inductor, to the input of the video driver (Q_{404}), which is an emitter follower. Thus no significant signal current will flow through it except that L_{109} will accomplish a small amount of additional ripple filtering in conjunction with the input capacitance of Q_{404}. Figure 6.20 shows the circuit divided up into three parts to facilitate the analysis. We will start at the right and move toward the left. We will use the fact that the Thevenin voltage is an AC short to draw this part of the circuit as shown in Fig. 6.21.

Since $C_{126} \ll C_{125}$, essentially all of the signal voltage is dropped across C_{126}, and C_{125} is an AC short. The Thevenin equivalent has no purpose relative to the signal but is simply used to add a DC component to it to facilitate DC coupling to Q_{404}. Thus we can redraw Fig. 6.20 in simplified form as Fig. 6.22. The circuitry to the right of the dashed line is a filter. We will analyze it in isolation and then investigate the effect of coupling it to the envelope detector.

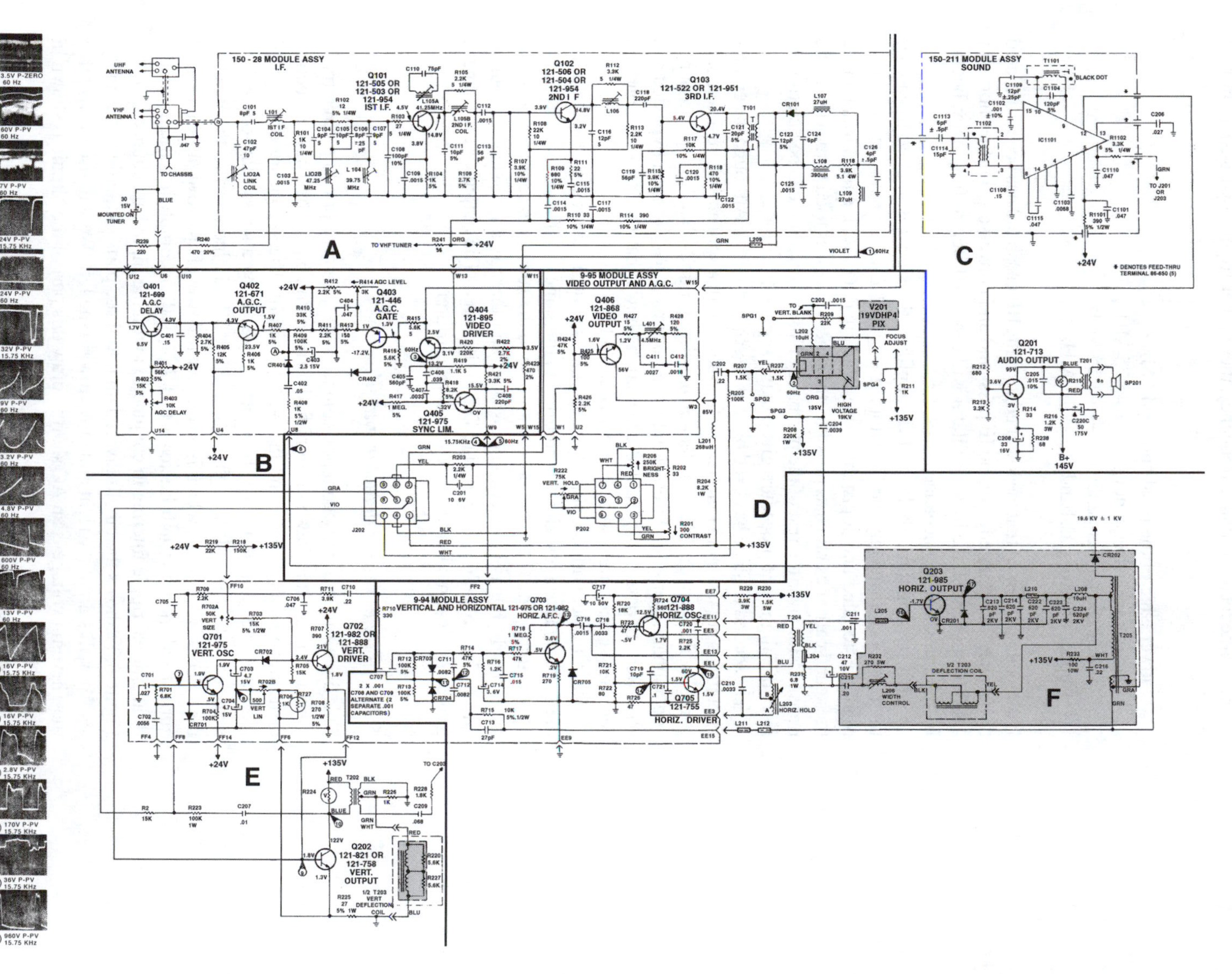

Fig. 6.19 Schematic diagram of the main chassis of the Zenith 19EB12 television receiver. An electronic version of this figure is available on the world-wide web at http://www.cup.org/titles/58/0521582075.html.

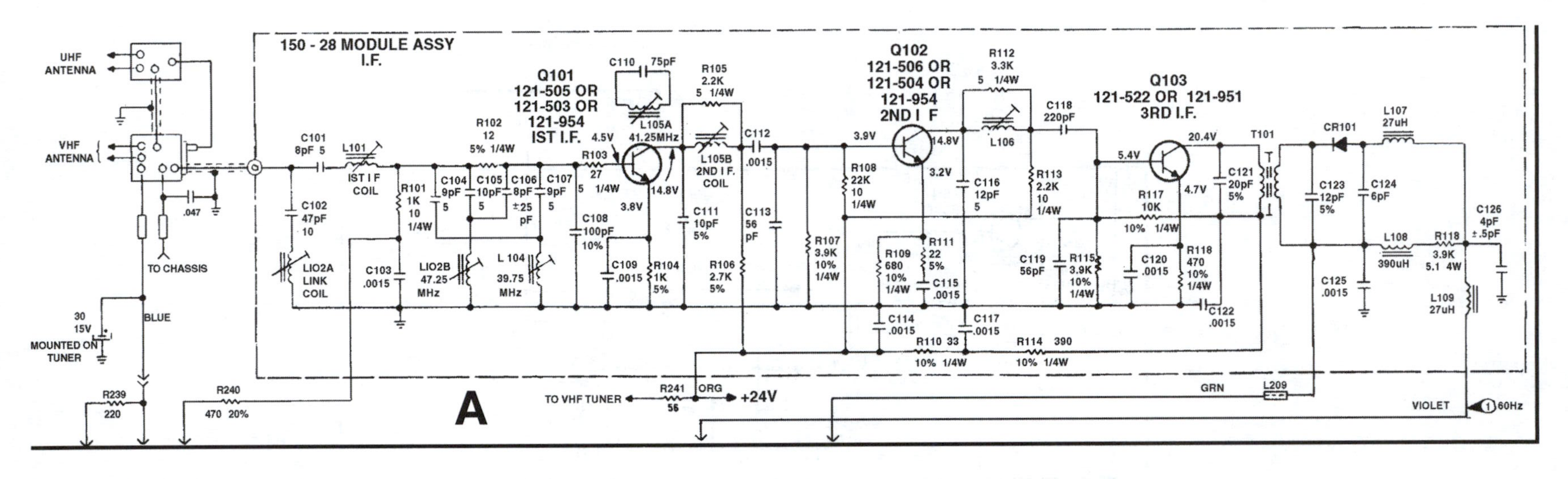

Fig. 6.19A If amplifier and video detector.

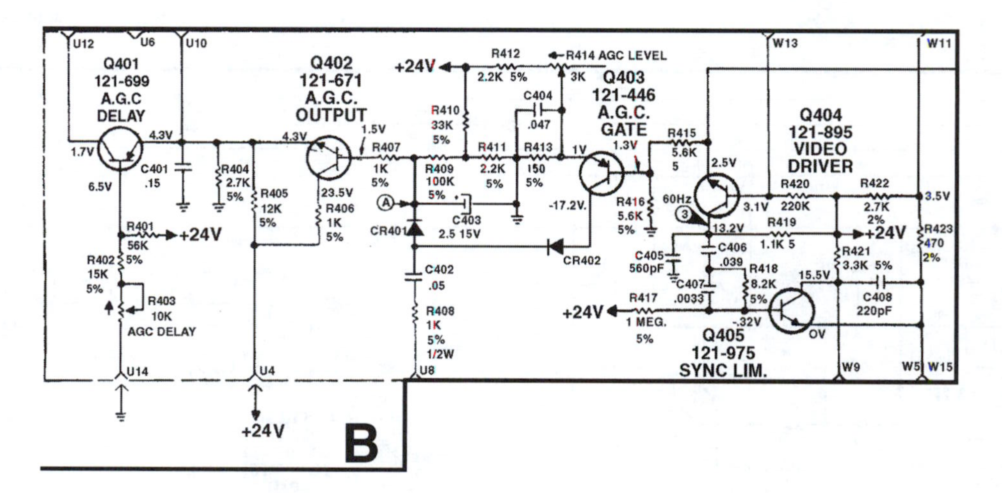

Fig. 6.19B AGC, sync and video driver.

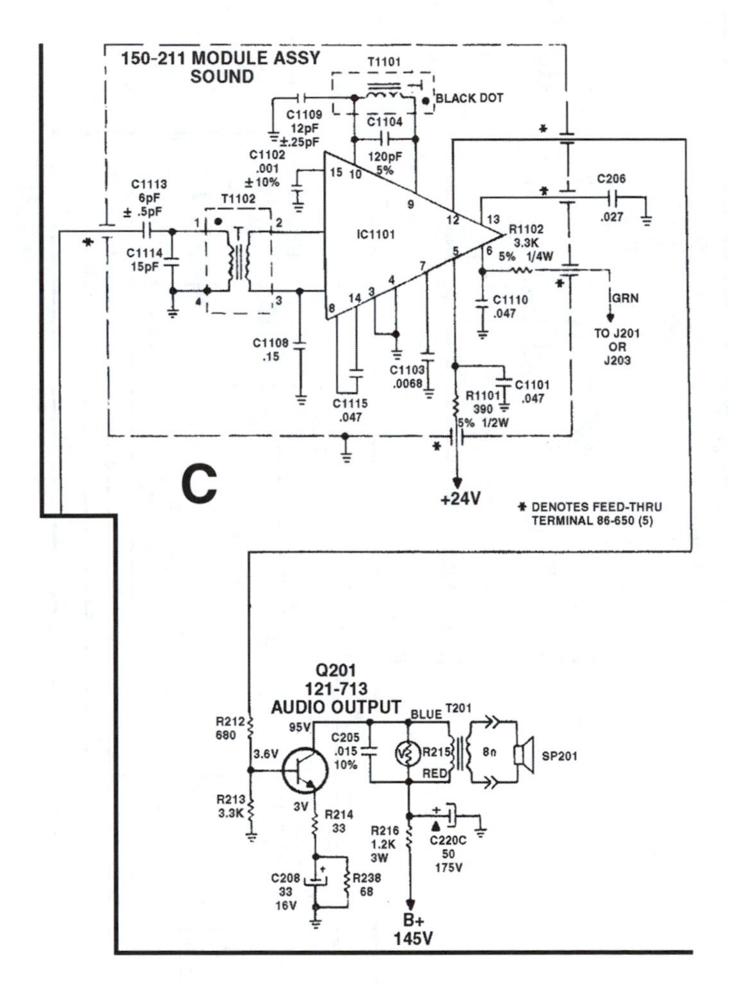

Fig. 6.19C Audio section.

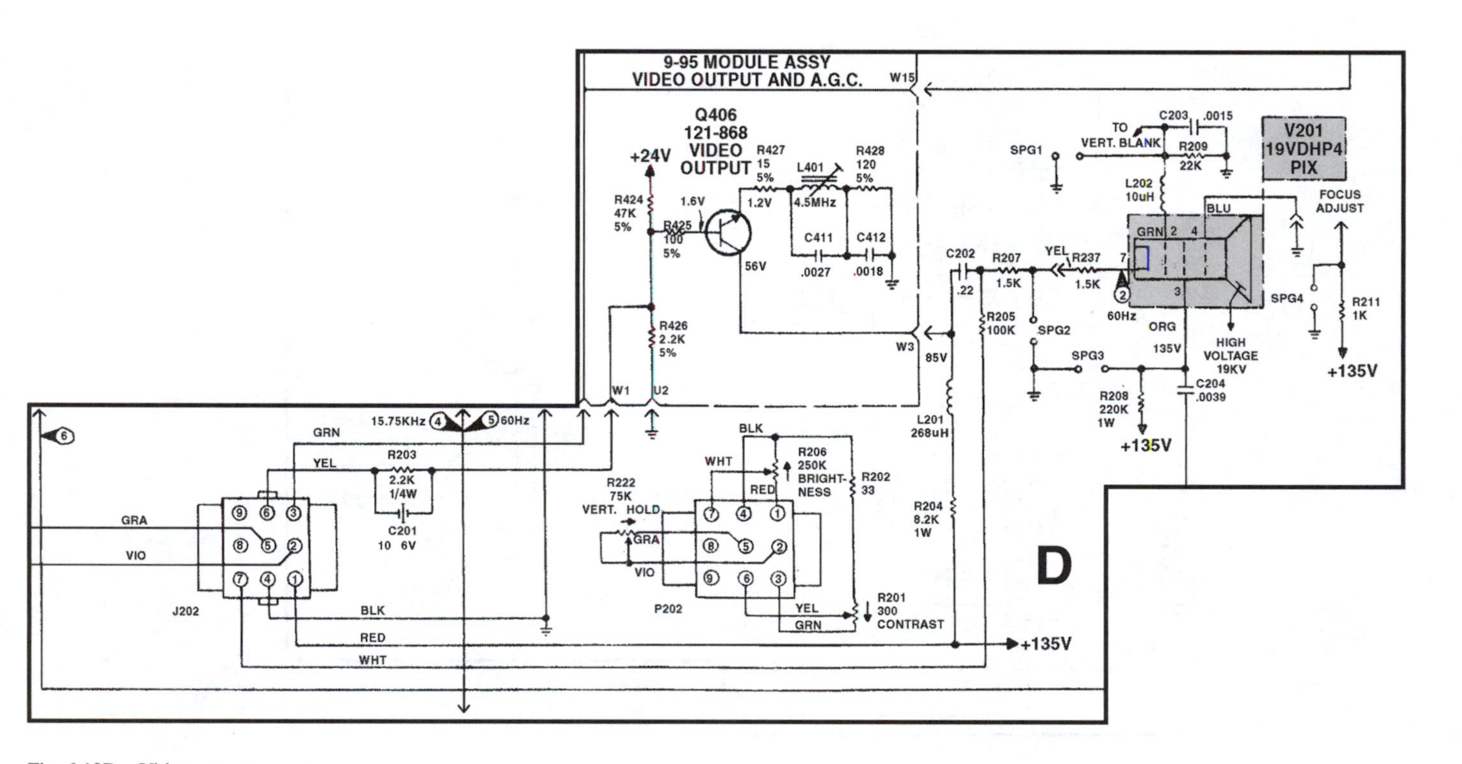

Fig. 6.19D Video output.

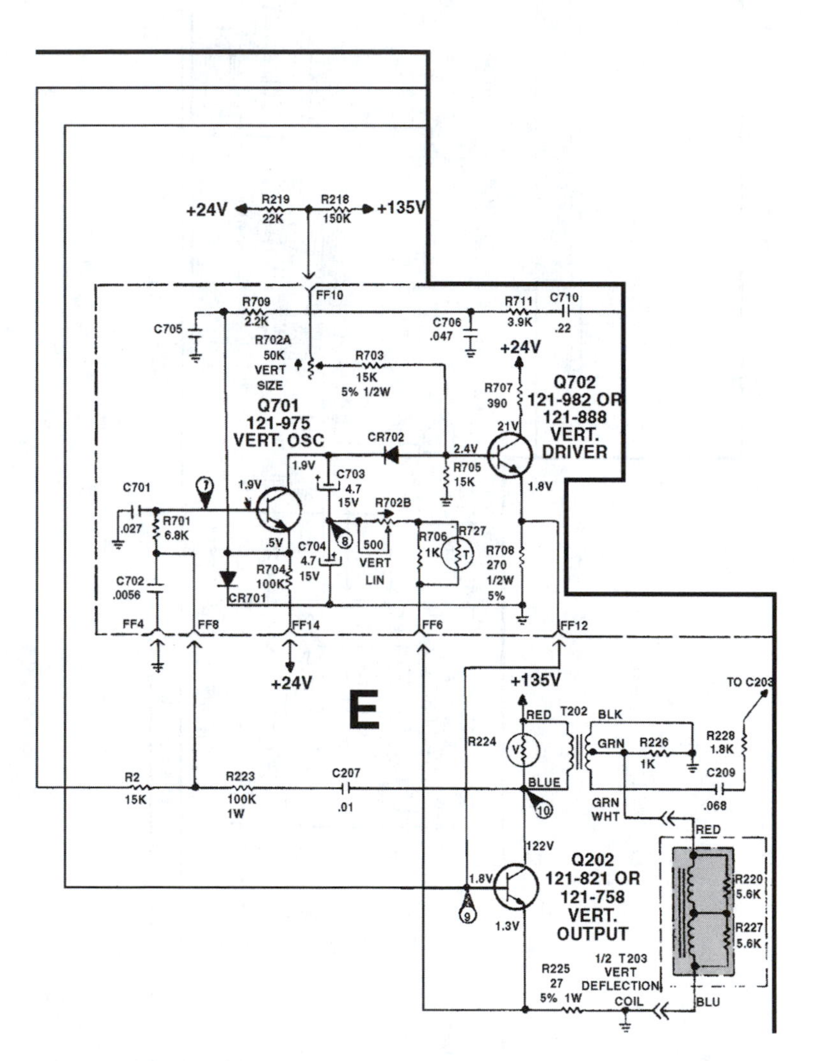

Fig. 6.19E Vertical section.

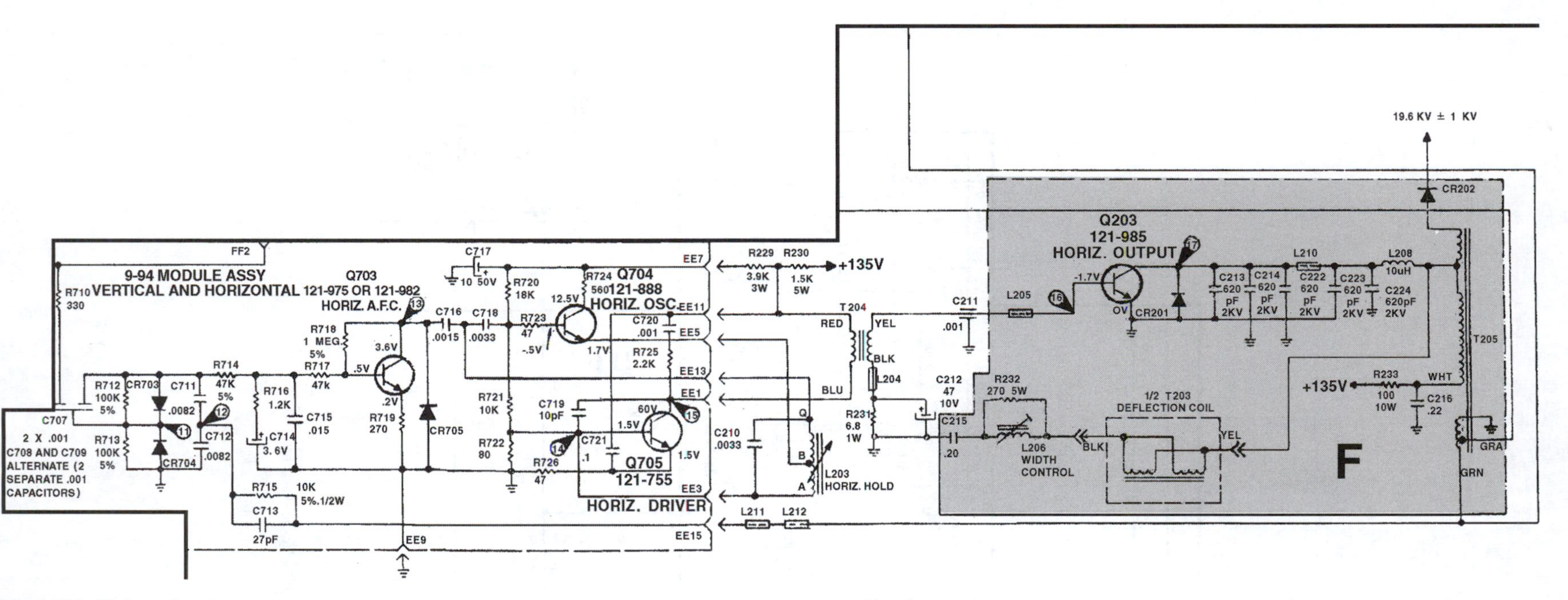

Fig. 6.19F Horizontal section.

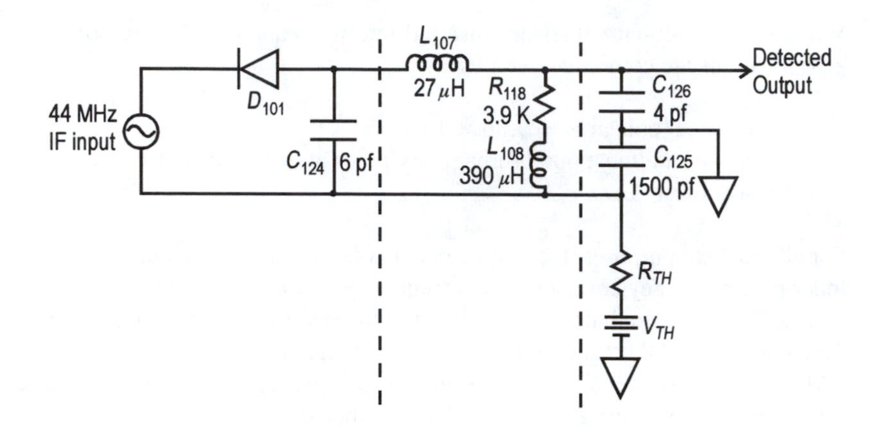

Fig. 6.20: The video detector circuitry of the Zenith 19EB12 television receiver.

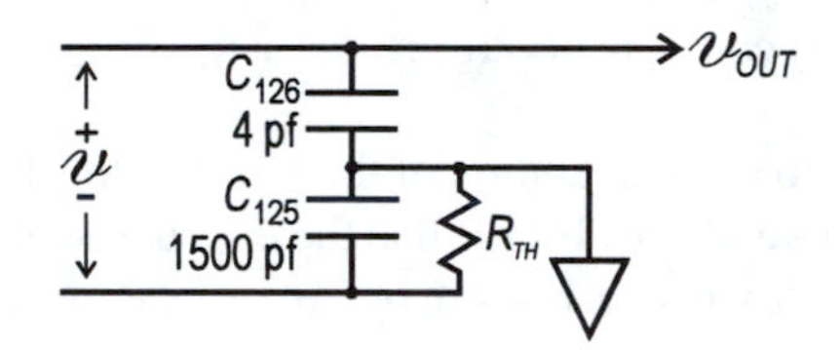

Fig. 6.21: The portion of Fig. 6.20 to the right of the rightmost dashed line after replacing the Thevenin voltage source by an AC short.

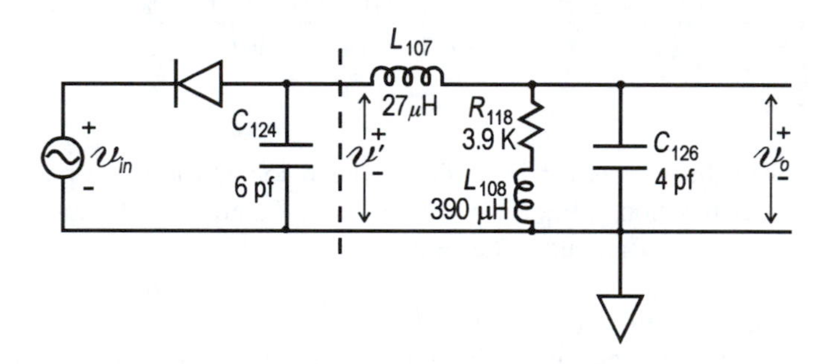

Fig. 6.22: Simplified version of Fig. 6.20 obtained by recognizing that C_{125} is an AC short.

The impedance of the R_{118}–L_{108}–C_{126} combination is

$$\mathbf{Z}(s) = \frac{(R_{118} + sL_{108})/sC_{126}}{R_{118} + sL_{108} + 1/sC_{126}} = \frac{R_{118} + sL_{108}}{s^2 L_{108} C_{126} + sC_{126} R_{118} + 1}. \tag{6.3}$$

From this we can set up another voltage division ratio to find $\mathbf{V}_o$ in terms of $\mathbf{V}'$:

$$\frac{\mathbf{V}_o}{\mathbf{V}'}(s) = \frac{\mathbf{Z}}{sL_{107} + \mathbf{Z}} = \frac{\frac{R_{118}+sL_{108}}{s^2 L_{108} C_{126} + sC_{126} R_{118} + 1}}{sL_{107} + \frac{R_{118}+sL_{108}}{s^2 L_{108} C_{126} + sC_{126} R_{118} + 1}}$$

$$= \frac{R_{118} + sL_{108}}{s^3 L_{107} L_{108} C_{126} + s^2 L_{107} C_{126} R_{118} + s(L_{107} + L_{108}) + R_{118}}. \tag{6.4}$$

When the component values are inserted into this equation and the pole and zero locations are computed, we find

(1) one real pole at $\omega = 9.36 \times 10^6$,
(2) a pair of almost purely imaginary poles at $\omega = 9.95 \times 10^7$, and
(3) a real zero at $\omega = 10^7$.

For all practical purposes, the real zero and pole cancel, leaving only the imaginary pole pair. They correspond to a frequency of about 15 MHz.

It will be recalled that the signal input to the envelope detector has a carrier frequency of 44 MHz and contains significant signal components as much as 4 MHz away from the carrier. The result we have just developed shows that the asymptotic response is flat out to 15 MHz and then drops off at −40 dB/decade, which will provide significant attenuation at 44 MHz. Owing to the very low damping ($\zeta = 0.0032$), there will be a large resonant peak at 15 MHz, but because no frequency components are present at this frequency, this peak should be unimportant.

We finally come to a consideration of the effect of this filter on the envelope detector itself. To this end, we observe that the impedance of the filter, as seen by C_{124}, is just $sL_{107} + \mathbf{Z}$, where $\mathbf{Z}$ is given by Eq. (6.3). Expressing this impedance as a ratio of polynomials in s, we get

$$\mathbf{Z}_{\text{in}}(s) = \frac{s^3 L_{107} L_{108} C_{126} + s^2 L_{107} C_{126} R_{118} + s(L_{107} + L_{108}) + R_{118}}{s^2 L_{108} C_{126} + s C_{126} R_{118} + 1}. \tag{6.5}$$

The zeros of this expression have already been solved for as the poles of the filter transfer function in Eq. (6.4). The poles of $\mathbf{Z}_{\text{in}}$ are solved for as the roots of the quadratic denominator. The solutions are:

(a) one real zero at $\omega = 9.36 \times 10^6$,
(b) a pair of almost purely imaginary zeros at $\omega = 9.95 \times 10^7$, and
(c) a pair of complex conjugate poles at $\omega = 2.53 \times 10^7$.

From these data we conclude that the asymptotic impedance plot is flat out to about 1.5 MHz, from whence it rises until about 4 MHz. The value of the impedance in the "low-frequency" region (up to 1.5 MHz) may be obtained by taking the limit of $\mathbf{Z}_{\text{in}}$ as $s \to 0$. The result obtained by so doing is $\mathbf{Z}_{\text{in}} = R_{118}$, which may be observed to be a lower limit on the envelope detector load.

It was observed in Section 1.5 that an envelope detector time constant should ideally satisfy the inequalities

$$\frac{1}{\omega_{\text{carrier}}} < RC < \frac{1}{\omega_{\text{mod}}}.$$

In our situation $R = 3.9\ \text{k}\Omega$, $C = 6$ pF, and so $RC = 23.4$ ns. For a 44 MHz carrier, $1/\omega_{\text{carrier}} = 3.6$ ns, nicely satisfying the left-hand inequality. The value of ω_{mod} is not as easy to pin down, since the video contains frequencies up to about 4 MHz. However, if the right-hand inequality is satisfied for this modulating frequency, it will certainly be satisfied for lower ones. The reciprocal

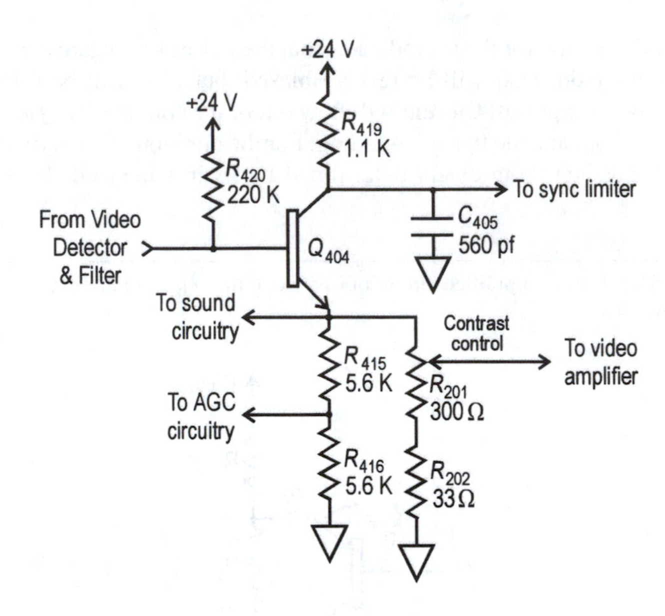

Fig. 6.23: The video driver stage of the Zenith 19EB12 television receiver. This stage is a distribution point from which signals are sent to four other main sections of the receiver.

of this radian frequency is 40.0 ns, which does indeed satisfy the right-hand inequality.

We thus conclude that the envelope detector will function properly with the load represented by the filter and that the filter will be effective in removing 44 MHz ripple.

6.3.4 The video driver

The video driver is an amplifier stage for the composite video, sync, and sound signals that leave the detector and filter. The schematic is shown in Fig. 6.23. As can be seen from the schematic, this stage is a distribution point from which the components of the composite signal go their various ways to be processed. All but the sync output are taken from the emitter circuit. The sync is taken from the collector. This allows the use of C_{405} to roll off the high-frequency response at the collector while retaining it at the emitter, where all of the outputs will need it. The rolloff frequency due to R_{419} and C_{405} will be at about 250 kHz, which is still high enough to allow the horizontal sync output to retain a fair semblance to a pulse, while attenuating the higher frequency video and sound components that are not desirable in the sync signal.

The contrast control is used as a video-signal voltage divider. This usage is analogous to the normal audio volume control configuration. In Figs. 6.19B and 6.19D, it can be seen that the emitter of Q_{404} is connected to the contrast control through J_{202}–P_{202}, a jack-plug combination that connects controls that are situated off the main chassis. Such provisions are ordinarily very helpful to the serviceman who must remove the chassis from the cabinet. R_{202} prevents the video signal from going to zero even when the contrast control is at its minimum setting.

6.3.5 Keyed AGC

The AGC circuitry can be redrawn for much greater clarity, as shown in Fig. 6.24. Point P is at ground potential in the absence of a flyback pulse.

Service information for the set indicates that the pulses are negative going with a 32 V amplitude. D_{401} will be reverse biased, but D_{402} will be forward biased during these pulses. Current will flow out of the collector of Q_{403} through D_{402}, C_{402}, R_{408}, and the flyback winding. For the duration of the flyback pulse, Q_{403} will conduct to an extent determined by its instantaneous base-emitter voltage.

Exercise 6.4 For the simplified circuit of Fig. E6.4, find i_E as a function of v_B, v_{EB}, R_u, and R_L.

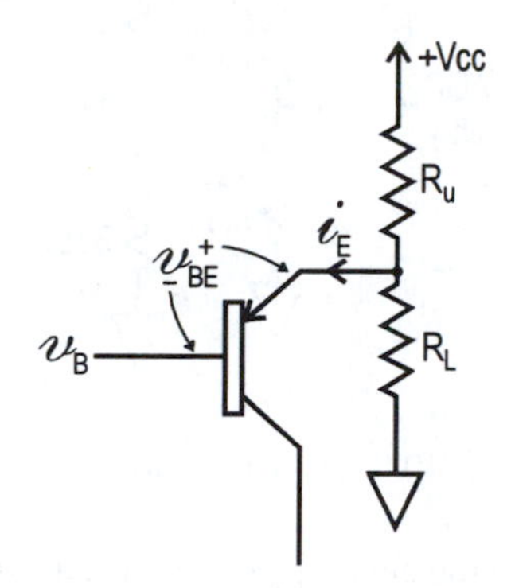

The signal applied to the base is the composite video and sync with inverted polarity, which means that the voltage drops during sync pulses. The emitter potential is derived from a variable voltage divider. The AGC level control should be set to hold Q_{403} in cutoff except on the sync pulses. In normal operation the horizontal sync pulses will coincide in time with the horizontal flyback pulses applied to the collector. Without such coincidence the transistor will not conduct, but this is unimportant, since the picture will not be synchronized if the pulses are not coincident in time. Thus it doesn't matter whether the gain is correct or not. Some other measure must be taken to restore sync. If, however, the pulses are coincident in time, the amount of conduction will be determined by the amplitude of the sync pulses applied to the base. The larger (more negative-going) these pulses are, the more heavily the transistor conducts. Thus, when operating properly, the transistor should always be in normal active mode during sync pulses and $i_C \approx i_E \propto$ sync pulse amplitude.

As the pulse of collector current is coupled through C_{402}, it acquires an increment of charge

$$\Delta Q = I_C \cdot T,$$

where T is the pulse duration and I_C is the magnitude of the collector current pulse of Q_{403}.

Q_{403} is called a gated amplifier in normal engineering terminology. In TV, this is more often called an *AGC keyer*. The enabling pulse is thus called a *gate pulse* or a *keying pulse*. When this pulse is gone, C_{402} discharges through D_{401} and R_{408} into C_{403} where it provides one component of the base bias for Q_{402}. The other component comes from the voltage divider R_{410} and R_{411} through R_{409}. The voltage from C_{402} adds to this. Thus this stage utilizes forward AGC, which is normal for keyed AGC circuitry.

Fig. 6.24: AGC keyer circuitry of the Zenith 19EB12 television receiver.

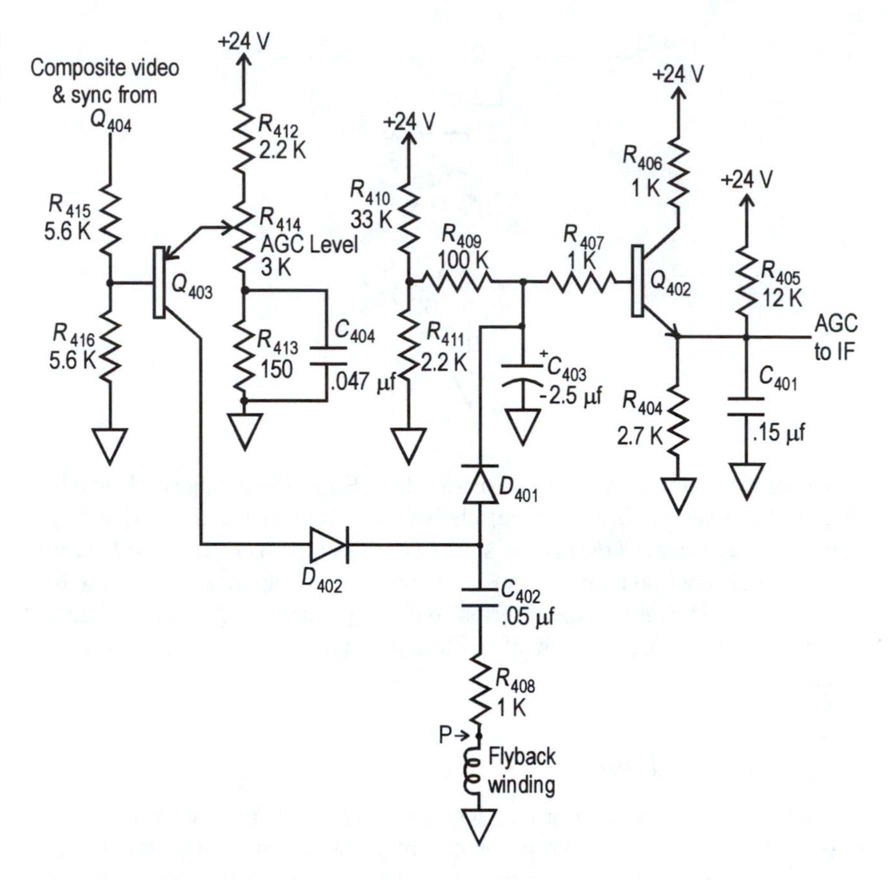

We now consider the operation of Q_{402}. In the absence of sync (as when the TV is tuned to an empty channel) there is never any AGC component of bias at the base of Q_{402}. The base potential set by the R_{410}–R_{411} pair is 1.5 V with a series resistance of 103 kΩ. The voltage divider connected to the emitter has a Thevenin voltage of 4.40 V and an effective resistance of 2.2 kΩ. Thus the base-emitter junction is reverse biased and Q_{402} is totally cut off. The potential on the AGC line is that from the emitter voltage divider, which we saw was equal to 4.4 V. Now let us assume that the circuit begins to receive some sync pulses. We will, for the moment, ignore C_{401}. It will be necessary for C_{402} to discharge into C_{403} a number of times before the base potential rises above the emitter potential and allows Q_{402} to conduct, but once it does conduct, charge will begin to be drained off of C_{403} between sync pulses and replenished during those pulses. The equilibrium voltage across C_{403} will be reached when the rates of charge loss and gain are equal, averaged over a line scan. This charging and discharging of C_{403} will cause the base voltage of Q_{402} to have a ripple at the horizontal frequency. It is the function of C_{401} to filter out this ripple. The output of this stage is the IF AGC voltage. It is the base bias voltage for Q_{101}, the first IF amplifier. This AGC voltage is also directly coupled to the emitter of Q_{401}, which has the misleading name of *AGC delay*. This is misleading because ordinarily delay refers to something that occurs later in time. In this case it refers to something that happens at a higher voltage. It might be more properly called RF AGC threshold. The schematic is shown in Fig. 6.25.

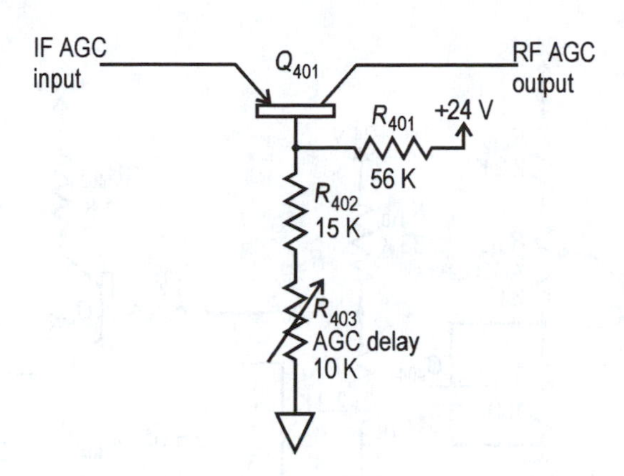

Fig. 6.25: The so-called delayed AGC circuit of the Zenith 19EB12 television receiver.

Operation is simplicity itself. Resistors R_{401}–R_{403} form an adjustable voltage divider that fixes the base voltage relative to ground. If the IF AGC voltage is high enough (which means the signal is high enough) to forward bias the base–emitter junction then the transistor conducts and applies AGC to the RF stage as well. Thus this stage receives AGC only when the gain-controlled IF amplifier is incapable by itself of holding the signal amplitude at a constant level.

6.3.6 *The sound section*

The signal from the emitter of the video driver (Q_{404}) is also coupled through C_{1113} and T_{1102} to IC_{1101}. From the circuitry connected to pins 9 and 10, it is apparent that this IC is a quadrature detector. Since no IF amplification or limiting is shown elsewhere, we conclude that this chip is of exactly the same type analyzed in Section 2.3.3; that is, the quadrature detector is preceded by several stages of IF preamplification, which have adequate gain to assure limiting in the worst case for which the chip is specified.

The audio output from pin 12 of IC_{1101} is fed to a conventional transformer-coupled class A power amplifier, which is distinctive only in the relatively high collector voltage at which it operates. C_{205} shunts out some of the highest frequencies to give a pleasing tone balance and R_{215}, which is a voltage-variable resistor, shunts current around the primary of T_{201} if Q_{201} shorts from collector to emitter.

Exercise 6.5

(a) What is the maximum output power available from this stage? Assume the DC primary resistance is 100 Ω.

(b) What must the transformer turns ratio be for maximum power output?

6.3.7 *The video amplifier*

The video output is taken from the emitter circuit of Q_{404}, the video driver, and directly coupled to Q_{406}, which is the video output stage. The circuitry is shown in Fig. 6.26. The signal from the contrast control is DC-coupled through R_{203} to assure that as a larger signal is fed to the video amplifier by moving the control

Fig. 6.26: The video amplifier stage of the Zenith 19EB12 television receiver.

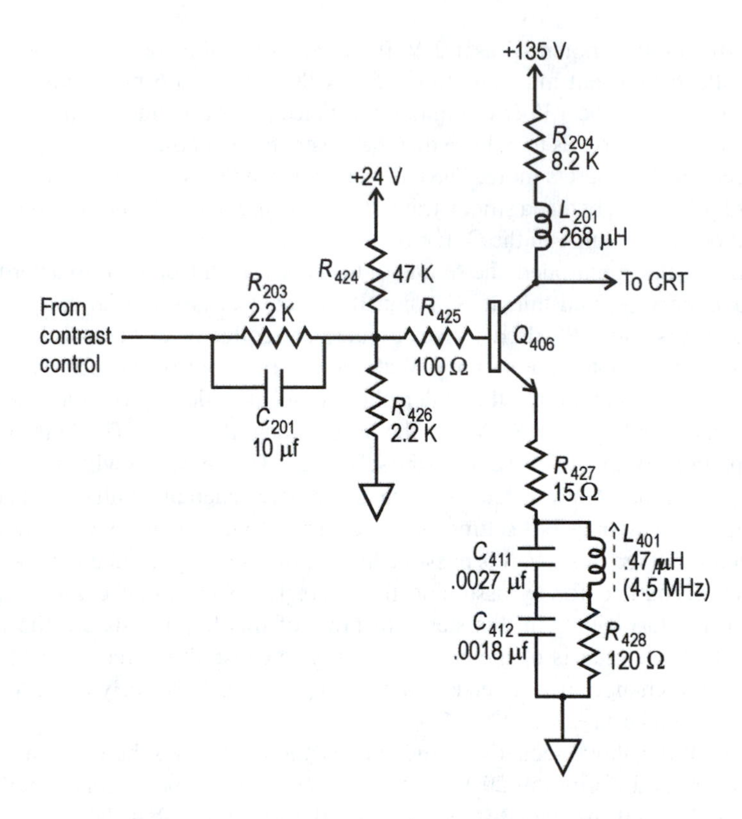

slider closer to the emitter of Q_{404}, the base bias of Q_{406} will also increase. This will occur because the contrast control divides both the AC and DC voltages at the emitter of Q_{404}. The increase of base bias as the signal increases will keep Q_{406} in the normal active mode. The capacitor across R_{203} acts as an AC short. The base bias is set by resistors R_{424} and R_{426}.

We recognize Q_{406} as a transconductance feedback amplifier. However, since both the collector and emitter impedances are complex, it is not immediately obvious what the frequency dependence of the voltage gain will be. It is clear, however, that C_{411} and L_{401} together comprise a parallel-resonant trap for the 4.5 MHz sound signal. If we assume this trap is lossless, it will appear as an open circuit at 4.5 MHz. This will cause the gain to drop to zero there.

To justify using the simple first-order feedback approximation for the gain of this stage, it is necessary to validate the assumption that the loop gain $|g_m \mathbf{Z}_{\mathrm{E\ min}}|$ will be much greater than unity, where $\mathbf{Z}_{\mathrm{E\ min}}$ is the minimum value of the emitter impedance over the amplifier's frequency range. First we must find the quiescent point, which presents us with some practical difficulties. These difficulties stem from errors in the manufacturer's schematic. The most glaring of these errors is seen by noting on Fig. 6.19D that the quiescent collector voltage is given as 56 V whereas if we follow this wire back through connector terminal W3, we see that the other end of this wire is shown as having a potential of 85 V. This is only the beginning of the difficulties. If we assume that the 85 V figure is correct, it means that there is a 50 V drop across the collector load impedance. Assuming negligible drop in L_{201}, this means that the collector current is $50\ \mathrm{V}/8.2\ \mathrm{k\Omega} \approx 6\ \mathrm{mA}$. This same current flowing through the emitter resistances would give a voltage drop of 0.81 volts, ignoring any resistance of L_{401}. But the schematic gives the emitter potential as 1.2 V. If we take the collector voltage as 56 V, the collector current must be (135 − 56) V/8.2 kΩ = 9.6 mA. This current flowing through the 135 Ω of emitter resistance gives a voltage drop of 1.3 V. This would dispose us to believe that the lower value of collector voltage is the correct one, but there's more. The base–emitter voltage is shown as only 0.4 V. It is highly unlikely that a silicon transistor with only 0.4 V on the B–E junction could conduct 10 mA in the C–E circuit.

As was just mentioned, the setting of the contrast control helps to determine the bias on Q_{406} and thus its Q point. If this control is set at its midpoint, a DC analysis using PSPICE® gives a quiescent collector current of 3.65 mA and a collector potential of 105 V. With the contrast control set such that there is 65 Ω resistance above the slider, the quiescent collector voltage is 85 V. The emitter voltage is 0.83 V and the base voltage is 1.65 V. The latter value compares favorably to the value on the schematic. This leaves us with the emitter voltage and the 56 V collector voltage as the only schematic values that seem substantially wrong. No setting of the contrast control can give a quiescent collector voltage of 56 V. It is possible that the discrepancy in schematic values can be accounted for by assuming that there is a substantial component of signal-developed bias in this stage. In favor of this hypothesis are the facts that: (1) The signal is of the correct polarity to cause this effect, and (2) the percentage change in I_{CQ} needed to make $V_{\text{CQ}} = 56$ V is fairly close to that needed to make $V_{\text{EQ}} = 1.2$ V.

Even if the above scenario is true, it does not justify the schematic showing two voltages differing by 29 V on opposite ends of a wire! We have perhaps overemphasized the bias discrepancies on the schematic, but the problem of schematic errors is not terribly uncommon, and we have here sought to demonstrate a critical approach to and reconciliation of such problems.

We got into this digression by trying to find the Q point of Q_{406}. If we accept the postulate that there is a large enough signal-developed bias to bring V_{CQ} down to 56 V, the collector current is 9.6 mA, giving $g_{\text{m}} \leq 0.384$ mho. As long as $\mathbf{Z}_{\text{E}}$ is greater than 26 Ω, $|g_{\text{m}}\mathbf{Z}_{\text{E}}|$ will be greater than 10, and we can therefore justify use of the first-order approximation for gain–namely $\mathbf{A}_{\text{G}} = -1/\mathbf{Z}_{\text{E}}(s)$ or $\mathbf{A}_{\text{v}} = -\mathbf{Z}_{\text{C}}(s)/\mathbf{Z}_{\text{E}}(s)$. It can be shown that if L_{401} has even as much as one ohm of resistance, this condition on emitter impedance will be met for all frequencies less than 8 MHz. Even so, the impedances are complicated enough to make the determination of the break frequencies challenging. Therefore, PSPICE® was used to perform an AC analysis of the video amplifier. The result is shown in Fig. 6.27. The form of the curve is surprisingly simple. The Bode breakpoints were computed to generate the dashed-line plot, and using a low-frequency transistor model, the circuit has three poles (including a complex-conjugate pair) and four zeros (also including a complex-conjugate pair due to the parallel-resonant trap in the emitter circuit). Yet the overall response is very close to that of a one-zero plot with the notch response superimposed on it. What we might call the dominant zero is set by the breakpoint of the parallel RC in the emitter circuit. The enhanced high-frequency response will sharpen picture detail.

The video output is taken to the CRT cathode (pin 7). The grid (pin 2) is held at ground potential by R_{209}. As the video signal swings positive, the grid-cathode bias gets increasingly negative, cutting the beam off, and making the screen black at that point in the scan.

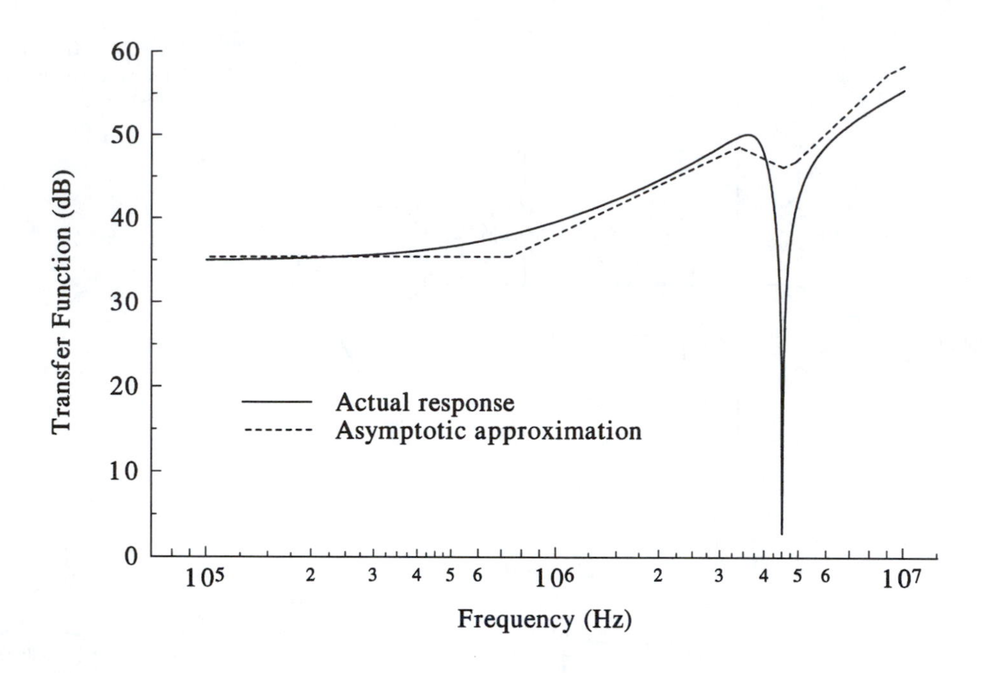

Fig. 6.27: Frequency response of the video amplifier shown in Fig. 6.26. The solid line represents the actual response, and the dashed line is the asymptotic approximation thereto, calculated from the break frequencies of the amplifier.

6.3.8 *The sync limiter*

In other sets this stage is called a *sync clipper* or, most commonly, a *sync separator*. It receives both vertical and horizontal sync pulses from the video driver stage, as well as sound and video. It is designed, however, to remain in or near cutoff until a sync pulse is received. Thus the video and sound are not present at its output. The sync pulses on the other hand, having a higher amplitude, are able to drive the stage not only into conduction, but into saturation. The schematic of this stage is shown in Fig. 6.28.

The positive-going sync pulse at the collector of Q_{404} may be approximated as a step input to the sync limiter stage (Q_{405}). Before this pulse arrives, the initial voltage on C_{406} may be shown to be 5.6 V. The input of Q_{405} will be approximated as an AC short. We want to find the current into the base in response to a voltage step input to the stage. Figure 6.29 shows the equivalent circuit that we will actually be analyzing. The transform of the transient part of the current may be shown to be

$$\mathbf{I}_b(s) = \frac{|V_{cc} - v_c(0)| C_{406}(1 + sC_{407}R_{418})}{s^2 C_{406}C_{407}R_{418}R_{419} + s(C_{406}R_{418} + C_{407}R_{418} + C_{406}R_{419}) + 1}. \tag{6.6}$$

This transforms back into the time domain to give

$$i_b(t) = 14.97\text{ mA} \cdot e^{-333518t} + 1.73\text{ mA} \cdot e^{-2689t}. \tag{6.7}$$

This result shows that there are two time constants of decay – a short one of about 3 μsec and a longer one of about 370 μsec. The coefficient of the slower exponential is so large (as base currents go) that the second term will hold the transistor in saturation long after the first term damps out, even though it is

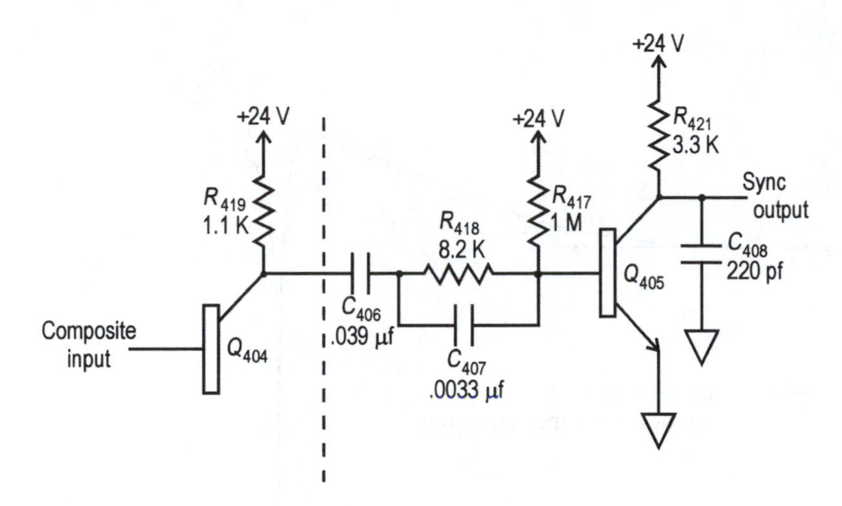

Fig. 6.28: The sync limiter stage of the Zenith 19EB12 television receiver.

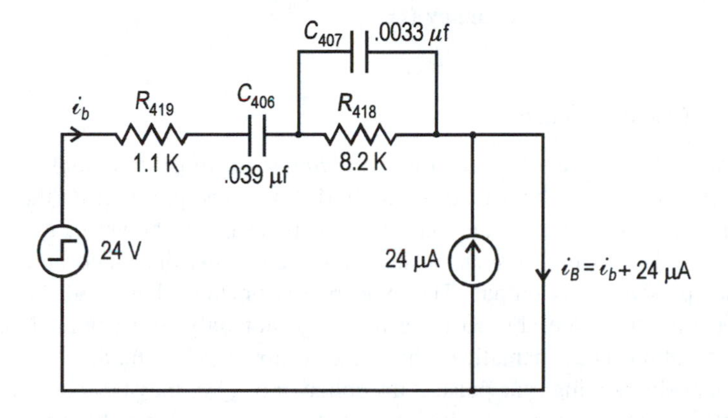

Fig. 6.29: Equivalent circuit used to analyze the schematic of Fig. 6.28.

initially much larger. The 370 μsec time constant is so long compared to the horizontal sync pulse ($\sim$5 μsec) that the base current essentially doesn't decay at all for the duration of the horizontal sync pulse. This assures us that Q_{405} will remain saturated during the pulse. When the input drops, the capacitive coupling causes the base voltage to drop negative, reverse biasing the B–E junction. There will thus be no base current. If we were to analyze this reverse-bias situation mathematically, we would have to use a different set of assumptions than those that led to Eq. (6.7). In particular, the transistor B–E junction would be an open circuit and capacitor recovery would occur by charging through R_{417}. But this is not necessary. We know that the negative excursion of v_{be} will hasten the switching off of Q_{405} by sweeping out base charge. For our purposes, we may assume that the transistor cuts off simultaneously with the drop of the input sync pulse.

The response to a vertical sync pulse is similar. Its duration is about 3 times the line scan period or about 190 μsec. Since the time constant is 370 μsec, the base current will have decayed to 1.06 mA. However, this value, multiplied by any reasonable value of β insures that throughout the vertical-sync pulse, Q_{405}, sufficient base drive exists to hold the transistor in saturation.

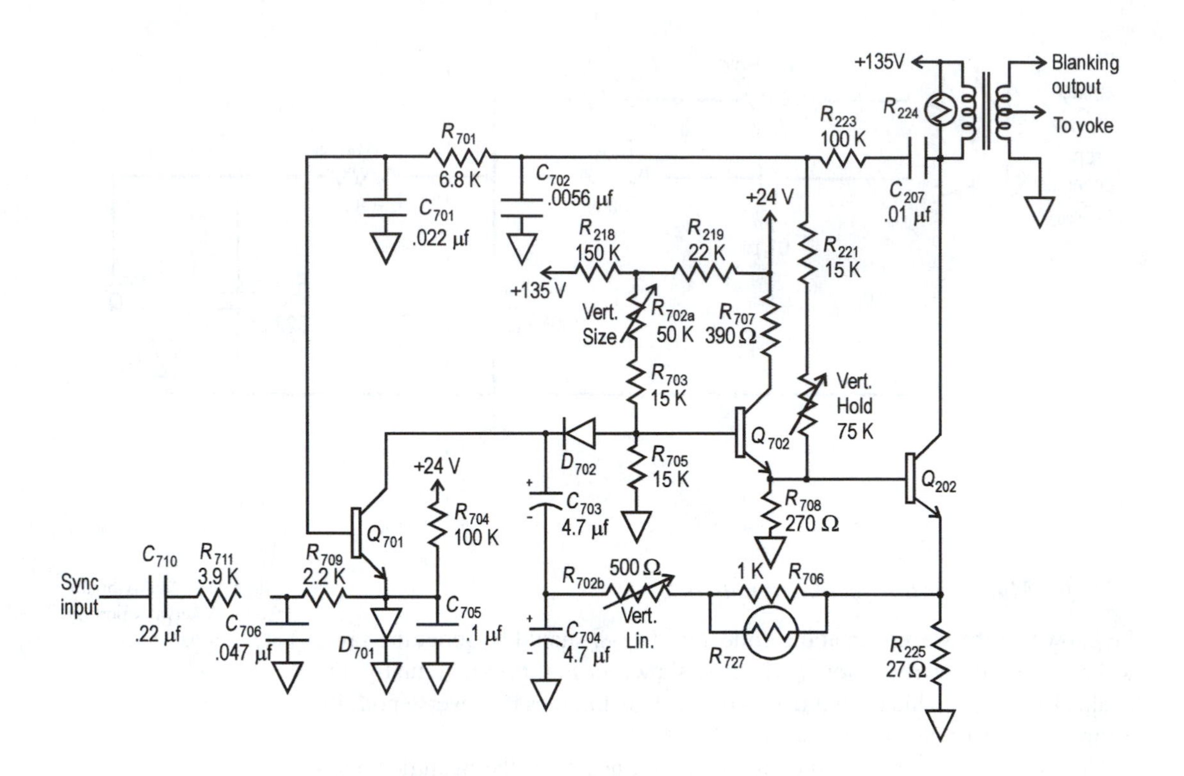

Fig. 6.30: Schematic diagram for the vertical oscillator and vertical output circuitry of the Zenith 19EB12 television receiver.

6.3.9 The vertical oscillator and output

The vertical oscillator and output circuit involves the transistors Q_{701}, Q_{702}, and Q_{202}. Its circuitry is shown in Fig. 6.30.

Our first observation is that the oscillator must work without sync input. This is evidenced by the fact that when a TV is tuned to a channel on which there is no signal, there is still vertical deflection of the beam. Thus we can ignore for the moment everything connected to the emitter of Q_{701}, except R_{704} and D_{701}. This allows us to examine the oscillator itself. Although there are many complications to be discussed, we first try to see the larger picture. In this case the most important step in the analysis is to recognize that the oscillator is an asymmetrical multivibrator with Q_{701} being one transistor and Q_{702} and Q_{202} (acting as a Darlington pair) being the other. The Darlington connection provides the high-power output needed to drive the vertical yoke as well as the low-power feedback signal back to the base of Q_{701}. We rightly expect, both from the complexity of the circuitry and from what we have already discussed about vertical sweep, that this multivibrator will not produce a square wave output voltage like the textbook multivibrators to which most of us have been exposed. The vertical size (called *height* in many sets) control simply varies the Q point and thus the gain of the Darlington pair. The remaining portion of the circuit will be considered under three headings:

(a) the multivibrator feedback network,
(b) the vertical linearity network, and
(c) the sync input network.

We shall consider these topics in the order listed.

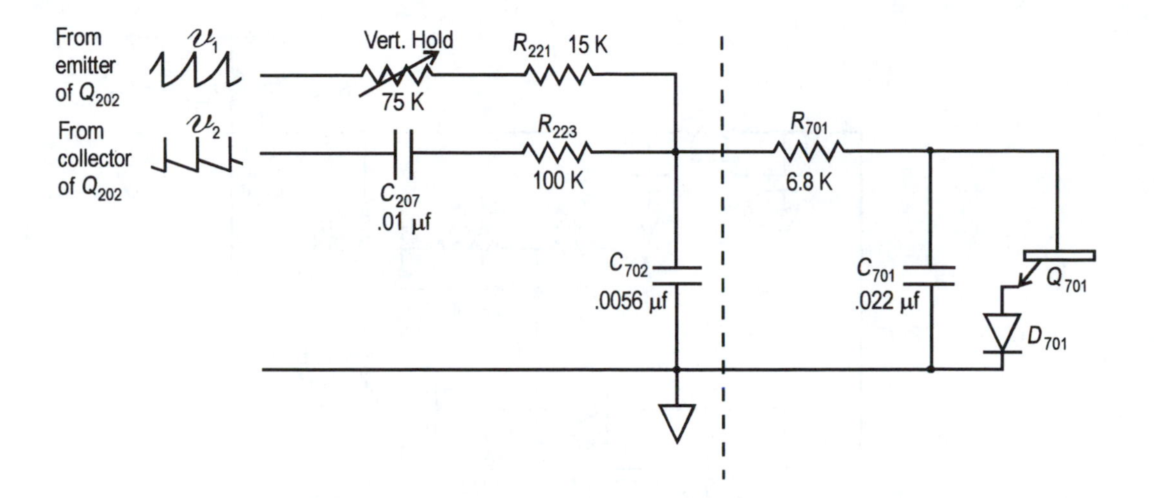

Fig. 6.31: Multivibrator feedback network portion of Fig. 6.30.

6.3.9.1 *Multivibrator feedback network*

We know that the waveform at the collector of Q_{202} should be about the same as that across the vertical yoke, which is shown in Fig. 6.11e. Actually, the feedback network, which is fed from Q_{202}'s collector, sees the reverse polarity of Fig. 6.11e. This network is shown in Fig. 6.31.

The two input waveforms shown are ascertained from the manufacturer's service data. The effects of the two sources will be treated by superposition. For the waveform from the collector of Q_{202}, the 0.01 μF (C_{207}) capacitor is critical. This waveform may itself be thought of as the sum of a sawtooth and a fast pulse. The responses of C_{207} to these two components are quite different. The pulse is fast enough that a substantial fraction of it is coupled through to the subsequent circuitry. The voltage across C_{701} has its positive excursion clamped by D_{701} and Q_{701}, but the resultant current pulse momentarily drives Q_{701} into saturation. The sawtooth component has a negative slope, so any output voltage it develops across C_{701} will be negative and hence will not be clamped. If the circuitry to the left of the dashed line is analyzed in isolation with v_1 shorted for superpositional analysis and v_2 being a negative-going ramp, we find that the voltage across C_{702} asymptotically approaches a negative constant. This implies that the current through C_{702} is zero and the voltage is developed across R_{221} and the vertical hold control. This in turn means that the current through them and through R_{223} and C_{207} as well is constant. A constant current through C_{207} can only result if the voltage across it is changing linearly with time. But this is exactly what happens when the right end of C_{207} is at a fixed (negative) potential and the left end is connected to a (negative-going) ramp. Since all of the time constants in this circuit portion are on the order of a millisecond or less, this constant potential across C_{702} is reached fairly early in the ~16 ms period of v_2. The value of the negative potential is

$$v_{C_{702}} = C_{207}(R_{221} + R_{\text{VH}}) \cdot \frac{dv_2}{dt}\,\text{(Ramp)}.$$

In this equation, $\frac{dv_2}{dt}$ (Ramp) < 0 is the slope of the ramp portion of v_2. Its value is about -16 V/msec. The significant point here is that this voltage is

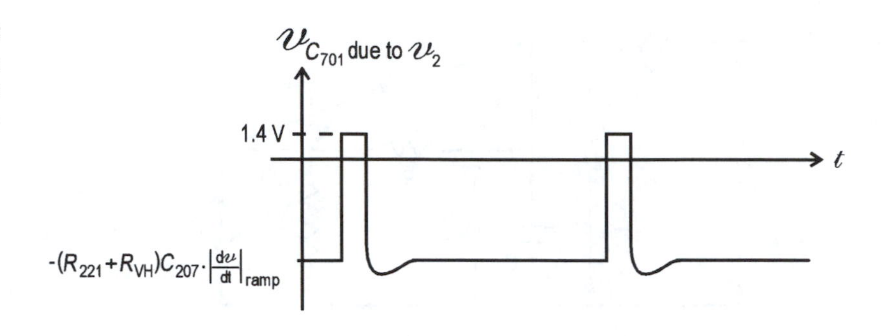

Fig. 6.32: Component of the signal across C_{701} in Fig. 6.31 due to feedback from the collector of Q_{202}.

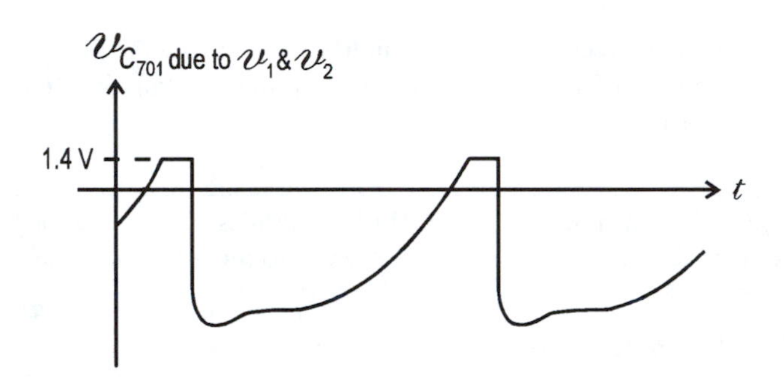

Fig. 6.33: Components of the signal across C_{701} in Fig. 6.31 due to feedback from the emitter of Q_{702} as well as from the collector of Q_{202}.

directly proportional to R_{VH}, the resistance setting of the vertical hold control. The importance of this fact will be apparent shortly. A little additional low-pass filtering is provided by R_{701} and C_{701}, but since the R_{701}–C_{701} time constant is only about $1/7$ ms, the voltage across C should be nearly the same as that across C_{702}. If we combine what we have learned about $v_{C_{701}}$ due to the two components of v_2, we are led to a picture resembling Fig. 6.32.

But there is also going to be a component of this voltage due to v_1 when v_2 is shorted out. Because Q_{202} has an output current ($g_m v$) that is basically a sawtooth, we expect the input voltage ($\sim v = v_1$) to be similar in form. Since v_1 is much smaller than v_2, and since we have noted that a sawtooth will not be well coupled by a 0.01 μF capacitor, we ignore the circuit branch containing C_{207} when calculating the contribution of v_1. Again, the short time constants due to $(R_{221} + R_{VH})$–C_{702} and R_{701}–C_{701} tell us that the voltages across these two capacitors will essentially follow v_1. If this component of $v_{C_{701}}$ is added to those shown in Fig. 6.32, the composite looks like Fig. 6.33.

This same voltage is applied to the base of Q_{701}. As it swings positive, Q_{701} turns on, discharging C_{703} and C_{704}. This drops the potential at the base of Q_{702} and thus the base potential of Q_{202}. The base drop raises the collector potential of Q_{202}. The increase is coupled back to the base of Q_{701} to accelerate the switching process. Thus regenerative switching takes place. Q_{701} saturates rapidly giving the retrace on the sawtooth. Two important observations on this sequence should be made:

(1) In comparing Figs. 6.32 and 6.33 we observe that the pulse component due to v_2 coincides in time with the retrace of v_1. This, of course, is not an accident. Figure 6.32 is perhaps a little misleading in that without the component due to v_1, the component due to v_2 would never be observed. It is the ramp component that causes the regenerative switching sequence that in turn generates the pulse component of v_2.

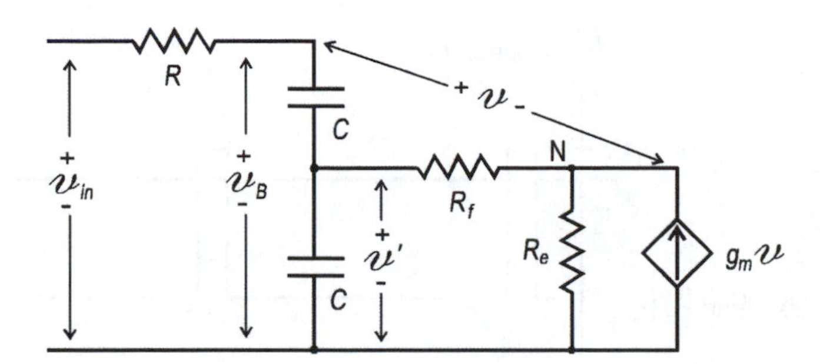

Fig. 6.34: The vertical-linearity portion of the vertical-oscillator circuitry shown in Fig. 6.30.

(2) The main effect of the vertical hold control is to set the DC component at the base of Q_{701}. As R_{VH} is varied, the free-running frequency of the vertical oscillator is varied.

Exercise 6.6 Explain the mechanism by which R_{VH} varies the frequency of the vertical oscillator in the absence of sync. Will an increase in R_{VH} raise or lower the frequency?

6.3.9.2 The vertical linearity network

Besides Q_{702} and Q_{202}, the vertical linearity network includes C_{703}, C_{704}, R_{706}, R_{727}, D_{702}, R_{225}, and the vertical linearity control. After Q_{701} has saturated and shorted out C_{703}–C_{704}, it cuts off because it has no more base drive (Fig. 6.30). Then C_{703}–C_{704} begins to charge again through D_{702}, R_{703}, the vertical size control, R_{218}, and R_{219}. If the vertical size control resistance is decreased, it will not only raise the sawtooth sweep amplitude by raising the Thevenin voltage, but the resultant increase in v_1 will also cause Q_{701} to switch earlier in the cycle, thus raising the sweep frequency. This undesirable interaction between the vertical adjustments was a universal problem with all of the vertical oscillator circuits used in older receivers.

Level shifting and a certain amount of shaping are accomplished by D_{702}. To whatever extent it can be considered a level shifter only, it will not appear in the AC equivalent of this circuitry. The AC equivalent of the circuitry to be analyzed is shown in Fig. 6.34.

Let us make the following observations:

(1) R is the Thevenin equivalent resistance of the DC source used to charge the capacitors. It incorporates R_{219}, R_{218}, R_{703}, R_{705}, and R_{702A}, the vertical size control. The subscript "A" after 702 refers to the fact that this is part of a dual control. Most often this takes the form of two controls mounted together with concentric shafts, with the inner shaft going through the front control and fastening to the rear one. As we shall see, the other control *ganged* with the vertical size is the vertical linearity.

(2) The amplitude of v_{in} in will be given by the Thevenin voltage. Since v_B is shorted by Q_{701} in the retrace part of the cycle, when Q_{701} cuts off, the capacitors start to charge just as if a voltage step had been applied at the input. This will be the input used in the analysis.

(3) The resistance r_π is not shown in the equivalent circuit because the current into node N from r_π is insignificant compared to the other

currents at that node. This is particularly true since the output stage is a Darlington.

(4) R_f is a composite of R_{706}, R_{727}, and R_{702B}, the vertical linearity control. R_{727} is a thermistor whose resistance will vary with temperature.

(5) The transconductance g_m in the equivalent circuit can be shown to be that of Q_{202} provided that g_m of $Q_{702} \times R_{708} \gg 1$. This inequality is strongly satisfied in this circuit. Thus the following three circuit equations apply:

$$\frac{\mathbf{V}_{\text{in}} - \mathbf{V}_{\text{B}}}{R} = sC(\mathbf{V}_{\text{B}} - \mathbf{V}') \qquad \text{(at the } \mathbf{V}_{\text{B}} \text{ node)},$$

$$sC(\mathbf{V}_{\text{B}} - \mathbf{V}') = sC\mathbf{V}' + \frac{[\mathbf{V}' - (\mathbf{V}_{\text{B}} - \mathbf{V})]}{R_f} \qquad \text{(at the } \mathbf{V}' \text{ node)},$$

$$\frac{\mathbf{V}' + (\mathbf{V}_{\text{B}} - \mathbf{V})}{R_f} + g_m\mathbf{V} = \frac{\mathbf{V}_{\text{B}} - \mathbf{V}}{R_e} \qquad \text{(at the N node)}.$$

After a great deal of algebraic manipulation, we find

$$\mathbf{V}(s) = |\mathcal{v}_{\text{in}}| \cdot [1 + sC(R_e + 2R_f)]/\{s[s^2C^2R(g_m R_e R_f + R_e + R_f) + sC\{R + 2(g_m R_e R_f + R_e + R_f)\} + (g_m R_e + 1)]\}. \tag{6.8}$$

The quadratic term in the denominator may be shown to have complex-conjugate roots for all of the parameter and component values achievable with the circuit. We may thus express the poles of the quadratic as

$$s_1 = \text{re} + j \cdot \text{im}$$

and

$$s_2 = \text{re} - j \cdot \text{im}.$$

The inverse transform is

$$\mathcal{v}(t) = \frac{|\mathcal{v}_{\text{in}}|(R_e + 2R_f)}{RC[g_m R_e R_f + R_e + R_f]} \cdot \left[\frac{\alpha}{(\text{re}^2 + \text{im}^2)}(1 - e^{re \cdot t} \cdot \cos(\text{im} \cdot t)) + \frac{e^{re \cdot t} \cdot \sin(\text{im} \cdot t)}{\text{im}} \cdot \left(1 + \frac{\alpha \cdot \text{re}}{(\text{re}^2 + \text{im}^2)}\right)\right], \tag{6.9}$$

where $\alpha \equiv 1/C(R_e + 2R_f)$ is numerically equal to the zero frequency of Eq. (6.8).

If we assume $g_m R_e R_f \gg R_e$, R_f, and R, then the denominator of (6.8) becomes

$$s g_m R_e \cdot [s^2 C^2 R R_f + 2sCR_f + 1].$$

The roots of the quadratic term are

$$s = \frac{-1 \pm j\sqrt{R/R_f - 1}}{RC},$$

so that

$$\text{re} = -1/RC \quad \text{and} \quad \text{im} = \sqrt{R/R_f - 1}/RC.$$

If these two definitions and that of α are used in (6.9), the result is

$$v(t) = \frac{v_{\text{in}}}{g_{\text{m}} R_{\text{e}}} \cdot \left[\left(1 - e^{-t/RC} \cdot \cos \frac{\sqrt{R/R_f - 1}}{RC} t \right) + \frac{1}{\sqrt{R/R_f - 1}} \cdot e^{-t/RC} \cdot \sin \frac{\sqrt{R/R_f - 1}}{RC} t \right].$$

It is important to find $v(t)$, since it is proportional to the collector current of Q_{202}, which in turn is proportional to the current in the vertical yoke, provided that T_{202} does not saturate.

It is therefore of interest to find that, depending on the choice of R_f, $v(t)$ can be made to rise linearly, with an upward curvature or with a downward curvature. This is exactly what a linearity control should be able to do. If the vertical size control is at midrange, the value of R_f for optimum linearity is 2,000 Ω. However, even if the thermistor, R_{727}, were an open circuit, R_f could not be greater than 1,500 Ω. Smaller values of $R_{702\text{A}}$ will reduce the Thevenin resistance R, which will lower the value of R_f giving best linearity. Moreover, a number of second-order effects will introduce nonlinearities of their own, but these may be compensated for, to a certain extent, by the vertical linearity control. These include D_{702} and the nonlinearities of the Darlington pair, which is not really being operated under small-signal conditions, but which should still be fairly linear since it has significant transconductance feedback due to R_{225}.

Physically, we may understand the circuit operation by imagining for the moment that $R_f \to \infty$. The two capacitors would charge through R with a simple exponential rise. Meanwhile, $g_{\text{m}} v$ flows through R_{e} giving a voltage drop $g_{\text{m}} R_{\text{e}} v$. By Kirchoff's voltage law, $v_{\text{B}} = v + g_{\text{m}} R_{\text{e}} v$. Therefore, as v_{B} rises, so do v and the voltage across R_{e}.

If we now let R_f become finite, but still $\gg R_{\text{e}}$, the voltage across R will change little, but current is now entering the lower capacitor through R_f as well as through R, causing it to charge more rapidly than if the rise were a simple exponential. The voltage across the lower C approaches $g_{\text{m}} R_{\text{e}} v (\approx v_{\text{B}})$. Thus since the junction between the capacitors has a rising potential, but one that is still less than v_{in}, the potential at the top of the upper capacitor (v_{B}) is increased. Thus v_{B} increases not only because of current flowing through R, but also because it is "pushed up" from below. This in turn increases v, which sustains the process of ramp generation.

6.3.9.3 The sync input network

The sync input network includes C_{710}, R_{711}, C_{706}, R_{709}, and C_{705}. It is essentially a DC blocking capacitor (C_{710}) followed by two sections of low-pass filtering. This serves to remove the horizontal sync pulses before they can be applied to the vertical oscillator. The pulses input to this network from the sync section are negative-going with an amplitude of 24 V. After passing through the low-pass

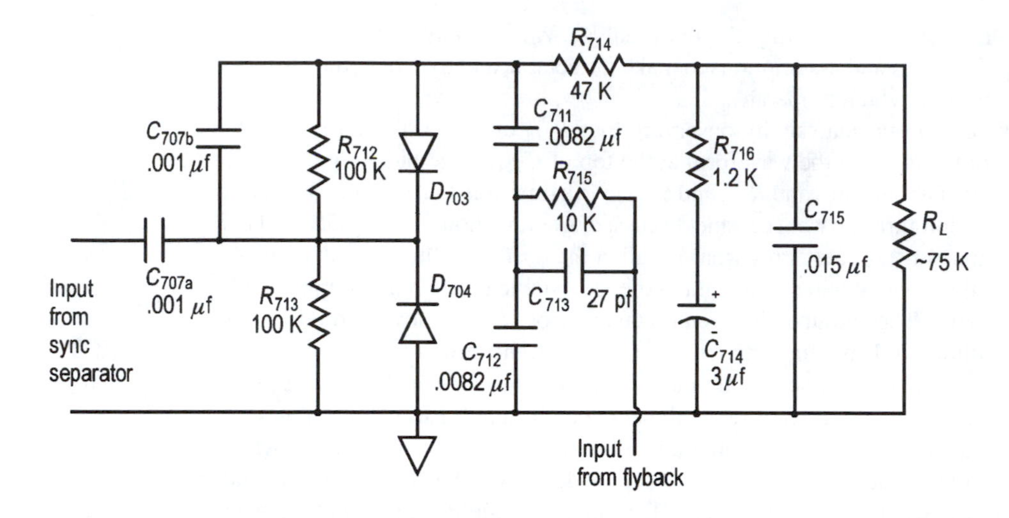

Fig. 6.35: The horizontal phase-detector circuitry of the Zenith 19EB12 television receiver.

network, the vertical pulse amplitude at the emitter of Q_{701} is about a third of this. But with sync input to Q_{701}, the retrace is not initiated by the rising voltage at the base of Q_{701} but by the falling sync pulses at its emitter. These will forward bias the B–E junction of Q_{701} and initiate the regenerative switching transition already discussed. Seen from this perspective, it should be apparent that the free-running period of the vertical oscillator must be longer than the synced period. This assures that the arrival of the sync pulse will occur before the ramp at the base of Q_{701} is large enough to initiate the regenerative-switching cycle.

6.3.10 The horizontal circuitry

The pulses from the sync limiter were, as we saw, passed through an LPF and used directly to trigger the vertical oscillator. We might therefore think that we would only need to high-pass filter the composite sync and apply it directly to the horizontal oscillator. Unfortunately, the situation is not this simple.

High-pass filtering passes not only the desired horizontal sync but undesired amplitude variations (noise) that are picked up at the TV antenna and passed right through the video detector. If the oscillator can be triggered on a sync pulse, it can be prematurely triggered by a noise pulse received at the sync input. Thus more sophisticated means of synchronizing the horizontal oscillator must be used. In this receiver, the horizontal circuitry consists of a phase detector, AFC, horizontal oscillator and driver, and the horizontal output stage. These will be examined in turn. Alternative approaches to the problem will be discussed in Section 6.3.11.

6.3.10.1 The phase detector

The output of the sync separator circuit goes to the horizontal phase detector as well as to the vertical oscillator. The function of the phase detector is to compare the phases of the horizontal-sync signal and the flyback pulse and to generate a DC correction voltage that will adjust and synchronize the horizontal oscillator to the flyback pulses. The phase detector circuitry is shown in Fig. 6.35. This appears to be an intimidating circuit to analyze. We shall have to make a number of observations and assumptions to make it tractable and to extract the physical principles on which it operates.

We begin our analysis by observing that R_{714} and everything to the right of it are for filtering the waveform at the top of C_{711}. C_{715} and R_L give a time constant of about 1 ms, and R_{716} and C_{714} give a time constant of about 3.6 ms. A similar configuration was seen and discussed in connection with Fig. 5.13b. This dual time constant performs what is called the *anti-hunt* function. In controls terminology, we would call it making the loop stable through the use of a lead-lag network. The output of the circuit is taken across R_L. Removing the elements of the filter still leaves us with ten components in the circuit to be analyzed.

The manufacturer's schematic shows a quasi-sawtooth waveform at the junction of C_{711} and C_{712}. Because the flyback input is a pulse waveform, this suggests that R_{715} and C_{712} form an integrating network. This is confirmed when we calculate that $R_{715} \cdot C_{712} = 82$ μsec, which is greater than the horizontal sweep period. On this account, we will remove R_{715} and C_{713} and replace them with a sawtooth source. The sawtooth-forming capacitor (C_{712}) remains in the circuit because it is also part of the path through which the input pulse current flows. This simplification reduces our circuit component count by two more, leaving eight.

We next note that C_{707a}, R_{713}, and D_{704} form a standard clamp. A clamp is ordinarily designed so that the resistance is large enough for it to be considered an open circuit. Therefore, we will attempt to treat the resistors as open circuits. We will, however, consider their effect as an exercise after we finish the first-pass analysis without them in the circuit. Removing the resistors reduces the component count to four capacitors and two diodes. The circuit also contains two "sources": a sync pulse input from the sync separator and the "sawtooth" formed by integrating the flyback pulse.

We next propose to use superposition to "turn off" each of these two sources in turn and find the effect of the other. The reader may object that superposition only applies to linear circuits and a circuit with diodes in it is not linear. The objection is valid, but we can circumvent it. In the first place, we will treat the diodes as ideal switches. If they are forward biased, they are considered short circuits, and if they are reverse biased, they are open circuits. If we then break up a cycle into times when the diodes are open and closed, the circuit becomes piecewise linear. Furthermore, as we have already noted, the diodes are apparently used as clamps, and we know that ideally a clamp does not change the shape of a waveform. It only changes its DC component.

Pursuant to this plan, we short the sync pulse input and find the response of the circuit to the sawtooth generated by R_{715} and C_{712}. For this portion of the analysis, C_{712} need not be considered beyond its role in the sawtooth formation. For this half of the superpositional analysis, we shall also consider C_{711} as an AC short, since it has a reactance less than $1/8$ of that of C_{707a} and C_{707b}. (A more-exact analysis than the one we will do shows that considering this capacitor a short for the sawtooth introduces an error of about 6% in our calculated response to that input.) With the sync input shorted, C_{707a} is in parallel with R_{713} and D_{704}, and the circuit is symmetrical with respect to the positive and negative half-cycles of the sawtooth. The DC component of the sawtooth that is intrinsic to its method of generation is removed by C_{711}, which is assumed to pass the AC without attenuation.

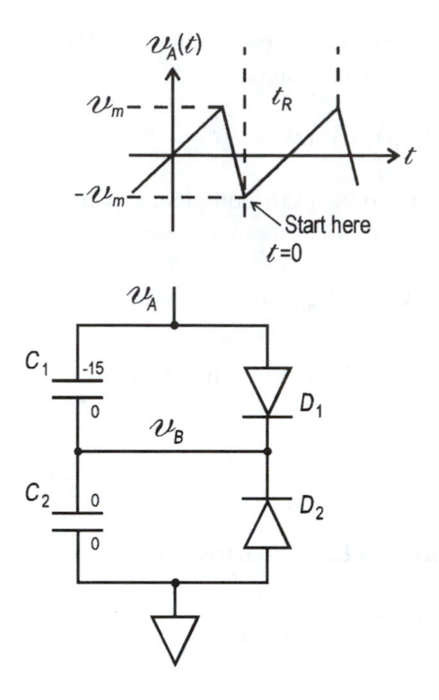

Fig. 6.36: Simplified subcircuit of Fig. 6.35 for determining the response to the sawtooth waveform formed by integration of the flyback pulses.

It should not be inferred that the presence of C_{711} in the circuit rules out there being any DC component at the anode of D_{703}. Quite the opposite is true. There must be DC here in general, because this is the phase detector output that goes to the filter and then on to the AFC stage. It only needs to be observed that this DC component results from processes transpiring in the phase detector proper and not from the generation of the sawtooth from the pulse. Figure 6.36 illustrates the essential circuitry whose sawtooth response we wish to find, but with simplified designators for the circuit components.

We proceed as follows: Assume that $v_A(t)$ has just reached a negative peak, and D_2 has conducted adequately to charge C_1 as shown. At this time C_2 is uncharged since we assumed that the diodes are ideal switches. Since $v_A(t)$ is reversing slope at $t = 0$, we will assume that both diodes are open and find the voltage response of the circuit; we then check to insure that the diodes *were* open.

Because the diodes are assumed to be open, the currents through the capacitors are equal. In terms of Laplace transforms, this yields

$$sC_1\mathbf{V}_1(s) - C_1v_1(0) = sC_2\mathbf{V}_2(s) - C_2v_2(0).$$

Since $v_B(0) = 0$, this simplifies to

$$\mathbf{V}_2(s) = \mathbf{V}_1(s) - \frac{v_1(0)}{s},$$

where we have defined $C \equiv C_1 = C_2$.

But by setting our zero of time where we did, $v_1(0) = -V_m$, allowing us to write

$$\mathbf{V}_2(s) = \mathbf{V}_1(s) + \frac{|V_m|}{s}. \tag{6.10}$$

We have here one equation in two unknowns. The other equation required is clearly the voltage loop equation:

$$\mathbf{V}_{\mathrm{A}}(s) = \mathbf{V}_1(s) + \mathbf{V}_2(s). \tag{6.11}$$

Since $\mathbf{V}_{\mathrm{A}}(s)$ is known to be a sawtooth, its rise can be expressed in the time domain as

$$v_{\mathrm{A}}(t) = -V_{\mathrm{m}} + 2V_{\mathrm{m}} \cdot \frac{t}{t_{\mathrm{R}}},$$

where t_{R} is the rise time of the sawtooth. The transform of the sawtooth rise is then

$$\mathbf{V}_{\mathrm{A}}(s) = \frac{-|V_{\mathrm{m}}|}{s} + \frac{2|V_{\mathrm{m}}|}{t_{\mathrm{R}}} \cdot \frac{1}{s^2}.$$

When this is inserted into Eq. (6.11), the result is

$$\frac{-|V_{\mathrm{m}}|}{s} + \frac{2|V_{\mathrm{m}}|}{t_{\mathrm{R}}} \cdot \frac{1}{s^2} = \mathbf{V}_1(s) + \mathbf{V}_2(s).$$

Finally, this equation is solved simultaneously with (6.10) to give $\mathbf{V}_1(s)$ and $\mathbf{V}_2(s)$, which are in turn transformed back into the time domain to give $v_1(t)$ and $v_2(t)$:

$$v_1(t) = V_{\mathrm{m}}\left(\frac{t}{t_{\mathrm{R}}} - 1\right) \quad (t \le t_{\mathrm{R}}), \tag{6.12a}$$

$$v_2(t) = V_{\mathrm{m}} \cdot \frac{t}{t_{\mathrm{R}}} \quad (t \le t_{\mathrm{R}}). \tag{6.12b}$$

These equations show that $v_1(t)$ rises linearly from an initial value of $-V_{\mathrm{m}}$ and reaches zero at the end of the sawtooth rise time. Similarly, $v_2(t)$ rises linearly from an initial value of zero to its final value of $+V_{\mathrm{m}}$. It should be obvious by referring back to Fig. 6.36 that these behaviors will justify our assumption that the diodes remain cut off at all times if we consider just the effect of the sawtooth.

Exercise 6.7

(a) Repeat the calculations just completed if the 100 kΩ resistors are not treated as open circuits.

(b) The calculations are straightforward, but you will find that late in the rise of the sawtooth, D_1 becomes forward biased. Given that t_{R} is 51 μsec, find the time at which this happens.

The finite value used for R and the resulting brief conduction of D_1 allows the phase detector as a whole to respond within one cycle to changes in the phase relationship between the sync and flyback pulses. The sawtooth response in our idealized circuit is summarized in Fig. 6.37.

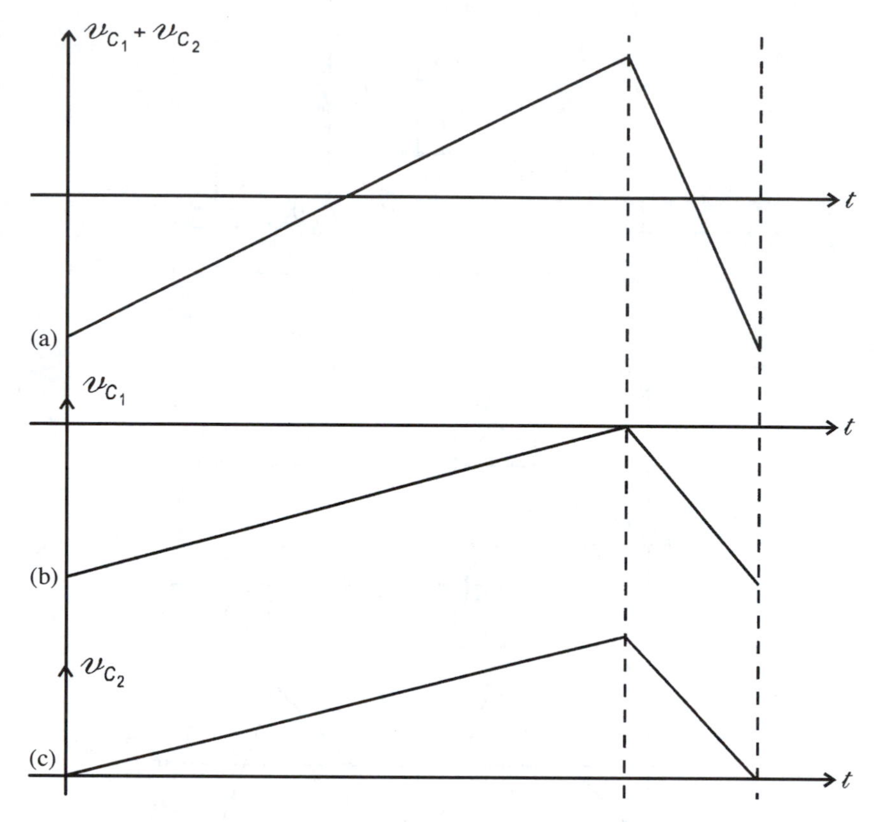

Fig. 6.37: The distribution of the sawtooth voltage between capacitors C_1 and C_2. (a) The sum of these two voltages is essentially that formed by the integration of the flyback pulses. (b) The voltage across C_1 always remains below zero. (c) The voltage across C_2 always remains above zero.

We now move on to the second half of our superpositional analysis. This involves setting the sawtooth "source" to zero and finding the effect of the input sync pulse. The circuit for this analysis is shown in Fig. 6.38. Subject to our assumptions, we can see that all of the capacitors are in series. Furthermore, since there is no resistance included in this simplified circuit, the capacitors will charge instantaneously both during the pulse and after its return to zero.

Exercise 6.8 Show that, for the assumptions made relative to the circuit of Fig. 6.38, during the pulse $v_1 = 0.45V_p$ with the polarity shown and $v_2 = 0.55V_p$ with polarity opposite to that shown. Of course, when the pulse returns to zero, so do both of the capacitor voltages.

Unfortunately, the polarities of v_1 and v_2 are both such as to forward bias their respective diodes. To say for sure if this is going to be a problem, we must first sum the separate solutions given by superposition to see if this will be true of the composite waveforms.

We thus need to synthesize our understanding of the circuit's responses to the sawtooth and sync pulse into a composite response. We proceed by noting that we would expect the sawtooth-forming pulse from the flyback and the sync pulse to be pretty-nearly coincident in time in a correctly operating receiver. Figure 6.39 shows a summation of the two components of v_1 and v_2, still under the assumption that the diodes are nonconducting. In constructing this composite we have used the fact that the service information shows the sawtooth

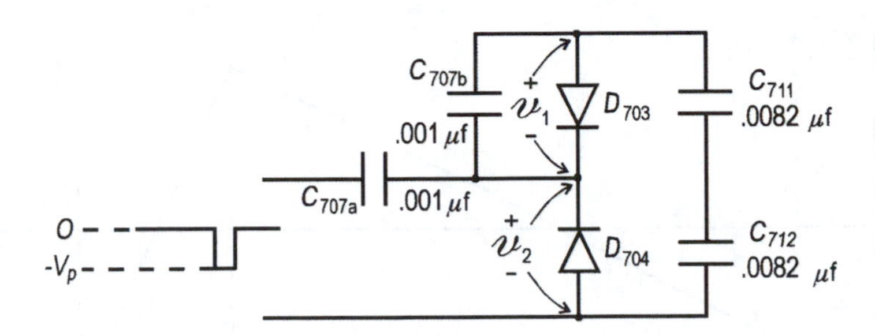

Fig. 6.38: Simplified subcircuit of Fig. 6.35 for determining the response to the horizontal sync pulses.

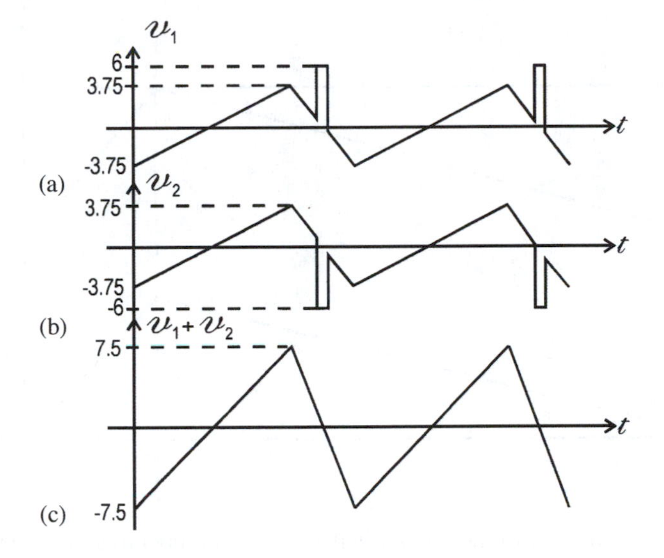

Fig. 6.39: Composite response of the horizontal phase detector of Fig. 6.35 to the sawtooth and horizontal-sync pulses assuming that the diodes never conduct. (a) The voltage across C_1. (b) The voltage across C_2. (c) The sum of these two capacitor voltages.

as having a peak-to-peak amplitude ($2V_\mathrm{m}$) of 15 V, which means our parameter V_m is 7.5 V. Similarly, the service information characterizes the sync pulse as dropping from 0 down to -12 V. Thus our parameter V_p is 12 V.

However, as this figure shows, we can no longer ignore the diodes, since the pulses attempt to drive v_1 above zero (forward biasing the upper diode) and also attempt to drive v_2 below zero (forward biasing the lower diode). How do we handle this? We recall that the elements now identified as C_2 and D_2 form a clamp. In fact, they clamp the negative excursion of v_2 to zero. Similarly, v_1 is clamped to zero by D_1. This causes a revision of Fig. 6.39 to Fig. 6.40.

The clamping action has produced no change in the sum $v_1 + v_2$. As we shall see, this is because the pulse coincides with the center of the retrace. If such were is not the case, then we would have a different story, as illustrated in Fig. 6.41, which shows only the retrace time slot. Note that in these figures, t_F is the fall time of the sawtooth and t_p is the time (measured from the beginning of the fall or retrace) on which the sync pulse is centered. These curves show that, due to the clamping action of the diodes, the sum $v_1 + v_2$ has acquired a DC component equal to

$$V_\mathrm{A}(\mathrm{DC}) = \frac{4V_\mathrm{m}t_\mathrm{p}}{t_\mathrm{F}} - 2V_\mathrm{m} = 2V_\mathrm{m} \cdot \left(\frac{2t_\mathrm{p}}{t_\mathrm{F}} - 1\right).$$

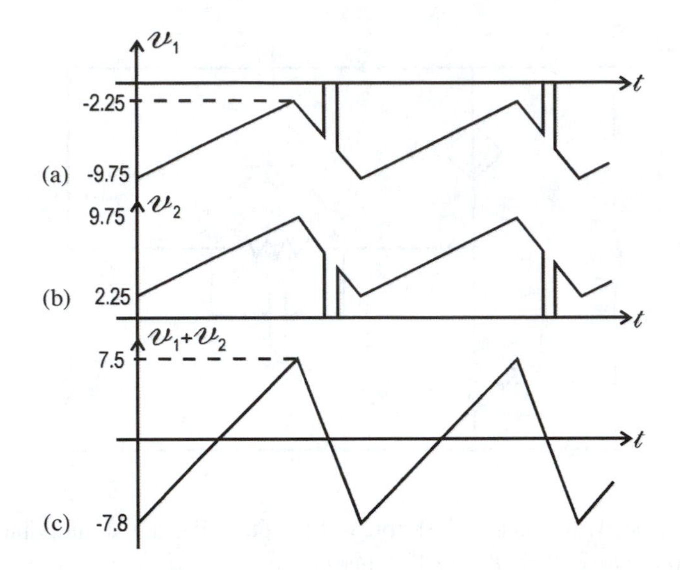

Fig. 6.40: Modification of the waveforms of Fig. 6.39 taking into account the clamping action of the diodes. (a) The voltage across C_1. (b) The voltage across C_2. (c) The sum of these two capacitor voltages.

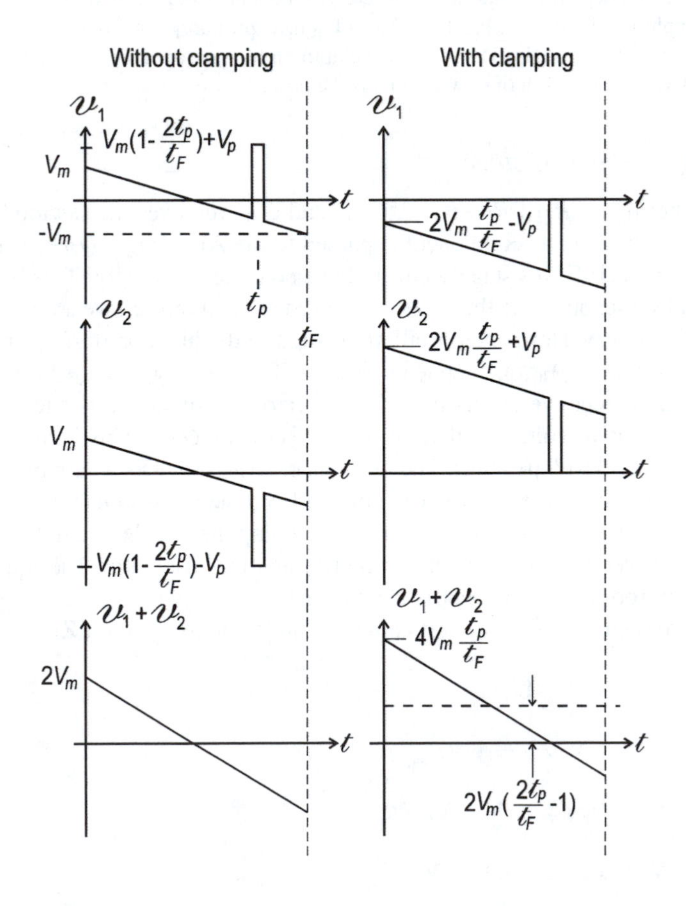

Fig. 6.41: Detail drawing of the retrace portion of Fig. 6.40 showing the effect in the right-hand column of varying the time during the retrace when the sync pulse arrives. If the sync pulse arrives past the midpoint of the retrace, the sawtooth voltage that is the sum of the capacitor voltages acquires a positive DC component. Without the diode clamping action, there is no DC component to the sawtooth regardless of when in the retrace the sync pulse arrives, as is seen by the waveforms in the left-hand column.

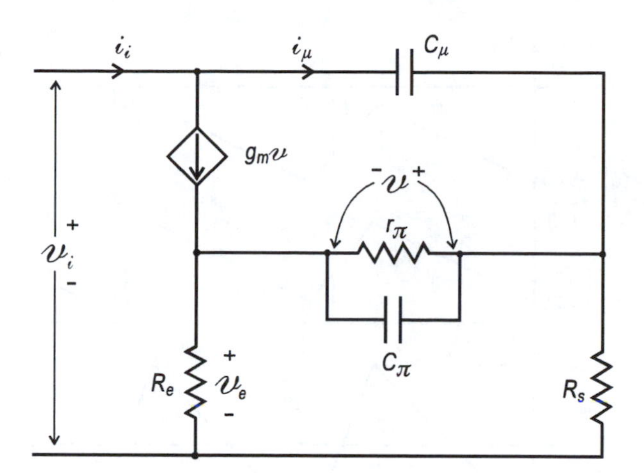

Fig. 6.42: Equivalent circuit of the AFC stage of the Zenith 19EB12 television receiver.

In accordance with what we just noted, if the pulse lies at the midpoint of the retrace ($t_p = t_F/2$) there is no DC component.

Exercise 6.9 This analysis has assumed that both v_1 and v_2 will clamp at the sync pulse level. Under these conditions, the DC component is independent of V_p, the sync pulse amplitude. If $|V_p| < 2V_m$, for values of t_p near zero or t_F, only one of the voltages will clamp on the sync tip. The other will clamp on the sawtooth. Derive an expression for $V_A(\text{DC})$ as a function of t_p when this is the case.

6.3.10.2 The AFC stage

After filtering by R_{714}, R_{715}, R_{716}, C_{714}, and C_{715} removes the sawtooth component of v_A, the DC component is passed to the AFC stage, Q_{703}. Referring back to Fig. 6.19F, this stage seems to have no collector supply! This is true in the usual sense, but since the stage is not used as an amplifier, we are not bound to usual circuitry. Horizontal oscillation is fed to the junction of C_{716} and C_{718} and is applied to Q_{703} and D_{705} in parallel. The polarity of D_{705} is such that it will clamp negative excursions of the horizontal voltage at its anode, while allowing positive voltage pulses at the collector of Q_{703}. This is an extreme case of signal-developed bias. The resistance, R_{718}, provides a base bias that is augmented or decreased by the DC voltage from the phase detector.

The output impedance of Q_{703} serves as a component in the oscillator circuit. We shall therefore analyze it before proceeding to the oscillator. The equivalent circuit appropriate to the task is shown in Fig. 6.42, where R_s is the source resistance seen by Q_{703}. A set of equations adequate to solve for $\mathbf{Z}(= \mathbf{V}_i/\mathbf{I}_i)$ is

$$\mathbf{I}_i = g_m\mathbf{V} + \mathbf{I}_\mu,$$

$$g_m\mathbf{V} + \mathbf{V}/\mathbf{Z}_\pi = \mathbf{V}_e/R_e$$

$$\mathbf{I}_\mu = \mathbf{V}/\mathbf{Z}_\pi + (\mathbf{V} + \mathbf{V}_e)/R_s,$$

$$\mathbf{V}_i = \mathbf{I}_\mu/sC_\mu + \mathbf{V} + \mathbf{V}_e,$$

where $\mathbf{Z}_\pi$ is the impedance of r_π and C_π in parallel.

Understandably, the expression for **Z** is quite complex. Fortunately, we can readily justify the use of four simplifying assumptions:

(1) $\beta \gg 1$,
(2) $R_e \ll R_s$,
(3) $g_m R_e \gg 1$, and
(4) $\omega \ll \omega_\beta$, where ω_β is the β cutoff frequency.

All of these assumptions should commend themselves to the reader.

After making these assumptions, the resultant expression for **Z** of the AFC output port becomes

$$\begin{aligned}\mathbf{Z} = \frac{\mathbf{V}_i}{\mathbf{I}_i} &= \frac{(1/\beta + R_e/R_s + j\omega C_\mu R_e)}{j\omega C_\mu} \\ &= R_e + \frac{(1/\beta + R_e/R_s)}{j\omega C_\mu} \\ &= R_e + \frac{1}{j\omega\left[\frac{C_\mu}{1/\beta + R_e/R_s}\right]} \\ &= R_e + \frac{1}{j\omega\left[\frac{\beta C_\mu}{1+\beta R_e/R_s}\right]}.\end{aligned}$$

The quantity in square brackets is the effective capacitance of the AFC device.

If we choose as typical values $C_\mu = 2$ pF and $\beta = 100$ we find an effective capacitance of 127 pF and a capacitive reactance of 80 kΩ at the horizontal oscillator frequency. This is so much larger than the resistive component that the latter can be neglected, and we can represent the AFC device to the oscillator simply as a high-Q capacitance of value

$$C_e \equiv \frac{\beta C_\mu}{1 + \beta R_e/R_s}. \tag{6.13}$$

Prior to proceeding to the oscillator, we observe that C_e is proportional to C_μ, which will vary with the quiescent collector–emitter voltage of Q_{703}, which in turn varies with the "DC" error voltage from the phase detector. It can, in fact, be shown that

$$\frac{dC_e}{dv_p} = -\beta C_\mu \left(\frac{R_{718}}{3V_{CEQ} R_{717}} \right).$$

Plugging the values $R_{718} = 1\,\text{M}\Omega$, $R_{719} = 47\,\text{k}\Omega$, $\beta = 100$, and $V_{CEQ} = 3.4$ V (from the schematic) into (6.13) yields

$$\frac{\Delta C_e}{C_e} = \frac{-\left(1 + \beta \cdot \dfrac{R_{719}}{R_{717}}\right) R_{718}}{3R_{717} \cdot V_{CEQ}} \cdot \Delta v_p = -3.28 \cdot \Delta v_p,$$

where v_p is the DC control voltage to this reactance stage. This should have a range in excess of $0 \pm .1$ V, resulting in a fractional variation in effective capacitance of ±33%.

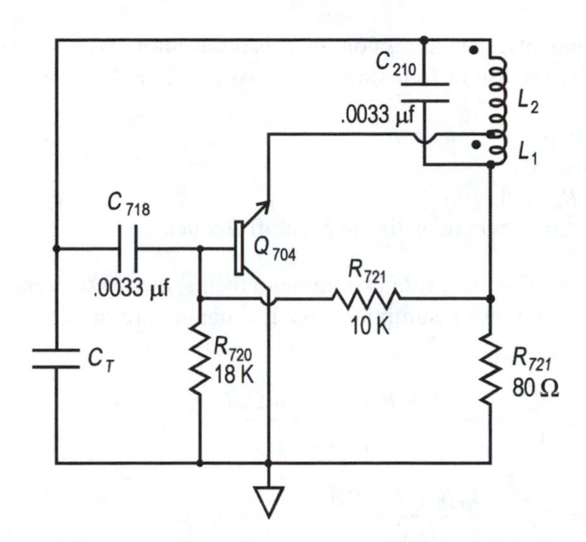

Fig. 6.43: Schematic diagram of the horizontal-oscillator stage of the Zenith 19EB12 television receiver.

6.3.10.3 *Horizontal Oscillator*

With this information in hand, we are now in a position to analyze the horizontal oscillator. The pertinent circuitry is redrawn in Fig. 6.43. The tapped coil identifies this as a Hartley oscillator of sorts, but its use in an emitter–follower circuit is unusual enough that we shall have to analyze the circuit for ourselves.

In drawing the oscillator in this form we have:

(1) Returned the base resistor to ground rather than to the positive supply. This is because it is held at AC ground potential by C_{717}.

(2) Eliminated the 47 Ω base series resistor, R_{723}. This resistor is probably used because the emitter follower can display an input impedance with a negative real part when it has a capacitive load. This could lead to oscillation at a frequency not determined by the feedback components. We will assume that at the designed frequency of oscillation, Q_{704} has a real input impedance appreciably greater than R_{723}.

(3) Used C_{T} to represent the series combination of C_{716} (1,500 pF) and C_{e}, which is the effective capacitance of the AFC stage.

(4) Represented L_{203} as two mutually coupled inductances, L_1 and L_2. We will assume perfect flux coupling between L_1 and L_2.

The feedback portion of the circuit is isolated and redrawn in Fig. 6.44.

Since the amplifier to which this network is applied is an emitter follower, we need to compute a voltage-transfer function, F. We anticipate that it will be greater than 1, so that we can have $|AF| = 1$. The circuit equations are

$$\mathbf{V}_{\mathrm{in}}(s) = sL_1\mathbf{I}_1 - sM\mathbf{I}_2 + \mathbf{V}_{\mathrm{A}},$$

$$\mathbf{I}_1 = \mathbf{V}_{\mathrm{A}}/R_{722} + (\mathbf{V}_{\mathrm{A}} - \mathbf{V}_{\mathrm{F}})/R_{721} + sC_{210}(\mathbf{V}_{\mathrm{A}} - \mathbf{V}_2),$$

$$(\mathbf{V}_{\mathrm{A}} - \mathbf{V}_{\mathrm{F}})/R_{721} = \mathbf{V}_{\mathrm{F}}/R_{720} + sC_{718}(\mathbf{V}_{\mathrm{F}} - \mathbf{V}_2),$$

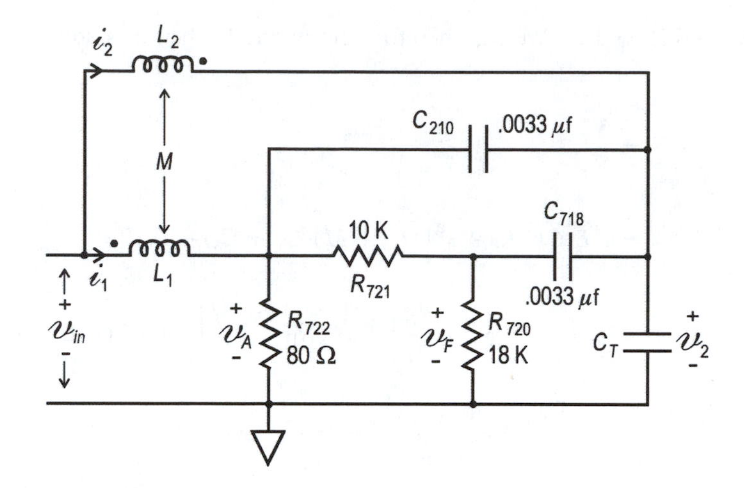

Fig. 6.44: Schematic diagram of the feedback network of the horizontal oscillator shown in Fig. 6.43.

$$\mathbf{V}_{\text{in}}(s) = sL_2\mathbf{I}_2 - sM\mathbf{I}_1 + \mathbf{V}_2,$$

$$\mathbf{I}_2 + sC_{210} \cdot (\mathbf{V}_{\text{A}} - \mathbf{V}_2) + sC_{718}(\mathbf{V}_{\text{F}} - \mathbf{V}_2) = sC_{\text{T}}\mathbf{V}_2.$$

We need the transfer function $\mathbf{V}_{\text{F}}/\mathbf{V}_{\text{in}}$, where $\mathbf{V}_{\text{in}}$ is input to the feedback network from the output of Q_{704}, and $\mathbf{V}_{\text{F}}$ is the voltage fed back to its base. We introduce the following shortened notations:

$$\begin{aligned}
R_{720} &\equiv R_{\text{B}}, \quad C_{210} \equiv C, \\
R_{721} &\equiv R_{\text{F}}, \quad C_{718} \equiv C_{\text{s}}, \\
R_{722} &\equiv R_{\text{s}}, \quad L_{\text{e}} \equiv L_1 + 2\sqrt{L_1 L_2} + L_2 = L_1 + 2M + L_2.
\end{aligned}$$

We can then express the voltage-transfer function as

$$\begin{aligned}
-\mathbf{F} = \frac{\mathbf{V}_{\text{F}}}{\mathbf{V}_{\text{in}}} &= s^3 L_{\text{e}} C C_{\text{s}} R_{\text{F}} + s^2\bigg[(L_2 + M)(C_{\text{T}} + C_{\text{s}}) + L_{\text{e}}C \\
&\quad + (L_1 + M)C_{\text{s}}\left(1 + \frac{R_{\text{F}}}{R_{\text{s}}}\right)\bigg] + sC_{\text{s}}R_{\text{F}} + 1 \\
&\quad \Bigg/ \Bigg\{ s^3 R_{\text{F}} C_{\text{s}}(L_{\text{e}}C + L_2 C_{\text{T}}) + s^2\bigg[\left(1 + \frac{R_{\text{F}}}{R_{\text{B}}}\right) \\
&\quad \times \{L_{\text{e}}C + L_2(C_{\text{s}} + C_{\text{T}})\} + L_1 C_{\text{s}}\left(1 + \frac{R_{\text{F}}}{R_{\text{s}}}\right) + 2MC_{\text{s}}\bigg] \\
&\quad + s\left[C_{\text{s}}R_{\text{F}} + L_1\left\{\frac{1}{R_{\text{B}}}\left(1 + \frac{R_f}{R_{\text{s}}}\right) + \frac{1}{R_{\text{s}}}\right\}\right] + \left(1 + \frac{R_{\text{F}}}{R_{\text{B}}}\right)\Bigg\}.
\end{aligned} \tag{6.14}$$

We can effect a small simplification by observing that $R_{\text{F}}/R_{\text{s}} = R_{721}/R_{722} =$

$10\ \text{k}\Omega/80\ \Omega = 125$. We can therefore throw out 1 when it is added to this ratio:

$$
\begin{aligned}
-\mathbf{F} &= \frac{\mathbf{V}_\mathrm{F}}{\mathbf{V}_\mathrm{in}} \\
&= s^3 L_\mathrm{e} C C_\mathrm{s} R_\mathrm{F} + s^2\Bigg[(L_2 + M)(C_\mathrm{T} + C_\mathrm{s}) + L_\mathrm{e} C \\
&\quad + (L_1 + M) C_\mathrm{s} \cdot \frac{R_\mathrm{F}}{R_\mathrm{s}}\Bigg] + s C_\mathrm{s} R_\mathrm{F} + 1 \Bigg/ \Bigg\{ s^3 R_\mathrm{F} C_\mathrm{s} (L_\mathrm{e} C + L_2 C_\mathrm{T}) \\
&\quad + s^2\Bigg[\left(1 + \frac{R_\mathrm{F}}{R_\mathrm{B}}\right)\{L_\mathrm{e} C + L_2(C_\mathrm{s} + C_\mathrm{T})\} + C_\mathrm{s}(2M \\
&\quad + L_1 R_\mathrm{F}/R_\mathrm{s})\Bigg] + s\left[C_\mathrm{s} R_\mathrm{F} + \frac{L_1}{R_\mathrm{s}}\left(1 + \frac{R_f}{R_\mathrm{B}}\right)\right] + \left(1 + \frac{R_\mathrm{F}}{R_\mathrm{B}}\right)\Bigg\}.
\end{aligned}
\tag{6.15}
$$

If we convert to complex rectangular form, we get

$$
\begin{aligned}
\frac{\mathbf{V}_\mathrm{F}}{\mathbf{V}_\mathrm{in}} &= \{1 - \omega^2[(L_2 + M)(C_\mathrm{T} + C_\mathrm{s}) + L_\mathrm{e} C + (L_1 + M) C_\mathrm{s} R_\mathrm{F}/R_\mathrm{s}]\} \\
&\quad + j C_\mathrm{s} R_\mathrm{F} \omega (1 - \omega^2 L_\mathrm{e} C) \Bigg/ \Bigg\{\Bigg\{\left(1 + \frac{R_\mathrm{F}}{R_\mathrm{B}}\right) \\
&\quad - \omega^2\Bigg[\left(1 + \frac{R_\mathrm{F}}{R_\mathrm{B}}\right)\{L_\mathrm{e} C + L_2(C_\mathrm{s} + C_\mathrm{T})\} + C_\mathrm{s}(2M + L_1 R_\mathrm{F}/R_\mathrm{s})\Bigg]\Bigg\} \\
&\quad + j C_\mathrm{s} R_\mathrm{F} \omega \left[1 + \frac{L_1(1 + R_\mathrm{F}/R_\mathrm{B})}{C_\mathrm{s} R_\mathrm{s} R_\mathrm{F}} - \omega^2(L_\mathrm{e} C + L_2 C_\mathrm{T})\right]\Bigg\}.
\end{aligned}
\tag{6.16}
$$

This equation is of the form

$$
\frac{\mathbf{V}_\mathrm{F}}{\mathbf{V}_\mathrm{in}} = \frac{a + jb}{c + jd} = \frac{(a + jb)(c + jd)}{c^2 + d^2} = \frac{(ac + bd) + j(bc - ad)}{c^2 + d^2}.
$$

Therefore a necessary condition for oscillation may be expressed as

$$
bc = da. \tag{6.17}
$$

Applying condition (6.17) to (6.16) yields a quadratic in ω^2. The positive root(s) of this equation are the frequency or frequencies for which the transfer function is purely real. These could be solved for readily if the numerical values of the inductances were known. As it stands, it would be very cumbersome and almost pointless to carry out an exact solution in terms of variable names. We can however obtain an approximate solution for the frequency of oscillation by carrying out the indicated operations and making some reasonable assumptions. Using (6.17), and using $C = C_s$, the condition for oscillation is

$$(1 - \omega_o^2 L_e C)\left\{\left(1 + \frac{R_F}{R_B}\right) - \omega_o^2\left[\left(1 + \frac{R_F}{R_B}\right)\{L_e C + L_2(C + C_T)\} + C(2M + L_1 R_F/R_s)\right]\right\} = \left[1 - \omega_o^2(L_e C + L_2 C_T) + \frac{L_1\left(1 + \frac{R_F}{R_B}\right)}{C R_s R_F}\right]\{1 - \omega_o^2[(L_2 + M)(C_T + C) + L_e C + (L_1 + M)C R_F/R_s]\}. \tag{6.18}$$

Note that in the circuit of Fig. 6.44, $C \gg C_T$. We will therefore initially assume that C_T is negligible. We do this only to "get a handle on" the inductance values in the circuit. Note also that the last term on both sides of Eq. (6.18) involves the ratio R_F/R_s, which, as noted previously, is equal to 125. Finally, by the assumption of perfect flux coupling, we may say that

$$M = nL_1,$$
$$L_2 = n^2 L_1,$$

and

$$L_e = (n + 1)^2 L_1,$$

where n is the turns ratio N_2/N_1.

Using all of these simplifications in the quadratic for ω_o^2 yields

$$\left[1 - \omega_o^2(n + 1)^2 L_1 C\right]\left\{1.56 - \omega_o^2\left[1.56\left\{(n + 1)^2 + n^2\right\}L_1 C + (2n + 125)L_1 C\right]\right\} = \left[1 - \omega_o^2(n + 1)^2 L_1 C + (1.56 L_1/2.64\ \mathrm{mH})\right] \times \left\{1 - \omega_o^2\left[L_1 C(n^2 + n) + (n + 1)^2 L_1 C + 125(n + 1)L_1 C\right]\right\}. \tag{6.19}$$

In writing this equation, we have used numerical values for all circuit components except C, L_1, and n. The outer curly brackets on both sides of this equation both have square brackets inside, and in both cases considerable simplification occurs if $n \ll 62$. On the left side it will always be true that

$$1.56\{2n^2 + 2n + 1\} + 125 \gg 2n.$$

On the right side

$$125(n + 1) \gg n(n + 1) + (n + 1)^2$$

$$125 \gg n + n + 1 = 2n + 1$$

$$124 \gg 2n$$

$$62 \gg n.$$

We will also define $x \equiv \omega_o^2 L_1 C$. Then

$$\begin{aligned}1.56[1 - x(n+1)^2]\{1 - x[(2n^2 + 2n + 1) + 80.13]\}\\ = [1 - x(n+1)^2 + 18.28x]\{1 - 125x(n+1)\}.\end{aligned}$$

Here we have an equation involving only n and x. For a given value of n, we will have a quadratic in x.

Exercise 6.10

(a) What sign for x will correspond to a real solution?

(b) Make a table of x vs. n for n varying from 2 to 4 in steps of 0.25.

(c) Add a column to the table that converts x to its corresponding value of L_1. What value will you use for ω_o?

(d) For $n = 3$, what is the corresponding value of L_1?

(e) What are the corresponding values of M, L_2, and L_e?

Based on these calculations, let us suppose that $n = 3$ and $L_1 = 2.5$ mH are a reasonable set of values for the oscillator coil.

Exercise 6.11 Return to Eq. 6.18, but now suppose that ω_o^2 is treated as an unknown, while all of the inductances are treated as known quantities based on the calculations of the previous exercise. Then the only unknown on the right-hand side of the equation is C_T. Calculate the fractional change in ω_o for a given fractional change in C_T.

We would like to go on and find $A_{v_{min}}$, but there is a problem in that the usual procedure in oscillator analysis is to solve for ω_o and then substitute it into the real part of F to satisfy the condition $|AF| = 1$. If we use the approximate value of ω_{osc} just derived, we should not expect to get very good results for $A_{v_{min}}$. This is because the oscillator stage is an emitter follower for which we already know $A_v < 1$. If we need any information beyond this about the required gain, it would be *how much* less than unity A_v can be and still allow oscillation. This is a second-order effect. Since Eq. (6.19) is the result of a first-order calculation, the second-order terms in it have already been discarded. This would make a calculation of $1 - A_v$ based on (6.19) pretty inaccurate.

Under most conditions, we would not need to pursue the value of $A_{v_{min}}$ beyond saying that it must be provided by an emitter follower. However, we will do so in this case because we have at our disposal a powerful technique from control theory that allows us to find the minimum gain for oscillation without knowing the resonant frequency. This technique is based on the Routh–Hurwitz criterion for stability of a feedback system. We shall use it not only to find $A_{v_{min}}$, but to illustrate the application of the technique to a "real-world" circuit.

We begin by setting up the loop gain and equating it to zero. This is called the characteristic equation:

$$1 + \mathbf{A}_v\mathbf{F} = 0,$$

where $\mathbf{A}_v$ is the open-loop amplifier gain.

As we have defined $\mathbf{F}$ throughout this book,

$$\mathbf{F} = -\mathbf{V}_F/\mathbf{V}_{in}$$

Table 6.3. *Routh–Hurwitz table for the horizontal oscillator*

s^3	$d_3 - A_v n_3$	$d_1 - A_v n_1$	0
s^2	$d_2 - A_v n_2$	$d_0 - A_v n_0$	0
s^1	$\frac{(d_2 - A_v n_2)(d_1 - A_v n_1) - (d_3 - A_v n_3)(d_0 - A_v n_0)}{d_2 - A_v n_2}$	0	0
s^0	$d_0 - A_v n_0$	0	0

so that

$$1 - \mathbf{A}_v \mathbf{V}_F / \mathbf{V}_{in} = 0. \tag{6.20}$$

We next further shorten the notation in (6.15) to

$$\frac{\mathbf{V}_F}{\mathbf{V}_{in}} = \frac{n_3 s^3 + n_2 s^2 + n_1 s + n_0}{d_3 s^3 + d_2 s^2 + d_1 s + d_0}, \tag{6.21}$$

where

$$n_3 \equiv L_e C C_s R_F,$$

$$n_2 \equiv (L_2 + M)(C_T + C_s) + L_e C + (L_1 + M) C_s R_F / R_s,$$

$$n_1 \equiv C_s R_F,$$

$$n_0 = 1,$$

$$d_3 \equiv C_s R_F (L_e C + L_2 C_T),$$

$$d_2 \equiv (1 + R_F/R_B)\{L_e C + L_2(C_s + C_T)\} + C_s(2M + L_1 R_F / R_s),$$

$$d_1 \equiv C_s R_F + (1 + R_F/R_B) L_1 / R_s, \text{ and}$$

$$d_0 \equiv 1 + R_F/R_B.$$

Putting (6.21) into (6.20) yields

$$1 - \frac{(A_v n_3 s^3 + A_v n_2 s^2 + A_v n_1 s + A_v n_0)}{d_3 s^3 + d_2 s^2 + d_1 s + d_0} = 0,$$

which means

$$(d_3 - A_v n_3) s^3 + (d_2 - A_v n_2) s^2 + (d_1 - A_v n_1) s + (d_0 - A_v n_0) = 0. \tag{6.22}$$

Next, the Routh–Hurwitz table is formed as seen in Table 6.3.

In order to get oscillation, we need a row with all zeros in it. If we assume that all of the coefficients of Eq. (6.22) are nonzero, the only way this can be

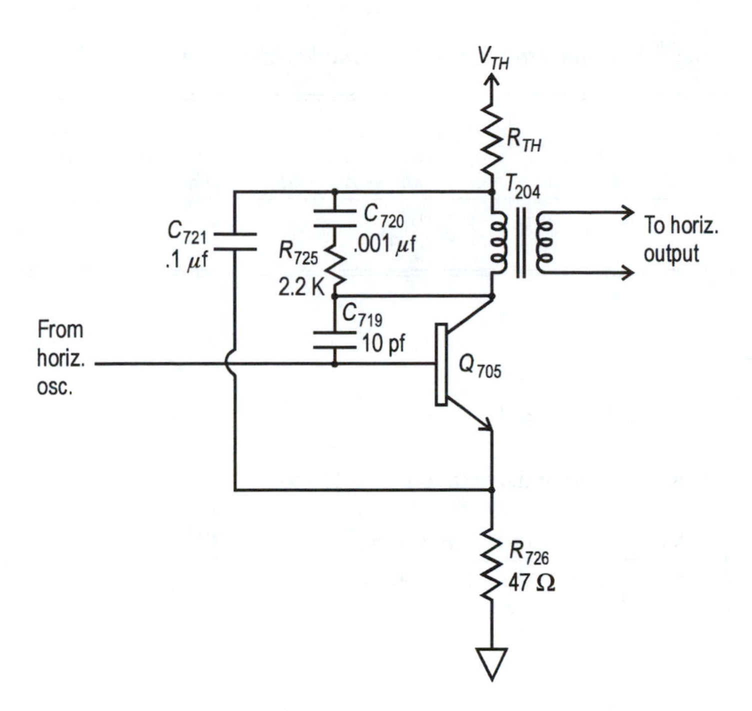

Fig. 6.45: Schematic diagram of the horizontal-driver stage of the Zenith 19EB12 television receiver.

satisfied is if the first entry in the s^1 row is zero. If the definitions of the d_i and n_i are used, we get

$$\begin{aligned}&\{(1 + R_F/R_B)[L_eC + L_2(C + C_T)] + C(2M + L_1R_F/R_s) - A_v\\&\quad\times [(L_2 + M)(C + C_T) + L_eC + (L_1 + M)CR_F/R_s]\}\\&\quad\times \{CR_F(1 - A_v) + (1 + R_F/R_B)L_1/R_s\} = CR_F\{L_eC(1 - A_v)\\&\quad+ L_2C_T\} \cdot \{(1 - A_v) + R_F/R_B\}. \qquad (6.23)\end{aligned}$$

Again we have one equation in one unknown, A_v, which is to be understood as the minimum gain that will allow oscillation. Inspection will confirm that this equation is also a quadratic that is readily solved numerically if L_1 and L_2 are known. Let us return to (6.23) and substitute into it the values of inductance from Exercise 6.10 as well as numerical values for all the resistors and capacitors. We will again ignore C_T, since it should be much less than C, and although its second-order effect on ω_o is important, its second-order effect on $A_{v_{min}}$ is not.

Performing this calculation yields the result $A_{v_{min}} = 0.18$. As we expected, the gain is less than unity and can thus be achieved with an emitter follower with lots of overdrive available.

6.3.10.4 Horizontal driver

The output from this oscillator is taken from across $R_s(=R_{722})$. Since this is a low impedance, loading by the next stage (Q_{705}, the horizontal driver) will not significantly affect the oscillator. The driver stage is shown in Fig. 6.45.

The Thevenin equivalent for the supply voltage (V_{TH} and R_{TH}) is determined by R_{229}, R_{230}, and the 135 V supply. Because there is no bias on the base, the

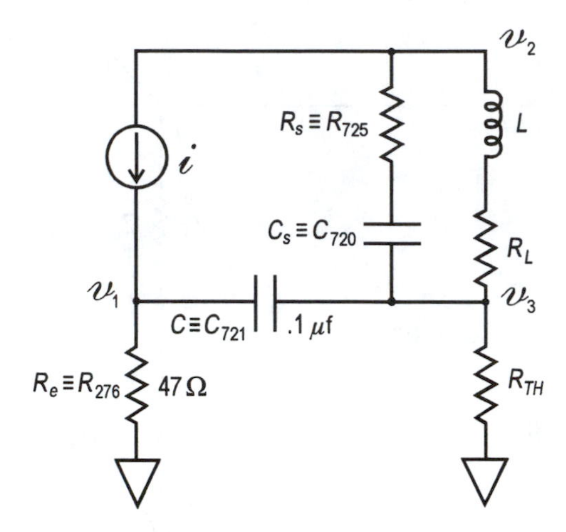

Fig. 6.46: Simplified AC equivalent circuit of the stage shown in Fig. 6.45.

stage most likely operates as a switch. The manufacturer's service information shows a waveform at the collector that is essentially a square wave, confirming this supposition. This stage differs from the usual switching stage in two significant respects:

(1) the addition of C_{720} and R_{725} across the primary of T_{204} and
(2) the addition of C_{721} between the emitter and the top of the primary of T_{204}.

Both of these components are necessary because of the inductive component of the impedance seen by the collector circuit. We will represent the transformer primary to the transistor as a load resistance, R_{L}, in series with a primary-referred leakage inductance, L. The load resistance seen by the transistor will be that seen by the secondary, transformed by the turns ratio squared. To keep the problem within reason, no other nonideality of the transformer is considered. The exact value of R_{L} is hard to pin down, since the load on the secondary is nonlinear. It seems reasonable to assume, however, that R_{L} is at least 10 times greater than R_{e}, so that the voltage across the load is much greater than that across the emitter resistance.

The circuit of Fig. 6.46 describes the situation as we have discussed it. The current source, i, represents the collector current flowing through the transistor. We have also returned the low end of R_{TH} to ground since we want to find the transient response to i, and not to the supply voltage. If we desire to know the total voltage at any point, we will have to add the steady-state component due to V_{TH}.

To analyze this circuit we could, of course, write four or five equations in as many unknowns and "turn the crank." But this is a good example of a problem that is a prime candidate for the application of the *Extra Element Theorem* (*EET*) (Middlebrook, 1989).

How many of us have not looked at a circuit and said, "That would be easy to analyze if it weren't for that one element"? The EET is an extremely powerful circuit-analysis technique that allows you to remove that "offending element" by either shorting or opening it and, after analyzing the resulting simpler circuit, perform two other simple calculations and combine the three results to get an

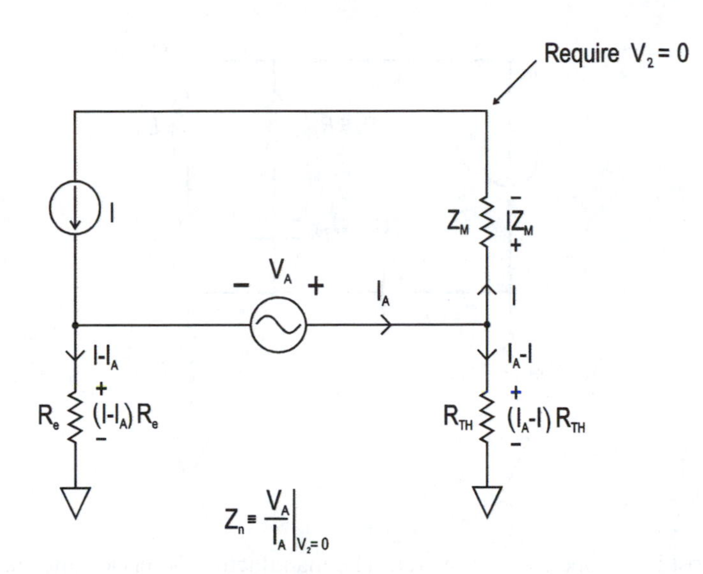

Fig. 6.47: Circuit for finding the null-double-injection impedance of the circuitry of Fig. 6.46.

exact expression for the transfer function desired. Let us see how it plays out in the circuit of Fig. 6.46. We seek the phasor transfer function $\mathbf{V}_2/\mathbf{I}$.

While in some cases it takes careful thought to decide which element is the best one to remove, in this case it is no contest. Let us remove C by making it an open circuit. Then

$$\frac{\mathbf{V}_2}{\mathbf{I}} = \mathbf{Z}_\mathrm{M} + R_\mathrm{TH}.$$

Here, $\mathbf{Z}_\mathrm{M}$ denotes the impedance of the RC branch in parallel with that of the LR branch. This result is one of the three needed to complete the solution.

Next we want to imagine that we look into the terminals where C *was* connected and find the driving-point impedance. This is calculated by turning off the *input variable*, which is defined as the one in the denominator of the desired transfer function. To turn off the current source requires that we open it. Obviously, the only impedance seen looking into the capacitor terminals is then

$$R_\mathrm{D} = R_\mathrm{E} + R_\mathrm{TH}.$$

Since R_E is 47 Ω and R_TH is 1.1 kΩ, it will obviously be a good approximation to represent R_D as simply R_TH. This is the second of the three calculations we need to finish the solution.

The third calculation sounds a little more intimidating. We need to calculate the *null-double-injection impedance*. In spite of the formidable-sounding name, it is often the easiest of the three to calculate. Though this is not the case for the present circuit, the calculation is still far less taxing than writing all the equations for Fig. 6.47 and solving them simultaneously. To do this calculation, we conceptually:

(a) Restore the source turned off in calculating R_D (or $\mathbf{Z}_\mathrm{D}$ in the more general case).

(b) Connect another source to the terminals where the extra element *used to be*. There are now two sources connected to the circuit (thus the words "double" and "injection" in the name of the impedance).
(c) "Tweak" this extra source until the output variable (the numerator of the desired transfer function) goes to zero (thus the word "null" in the name of the impedance).
(d) Take the ratio of the voltage of the added source divided by the current flowing through it to find the null-double-injection impedance, R_n (or $\mathbf{Z}_n$ in the more general case).

To see how this strategy works out, refer to Fig. 6.47, which shows the added source. To make $\mathbf{V}_2 = 0$, the voltage drops across $\mathbf{Z}_M$ and R_{TH} must be equal and opposite. Since the current through $\mathbf{Z}_M$ is $\mathbf{I}$ and the current through R_{TH} is $\mathbf{I}_A - \mathbf{I}$, this is equivalent to a requirement that

$$\mathbf{I}\mathbf{Z}_M = (\mathbf{I}_A - \mathbf{I})R_{TH} \quad \text{or} \quad \mathbf{I}(\mathbf{Z}_M + R_{TH}) = \mathbf{I}_A R_{TH}.$$

Similarly, the lower voltage loop yields

$$\mathbf{V}_A = (\mathbf{I}_A - \mathbf{I})(R_{TH} + R_E).$$

Eliminating $\mathbf{I}$ between these two equations yields the ratio of voltage to current for the added source:

$$\mathbf{Z}_n = \frac{\mathbf{V}_A}{\mathbf{I}_A} = \frac{\mathbf{Z}_M}{R_{TH} + \mathbf{Z}_M}(R_{TH} + R_E) \cong \frac{\mathbf{Z}_M R_{TH}}{R_{TH} + \mathbf{Z}_M} = \mathbf{Z}_M \parallel R_{TH}.$$

This is the third result we need. Now it is simply a matter of putting the pieces together. The proper relationship is given by Eq. (6.24):

$$\left(\frac{\mathbf{V}_2}{\mathbf{I}}\right)' = \left(\frac{\mathbf{V}_2}{\mathbf{I}}\right) \cdot \frac{1 + \mathbf{Z}_n/\mathbf{Z}_x}{1 + \mathbf{Z}_d/\mathbf{Z}_x} = (\mathbf{Z}_M + R_{TH}) \cdot \frac{1 + \frac{R_{TH} \parallel \mathbf{Z}_M}{1/sC}}{1 + \frac{R_{TH}}{1/sC}}$$

$$= (\mathbf{Z}_M + R_{TH}) \frac{1 + sC(R_{TH} \parallel \mathbf{Z}_M)}{1 + sCR_{TH}}. \qquad (6.24)$$

In this equation, the unprimed transfer function refers to the circuit with the extra element removed. The primed transfer function refers to the whole circuit with the extra element restored. $\mathbf{Z}_x$ is the impedance of the element we removed – in this case, the capacitor C.

Before we go on to the completion of this analysis, we need to make a couple of clarifications. We mentioned that the troublesome element could be removed by opening or shorting it. If it had been removed by shorting, Eq. (6.24) would have had to be modified slightly. In that case, the ratios of impedances in both the numerator and denominator have to be inverted. Also note that the extra element must be an impedance, even if it is composed of several elements. It cannot be a source. However, if there is an impedance in series with a current source, and we let that impedance go to the open-circuit condition, the source is removed also. The dual is true for an impedance in parallel with a voltage source.

Further algebraic simplification reduces the result of (6.24) to

$$\left(\frac{\mathbf{V}_2}{\mathbf{I}}\right)' = \mathbf{Z}_\mathrm{M} + \frac{R_\mathrm{TH}}{1 + sCR_\mathrm{TH}}.$$

We wish to find the step response, which should closely approximate the turn-on and turn-off transients of the transistor. Accordingly, we set

$$\mathbf{V}_2(s) = \mathbf{I}(s)\left[\mathbf{Z}_\mathrm{M} + \frac{R_\mathrm{TH}}{1 + sCR_\mathrm{TH}}\right].$$

The step response will be found by inputting a current step. Accordingly, we make $\mathbf{I}(s) = |I_\mathrm{o}|/s$. When this is inserted into the above equation, we obtain

$$\mathbf{V}_2(s) = \frac{|I_\mathrm{o}|}{s} \cdot \left[\mathbf{Z}_\mathrm{M} + \frac{R_\mathrm{TH}}{1 + sCR_\mathrm{TH}}\right]. \tag{6.25}$$

Expanding $\mathbf{Z}_\mathrm{M}$ yields

$$\mathbf{Z}_\mathrm{M} = \frac{s^2LC_\mathrm{s}R_\mathrm{s} + s(L + C_\mathrm{s}R_\mathrm{s}R_\mathrm{L}) + R_\mathrm{L}}{s^2LC_\mathrm{s} + sC_\mathrm{s}(R_\mathrm{L} + R_\mathrm{s}) + 1}. \tag{6.26}$$

Let us substitute (6.26) into (6.25), so that we can take the inverse Laplace transform and obtain the time-domain response. The result is

$$\mathbf{V}_2(s) = \frac{|I_\mathrm{o}|}{s} \cdot \left[\frac{s^2LC_\mathrm{s}R_\mathrm{s} + s(L + C_\mathrm{s}R_\mathrm{s}R_\mathrm{L}) + R_\mathrm{L}}{s^2LC_\mathrm{s} + sC_\mathrm{s}(R_\mathrm{L} + R_\mathrm{s}) + 1} + \frac{R_\mathrm{TH}}{1 + sCR_\mathrm{TH}}\right]. \tag{6.27}$$

It might be thought strange that R_E does not appear in this expression. This is a consequence of the fact that R_E only appears in the equations in series with R_TH, which, we argued, is by far the larger of the two. A check of the schematic will verify that these two resistors are, in fact, in series. Despite their grounded common connection, current does not flow into or out of ground.

Since the first term on the right-hand-side of (6.27) is second order, the form of the time-domain response will depend on the type of damping. The C_s–R_s network is a *snubbing* network, whose purpose is to absorb the stored inductive energy when the transistor turns off. As such, it is reasonable to assume that these two components have been chosen to give near critical damping.

Exercise 6.12

(a) If R_s and C_s have been selected for critical damping, what value of L does the transformer exhibit?

(b) The manufacturer's service information shows that the waveform is slightly underdamped. Will this mean that L is larger or smaller than your calculated value?

The calculation performed in Exercise 6.12 is a very meaningful one. If we did not know the value of the inductor and it had to be replaced, the set would function satisfactorily with a 2 mH inductor. If this value for L and the other

Fig. 6.48: Schematic diagram of the horizontal output stage of the Zenith 19EB12 television receiver.

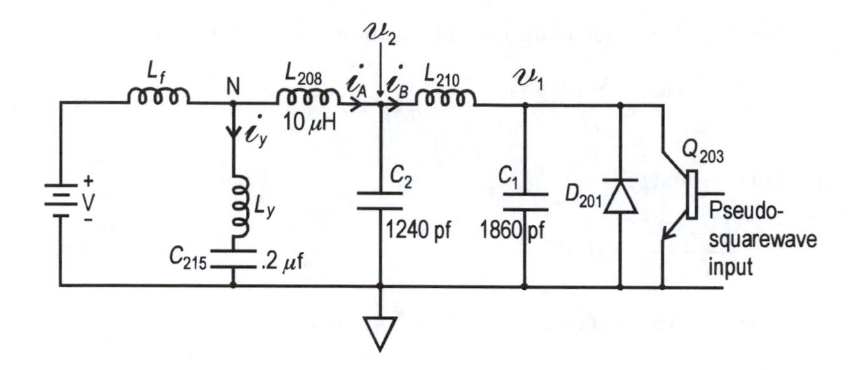

numerical values are inserted into (6.27), we find that there are three terms in the time-domain solution. The first term is a rapidly damped sinusoid. This is certainly in keeping with the nature of the response we see at the rising edge of each square pulse. The second term is a constant due to the flow of I through R_L. The third term is a rising exponential due to the charging of C, which is effectively in parallel with R_{TH}. This rise is quite slow compared to the width of the pulses and accounts for the slightly rising response shown in the service data.

Exercise 6.13

(a) Starting with Eq. (6.27), find the initial value of dv_2/dt. If this value remained constant, how long would it take for v_2 to swing between V_{TH} and zero?

(b) How does this compare to the time of conduction (which is approximately 0.5 × horizontal sweep period)?

6.3.10.5 Horizontal output stage

The pseudo-square-wave output from this stage is coupled through T_{204} to the base of the horizontal output stage, Q_{203}. The essential elements of this stage's schematic are shown in Fig. 6.48. In this schematic:

V is the 135 V supply. We have ignored the 100 Ω resistor, R_{233}.

L_y is the inductance of the horizontal yoke and the width coil in series with it. The effect of R_{232} in shunt with the width coil is ignored. Furthermore the 0.2 μF series capacitor, C_{215}, will be considered an AC short. It only serves to prevent the inductors from shorting V.

L_f is the inductance of the flyback primary. C_1 and C_2 are both formed from the parallel combination of several capacitors. This sort of thing is often done when a manufacturer can get a good buy on a particular capacitor value.

L_{210} is simply a ferrite bead slipped over the wire. Its exact inductance value will depend on the geometry and permeability of the bead, but 0.1 μH is probably of the right order of magnitude.

There are many features of this circuit that we could consider, but we will restrict ourselves to two of them.

We shall consider first the form of the current, i_y, when the transistor is saturated. We know that this should be a ramp in order to give linear deflection. It is not physically obvious, however, that this will be the case.

When Q_{203} is saturated, the current equation at node N is

$$\frac{\mathbf{V}-\mathbf{V}_{\mathrm{N}}}{sL_f}=\frac{\mathbf{V}_{\mathrm{N}}}{sL_{\mathrm{y}}}+\frac{\mathbf{V}_{\mathrm{N}}-\mathbf{V}_2}{sL_{208}}=\mathbf{I}_{\mathrm{y}}+\mathbf{I}_{\mathrm{A}} \tag{6.28a}$$

and at the v_2 node is

$$\frac{\mathbf{V}_{\mathrm{N}}-\mathbf{V}_2}{sL_{208}}=sC_2\mathbf{V}_2+\frac{\mathbf{V}_2}{sL_{210}}. \tag{6.28b}$$

When these two equations are solved for $\mathbf{V}_{\mathrm{N}}$, the result is

$$\frac{\mathbf{V}_{\mathrm{N}}}{\mathbf{V}}(s)=\frac{s^2L_{208}C_2+\dfrac{L_{208}}{L_{210}}+1}{s^2L_{208}C_2\left(1+\dfrac{L_f}{L_{\mathrm{y}}}+\dfrac{L_f}{L_{208}}\right)+\left(1+\dfrac{L_f}{L_{\mathrm{y}}}\right)\left(\dfrac{L_{208}}{L_{210}}+1\right)+\dfrac{L_f}{L_{210}}}. \tag{6.29}$$

The current through the yoke is then

$$\mathbf{I}_{\mathrm{y}}(s)=\frac{\mathbf{V}_{\mathrm{N}}(s)}{sL_{\mathrm{y}}}. \tag{6.30}$$

Therefore

$$\mathbf{I}_{\mathrm{y}}(s)=\mathbf{V}(s)\left[s^2L_{208}C_2+\frac{L_{208}}{L_{210}}+1\right]\Big/\{s[s^2L_{208}C_2(L_{\mathrm{y}}+L_f+L_fL_{\mathrm{y}}/L_{208})+(L_{\mathrm{y}}+L_f)(L_{208}/L_{210}+1)+L_fL_{\mathrm{y}}/L_{210}]\},$$

which can be reduced to

$$\mathbf{I}_{\mathrm{y}}(s)=\frac{\dfrac{\mathbf{V}(s)}{(L_{\mathrm{y}}+L_f)}\left[s^2L_{208}C_2+\dfrac{L_{208}}{L_{210}}+1\right]}{s\left[s^2(L_{208}+L_{\mathrm{TH}})C_2+\left(\dfrac{L_{208}}{L_{210}}+1\right)+\dfrac{L_{\mathrm{TH}}}{L_{210}}\right]},$$

where we have defined

$$L_{\mathrm{TH}}\equiv\frac{L_{\mathrm{y}}L_f}{L_{\mathrm{y}}+L_f}$$

as the "Thevenin inductance," since it is the value of the Thevenin impedance of the source which incorporates $\mathbf{V}(s)$.

We now utilize the following inequalities relative to the inductances:

$$L_{210}\ll L_{208}\ll L_{\mathrm{y}},L_f,$$

since L_{y} and L_f should be of the order of millihenries. Thus

$$\mathbf{I}_{\mathrm{y}}(s)\cong\frac{\dfrac{\mathbf{V}(s)}{(L_{\mathrm{y}}+L_f)}\left[s^2L_{208}C_2+\dfrac{L_{208}}{L_{210}}\right]}{s[s^2L_{\mathrm{TH}}C_2+L_{\mathrm{TH}}/L_{210}]}.$$

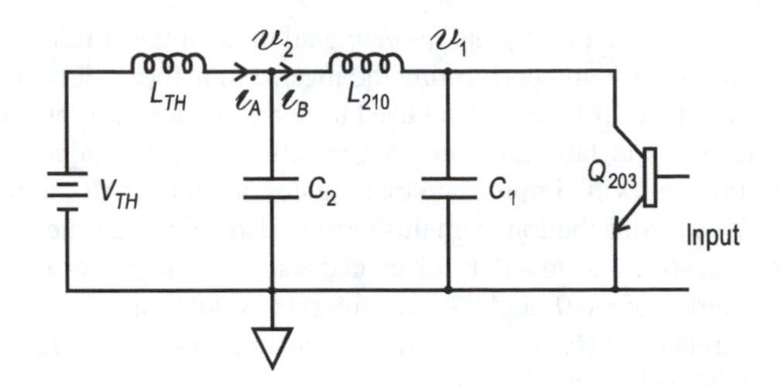

Fig. 6.49: Simplification of the circuit of Fig. 6.48 suitable for calculating the nature of the pulses at the collector of Q_{203}.

Factoring out common terms, we get

$$\mathbf{I}_y(s) = \frac{\mathbf{V}(s)L_{208}}{sL_{\text{TH}}(L_y + L_f)} = \frac{\mathbf{V}(s)L_{208}}{sL_yL_f}.$$

If we use $\mathbf{V}(s) = |V|/s$, we have

$$\mathbf{I}_y(s) = \frac{|V|L_{208}}{\cdot L_yL_f} \cdot \frac{1}{s^2},$$

which transforms into the time domain as

$$i_y(t) = \frac{|V|L_{208}}{L_yL_f}t,$$

which is the required ramp form.

The second aspect of the circuit that we wish to examine is the waveform of v_1. The service information shows narrow (~9 μsec) high-amplitude (~960 V) pulses at this point. Physically, these arise when the transistor goes into cutoff and the stored energy in the inductors and capacitors is redistributed.

For this analysis it will be advantageous to begin by incorporating v, L_F, L_y, and L_{208} into a Thevenin equivalent source to begin. We will also omit the diode, D_{201}, in our analysis and then reckon its effect by a physical argument from our results. Figure 6.48 then simplifies to Fig. 6.49.

Since we want to examine the pulses, our analysis will assume the transistor is an open circuit. However, since the pulse phenomenon is dependent upon stored energies in the reactances, the circuit formulation must include "initial values" of the capacitor voltages and inductor currents (i.e., those present just before the transistor switches from saturation to cutoff). We will write the circuit equations for $v_1(0) = 0$, since this is assured by the saturated transistor. The transformation equations are

$$\mathbf{V}_{\text{TH}} = L_{\text{TH}}[s\mathbf{I}_{\text{A}} - i_{\text{A}}(0)] + \mathbf{V}_2,$$
$$\mathbf{I}_{\text{A}} = C_2[s\mathbf{V}_2 - v_2(0)] + \mathbf{I}_{\text{B}},$$
$$\mathbf{V}_2 = L_{210}[s\mathbf{I}_{\text{B}} - i_{\text{B}}(0)] + \mathbf{V}_1,$$
$$\mathbf{I}_{\text{B}} = sC_1\mathbf{V}_1.$$

The transform, $\mathbf{V}_1(s)$, can be solved for from these equations to give

$$\mathbf{V}_1(s) = \frac{\mathbf{V}_{\text{TH}}(s) + sC_2L_{\text{TH}}v_2(0) + L_{\text{TH}}i_{\text{A}}(0) + [1 + s^2L_{\text{TH}}C_2]L_{210}i_{\text{B}}(0)}{s^4L_{\text{TH}}L_{210}C_1C_2 + s^2(L_{\text{TH}}C_2 + L_{\text{TH}}C_1 + L_{210}C_1) + 1}. \tag{6.31}$$

At this point, we are faced with evaluating all three of the "initial condition" terms in the numerator and performing the inverse transforms. The initial conditions are evaluated from the model used to find the yoke current. The whole process is a long and laborious one. Since little would be gained by going through all this, we will simply state that the dominant term in the numerator is $L_{TH}\dot{\imath}_A(0)$. The contribution from this term is also about 12 times as large as that due to V_{TH}, so that term will be discarded also. We will perform the inverse transform in terms of $\dot{\imath}_A(0)$ and then consider its evaluation.

The denominator of (6.31) is a quadratic equation in s^2. Since $L_{210} \ll L_{TH}$, the roots can be well approximated as

$$s^2 = -1/L_{TH}(C_1 + C_2) \equiv -1/L_{TH}C_p,$$

$$s^2 = -(C_1 + C_2)/L_{210}C_1C_2 \equiv -1/L_{210}C_s,$$

where C_p and C_s are the equivalent parallel and series capacitances of C_1 and C_2.

Accordingly, we seek the inverse transform of

$$\mathbf{V}_1(s) = \frac{\dot{\imath}_A(0)}{L_{210}C_1C_2(s^2 + 1/L_{TH}C_p)(s^2 + 1/L_{210}C_s)},$$

which may be inverse transformed as the integral of the tabulated transform pair

$$\frac{s}{(s^2 + a^2)(s^2 + b^2)} \Leftrightarrow \frac{\cos at - \cos bt}{b^2 - a^2}.$$

Accordingly,

$$v_1(t) = \frac{\dot{\imath}_A(0)}{L_{210}C_1C_2} \cdot \left[\frac{\sin \dfrac{t}{\sqrt{L_{TH}C_p}}}{\dfrac{1}{\sqrt{L_{TH}C_p}}} - \frac{\sin \dfrac{t}{\sqrt{L_{210}C_s}}}{\dfrac{1}{\sqrt{L_{210}C_s}}} \right]$$

$$\cdot \frac{1}{1/L_{210}C_s - 1/L_{TH}C_p}.$$

Since $C_s < C_p$ and $L_{210} \ll L_{TH}$, some terms will be negligible. Using these inequalities reduces the previous equation to

$$v_1(t) = \frac{\dot{\imath}_A(0)}{C_1C_2}\sqrt{L_{TH}C_p} \cdot C_s \cdot \sin \frac{t}{\sqrt{L_{TH}C_p}}.$$

Since $C_s \equiv C_1C_2/(C_1 + C_2)$ and $C_p \equiv C_1 + C_2$

$$v_1(t) = \dot{\imath}_A(0) \cdot \sqrt{\frac{L_{TH}}{(C_1 + C_2)}} \cdot \sin \frac{t}{\sqrt{L_{TH}(C_1 + C_2)}}. \tag{6.32}$$

The ratio $\sqrt{L_{TH}/(C_1 + C_2)}$ is the effective resistance through which $\dot{\imath}_A(0)$ flows. Using $C_1 + C_2 = 3{,}100$ pF and assigning L_{TH} a typical value of 3 mH, this resistance comes to about 1.0 kΩ.

We now need to go back to Eqs. (6.28a), (6.29), and (6.30) to find $i_A(t)$ when the transistor is saturated.

Exercise 6.14 Again assuming $L_{210} \ll L_{208} \ll L_{TH}$, show from (6.28a), (6 29), and (6.30) that $i_A(t)$ is also a ramp given by

$$i_A(t) = \frac{|V|}{L_f} t.$$

This result applies while the transistor is saturated, which lasts for about 35 μsec, according to the waveform in the service information. If we assume $L_f = 5$ mH,

$$i_A(35\ \mu\text{sec}) \rightarrow i_A(0) = \frac{135}{.005} \times 35 \times 10^{-6} \approx 1\ \text{A}.$$

If this typical value is inserted into (6.32), we get

$$v_1(t) = 1000 \cdot \sin \frac{t}{\sqrt{L_{TH}(C_1 + C_2)}}.$$

This amplitude should be compared with the actual value of 960 V mentioned previously. Not only is it clearly of the right order of magnitude, but it is remarkably close, considering that the values used for L_F and L_y were educated guesses.

Likewise, the time at which the first positive half-cycle is completed is found by setting

$$\pi = \frac{t_w}{\sqrt{L_{TH}(C_1 + C_2)}}.$$

Again using $L_{TH} = 3$ mH and $C_1 + C_2 = 3{,}100$ pF, we find $t_w = 9.6\ \mu$sec. This again represents remarkably good agreement with the actual value of 9 μsec, cited previously.

When this voltage reverses, the negative half-cycle is shorted out by the diode across Q_{203} making v_1 remain at zero rather than allowing it to swing negative. Hence this is the damper diode.

It is left as an exercise to the reader to find the behavior of V_N when the transistor goes into cutoff. Physically, we expect a large pulse here also. This pulse is transformed up by the flyback transformer, T_{205}, and rectified by diode, D_{202}, to provide high DC voltage to the picture tube. The CRT has conductive coatings over the inside and outside of the bell. The outer coating is grounded, and the inner is connected to the high voltage. Thus the glass serves as the dielectric in a very large filter capacitor. Because the ripple frequency is so high, the filtering is excellent.

In the vicinity of the CRT are a number of spark gaps that protect external circuitry if the high voltage in the CRT should internally arc over to another electrode.

6.3.11 Variations

Other monochrome sets of this vintage, although performing the same circuit functions, often used different circuitry to do so.

In the tuners, channel switching was usually accomplished either by wafer switches or turrets. The tuner in the Zenith 19EB12 is quite unusual in that the local oscillator switching is of the turret type, and all the rest of the channel switching is done with wafer switches. Both types of tuners have been rendered obsolete by electronic tuning as discussed in Sections 3.3 and 3.4.

The tuned circuits coming into the IF strip, as well as the interstage coupling circuits, can be quite different from those in this set, which has a particularly elegant input filter.

The video detector is, as has been mentioned, unorthodox not only in not having the lower end of T_{101}'s secondary grounded (to facilitate AGC takeoff) but also in the number of inductors used in the filter.

Other sets may use simpler AGC circuitry than the keyed circuitry used in the Zenith. Color sets have almost all used keyed AGC, but only the better monochrome sets did so when discrete circuitry was dominant. Simpler AGC circuitry is similar in principle to the AVC we saw used on radios.

In the video amplifier, the contrast control is often found in the emitter circuit. Contrast change can be accomplished in one of two ways. The first way is to have the resistance in the emitter circuit variable, controlling the negative feedback and thus the stage gain. The other way is to have a potentiometer connected as a fixed emitter resistor, but to take the output video signal from the slider of it, as was done in this set.

The brightness control varies the DC bias on the CRT cathode. In some sets it varies the control grid's DC potential. In fact it is not uncommon to find circuits where the video is input to the control grid and the cathode has a fixed potential. The brightness control can be connected to either grid or cathode in this case also.

As has been pointed out, the sound can be taken off before the video detector and passed through a separate AM detector prior to the sound IF amplifier, or it can be taken off right at the output of the video detector or at some point within the video amplifier, as is the case here, where it is taken from the output of the video driver.

The sound detector was, in most sets of this vintage, a ratio detector or an integrated quadrature detector, as in our set.

Several different configurations were used for the vertical oscillator, but the two most common were the blocking oscillator and the multivibrator. Within these main divisions, there are numerous variations. It was noted that this set features a very asymmetrical multivibrator. Although there will always be some degree of asymmetry, the circuit used in this set is particularly asymmetrical.

There is even more variety in the horizontal-sweep section of discrete TV receivers due to different forms of the phase detector, the AFC, and the oscillator itself. Here also, multivibrators and blocking oscillators were frequently used.

In addition to the phase detector used in this set, there is a variety that requires the use of a phase splitter (often called sync splitter) to drive it. An example of this circuitry is shown in Fig. 6.50.

In addition to circuitry in which the AFC stage is capacitive, there are also versions where the AFC stage looks inductive.

Many receivers do not have a reactance transistor at all, but the DC voltage from the phase detector is filtered and applied directly to the horizontal oscillator. Where an emitter-coupled Hartley is used as the horizontal oscillator (as

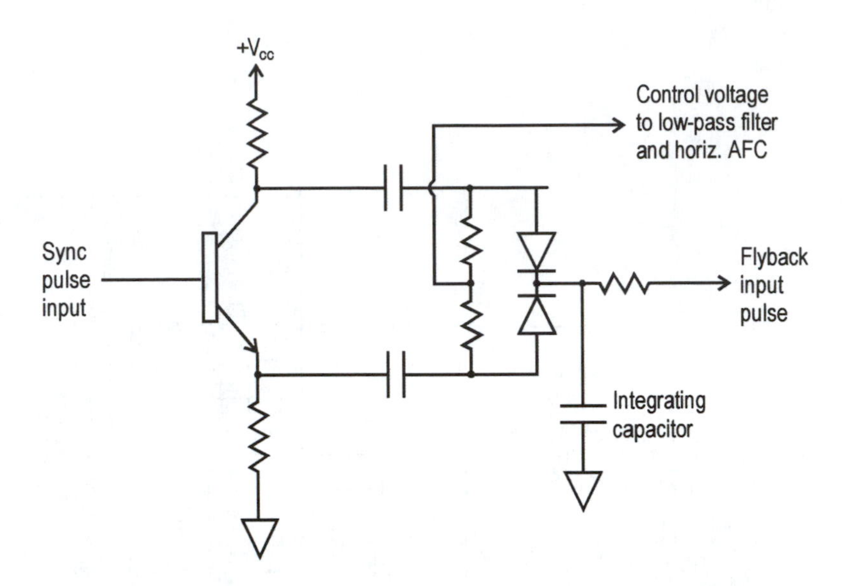

Fig. 6.50: Schematic diagram of a typical discrete sync splitter.

in our circuit), it is more satisfactory to use a reactance stage. This is because the configuration oscillates so stably that the frequency does not depend on any transistor parameter to the first order. This frequency independence can be confirmed for our circuit by referring back to Eq. (6.19).

To the multivibrator and blocking oscillator we add a third commonly used horizontal oscillator configuration – the sinusoidal oscillator. The former two configurations are generally fed to horizontal-output stages that operate linearly, rather than those that use the switching approach found in this set.

We next update our understanding of monochrome TV circuitry by moving from the discrete design we have been studying to a more modern, integrated version.

6.4 Analysis of an integrated monochrome receiver

The Zenith model 5NB4X, a small 5″ portable set, is highly integrated. As may be seen from its schematic in Fig. 6.51, it basically has three integrated circuits and six transistors (plus those in the tuners, which are not shown on this schematic). The transistors are distributed as follows:

(1) Q_{601}–Q_{603} in the vertical output amplifier,
(2) Q_{801} in the video output amplifier, and
(3) Q_{501}–Q_{502} in the horizontal output amplifier.

Observe that all of these transistors are used in high-power stages. One IC handles the last power stage – the audio. All other functions of the TV are handled by the remaining two ICs. The largest chip is a type manufactured by Motorola who call it the Monomax™ TV Subsystem. The chip number is MC13001. A block diagram of this chip, shown in Fig. 6.52, shows the high level of integration.

In Fig. 6.51A, switch S_{202} selects between the built-in whip antenna of the TV and an external 300 Ω dipole. If the latter is selected, the antenna signal

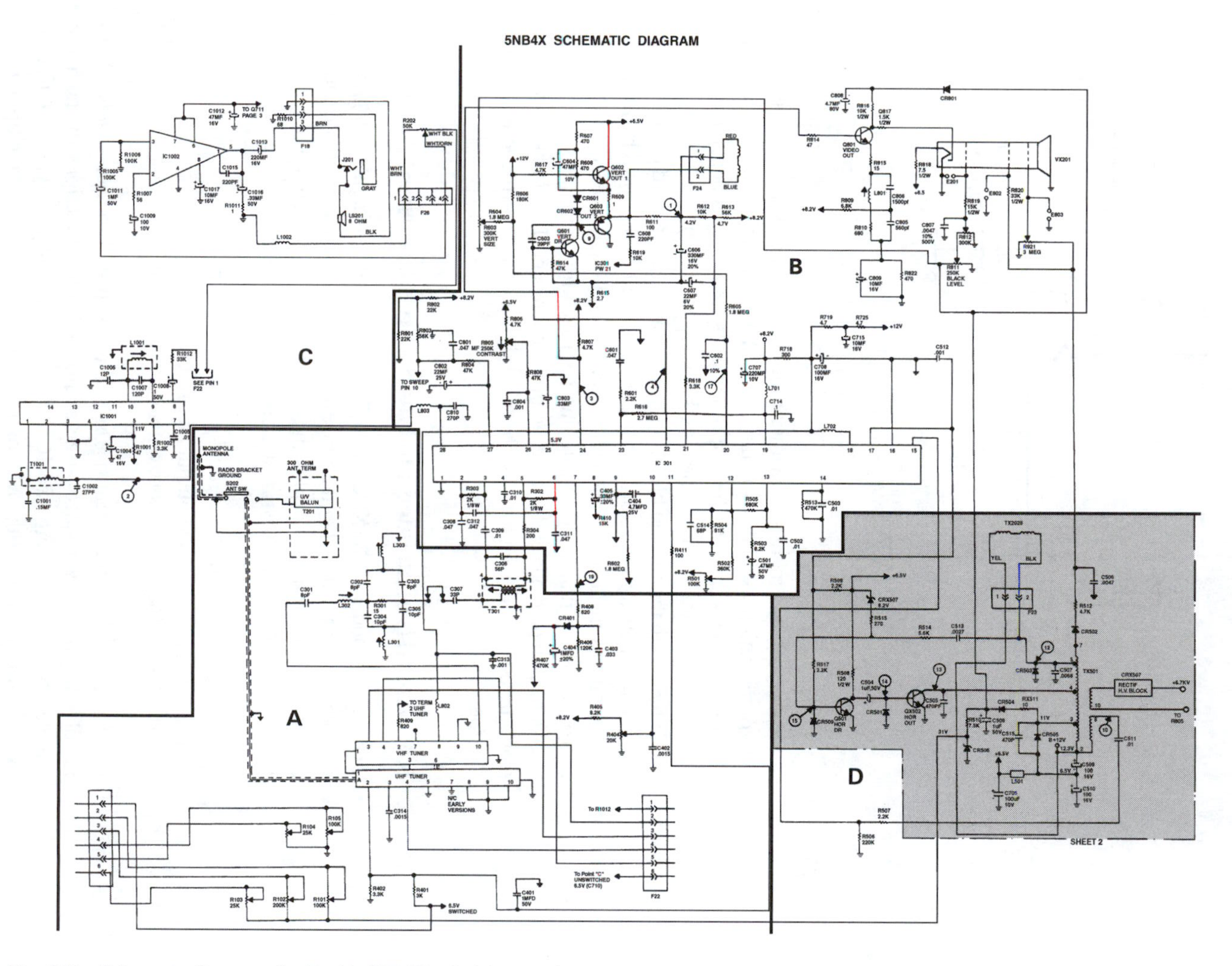

Fig. 6.51 Schematic diagram of a Zenith 5NB4X television receiver. An electronic version of this figure is available on the world-wide web at http://www.cup.org/titles/58/0521582075.html.

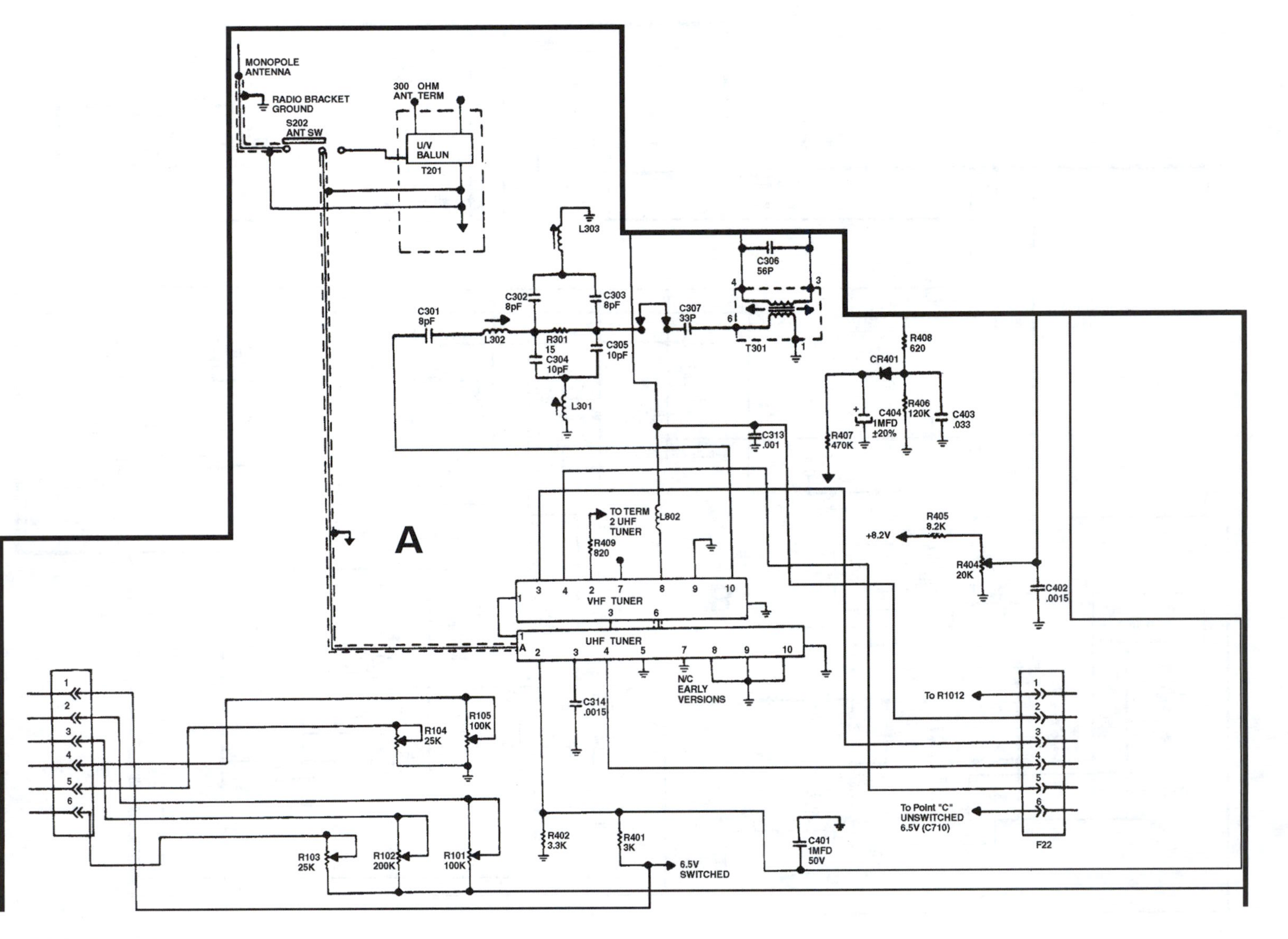

Fig. 6.51A Front end.

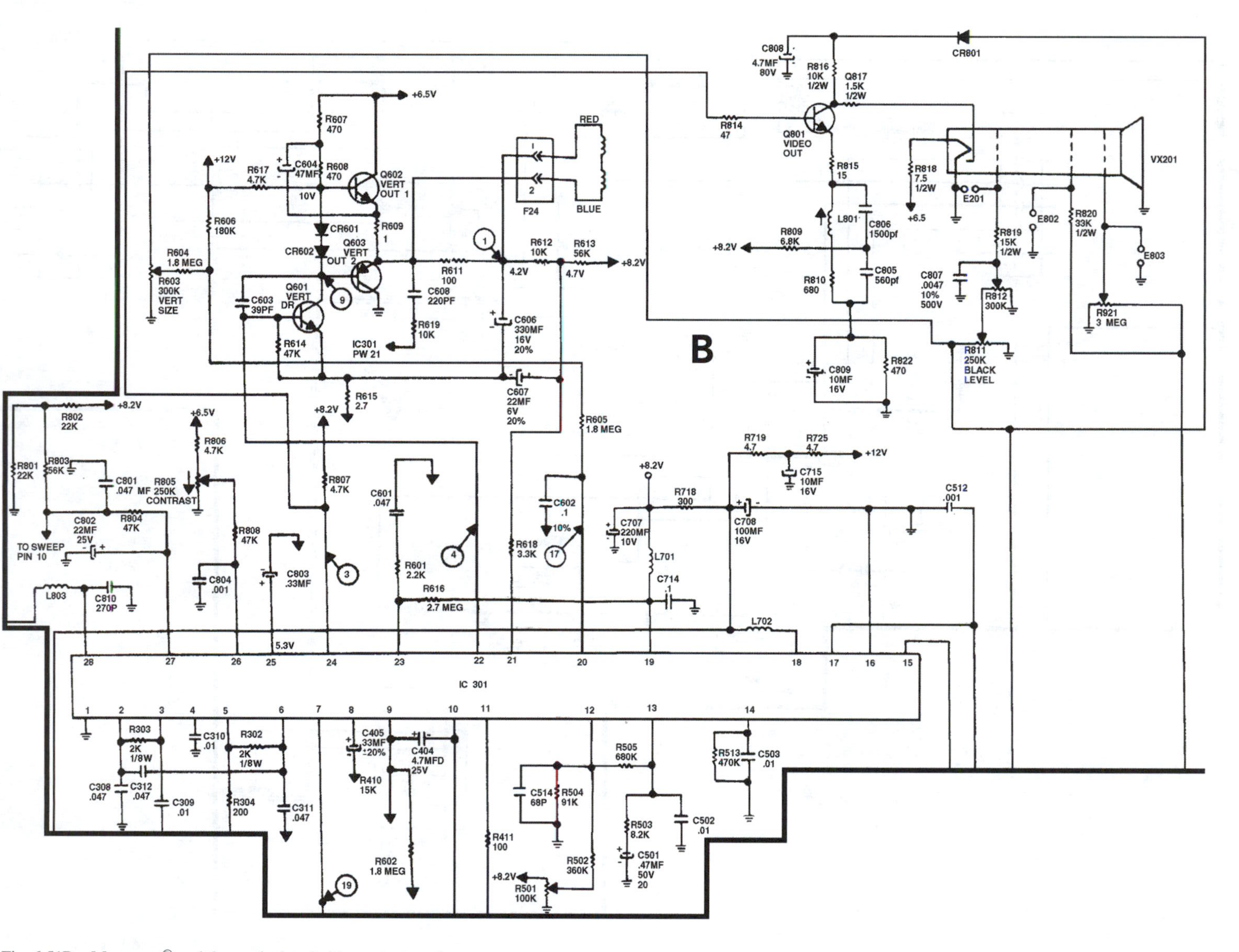

Fig. 6.51B Monomax® and the vertical and video output sections.

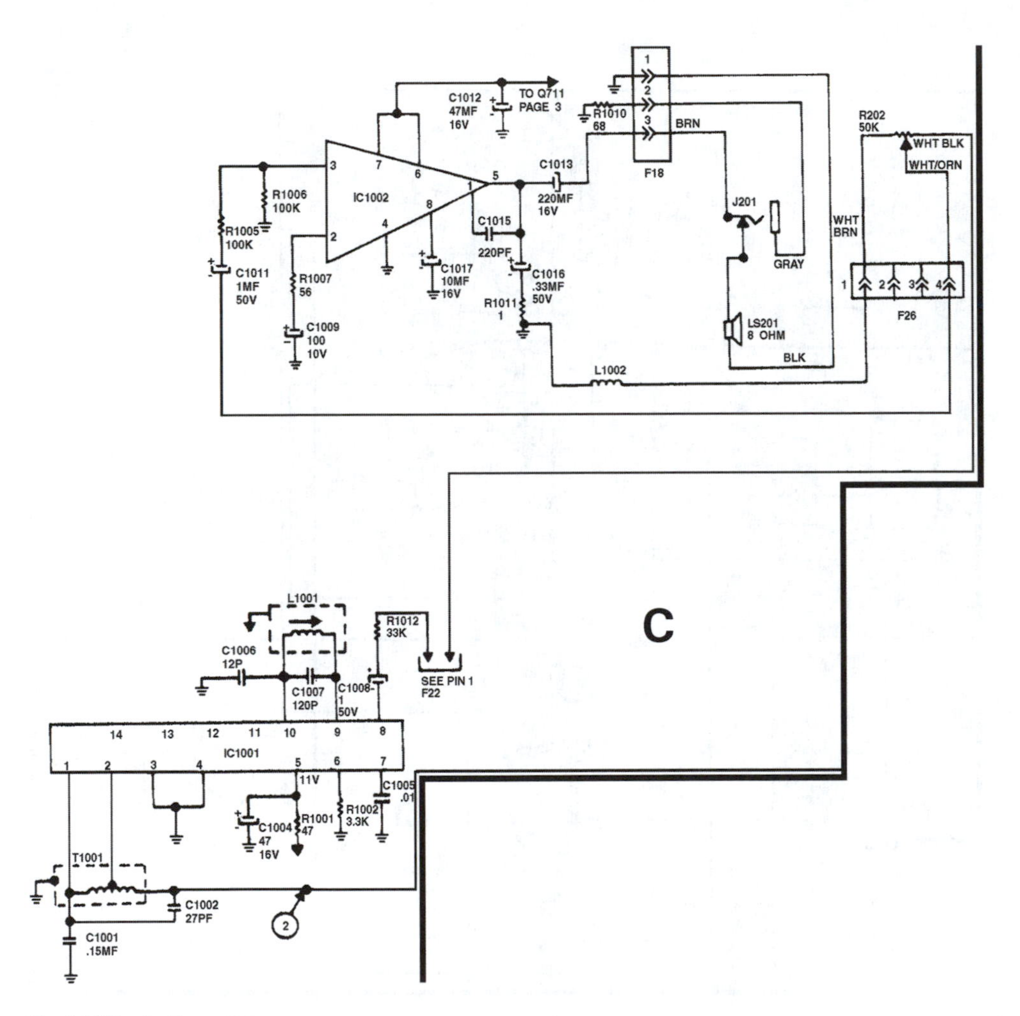

Fig. 6.51C Audio sections.

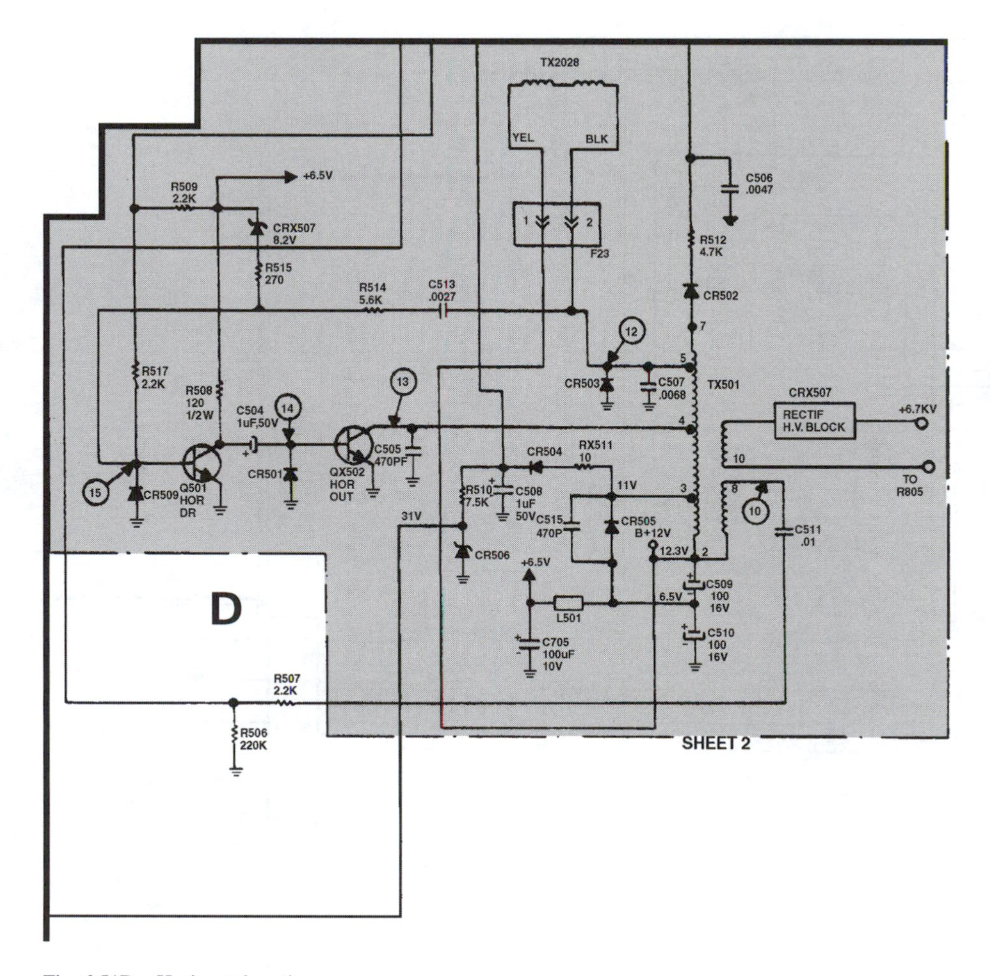

Fig. 6.51D Horizontal section.

passes through a balun transformer, T_{201}. V_{CC} is applied at pin 8 of the VHF tuner. The IF output comes from pin 10 of this tuner and passes through a bridged double-tee filter very similar in form and function to that in the discrete Zenith monochrome receiver just analyzed, and hence to T_{301}, which is a video IF transformer.

The output of T_{301} is differentially balanced above ground and is applied to the IF input of the Monomax™ chip. The chip provides all necessary IF amplification as well as video detection. The chip video detector is a full-wave envelope detector and features nonlinearity compensation, achieved by incorporating a similar nonlinear element in the feedback loop of the detector output amplifier. As shown in Fig. 6.52, and as was the case for the discrete TV analyzed, four outputs are derived from the video detector output:

(1) The sound output goes to a sound IF-quadrature detector chip and from there through the volume control, R_{202}, to the audio output IC, which drives a speaker. The quadrature chip is a Motorola MC1358 type, and the audio chip appears to be a National LM386.

(2) The video output passes first through a contrast (video gain) control. This has a gain that varies by a factor of 14 as the DC voltage on pin 26 is varied by the contrast control. This contrast range runs between the white and black reference levels. This section also includes a beam-current limiter that automatically reduces the contrast if the beam current is too large. A clamping circuit is included that holds the black level of the video signal at or near the voltage required to produce a black screen despite changes in the contrast setting, temperature, and power supply voltage. It is this circuitry that makes the beam limiter effective. At the black level, the beam current is essentially zero. Since this level is fixed, increasing contrast consists solely of making the white whiter. This, of course, entails an increase in beam current. Thus, when the beam limiter lowers the contrast, it makes the white less white and thus lowers the beam current.
All of the video circuitry discussed thus far is in the "Video Process(or)" block of Fig. 6.52. The subsequent "Blank Buffer" block performs a buffering of the video signal and adds vertical-blanking pulses. The output of this stage is taken from pin 24 of the Monomax™ to the video output amplifier.

(3) The AGC is of the gated or keyed variety as was the case for the discrete design previously analyzed. The keying pulses come into the chip at pin 15 and are buffered before being applied to the AGC circuitry and the horizontal phase detector #2. The AGC block has separate outputs for controlling the RF and IF stages as well. The RF AGC in our receiver is taken from pin 11 of the chip to pin 2 of both the UHF and VHF tuners. A variable resistive voltage divider connected to pin 10 of the chip sets the threshold for the "delayed" AGC. The chip also outputs a "feedforward AGC" on pin 9, which is AC-coupled to the AGC threshold line by C_{406}. This provides additional AGC range during transient strong-signal conditions without modifying the time constants or endangering the stability of the main AGC system. Since

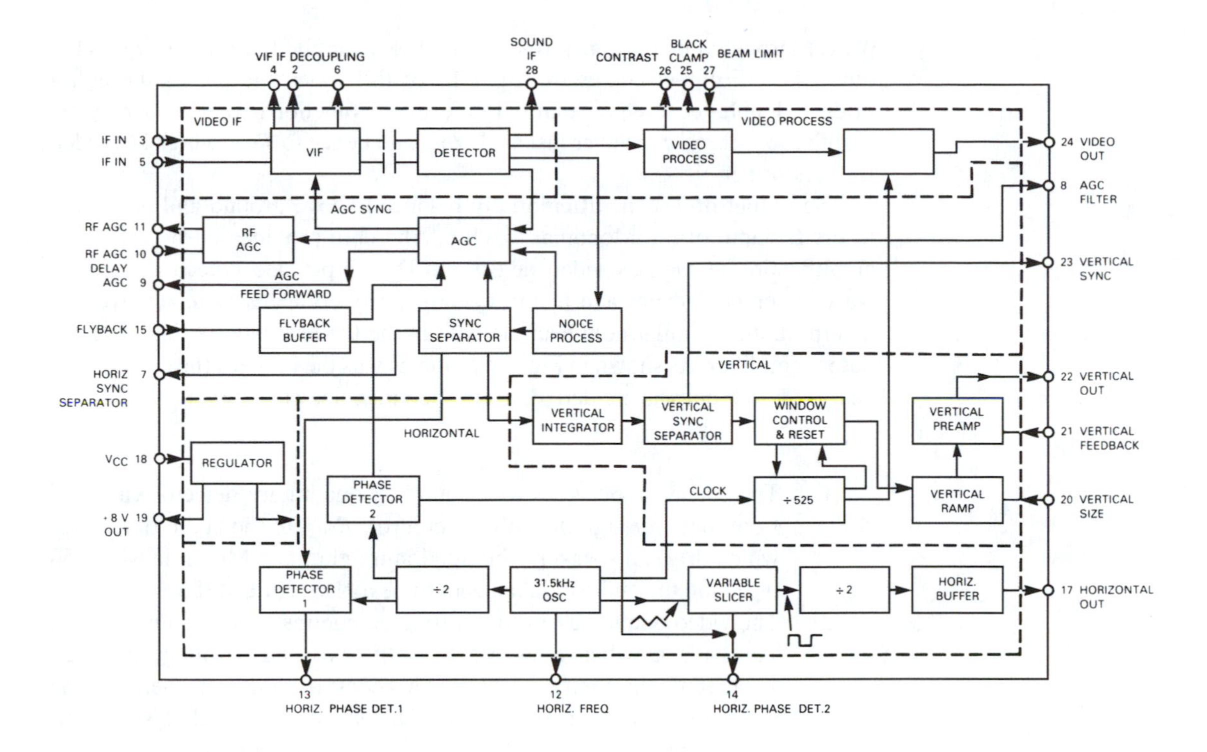

Fig. 6.52: Block diagram of the Motorola Monomax^{TM} integrated circuit, which is the heart of the Zenith 5NB4X television receiver.

the video IF amplifier is "on board," the AGC connection to it is internal to the chip.

(4) The detected video is also fed to the sync section after passing through a block called the "Noise Process(or)." The signal out of the video detector is inverted and hence has negative-going sync pulses. When the circuitry senses a negative voltage 1.4 V or more below that of the sync tips, it is assumed to be noise and is used to generate a pulse which blanks the video signal to the AGC and sync circuits. The use of this noise-blanked video eliminates picture instability induced by noise.

The components connected to the Monomax^{TM} pin 7 determine the clipping level and time constant. Good separation of horizontal sync demands a relatively short time constant, which is provided by R_{406} and C_{403}. However, this is too short for good vertical-sync separation. Diode CR_{401} connects a longer time constant (R_{407} and C_{404}) to pin 7 when it conducts. The horizontal-sync pulses are short enough that they do not raise the potential of C_{403} enough to cause significant conduction of the diode. Vertical-sync pulses are much longer and will allow the diode to conduct, thus activating the longer time constant and allowing good vertical-sync separation. The separated horizontal and vertical sync pulses are fed to their respective sweep sections for further processing.

The horizontal-sweep circuitry is primarily two interlocked PLLs, one of which secures frequency-lock to the sync pulses, and one of which secures phase-lock between the horizontal-amplifier drive signal and the flyback pulse (i.e., it serves as the phase detector).

The oscillator is a VCO with a triangle-wave output running at a frequency of 31.5 kHz. This is squared up and divided by two to give a frequency of

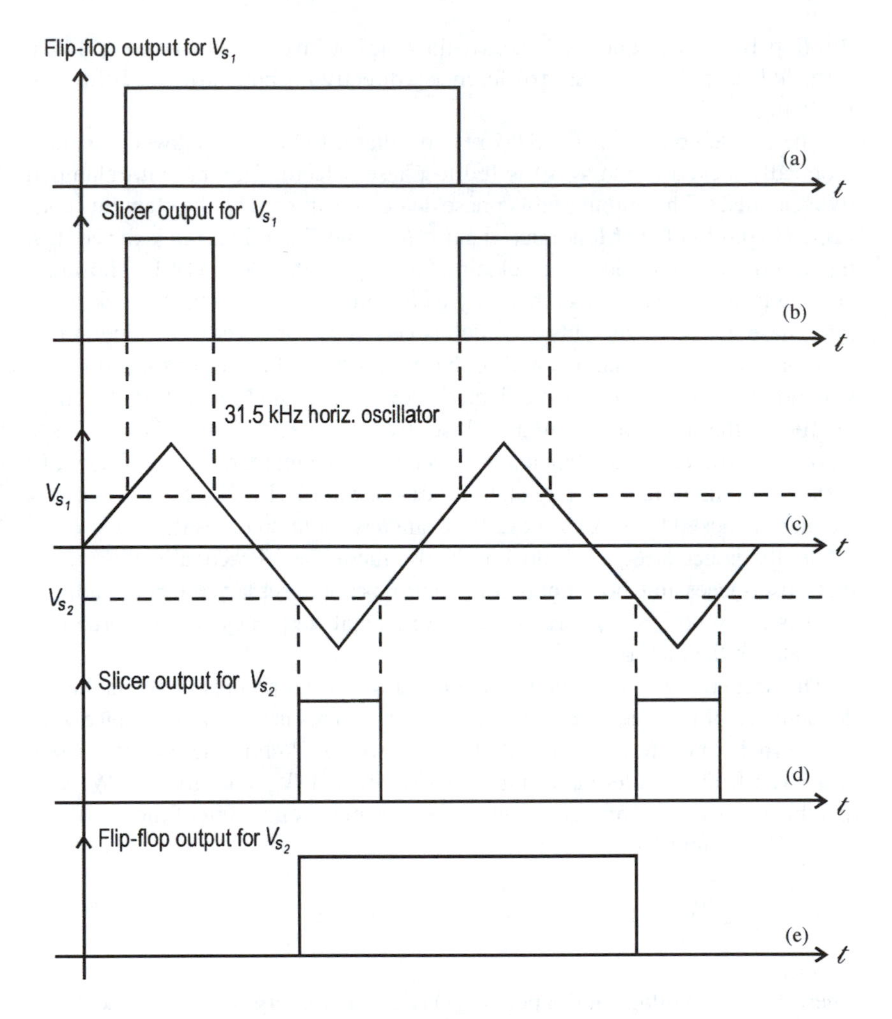

Fig. 6.53: Waveforms illustrating the operation of the horizontal phase-lock circuitry in the Zenith 5NB4X television receiver. These waveforms illustrate the variation in the phase of the output of a flip-flop due to variation of the slicing level. V_{s1} and V_{s2} are two different slicing levels. The slicing voltage is the output of a phase detector that compares the flyback pulse to the output of the horizontal oscillator after the latter is passed through a different flip-flop.

15.75 kHz. This frequency is compared to that of the sync pulses in the phase detector of PLL #1. The output of the phase detector is then fed to C_{502} and C_{501} in series with R_{503}. This network comprises the loop filter. Further filtering is provided by R_{505}, R_{504}, and C_{514}, which all connect to Monomax^{TM} pin 12, the control voltage input to the VCO.

Exercise 6.15 If R_{501} were varied, what effect might it have on the picture?

This loop assures that the horizontal square wave derived from the VCO will exactly match the frequency of the horizontal-sync pulses. The VCO output also goes into a separate slicer (or clipper) with a variable slice level. Varying the slice level will vary the phase of the clipped triangle wave. The waveform out of the slicer will, in general, be asymmetrical. Running it through a $\div 2$ (flip-flop) restores the symmetry and cuts the frequency back down to the 15.75 kHz needed to drive the horizontal-output circuitry. This is shown in Fig. 6.53, which should be "read" both upward and downward from the middle waveform.

It has been assumed in drawing Fig. 6.53 that the flip-flop is positive-edge triggered. The bottom and top waveforms are two possible outputs from the flip-flop. Both are frequency-locked to the sync but have different phases relative to the flyback pulse. The phase difference is directly attributable to the difference in slicing level.

The divided output of the 31.5 kHz oscillator (which we showed was harmonically locked to the sync) is fed to phase detector 2 along with clamped flyback pulses. The output of this phase detector is filtered by the elements connected to pin 14 of the Monomax™ chip (R_{513} and C_{503}). This pin is the control input of the variable slicer, thus closing the loop for the second PLL. The slicer voltage level will vary in such a way as to force the horizontal flyback pulse into phase with the sync pulses, which is necessary for keyed AGC operation.

It has been noted that a sound IC-based design will not only simplify construction but will also provide enhanced performance and reliability. A pair of features of the horizontal oscillator illustrates this. The first feature is that it uses an on-chip silicon nitride capacitor with a very low temperature coefficient to yield an oscillator temperature stability of less than 1 Hz/°C. This oscillator is frequency-locked by a PLL whose lock range is about 20 times the worst-case, thermally induced frequency drift of the oscillator itself. Practically, this means that a user-accessible horizontal hold control is not necessary. The only adjustment is a trimpot (R_{501}) used to set the nominal frequency of the horizontal oscillator at the factory.

The second high-performance feature of the horizontal oscillator is the independence of the frequency on supply voltage. This is achieved because with no external connections to pin 12 of the Monomax™ chip, the potential there is derived from an internal voltage divider between V_{cc} and ground. We will call this voltage V_D. This same voltage sets the upper end point of the oscillator ramp. The current fed into pin 12 is

$$I = \frac{V_p - V_D}{R_{502}},$$

where V_p is the voltage at the pot (R_{501}) slider. The charging time of the oscillator's internal capacitance is given by $T = CV_D/I$, from which we get

$$T = \frac{C \cdot R_{502} \cdot V_D}{V_p - V_D}.$$

Since both V_D and V_p are derived from voltage dividers across the same power supply, we can write

$$V_D = F_1 \cdot V_{cc} \quad \text{and} \quad V_p = F_2 \cdot V_{cc},$$

where F_1 and F_2 are constants set by resistances. Thus the frequency of oscillation is given by

$$f = \frac{1}{T} = \frac{F_2 \cdot V_{cc} - F_1 \cdot V_{cc}}{C \cdot R_{502} \cdot F_1 \cdot V_{cc}} = \frac{F_2 - F_1}{F_1 \cdot C \cdot R_{502}}.$$

Observe that the frequency does not depend on the supply voltage. In addition to the current feeding pin 12 through R_{502}, another component (the error signal) is summed in through R_{505} from the filtered output of phase detector 1, allowing

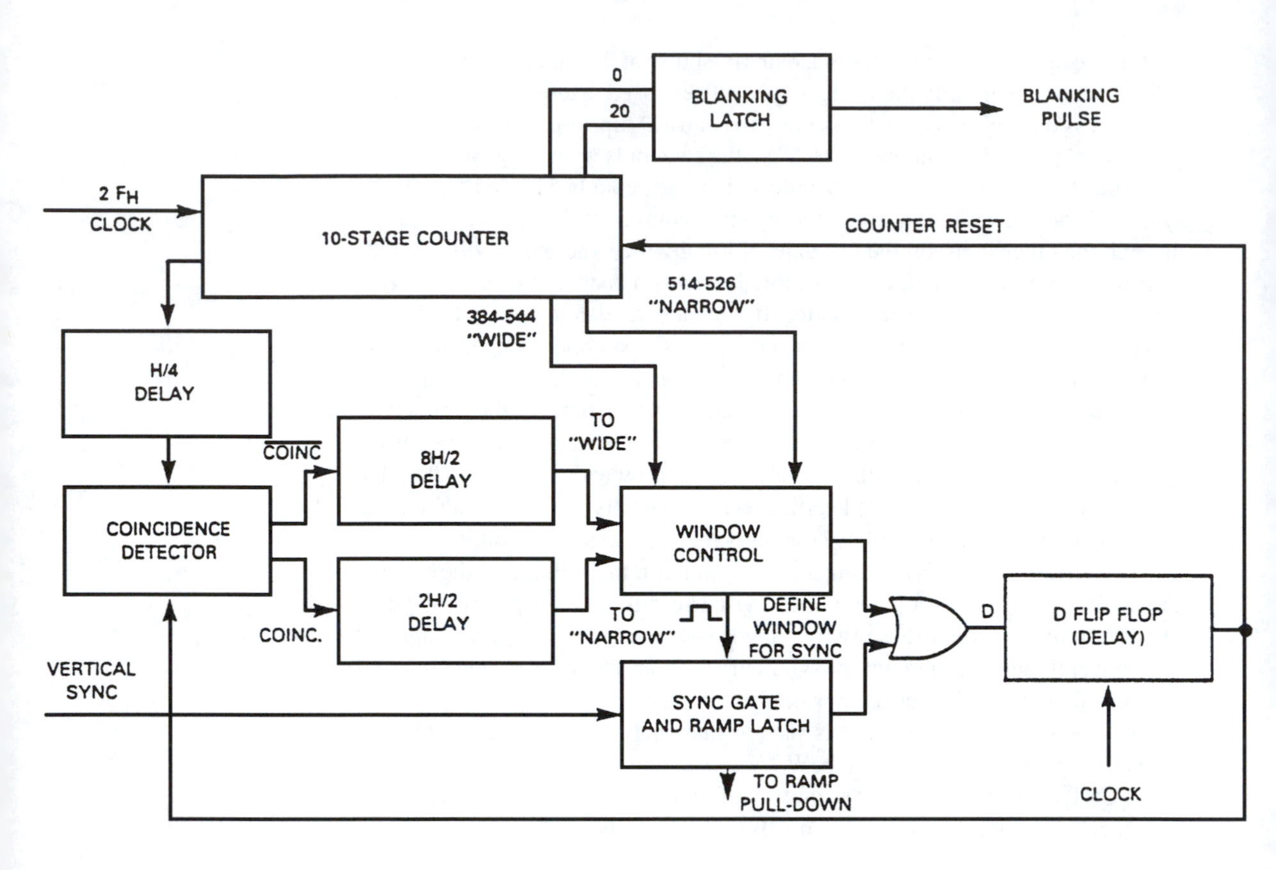

Fig. 6.54: Detailed block diagram of the portion of the Monomax™ chip that is responsible for both maintaining sync and minimizing noise immunity through the creation of a variable-width "window" for the vertical sync pulse.

it to control the frequency. The output of the horizontal oscillator is fed to a two-stage power amplifier prior to being applied to the flyback and horizontal yoke. R_{509} is the load resistor for the Monomax™ horizontal output. The signal developed across it is DC-coupled to the base of the driver transistor (Q_{501}) through R_{517}. The yoke voltage is capacitively coupled back into the base circuit of Q_{501} to cut it off during retrace. This coupling would allow the base voltage to go negative enough to damage the transistor if it were not for D_{509}, which clamps the negative voltage at a level safe for the B–E junction. The driver stage acts as a switch to turn the horizontal output stage (Q_{502}) off and on. When it is on, the collector current begins to ramp up at a rate determined by the inductance between terminals 2 and 4 of the flyback. When the base drive is removed, Q_{502} tries to cut off, but the inductance tries to sustain the current and, in so doing, generates the high-voltage spikes that are rectified to supply the high voltage.

The boost voltage is provided by the damper diode, D_{502}, and is used only for the screen and focus grids of the CRT.

The vertical-sweep circuitry is quite unorthodox by "pre-IC" standards, but it is typical of the circuitry currently used to give superior performance and to minimize the adjustments the consumer needs to make. It makes use of the excellent frequency stability of the horizontal oscillator to provide highly stable vertical sweep as well. A simplified block diagram is shown in Fig. 6.54.

The 31.5 kHz oscillator is fed to a frequency-divider chain that not only decodes the 525 count (31.5 kHz/525 = 60 Hz) but several other counts as well. For example, a latch initiates a blanking pulse at 0 count and terminates it at the 20 count. Normally the vertical-sync pulse is used to reset this counter, but it must pass through a sync gate, an OR gate, and a D flip-flop to do so. The sync gate is enabled by a "window control," so the sync pulse can only pass through when the counter is in the time "window" between counts 514 and 526. If the sync has not arrived in that time, the window control itself resets the counter through the other input on the OR gate. Normally the sync pulse will coincide with the 525 count. Whether or not this is true (on each cycle of vertical) is determined by a coincidence detector. If coincidence is not detected for eight successive cycles, the window is "opened wider" to examine a longer time span for the arrival of the sync pulse. When the pulse is found the circuitry quickly relocks to it. After two cycles where coincidence is achieved, the window is narrowed down again. This scheme normally gives a narrow lock range for good noise immunity, but it allows fast acquisition by widening the lock range when necessary. The same gated vertical sync that resets the counter also resets the vertical ramp generator whose output is the vertical sweep voltage.

The vertical driver, Q_{601}, compares this output ramp voltage to the voltage across R_{615}, which is proportional to the current through the vertical yoke. The difference between these two voltages is amplified by Q_{601} and applied to a conventional complementary-symmetry emitter–follower stage (Q_{602} and Q_{603}), which drives the vertical yoke.

Even beyond integrated designs such as that analyzed here, we have begun to see shirt pocket and even wrist B/W TV sets. If an LCD display is used rather than a CRT, the scan signals are of quite a different character, it being necessary to sequentially address the various pixels on the display.

References

Fink, D.G. 1957. Television Standards. In *Television Engineering Handbook*, ed. D.G. Fink, p. 2–20. New York: McGraw-Hill.

Middlebrook, D. 1989. Null Double Injection and the Extra Element Theorem. *IEEE Transactions on Education*, 32, #3: 167–80.

Neuhauser, R.G., Cope, A.D. 1992. Imaging Devices. In *Television Engineering Handbook*, ed. K. Blair Benson and Jerry Whitaker, Chapter 11. New York: McGraw-Hill.

Schlesinger, K. 1957. Scanning, Deflection and Color Registration. In *Television Engineering Handbook*, ed. D.G. Fink, p. 6–12. New York: McGraw-Hill.

7

Color TV

The main issue at the inception of color TV broadcasting was again one of compatibility. The requirement that the color signal be usable by black-and-white TV sets led to a system in which additional information was broadcast along with the previously used signals. This additional information was simply not processed by monochrome receivers. Compatibility also requires that a color set be able to receive a monochrome picture properly.

To understand the characteristics that the color signal must have, it will help to digress and take a look at the color cathode-ray tube (CRT).

7.1 The color CRT

Even though the predominant color reproduction method used today is pretty much a "brute force" solution, it is at the same time, a testimonial to modern manufacturing techniques that it can be mass produced with the high degree of precision required to make it work.

In the back of the neck of the CRT, arranged as shown in Fig. 7.1, are three electron guns instead of the single on-axis gun used in a B/W set. Even though these guns obviously don't shoot electrons of various colors, they are identified by the color they will cause on the screen. Instead of having a single (white) phosphor as in monochrome TV, the color CRT has three different phosphor colors on the inside face of the CRT. They are, in most cases, arranged in a dot pattern as shown in Fig. 7.2. The sets of dots enclosed in the triangular outlines include one of each color and are called *triads*. On a 19″ (diagonal measurement) color CRT, the dots are about .38 mm in diameter. Between the dots and the gun assembly and close to the dots lies a sheet of metal called the *shadow mask*. There are holes in the shadow mask, positioned over the center of each triad of dots. The function of this mask is illustrated by Fig. 7.3, which shows only two guns and two dots. It can be readily perceived how the concept could be extended to three of each if the third spatial dimension were added.

Note that although the vertical scale shown is the same for the dots and the guns, it is not the same as the horizontal scale. To the extent to which the shadow

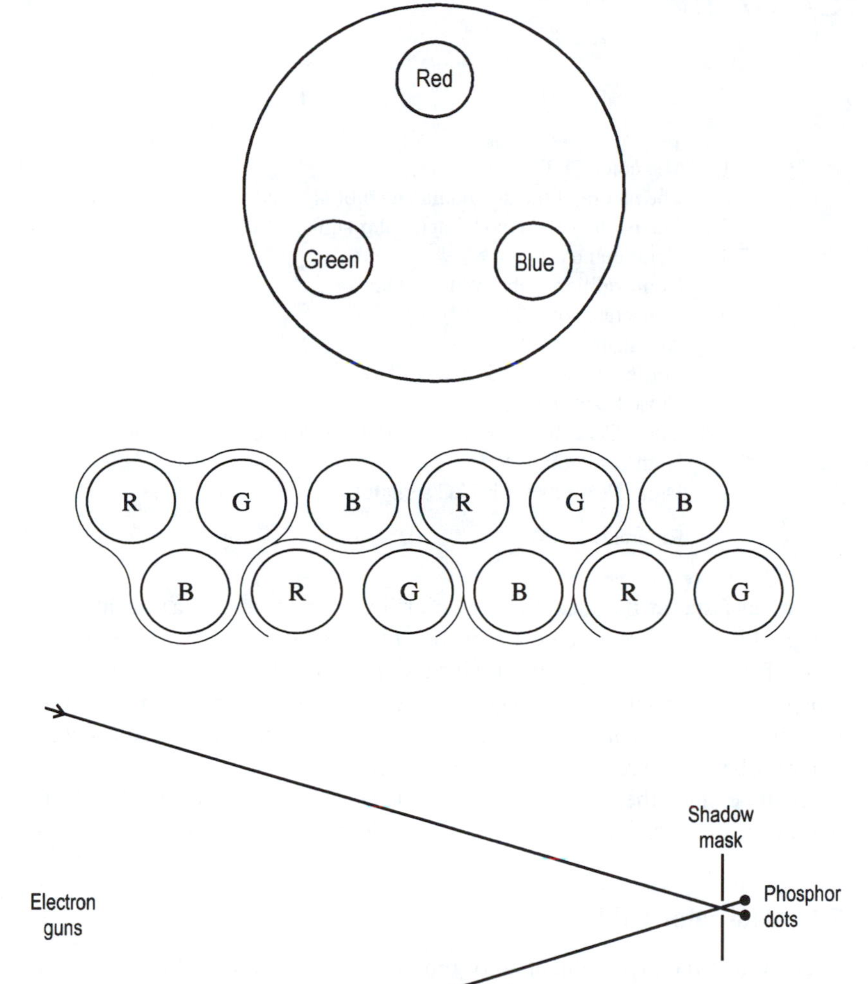

Fig. 7.1: A cross-sectional view of the neck of a color CRT showing the placement of the electron guns for the three receiver primary colors.

Fig. 7.2: A magnified view of a small portion of the face of a color CRT showing the phosphor dot triads.

Fig. 7.3: An illustration of how the shadow mask allows an electron beam to access only one phosphor dot in a triad.

mask is properly fabricated and positioned in manufacture, it is impossible for electrons from the upper gun to reach the upper dots. Thus ideally each gun activates only one color of phosphor dot. Not only must a great deal of care be taken in manufacture, but a lot of circuitry is necessary to make results even approach the ideal. This circuitry will be discussed later.

7.2 The nature of the demodulated color signals

For now, the information we need out of all this is that the color TV set must generate separate signals to drive the red, blue, and green guns of the CRT. This requirement, coupled with that of compatibility, led designers to develop a system wherein the color information is transmitted as *color minus luminance*. The signal we called video in a monochrome set is processed in exactly the same way in a color set but is called *luminance* in recognition of the fact that it determines the point-by-point illumination level of the picture. This might lead us to believe that we would have to process not only the luminance (abbreviated as Y) but three color difference signals:

$$R - Y,$$
$$B - Y,$$

and

$$G - Y.$$

Although this approach was taken in some very early receivers, it is not necessary, because the Y signal itself is constituted in the TV camera as a linear combination of the R, B, and G signals. The relative weighting of these signals to give the best monochrome picture has been found to be

$$Y = .30R + .59G + .11B. \tag{7.1}$$

Thus if $R - Y$, $B - Y$, and Y are recovered in the receiver, they can be combined to give $G - Y$:

$$G - Y = \alpha(R - Y) + \beta(B - Y) + (\gamma - 1)Y, \tag{7.2}$$

where α, β, and γ are constants to be determined.

Exercise 7.1 Combine (7.1) and (7.2) to eliminate Y. The fact that R, G, and B are all independent quantities says we can equate their coefficients individually to zero, yielding three equations in the three constants, α, β, and γ. Solve for these constants and show that

$$\alpha = -.508,$$

$$\beta = -.186,$$

$$\gamma = 1.000.$$

Using the results of Exercise 7.1 allows us to write Eq. (7.2) as

$$G - Y = -.508(R - Y) - .186(B - Y). \tag{7.3}$$

This result shows that Y itself is not even needed to recover $G - Y$. Only the $R - Y$ and $B - Y$ signals are needed.

7.3 The nature of the modulated color signal

When we consider the constraint that all of the color information must be put into the standard 6 MHz channel bandwidth along with the monochrome and sound information already there, it is clear that some powerful measures must be resorted to in order to accomplish this. The successful implementation of color TV requires two such measures: (1) the encoding of the color signals to provide just enough but no more information than human vision requires, and (2) the exact placement of the color information in the spectrum to give the best all-around performance.

7.3.1 Encoding of the color signals

Experiments on human perception of color have shown that color is not perceived on very small objects, and that medium-sized objects appear to be satisfactorily represented if they are perceived as being a "warm" color or a "cool" color. The best choice is an orange–cyan (cyan is blue–green) combination. Large objects require the full spectrum of colors to be perceived correctly. This information is likely to evoke a certain amount of skepticism in some readers. Let us look at the matter further to see if we can make it appear more plausible.

We might first ask what defines "small," "medium," and "large." "Small" will be defined for our purpose as 1/150 of the screen width, which is just over .5% of the screen diagonal dimension. Thus for a 19″ TV, dimensions less than about .1″ are considered small and are rendered in monochrome. This is most readily observed by noticing a thin colored vertical object in the background of a TV scene (a picture frame, for example). If it is far enough back to be "small" it will appear black or white. The color will show up if the camera moves in on it. "Medium" objects range from about .5% to 1.5% of the screen diagonal measurement, or up to about .3″ for a 19″ TV. Objects in this size range will be reproduced with a color somewhere between saturated reddish–orange and saturated blue–green. Between these extremes lie a less-saturated orange that is actually closer to red in hue, a wide range of pinks, whites, and a full range of blue–greens from the lightest to the most saturated.

Although it is theoretically possible to transmit only $R - Y$ and $B - Y$ and thereby correctly reproduce all colors within the range allowed by the phosphors, it is more advantageous for bandwidth preservation purposes to instead transmit two other signals that are linear combinations of these. They are constituted as

$$I = .74(R - Y) - .27(B - Y)$$

and

$$Q = .48(R - Y) + .41(B - Y).$$

This choice of signals causes the I signal to lie along the orange–cyan axis. These two signals are quadrature amplitude modulated (QAM) onto a 3.58 MHz color carrier in a balanced modulator, which achieves suppression of the carrier. This choice of the chroma subcarrier frequency will be discussed in the next section. QAM was discussed in Section 1.9 where we saw a slight variant of it used in the Motorola AM stereo system. The two quadrature chroma components combine in the camera as shown in Fig. 7.4.

Just as we saw was the case with the $L - R$ channel in FM stereo, suppression of the subcarrier at the transmitter requires its regeneration at the receiver. In color TV, the method used is different, however. There are two reasons for

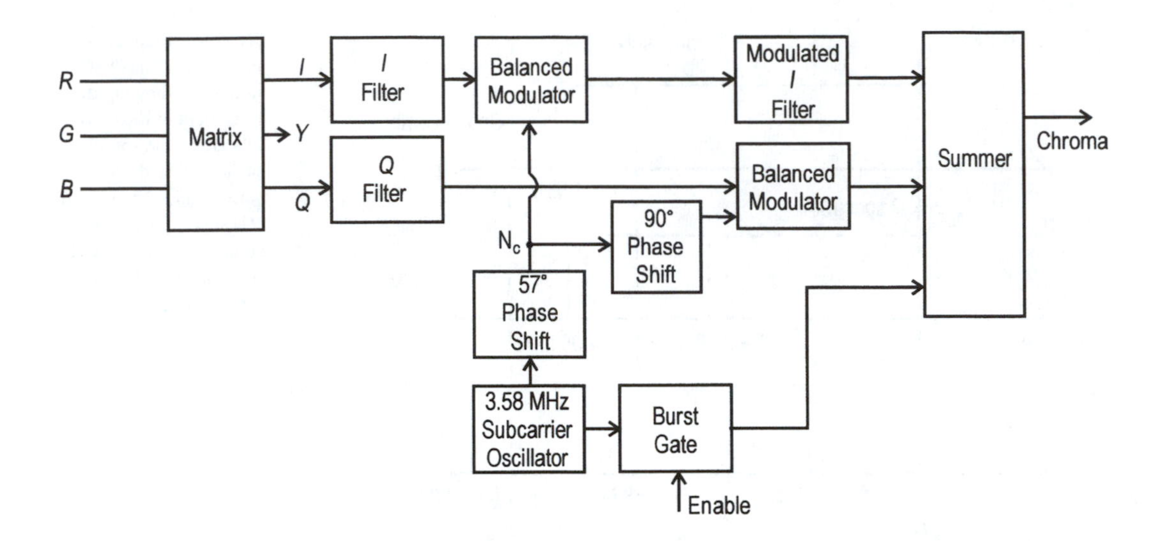

Fig. 7.4: A block diagram showing how an NTSC chroma signal is generated from R, G, and B signals.

suppressing the subcarrier. The first is that the chroma frequency will combine with the sound carrier at 4.5 MHz in any circuit nonlinearity and will generate a .92 MHz beat note that shows up as waviness in the picture. The second is that if no subcarrier is transmitted, the entire chroma signal disappears for monochrome transmission. Under these conditions, any power transmitted as chroma subcarrier would be totally wasted.

Carrier regeneration begins when the burst gate is briefly enabled during the "back porch" of the horizontal-blanking pulse. A short burst of the subcarrier is thus sent during the blanking time between horizontal scan lines. This prevents it from ever being seen on the screen. Referring back to Fig. 6.6, we see that the "back porch" of the blanking pulse is at least .05 H (=3.18 μsec) in duration. FCC broadcast standards call for a burst of at least 8 cycles of the suppressed chroma subcarrier to be superimposed on the back porch. It is the task of subsequent circuitry to take these bursts and use them to produce a continuous wave (CW) sinusoid locked in phase and frequency to the oscillator in the camera producing the bursts. This CW oscillator in the TV receiver is generally called the *burst oscillator*. This name is slightly misleading, however, since it doesn't oscillate in bursts, but rather produces a continuous wave *locked* to the bursts.

Circuitry to accomplish this task will be examined in Section 7.6.3. This burst constitutes the zero reference for phase of the chroma signal. The 57° phase shift common to both the I and Q signals is the amount by which I differs from the burst. Q, of course, is an additional 90° removed in phase.

If we think of $\mathrm{N_C}$ in Fig. 7.4 as the actual carrier input to the QAM modulator, we see that the modulated I signal will be *In phase* with the carrier at $\mathrm{N_C}$. This is why it is called the I component. Likewise the Q component is so named because it is in *Quadrature* with the carrier at $\mathrm{N_C}$.

These two components each have another name that is an important descriptor of their nature. The I signal is also called the wideband component and the Q signal is called the narrow-band component. The latter component is needed to reproduce large colored areas (>.3″ on a 19″ TV). We ascertain that frequency

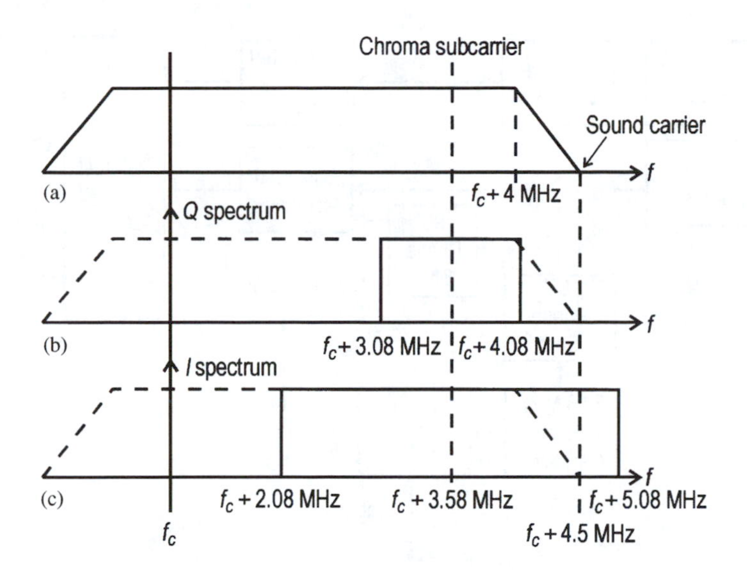

Fig. 7.5: Frequency spectrum of an NTSC color signal: (a) showing the placement of the chroma subcarrier and sound carrier, (b) with the narrowband chroma signal (Q) superimposed upon it, and (c) with the wideband chroma signal (I) superimposed upon it.

range by recalling that a horizontal scan time (exclusive of the blanking time) is about 50 μsec. For a 19″ CRT, the screen width is $.8 \times 19'' = 15.2''$. The horizontal scan will thus traverse an object .3″ wide in

$$\frac{.3}{15.2} \times 50\,\mu\text{sec} = 1\,\mu\text{sec}.$$

This time corresponds to a half-cycle of .5 MHz frequency, which is the upper frequency limit for the Q component. Since it should be possible to have the entire screen one color, the lower frequency limit for Q is DC. Thus the filter for the Q component should be a low-pass type with a cutoff frequency of .5 MHz.

Exercise 7.2 Based only on information given in this chapter, what should be the cutoff frequency for the modulated I component?

The answer, of course, is 1.5 MHz, since the smallest object for which I is used is 1/3 of that for Q. It will also be necessary for the I filter to extend down to DC because both I and Q are needed to reproduce the full spectrum of colors.

There is another complication that arises in considering the I filter, however. This can be appreciated by referring back to Fig. 6.4b, which is repeated here with the color signals added as Fig. 7.5.

While the Q spectrum fits pretty well into the passband of Fig. 7.5a, the I spectrum not only does not do so, but it actually overlaps not only the sound carrier, but even the edge of the channel. This clearly will not do. The problem is circumvented by following the I modulator with a vestigial sideband filter (Section 6.1.2.1) that truncates the upper sideband to .5 MHz, while allowing the lower one to remain at 1.5 MHz relative to the chroma subcarrier.

One might well wonder why the color information, extending down to within 2 MHz of the main carrier (f_c), does not interfere with the luminance information, which extends up to 4 MHz above the main carrier, since there is an overlap region 2 MHz wide. There are two reasons for this. The first is that the luminance spectrum falls off sharply with frequency, but the more important reason is called *interleaving*. It is discussed next.

7.3.2 Placement of the color information in the spectrum

The frequency of the color subcarrier is chosen on the basis of three criteria:

(a) It should be as high as possible, consistent with its fitting into the channel allocation. This will help to minimize interference with the video signal.
(b) Its harmonics should not coincide with those of the video signal. Since, as shown in Fig. 6.7, the basic periodicity of the video signal is that of the horizontal scan frequency, this requirement is best satisfied in practice by making the chroma subcarrier frequency equal to an odd multiple of half the horizontal sweep frequency.
(c) The beat frequency between the sound carrier and the chroma subcarrier will be in the video passband. It can give rise to sound interference called *sound bars* in the picture. Again this interference will be minimized if this beat frequency is at an odd multiple of half the horizontal sweep frequency. These last two conditions may be stated mathematically as

$$f_{\mathrm{sc}} = \frac{f_{\mathrm{H}}}{2} \cdot (2n + 1)$$

and

$$4.5\ \mathrm{MHz} - f_{\mathrm{sc}} = \frac{f_{\mathrm{H}}}{2} \cdot (2m + 1),$$

from which we get

$$4.5\ \mathrm{MHz} = \frac{f_{\mathrm{H}}}{2}(2n + 1 + 2m + 1) = f_{\mathrm{H}}(n + m + 1).$$

If we use the monochrome value $f_{\mathrm{H}} = 15{,}750$ Hz, we get

$$n + m + 1 = 285.71.$$

But since n and m were assumed to be integers, this is not possible. The problem is resolved by adjusting f_{H} slightly for color transmission. If we set

$$n + m + 1 = 286 = \frac{4.5\ \mathrm{MHz}}{f_{\mathrm{H}}}$$

we find $f_{\mathrm{H}} = 15{,}734.26$ Hz, which is still within the FCC specifications, and $n + m = 285$.

Any integer values of n and m satisfying this equation will meet requirements (b) and (c) and will give "strange" values for f_{sc}, that is, values that are not some

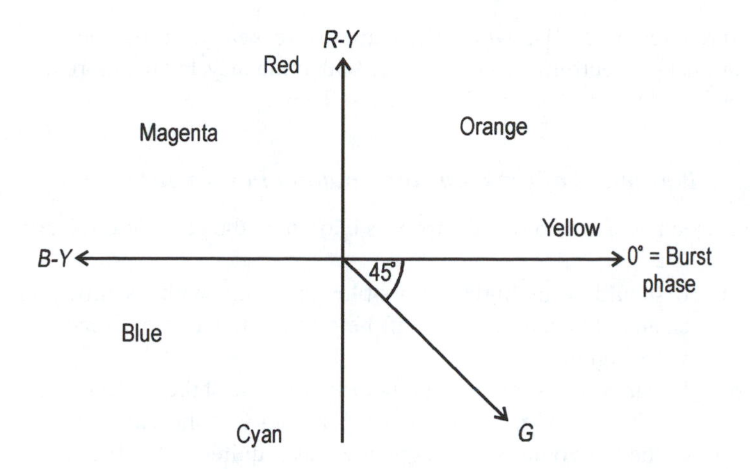

Fig. 7.6: A diagram showing how chroma phase (relative to the burst) encodes color according to the NTSC standard. This diagram should be compared to Fig. 7.9.

"nice round number." We must resort to requirement (a) to narrow the choice further. Since the upper sidebands of the Q signal extended .5 MHz above the chroma subcarrier, the requirement that they not reach up to the lower sidebands of the sound signal requires $n < 253$. Of course this limit would not even be approached in a real system since some "guard band" of frequencies would be left between the chroma and sound sidebands to account for the nonideality of filters. The actual value of n used is 227, making $m = 58$ and $f_{sc} = 3.579545$ MHz. We might note in concluding this section that the ratio of horizontal to vertical scan frequencies remains at 525/2 in color transmission. Thus the lowering of f_H to 15,734.26 Hz dictates a lowering of the vertical scan rate to 59.94 Hz.

7.4 Color demodulation

If the phase of the reconstructed subcarrier is correct, any color can be closely approximated by having the proper mix of the primary color components. The color TV standards used in the United States are those of the NTSC (National Television System Committee). The NTSC color encoding is shown in Fig. 7.6. That the placement of G is correct can be seen as follows: Suppose the TV screen is pure green. Then $R = 0$ and $B = 0$. But $Y = .59G + .30R + .11B = .59G$ in this case. Thus $R - Y = -.59G$ and $B - Y = -.59G$. It is easily verified that these two components will give a resultant at an angle of 45° from the burst in the fourth quadrant. This procedure also shows us that vectorial combination of these phasors gives a qualitative indication of the color resulting from mixing two primaries.

Exercise 7.3 What color will be produced by mixing red and green while the blue gun is off?

The result of this exercise is contrary to what most of us might expect, because what we have here is additive mixing rather than subtractive mixing. If we see a piece of cloth that appears red, it is because red is the only component of incident white light that is not absorbed by the cloth. If we dye a cloth in a mixture of red and green dyes, the red dye absorbs the green light and the

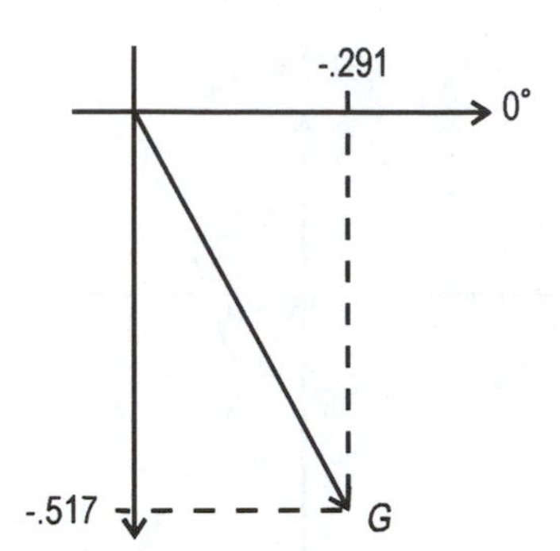

Fig. 7.7: Phasorial computation of the position of the G signal in an overload-compensated NTSC system.

green absorbs the red. Obviously other colors are absorbed also, and the cloth will probably appear brown or some other shade of low chromaticity. This is subtractive mixing. Additive mixing is most readily illustrated by overlapping spotlights of different colors. Red and green will produce yellow.

At this point we introduce the first of several complications in this nice simple picture. To avoid overmodulation, both $R-Y$ and $B-Y$ are scaled down by different factors, $R-Y$ by a factor of .877 and $B-Y$ by a factor of .493 (Liff, 1979). This scaling will modify the phases of the encoded colors on the NTSC chart. To illustrate this, let us repeat the previous calculation for the angle of G. As Fig. 7.7 shows, we now have a component along $R-Y$ of $.877 \times (-.59G) = -.517G$ and a component along $B-Y$ of $.493 \times (-.59G) = -.291G$, from which $\theta = -60.7^\circ$. We can, in like manner, find that R has an angle of 76.5° and B lies at -167.6°.

We now know the angular positions of R, G, B, $R-Y$, and $B-Y$. Finding $G-Y$ is a bit trickier. We have noted that at the transmitter $R-Y$ and $B-Y$ are modified to $.877(R-Y)$ and $.493(B-Y)$, respectively. If we know $(R-Y)$ and $(B-Y)$, then $(G-Y)$ is found from (7.3):

$$G - Y = -.508(R - Y) - .186(B - Y).$$

But since the receiver gets weighted versions of the two quadrature components, this becomes

$$\begin{aligned} G - Y &= \frac{-.508}{.877}[.877(R - Y)] - \frac{.186}{.493}[.493(B - Y)] \\ &= -.579[.877(R - Y)] - .377[.493(B - Y)]. \end{aligned}$$

The first term on the right is the component along $R-Y$; the second is the component along $B-Y$. Thus we have the situation depicted in Fig. 7.8.

The revised NTSC color chart for the overload-compensated system can thus be drawn as shown in Fig. 7.9, incorporating the calculations just completed and corresponding calculations for I and Q. It should be understood that this diagram as a whole has meaning only in the context of the chroma signal at the receiver, since $G-Y$ has no existence elsewhere. Since no information concerning relative amplitudes of the components is intended to be conveyed by this diagram, the scaling factors of .877 and .493 have been omitted from the quadrature components.

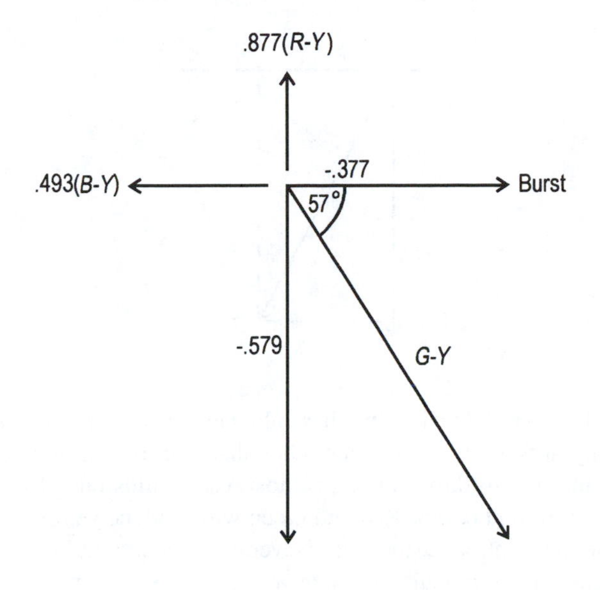

Fig. 7.8: Phasorial computation of the position of the G–Y signal in an overload-compensated NTSC system.

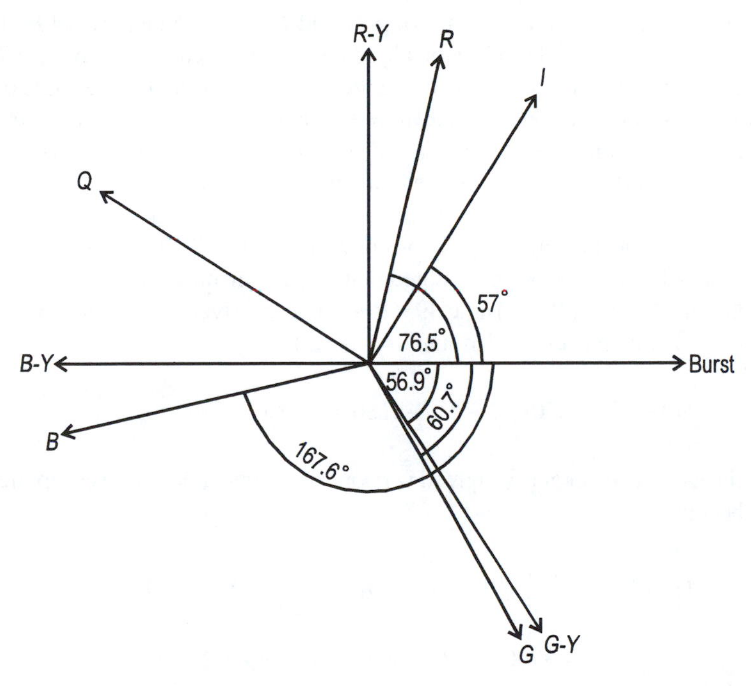

Fig. 7.9: A diagram showing how chroma phase (relative to the burst) encodes color in an overload-compensated system according to the NTSC standard. The Q and I phasors are also shown.

Color demodulation is achieved by multiplying each color component by a sinusoid with the same frequency as the burst and with a fixed phase relationship to it. For example, to recover $B - Y$:

$$\begin{aligned} v_o &= \cos(\omega_{\rm B} t) \cdot (B - Y) \cdot \cos(\omega_{\rm B} t + \theta) \\ &= (B - Y)[\cos^2(\omega_{\rm B} t) \cos\theta - \cos(\omega_{\rm B} t) \cdot \sin(\omega_{\rm B} t) \cdot \sin\theta]. \end{aligned}$$

After low-pass filtering, the output is

$$v_o = .5(B - Y) \cdot \cos\theta. \tag{7.4}$$

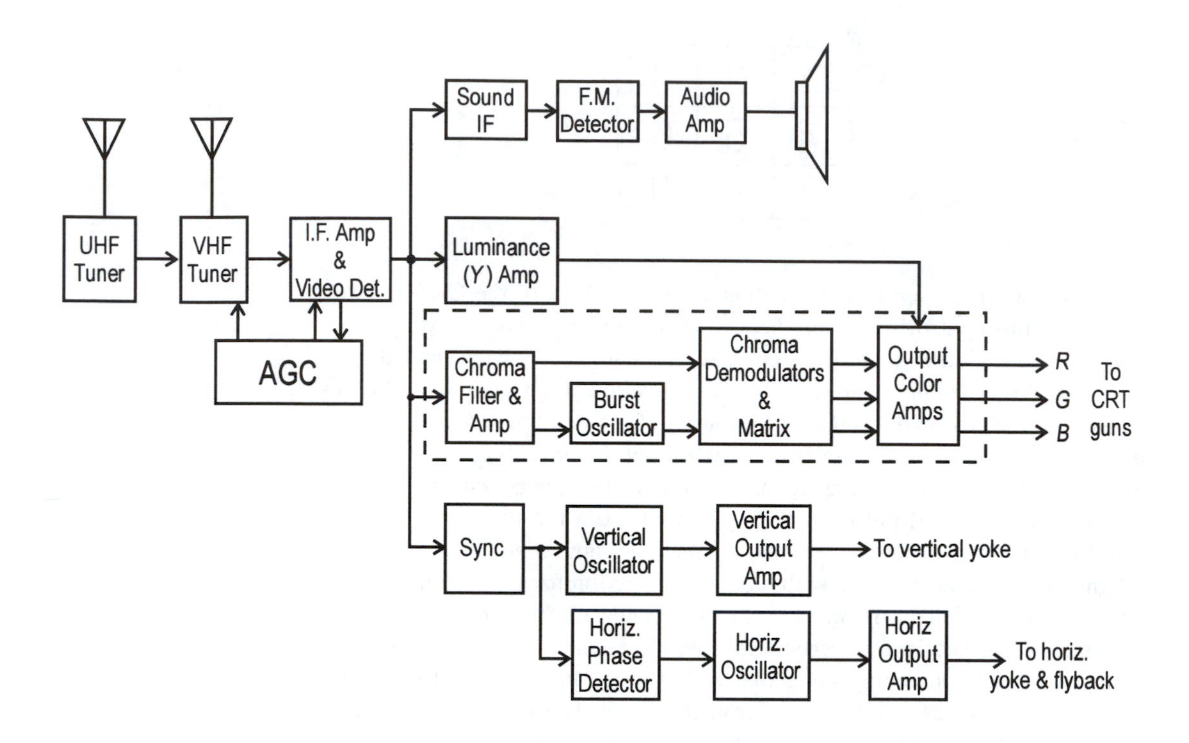

Fig. 7.10: Block diagram of a typical color television set.

In the above equations, θ is the phase difference between $B - Y$ and the burst. It should not be inferred from (7.4) that $B - Y$ and the burst should be in phase. In fact, $B - Y$ is about 180° out of phase with the burst. In other words, θ in (7.4) is about 180°. TV receivers are not necessarily designed to force θ to equal exactly 180°, since they have a control labeled "Tint" or "Hue" that allows θ to be varied while maintaining a quadrature relationship between the sinusoids that will be used for demodulating $R - Y$ and $B - Y$. The tint control is adjusted to give the proper flesh tones, since errors there are most obvious and since all colors will be correct if any one is, provided the rest of the set is properly designed and adjusted.

Figure 7.10 shows a block diagram of a color TV with the special color circuitry just discussed enclosed in dashed lines. It will be observed that the rest of the circuitry is the same as for the monochrome set.

Before proceeding to the analysis of a color receiver, we will discuss some of the adjustments and circuitry that are not shown in Fig. 7.10 because they exist only to compensate for nonidealities.

7.5 Nonidealities of the color system

The first adjustment is to equalize all three electron guns in the CRT by setting them at the same threshold so any video signal will turn them on. These are called *kine bias adjustments*. The name comes from the fact that the CRT used to be known as a *kinescope* a long time ago. The next adjustment is *purity*. For this adjustment, the green and blue guns are turned off and the yoke position and the purity rings on the yoke are adjusted to give a pure red screen.

This brings us to the convergence procedure. We imagine the face of the CRT to be divided up into nine roughly equal rectangles, as shown in Fig. 7.11.

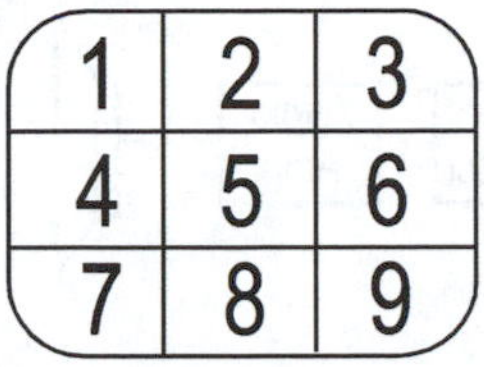

1	2	3
4	5	6
7	8	9

Fig. 7.11: A view of the front of a color CRT showing the convergence zones.

Static convergence magnets are mounted on the neck of the CRT and may be adjusted to control where the beams hit the screen. Static convergence is done for zone 5–first to bring green and red into convergence and then to bring the blue beam into convergence with the yellow to give white. Ordinarily convergence is accomplished using a pattern of dots on the screen, since this makes it very easy to see color fringing when out of convergence.

The next phase of convergence is dynamic. This is necessary because, as the beam is deflected out of zone 5, electrons from the three guns have to travel significantly different distances to reach the same triad. Having a longer "flight path" means that even with the same deflection impulse from the yoke, an electron will deflect more. This is compensated for by two electromagnet windings on each of the convergence magnets. These are driven from the sweep voltages, giving converging fields that will bring all three beams to the same triad, even at distances well removed from the center. The adjustments for these electromagnets are made on a formidable array of about twelve potentiometers and coils called the convergence panel. These controls affect the convergence in zones 2, 4, 6, and 8. No provision is made for converging the corner zones. The quality of the convergence there will be wholly dependent on the manufacturing quality if the rest of the CRT is properly converged.

The final nonideality that must be compensated for is the pincushion distortion as shown in Fig. 6.10. The approach to its elimination is based on the fact that the curved lines have the form of half-sinusoids. For instance, to straighten out the bottom line, we could add a properly phased sinusoid at the horizontal sweep frequency to the vertical sweep signal. The amplitude of the correction sinusoid is decreased toward the center of the screen because saturable reactors whose inductance is controlled by the vertical sweep are used to generate it. Similar circuitry generates correction terms at the top, left, and right.

7.6 A discrete color TV

Because so much of a color set is the same as a monochrome set, we will not do a stage-by-stage analysis of an entire discrete set but rather will analyze only the chroma circuitry of a Sony KV-1200U in detail and then proceed to a more-modern integrated receiver.

The 17 transistors incorporated in the circuitry of Fig. 7.12 are just over a third of the total in the set. There are also two tubes: the CRT and the high-voltage rectifier tube.

7.6.1 First bandpass amplifier and ACC

After video detection, the composite signal passes through one stage of video amplification (not shown). Output is taken from the emitter of this stage to the

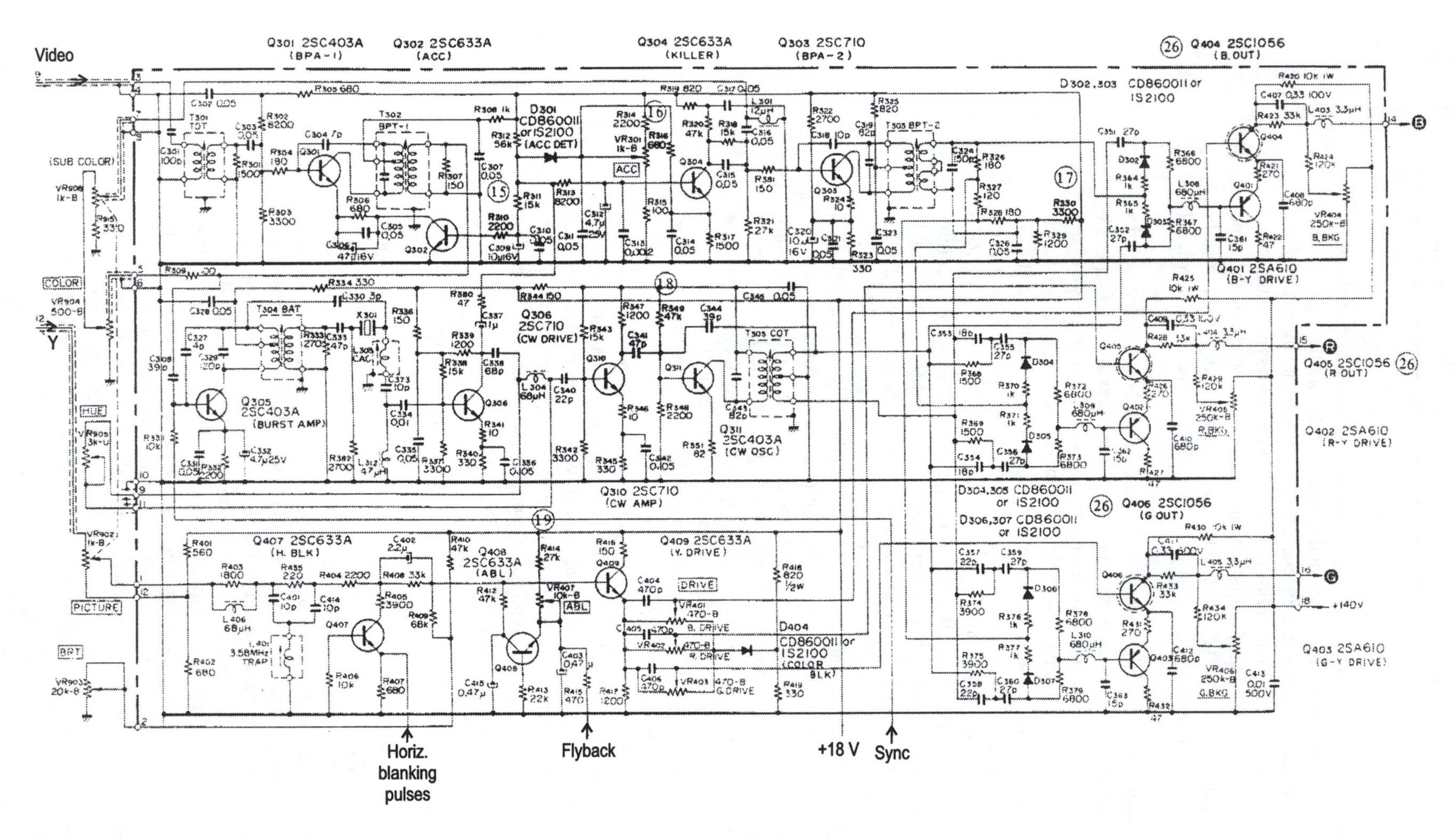

Fig. 7.12: A partial schematic diagram of a Sony KV1200U discrete television receiver showing the chroma-processing circuitry.

luminance (Y) amplifier (Q_{407} and Q_{409}) and to the chroma circuitry (Q_{301}–Q_{306} and Q_{310}–Q_{311}). In this set, sound is taken off before the video detector and passes through its own envelope detector (to translate the sound from 41.25 MHz down to 4.5 MHz) prior to FM detection. The sync is taken off after two stages of luminance amplification. Neither the sound nor the sync takeoff is shown in Fig. 7.12. Thus only Y and chroma are driven directly by the video amplifier output, which enters Fig. 7.12 at the upper left-hand corner.

Entering the chroma section, the signal first encounters T_{301}, which is called a *takeoff transformer*. In conjunction with C_{301}, this attenuates the low-frequency components in the video signal. The secondary of this transformer feeds the base circuitry of Q_{301}, which is the first chroma bandpass amplifier. Because this is a tuned amplifier, both mismatching and neutralization are used to stabilize it. Although its base bias circuit is conventional, the emitter circuit has transistor Q_{302} in series with it. This device provides automatic color control (ACC) by varying the gain of Q_{301}. We may think of Q_{301} as having two emitter resistances. The fixed one (R_{306}) is bypassed by C_{305} and C_{306}. The variable one (Q_{302}) is unbypassed. As this device's effective resistance is varied by its base potential, it varies the color signal output from Q_{301}. In the absence of a color signal, the base bias on Q_{302} is derived from the $+V_{cc}$ line through R_{312} and R_{311}. This is adequate to saturate Q_{302}, so Q_{301} runs at full gain.

The first color amplifier passes not only the chroma but also the burst, since both are at 3.58 MHz. Thus the signal out of Q_{301} is routed to the "burst oscillator" as well as on to the second color amplifier. The amplitude of the burst will be proportional to the amplitude of the received signal. Thus a sinusoid from the collector circuit of Q_{306} is fed to the anode of D_{301}, the ACC detector, and will affect the base bias on Q_{302}.

The cathode of D_{301} is biased from V_{cc} by a voltage divider composed of R_{314}, R_{315}, and VR_{301B}. The latter resistor is adjusted to place the diode at the threshold of conduction with no signal applied. When the 3.58 MHz sinusoid from the "burst oscillator" is applied to the anode side, the positive peaks are rectified and filtered by C_{311}, making the cathode voltage more positive. This voltage is applied to the emitter of Q_{304} through R_{316}.

The negative-going half-cycles which do not pass through D_{301} will lower the anode potential of D_{301}. The negative DC component is extracted by the LPF comprised of C_{309}, C_{310}, and R_{311} and is applied to the base of Q_{302}. It goes through a separate LPF consisting of C_{312}, C_{313}, and R_{313} and is then applied to the base of Q_{304}.

To review, the ACC system functions as follows: As the received color-signal amplitude increases, the burst-related sinusoid amplitude does also. The negative half-cycles of this are rectified, filtered, and applied to the base of the ACC transistor, reducing the positive bias there. This reduces the conduction of Q_{302}, raising the unbypassed emitter resistance of Q_{301}, which reduces its gain and thus reduces the signal delivered to the burst oscillator circuitry. This, as we shall see, directly reduces the amplitude of the sinusoid whose increase conceptually started this sequence of events. In some receivers, the color signal itself, rather than the burst-related sinusoid, is used to gauge the signal strength. This is an inferior system, however, since the chroma signal strength will depend not only on the RF signal strength, but also on the specific scene being transmitted.

7.6.2 Color killer and second bandpass amplifier

It was noted in the previous section that in the absence of a color signal, Q_{301} runs at full gain. Because color reproduction requires a stronger signal than monochrome, weak noisy pictures, which might be satisfactory for monochrome viewing, may look like they are full of colored confetti unless countermeasures are taken. It is the purpose of the *color killer* to change the picture to monochrome if the signal is too small to allow good color pictures.

It was noted that the effect of the burst-related sinusoid in conjunction with the ACC rectifier was to raise the emitter potential of Q_{304} and lower its base potential. Both of these changes will tend to cut it off, making its collector appear as an open circuit. The signal from the first bandpass (color) amplifier output is applied to the top of VR_{904B}, which acts as a variable voltage divider. The output of this goes to VR_{908B}, which serves as yet another voltage divider. This one seems to be a service adjustment to set the proper range of the front panel control VR_{904B}. This is identified as the *color control*, and its effect on the color signal is exactly analogous to the effect of a volume control on sound. The effect of varying this potentiometer is to make the colors more or less intense. The output of VR_{908B} is applied to the collector of the color killer transistor, Q_{304}, and through R_{381} to the base of the second bandpass (color) amplifier. In saying this, we recognize that the .05 μF capacitors scattered throughout the chroma section may be considered as AC shorts, since their impedance is less than 1 Ω at 3.58 MHz.

As long as the collector of Q_{304} can be considered as open, the chroma signal passes into the base of Q_{303}, but if the signal amplitude drops, it will have the opposite effect to that previously discussed on the bias of Q_{304}. The emitter potential will fall and the base potential will rise. This will result in Q_{304} going into saturation if the decline in signal strength is severe enough. When Q_{304} saturates, it grounds out the signal from the first color amplifier and the input of the second, thus killing the color. The dual effect of varying the potential on both the base and emitter of Q_{304} means that the transition from cutoff to saturation occurs over a very narrow range of signal amplitudes, as it should. Ideally, it should function bistably.

The 12 μH inductor (L_{301}) in the collector circuit of Q_{304}, if it were ideal, and if the effective source impedance feeding it were large, would introduce a 90° phase lead into the chroma signal. Whereas the coil might qualify as nearly ideal, the source resistance is relatively low and somewhat dependent on the settings of VR_{904} and VR_{908}. Thus the phase shift will be significantly smaller than 90°. This, as we shall see, is just what is needed for proper operation.

When enabled by the color killer, the signal passes to the second color amplifier, Q_{303}. This stage is quite similar in form and function to Q_{301} except for the absence of gain-control capability in the emitter circuit. It does have one distinctive feature of its own, however. That is R_{322}, which connects from V_{cc} to the junction of R_{323} and R_{324} in the emitter circuit. This resistance can have no AC function, since both its lower and upper ends are at AC ground potential. Therefore it must be there for bias purposes.

Exercise 7.4 Find the quiescent collector current of Q_{303} when the color killer is cut off. Assume $\beta = 100$. Assume that the DC drop across the decoupling resistor, R_{319}, is negligible and that $V_{cc} = 18$ V.

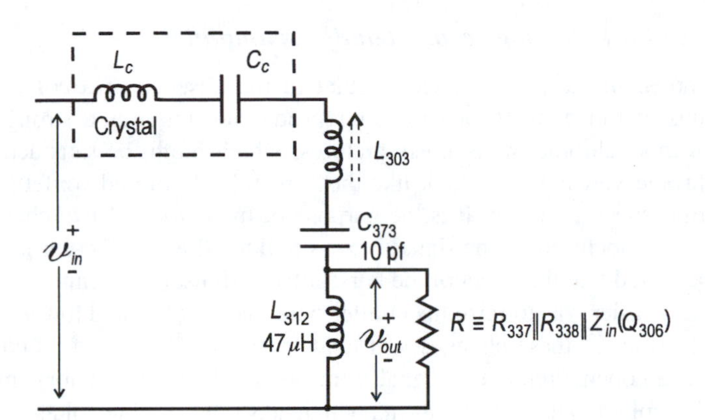

Fig. 7.13: Schematic diagram of the ringing network at the output of the burst amplifier in the Sony KV1200U.

By pulling the emitter up to a potential closer to the Thevenin voltage of the base circuit, R_{322} actually lowers the base–emitter drive and thus the collector current.

7.6.3 *Burst circuitry*

We now leave the chroma signal to examine the burst-processing circuitry. We have already observed that the base of the burst amplifier is driven by the output of the first color amplifier. From the schematic of Fig. 7.12, we can see that sync is also fed to the base. The sync source is called the *sync separator* (Q_{309}), but this stage is not shown in Fig. 7.12. The burst amplifier is well-named: Although it receives all signal components at the color subcarrier frequency, it amplifies only the burst. This is because the base of Q_{305} has no quiescent bias. The transistor goes into conduction only when the horizontal-blanking pulse is present. Since the burst occurs during the latter part of this pulse, it is there at the right time to be amplified.

In modern sets, the gating or enabling pulse for the burst amplifier is most often derived from a separate winding on the flyback. This allows the sync clipping level to be set for best noise immunity without regard for the requirements of the burst processing circuitry.

The output of the burst amplifier is fed to an RLC network in which the crystal is embedded. To ascertain the purpose of this network, we represent the crystal itself with an equivalent circuit of a series LC. In these terms, we can draw the network to be analyzed as we have done in Fig. 7.13.

By a straightforward application of the voltage divider, we have

$$\frac{\mathbf{V}_{\text{out}}}{\mathbf{V}_{\text{in}}} = s^2 L_{312} R C_c C_{373} / \{ s^3 L_{312}(L_c + L_{303}) C_c C_{373} + s^2 (L_c + L_{303} + L_{312}) R C_c C_{373} + s L_{312}(C_c + C_{373}) + R(C_c + C_{373}) \}.$$

By defining

$$C_s \equiv C_c C_{373} / (C_c + C_{373})$$

the above equation becomes

$$\frac{\mathbf{V}_{\text{out}}}{\mathbf{V}_{\text{in}}} = \frac{s^2 L_{312} R C_s}{s^3 L_{312}(L_c + L_{303})C_s + s^2(L_c + L_{303} + L_{312})RC_s + sL_{312} + R}.$$

The resonant frequency will be that for which the denominator is pure real. This will be achieved when terms with odd powers of s sum to zero:

$$s^3 L_{312}(L_c + L_{303})C_s + sL_{312} = 0,$$

$$s^2(L_c + L_{303})C_s + 1 = 0,$$

$$\omega^2(L_c + L_{303})C_s = 1,$$

$$\omega_{\text{Res}} = \sqrt{\frac{1}{(L_c + L_{303})C_s}}. \tag{7.5}$$

Since $C_c \ll C_{373}$, $C_s \approx C_c$ and so

$$\omega_{\text{Res}} \cong \sqrt{\frac{1}{C_c(L_c + L_{303})}}.$$

L_{303} is adjustable, as indicated in Fig. 7.13, and its adjustment allows one to "pull" the resonant frequency of the crystal slightly. Therefore if this coil is properly adjusted, the bursts will excite the crystal to "ring" at this resonant frequency. Since the crystal Q is very high, this ringing dies out quite slowly. In fact, a scope trace of the voltage across R shows that it exhibits less than 5% damping before another burst arrives. The ringing voltage is applied to a totally conventional amplifier stage, Q_{306}. One of the two outputs from the collector of this stage is the "burst-related sinusoid" input to the chroma section as mentioned in Sections 7.6.1 and 7.6.2. The other goes to a variable-phase-shift network shown in Fig. 7.14. Here $R \equiv R_{342} \| R_{343} \| Z_{\text{in}}(Q_{310})$. There is no obvious reason for the use of two capacitors in series unless it is to minimize the effect of some stray capacitance not shown in the schematic. Again the voltage divider yields

$$\frac{\mathbf{V}_{\text{out}}}{\mathbf{V}_{\text{in}}}(s) = \frac{s^2 L_{304} CR + sCRR_{905}}{s^2 L_{304} C(R + R_{905}) + s(L_{304} + CRR_{905}) + R_{905}},$$

where C is the series combination of C_{338} and C_{340}. In this case we are interested in knowing the phase difference between input and output.

$$\frac{\mathbf{V}_{\text{out}}}{\mathbf{V}_{\text{in}}} = \frac{-\omega^2 L_{304} CR + j\omega CRR_{905}}{[R_{905} - \omega^2 L_{304} C(R + R_{905})] + j\omega(L_{304} + CRR_{905})}.$$

From this we can write

$$\phi = -\arctan\frac{R_{905}}{\omega L_{304}} - \arctan\frac{(L_{304} + CRR_{905})\omega}{[R_{905} - \omega^2 L_{304}(R + R_{905})C]}.$$

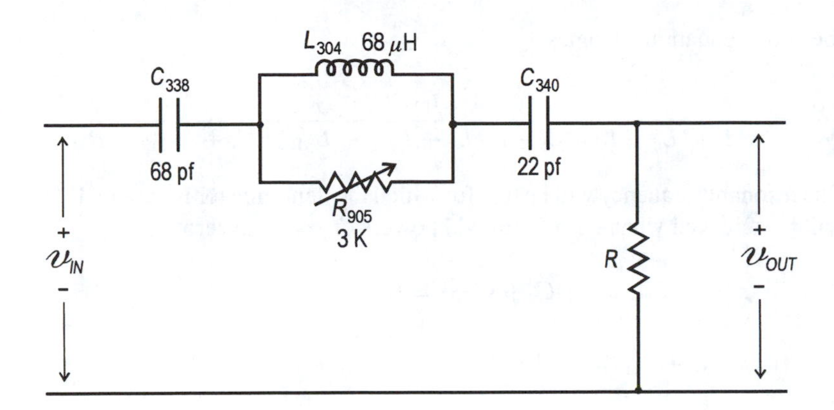

Fig. 7.14: Schematic diagram of the tint-control network in the Sony KV1200U.

By trigonometric combination, this equation can be expressed as

$$\phi = \arctan \frac{\left[R_{905}^2(1 - \omega^2 L_{304} C) + L_{304}^2 \omega^2\right]}{C\omega\left[R R_{905}^2 + L_{304}^2 \omega^2 (R + R_{905})\right]}.$$

Inserting the component values reduces this to

$$\phi = \arctan \left[\frac{.428 R_K^2 + 2.34}{.349 R_K^2 + .875 R_K + .816}\right],$$

where R_K is the value of R_{905} in kilohms. For the limiting values of $R_K = 0$ and $R_K = 3$, the phase shifts are 70.8° and 43.3°, respectively. This variation is produced by R_{905}, which is the *tint* control or *hue* control as Sony called it. This control varies the phase of the CW subcarrier relative to the chroma information and will thus vary the demodulated colors about their nominal tint.

The load resistance seen by this phase shift network is the input resistance of the Q_{310} stage. This stage is a conventional amplifier that feeds a fixed-phase-shift network composed of C_{341} and the input impedance of the Q_{311} stage. This network introduces an additional phase shift of about 24°, making the total range vary between 67° and 95°. The Q_{311} stage is a most unusual one. Sony calls it the CW oscillator, and indeed it is an oscillator. In fact, the change of one part value would allow the stage to oscillate at the resonant frequency of T_{305}'s primary even if isolated from the rest of the circuit. That part is R_{348} in the base bias network. As the circuit is designed, R_{348} is so low that the quiescent collector current is between 20 and 30 microamps. This lowers the loop gain enough to prevent oscillation until a ringing pulse arrives at the base and drives it into conduction. Thus it is only necessary to tune the primary of T_{305} reasonably close to 3.58 MHz to have it output a waveform of the exact same frequency as the burst and having a fairly good sinusoidal waveform.

With the introduction of integrated circuits into TV, a more elegant and precise method of converting the burst to a continuous wave was introduced. In this method, an integrated-circuit oscillator capable of producing a high-quality sinusoid uses a voltage-variable capacitance (VVC) as one of its frequency-determining elements. The burst and the output of this oscillator are both fed to a phase detector, whose DC output varies the voltage across the VVC in such a direction as to secure frequency and phase lock. It uses the phase-locked-loop principle, but is designed to give a sinusoidal output.

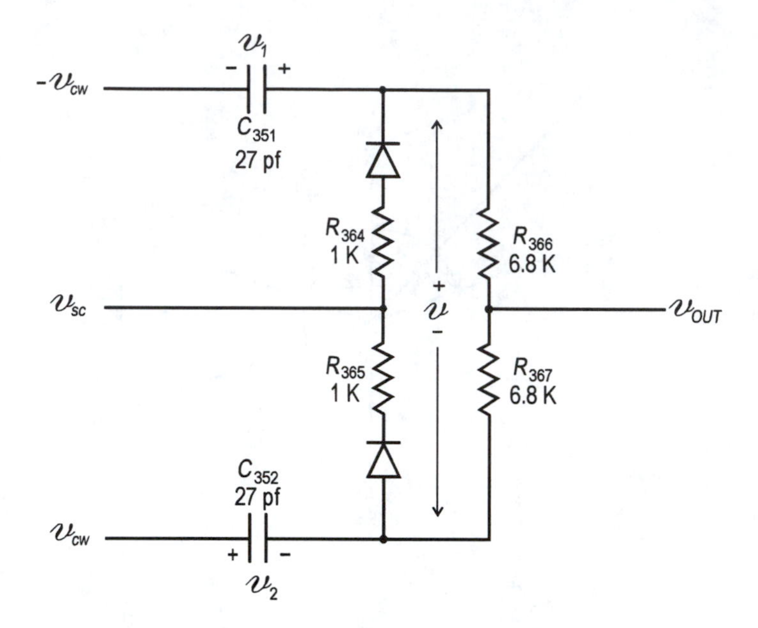

Fig. 7.15: Schematic diagram of the "Blue" phase detector in the Sony KV1200U.

7.6.4 Color demodulation circuitry

The outputs of the chroma section and the "burst oscillator" are brought together into circuitry that is evidently a phase detector. The blue phase detector is simplest and is shown in Fig. 7.15, where v_{sc} is the phase-shifted chroma from T_{303}. The push-pull outputs from the secondary of T_{305} are 180° out of phase with each other relative to the grounded center tap. We have defined the voltage on the upper half of the secondary as v_{cw}, which makes that on the lower half equal to $-v_{cw}$.

To analyze this circuit, we assume first that the diodes are ideal. Each diode will be assumed to charge its associated capacitor to a DC voltage equal to the amplitude of the phasor voltage difference applied to its loop. Thus we have

$$v_1 = |v_{cw} + v_{sc}| \text{ for the upper loop}$$

and

$$v_2 = |v_{cw} - v_{sc}| \text{ for the lower loop.}$$

This result's validity is dependent in part on the fact that the charging time constant (27 pF $\times$1 kΩ = 27 ns) is much less than the period of the 3.58 MHz signal (280 ns).

We now have expressions for the "DC" components v_1 and v_2. As we have done previously, we put "DC" in quotes to indicate that it varies slowly compared to the 3.58 MHz voltages. Since v_{cw} and $-v_{cw}$ are at the same DC potential, v must also be a "DC" term, and we can write an equation for the outside loop:

$$-v_1 + v - v_2 = 0$$

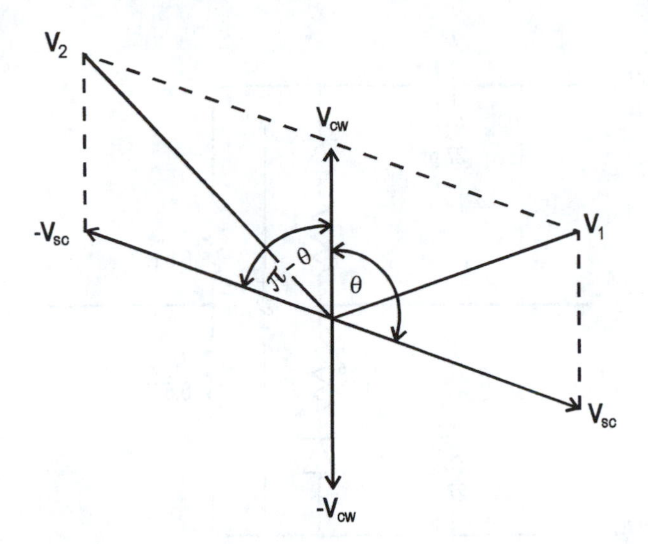

Fig. 7.16: Phasor diagram showing the summing of the CW 3.58 MHz signal (reconstructed from the burst) with the chroma signal.

or

$$v = v_1 + v_2.$$

For the same reason

$$\begin{aligned} 0 &= -v_1 + v/2 + v_{\text{out}} \\ &= -v_1 + (v_1 + v_2)/2 + v_{\text{out}} \end{aligned}$$

and thus

$$v_{\text{out}} = (v_2 - v_1)/2. \tag{7.6}$$

We next need to find v_1 and v_2 in terms of the phasor voltages that produce them. This can be done by using the law of cosines. The phasor diagram to be analyzed is shown in Fig. 7.16. We have

$$\begin{aligned} v_1 &= \sqrt{|v_{\text{cw}}|^2 + |v_{\text{sc}}|^2 - 2|v_{\text{cw}}| \cdot |v_{\text{sc}}| \cdot \cos(\pi - \theta)} \\ &= \sqrt{|v_{\text{cw}}|^2 + |v_{\text{sc}}|^2 + 2|v_{\text{cw}}| \cdot |v_{\text{sc}}| \cdot \cos\theta} \end{aligned} \tag{7.7a}$$

and

$$v_2 = \sqrt{|v_{\text{cw}}|^2 + |v_{\text{sc}}|^2 - 2|v_{\text{cw}}| \cdot |v_{\text{sc}}| \cdot \cos\theta}. \tag{7.7b}$$

The service information on this TV shows that $|v_{\text{cw}}| = 16.5\ \text{V}_{\text{pk-pk}}$ and $|v_{\text{sc}}| = 3.0\ \text{V}_{\text{pk-pk}}$. Since $|v_{\text{cw}}|^2 \gg |v_{\text{sc}}|^2$, Equation (7.7a) may be simplified by binomial expansion of the square root:

$$\begin{aligned} v_1 &\cong |v_{\text{cw}}| \sqrt{1 + 2 \cdot \frac{|v_{\text{sc}}|}{|v_{\text{cw}}|} \cdot \cos\theta} \approx |v_{\text{cw}}| \left(1 + \frac{|v_{\text{sc}}|}{|v_{\text{cw}}|} \cdot \cos\theta\right) \\ &\approx |v_{\text{cw}}| + |v_{\text{sc}}| \cos\theta. \end{aligned}$$

Likewise

$$v_2 \approx |v_{\text{cw}}| - |v_{\text{sc}}| \cos\theta.$$

Inserting these results into (7.6) gives

$$v_{\text{out}} \cong -|v_{\text{sc}}| \cos\theta. \tag{7.8}$$

Having derived this important result, it is necessary to point out that the values of R_{366} and R_{367} used in this design are unusually low. This means that there will be significant discharge of the capacitors during a cycle of 3.58 MHz, lowering the voltage across them and thus the output voltage as well. Although this will introduce more ripple in the output, it would not be expected to change the functional dependence of v_{out} on θ.

Before we look at the red and green phase detectors, we need to address the question of what phase detectors have to do with color demodulation. It was mentioned in conjunction with Eq. (7.4) that the $B - Y$ signal could be recovered by multiplying v_o by $\cos(\omega t + \omega)$. We later noted that ideally $\theta = \pi$. Let us now generalize this to consider synchronous demodulation of a chroma signal shifted by an angle θ relative to the CW signal, assumed to be $\cos(\omega t)$:

$$v_{\text{demod}} = \sqrt{(R - Y)^2 + (B - Y)^2} \cdot \cos(\omega t - \theta) \cdot \cos\omega t.$$

Expansion of the first cosine term yields

$$\begin{aligned} v_{\text{demod}} &= \sqrt{(R - Y)^2 + (B - Y)^2} \cdot [\cos\omega t \cdot \cos\theta + \sin\omega t \cdot \sin\theta] \cdot \cos\omega t \\ &= \sqrt{(R - Y)^2 + (B - Y)^2} \cdot [\cos^2\omega t \cdot \cos\theta + \cos\omega t \cdot \sin\omega t \cdot \sin\theta] \\ &= \sqrt{(R-Y)^2 + (B-Y)^2} \cdot \left[\left(\frac{1 + \cos 2\omega t}{2}\right) \cdot \cos\theta + \left(\frac{\sin 2\omega t}{2}\right) \cdot \sin\theta\right]. \end{aligned}$$

This expression shows three terms in the square brackets. The first is a "DC" term. The latter two are at a frequency of 2ω. These are readily removed by low-pass filtering, leaving

$$v_{\text{demod}} = \frac{\sqrt{(R - Y)^2 + (B - Y)^2}}{2} \cdot \cos\theta.$$

This equation is exactly analogous to Eq. (7.8) and shows that, in this application, the phase detector is a simple, inexpensive, and adequate substitute for a balanced demodulator.

Having established this result, we return to the red and green phase detectors, which each have four circuit components in addition to those used in the blue phase detector. The red and green detectors are topologically identical and have the form shown in Fig. 7.17. The circuitry to the right of the dashed line is identical to that of Fig. 7.16. That to the left is used to introduce an additional phase shift in the CW carrier prior to applying it to the phase detector.

Exercise 7.5 Show that if loading by the phase detector portion of Fig. 7.16 is neglected, $\mathbf{V}_{\text{S}}/\mathbf{V}_{\text{cw}} = 1\angle{-2}\arctan R_{\text{S}}C_{\text{S}}\omega$.

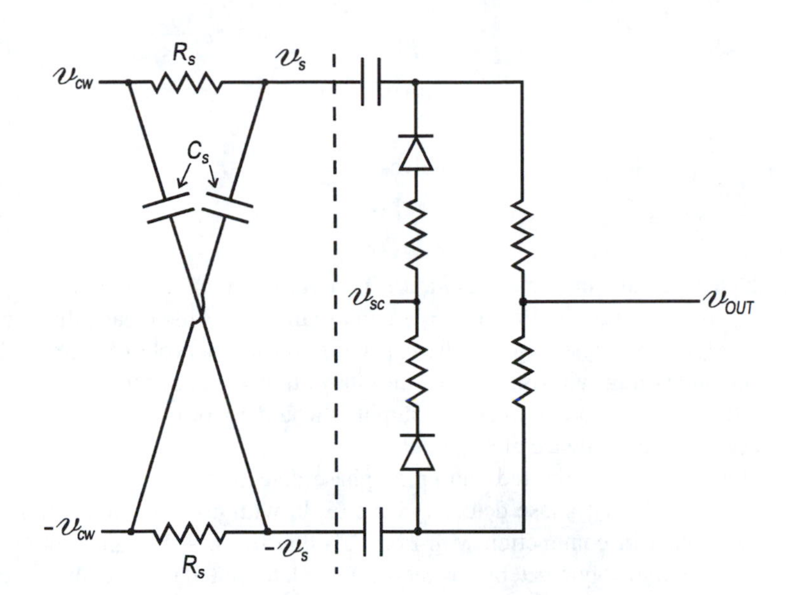

Fig. 7.17: Schematic diagram of the "Red" and "Green" phase detectors in the Sony KV1200U.

For the red phase shifter $R_s = 1.5\ \text{k}\Omega$ and $C_s = 18$ pF; so the phase shift is $-63°$. For green, $R_s = 3.9\ \text{k}\Omega$ and $C_s = 22$ pF and the phase shift is $-125°$.

Before proceeding to the mathematics of color formation, we note in Fig. 7.12 that across the chroma output there is a resistive divider comprised of R_{326}, R_{327}, and R_{328}. The right side of R_{328} is returned to AC ground by C_{326}. The DC voltage here is derived from the R_{329}–R_{330} voltage divider and is used to bias the diode pairs in the phase detectors (probably close to the threshold of conduction). The R_{326}–R_{327}–R_{328} divider sets the relative chroma values delivered to the color phase detectors. The blue receives v_{sc}, that to the red is $(5/8)\, v_{sc}$, and that to the green is $(3/8)\, v_{sc}$. Then using (7.8), we may write the equations for the magnitudes of the three color components out of the demodulators:

$$B - Y = -|v_{sc}| \cos(\theta_{sc} - \theta_{cw}), \tag{7.9a}$$

$$R - Y = \tfrac{5}{8} |v_{sc}| \cos(\theta_{sc} - \theta_{cw} + \theta_{RS}), \tag{7.9b}$$

and

$$G - Y = -\tfrac{3}{8} |v_{sc}| \cos(\theta_{sc} - \theta_{cw} + \theta_{GS}), \tag{7.9c}$$

where θ_{RS} and θ_{GS} are the additional phase shifts introduced by the left-hand side of Fig. 7.17. $R - Y$ has the opposite sign to the other two outputs because the connections from T_{305} are opposite.

We might imagine θ_{sc} to be the phase of the transmitted chroma signal and θ_{cw} to be the phase of the burst. In fact, however, both v_{sc} and v_{cw} are substantially shifted relative to their transmitted counterparts. The origins of these phase shifts have already been discussed. The shift of v_{cw} originates in the tint control network between Q_{306} and Q_{310} and in the RC network between Q_{310} and Q_{311}. The total shift due to these sources was shown to lie between 67° and 95°, depending on the setting of the tint control. The shift of v_{sc} is primarily due to L_{301} and will be something between 60° and 90°. If both of these shifts are made equal by the tint control, the phase difference between θ_{sc} and θ_{cw}

Table 7.1. *Theoretical response of the color demodulator circuitry in the Sony KV-1200U television receiver*

Transmitted color signal	θ_c	Output of $B-Y$ demodulator from Eq. (7.10a)	Output of $R-Y$ demodulator from Eq. (7.10b)	Output of $G-Y$ demodulator from Eq. (7.10c)	Normalized phasor resultant of color signals	Normalized ideal resultant of color signals
R–Y	90°	0	$.557V_{sc}$	$-.307V_{sc}$	$.83\angle 102°$	$1\angle 90°$
G–Y	−56°	$-.559V_{sc}$	$-.303V_{sc}$	$.375V_{sc}$	$.98\angle -39°$	$1\angle -56°$
B–Y	180°	V_{sc}	$-.284V_{sc}$	$-.215V_{sc}$	$1.13\angle 185°$	$1\angle 180°$

should be the same as that between the transmitted chroma and burst signals. Then if the decoding is correct, color reproduction will be correct. We will now examine the remainder of the decoding process. We shall denote $\theta_{sc} - \theta_{cw}$ as θ_c, the color phase angle. Equations (7.9) establish the $B - Y$ signal as a sort of de facto phase reference, relative to which θ_{RS} and θ_{GS} are measured (relative to the $R - Y$ and $G - Y$ signals respectively).
Using this notation in (7.9) yields

$$B - Y = -|\mathcal{V}_{sc}| \cos \theta_c, \tag{7.10a}$$

$$R - Y = \tfrac{5}{8}|\mathcal{V}_{sc}| \cos(\theta_c - 63°), \tag{7.10b}$$

and

$$G - Y = -\tfrac{3}{8}|\mathcal{V}_{sc}| \cos(\theta_c - 125°). \tag{7.10c}$$

We know that component tolerance and other nonidealities will make the color demodulation less than ideal, but at this point we neither have a feel for the magnitude of errors inherent in this particular decoding system or of the maximum tolerable error for producing a quality picture. To get a feeling for the errors produced by this circuit, we consider the response of the demodulators for three different values of chroma signal–namely along the three $\mathcal{C} - Y$ axes, where $\mathcal{C}$ may be R, G, or B. The results of the calculations are shown in Table 7.1. The table shows angular errors of up to 17° and amplitude errors of up to 17%. The NTSC specification calls for an absolute angular error of no more than 10° and a percent amplitude error of no more than 20% (Reference Data for Radio Engineers, 1956). We thus see that the amplitude spec is satisfied, but apparently the angular one is not. In this case, appearances are deceiving, however. If we examine the actual angles relative to the ideal, we see that in each case the actual angles are larger than the ideal. If we could reduce all angles by 11° we would have the situation shown in Table 7.2.

This set of decoded signals satisfies the NTSC requirements on both amplitude and phase. Of course, the control that is capable of introducing an 11° phase shift between the shifted chroma and the shifted burst is the tint control. This phase shift will not translate directly to a shift in the phase of the decoded color on the NTSC color chart (Fig. 7.9). This is because, as Table 7.1 shows, the actual decoded color is not just a function of the one demodulator that should ideally be outputting it. There are also smaller components of the

Table 7.2. *The effect of tint control variation on the demodulated color picture*

Color signal (transmitted)	Actual decoded color signal	Ideal decoded color signal
$R–Y$	$.83\angle 91°$	$1\angle 90°$
$G–Y$	$.98\angle -50°$	$1\angle -56°$
$B–Y$	$1.13\angle 174°$	$1\angle 180°$

screen color from the outputs of the other two demodulators in general. Nevertheless there is every reason to expect that the phase shift introduced by the tint control will be closely reflected in the phase angle of the recovered color signal.

Exercise 7.6 Referring to Eqs. (7.9), suppose the tint control is used to increase θ_{cw} by 11°. Generate a table analogous to Table 7.1 in this case. You may find it a useful computational aid to treat the color components as phasors that can be resolved into "real" components along the burst phase and "imaginary" components along the $R - Y$ axis. These "phasors" can then be added in the usual way to find the resultant in polar form.

It is also pertinent to point out that the equations we have presented and those we have derived are predicated on a CRT with phosphors called the NTSC *receiver primaries*. When newer and better phosphors are developed, it becomes necessary to adjust the color decoder amplitudes and phases. A matrix method of doing so has been developed and is the subject of a classic paper in color TV (Parker, 1966).

7.6.5 Color output amplifiers

Each of the phase detector outputs is fed to a color output amplifier of the type shown in Fig. 7.18. The LC filter on the input to the lower transistor removes any high-frequency ripple on the output of the phase detectors and also band-limits the $\mathcal{C} - Y$ signal to remove frequency components that carry no information. The lower transistor functions as an emitter follower to buffer the $\mathcal{C} - Y$ signal to the emitter circuit of the upper transistor. Again recall that $\mathcal{C}$ stands for one of the NTSC primary colors, R, G, or B.

Exercise 7.7

(a) Use the equivalent circuit shown to find v_o at low frequencies as a function of $\mathcal{C}$ and Y.

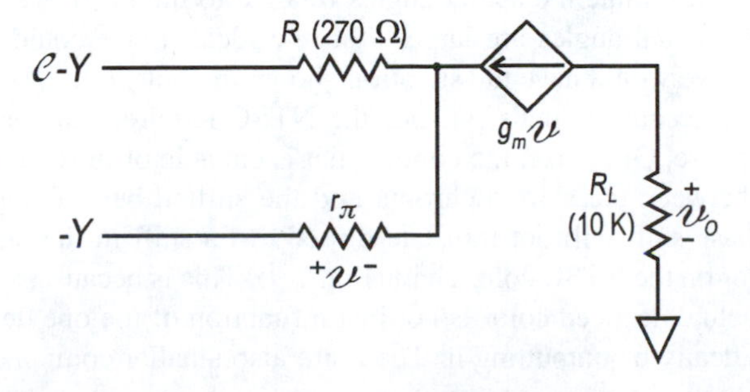

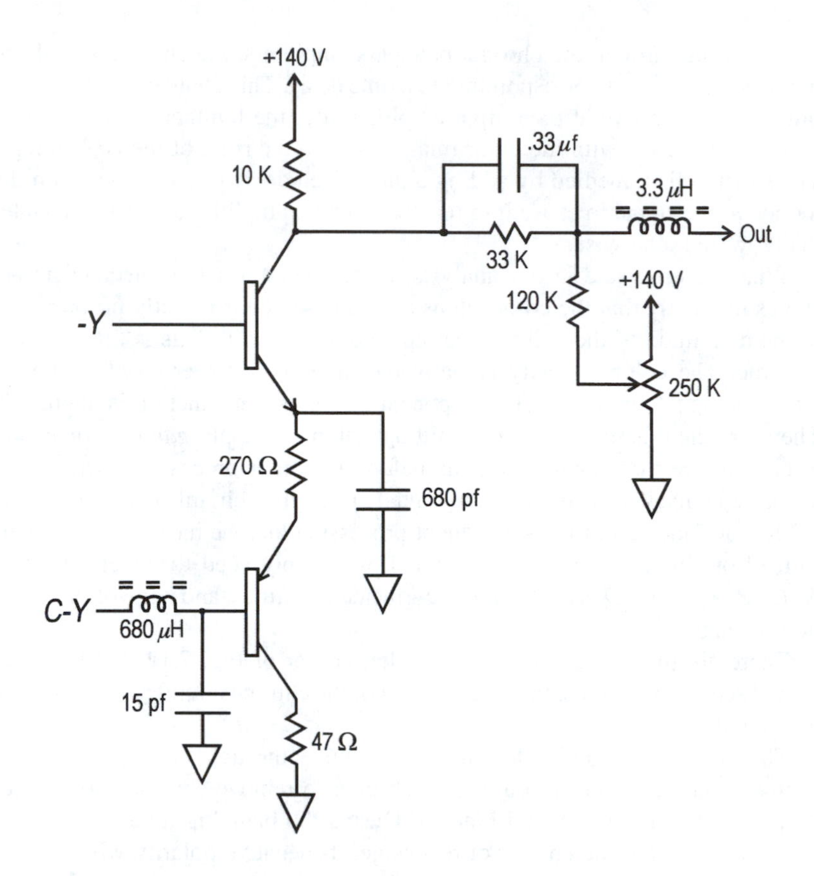

Fig. 7.18: Schematic diagram of one of the color output amplifiers in the Sony KV1200U. It is in this stage that the luminance and the chroma signals are combined to recreate the pure color signals that drive the CRT.

(b) The result of part (a) is oversimplified in that $C - Y$ is band-limited and $-Y$ is not. Set $Y = Y_{\text{lf}} + Y_{\text{hf}}$. Then $C - Y$ should be $C - Y_{\text{lf}}$ and $-Y = -Y_{\text{lf}} - Y_{\text{hf}}$. Find the revised expression for v_{o} in this case and discuss its physical significance.

(c) Suppose the color killer is activated because of a low-amplitude color signal. Find v_{o} in this case and discuss your result physically.

Just as the inputs to this stage are DC-coupled from the phase detectors, so the outputs are direct-coupled to the CRT cathodes. This extends frequency response down to DC, which makes it possible to reproduce large color areas on the screen. The 33 kΩ resistor in Fig. 7.18 is bypassed with a .33 μF capacitor so the AC signal will not be attenuated.

The potentiometer in the collector circuit is effectively a brightness control for each primary color. The $-Y$ signal is also DC coupled to the output stage, and there are separate controls to vary this direct-coupled signal to each of the color output amplifiers. These controls are in the emitter circuit of Q_{409}, and they effectively set the video signal threshold level for conduction in the output stage.

7.6.6 Loose ends

An examination of the connections from the color output amplifiers to the CRT shows that as C gets larger, the CRT gun moves toward cutoff. This we take to mean that the outputs are actually $\alpha(-C)$, rather than to C. This is what we would expect if the inputs to this stage both had their signs reversed to $+Y$ and $-(C - Y)$, which is the actual situation.

In passing through the chroma bandpass amplifiers, the chroma signal undergoes a phase shift corresponding to a time delay. This could cause the $C - Y$ information to arrive at the output amplifier after the luminance information, giving two images with the color image offset to the right of the B/W image. The situation is remedied by adding a distributed delay line in series with the monochrome signal to cause it to reach the output amplifier about 1 μsec later than it would otherwise.

What we have said in our analysis of the color formation mechanism assumes implicitly that the color intensity on the screen is directly proportional to the amplitude of the color signal applied to the CRT. This is far from being true. The color intensity actually increases as a power-law function of the color signal amplitude. The exponent of this power function is about 2.2. Therefore the practice is for transmitting stations to apply *gamma correction* to each of the three camera signals before any other processing. This is essentially equivalent to passing the signals through a circuit that extracts the 2.2th root. Since all of the subsequent processing that we have studied is performed on this gamma-corrected signal, we simply need to understand that R, G, B, Y, I, and Q are all gamma-corrected quantities, and none of the equations change.

There are three stages in the lower left corner of Fig. 7.12 that have not been discussed yet. Since they are not part of the chroma circuitry, we will only mention them briefly.

The luminance signal is fed into this circuitry through VR_{902B}, which is a contrast control. It goes through a notch filter to remove chroma and reaches Q_{407}, which is the horizontal blanker. During the blanking time, the flyback pulse is coupled to the emitter of this stage. Its negative polarity will forward bias the B–E junction, and Q_{407} will conduct, pulling the collector down toward ground and thus shunting out the Y signal.

The ABL (Automatic Brightness Limiter – Q_{408}) limits the maximum current the CRT can draw from the HV supply. The emitter potential is set by VR_{407B} and a negative DC voltage proportional to the beam current. This voltage is ripple filtered by R_{415} and C_{403}. As the HV supply current rises, the emitter potential of Q_{408} falls until the bias set by VR_{407B} is overcome by the HV supply current component and goes negative by about .6 volts. At this point, Q_{408} conducts, lowering the base bias on Q_{409}. This is just an emitter follower stage of amplification for the Y signal. Since it is DC-coupled all the way to the CRT, the reduced base drive will cause the base potentials of the upper transistors in the color output amplifiers to drop. This will raise their collector potentials, which will make the CRT cathodes more negative relative to the grid and will thus reduce the current demand on the HV power supply.

7.7 An integrated color TV

We now proceed to the analysis of a modern integrated set based on the RCA CTC136 chassis. As is common in TV, several models will use the same basic chassis, but with different added modules to allow a choice of capabilities. The RCA trade name in TV is now owned by Thomson CSF, the French electronics giant. Nevertheless, due to the enormous contributions RCA made to the development of TV, we have chosen a set bearing the RCA name and manifesting the RCA design philosophy as our set to analyze.

The discrete Sony set just examined had forty-six transistors. A direct comparison with the RCA is difficult since the latter is totally modular in concept as well as construction. One main board contains all of the IF, video, chroma, sound, sync, and deflection circuitry except for power stages. In order to make a functioning set, it is necessary to add a tuning module, an interface board, user controls, CRT, and yoke, in addition to the power output stages.

The tuning circuit options include: (1) mechanical tuners, (2) wired remote for hospital use, (3) wireless (infrared) remote that allows power on-off, channel up-down and volume up-down, and (4) a full-function wireless remote with keyboard selection of channels.

The interface board will be one of four modules suited to the tuner option the buyer selects.

In addition to these required modules, there are some options. These include boards for AM–FM reception (primarily for hospitals and motels), TV stereo sound and an autoprogram module. When a remote is used to scan up or down through the channels, it increments to the next channel in numerical sequence. However, in most localities, the majority of channel numbers will not deliver a useful viewing signal without cable service. The autoprogram module will skip over channels that do not provide a strong enough signal to give a good picture. As new channels are added in a viewing area, the autoprogram circuit will automatically include them in its scheduled tour of "channel land." Because this requires a lot of circuitry for a relatively insignificant circuit function, we will not look at the autoprogram circuitry in our analysis.

Finally, a given choice of tuning mechanism may require one or more outboard modules. For example, remote infrared (IR) tuning will require a module to receive the IR, convert it to an electrical signal, and amplify it.

The general block diagram of this class of TVs, shown in Fig. 7.19, illustrates much of this, and more besides. The large assembly enclosed in dashed lines is a single IC! The rest of the circuit includes the tuner, three filters including the exotic SAW filter to be discussed later, several power output stages, and power supply circuitry. Thus, conceptually, this TV has much in common with the Monomax®-based monochrome set analyzed in Chapter 6. Both have:

(1) the great majority of their circuitry on a single chip,
(2) power output functions handled by discrete transistors, and
(3) separate tuner circuitry.

The differences between these integrated sets (besides the obvious fact that one has color and the other doesn't) are primarily based on the higher level of conveniences generally found in color sets.

We will be analyzing a set with a wireless remote control that is keyboard programmable but is without stereo sound or a built-in radio receiver. As mentioned previously, we will not include the autoprogram circuit in our analysis, even though a receiver otherwise configured as we have described would have this capability. Figure 7.20 shows how these functions are partitioned among the modules. Having selected this configuration, we can now determine the device-level complexity as shown in Table 7.3.

We will not analyze this circuit in quite the detail devoted to previous circuits, both because information on the ICs is difficult to find, and because most of the circuit functions and means of implementing them are already known

Table 7.3. *Active device count in various modules of the RCA CTC-136 TV receiver*

Module	ICs	Transistors
IR transmitter	1	1
IR receiver	0	4
Tuning unit: RF section	1	8
Tuning unit: control section	2	7
Interface board	0	5
Main circuit board	2	15
CRT drive borad	0	3
Degaussing circuit	0	1

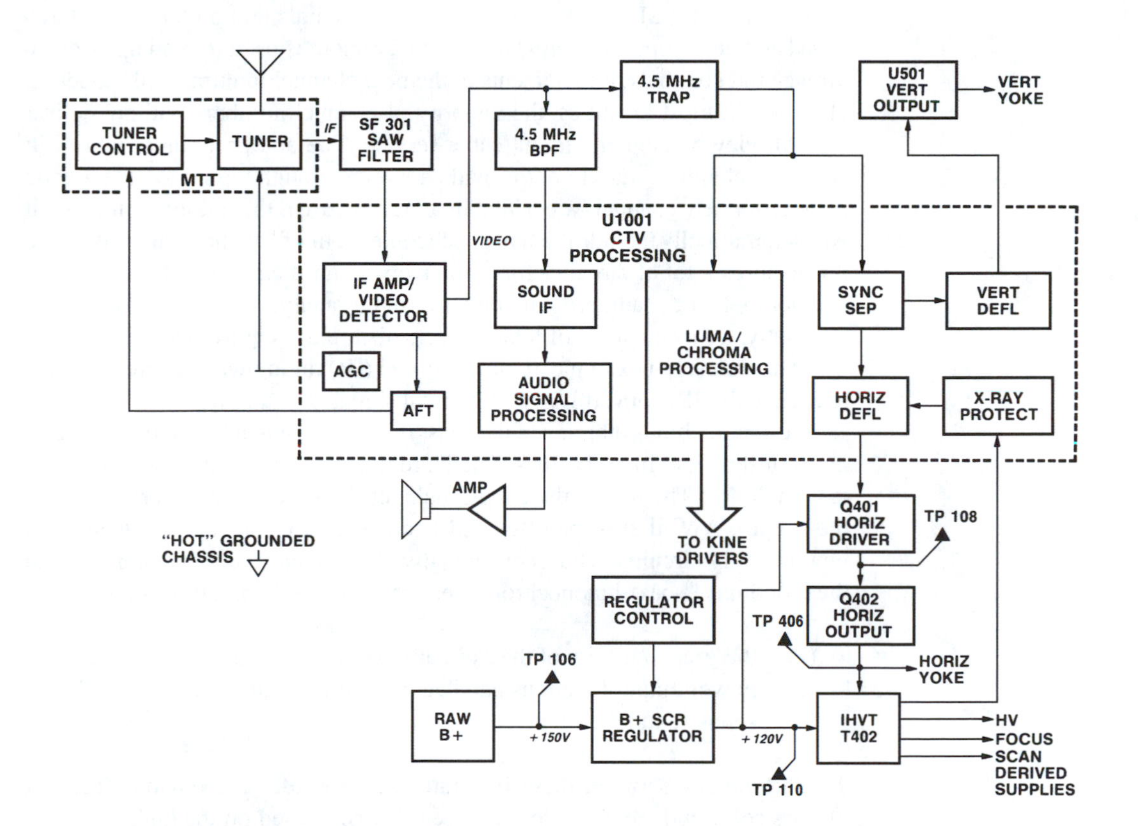

Fig. 7.19: Block diagram of the RCA CTC-136 color television receiver.

to us. Where this is not the case, we will devote more detail to the circuit description. We shall study the modules in the same order in which they are listed in Table 7.3.

7.7.1 IR transmitter

The schematic of this module is shown in Fig. 7.21. This is the user's hand-held unit. The transmitter consists, almost entirely, of the keyboard and the IC. This IC has a clock derived from a crystal, Y_{1001}. It decodes the keyboard and, for each key pressed, outputs a unique sequence of pulses at pin 9. This pulse train

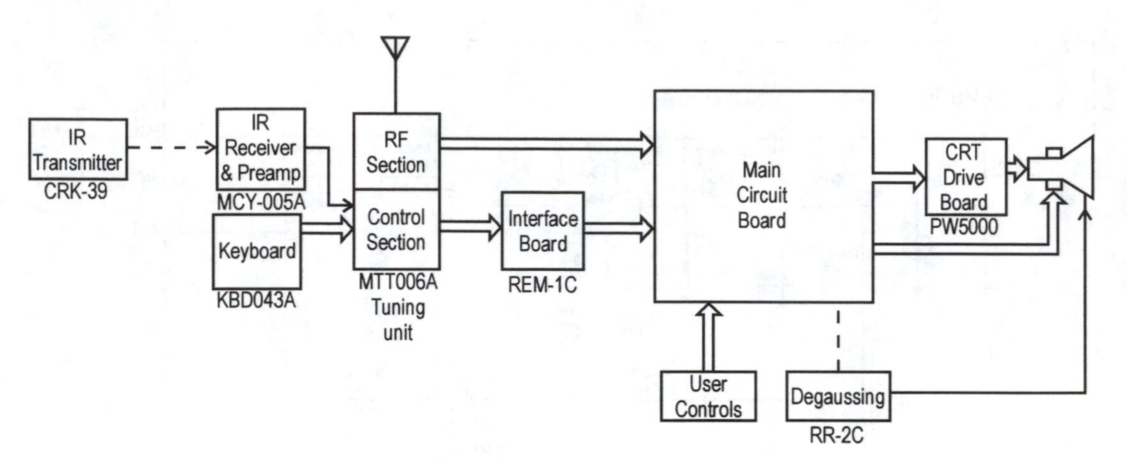

Fig. 7.20: Modular diagram of the RCA CTC-136 color television receiver.

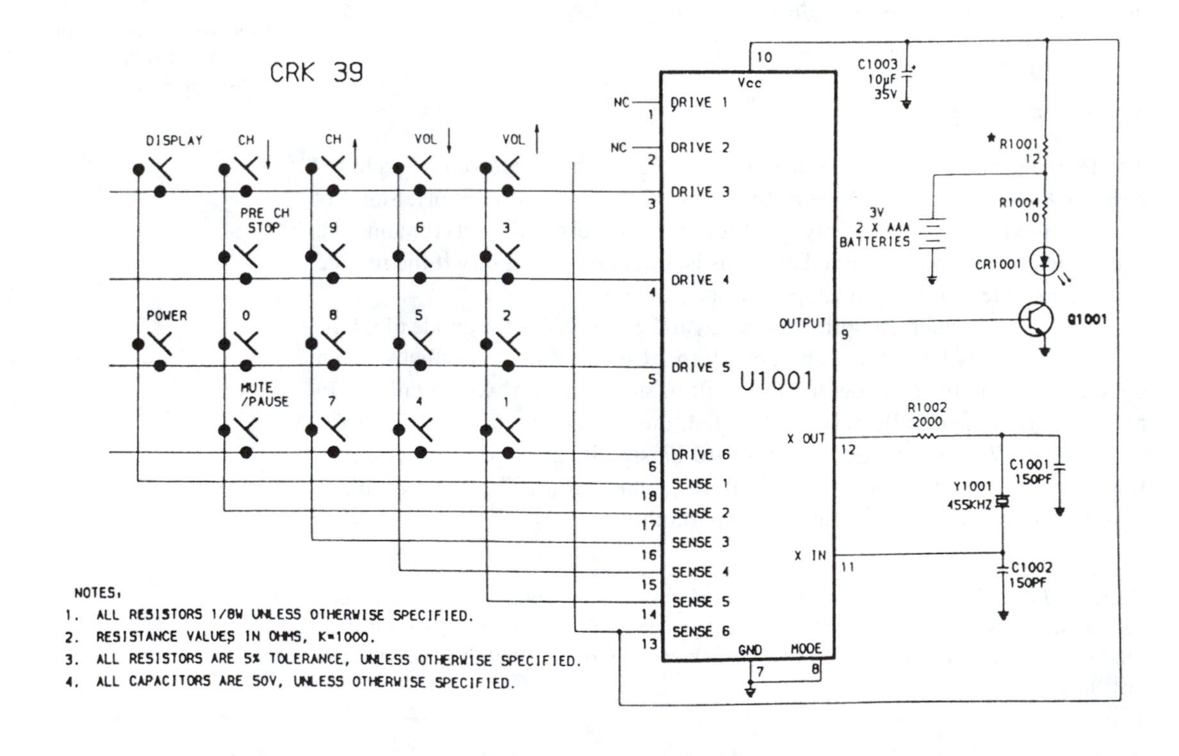

Fig. 7.21: Schematic diagram of the infrared (IR) remote-control transmitter for the RCA CTC-136 color television receiver.

is fed to Q_{1001}, which acts as a driver for the IR LED, CR_{1001}. The entire unit is powered by two miniature penlight batteries, which should give nearly a year's use. Since the IC is powered at all times, we are probably safe in assuming that it is fabricated in CMOS. The LED is not passed through a lens because it naturally has a beam divergence of about 30°, and this translates more or less directly to the allowable angular error in direct aiming. It is sometimes possible to activate the TV by aiming further away from the set because reflections of this divergent beam reach the receiver, which is quite sensitive and thus able to

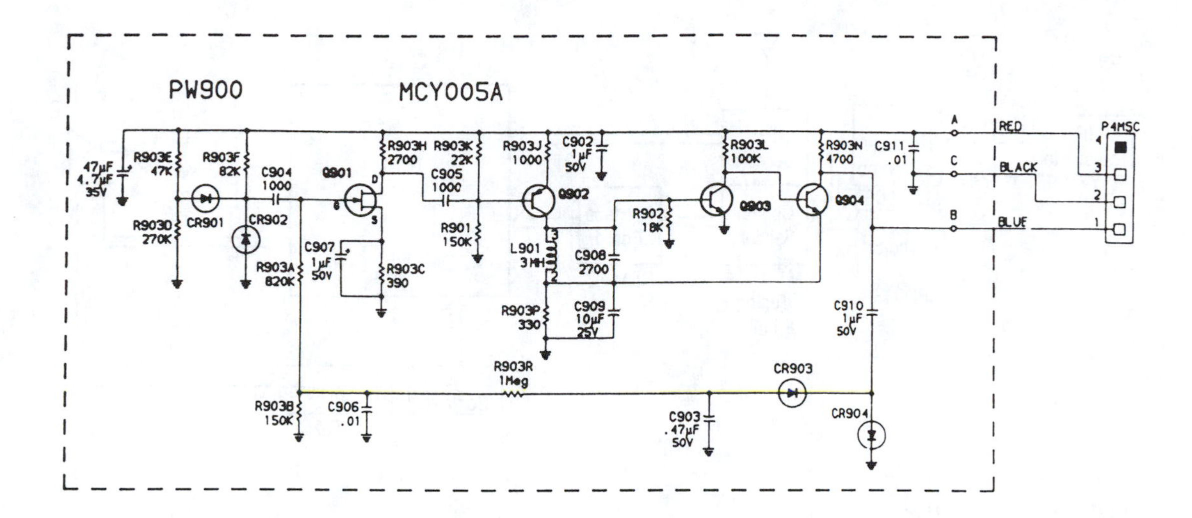

Fig. 7.22: Schematic diagram of the IR remote-control receiver for the RCA CTC-136 color television receiver.

respond to them. Good IR reflectors may not necessarily be those things that are good reflectors of visible light.

7.7.2 *IR receiver*

The IR receiver schematic is shown in Fig. 7.22. Light is received by a photodiode, CR_{902}, which is held in reverse bias by an 11 V standby supply. Standby refers to the fact that this supply (among several others in the set) remains on, even when main power is turned off. This is, of course, necessary if the remote is to be capable of turning main power on.

The voltage divider, comprised of R_{903D} and R_{903E}, holds the anode of CR_{901} at about 9.4 volts DC. In the absence of an IR pulse, CR_{902} is effectively an open circuit, and the cathode of CR_{901} will be at about 11 V, which holds it in reverse bias. When an IR pulse is detected, the photodiode conducts, and the potential at CR_{901}'s cathode falls, forward biasing it, and allowing current to flow to ground through CR_{902} from both R_{903F} and CR_{101}. The effect of this is to reduce the drop in voltage across the photodiode.

Exercise 7.8 In this diagram we have modeled the reverse-biased photodiode as a current sink. Assuming its saturation current is negligible, this is exactly what it is. Find $v(I_L)$ for the circuit as shown and when the circuit to the left of the dashed line is disconnected.

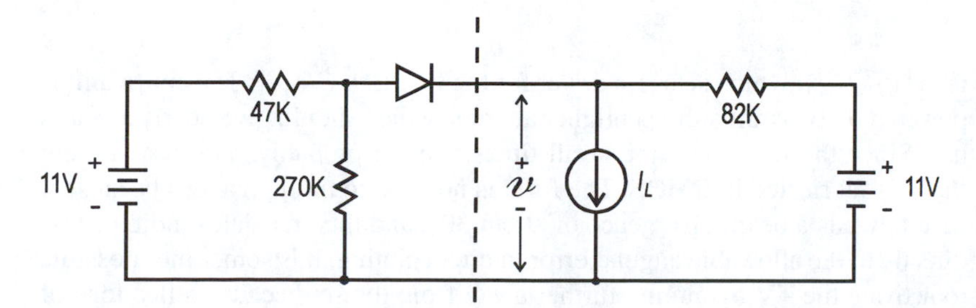

At first it seems counter-productive to deliberately reduce the sensitivity of the circuit to the IR pulse. The reason for doing so can be seen by looking onward in the circuit. The photodiode signal is capacitively coupled to the gate of a JFET. It is important that this gate never become forward biased, because if it does, the FET may require a time to recover that is longer than the duration of the IR pulses, and some may be lost and lead to improper decoding. As we shall see, other measures are taken to prevent this as well.

The output of the FET is taken to a direct-coupled three-stage amplifier with DC feedback from the emitter of Q_{904} into the collector circuit of Q_{902}. The bypass capacitor, C_{909}, prevents any AC feedback between these stages.

Exercise 7.9

(a) Assuming all transistors are in normal active mode and have $\beta = 100$ and assuming $V_{BE} = .6$ V, find the DC quiescent point for all three transistors.

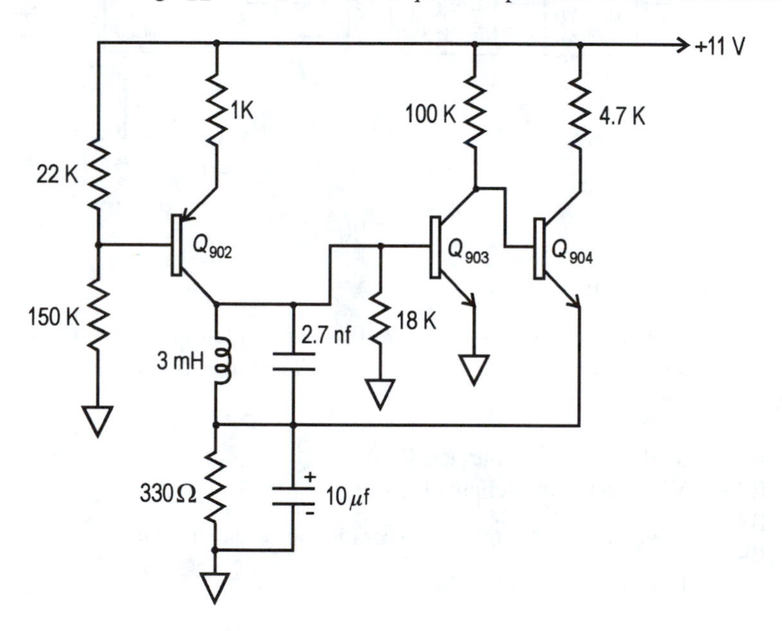

(b) Find the resonant frequency of the collector load of Q_{902}.

(c) Estimate the gain of Q_{902} at resonance, including the loading by Q_{903}.

(d) Estimate the gain of Q_{903} when loaded by Q_{904}.

(e) Estimate the gain of Q_{904}, assuming the 4.7 kΩ collector resistor is its only load.

The output of this entire amplifier is applied to a network that connects back to the gate of the FET. Before looking on to the next paragraph, determine the purpose of this network.

The first four elements encountered in the network, C_{910}, CR_{904}, CR_{903}, and C_{903} are part of a voltage doubler that outputs a negative voltage proportional to the signal amplitude at the output of the preamp. This DC voltage is further ripple-filtered by R_{903R} and C_{906}. The voltage divider made up of R_{903R} and R_{903B} applies part of this voltage to the FET gate through R_{903A}. This feedback system exists to make the gate more negative when the IR signal gets stronger, and keep the FET gate reverse biased.

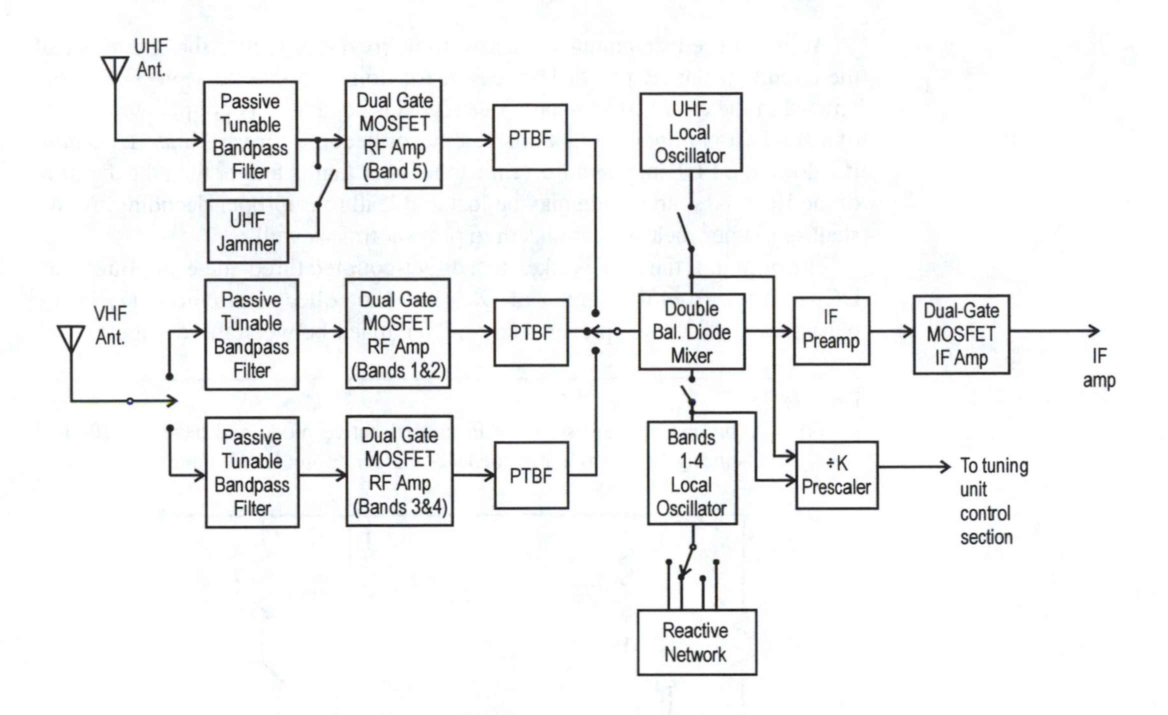

Fig. 7.23: Device-level block diagram of the electronic tuner of the RCA CTC-136 color television receiver.

7.7.3 *Tuning unit*

As indicated in Table 7.3, and in Fig. 7.20, the tuning unit is composed of two sections – an RF section and a control section. The tuner is designed to receive five bands as follows:

B1 VHF low band (channels 2–6)
B2 VHF high band (channels 7–13)
B3, B4 } cable bands (channels not specified in service literature)
B5 UHF band (channels 14–70)

7.7.3.1 *RF section*

As is typical of all electronic tuners, this one makes heavy use of switching diodes to accomplish band switching. An example of this will be shown shortly. First, however, the reader's attention is directed to Fig. 7.23, which shows a device-level partitioning of the RF section of the tuning module.

Relative to this figure, the following observations are important:

- All switches shown are "gangs" of the electronic bandswitch.
- Not all "gangs" are shown. The circuit contains a total of twenty-six bandswitching diodes. Those shown are the most important ones, conceptually.
- The tunable bandpass filters are tuned by varicaps. The filters on the inputs to the RF amplifiers each contain one, and those on the outputs of the RF amplifiers each contain two.

- The UHF local oscillator contains one varicap, and the local oscillator for bands 1–4 contains two.
- All twelve of these varicaps are varied simultaneously by one line from the control section. This indicates good design. In fact, only seven lines connect the RF and control sections of the tuning module; the B1–B5 bandswitch enable lines (digital), the varicap control voltage (analog), and the divided-down local oscillator signal (digital), which is an output from the RF section.
- The UHF jammer is a switching transistor which grounds out the varicap control voltage delivered to the input bandpass filter of the UHF stage when B5 is not enabled.
- The double-balanced diode mixer is used because it is less expensive than a UHF transistor capable of performing the same function.

Most receivers utilize a separate transistor mixer for lower frequencies, but the balanced diode mixer is excellent in terms of both linearity and frequency response. Its only drawback is a low-amplitude output. Since the mixer is followed by three discrete IF amplifier stages, plus further amplification in the main IC, there should be plenty of gain to offset the low-level mixer output.

Exercise 7.10 Two of the three discrete IF amplifier stages are found in the RF portion of the MTT-006A tuning system we have been examining. These two stages are shown below.

Because the service literature fails to show the values of any of the inductors in this circuit, a detailed analysis is not possible. Certain features of the circuit are, nonetheless, explicable.

(a) The first stage (bipolar) is obviously a tuned amplifier, yet it contains no mismatching or neutralization circuitry. How is this possible?
(b) What two types of feedback does this stage have?
(c) Why is a dual-gate MOSFET used for the second stage rather than a single-gate one (two reasons)?
(d) What will happen to the bias on G2 when the UHF band is selected?

The inductors in this circuit (besides the one in the tuned load of the first stage) are used to shape the frequency response. The need for this was discussed in Section 6.3.2. However, most of the frequency shaping is accomplished by a SAW filter (which is discussed in Section 7.7.5.1) and by an inductor (L_{302}), which is in parallel with its output port.

The local oscillator for bands 1–4 is shown in Fig. 7.24. All diodes shown with a switch within the circle are switching diodes.

Their bias states are shown in Table 7.4. We will consider the case where the B1 bandswitching line is high and the others are low.

With the switching diodes removed from the circuit in favor of opens and shorts, the circuit reduces to Fig. 7.25. In drawing this diagram, we have combined in series or parallel whatever elements are so connected with the diodes in the states shown in Table 7.4. Note that the positive voltage on the B1 enable-line not only switches the diodes but also goes through a two-stage decoupling network (200 Ω/940 pF and 150 Ω/470 pF) to become the V_{cc} supply for the oscillator. This assures that it will not even run when Q_{103} is serving as the local oscillator for UHF. To proceed further in the analysis of this circuit, we

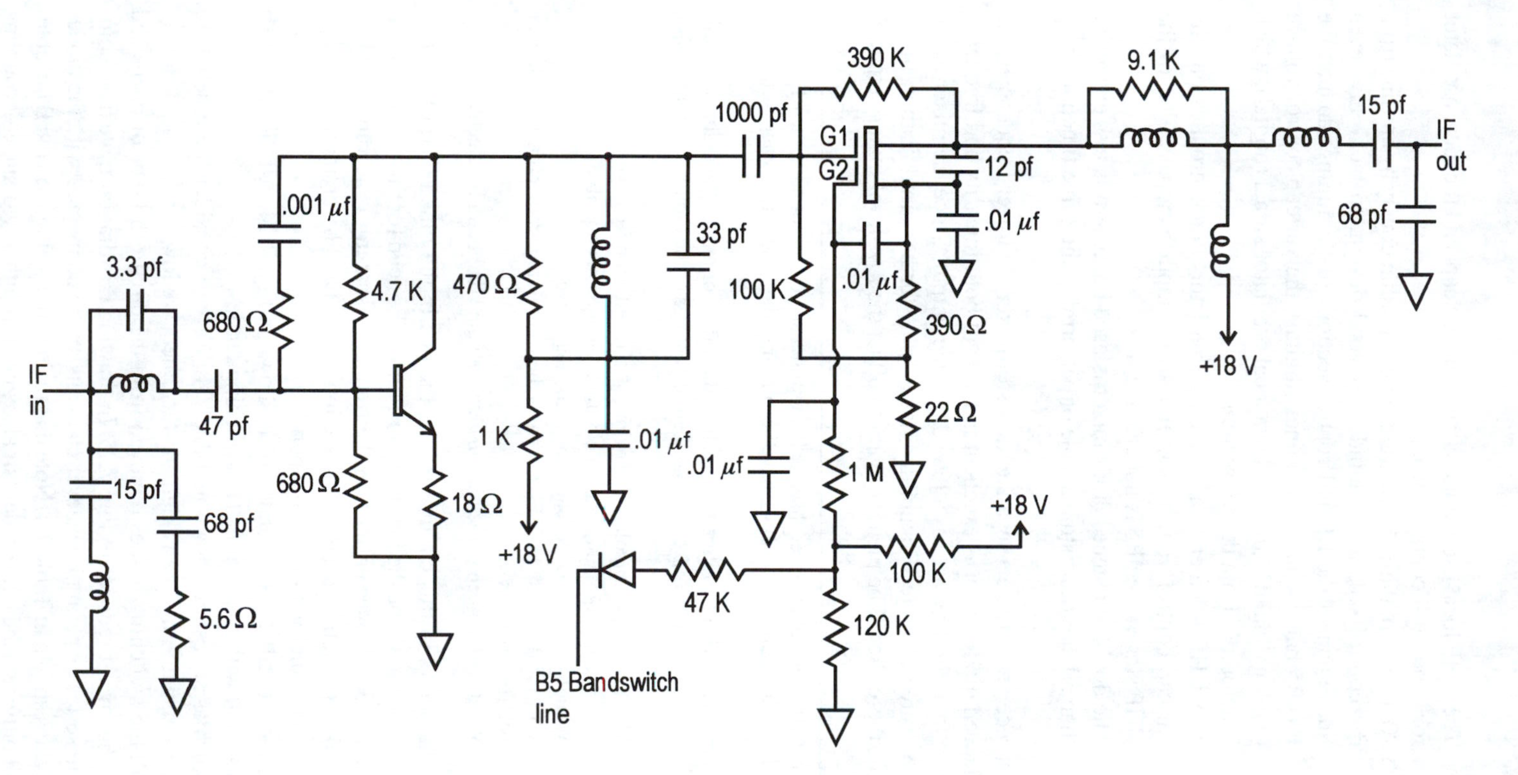
IF
in
3.3 pf
47 pf
15 pf
68 pf
5.6 Ω
680 Ω
.001 μf
4.7 K
680 Ω
18 Ω
470 Ω
1 K
+18 V
.01 μf
33 pf
1000 pf
390 K
G1
G2
100 K
.01 μf
12 pf
.01 μf
390 Ω
22 Ω
.01 μf
1 M
+18 V
100 K
47 K
120 K
B5 Bandswitch
line
9.1 K
+18 V
15 pf
68 pf
IF
out

Fig. E7.10.

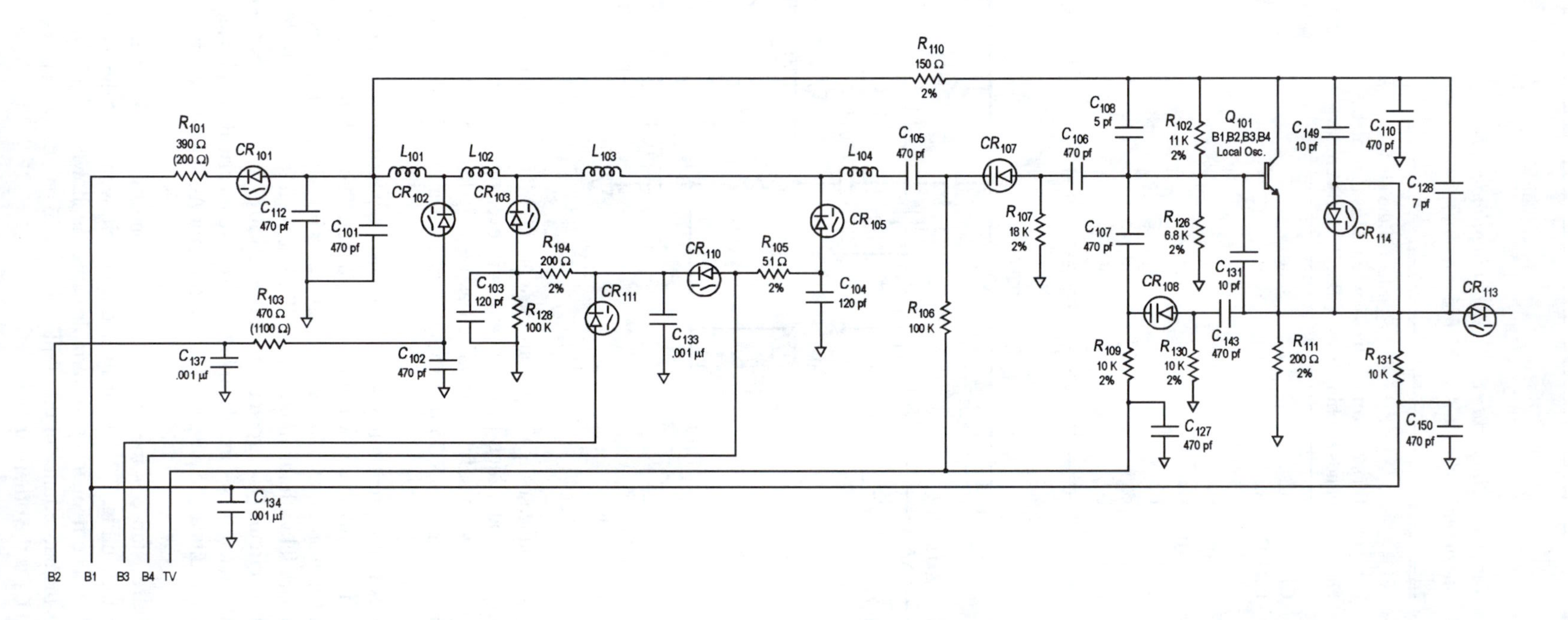

Fig. 7.24: Schematic diagram of the local oscillator portion of Fig. 7.23. All of the diodes except the varicaps are switching diodes.

Table 7.4. *Bias state of the switching diodes in the tuner of the RCA CTC-136*

Diode	Bias determined by bandswitching lines	State of the diode when B1 is high and B2, B3 & B4 are low
CR101	B1	short
CR102	B2	open
CR103	B3 or B4	open
CR105	B4	open
CR110	B4	open
CR111	B3	open
CR113	B1	short
CR114	B1	short

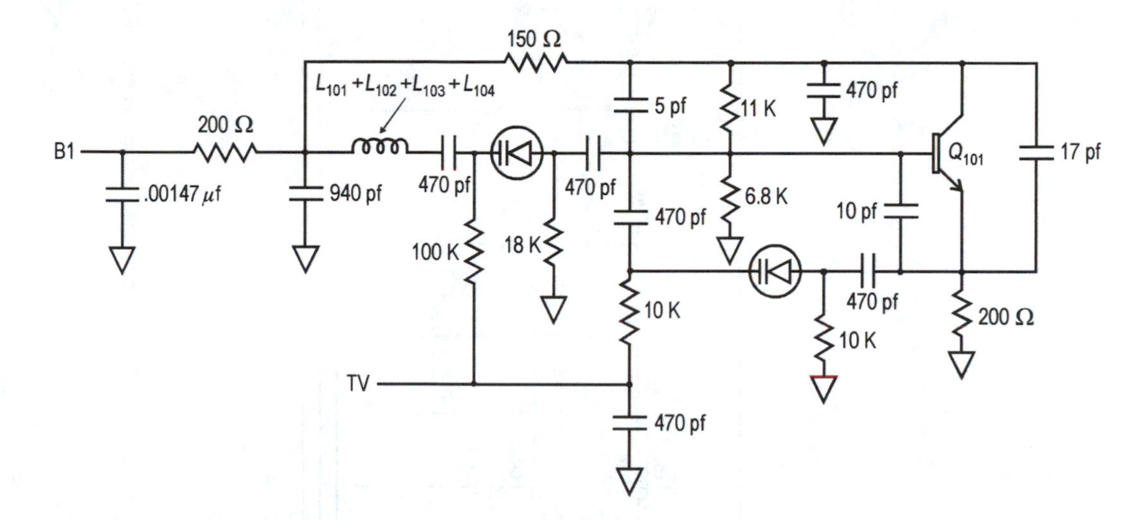

Fig. 7.25: A simplified version of Fig. 7.24 obtained by replacing each switching diode by an open or short as is appropriate for Band 1 operation.

calculate the impedance of a 470 pF capacitor at 100 MHz, which is near the minimum frequency at which this oscillator will ever operate, and find that it is 3 Ω. It can thus be safely considered as an AC short. Shorting out all of these capacitors in Fig. 7.25 in a straightforward way yields the circuit of Fig. 7.26.

It is obvious that a great many simplifications are possible on this circuit. Only three of these involve discarding a component. The 10 k resistor in the emitter circuit is negligible relative to the 200 Ω with which it is in parallel. The 100 k in parallel with the inductance will be negligible to the same degree that the resonant circuit Q is greater than one. The 150 Ω resistor has both ends AC-grounded. When these and the more obvious geometrical simplifications are made, the circuit can be put into standard form for AC analysis as shown in Fig. 7.27.

The 2.54 k resistance is the parallel combination of the 6.8 k, 11 k, 10 k, and 18 k resistors in Fig. 7.25. To analyze this circuit, we extract the feedback network and label components with variable names as shown in Fig. 7.28. Here C_2 represents the capacitance of the 10 pF capacitor in parallel with one of the varicaps, and C_4 is the other varicap.

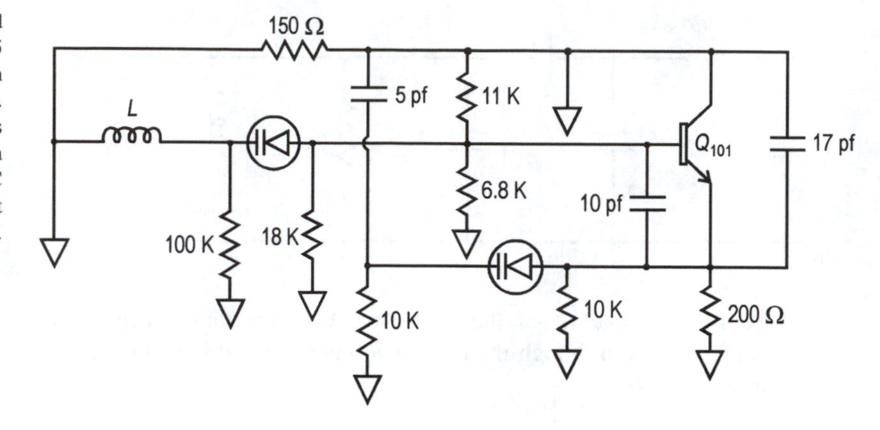

Fig. 7.26: A simplified version of Fig. 7.25 obtained by replacing each 470 pF capacitor by a short. This simplification means that the resulting circuit can only be used for AC analysis from this point forward.

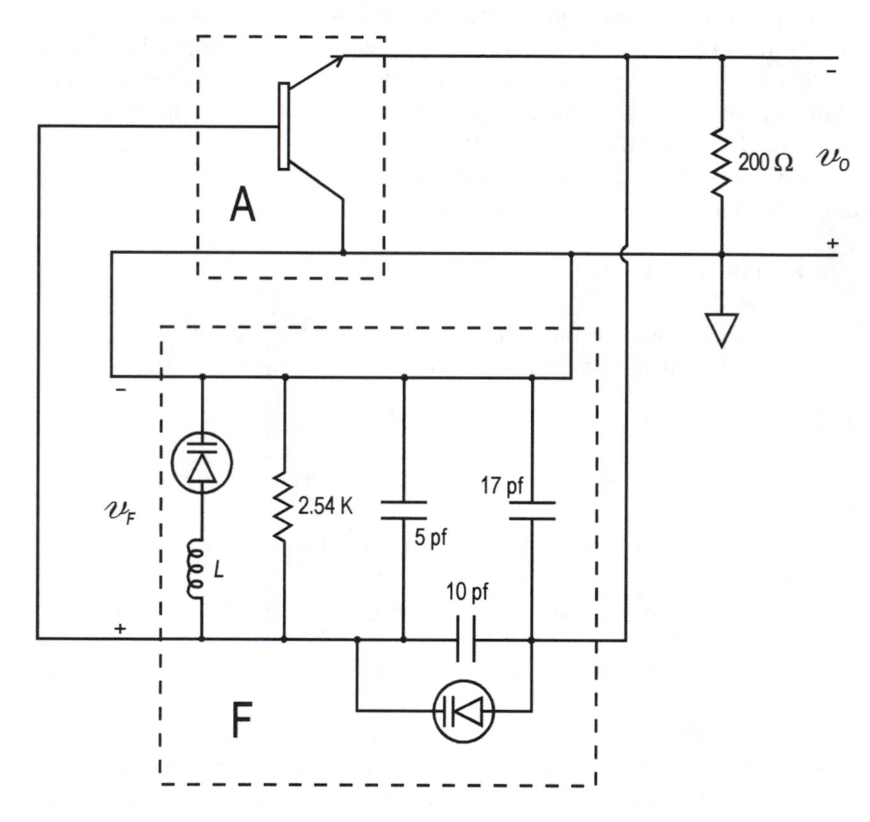

Fig. 7.27: A schematic diagram of the local oscillator of Fig. 7.26 in standard feedback form. The feedback network is obtained from Fig. 7.26 by combining elements in parallel as appropriate.

Analysis of the transfer function of Fig. 7.28 with $R_1 = 0$ yields no circumstances under which the Barkhausen criterion is satisfied. If we were to believe this, we would conclude that the circuit will not oscillate. Since it is part of a working TV, however, it must oscillate. In reevaluating our analysis, we consider the following possible explanations:

- Perhaps the loading of the output of F by the input of A is not negligible, even though the emitter follower used for A should have high input impedance.
- Perhaps the amplifier is not operating in its midband range where its capacitive effects can be ignored.

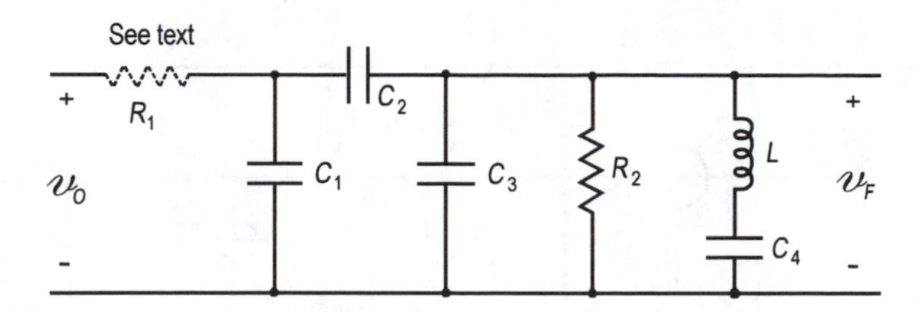

Fig. 7.28: A schematic diagram of the feedback network of Fig. 7.27 used to find its transfer function.

- Perhaps the loading of the output of A by the input of F is not negligible, even though the emitter follower should have low output impedance.

We provisionally reject the second possibility because it is generally bad design to make a circuit's functionality depend on a nonideality. This leaves the first and third possibilities. The third is the more probable since if the amplifier output impedance actually were negligibly small, C_1 would do nothing. Accordingly, we add the resistance R_1 (shown in dashed lines) to represent the output impedance of A. It will be of the order of $1/g_{\mathrm{m}}$ – in the range of tens of ohms. The analysis is carried forward as usual:

- Find $\mathbf{F} = \mathbf{V}_{\mathrm{F}}/\mathbf{V}_{\mathrm{o}}$
- Set $\mathbf{AF} = -1$. This is a complex equation.
- Equate real parts on both sides of this equation.
- Equate imaginary parts on both sides of this equation.

The results are:

$$\omega_{\mathrm{o}}^4 L(C_1C_2 + C_1C_3 + C_2C_3)C_4R_1R_2 - \omega_{\mathrm{o}}^2[LC_4 + R_1R_2(C_1C_2 + C_1C_3 + C_2C_3 + C_1C_4 + C_2C_4)] + 1 = 0, \tag{7.11a}$$

$$A_{\mathrm{min}} = -\omega_{\mathrm{o}}^2 LC_4[R_1(C_1 + C_2) + R_2(C_2 + C_3)] + [R_1(C_1 + C_2) + R_2(C_2 + C_3 + C_4)] / \{C_2R_2(1 - \omega_{\mathrm{o}}^2 LC_4)\}. \tag{7.11b}$$

The first equation is a quadratic in ω_{o}^2 and can thus in theory be solved in closed form. However, the answer is so voluminous as to preclude any possibility of extracting physical significance. There is, however, one special case in which something useful can be salvaged from this analysis. It is often true in a quadratic describing a physical system that $b^2 \gg 4ac$. When this is true, the two roots of the quadratic can be approximated as $-c/b$ and $-b/a$. Using the typical values:

$$\begin{aligned} C_1 &= 17\,\mathrm{pF} & R_1 &= 50\,\Omega \\ C_2 &= 14\,\mathrm{pF} & R_2 &= 2540\,\Omega \\ C_3 &= 5\,\mathrm{pF} & L &= .22\,\mu\mathrm{H} \end{aligned}$$

it is found that $b^2 > 80\,ac$, so it is a good approximation to use these simplified expressions for the roots. The two calculated roots are thus

$$\omega_o^2 = \frac{1}{R_1 R_2(C_1C_2 + C_1C_3 + C_2C_3 + C_1C_4 + C_2C_4)}$$

and

$$\omega_o^2 = (1/L)(1/C_s + 1/C_3 + 1/C_4),$$

where $C_s \equiv C_1C_2/(C_1 + C_2)$ is the effective series capacitance of C_1 and C_2. Now the problem becomes one of finding which of these roots gives the actual oscillation frequency. We are immediately inclined toward the latter because it shows L dependence. If ω_o does not depend on L, the inductance has no purpose in the circuit. This inclination becomes certainty, however, when we plug the first expression for ω_o^2 into (7.11b) and find that it gives $A_{min} > 1$. Because the emitter follower cannot provide this gain, the circuit cannot oscillate at the frequency given by the first root, and thus the second one is correct. Inserting the correct expression

$$\omega_o^2 = (1/L)(1/C_s + 1/C_3 + 1/C_4)$$

into (7.11b), we get

$$A_{min} = \frac{R_1(C_1 + C_2)^2 + R_2C_2^2}{R_2C_2(C_1 + C_2)}.$$

Using $R_1 = 1/g_m$ and defining $f \equiv C_2/(C_1 + C_2)$, this becomes

$$A_{min} = (1/g_m R_2 f) + f.$$

Exercise 7.11 Since C_2 is 10 pF in parallel with a varicap, is it possible that variation in the varicap capacitance can raise A_{min} above unity and hence cause oscillation to cease? Find the range of f and thus the range of C_2 for which oscillation is possible.

The upshot of all these calculations is that analysis of the very complex oscillator circuit shown in Fig. 7.24 can actually be reduced down to analyzing the moderately tractable form of Fig. 7.27, and from that analysis we are able to extract physically reasonable expressions for the frequency of oscillation and the minimum gain for oscillation.

Exercise 7.12 Since both varicaps in the local oscillator are identical, and since they are driven by the same DC control voltage, they will presumably have the same capacitance. For $L = .22\,\mu$H, $C_1 = 17$ pF, $C_2 = 10\,\text{pF} + C_v$, $C_3 = 5$ pF, and $C_4 = C_v$, plot ω_o vs. C_v for $.1 < C_v < 1{,}000$ pF.

7.7.3.2 Control section

Referring back to Fig. 7.20, we see where this module fits into the overall scheme of things. Figure 7.29 shows the schematic.

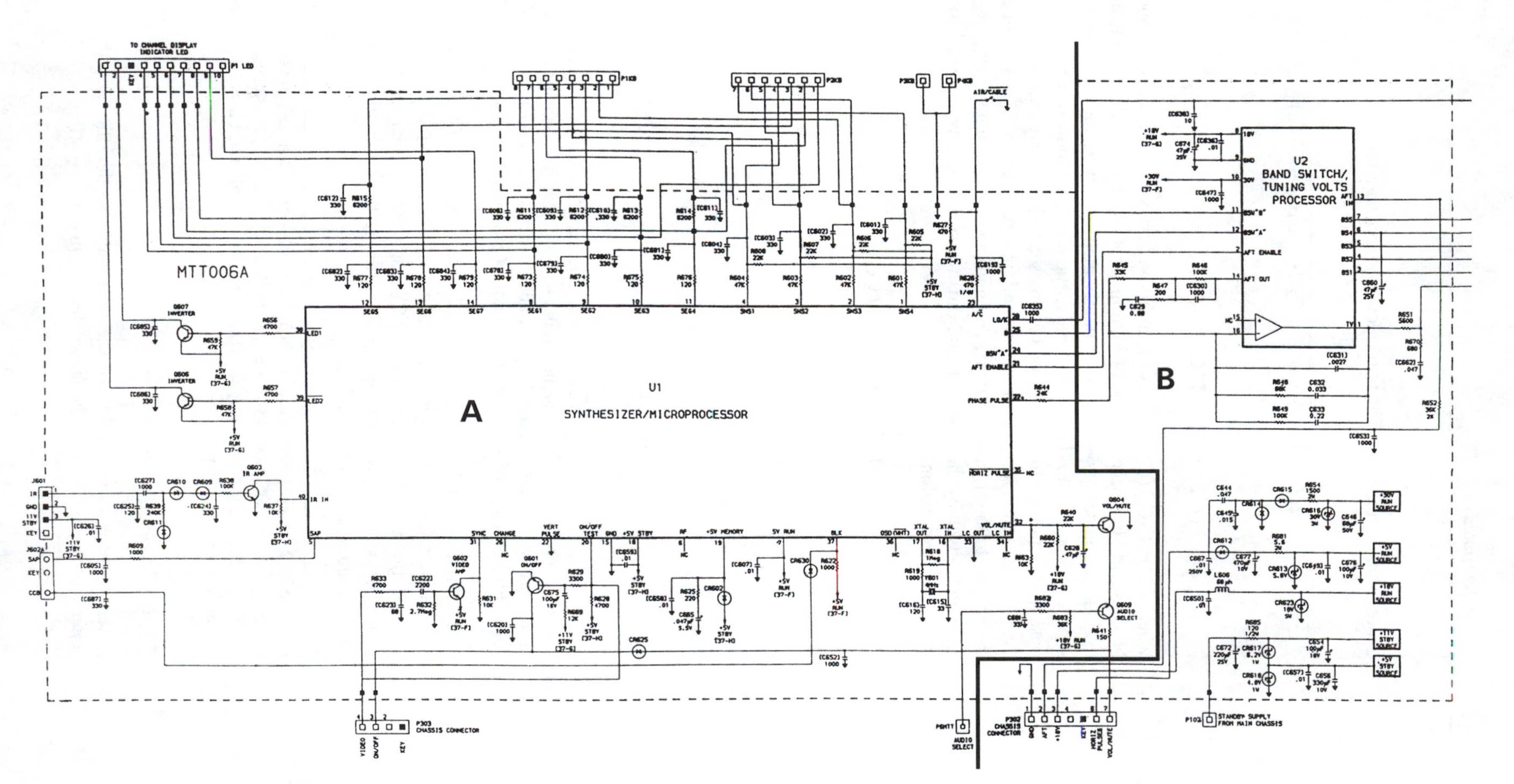

Fig. 7.29: A schematic diagram of the control section of the tuning unit in the RCA CTC-136 color television receiver. An electronic version of this figure is available on the world-wide web at http://www.cup.org/titles/58/0521582075.html.

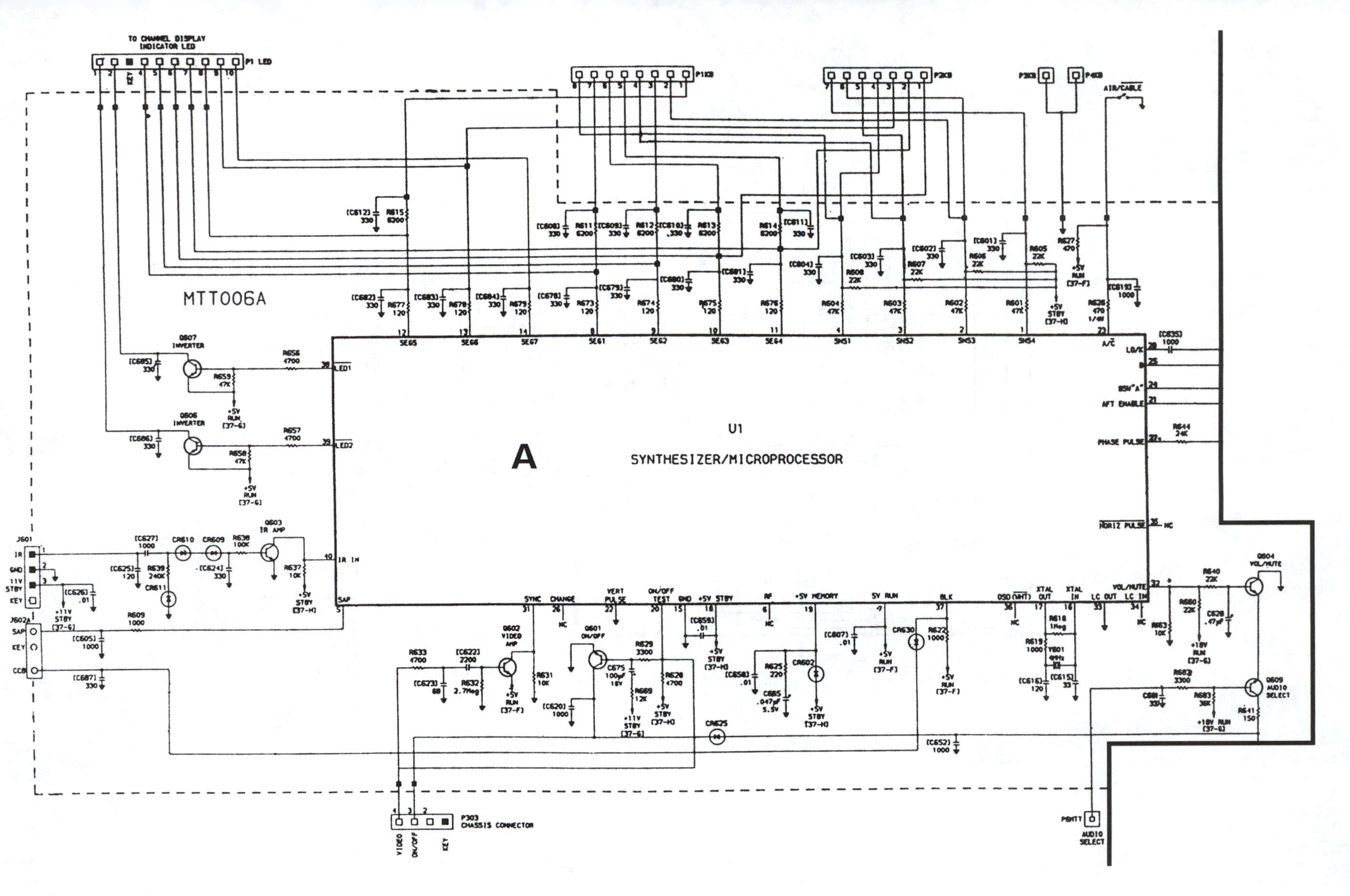

Fig. 7.29A: Synthesizer/microprocessor.

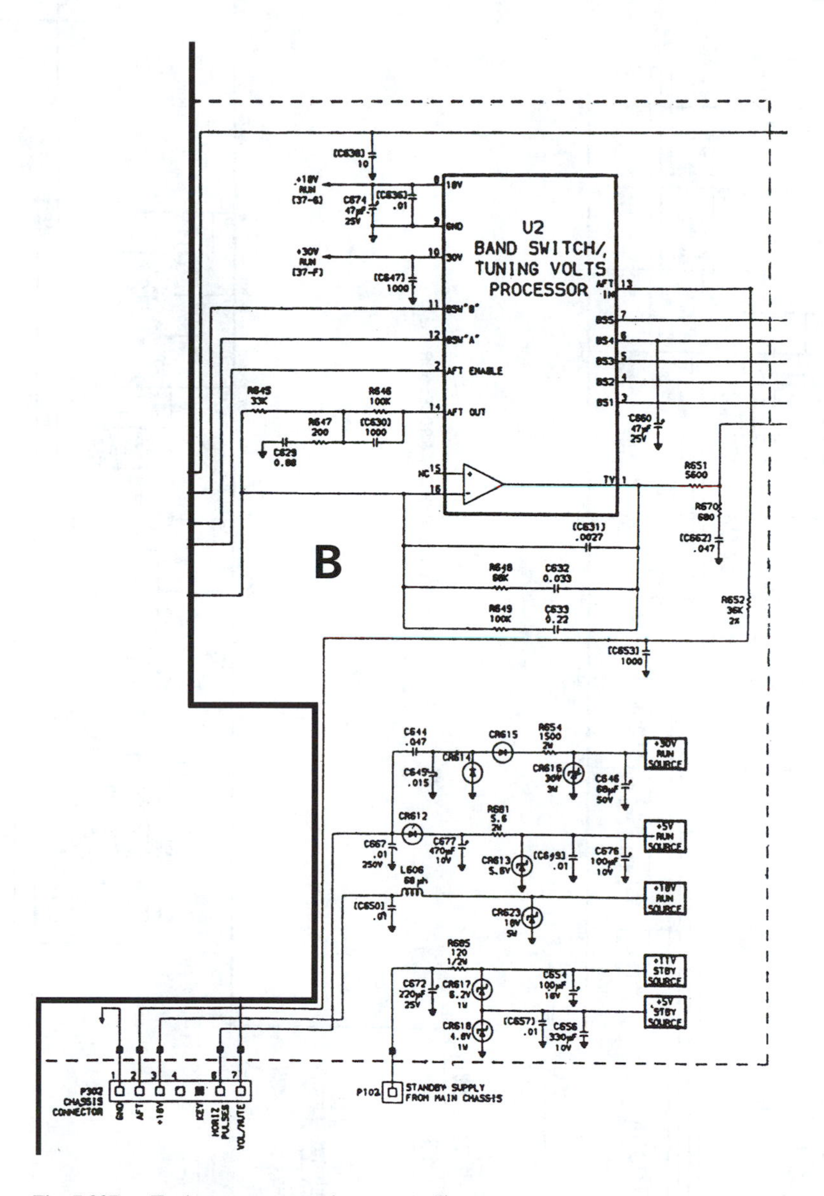

Fig. 7.29B: Tuning processor and power supplies.

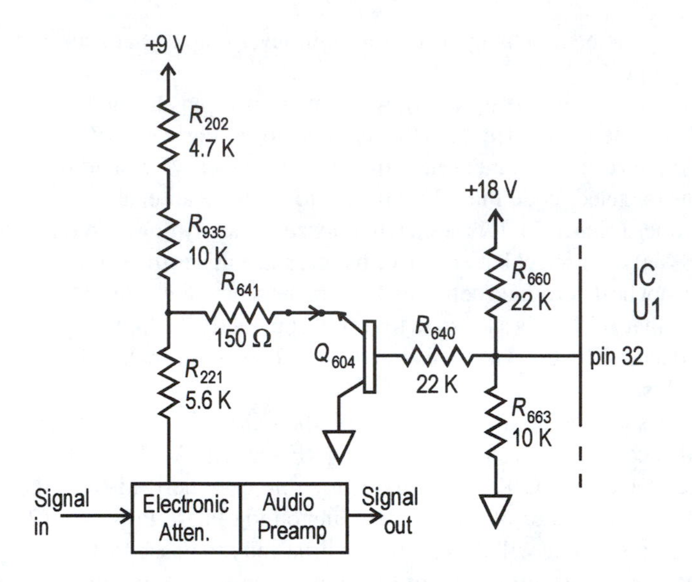

Fig. 7.30: A schematic diagram showing the electronic volume-control function of the RCA CTC-136 color television receiver.

We shall start our tour of U1 in the lower right-hand corner, where we see the crystal that generates the chip clock. The "LC Out" pin next to it is for a clock oscillator for OSD (On-Screen Display of time and channel) in sets with this feature. There is also an "LC In" pin that has no connection in this set.

Turning the corner and heading north, we find the volume control pin. In response to signals from the remote or the keyboard on the TV itself, U1 outputs a voltage to the base of Q_{604}. The audio select transistor (Q_{609}) is normally held in saturation by the large base current that flows through R_{683}, the B–E junction of Q_{609}, into the emitter and out the collector of Q_{604}. The magnitude of this current is determined primarily by the output from pin 32 of U1 and secondarily by R_{660} and R_{663}, which are connected to that pin. This may be seen a bit more clearly if we pull onto one diagram all of the circuitry that is connected with the volume control function, as is done in Fig. 7.30.

The shorted switch in this diagram is Q_{609}. This transistor is activated by the Audio Select pin (in receivers that include a radio) when the user only wants to listen. Since the configuration we are analyzing has no radio, this pin is never activated, and the transistor remains a closed switch.

It cannot be determined from this diagram whether the attenuator sources or sinks current, or even if it is voltage controlled. It is clear, however, that whichever of these is true, when U1 changes the drive to Q_{604}, the voltage and current applied to the attenuator will change, changing the volume. Both the attenuator and the audio preamp are part of the huge chip on the main chassis module. We shall call this chip MAGNUM, rather than the nondescript 176854 assigned to it in the RCA service literature.

We note in Fig. 7.29A that the 18 V supply shown in Fig. 7.30 is identified as "RUN". This is in contrast to the standby supplies we have already discussed. A RUN supply is only on after power has been turned on.

As we resume our northward trek, we come to a pin labeled "Phase Pulse." We shall defer discussion of this pin until we cover U2. The next pin is "AFT Enable" (pin 21). This allows automatic fine tuning (AFT) to be disabled while the channel is being changed. It is a logic-level output that acts on the AFT circuitry in U2.

Again resuming our trip north, we come to two pins labeled "BSW A" and "BSW B." These are select pins that connect to similarly identified pins on U2. Based on the user's channel selection, U1 determines what band is desired and generates a select code that U2 will use to produce an enable on the B1–B5 output lines (pins 3–7). It is somewhat obscure exactly how two lines (BSW A and BSW B) can select between five bands, since this is the sort of information that can not be determined definitively from the schematic alone. However, RCA explains that BSW A is a tri-state line, and apparently the high-impedance state is also decoded by U2. We have already seen how these enable signals are used in the RF section.

Our last stop going north is at pin 29, the LO/K pin. This is the frequency-divided local-oscillator signal from the RF section. A PLL in U1 compares the frequencies of LO/K and a frequency reference derived from the crystal. The phase comparator of the PLL outputs on the Phase Pulse pin (27) pulses whose average value will depend on the proximity to phase lock. These will be low-pass filtered to supply one of two components of the tuning voltage (TV). As we have already seen, TV is used to tune the varicaps in the RF section, particularly those in the local oscillator, thus closing the control loop.

As we now reach the northern boundary of U1 and turn west, we first see a pin labeled "A/C" for Air/Cable. This, of course, will be activated by the user's choice of signal source.

Continuing our stroll westward, we come to four lines labeled "SNS." The one marked SNS4 is not used, but the other three go to the keyboard on the TV. Apparently the SNS pins, which are normally pulled up by the 5 V standby supply, are grounded out when a key to which they are connected is pressed.

After SNS pins, we encounter seven SEG pins. The name suggests the word segment, as in seven-segment LED display. We see, in fact, that they do go to the channel display. However, some of them are multiplexed and do double duty. The SNS lines decode the columns of the keyboard, and SEG lines 1–5 decode the rows. Figure 7.31 shows the keyboard layout.

We now round the corner and head south, first encountering two drive pins for the two-digit seven-segment display. This will display the channel number selected. A set with the OSD option would use a different tuning module – one that would display the channel number on the screen. The clock module would also display on the screen if that option were chosen. The display drivers in this set are PNP transistors that feed the anodes of common-anode displays.

Proceeding south, we next stop and consider the IR input. This, of course, is the output from the IR detector/preamp module of Fig. 7.22. Since this signal is capacitively coupled by C_{627}, it will have both positive and negative excursions at the anode of CR_{610}. This diode and the following one do level shifting for the input of the IR amplifier. The half-cycle that cannot pass through these two series diodes is handled by CR_{611} and R_{639}, so that C_{627} does not charge up and block the passage of any further signal. The U1 chip decodes these IR signals and thus determines the action called for from the remote. It then outputs to the appropriate pin(s) to accomplish it.

As we enter the final leg of our walking tour, we reach the southern border and turn east. Although it is not used or even shown in Fig. 7.29, there is a pin

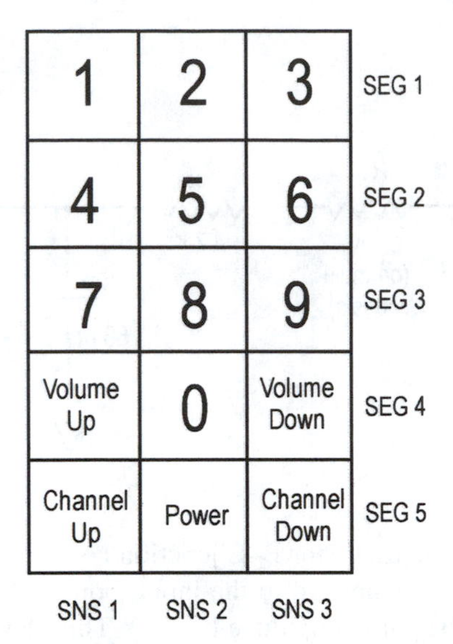

Fig. 7.31: The control keyboard layout of the RCA CTC-136 color television receiver.

on U1 called SAP. This SAP input is used in sets that are designed to receive a Second Audio Program (provided that one is transmitted). This audio program is transmitted in quadrature with normal TV sound. The next sight to meet our eyes is the sync input. The sync signal applied to this pin is derived from the video signal applied to the base of Q_{602}. This is a PNP transistor, which allows it to drive a collector load that is referenced to DC ground. Its load consists of R_{631} in parallel with the input impedance of pin 31 of the chip. A cursory look at the frequency-selective elements in the base circuit of Q_{602} shows an LPF section followed by a high-pass section. This might seem contradictory until we realize that C_{622} is necessary as a DC blocking capacitor and R_{632} is necessary as a base bias resistor. We therefore direct our attention to determining the frequency-dependent effects of these components, given their magnitudes and the signals they will see. The quiescent base current will be about 2 μA, presumably giving a collector current of about .2 mA. This, flowing through the 10 kΩ collector resistor, will give a drop of about 2 V, or roughly half the collector supply voltage. The video signal applied to the base will swing above and below zero, since it is AC coupled through C_{622}. Although the rest of the TV schematic does not give the video amplitude applied to the base circuit of Q_{602}, it may be calculated to be about 3 $V_{\text{pk-pk}}$ from other information that is given. Thus Q_{602} will not be operating in the small signal region and consequently, a transient time domain analysis will be required.

The relevant circuitry is shown in Fig. 7.32. We will consider first what this circuit does with a positive-going vertical-sync pulse, which is a portion of the video signal applied to this circuit. This signal also contains a large DC component, but this will be insignificant after a cycle or two because C_{622} will charge up to the average voltage difference across it between the vertical-sync pulses. Furthermore, the time constants of the vertical pulse are so long that C_{623} cannot have any effect in its processing by the stage. We shall see that C_{623} "does its thing" with the horizontal-sync pulse. We shall, therefore, ignore it for the present.

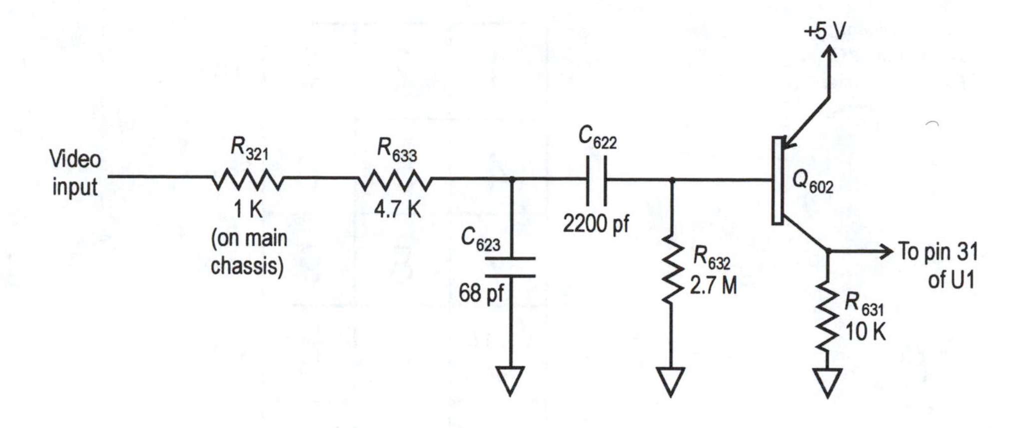

Fig. 7.32: The sync-processing circuit of the RCA CTC-136 color television receiver. The output of this stage is input to the control microprocessor of Fig. 7.29.

Between pulses, the transistor B–E junction conducts through R_{632}. When a positive-going pulse is applied to the input, both ends of C_{622} jump up by a voltage equal to the pulse amplitude (~3 V). This elevates the base voltage to about 7.4 V and drives the transistor into cutoff, which causes the collector voltage to drop to zero. With the pulse still high, C_{622} starts to charge through R_{632}, bringing the base potential back down toward its nominal 4.4 V value. But the R_{632}–C_{622} time constant is so long (~6 ms) compared to the vertical sync pulse width (~.2 ms) that there is only time for v_{Base} to charge back to about 7 V, before the sync pulse drops back down by 3 V. This voltage step is also coupled through C_{622}, but as it attempts to pull the base down by the same 3 V, it heavily forward biases the B–E junction, which clamps the lower voltage excursion at about 4.4 V. The transistor is saturated, and the collector voltage jumps to 5 V.

Now C_{622} starts to discharge. The discharge path is through the B–E junction, R_{633}, R_{321}, and the pulse source. This gives a discharge time constant of about 13 μsec. In the initial stages of this discharge, when the current is large, it is adequate to hold the transistor in saturation. As the base current decays, the transistor comes out of saturation, and the collector voltage starts decaying from its previous 5 V value. This decay continues until the base current drops to the roughly 2 μA that flows through R_{632}. Figure 7.33 shows these waveforms.

Exercise 7.13 Calculate the time duration for which $v_{\text{Collector}} = 5\,\text{V}$ after the sync pulse, assuming the transistor β is 100.

When we repeat this analysis for a horizontal sync pulse, we find that the pulse duration is so small that the decay of base voltage shown in Fig. 7.33b cannot occur. Hence when the sync pulse drops back to its low level, it does not drive the transistor into saturation, so the collector voltage just returns to its quiescent value. This reasoning would lead us to expect the waveforms shown in solid lines in Fig. 7.34.

The dashed line in Fig. 7.34c reveals three effects not predicted in the "hand waving" discussion just completed:

- There is a negative-going spike at the conclusion of the sync pulse. This is due to the negative transition on the base being coupled through to the collector by C_μ, which connects base and collector internal to

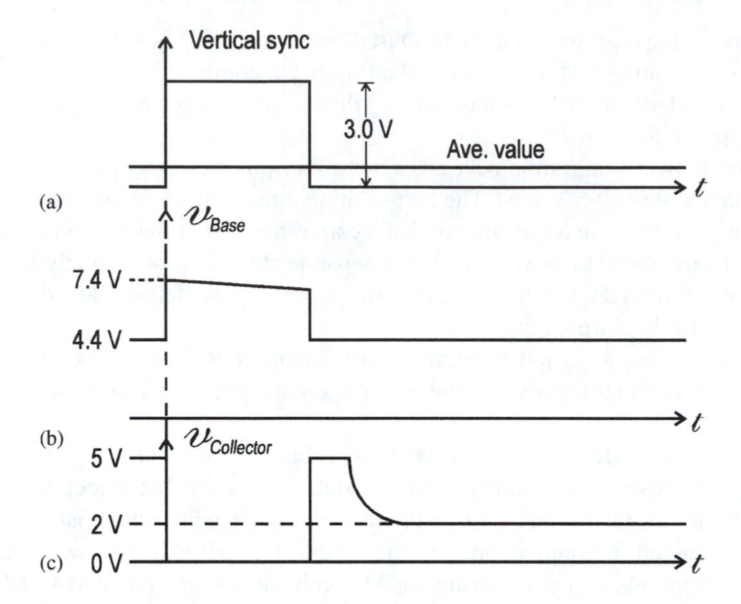

Fig. 7.33: Timing diagram showing the response of the circuit of Fig. 7.32 to a vertical sync pulse. (a) The input sync pulse. (b) The base voltage. (c) The collector voltage.

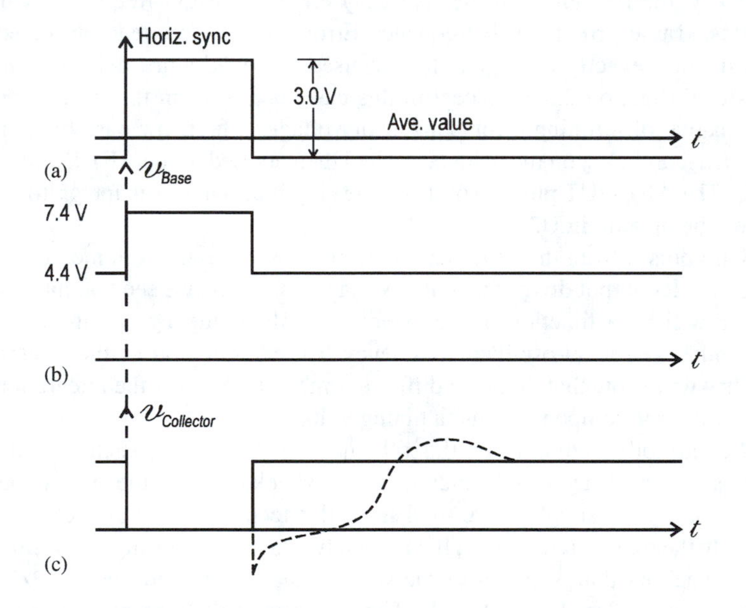

Fig. 7.34: Timing diagram showing the response of the circuit of Fig. 7.32 to a horizontal sync pulse. (a) The input sync pulse. (b) The base voltage. (c) The collector voltage.

the terminals of the transistor. The 68 pF capacitor (C_{623}) exists only to suppress this spike. Without this capacitor, the pulse amplitude is about -1 V, and it poses a definite hazard to U1. With C_{623} in circuit, the amplitude is reduced to about $-.3$ V, which is presumably safe for the chip.

- There is a widening of the time that the collector voltage is low. This is a standard delay-time phenomenon.
- When the collector voltage finally does rise back to its nominal value, it slightly overshoots and then returns. This is the very low-level counterpart of the effect that causes the large spike with a vertical sync pulse. The effect here is inadequate to reach saturation at all.

Proceeding east, we come to the On/Off pin (pin 20). This is an output from the μP, indicating that it has received a Power-On command. This signal goes to the interface board. The method by which this powers up the TV will be discussed in Section 7.7.4.

Our eastward migration next takes us to four power supply pins – namely ground and three +5 V pins. The first of these three (pin 18) is standby power for the μP. It must be on standby so that it can respond to a Power-On command from the remote. The next (pin 19) is a separate standby power supply for the memory in the μP. The third is the main power supply delivered to the chip after the TV is powered up.

BLK is a blanking pulse output from U1 that, after further processing on the REM-1C board, provides the OSD blanking pulses for sets using this capability.

A short walk further to the east takes us back to our starting point. This pretty well covers the circuitry in this module except for that associated with U2, the Bandswitch/Tuning Volts Processor. Even much of this has already been discussed in conjunction with the Synthesizer/μP chip. We will briefly discuss the remaining pins. An analog AFT voltage is generated in MAGNUM at the output of the IF amplifier. If the local oscillator is at exactly the correct frequency, the IF signal will be also. Any error in the LO frequency will be manifested as an error in the IF frequency. Errors in the IF frequency are detected by a method exactly analogous to that used for quadrature detection of an FM signal (Section 2.3.3), except in this case, output from the AFT detector is indicative of a tuning error. This error voltage is first low-pass filtered by C_{315}, L_{312}, and C_{326} on the main board and then applied to the AFT IN pin (13) of U2. The AFT OUT pin (14) of this same chip is applied to an integrator built around the op amp in U2.

Phase pulses from the PLL phase detector in U1 are also summed into this integrator. Its output drives the tuning voltage line. Thus we see that the tuning voltage will be a function of the error in the IF frequency and the error in the synthesized local-oscillator frequency. An examination of the integrator circuitry will show that there are differing time constants for the integration of these two error components into a tuning voltage.

The only other circuitry on this schematic is that of five standby and run power supplies. A dedicated winding on the flyback provides the energy source for two of the run supplies. We shall study this technique in more detail when we study the main circuit board. It has the advantage of requiring much smaller filter capacitors than supplies of the same capacity run from the 60 Hz line, since the "ripple" is close to 16 kHz. The standby supplies are derived from the AC line and can not be switched off.

7.7.4 Interface board

Referring back to Fig. 7.20, we see that only the REM-1C board remains between us and the main circuit board. Figure 7.35 shows its schematic. Besides providing some "pass-through" connections, it comprises three main sections.

The first of these consists of Q_{906}, Q_{907}, and their associated circuitry. This portion of the schematic is so defective that it is impossible to ascertain its function. For example:

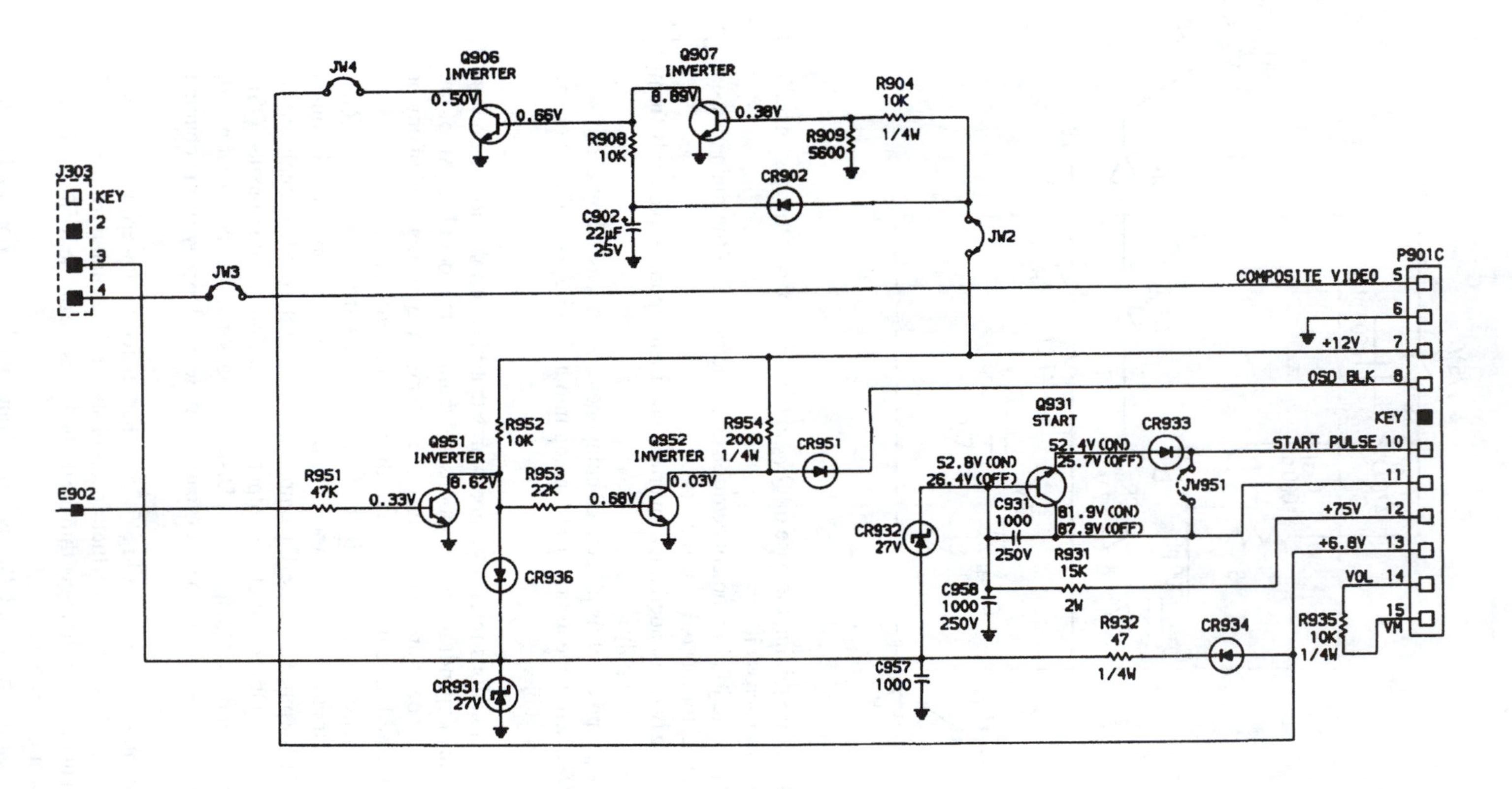

Fig. 7.35: Schematic diagram of the tuner interface board in the RCA CTC-136 color television receiver.

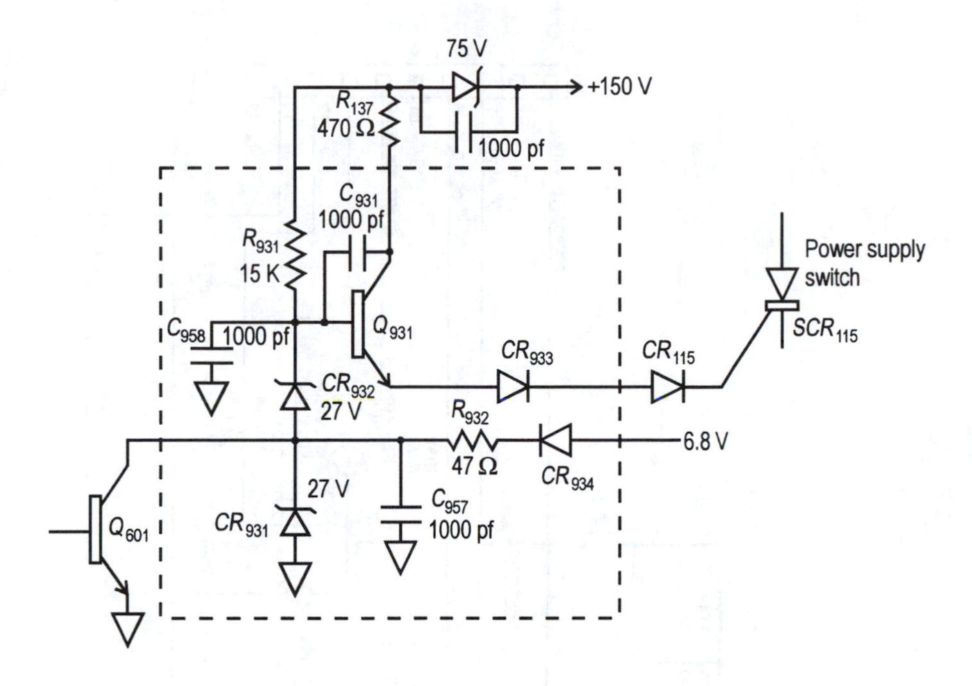

Fig. 7.36: Schematic diagram of the power turn-on circuit in the RCA CTC-136 color television receiver.

- The wire joining the base of Q_{906} to the collector of Q_{907} has an 8.23 V drop across it.
- The wire joining the collector of Q_{906} and pin 13 of the main chassis connector has 6.3 V across it.
- These two cascaded logic inverters have their inputs derived only from the 12 V supply.
- Q_{907} and all of its associated components are bypassed by CR_{902}.
- Q_{907} will have almost 1 mA going into its base. A V_{BE} of .38 volts is not credible.

The "JW" notations refer to jumper wires that may or may not be present in a given unit, depending on the options and features its model has. All of the jumpers seem to be connected for the circuit shown, and the removal of any or all of them would raise more questions than it would answer.

The second circuit section in Fig. 7.35 is that comprised of Q_{951}, Q_{952}, CR_{936}, CR_{951}, and their associated resistors. The drive for this section, if used, comes from the BLK (pin 37 of the U1 chip in the MTT-006 module analyzed in Section 7.7.3.2). Observe that the output of this section is a line labeled OSD BLK, which stands for "On-Screen (Display) Blanking." As has been noted, the set we are considering does not feature On-Screen Display of the channel and time.

The remaining circuitry in Fig. 7.35 is that related to powering on the set upon command. The circuitry which accomplishes this is isolated from Fig. 7.35 and shown in Fig. 7.36. Items within the dashed lines are on the REM-1C board. The others are elsewhere.

When Q_{601} is saturated, CR_{931} is shorted out, and the base potential of Q_{931} is too low to turn on the SCR. Saturation of Q_{601} is maintained when the set is off because the 5 V standby supply, which is input to pin 18 of U1 on the MTT-006A board, appears on pin 20 of the same chip. When this pin is brought

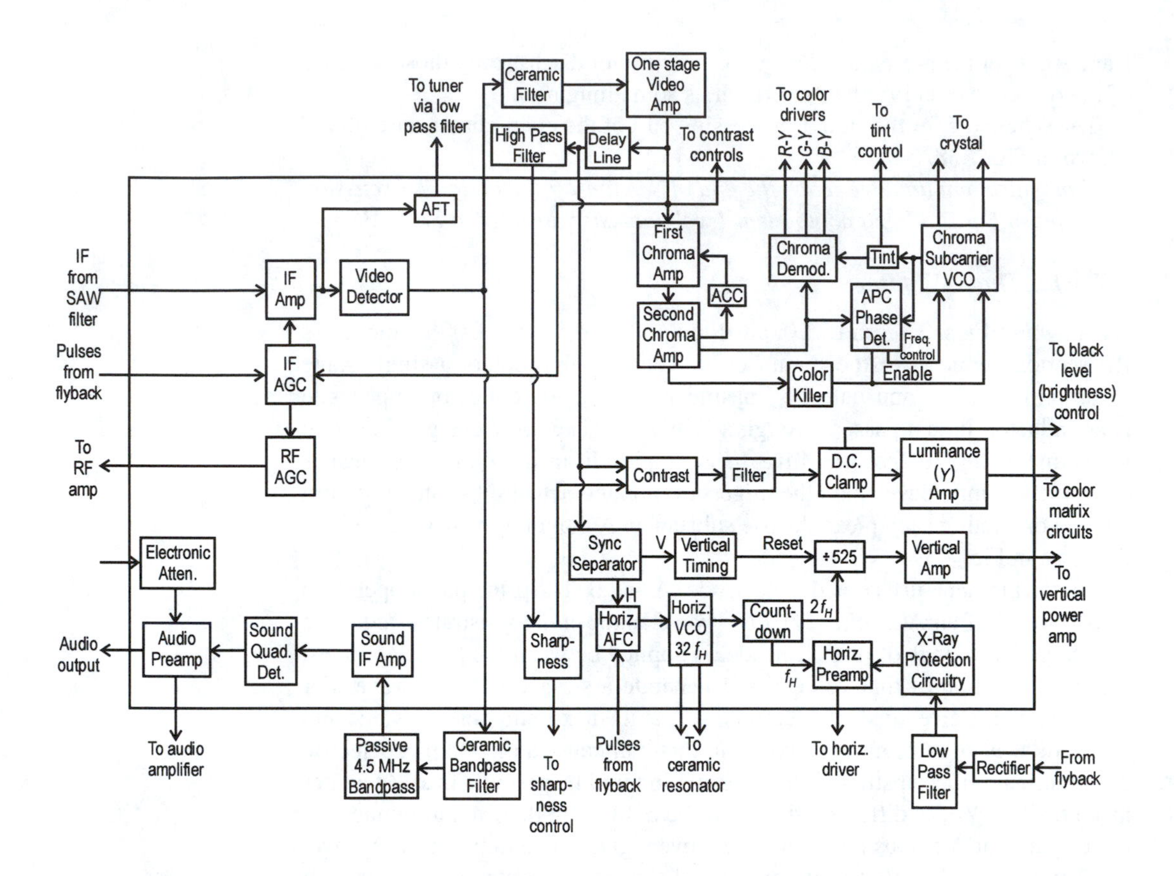

Fig. 7.37: Block diagram of the functions incorporated in the MAGNUM IC in the RCA CTC-136 color television receiver.

low by a signal from the IR remote, Q_{601} is cut off and CR_{931} "Zenes." This produces a rapid 27 V rise in the base potential of Q_{931}, which is switched to the SCR gate by Q_{931} and turns it on.

From Fig. 7.35, we would conclude that when Q_{601} is saturated, the 6.8 V supply would be heavily loaded by the 47 Ω resistor. This is not the case, however. This 6.8 V supply is unique among the several supplies in the TV in that it acts as a sort of "semi-standby" supply. It actually consists of a standby and a run supply in parallel. The standby one has a high internal source resistance, and the loading by CR_{934} and R_{932} suffices to pull it down to about a volt, instead of 6.8 V as the schematic suggests. When Q_{601} goes into cutoff (assume that this happens instantaneously), C_{957} charges through R_{931} and CR_{932} toward $150-75-27 = 48$ V. It reaches 6 V in 2 μsec and 27 V in 12 μsec. This sequence not only brings the "standby" part of the 6.8 V source on line, but enables the rest of the power-on sequence, which activates the "Run" part of the 6.8 V supply.

7.7.5 Main chassis

We move now to the last major module in the TV. It contains far more circuitry than any of the others. A great deal of this circuitry resides in the single chip that we have dubbed MAGNUM. A block diagram of this chip is shown in Fig. 7.37.

Most of the remaining components on this board are support parts for MAGNUM. We will, therefore, use the innards of MAGNUM as our analysis framework for the present, and we will show and/or discuss only those sections of the outboard circuitry that will teach us something new.

The schematic for the signal-processing part of the main-chassis circuitry is shown in Fig. 7.38.

From this point until we reach the end of Section 7.7, block-level discussion will refer to Fig. 7.37, but component-level analysis will refer to Fig. 7.38.

7.7.5.1 The SAW filter

As shown in Figs. 7.17 and 7.20, an IF output from the MTT-006A module's RF section connects to the main board. There, it first enters a single-stage IF amplifier that is unusual only inasmuch that its AC collector impedance is an inductor. It next passes through a *SAW filter*. The SAW designation is an acronym for *Surface Acoustic Wave*. In its simplest form, a SAW device consists of two sets of interleaved metallic fingers evaporated onto a substrate of lithium niobate or another such piezoelectric substance. A diagram of such a structure is shown in Fig. 7.39.

For the moment, just consider electrodes A and B. The principle of operation is as follows: If an AC voltage, whose wavelength in the substrate is equal to the "finger" spacings of one electrode, is applied between the two electrodes, the time variation of that voltage will generate a surface acoustic wave. For instance, if, at a given instant, electrode A is at its maximum positive value and electrode B is at its negative maximum, this will cause a certain piezoelectric deformation of the substrate. This deformation will be periodic in x. At a later instant, when V_{AB} is different, there will be a different deformation under the fingers, and the previous one will have moved, giving rise to the surface wave.

This generated wave travels on toward the C and D electrodes. When it passes under them, a voltage will be induced between C and D. If the finger structure is the same as that of A and B, the wave crests will all lie under the "fingers" of one electrode at a given instant, and a relatively large voltage will be generated. The amplitude response can be extensively tailored by varying the substrate material, the finger spacing, progression of finger length, finger width, and by the omission of some fingers. Since the surface waves propagate in both directions from the source transducer, and since only the one going toward the receiving transducer produces any output, a double finger structure is used to suppress the unwanted wave. Gerard (1978) gives a good overview of the design principles for these filters.

In this receiver, the SAW filter is a bandpass for the video IF signal whose spectrum is 6 MHz wide and whose center frequency is 44 MHz. The SAW output enters the IF amplifier in MAGNUM.

7.7.5.2 AGC

Since the RF AGC is driven by the IF AGC circuit, we are probably safe in assuming that this is delayed AGC.

7.7.5.3 AFT

The output of the IF amplifier, as shown in Fig. 7.37, goes to the video detector. Recall that this is an AM detector. At the same time, however, the video is

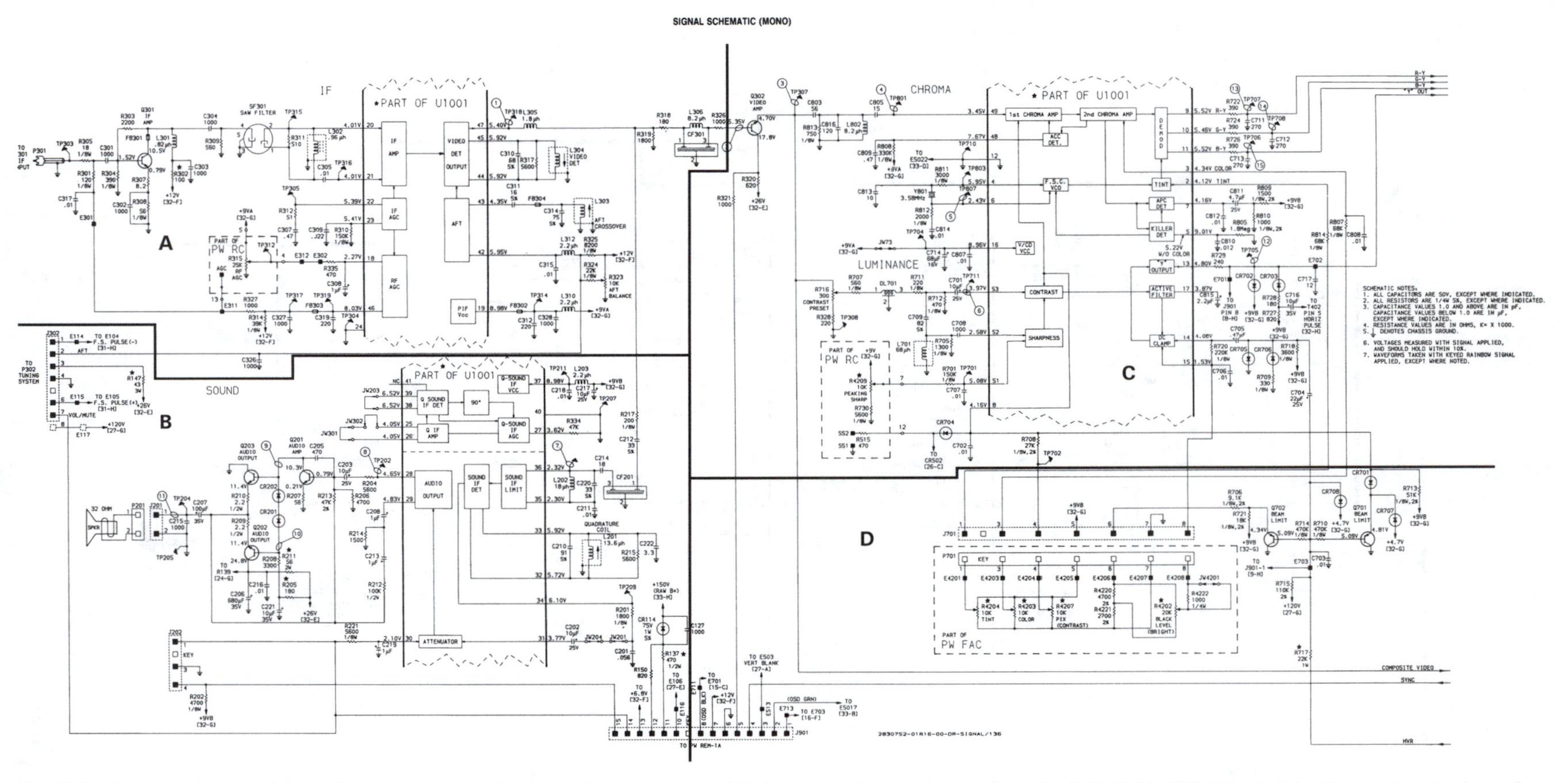

Fig. 7.38: Schematic diagram of the portion of the main board which performs processing of the luminance, chroma, and sound circuitry in the RCA CTC-136 color television receiver. An electronic version of this figure is available on the world-wide web at http://www.cup.org/titles/58/0521582075.html.

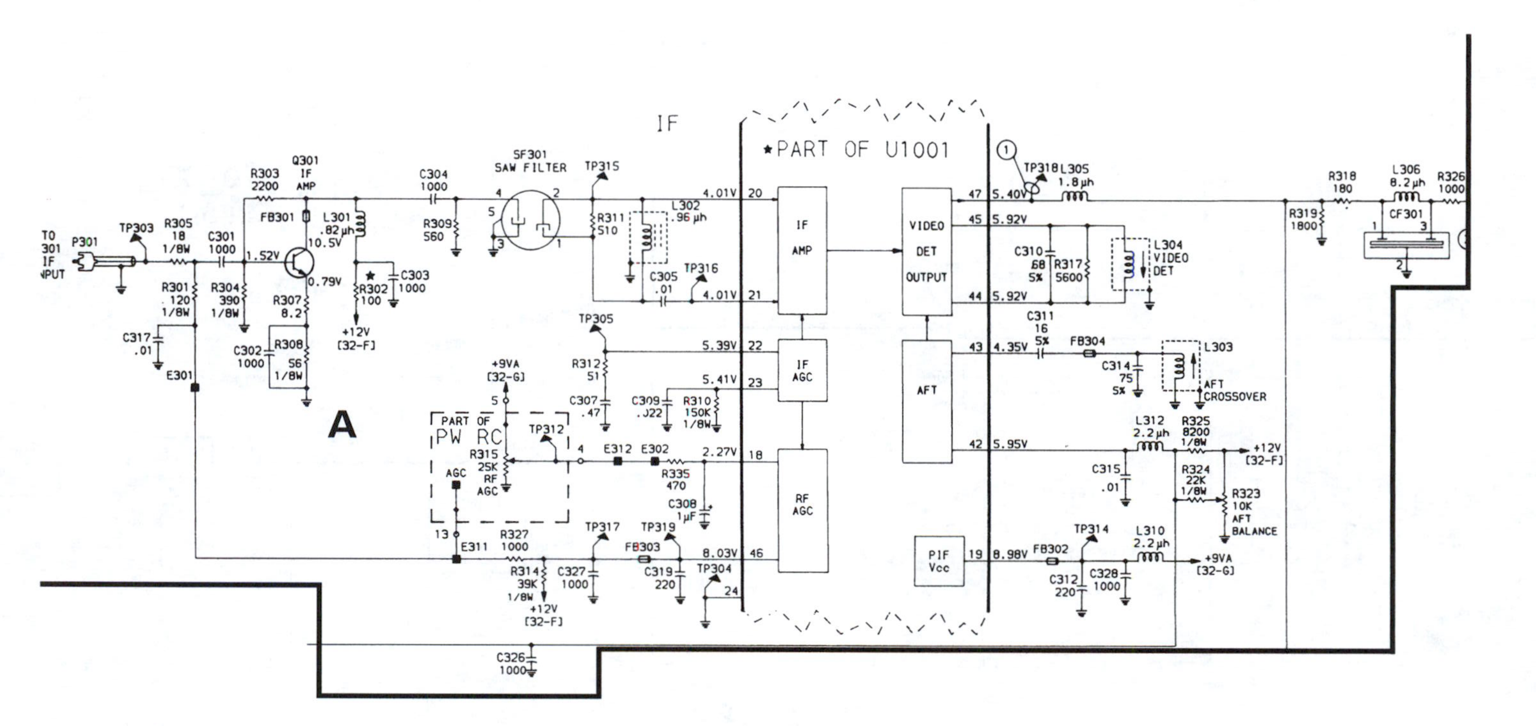

Fig. 7.38A: IF and video detector sections.

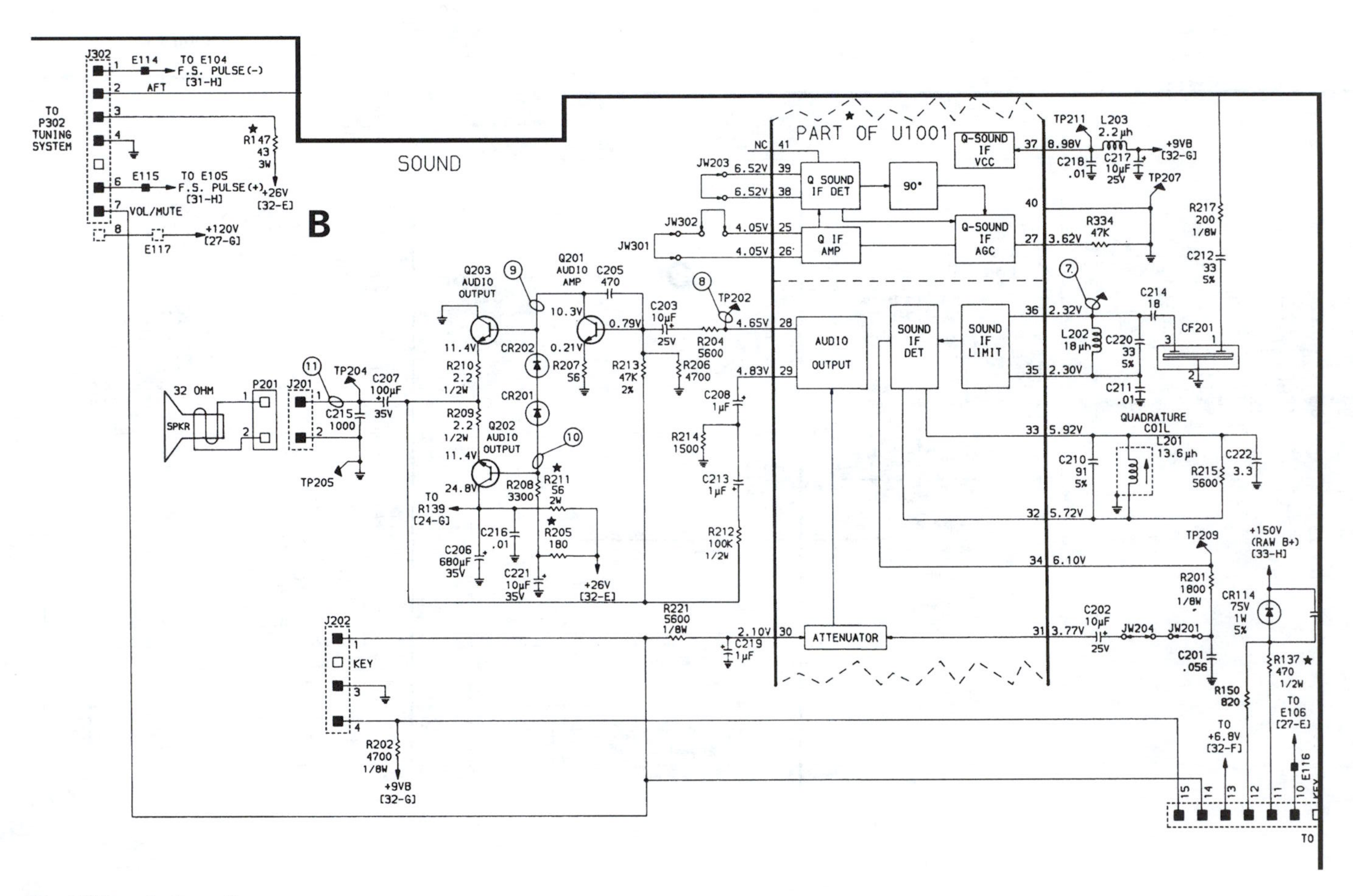

Fig. 7.38B: Audio section.

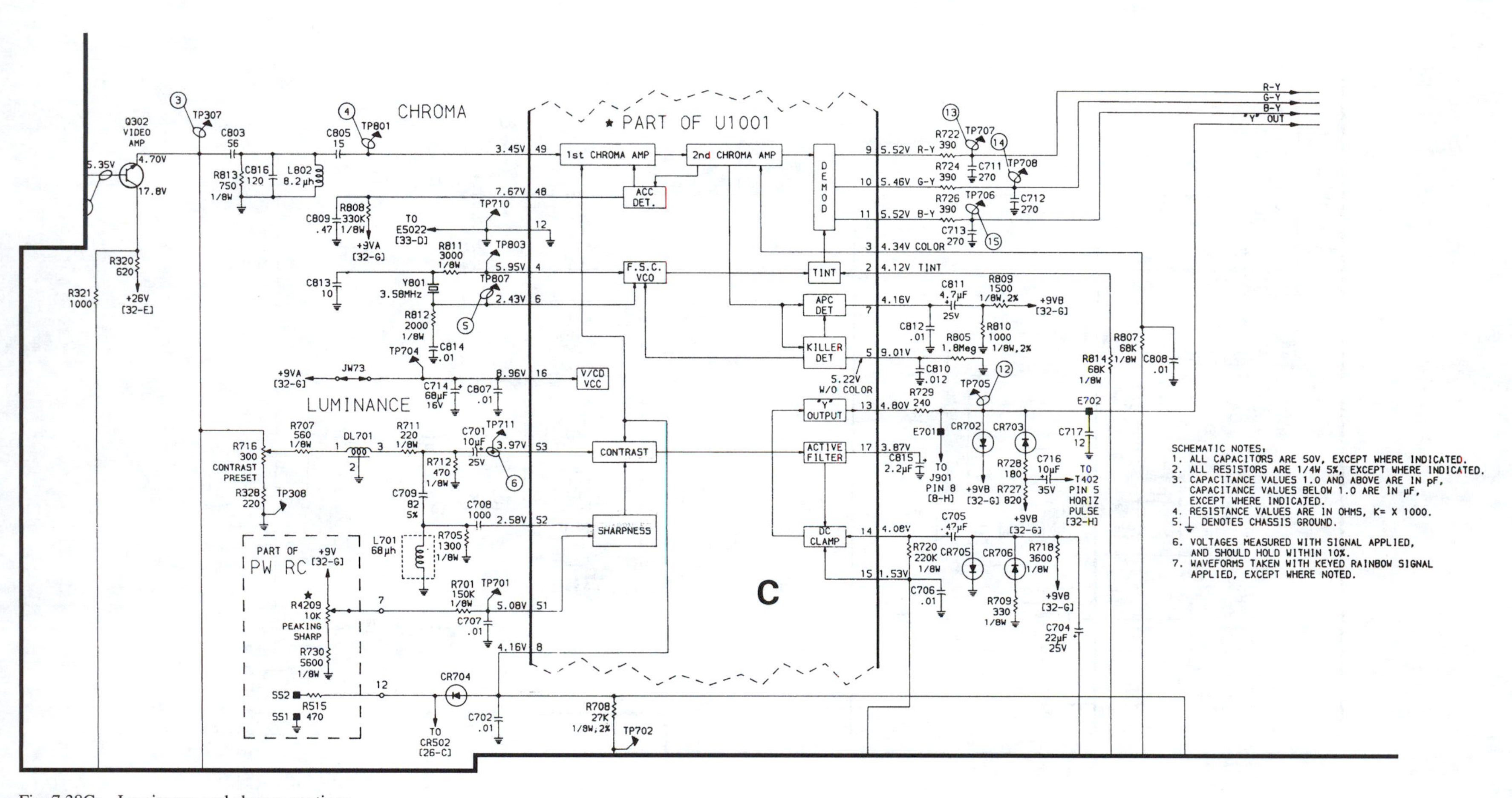

Fig. 7.38C: Luminance and chroma sections.

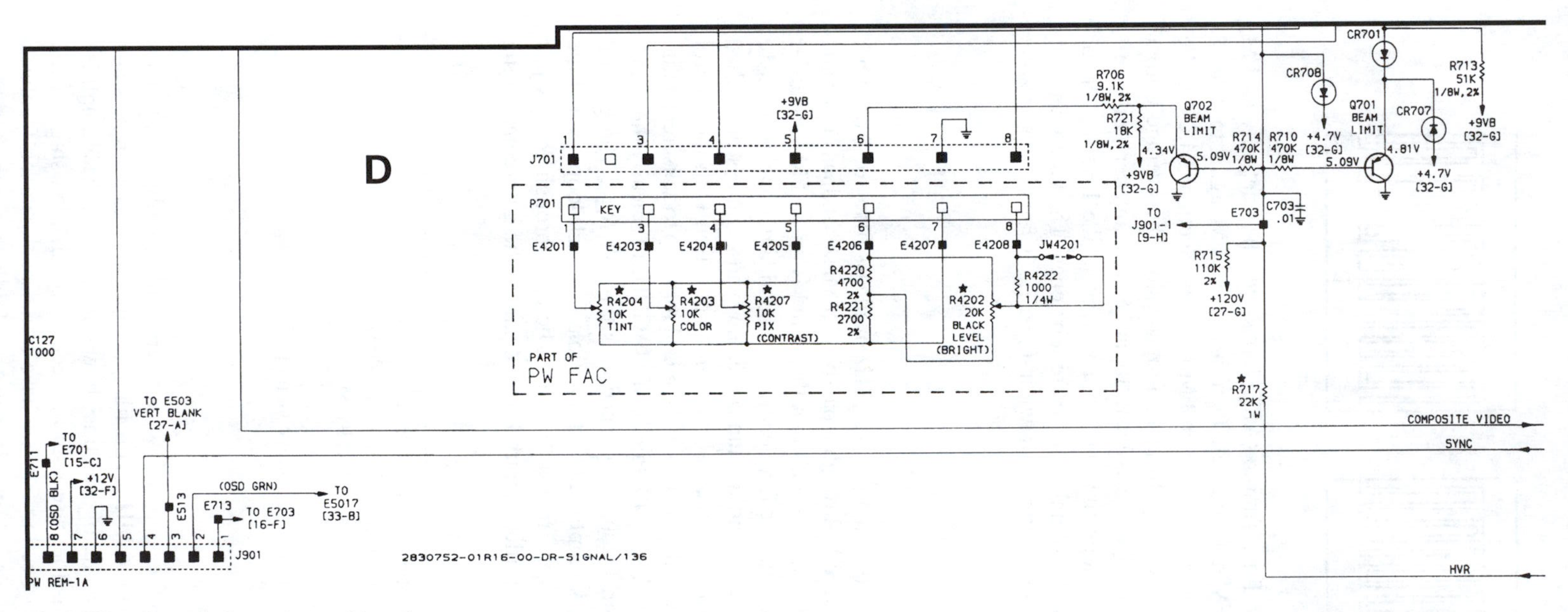

Fig. 7.38D: Beam limiter and control board.

Fig. 7.39: Schematic representation of the structure of a SAW filter.

applied to an Automatic Fine Tuning (AFT) block, which is actually an FM detector. How does an FM detector help us if there is no FM signal to detect? It works exactly the same as AFC on an FM receiver (Section 2.4.2). If the IF center frequency moves away from its proper value, the average output voltage of the FM detector (which is basically an $f-v$ converter) will shift also. This voltage will change the bias on a varicap in the local oscillator circuit and thus force the IF signal back where it belongs. The video detector output, which will have the spectrum shown in Figs. 7.5 is fed to two ceramic filters.

7.7.5.4 *Sound*

One of the ceramic filters (CF_{201}) is a 4.5 MHz bandpass filter for the sound. Its output signal then proceeds through a passive 4.5 MHz bandpass filter and is then fed to a sound IF stage. Recall that the sound is on an FM subcarrier 4.5 MHz above the video carrier frequency. The sound signal has its own IF amplifier/quadrature detector and an audio preamp whose gain is controlled by the electronic attenuator discussed briefly in conjunction with the control section of the MTT-006A. It then leaves MAGNUM for a trip to a discrete audio power amplifier, shown in Fig. 7.40. Pin 28 of MAGNUM is clearly the signal output lead, and the main feedback connection is through R_{206}. The general rule for negative-feedback amplifiers with multiple feedback loops is that the outermost loop determines the first-order gain. The path through R_{212} is just part of the load on the amplifier output. It is not part of the gain-determining circuitry of Fig. 7.40.

Exercise 7.14 For the amplifier of Fig. 7.40:

(a) What kind of negative feedback amplifier do we have?
(b) What will its input impedance ideally be?
(c) What does R_{206} accomplish?
(d) What would be the midband voltage gain of this audio amplifier?

7.7.5.5 *Video*

The other ceramic filter (CF_{301}) is a low-pass type with its corner set to pass all frequencies below about 4 MHz. This filter's output goes to a single-stage video amplifier, the schematic of which is shown in Fig. 7.41. The amplifier has three outputs. As was the case with the record equalization stage of the Marantz tape recorder in Section 5.4, we have a transconductance feedback amplifier here with a complex emitter impedance. Unlike that amplifier, however, this one provides outputs from both the collector and emitter. It would

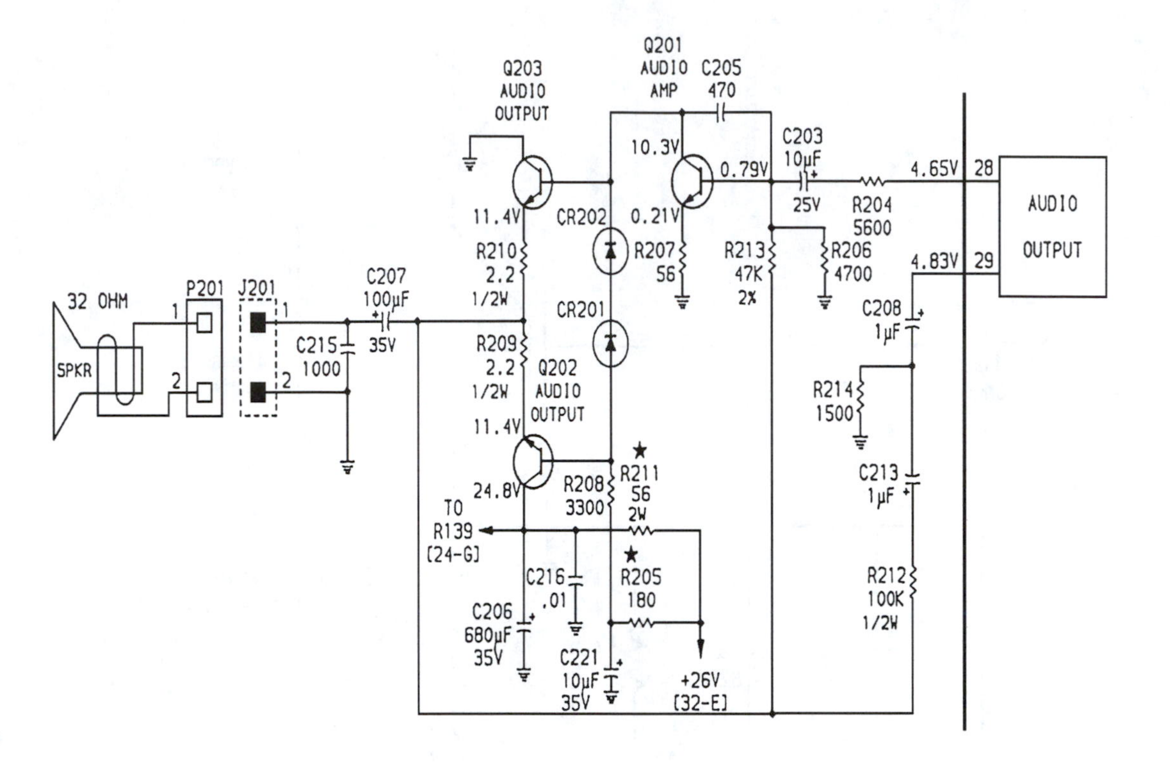

Fig. 7.40: Schematic diagram of the audio amplifier circuitry in the RCA CTC-136 color television receiver.

be an easy matter to write expressions for the feedback gains of this stage, but the resulting expressions (particularly for the chroma output) are cumbersome and difficult to interpret physically. We choose, therefore, to analyze the circuit using PSPICE®, obtaining the following results:

- The chroma output, as we should suspect, is a bandpass function that peaks at 4 MHz where it reaches 0 dB. This response is also reasonable since it is an emitter–follower output. At frequencies below 4 MHz, the response falls off rapidly, reaching −35 dB at 1 MHz. Above 4 MHz, it falls more slowly, reaching −8.4 dB at 10 MHz. Of course, the upper end fall-off is not so critical, since virtually all of the chroma is below 4 MHz.
- The collector output (Sync) reflects the strongly frequency-dependent emitter impedance, since the feedback gain is $-\mathbf{Z}_c/\mathbf{Z}_e$. It shows a flat gain of about .5 dB out to 1 MHz, followed by a gentle rise of gain to 6.5 dB at 4 MHz. Then there is a drop in gain back to 3 dB at 6 MHz followed by a steady rise in gain out past 100 MHz as C_{803} and C_{816} start to act as an emitter bypass, which removes the feedback and allows the gain to rise greatly. Again, this frequency range is of no practical interest.
- The luminance output is taken directly from the emitter and shows a frequency response that is flat to better than .1 dB out beyond 10 MHz. The R_{716} control in the emitter circuit is called the Contrast Preset control. It is a service adjustment that sets the range over which the user's contrast control will work.

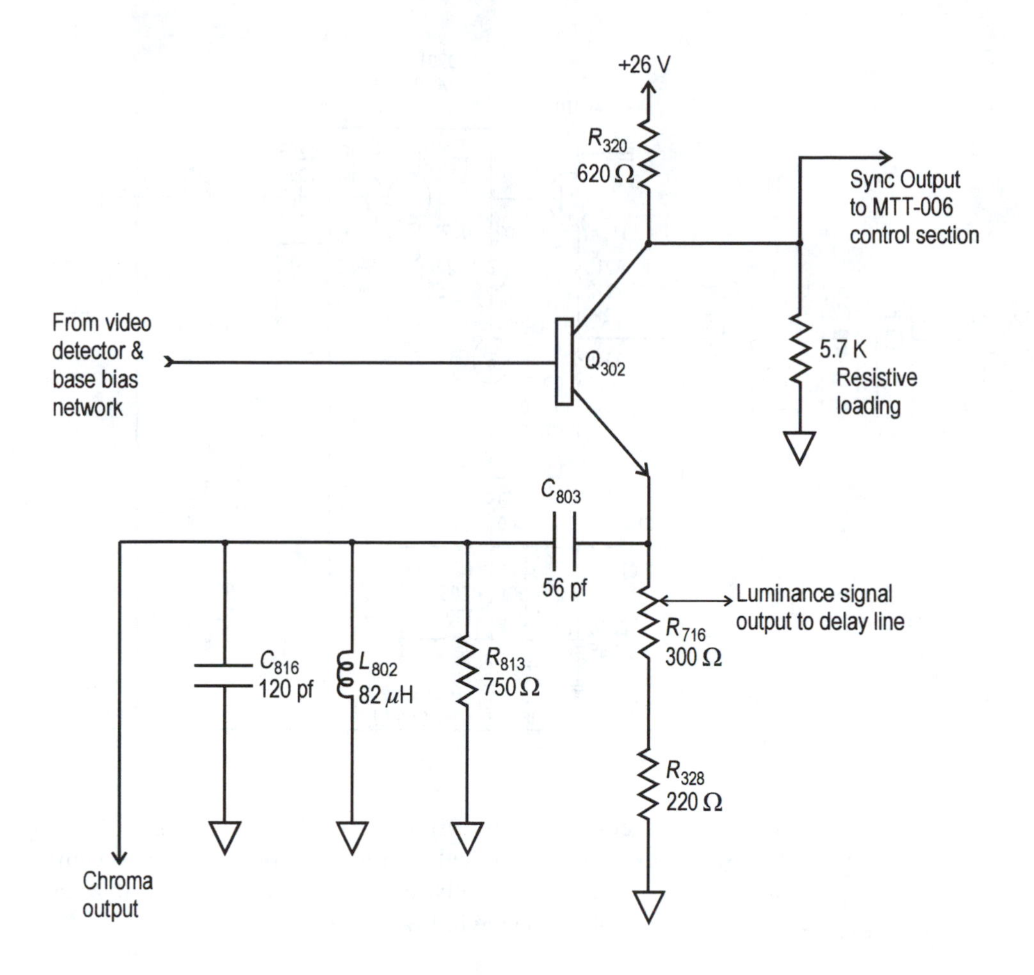

Fig. 7.41: Schematic diagram of the video amplifier stage in the RCA CTC-136 color television receiver. It should be observed that this stage acts as a distribution point to feed video (or some modification thereof) to the sync, luminance, and chroma circuits.

7.7.5.6 Chroma

This output from the stage just discussed goes to a two-stage chroma amplifier with AGC. As previously noted, in the chroma circuit this is called ACC for Automatic Color Control. To keep the circuit from leveling out color intensity changes that belong in the picture, the burst amplitude is servoed to a reference signal set by the user color control. The burst amplitude will be directly proportional to the strength of the received signal before this operation and will be constant afterward if it was originally within the control range of the ACC circuit. Outputs from the second stage of chroma amplification go to the chroma demodulator and the color killer. The functions of both have been discussed previously in more detail. The color killer senses when the color signal is so weak that ACC cannot bring it up to a usable level and kills the "burst oscillator" when this is the case. This in turn disables the chroma demodulator and APC phase detector, eliminating all color from the picture. This mechanism of color killing is in contrast to that used in the discrete Sony color TV analyzed earlier in this chapter. In that set, the color is killed by switching in an AC-grounding path for the input signal to the second chroma amplifier.

The block labeled as a chroma demodulator not only demodulates the chroma along two quadrature carrier phases but combines the resulting signals in resistive matrices to generate separate $R - Y$, $B - Y$, and $G - Y$ signals.

Fig. 7.42: Block diagram of the color phase-control loop in the RCA CTC-136 color television receiver.

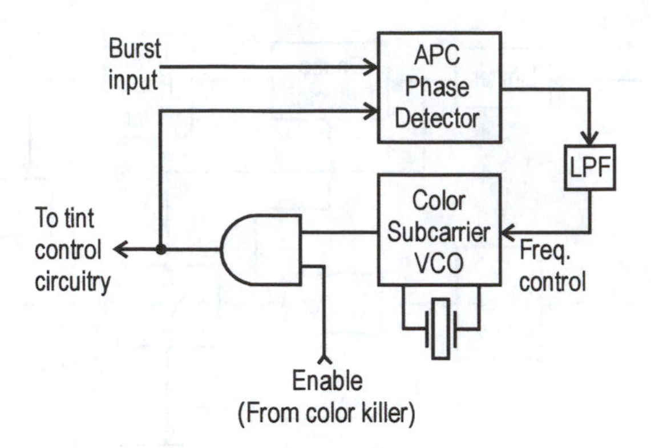

The APC Phase Detector, the Chroma Subcarrier VCO and the crystal (Y_{801}) comprise a PLL. This combination utilizes a different technique for recovering the color subcarrier than was used in the discrete Sony set. The chroma signal, which contains the color burst, is input to the APC phase detector. Figure 7.42 shows this circuitry in as standard a form as possible. The PLL action causes the VCO to lock onto the burst frequency. The time constant of the LPF is long enough to hold the VCO on frequency until the next burst. The LPF is not shown in Fig. 7.37 because it is an *RC* network outboard to MAGNUM. The mechanism by which the APC output varies the crystal-controlled VCO is not given, but it may well be a reactance-type circuit.

The output of the chroma subcarrier VCO will be a CW signal of 3.58 MHz that is phase locked to the one that was used in the camera that created the color signal. Being phase locked, it can at worst be different by an added constant phase. This VCO output next passes through the tint control circuitry, which shifts its phase by a constant amount, the amount being determined by the tint control setting. This allows the tint to be set to exactly what the camera saw. The correct setting will be recognized when flesh tones are correct. This is all very well and fine as long as the received signal continues to come from the same channel and camera for which you set the tint control.

All receivers now incorporate some kind of Automatic Tint Control (ATC) to minimize the difference in tint between signal sources. The earliest ATC circuits were called static ATC. They utilized the fact that the flesh tones lie close to the I color vector. These circuits attenuate the Q vector so that the whole color vector lies closer to the I vector. Although this measure does move errant flesh tones closer to what they should be, it also distorts other colors. While viewers are most particular about flesh tones, they are next most critical of sky and vegetation colors. These are precisely the colors that are most distorted by static ATC. These considerations have led to the development of dynamic ATC circuits, which are universally used now. These activate tint correction only for color signals that are within a certain phase difference from flesh tones.

RCA has used a circuit that modifies tints of signals that are within $+90°$ of flesh tones by pulling them toward the flesh tones. The remaining 180° range of the color/phase diagram is unaffected. As seen in Fig. 7.6, this process leaves what are sometimes called the cool colors roughly unaffected. Quasar ran tests that suggested that a much smaller range of color phases could have flesh-tone correction applied and do the job as well (± 30–$35°$). This, of course, has the

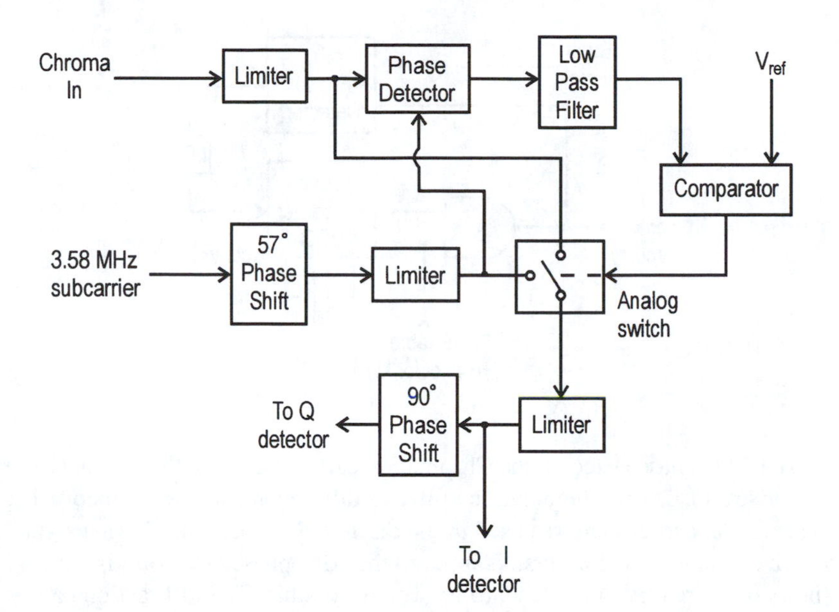

Fig. 7.43: Block diagram of the automatic tint control circuitry used in some Quasar television receivers.

advantage of leaving other colors less affected. The curious thing is that the RCA system only needs the addition of an adjustable DC voltage at one point in the circuit to smoothly reduce the range of color phase angles for which correction will be applied, but they have apparently never done so. A block diagram of the Quasar system is shown in Fig. 7.43. The 57° phase shift applied to the subcarrier rotates it to line up in phase with the I vector. This shifted subcarrier and the chroma are both amplitude limited and fed to the phase detector. Limiting assures that the phase detector output will depend only on the phase difference between its inputs ($\cos\theta$) and not on their amplitudes. When the chroma is lined up in phase with I (flesh tones), the output of the phase detector will be maximized, and it will fall off for phases on either side of flesh tones. Since the chroma and subcarrier are locked in frequency, the phase detector output will include a "DC" term that varies as the color changes from point to point. This component is extracted by the LPF and compared to an adjustable reference voltage. The comparator controls an analog switch. If the chroma is in the flesh-tone range, the phase detector output will be higher than the reference, and the comparator will toggle the analog switch to the left, passing through the phase-shifted subcarrier in place of the chroma. As this goes to detection, it will naturally be decoded as flesh tone. If the chroma is not within the flesh-tone range, the phase detector output will be below the reference, and the analog switch will patch through the chroma itself. When this is decoded, the colors will be true. This system is described in detail in the literature (Rzeszewski, 1975). The chroma proceeds off board to the color driver stages.

Exercise 7.15 If the reference voltage in Fig. 7.43 is raised, what will happen to the range of angles decoded as flesh tones?

7.7.5.7 Luminance

The video signal out of Q_{302} is taken from the slider of R_{716}, which is a preset control for contrast. This signal next passes through a delay line and is thus

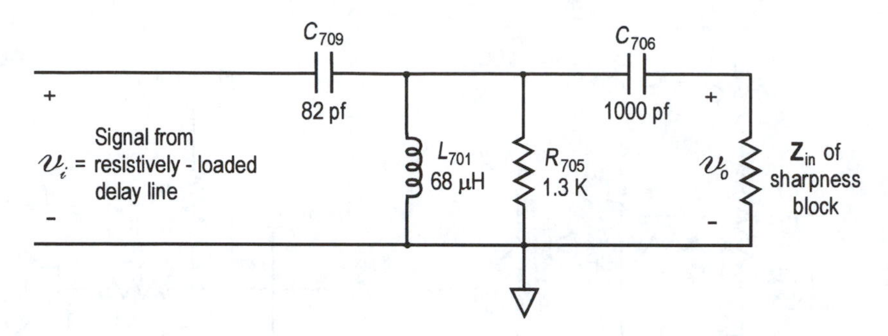

Fig. 7.44: Schematic diagram of the high-pass filter used in the sharpness-control circuitry of the RCA CTC-136 color television receiver.

identified as the luminance. From the delay line, it takes two paths. One is through the sharpness control circuitry, and the second through the user contrast control circuitry.

The sharpness control circuit is the simplest and since its output is one of the inputs to the contrast circuit, we will consider it first. The luminance goes through a high-pass filter before entering the sharpness block of MAGNUM. We show this filter in Fig. 7.44.

The transfer function for this network is

$$\frac{\mathbf{V}_o}{\mathbf{V}_i} = s^3 L_{701} C_{706} C_{709} R_{705} \mathbf{Z}_{in} / \{ s^3 L_{701} C_{706} C_{709} R_{705} \mathbf{Z}_{in} + s^2 L_{701} [(C_{706} + C_{709}) R_{705} + C_{706} \mathbf{Z}_{in}] + s[C_{706} R_{705} \mathbf{Z}_{in} + L_{701}] + R_{705} \}.$$

We thus have a third-order, high-pass filter. The exact break frequencies will depend on $\mathbf{Z}_{in}$, which is not known. However, the effect of $\mathbf{Z}_{in}$ is actually quite small as long as it is over 1 kΩ. For $\mathbf{Z}_{in} > 1$ kΩ, $\mathbf{V}_o/\mathbf{V}_i$ exhibits a pair of complex-conjugate poles at about 2 MHz and a real pole at a much lower frequency. This means that for amplitude purposes, it is effectively a second-order filter in the way the gain falls off. Basically, the low-frequency pole is due to C_{706}, which is an AC short at frequencies in or near the passband. The conjugate poles cause a minor resonant peak in the response near 2 MHz that gives a transmission factor of as much as 3 dB relative to the input. Both this peak and the overall high-pass response will provide the sharpness block with a version of the luminance that has high frequencies strongly emphasized.

The user Sharpness Control simply puts a varying DC voltage onto one input of the sharpness block. This is almost surely the gain-control input of a voltage-controlled amplifier. The output of this amplifier will thus be an amplified version of the high-frequency video components of the luminance signal as selected by the high-pass filter of Fig. 7.44.

The contrast block is much more than its name implies. Among its additional duties is the summing of the boosted high frequencies with the normal luminance signal. The greater the high-frequency content of the luminance, the greater the sharpness. Of course, the high-frequency end of the spectrum contributes a disproportionate amount of noise also, so too high a sharpness setting tends to give a noisy picture. The noise shows up differently than in a picture where the received signal is too weak. Too great a high-frequency component gives small and sharply defined flecks that come and go in the picture without obscuring it, as well as a picture with fine, moving grains.

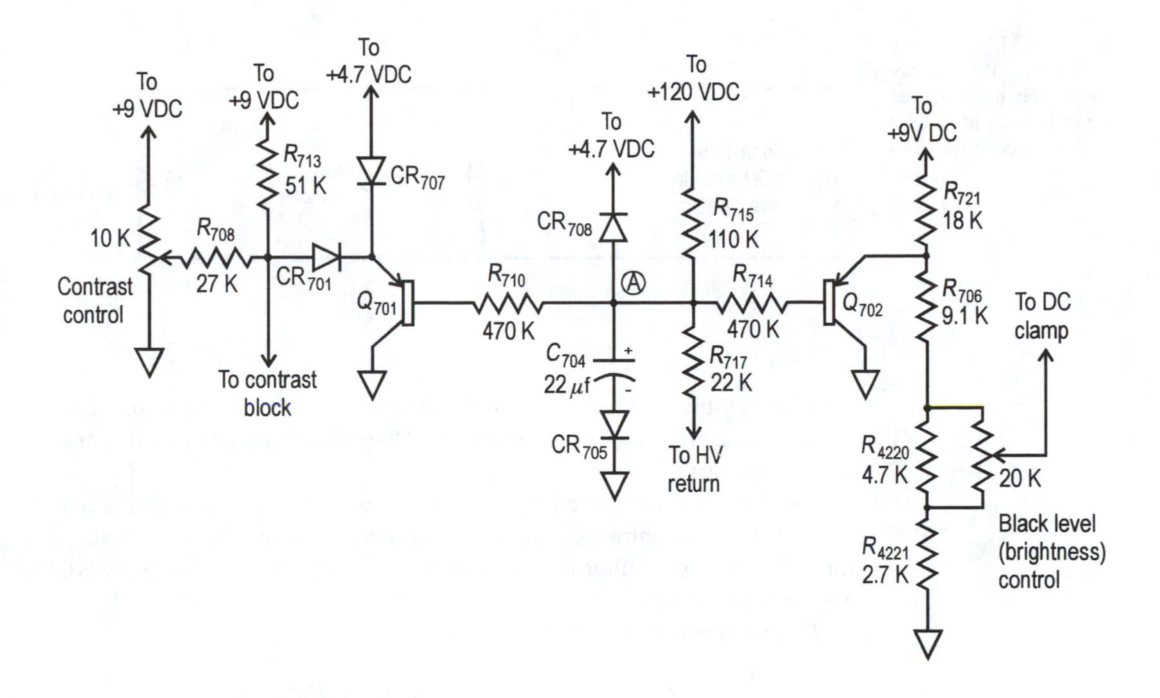

Fig. 7.45: Schematic diagram of the beam-current limiter in the RCA CTC-136 color television receiver.

The circuitry in the right-hand portion of Fig. 7.38D is the beam limiter. This set contains a two-stage beam current limiter. Its function is to prevent damage to the CRT. We analyze it with the luminance circuitry because it operates by overriding the contrast and black level controls. The relevant circuitry is shown schematically in Fig. 7.45.

The current drawn by the electron guns in the CRT comes from the +120 VDC supply through R_{715} and R_{717} and into the low end of the HV winding of the flyback. The flyback, of course, boosts the potential enormously, but the current comes from the 120 V supply.

In normal circumstances, both Q_{701} and Q_{702} are in cutoff. As long as this is the case, CR_{701} is an open circuit and the contrast control to the left functions normally. As the beam current rises, there is more drop across R_{715}, and the potential at node A falls. Any AC component of beam current is removed by C_{704} and CR_{705}. Until the beam current increases enough to pull down the voltage at node A, this voltage is clamped at about 5.4 volts by CR_{708}.

Exercise 7.16 In Fig. 7.44, determine the approximate beam current for which the node A voltage will drop below its clamp level.

When the node A voltage is pulled down by excessive beam current, the first response of the circuit is that Q_{701} conducts enough to forward bias CR_{701}. It pulls down the potential of the contrast control pin of MAGNUM, thus overriding the setting of the contrast control and reducing the contrast. This means that the bright parts of the picture will be less bright and can, therefore, be rendered with less beam current. The emitter voltage of Q_{701} at this point will depend somewhat on the setting of the contrast control, but should be near 5 volts. Further increases in beam current will continue to turn on Q_{701} more heavily until it is fully saturated, thus clamping the contrast control voltage to about .75 V through CR_{701}. If the beam current were to rise by just 20 μA above the value at which contrast limiting starts, not only is Q_{701} fully saturated, but Q_{702} would go from fully off to fully saturated. This would lower the brightness control voltage to zero, drastically curtailing beam current.

The signal out of the contrast block passes through an active filter whose type and purpose are not specified. Since the luminance signal has a bandwidth that ideally extends from DC to over 4 MHz, it may have the purpose of rolling off frequencies above 4 MHz (low pass) or it may be a notch filter to remove color at 3.58 MHz. The problem with both of these possibilities is that the frequencies are quite high by active filter standards at the time the set was designed – even with high quality "stand-alone" op amps. Here we have the filter circuitry as just a small part of a very large chip.

In any case, the filtered luminance signal is passed through a DC clamp similar to that found in the integrated Zenith monochrome receiver of Section 6.4. A user black-level control sets the clamp voltage to a value that will define the darkest parts of the picture. This black level will remain the same even if the received signal strength varies. The clamped luminance then receives a final stage of amplification before exiting MAGNUM at pin 13 and passing on to the color decoding circuits.

We have now reached the stage of the analysis where it will be necessary to do a lot of jumping back and forth between the main circuit board and other parts of the receiver. We will keep the circuit board as our point of reference for analysis.

The luminance and chroma from this board proceed to three color driver stages that are nearly identical. These signals enter at the upper right-hand corner of Fig. 7.46. This figure also incorporates the remainder of MAGNUM and the deflection and power supply circuitry. In these drivers, the $R - Y$, $B - Y$, and $G - Y$ are reunited with Y to produce the individual color drive signals. As was the case with the discrete Sony color set, the Y signal (actually $-Y$) is applied to the emitters of all three color drivers, and each has a $\mathcal{C} - Y$ ($\mathcal{C}$olor $-Y$) signal applied to its base. Figure 7.47 shows the configuration. The main observation to be made relative to this schematic is that the green drive control governs the amount of $-Y$ fed to the emitter of Q_{5005}. This is necessary to make v_{BE} of this stage strictly proportional to G, with no dependence on Y as such.

The green bias is, as we shall see, directly coupled to the green power-output stage, which in turn is directly coupled to the CRT. Thus, adjusting R_{5034} will change the Q point of the output stage and, thereby, the green beam bias. This will adjust the intensity of the green beam in the absence of signal. All three beam bias controls must be adjusted correctly to balance the three beam intensities.

The outputs of the color driver stages are coupled to the output stages, one of which is shown in Fig. 7.48. We note that the output stages are in the common-base configuration, which is fairly unusual. Furthermore, collector current for all three output stages is drawn through R_{5008} and L_{5001} in parallel. This is investigated in Exercise 7.17.

Exercise 7.17 Ignoring the effects of the resistors to the right of the dashed line in Fig. 7.47, find the voltage gain of this stage. It will be necessary to include a Thevenin source resistance to model the output of the driver stage. Make and state whatever simplifying assumptions are appropriate.

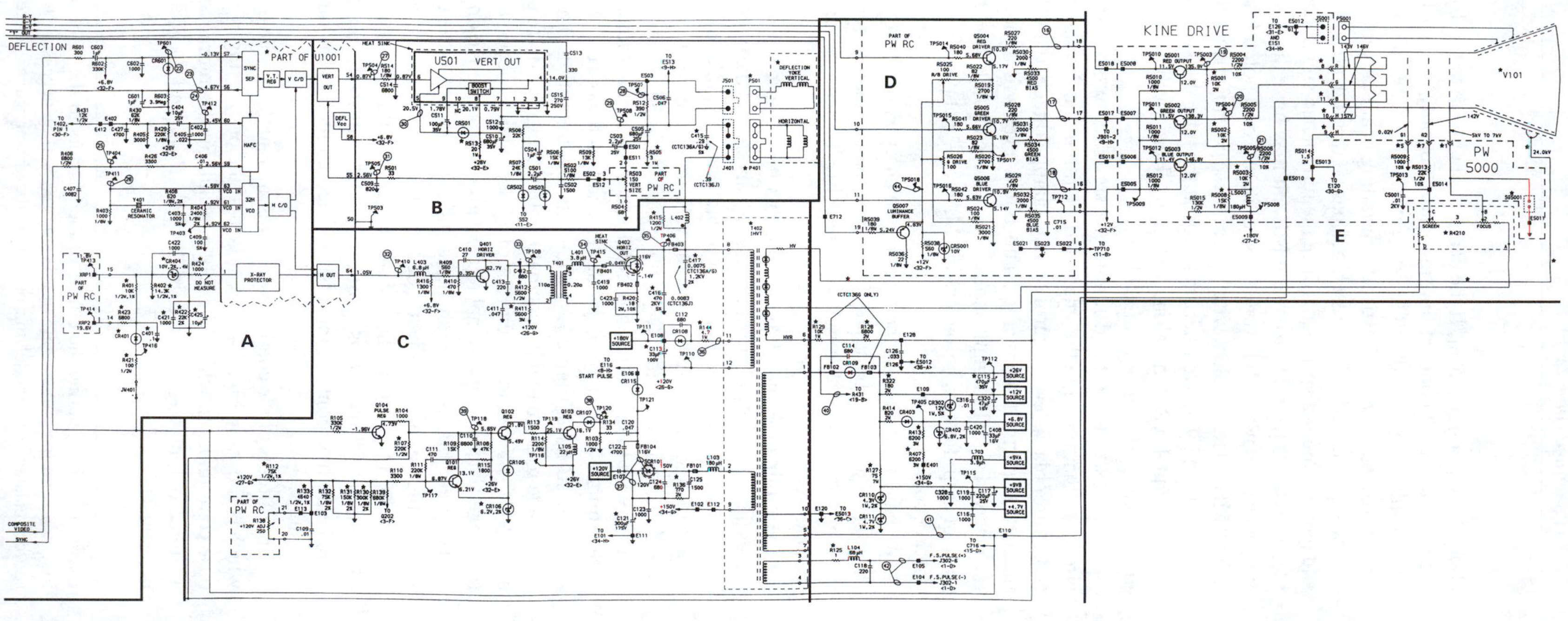

Fig. 7.46: Schematic diagram of the portion of the main Board which includes the deflection circuitry, the power supplies, and the color output circuitry in the RCA CTC-136 color television receiver. An electronic version of this figure is available on the world-wide web at http://www.cup.org/titles/58/0521582075.html.

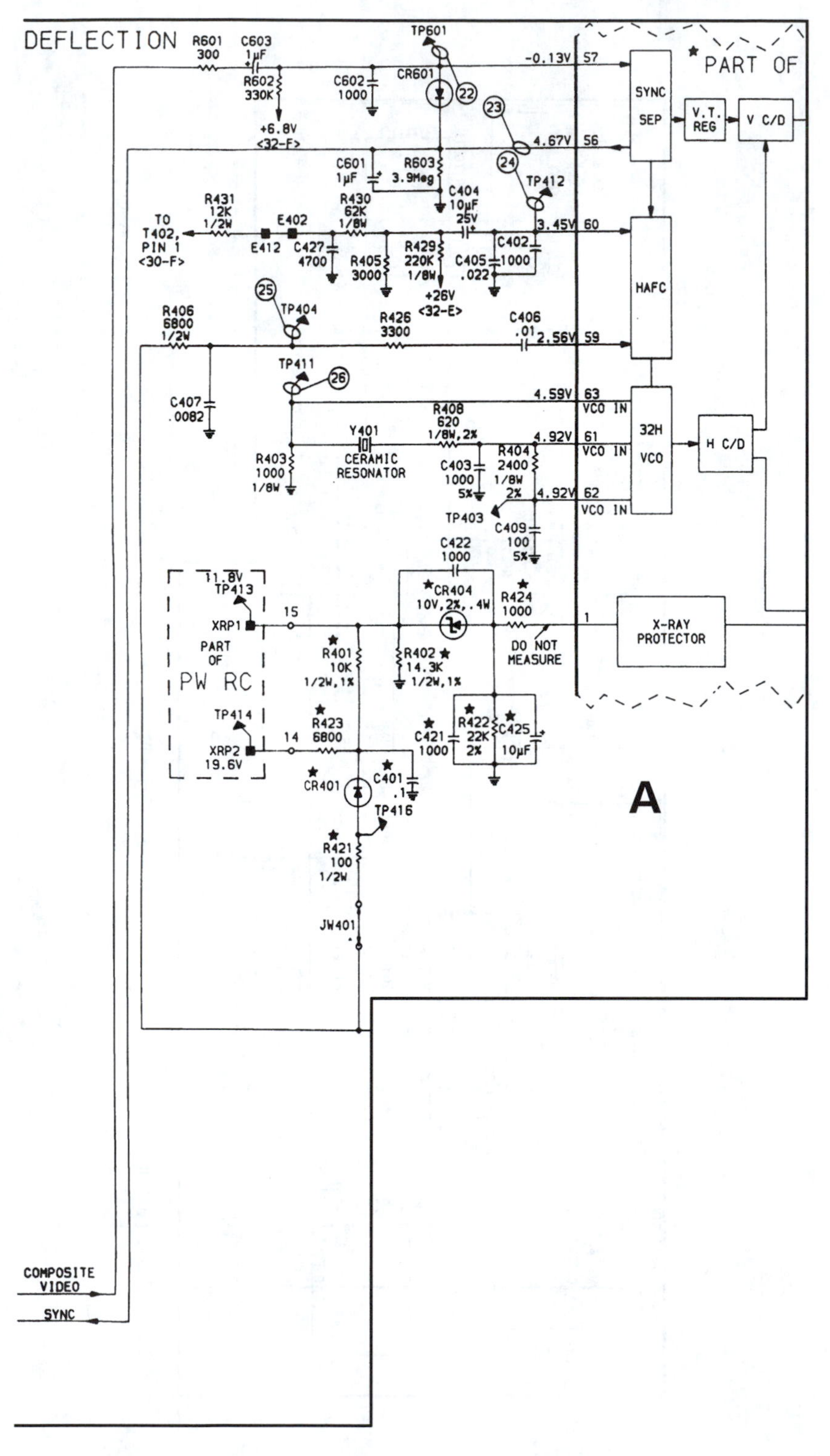

Fig. 7.46A: Horizontal and vertical sections.

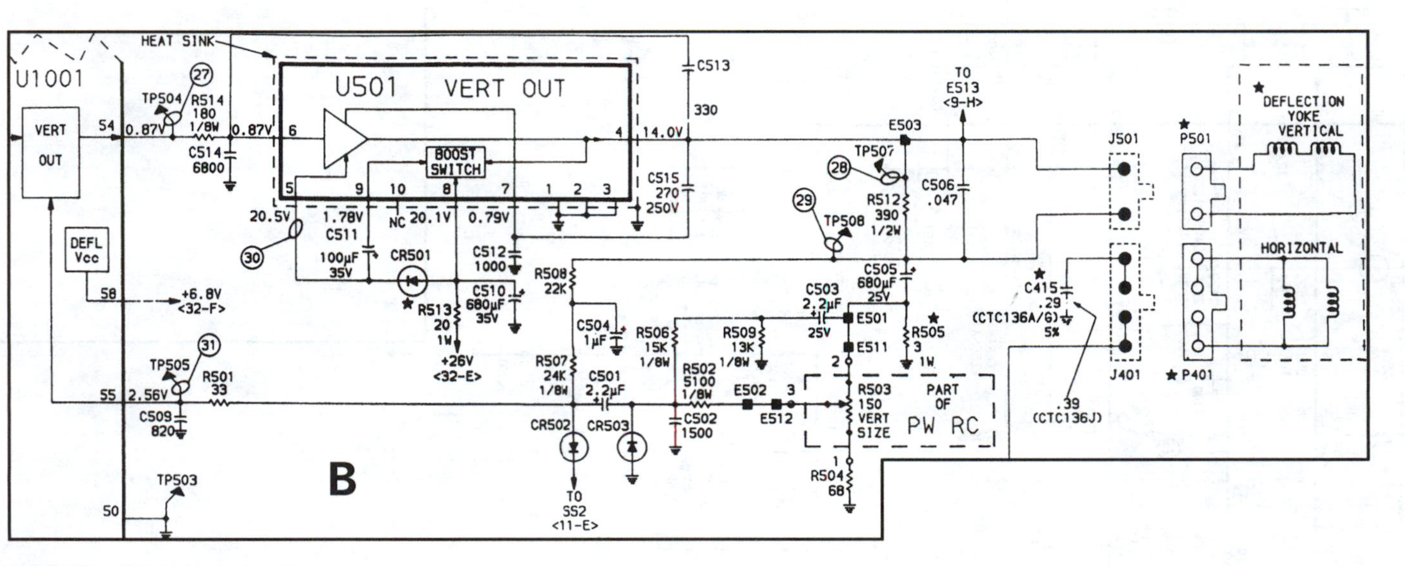

Fig. 7.46B: Vertical output section.

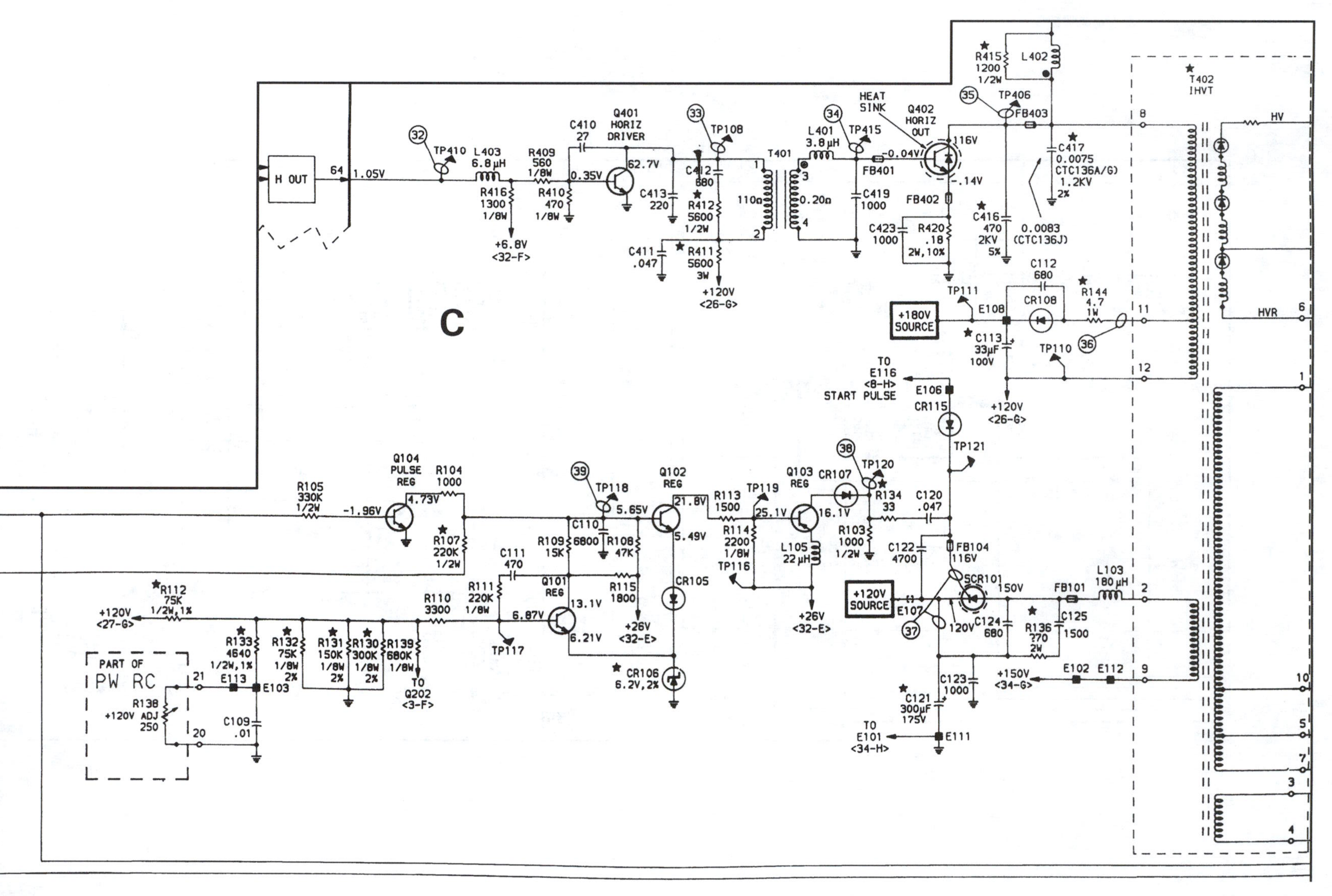

Fig. 7.46C: 120 V regulator and horizontal output section.

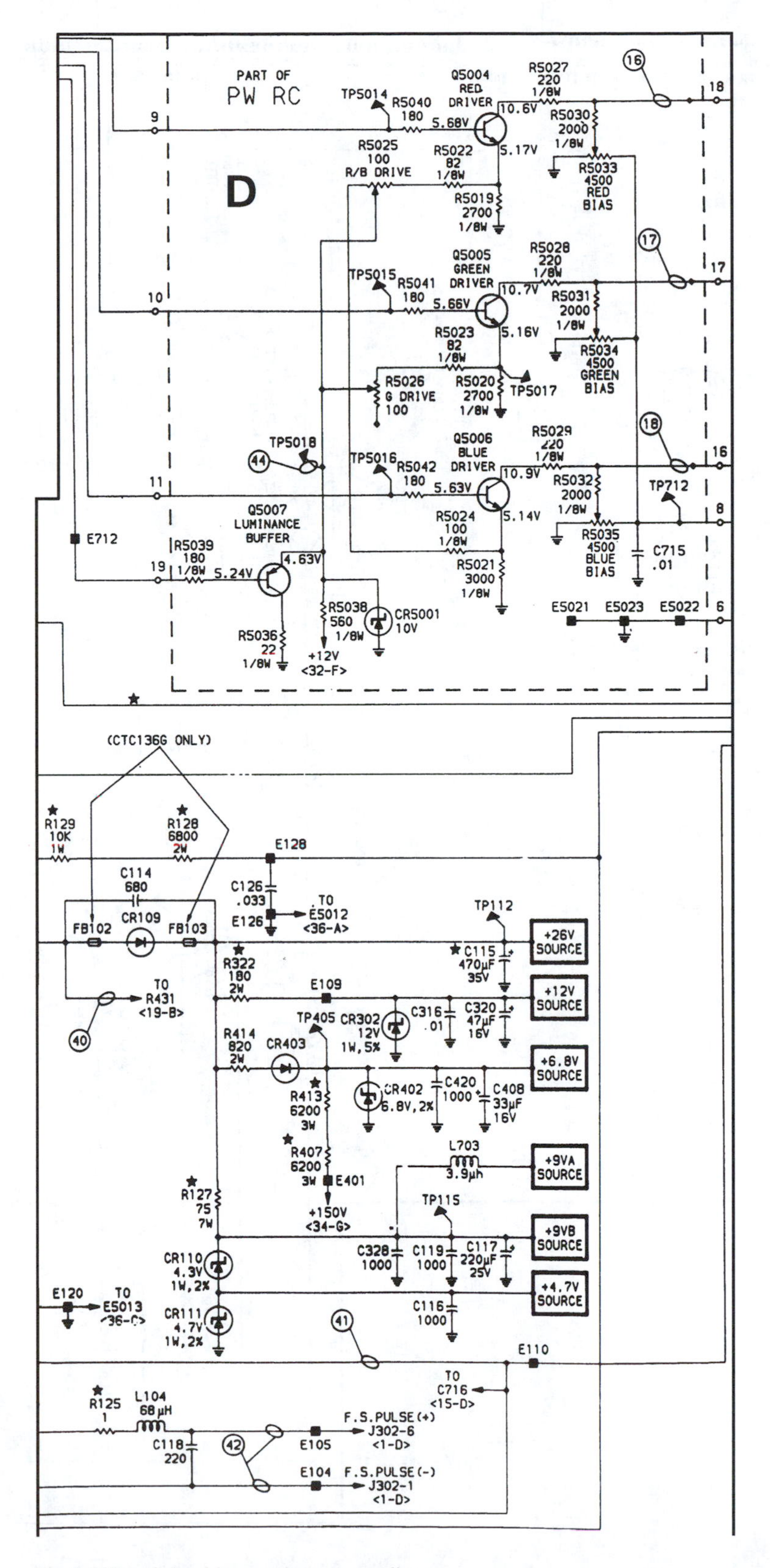

Fig. 7.46D: Color drivers and power supplies.

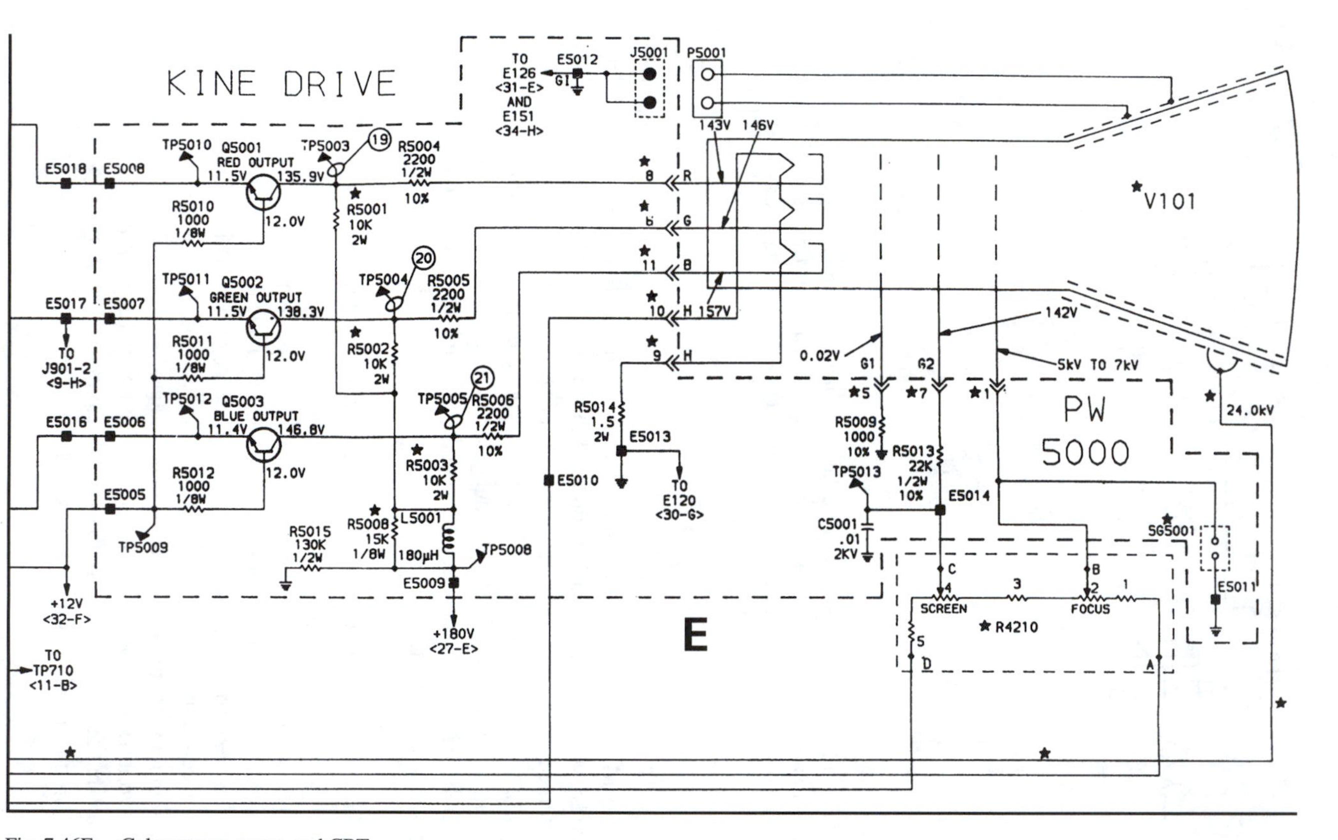

Fig. 7.46E: Color output stages and CRT.

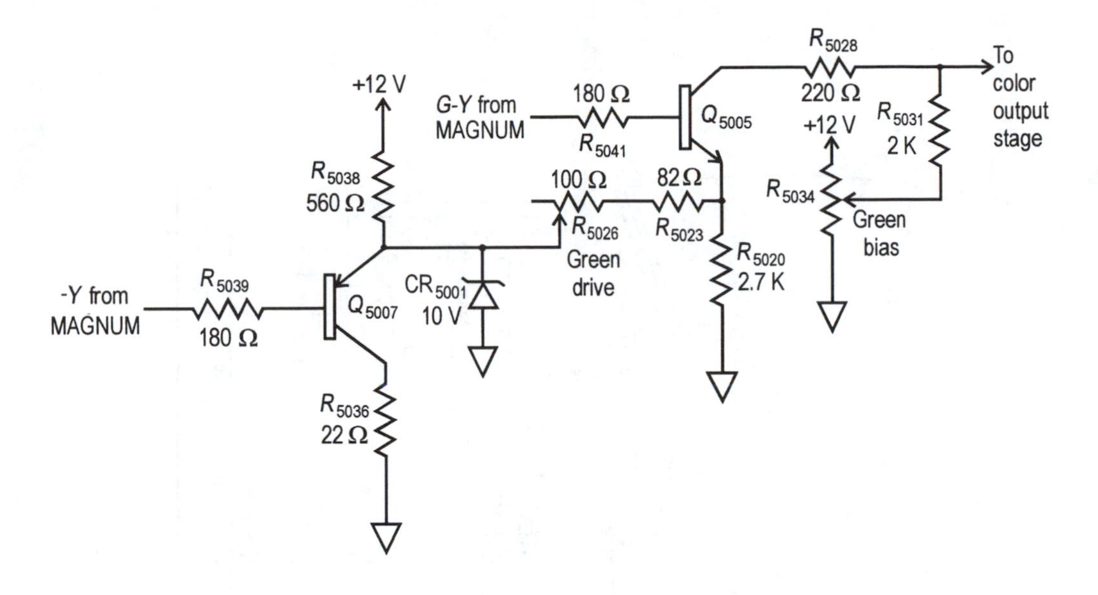

Fig. 7.47: Schematic diagram of one of the color driver amplifiers in the RCA CTC-136 color television receiver. It is in this stage that the luminance and the chroma signals are combined to recreate the pure color signals that drive the CRT.

Exercise 7.18 Also in Fig. 7.47, extend your calculation to include the identical circuits for R and B. Express your overall result for v_{out} as a constant term times the green drive voltage plus a frequency-dependent term that depends on all three drive voltages.

7.7.5.8 Horizontal sweep

We now return to MAGNUM and Associates on the main board where the composite video arrives next at the sync separator. The exact nature of this stage cannot be ascertained from the schematic, but as usual it provides horizontal and vertical sync pulses. The vertical and horizontal countdown circuits look quite similar to those in the Zenith 5NB4X analyzed in Section 6.4. Though the block diagram here is not detailed enough to make a complete comparison, these differences can be discerned:

- The horizontal VCO here runs at 32 times the horizontal sweep frequency ($32 f_{\text{H}}$) as compared to $2 f_{\text{H}}$ in the Zenith.
- The horizontal VCO frequency is controlled by a ceramic resonator. A oscillator based on a ceramic resonator can readily have its frequency "pulled" by external capacitance. We therefore postulate with a fair degree of certainty that the AFC stage must contain both a phase detector and a voltage-controlled reactance stage.
- The horizontal sweep output exits the chip through a voltage-controlled amplifier that cuts down the sweep signal if required by the X-ray protection circuitry. This is shown in Fig. 7.49.

7.7.5.9 X-ray protection

In the absence of flyback pulses, the 20 V supply sees a voltage divider consisting of the 6.8 k, 10 k, and 14.3 k resistors. This places a voltage of about 9.2 V at the cathode of the 10 V Zener diode. When the effect of the flyback

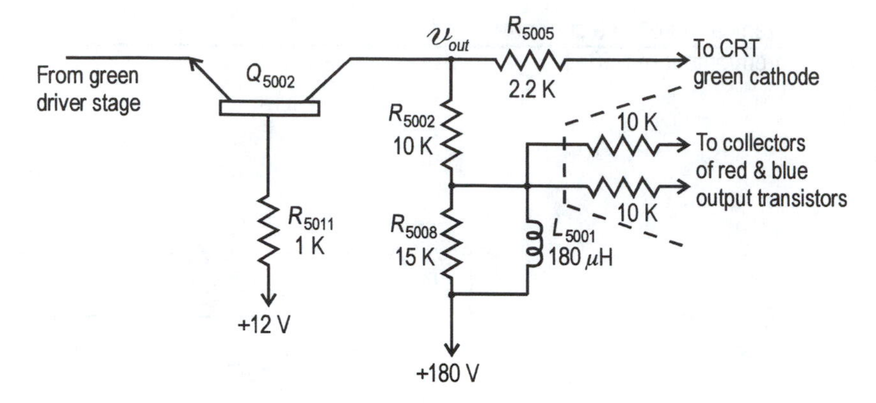

Fig. 7.48: Schematic diagram of one of the color output amplifiers in the RCA CTC-136 color television receiver.

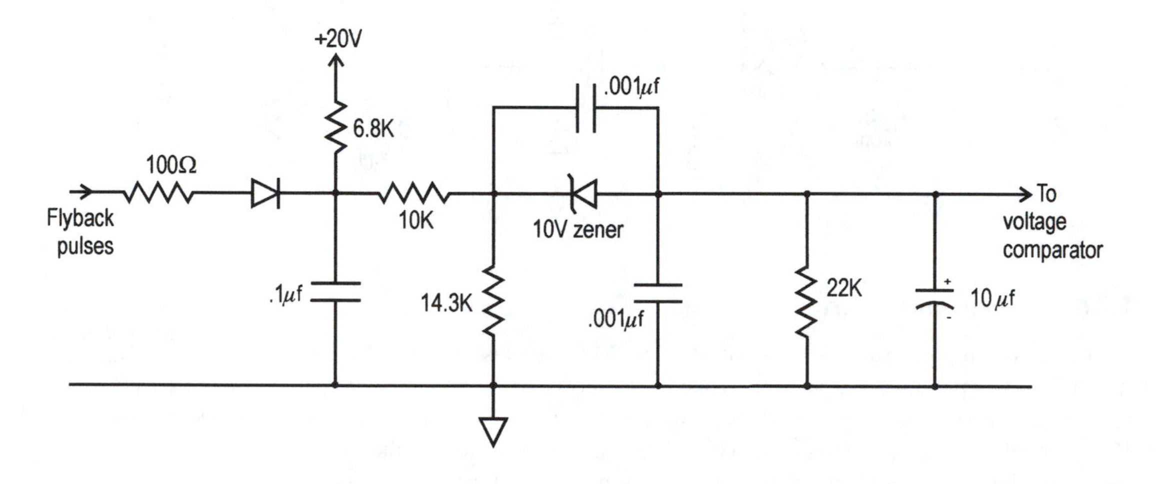

Fig. 7.49: Schematic diagram of the X-ray protection circuit in the RCA CTC-136 color television receiver.

pulses is superimposed, we see that they will be rectified by the diode and will charge the .1 μF filter capacitor. The 100 Ω resistor in series with the diode will limit the peak charging current through the diode. The nominal amplitude of the flyback pulses input to the circuit is about 28 V. They will thus be attempting to charge the filter capacitor to this value. However, as the voltage on this capacitor rises, the rectifier diode conducts for a smaller part of each pulse and so charges more slowly. Furthermore, after the capacitor voltage reaches the voltage set by the 20 V supply and its divider (about 15.6 V), the current from the 20 V supply decreases. Nevertheless, the capacitor will continue to charge slowly due to the rectified pulses until the voltage across it reaches about 17 volts. At this point, the discharge through the resistive divider and the charge due to the pulses and the 20 V supply are about equal. When the filter capacitor voltage reaches 17, the cathode of the Zener is at a potential of 10 V, i.e., at the threshold of breakdown. If the output of the horizontal oscillator were to rise, the resulting rise in high voltage would present an X-ray hazard. However, the increased output of the horizontal oscillator would also increase the amplitude of the flyback pulses input to this circuit. This in turn would raise the cathode of the Zener enough to cause it to break down. When it does, DC voltage will appear across the filter network connected to the Zener's anode. This DC voltage is sensed by the X-ray protection circuitry, which lowers the gain of the horizontal sweep amplifier to cut back the high voltage.

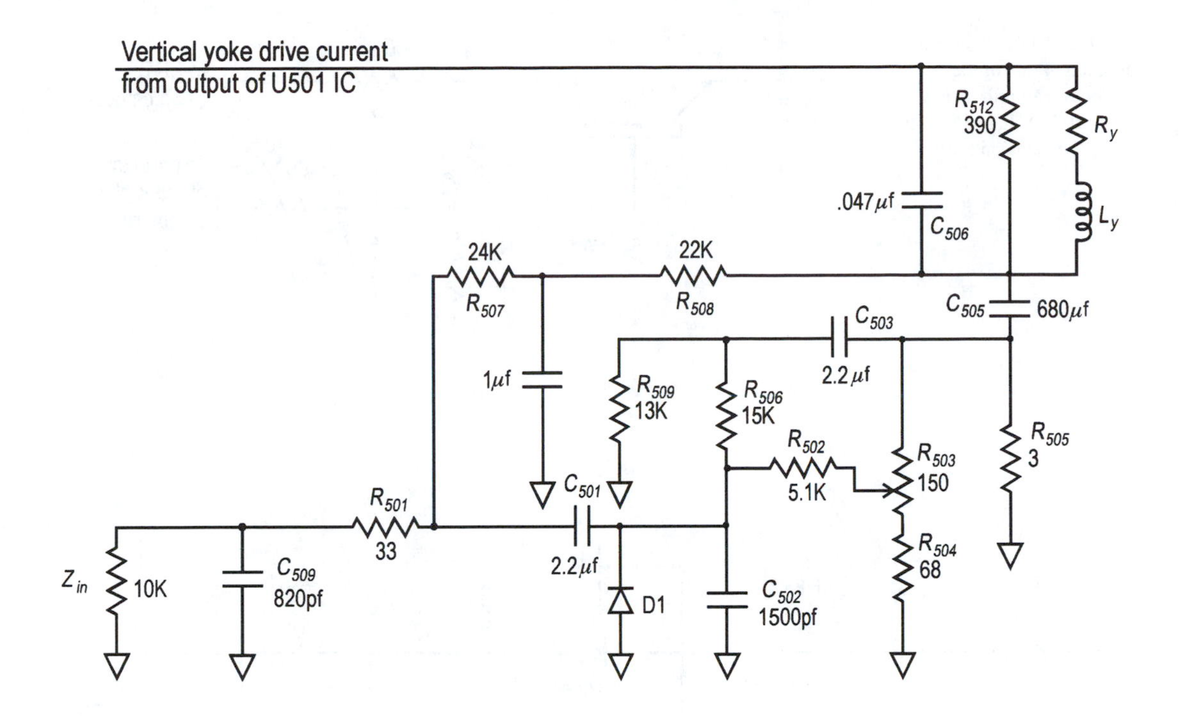

Fig. 7.50: Schematic diagram of the vertical yoke and feedback circuitry in the RCA CTC-136 color television receiver.

7.7.6 *Vertical output circuitry*

We now leave the congenial confines of MAGNUM for the last time and follow its vertical sweep voltage output to the yoke. It is a short trip! The sweep voltage is applied to a vertical-output IC through a 180 Ω resistor. The output of this IC is capable of driving the vertical yoke directly. The low side of the yoke, instead of being returned to ground directly, is grounded through the series combination of a 680 μF capacitor and a 3 Ω resistor. The yoke current flowing through the resistive part of this impedance causes a small voltage drop that is used to generate a feedback signal, which is fed back to the vertical output IC. The yoke and feedback circuitry are shown in Fig. 7.50.

Vertical blanking pulses are taken from the high side of the vertical yoke. The reason for this is clear if we look back to the yoke voltage in Fig. 6.11e. The pulses occur at the correct time (between frames) to accomplish blanking. In this RCA set, the voltage across the vertical yoke is almost identical to Fig. 6.11e but is inverted.

From Fig. 7.50 we wish to find the yoke inductance and resistance, the amplitude of the yoke sawtooth current, and the nature of the signal fed back to MAGNUM. All but the last of these may be determined with reasonable accuracy using the much simpler circuit of Fig. 7.51. We are able to eliminate the rest of the circuitry to find the yoke parameters because the impedances of the eliminated circuitry are expected to be much larger than those of the components shown in Fig. 7.51. We shall have to check this when our calculation is complete.

Exercise 7.19 Assume that $i_y(t) = I_M(1 - 2t/t_t)$, where t_t is the trace time of the sawtooth sweep current and I_m is its amplitude. The Laplace transform of a cycle is then $\mathbf{I}_y(s) = I_m(1/s - 2s^2 t_t)$. Since v_B is only about a tenth of v_A in peak-to-peak value, we may assume for a first-pass calculation that the voltage across the entire circuit of Fig. 7.50 comes from $i_y(t)$ flowing through the yoke. Using this assumption:

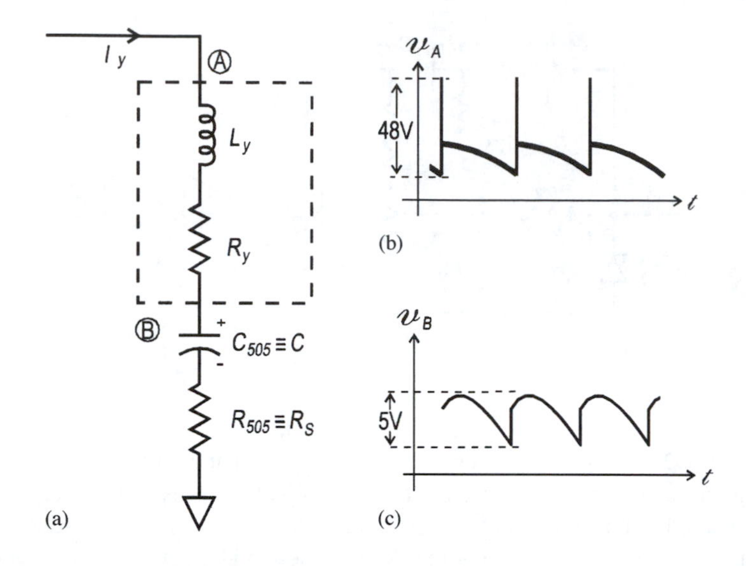

Fig. 7.51: (a) Schematic diagram of the simplified yoke circuit that will be used to estimate component values in the vertical yoke circuit of Fig. 7.50. (b) The voltage at node A of this diagram. (c) The voltage at node B. These waveforms come from the service information.

(a) Find $I_{\rm m}(R_{\rm y} + R_{\rm s})$ by comparison to $v_{\rm A}(t)$ during the trace cycle.
(b) Find $v_{\rm B}(t)$ during the trace cycle in terms of $I_{\rm m}$, $R_{\rm s}$, C, and $t_{\rm t}$.
(c) Find the peak-to-peak amplitude of $v_{\rm B}(t)$ in terms of these same quantities.
(d) Knowing the experimental value of the peak-to-peak value of $v_{\rm B}(t)$, find $I_{\rm m}$, using the result of part (c).
(e) Find $R_{\rm y}$.

All of these results come from considering only the trace cycle. The pulse portion of $v_{\rm A}(t)$ results from the yoke inductance.

(f) Use the amplitude of the $v_{\rm A}(t)$ pulse to find $L_{\rm y}$. Note that this will require you to assume something about the nature of the sawtooth retrace. Be advised that the simplest assumption is not the best! State whatever assumption you make.

Our analysis shows $R_{\rm y} \approx 13\,\Omega$, $L_{\rm y} = 28\,{\rm mH}$, and $I_{\rm m} = .54$ A. For these values, an analysis of the entire circuit using PSPICE® and assuming a linear retrace for the sawtooth shows $v_{\rm A}(t)$ with a peak-to peak-amplitude of 45 volts and $v_{\rm B}(t)$ with a peak-to-peak amplitude of 4.55 volts. These compare favorably with the values shown in Fig. 7.50. Every feature of the experimentally observed waveforms is reproduced in the computer output except that the pulses in $v_{\rm A}(t)$ have tops that slope up sharply. This is a consequence of having assumed a linear retrace and our having omitted the spike supression components R_{512} and C_{506}.

We must now, as promised, compare the impedances of these elements to those that were ignored by comparison. At 60 Hz a 28 mH inductance has a reactance of over 10 Ω. Since the sawtooth has higher-frequency components also, the reactance of $L_{\rm y}$ will be even larger for these. This impedance is in series with a 13 Ω resistance. In comparison to this series connection of impedances, we have ignored the series connection of $R_{\rm s}(=3\,\Omega)$ and C, whose 680 μF capacitance has a reactance of 4 Ω at 60 Hz and less at the harmonic frequencies. This is the least justifiable of the simplifying assumptions made in the course of the analysis, but considering the good agreement with the computer analysis, it still seems not to be too bad an assumption. The series combination of L_y and R_y is in parallel with R_{512} (390 Ω) and C_{506} (.047 μF). The resistance is obviously much larger than the impedances of L_y and R_y. The capacitor has an impedance of over 5 kΩ at 60 Hz and less at the harmonic frequencies. However, since the harmonics of a sawtooth fall off as $1/n^2$, no significant part of $\mathbf{I}_{\rm y}$ flows through it. Then, the reader might ask, why is it there? It is there (as

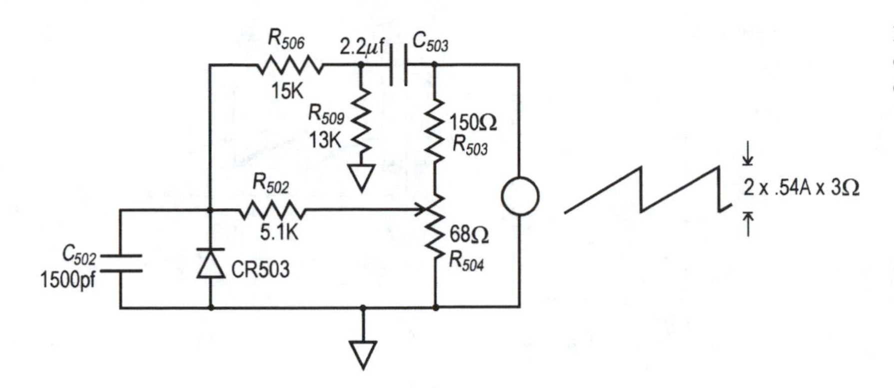

Fig. 7.52: Sawtooth clamp circuit portion of the circuitry in Fig. 7.50.

is the 390 Ω resistor) to suppress the voltage spikes that would otherwise occur when the sawtooth current through the vertical yoke begins its retrace.

It should also be clear that the rest of the circuitry connected to nodes B and C will not cause significant loading. Rather, these nodes act as voltage sources for the rest of the circuitry. Examining this remaining circuitry, we recognize that the voltage across the 3 Ω resistor we have called R_s will be a sawtooth by virtue of the yoke current flowing through it. The 150 Ω vertical size pot (R_{503}) and the 68 Ω R_{504} in series with it constitute a variable voltage divider, which will send part of this sawtooth voltage onward. This portion of the circuitry is shown in Fig. 7.52. C_{503} is effectively an AC short to the sawtooth waveform. Therefore, it places R_{509} in parallel with the voltage divider for AC purposes. Its resistance is so much larger than that of the divider that it can have no AC purpose.

The "source" voltage shown has no DC component. The diode and C_{502} combine with the Thevenin resistance of R_{502} (5.1 k) and R_{506} (15 k) in parallel to form a clamp that holds the negative excursion of the sawtooth at about $-.7$ volts with the expected rounding off. If it were not for this rounding of the negative peaks, they would be at -1.2 V. The positive peaks are at $+1.2$ V with the vertical size pot set at its center position. This means that the undistorted wave would have a peak-to-peak amplitude of $\sim$2.4 volts, which happens to be the Thevenin voltage for this setting of the pot.

Furthermore, this means that, but for distortion, the DC component for this pot setting would be zero. If the diode voltage has no DC component, neither will the voltage on the left side of C_{503}. Much the same thing is true for other pot settings. If there is no DC voltage on either side of C_{503}, it can be replaced by a short and R_{509}, the 13 kΩ resistor, can be eliminated. Computer analysis confirms these conclusions. Why are the components put in the circuit then? Simply, because in general the sweep voltage is not a "nice" sawtooth, but rather a relatively "ugly" one that has deliberately-introduced nonlinearities to compensate for pincushion and other distortion mechanisms. The blocking capacitor, which is unnecessary in the theoretical case, is needed in the practical case, where the clamp diode may have a nonnegligible average voltage across it.

We now direct our attention to the *RC* network connected to node B in Fig. 7.50, proceeding left. This network is shown in Fig. 7.53.

This is simply an integrating network. The waveform at B, which is being integrated, has a DC component in excess of 1.5 V, so this causes the integrated waveform to be "riding a ramp" until the DC voltage across C_{504} gets large

Fig. 7.53: Integrating network portion of the circuitry in Fig. 7.50.

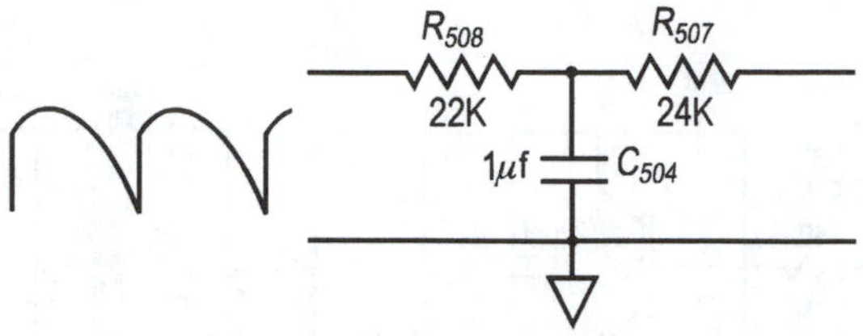

enough that the rate of charge loss through R_{507} (24 k) is equal to the rate of charge being integrated onto the capacitor through R_{508} (22 k). It would appear from Fig. 7.50 that there is no DC path by which charge can leak off through R_{507}, but where the feedback goes back into MAGNUM, the chip will have a resistance to ground. The calculations cited have been done with an assumed input resistance of 10 kΩ. Lower resistances (down to 100 Ω) have also been analyzed. The waveforms are generally the same in shape but lower in amplitude for lower input resistances.

Capacitor C_{501} of Fig. 7.50 is also an AC short. In this case, it is necessary for DC blocking even for a pure sawtooth current. This is true because although we have shown that the average voltage across the diode clamp is very nearly zero for a sawtooth yoke current, there will be a DC component of voltage at the left side of this capacitor because of the DC component of current out of the integrator.

To summarize the nature of the signal fed back to the vertical oscillator part of MAGNUM, we see that it has the following characteristics:

- It is basically a sawtooth which is in phase with and proportional to the vertical yoke current.
- Its amplitude will vary with the setting of the vertical size pot, R_{503}.
- It will contain a +DC component from the integrator section (R_{507}, R_{508}, and C_{504}). This component will vary linearly with sweep amplitude and indirectly with the setting of R_{503}. As this pot is varied, the signal fed back to the vertical oscillator causes its output to vary, thus causing sweep amplitude to vary.
- We can say no more than this about the mechanism of changing sweep amplitude without knowing more about how MAGNUM uses this feedback signal.

7.7.7 Horizontal output circuitry

An output from MAGNUM provides the input to the horizontal driver as well. Figure 7.54 shows the horizontal circuitry.

The input voltage, v_{in}, is a nearly symmetrical square wave from MAGNUM. The resistive network of 1.3 kΩ, 560 Ω, and 470 Ω in conjunction with the 6.8 V supply biases the base of Q_{401} to such a point that v_{in} is able to swing the stage output between saturation and cutoff.

In actuality, little more needs to be said about either the driver or the output stage, since both are very similar to the corresponding stages of the Zenith 19EB12 monochrome set studied in Sections 6.3.10.4 and 6.3.10.5. In fact, by comparing Figs. 7.54 and 6.45, we see that the circuits are topologically the same except:

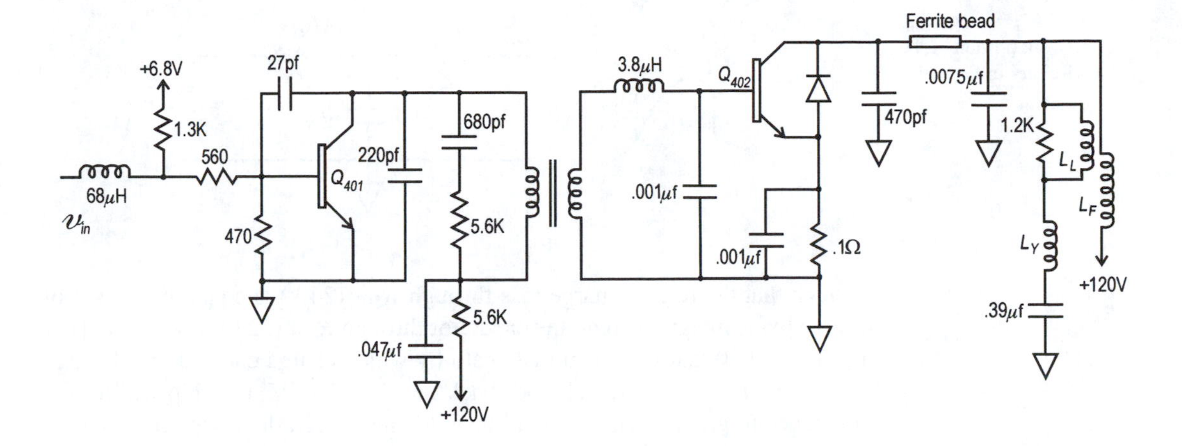

Fig. 7.54: Horizontal driver and output stages in the RCA CTC-136 color television receiver.

- Figure 6.45 has an additional resistor (R_{726}), whose effect we showed to be small.
- Figure 7.54 has an additional capacitor (220 pF) from the collector of Q_{401} to ground. Its effect is to slow the rise of the collector voltage, reducing overshoot slightly.

The horizontal output stage in Fig. 7.54 is even more like that in Fig. 6.48 than the corresponding driver stages are. In fact, if L_{208} in Fig. 6.48 were shorted out, the circuits would be topologically identical. The only question then is whether it will still work without that inductor.

Exercise 7.20 Starting with the most general equation for $\mathbf{I}_y(s)$ in Chapter 6, set $L_{208} = 0$ and solve for $i_y(t)$. Under what conditions will it give a ramp response similar to what was obtained with L_{208} in circuit? Is this condition ever likely to be achieved in practice?

7.7.8 Power-supply circuitry

The type of power supply used in this receiver might be considered quite unorthodox by most electronic engineers outside of the consumer electronics industry, but it has become common in modern TV sets. It is, first of all, a switching supply, but this is not unusual in itself. Its uniqueness stems from the fact that all of the "run" power supplies we have already discussed are driven from windings on the flyback. This power supply circuitry is shown in Fig. 7.55.

The standby supply in this set is an unregulated 150 VDC supply derived directly from the line and followed by a bridge rectifier. When the turn-on command is received from the console or the remote, the MTT-006A outputs a start pulse as discussed in Section 7.7.3.2. This pulse is applied to the turn-on circuitry in the interface board, which in turn fires the SCR, as discussed in Section 7.7.4. The circuit in Fig. 7.55 takes up where we left off there. In the lower right-hand corner of the schematic, we can see both the SCR and the previously discussed input, which is applied to its gate to start the power-up procedure.

When the SCR has been triggered, it allows current to flow from the unregulated 150 V standby supply to charge C_{121}. This voltage across C_{121} is the 120 V

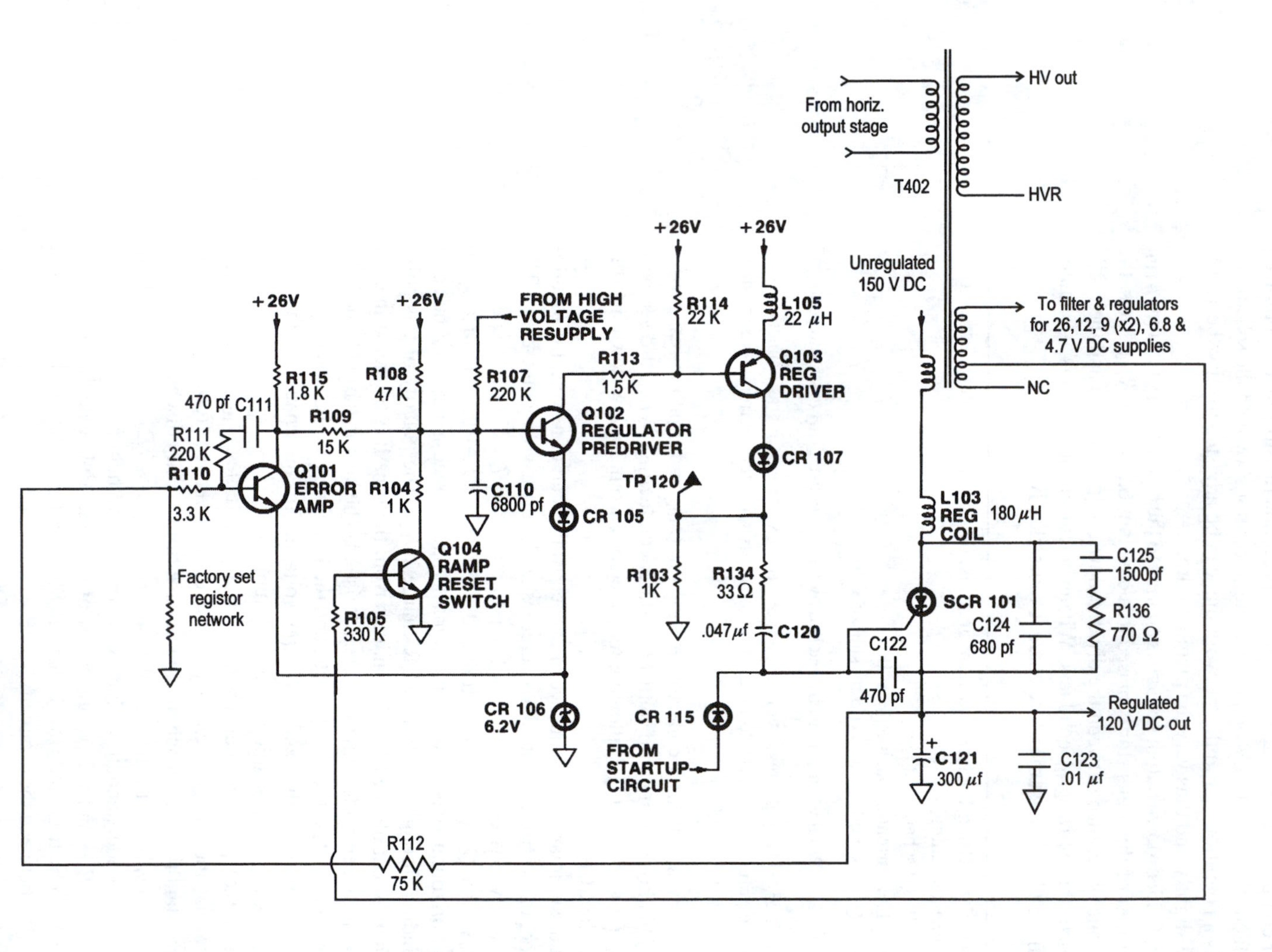

Fig. 7.55: Schematic diagram of the 120 V regulator in the RCA CTC-136 color television receiver.

output of the discrete regulator made up of all the transistors in the schematic. The output is fed back to the comparison input of the regulator through a voltage divider comprised of R_{112} and a network of precision resistors that is optimized at the factory to give 120 V output. Although it is not shown on the schematic, there is a potentiometer in this network that can be adjusted to vary the 120 V supply slightly and thus vary the picture width.

The divided-down output voltage is applied to the base of Q_{101} through R_{110}. The emitter of this transistor is biased to a constant 6.2 V by CR_{106}. Thus, the greater the difference between the divided-down output voltage and the 6.2 V reference, the more Q_{101} conducts. Whatever current it conducts is shunted away from C_{110}, which therefore charges more slowly.

Exercise 7.21 Using Laplace transform analysis, determine the voltage across C_{110} versus time, assuming:

(1) It is discharged at $t = 0$.
(2) The current through $R_{107} = 0$.
(3) Q_{101} is modeled as a current sink of value I.

Exercise 7.22 What voltage must be reached by C_{110} to cause Q_{102} to start to conduct?

Exercise 7.23 Using the results of the previous two exercises, find (as a function of I) the time required for the pseudo-ramp voltage across C_{110} to rise from zero to where Q_{102} starts conducting.

R_{114} serves both as a collector resistor for Q_{102} and a pullup resistor for Q_{103}. This holds it in cutoff until Q_{102} starts conducting. When it does, base current flows out of Q_{103}, allowing its collector current to flow down into the gate of the SCR.

Let us summarize the sequence of events. At the start of a cycle, the pseudo-ramp begins to rise at a rate that decreases as the error voltage increases. The more slowly it rises, the longer it takes to reach the 7.4 V or so required to put Q_{102} and Q_{103} into conduction and fire the SCR. At the end of a cycle of the horizontal sweep (fed to the flyback), the retrace generates a pulse in the winding through which the SCR charges C_{121}. The polarity of this pulse is such as to cut off the SCR. It conducts from the time it is triggered in each horizontal sweep cycle until the cycle is finished. The later in the cycle the SCR is triggered, the less time it conducts, and the less time C_{121} has to charge. The net effect is to lower the regulated output voltage. This in turn reduces the drive to the error amp, maintaining the equilibrium.

The retrace pulse that cuts off the SCR also saturates Q_{104} and discharges C_{110} so that the charge cycle can start over. We can see from the discharge time constant that the capacitor will not have time to fully discharge. This will somewhat invalidate the numerical results of Exercises 7.21 and 7.23, but not the physical understanding we have developed from them.

Diodes CR_{107} and CR_{115} form a sort of discrete AND gate that allows the SCR gate to be triggered from either the Start circuit or the regulator without interaction between those circuits.

Resistor R_{103} allows Q_{103} to have a DC path to ground for the collector current but is large enough that no significant part of the current pulse from Q_{103} will flow through it as opposed to the path through R_{134} and C_{120}.

The *RC* elements across the SCR constitute a "snubber" network that combats rate-effect turn-on and inductive transient damage to the SCR.

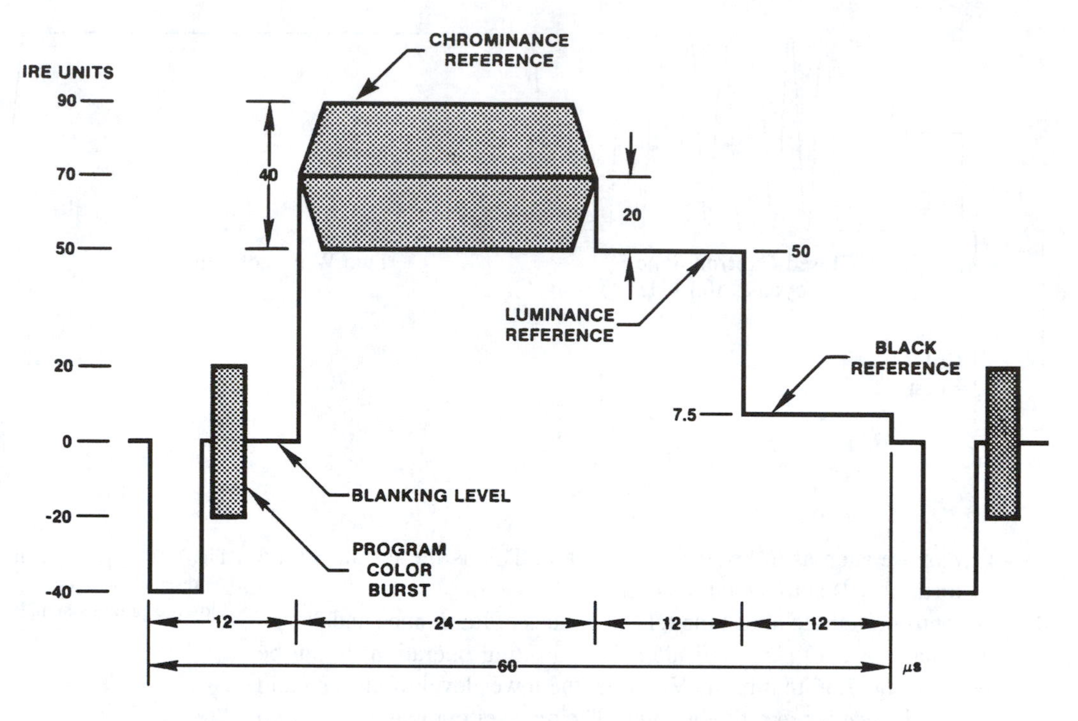

Fig. 7.56: Time-domain representation of a VIR signal.

7.8 Vertical Interval Reference (VIR)

VIR refers to a system in which the TV station transmits information as to the average luminance level, the average chroma level, a black-level reference, and a color-burst reference. On a receiver equipped to utilize this information, it can be used to automatically and continuously adjust the picture to be nearly exactly like the one viewed by the camera.

We have alluded previously to the vertical blanking bar that can be seen when the picture is rolled vertically. This bar occupies the first 21 lines in both fields of each frame, and since the beam is blanked out, information of all sorts can be sent during this time. In 1975, the FCC designated the 19th line of each frame field as being set aside for VIR. In 1994, however, the FCC designated line 19 for the GCR system (Section 7.10) and stated that VIR may now be transmitted on any line between 10 and 16. Many stations have been equipped to transmit VIR, but to receive the benefits of it, the receiver must be equipped to process the VIR signals. The basic form of this signal is shown in Fig. 7.56.

The VIR receiver circuitry consists of four parts:

(1) A circuit to recognize when VIR is being transmitted.
(2) A circuit to count the lines of each frame field.
(3) A control circuit to null out the decoded $R - Y$ during the chroma reference interval. Since the chroma reference is $B - Y$, any output from the $R - Y$ demodulator during this interval is indicative of a phase error in the CW subcarrier being used for demodulation. Any demodulated $R - Y$ becomes an error signal that is used to shift the

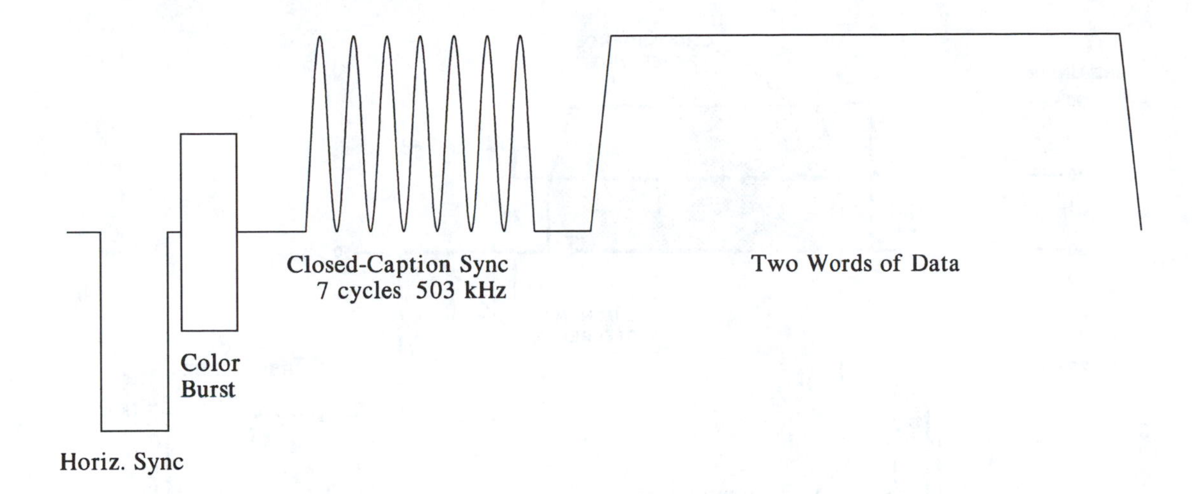

Fig. 7.57: Time-domain representation of a closed-captioning signal.

CW phase such as to bring $R - Y$ to zero. This is the ultimate (and distortionless!) automatic tint control.

(4) A control circuit to adjust the relative values of luminance and $B - Y$ chroma to secure perfect B after the matrixing operation. It can be seen in Fig. 7.56 that in the VIR line, the lower level of chroma and the luminance are exactly the same. The receiver compares these two during the VIR line (the luminance is clamped to the black level), and if they are not equal, the error signal changes the gain of the chroma to make them equal.

There is, of course, automatic bypassing of the receiver VIR system if the reference signals are not being transmitted. Furthermore, user controls of color and tint can still be provided to indulge viewer preferences. These are connected at all times except during the line where VIR is being transmitted. The VIR system thus assures that the center of the range of these preference controls will always be correct. A more thorough exposition of the VIR system may be found in the literature (see, for example, Neal, 1976).

7.9 Closed captioning

This is the most widely used form of what is called videotext or teletext. It places dialog, narration, and nonspoken cues in text form on the screen of sets equipped with a caption decoder.

Line 21 of the video frame has been allocated to teletext services. The format for a line of captioning data is shown in Fig. 7.57. The data bytes contain seven bits of data plus one parity bit. Characters are represented by their ASCII code, but there are also 55 non-ASCII characters that have significance specific to the line 21 system. A data rate of two bytes/frame at 30 frames/second means that only 60 bytes/second can be sent. Since each byte will encode a letter, punctuation mark or control code, the sending rate for text will be less than 60 letters/second, which averages about 12 words/second, or 720 words/minute. This is faster than most people can read, even without taking time to look at the picture! Nevertheless, in electronic terms, it is extremely slow. The other side of the coin is that it should also be very immune to noise. It can, in theory,

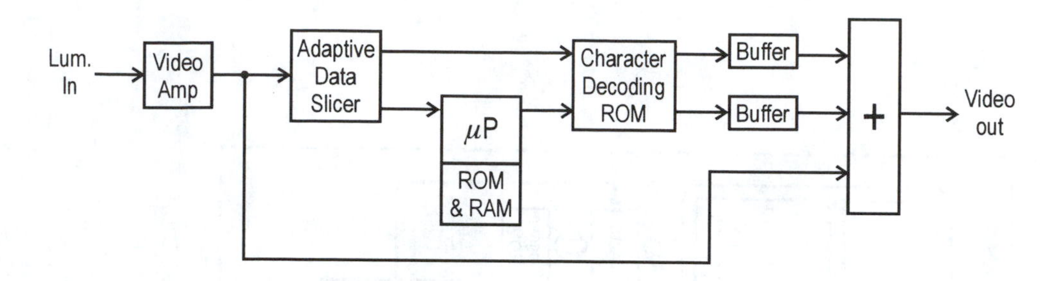

Fig. 7.58: Block diagram of the caption-extraction circuitry developed by the National Captioning Institute.

deliver clean captions even when the noise is so severe that the picture itself is obliterated. In practice, it does not always work out this way.

A block diagram of the decoder is shown in Fig. 7.58. The μP takes care of recognizing valid data, sending it to character decoding ROM if appropriate, interpreting and acting on control codes, and phasing the teletext properly with respect to the luminance. The adaptive slicer adapts to different signal and noise levels to secure as wide a range of operating conditions as possible. It and the μP and character ROM are part of a module developed at the National Captioning Institute, a nonprofit organization with the sole purpose of promoting captioning. More information on the system can be obtained by contacting them at 5203 Leesburg Pike, Falls Church, VA 22041.

When the inclusion of captioning capability was mandated by the FCC for all TV receivers over 13″, a monolithic IC caption decoder became an economic necessity. Such a chip (SAA5252) was introduced by Philips in 1992.

7.10 Ghost Cancellation Reference (GCR) system

The GCR system was a joint development between Philips Consumer Electronics (owners of the Magnavox trade name), Zoran, and Mitsubishi. It involves the transmission of a reference signal during line 19 of each field of a TV scan. In the time domain, this reference looks like a burst whose frequency is increasing. The duration of this "burst" is somewhere between 30 and 35 μsec. It replaces video during line 19, since that line won't show on the screen anyway. In a receiver equipped to handle it, the received signal is processed to extract the GCR, and this is compared to a reference stored in ROM. Based on a comparison between the received GCR signal and the reference, the coefficients of a 576-tap digital filter are calculated to a precision of 11 bits by an ALU under ROM program control. The digital filter has both FIR and IIR sections. The filter thus calculated and implemented is capable of suppressing the ghost relative to the desired signal by as much as 48 dB.

7.10.1 Implementation

The system has been implemented with only three chips – one primarily analog, one digital, and an A/D converter. The ideal arrangement is to have the circuitry built into new TV receivers, but it is also available as a set-top box to retrofit it into existing TV–VCR systems.

Figure 7.59 shows a block diagram of the GCR system as it is apportioned among the chips. Both of the LPF blocks are seventh-order elliptic filters, which introduce a 6 dB loss even in the passband.

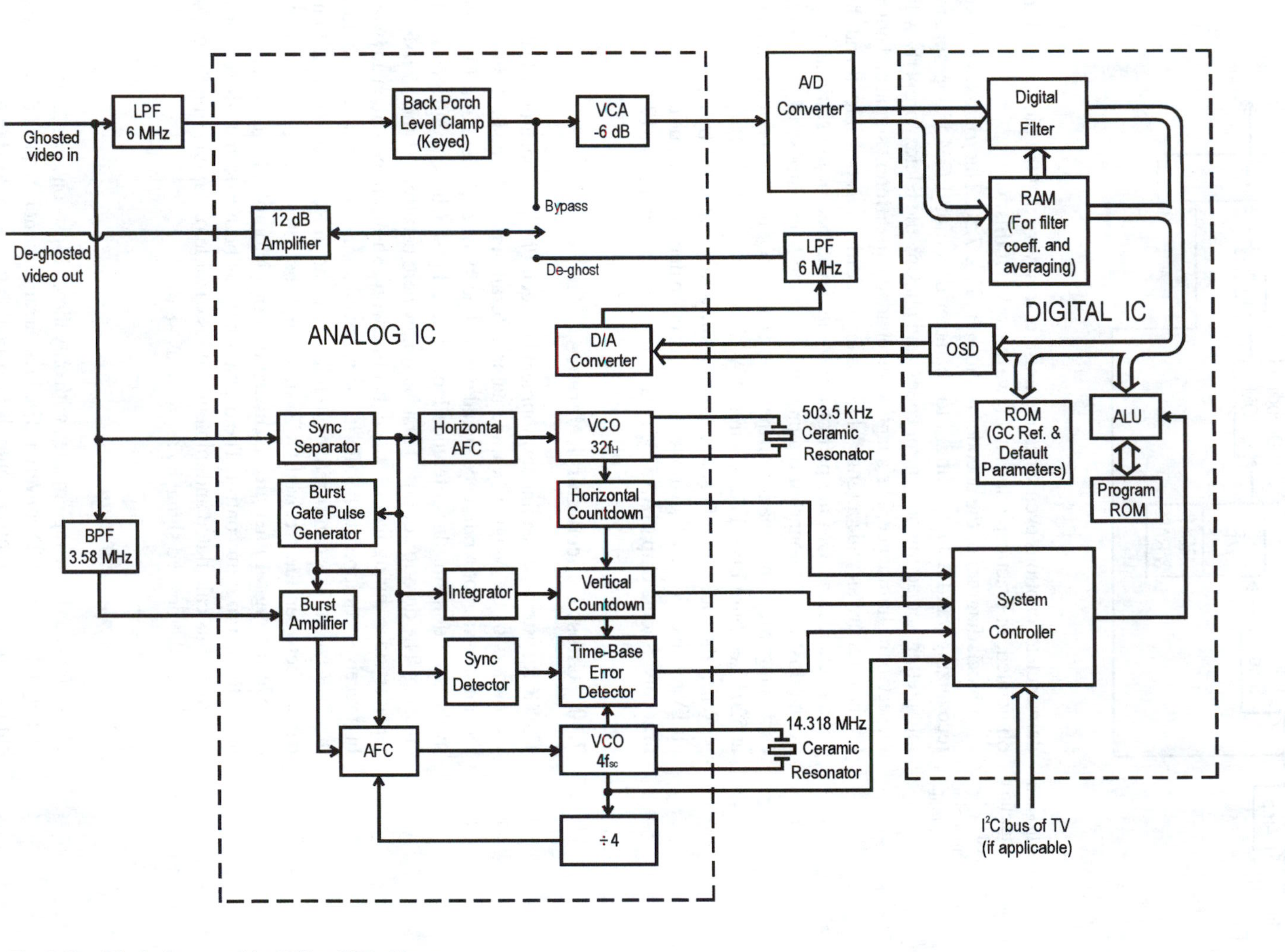

Fig. 7.59: Block diagram of the Phillps GCR chip set.

In the analog IC, the VCA has a gain of less than unity to ensure that the input to the A/D converter never exceeds the reference even in the worst case. The bypass switch can be operated either automatically or manually. It will be in the bypass position if no GCR information is detected in the received signal. There is also a demo mode in which the upper half of the screen is deghosted and the lower half is not. The analog switch toggles between the Bypass and Deghost positions during each vertical scan. In the Bypass position, the 12 dB amplifier restores the 6 dB lost in the input LPF and the 6 dB lost in the VCA. In the Deghost position, the D/A converter has an adjustable reference, which essentially controls its gain and allows the deghosted output signal to be set equal in amplitude to the input.

The lower portion of the analog IC should look very familiar by now. The circuitry is almost identical to that used in the RCA TV except for the time-base error detector. The purpose of this block is to detect changes of channel and when the video is coming from a VCR. Detection of channel change is important, because the algorithm changes filter parameters to their default values and thus facilitates rapid acquisition of the correct filter parameters for the new channel. When the video is determined to be coming from a VCR, a modified GCR reference signal is used for comparison, to account for the reduced bandwidth of a videotaped signal relative to a broadcast signal.

We need say little about the digital IC except to elaborate a bit on the RAM. Figure 7.59 shows that this RAM is used for storing the dynamically changing filter coefficients and for averaging. Since the 19th line will, like all other lines, have horizontal sync and blanking and burst pulses in addition to the desired GCR signals, it is desired to remove everything but the GCR. This is achieved by a combination of phase reversal and averaging. At the transmitter, the polarity of successive GCR "bursts" is phase inverted according to this pattern: $+ - + - / - + - +$. The pattern repeats every eight fields. In the averaging process, this phase inversion is undone, and the horizontal and burst pulses are cancelled, as well as any DC component that may be present. However, since the GCR was recorded with alternating phase, it is not cancelled but doubled. Ordinarily, the GCR lines of sixteen fields are averaged, and this is adequate to give a very clean GCR signal. Under unfavorable conditions, the averaging can be carried out over many more fields to clean up the GCR signal, but always some multiple of eight, since this is the periodicity of the alternating phase pattern.

The ghost-filtered video is combined with On-Screen Display characters to inform the user of the status of the operations. This OSD-bearing video is then converted back to an analog signal, restored to its original amplitude, and fed to the video input of the TV.

We might also mention that the chip set is capable of working on PAL as well as NTSC systems.

The system is so effective that in the three years between its FCC approval in October of 1992 and the end of 1995, it has been installed on practically all TV transmitters in the United States. It is now well on its way to being adopted by a large number of foreign countries as well.

For more details, the interested reader is referred to a journal article by the inventors and developers of the system (Johnson et al., 1994).

7.11 High-resolution TV

Research is going on in several different directions to improve the resolution or definition of TV. These can be separated into two categories called IDTV (Intermediate Definition TV) and HDTV (High-Definition TV). The technologies used are totally different.

7.11.1 *Intermediate-definition TV*

IDTV achieves an improved picture by eliminating interlace and its attendant picture degradation. This is achieved by storage, interpolation, and accelerated scanning. The IDTV set scans an entire frame of 525 lines in $1/60$ second. Where does the information come from for the field that is not present during the scan? There is enough RAM in the set (2.5 MBits) to store an entire field. There is a separate memory that stores just the previous line of the field being scanned. Let us suppose that the even field is the one stored in its entirety. Then as a given pixel in the odd field is being scanned, three values are correlated. They are:

- the signal for the odd-field pixel being scanned,
- the stored signal from the previous line of the odd field that is just above the odd field pixel being scanned, and
- the signal from the stored even-field pixel that lies between the first two just mentioned.

These three values go into a DSP circuit that outputs an interpolated value for the signal that should appear at the time and place where the beam is scanning. If the interpolation is made from the previous line, it significantly improves resolution for moving objects but causes vertical jitter on stationary objects. If the interpolation is made from the previous field, it gives substantial improvement in resolution for stationary objects but causes serrated edges on moving objects. The Thomson CSF chassis CTC-170 provides "The best of both worlds" by using a system called *motion adaptive processing*. In this system, the chroma and the high-frequency part of the luminance always use the data from the stored line. But the low-frequency luminance may use either one or the other stored value depending on whether the picture is primarily moving or stationary! It makes this determination by looking at the luminance spectrum around 2 MHz. This is the portion of the spectrum most indicative of motion.

7.11.2 *High-definition TV*

After more than ten years of "Preliminaries," HDTV remains in limbo. In the early days, there was a great deal of attention paid to making any HDTV system that might be adopted compatible with present NTSC receivers. This is no longer the case. In the interest of getting HDTV moving, all of the entities who proposed HDTV systems of their own have abandoned them in favor of a composite system worked out by what has been called "The Grand Alliance." The parties to this alliance are (in alphabetical order) AT&T, General Instrument, MIT, Philips, Sarnoff Research Center, Thomson CSF, and Zenith. The technical specifications of the system have been published (Hopkins, 1994).

The Grand Alliance system is an all-digital one. The advantages of the digital system in terms of bandwidth requirements and noise immunity (among others) were deemed to be of greater importance than maintaining compatibility with existing receivers. In addition to vastly improved pictures, this system also offers theater-quality sound. The system underwent technical evaluation and field tests by the FCC Advisory Committee, who recommended its adoption to the full FCC in November of 1995. As of this writing ten months later, the FCC still has not ruled on the recommendation. The officially cited reasons are (Eckert, 1996):

(1) There is concern that a government standard will "inhibit technological advances." This rings a little hollow when we have the examples of how the lack of such a standard has almost killed AM Stereo and how the imposition of such a standard led to an explosive growth of FM stereo.
(2) There has also been concern expressed that the Grand Alliance system is not sufficiently compatible with computers. This objection is amazing, as at least two of the seven parties to the Grand Alliance are computer manufacturers and/or marketers, and compatibility with computers was one of the criteria specifically addressed in the system design.
(3) Many owners of existing stations are understandably not excited about having all of their equipment obsoleted and/or perhaps losing their spectral allocation to the new system.

For these reasons, HDTV continues to languish in America after more than ten years of FCC inaction. There is some positive news, however, in that Lucent Technologies and Mitsubishi Electric Corp. have agreed to jointly develop a chipset for the reception of the proposed Grand Alliance system. They expect to have samples available by the first quarter of 1998.

The proposal for the Grand Alliance system involves having the FCC allocate new spectral space for digital TV transmission until HDTV becomes dominant, and then reclaiming the spectrum now allocated to analog transmission and making it available for further digital broadcasts. This would ultimately force owners of analog receivers to either scrap them and buy a digital receiver or lose TV access.

Japan is already transmitting in their own HDTV system called Hi-Vision, and all of the major Japanese TV makers are marketing sets that can receive it.

7.12 Digital TV sets for NTSC signals

By "Digital TV," we refer here to one in which most or all of the signal processing is done digitally, as opposed to one that is designed to receive digital TV broadcasts.

As early as 1981, ITT Semiconductors of West Germany had produced lab prototypes of a chip set for digital TV and demonstrated a TV based on them. Toshiba has also played a leading role in the development of this technology. Considering all the advantages of digital TV, it seems somewhat remarkable that there has not been a much more complete swing over to digital TV. The industry has, however, taken a more cautious approach to the "Digitalization" of analog TV. We shall look at three topics in the pages that follow: What is the state-of-the-art capability in DSP as applied to TV? What is the level of DSP currently found in TV? And what are the advantages of digital TV?

7.12.1 Current DSP capability

Conceptually, at least, the TV signal picked up at the antenna could be A/D converted, and DSP started there. Practically, of course, this is absurd both because of the very high frequencies encountered there (up to 806 MHz for UHF) and the extremely low signal amplitudes (10–100 μV). This pretty well dictates that tuners continue to process the RF signals in an analog manner. However, we cannot overlook the fact that all "High-end" TV sets as well as most of the mid-range ones now incorporate digital tuning technology or quartz-synthesized tuning. This was, in fact, one of the first inroads of digital technology into TV receivers.

Since we must abandon the RF signal to its analog origins, we might next ask about the feasibility of digitizing the IF signal out of the tuner. Because this signal lies roughly between 40–45 MHz and has been substantially amplified in the tuner, it is technologically feasible to digitize it. It is not economically feasible at the present time, however, since an A/D converter capable of doing the job is still quite expensive.

We thus conclude, as have the chip manufacturers, that for now DSP must begin at the output of the video detector.

7.12.1.1 A/D converter

The A/D converter is the first block to which the video signal goes. The highest-frequency components here in an NTSC system are below 5 MHz. By the Nyquist criterion, this would require sampling at 10 MHz or higher. A sampling frequency of $4 \times f_{sc}$ has been chosen, where f_{sc} is the frequency of the color subcarrier:

$$4 \times 3.58 = 14.32 \text{ MHz}.$$

Not only does this value give a nice margin above the Nyquist requirement, but it also aids in color processing, as we shall see. This sampling frequency is also used as the clock for all DSP operations.

It has been established that the digital encoding of the video signal requires eight bits for a quality picture. ITT utilizes a flash converter (so named for its speed). The n-bit flash converter needs 2^n comparators. To minimize cost, ITT developed a clever technique of using a seven-bit flash converter (which cuts the comparator count in half) and then "dithering" the A/D converter's reference voltage up and down by $1/2$ LSB during alternate horizontal line scans. The lines are close enough that the eye averages the two video levels so written to the screen and gives a picture that is subjectively equivalent to that given by an eight-bit converter without reference dither. The latter system is used by Toshiba, but without recourse to flash conversion. Presumably, the state of the A/D art improved after ITT's pioneering design to the point where non-flash converters could do the job.

7.12.1.2 Luminance

The output of the A/D is passed through comb filters to separate the Y and chroma signals. The comb filter is discussed in Section 8.4.1.2. The interleaved nature of these signals in the frequency domain, discussed earlier in this chapter, makes them ideal candidates for separation by the comb filter. The A/D output goes to two other circuits as well:

(1) A black level clamp. This clamps the black level of the digitized signal to a fixed reference. The purpose of this is to center the video signal in the dynamic range of the A/D input to give as broad an encoding range as possible.
(2) The sync separator. This circuit performs the same function as its analog counterpart. We shall trace its outputs when we get to the deflection system.

The separated luminance signal will generally go through further circuit blocks that will typically include various sharpness controls and automatic gain (contrast) controls. Toshiba utilizes what they call *contour correction.* A sharp vertical line on a picture will cause an abrupt rise in the video signal. If it is desired to increase the line sharpness, it is necessary to raise the gain of the high-frequency part of the video signal. This, however, will result in overshoot and undershoot in general. The contour correction cleans up the transitions, while preserving the improved rise time of the video signal. Although this effect exists on sharp horizontal lines also, it is not as noticeable. Nevertheless, Toshiba provides both horizontal and vertical contour correction.

7.12.1.3 Chroma

The chroma circuitry is more complex than that of the luminance. The chroma is first passed through an ACC stage, which holds the burst amplitude at a reference value and thus holds the color signal available to subsequent circuitry at a fixed level. If the ACC circuitry cannot bring the burst amplitude up to the reference level, the color killer is activated.

The ITT chip performs synchronous chroma demodulation digitally. Synchronous demodulation requires a color subcarrier in the receiver in phase with that sent from the transmitter. The phase of the subcarrier in the transmitter is sent as the phase of the burst. Therefore, the receiver contains a digital PLL that phase locks the system clock to the burst, even though the system clock is four times higher in frequency. With lock achieved, demodulation can take place, and it produces $R - Y$ and $B - Y$ signals as in analog processing. The digital versions of Y, $R - Y$, and $B - Y$ are D/A converted and then analog matrixed to extract the R, G, and B signals.

The Toshiba approach to digital color demodulation is an elegant matrix method (Suzuki, Kudo, Nakagawa, Yoshimoto, and Namiaka, 1984) incorporating the following relevant quantities in the overall block transfer function:

- the CRT phosphor colors and sensitivity,
- contrast control setting,
- color intensity control setting,

- automatic color control circuit output, and
- tint control setting.

This matrix process outputs digital R, G, and B signals that do not have to go through a separate matrix operation. They are individually D/A converted and fed to the CRT.

The ITT chip-set design also features automatic adjustment of maximum beam current and minimum beam current for each of the three guns. The maximum current must be controlled to prevent damage to the phosphors and the minimum beam currents must be balanced and at their threshold in the absence of a video signal. Then if there is no color signal, the screen will show an untinted black and white picture.

7.12.1.4 Sync and deflection

The sync and deflection circuitry is similar in the ITT and Toshiba chip sets. The digitized video is also fed to the sync section. Since it has been black-level clamped, the excursion of the sync pulses is easily determined. They will extend from the black level to the positive video peaks. The separation level for sync is set between these values. In both cases, the timing constants (width and period) of the horizontal sync pulses are averaged over several cycles, and the sync pulse used to trigger the scan generators is this averaged pulse. This serves to make the set much more immune to interference and noise pulses. In the Toshiba sets, the same thing is done with the vertical-sync pulses. In the ITT set, the "vertical oscillator" is a counter whose input is the horizontal sweep frequency. The counter is reset by the vertical-sync pulse, gated through a variable "time window" in a manner very similar to that used in the integrated monochrome set of Section 6.4.

The major networks transmit a signal in which both the horizontal and vertical sweep frequencies are submultiples of the color burst signal. This affords the ultimate in picture stability, if the set's sync section is equipped to handle it. The ITT chip set is so equipped. It can automatically determine when all three signals are and are not phase locked and switch its processing system to the optimum in either case.

Both chip sets also feature digitally generated pincushion correction signals which are summed into the scan signals to account for the nonsphericity of the CRT face. This method eliminates the pincushion transformers and wave-shaping components required in analog receivers. Both sets are also designed with pulse-width modulators as the link between the deflection generators and the yoke windings. These provide D/A conversion, high power efficiency, and low cost.

7.12.1.5 Audio

The output of the video detector also contains the audio carrier at 4.5 MHz. This output is filtered to extract the sound carrier and is FM demodulated to recover the sound signal (or signals in the case of stereo transmission). These signal(s) plus any pilot signal present are digitized in three different channels by a multiplexed A/D converter and undergo extensive DSP. The digital operations performed include de-emphasis, tone and loudness controls, volume and balance controls, as well (in the ITT set) as a pseudo-stereo simulation capability for programs without a stereo channel. The L and R signals are also output through pulse-width modulators, which are designed to interface with class D audio power amplifiers.

7.12.1.6 Microprocessor

As we shall see, the microprocessor (μP) is the key component that makes a set with DSP both more flexible and more integrated in function. The μP controls the conversion processes, as well as performing the calculations for many of the DSP operations. In Section 7.12.3, we will see another extremely important function of the μP.

7.12.2 Current DSP usage in TV

Given the high level of performance that DSP receivers have demonstrated, it is somewhat surprising that some comprehensive DSP chip set has not been incorporated into current receiver designs. To what can we attribute this conservatism? For one thing, the cost of a set using the digital chips has always been higher. It is unquestionably true that this is not intrinsic to the technology. If produced in numbers comparable to analog sets, the economies of scale in the chip production as well as the large number of "outboard" components that would be eliminated by the conversion from analog would make the digital set less expensive.

In any case, however, receiver manufacturers have chosen the gradual approach to digital. The following digital techniques are some of those that have been incorporated into otherwise analog receivers since the advent of the first digital TV chip sets:

- comb filters,
- horizontal and vertical countdown circuits,
- double scan capability (as discussed in Section 7.11),
- picture-in-picture,
- variable "time window" syncing for maximum noise immunity,
- class D audio amplifiers, and
- quartz-synthesized tuning.

There is clearly much more that can be done to bring DSP into the average TV.

7.12.3 Advantages of digital TV

Besides the higher level of performance and the eventual lower price tag, DSP has other capabilities and advantages that stretch the imagination. These include:

(a) Greater reliability. The fewer components there are in a set, the fewer interconnections there are. In a well-built printed circuit set, this translates to fewer malfunctions.
(b) Automated alignment. This is perhaps the main payoff of digital TV. Receiver fabrication is now highly automated, requiring a minimum of

human intervention. This is important in holding down cost, because the reduction of labor costs and the economies of integration have been responsibe for giving us a much better TV than was available twenty years ago at a markedly reduced cost in constant dollars. The digital chip sets need only have the alignment parameters made available in a ROM, and the μP will perform the alignment of all the parts of the set that are digital. Furthermore, as components age and drift, the μP continually realigns the receiver to compensate for it.

(c) Compatibility with different systems. The NTSC system used in the United States is one of three systems used throughout the world. The other two are PAL and SECAM. PAL is most like NTSC with different parameters. SECAM encodes some information differently altogether. It is entirely possible to make a DSP TV that not only recognizes which signal format is being transmitted but switches its own circuitry to decode it properly.

(d) Noise reduction processing can be incorporated. Some of this is discussed in Chapter 8.

(e) Double-scan capability, as in the Toshiba chip set. This allows higher resolution as described in Section 7.11, but of even greater importance, it reduces flicker. This is particularly important in Europe where the vertical sweep frequency is 50 Hz.

(f) Freeze-frame capability can be incorporated.

(g) Zoom capability can also be incorporated.

The future is indeed bright for digital TV. The main question now is "When will its future arrive?"

References

Eckert, R., 1996. Private communication.

Gerard, H. 1978. Principles of Surface Wave Filter Design. In *Acoustic Surface Waves*, ed. A.A. Oliner, pp. 61–96. Berlin: Springer-Verlag.

Hopkins, R. 1994. Digital Terrestrial HDTV for North America: The Grand Alliance HDTV System. *IEEE Transactions on Consumer Electronics*, CE-40 (3): 185–198.

Johnson, L., McNay, S., Hill, R., Greene, D., Manor, Y., Horowitz, S., Casey, R., Murakami, K., Kaku, K., Koo, D. 1994. Low Cost Stand-Alone Ghost Cancellation System. *IEEE Transactions on Consumer Electronics*, CE-40 (3): 632–639.

Liff, A. A. 1979. *Color and Black & White Television Theory and Servicing*, pp. 78–81. Englewood Cliffs, NJ: Prentice-Hall.

Neal, C.B. 1976. Improving Television Color Uniformity Through Use of the VIR Signal. *IEEE Transactions on Consumer Electronics*, CE-22 (3): 230–237.

Reference Data for Radio Engineers, 1968. 5th ed., p. 28–16. Howard W. Sams and Co.

Rzeszewski, T. 1975. A Novel Automatic Hue Control System. *IEEE Transactions on Consumer Electronics*, CE-21 (2): 155–163.

Suzuki, S., Kudo, Y., Nakagawa, N., Yoshimoto, A., Namiaka, T. 1984. High Picture Quality Digital TV for NTSC and PAL Systems. *IEEE Transactions on Consumer Electronics*, CE-30 (3): 213–219.

8

Video cassette recorders

Although recorders capable of recording a TV signal on magnetic tape have been available since the early 1950s and videocassette recorders have been since the early 1970s, they never became a consumer item until 1975, when Sony introduced the Betamax VCR. Earlier units were generally too complex, too expensive, and too voracious in their appetite for tape. The first Betamax units could record for one hour on a cassette, making them almost as easy to use as an audio tape recorder. Two years later, Matsushita introduced the VHS (Video Home System) VCR, which boasted recording times of two or four hours on a somewhat larger cassette. The Betamax system was then modified to allow two hours of recording. Subsequently VHS achieved a six-hour record time and Beta got up to three hours. For both machines, the longer record time yielded slightly inferior picture quality relative to the shorter record times. Although both systems used the same general principles in their electronics, they were not compatible either mechanically or electrically. If both systems were played at their fastest speed (one-hour play time for Beta; two hours for VHS), the Beta probably produced a slightly better picture quality, but if both were operated in the two-hour mode, the VHS picture was generally better. The fundamental reason for this is that the Beta cassette is smaller and simply cannot hold as much tape as the VHS cassette does. Thus the Beta achieved the two- and three-hour modes by slowing down the tape speed. This hurt high-frequency response and so picture details were lost.

Although there were a number of more esoteric differences between the Beta and VHS systems such as tape tension (Beta lower in record and playback; VHS lower in fast forward and reverse), it can hardly be maintained that the chief issue in the minds of most consumers was anything but the longer record time of the VHS. This led to the situation in 1985 that all manufacturers of the Beta system except Sony had either switched to the VHS system or were "hedging their bets" by manufacturing both VHS and Beta systems. The dominance of VHS over Beta is now total. Virtually no tapes are produced in Beta format any longer. For these reasons, we will direct our attention exclusively to the VHS system.

It was stated in the introduction to this book that no attempt would be made to cover the mechanical portions of consumer electronic equipment except to the degree necessary to understand the electronic part. In the VCR we have not only some purely mechanical systems, but also some servo systems, which can scarcely be separated into electronic and mechanical parts. The same thing will be seen in Chapter 9, where we will discuss digital discs. We will therefore cover the servo systems as a unit.

8.1 Comparisons to audio tape recording

Since we analyzed an audio tape recorder in some detail in Chapter 5, it would obviously aid our understanding of the VCR if the material there were all directly applicable in the present context. This is the case as regards the audio portion of the program material, which is recorded on its own track with a bias oscillator frequency of 65 kHz. With respect to the video and chroma signals, however, only some fundamental tape principles are applicable; the circuitry is totally different. Let us see why this is necessary.

We noted in Chapter 4 that the frequency at which the response of a tape head begins to fall off is

$$f = v/2x_{\mathrm{g}}. \tag{8.1}$$

We recognize that the upper frequency limit of an audio tape recorder (~20 kHz) is wholly inadequate to record a video signal, which has frequency components up to 4 MHz. If we were to take the simple-minded approach that we need only increase the tape speed by a factor of 200 to raise the frequency response by the same factor, we would require a mile of tape for every three minutes of recording. This obviously won't do! Alternately, we might consider narrowing the head gap to raise the frequency response. There are obviously limits to this approach due to both the precision of the manufacturing process and considerations of widening of the gap by wear. Good audio heads have gaps on the order of a micrometer. The VHS head gap is typically .3 μm. This obviously will not be adequate to raise the frequency response to the desired level. The current solution to this problem is known as helical scanning. It was first demonstrated by Ampex in 1956 for broadcast video tape recorders and by Toshiba in 1959 for home units. The technique is based on the fact that the velocity (v) in Eq. (8.1) is the relative velocity between head and tape. If it is desired to move the tape more slowly in order to reduce tape consumption, high-frequency response can still be achieved if the heads are moved rapidly. The exact method used to accomplish this will be discussed later. For now it will suffice to observe that whereas the linear tape speed is only about 3 cm/sec, the relative speed between the heads and the tape is 5,800 cm/sec, giving a frequency response more than adequate for video recording.

8.2 Encoding

8.2.1 Luminance

Another problem arises in the VCR that we did not have to consider for audio. In Chapter 5, we observed that the total dynamic range of magnetic tape was of the order to 50–60 dB between saturation and noise. On the low-frequency end of the playback head's frequency response, there is a linear rise of 6 dB/octave. It will rise 50–60 dB in 8–10 octaves. It is thus practically impossible to have a functioning tape system that can record over a frequency range of more than 10 octaves. In audio, there are 10 octaves between 20 Hz and 20 kHz, so good design can approach this ideal. To record a luminance signal ranging from 10 Hz to 4 MHz, however, requires 19 octaves of bandwidth on the part of the tape system.

This problem is solved by NBFM modulation of the luminance signal onto a 3.9 MHz carrier. The frequency deviation of the modulated signal is ±.5 MHz about the 3.9 MHz center frequency. An additional advantage of this method is that even though the strong dependence of the playback signal with frequency will introduce amplitude variation in playback, this variation is easily removed by limiting prior to demodulation.

Exercise 8.1 The phasor diagram of a normal NBFM signal is as shown (Haykin, 1983):

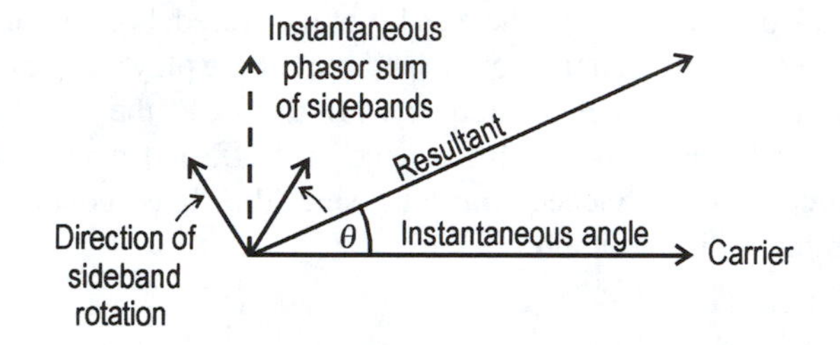

(a) Under what conditions on the sidebands will θ be a maximum?
(b) Under what conditions will θ be a negative maximum?
(c) Under what conditions will the amplitude of the resultant be essentially the same as that of the carrier?
(d) If these amplitudes are not the same, how will it show up on a scope?
(e) Suppose this FM wave were to pass through a network that would increase the magnitude of one sideband by an additive factor of δV and decrease the other by the same amount. Compare the maximum positive and negative angles to those obtained when the sidebands were equal. How would the scope display differ from that of part (c)?
(f) Repeat part (e), but assume that the network has a response over the passband of the NBFM signal that rises at a slope of 20 dB/decade. The carrier is at 3.9 MHz, and the maximum deviation of the sidebands is ±.5 MHz

There is, as we see next, an additional advantage to FM modulating the luminance.

8.2.2 *Chroma*

Since the luminance carrier is at 3.9 MHz and has sidebands extending out ±.5 MHz, it probably would not do to leave the 3.58 MHz chroma signal unaltered, since it would most likely not be separable in frequency from the modulated luminance. In the VHS system, the chroma is mixed with a 4.209 MHz carrier to heterodyne it down to a center frequency of 629.37 kHz, which is exactly 40 times the horizontal sweep frequency for color TV. The resulting chroma is still a QAM signal. The relatively large frequency difference between the luminance and chroma after encoding certainly facilitates their separation on playback, but there is another, equally significant benefit. Because the encoded chroma still carries amplitude modulation, it must be recorded in a way that preserves the amplitude information. We saw in Chapter 5 that in terms of current practice, this calls for the use of AC bias at a frequency well above that of the encoded chroma. But the encoded luminance is just such a signal! In other words, the simple act of summing the encoded luminance and chroma signals produces a composite that is suitable for direct application to the recording head. The FM-encoded luminance signal thus doubles as bias for the chroma.

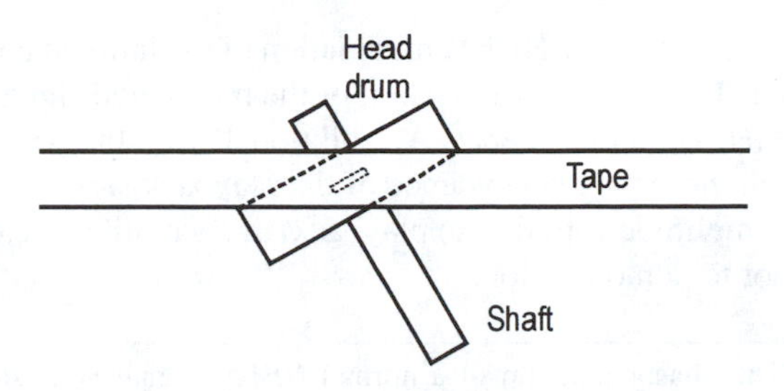

Fig. 8.1: A diagram showing the relative positions of the videotape and the head drum in a VHS VCR. The rotation of the head drum creates slanted tracks on the tape.

In playback of an audio tape, the bias is not recovered, both because there is no use for it and because the frequency spectrum of the playback response rolls off well below the bias oscillator frequency. In the VCR, the "bias" frequency range is primarily below the high frequency rolloff, and it not only *can* be separately recovered, but indeed it *must* be, since the "bias" contains all of the encoded luminance information.

8.2.3 Sync

An NTSC signal contains luminance, chroma, and sync all together. As this information is all to be stored on magnetic tape, and chroma and luminance have been recorded separately, it will be necessary for the synchronizing information to be recorded also. Vertical sync information is recorded with a separate head and on a separate tape track from the picture information. It is not called the sync track, because the sync information is recorded in a modified form. The vertical sync pulses are frequency-divided by two to produce a 30 Hz control signal that is, of course, phase locked to the vertical sync. Horizontal sync is recorded with the luminance. The track containing the 30 Hz pulses is called the *control track*. The reasons for using this approach will be discussed later.

Exercise 8.2 Would you expect to find AC bias used to record the control track? Explain.

8.2.4 Audio

The audio information is also recorded by a separate head on a separate track. The audio and control heads are, in fact, stacked and record their tracks after the picture information has been recorded. The audio track is, as has been noted, wholly conventional in its recording.

8.2.5 The helical scan

As has been mentioned, the key to recording with a reasonable amount of tape is the helical scan. In its simplest form, this is achieved by placing two heads 180° apart on a rotating cylinder. The tape is led past this rotating cylinder at a substantial angle to the axis of rotation, as shown in Fig. 8.1. The dashed lines on the tape show the head gap under the tape and the one band of recorded tape produced by this one pass of the tape head. We see that the recorded band or track will be diagonal relative to the edges of the tape. Since the other head is on the other side of the cylinder, the rotation of the cylinder will bring that head to

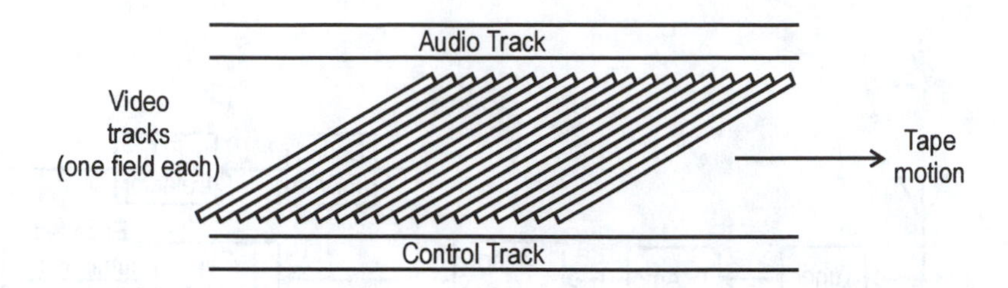

Fig. 8.2: A segment of VHS videotape showing the locations on the tape where various signals are recorded.

the same position shown in a short time and will produce another diagonal track parallel to the one shown. The tape transport rate should be such as to allow this new track to lie right beside the previous one. The process continually repeats. It may be objected that some of the information from one track is bound to get picked up when an adjacent track is "read" on playback. This is absolutely correct. The measures for dealing with this and many other nonidealities will be discussed after we have covered all of the basics of operation.

The cylinder rotates at 1,800 RPM = 30 RPS. Thus every $1/30$ of a second, each of the two heads will make one pass over the tape, and each head will contact the tape for $1/60$ sec. If these figures sound familiar, they should. NTSC standards call out $1/30$ second as the time to scan one frame and $1/60$ sec as the time to scan each of the two interlaced fields that make up a frame. We see, therefore, that each recorded track on the tape contains one field.

We are now in a somewhat better position to appreciate why the control track contains 30 Hz pulses. Since the rotation of the head cylinder should be synchronized to the vertical sweep rate, and since it nominally rotates at 30 RPS, it is a relatively simple matter to put a magnet on the cylinder and a pickup just below it that will generate one pulse with each rotation. This pulse should ideally be locked to the frequency-halved vertical sync pulse in both frequency and phase. Someone might ask if it wouldn't be possible to put two magnets on the head cylinder to raise the frequency from the magnetic pickup to 60 Hz so the sync frequency need not be divided down. This would surely be conceptually possible, but the practical difficulty is that if pulses generated by the two magnets were not exactly evenly spaced in time, there would be intolerable jitter in the picture.

Of course, a servo is necessary to achieve frequency and phase lock. This will be examined later.

Because the heads rotate, feeding signal to them or picking signal from them presents an interesting challenge. An arrangement of commutator rings and brushes might conceivably work, but it is both too noisy in the electrical sense and too mechanically unreliable to make it of much value. What is actually done is to couple the heads electrically to magnetic coils, in the rotating cylinder. These pass by stationary coils, which induce a signal in the rotating coils (on record) or which receive a signal from the rotating coils (on playback). This arrangement is called a *rotating transformer*, and it works quite satisfactorily.

We conclude our present discussion of helical scan with Fig. 8.2, which shows a segment of tape and the positions in which the various signals are recorded.

Exercise 8.3 Using only numerical information that has already been given in this chapter, and assuming that the video tracks occupy 10 mm of the 12.65 mm tape width, find the approximate angle that the video tracks make with the edge of the tape.

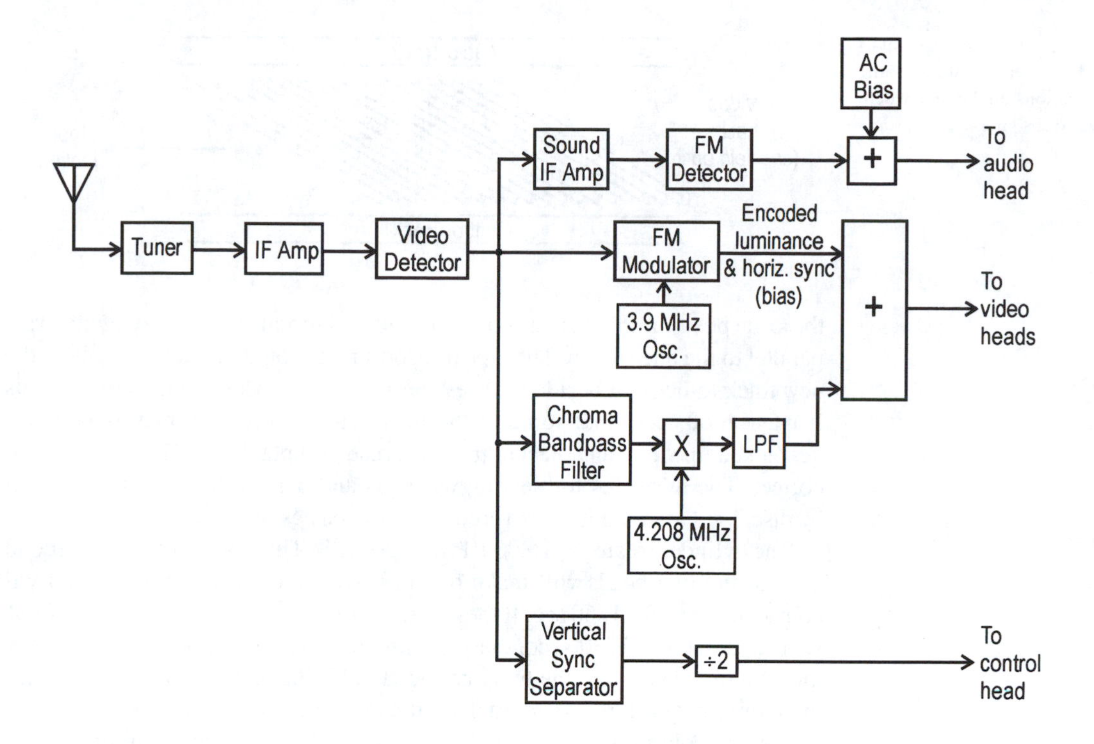

Fig. 8.3: Block diagram of the record circuitry of the conceptual VCR.

8.3 The conceptual VCR

It is somewhat overwhelming to see even the block diagram of a "real-life" VCR and try to understand the functions of the blocks. To break the reader in gently, we will start off with what we might call a conceptual VCR diagram. It might be thought of as the way a bright technician with no experience in actual engineering design might have conceived of the system if he were to try to propose it in block diagram form. It must be stated at the outset that as drawn, this system *will not work*. It is devoid of both electromechanical and purely electrical control systems as well as all of the circuitry that overcomes the many nonidealities. As we understand this conceptual VCR, however, we will be in a much better position to see the practical problems that it entails and to appreciate the measures that are incorporated in an actual recorder to solve those problems. We will see that the problems are so many and so serious that, before the advent of large-scale ICs, the consumer VCR would have been an impossibility, both from the standpoints of initial cost and reliability.

8.3.1 The record mode

All of the foregoing should fairly well commend itself to the reader. Figure 8.3 summarizes the circuit functions we have deduced. The circuitry up through the video detector and from there up through the sound FM detector is exactly like that of a TV.

The chroma bandpass filter is also familiar from color TV circuitry. It feeds circuitry that encodes the chroma information as described in Section 8.2.2.

What we have called the vertical sync separator is actually a standard TV sync separator followed by an LPF.

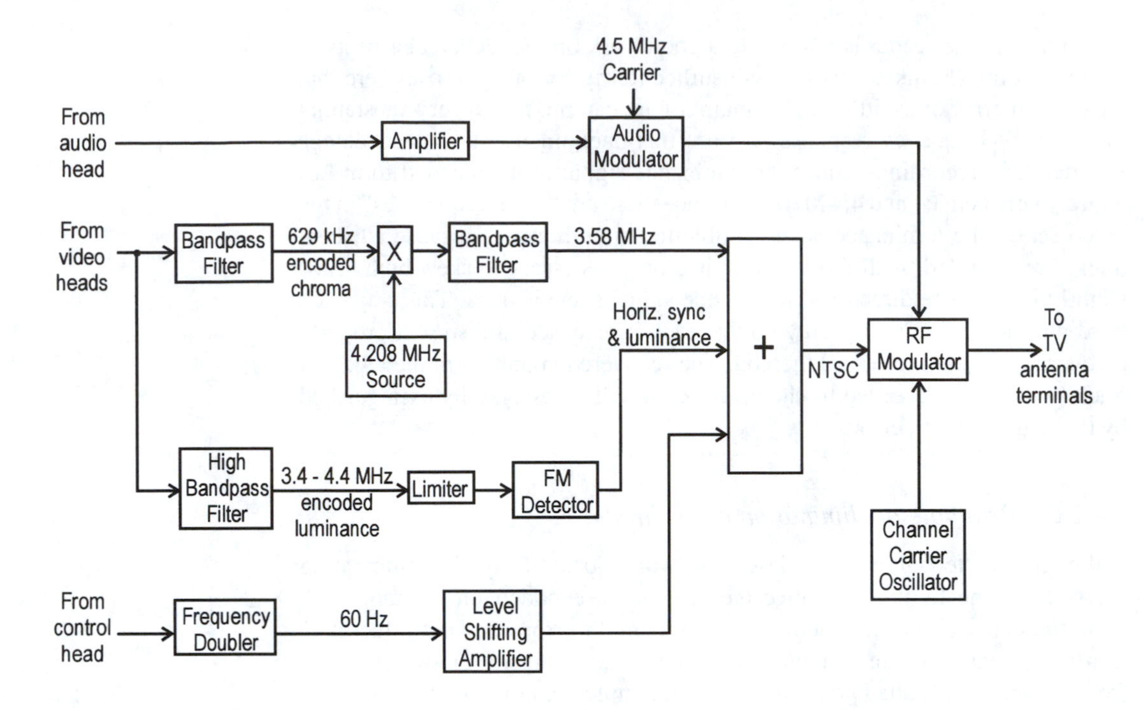

Fig. 8.4: Block diagram of the playback circuitry of the conceptual VCR.

8.3.2 The playback mode

Just as the VCR must convert an NTSC signal to a different format before recording, on playback, the recovered signal must be converted back to NTSC before being output to the RF modulator for application to the TV antenna terminals. As shown in Fig. 8.4, the signal from the video heads (after amplification) is sent to two bandpass filters to separate out the encoded luminance and chroma information. The chroma is heterodyned back up to 3.58 MHz. The bandpass filter after the multiplier selects the lower sideband, which centers at 3.58 MHz. Since the luminance was FM modulated, it is limited to remove any AM information and then detected to reproduce the horizontal sync and luminance. The vertical sync is reconstructed from the control track. This frequency is obtained by frequency doubling the 30 Hz control signals.

8.4 Nonidealities and their solutions

8.4.1 Crosstalk

Since we have already mentioned the problem of reading part of the wrong picture track (field information) on playback, let us address this problem first. Two techniques are used. One eliminates crosstalk in the luminance, and one eliminates it from the chroma.

8.4.1.1 Principle of elimination from luminance

There exists a nuisance problem in audio tape recording called *azimuth error*. Its effect is to dramatically reduce high-frequency response if the head gap is not exactly aligned with the recorded magnetization. For an explanation of this phenomenon, the reader is referred to a good book on the technical aspects of tape recording (Lemke, 1988). It will suffice to say for our purposes here that an azimuth error of as little as 10 minutes of arc can produce a very substantial loss of high-frequency response. We use this phenomenon to our advantage in video tape recordings. Since the luminance signal in its encoded form lies entirely between 3.4 and 4.4 MHz, and since these are "high frequencies" in the video sense, the luminance is susceptible to azimuth error. Because adjacent tracks are recorded by different heads, it is only necessary to skew both heads slightly in opposite directions. In practice a $\pm 6^\circ$ skew is used. This way each head will be perfectly aligned with the track it recorded and so will properly recover the signal impressed thereon. The recovered signal from the adjacent track (which was recorded by the other head) will be essentially extinguished by the large azimuth error.

8.4.1.2 Principle of elimination from chroma

Although the technique just described works beautifully for eliminating crosstalk in the luminance, since the chroma as encoded on the tape is at "low frequencies" in the video sense, the azimuth error does not affect it to nearly as great a degree. Another technique is necessary to remove crosstalk from chroma. It is called *phase inversion*. Its implementation requires a cyclical switching of the phase of the 629 kHz chroma signal. This is accomplished in the heterodyning process during the Record operation. We shall show the principle first and then the circuitry used to implement it.

The encoded chroma signal is, as has been observed, centered around a frequency of 629.37 kHz, which is exactly 40 times the horizontal sweep frequency. Thus during the recording of each horizontal line, there will be 40 cycles of the translated chroma carrier. At the end of each scan line, the phase of the 629 kHz carrier is rotated 90° relative to what it was for the previous line. This continues over the entire track, so that every 4th line recorded on that track is recorded with the same phase of carrier.

When the other head records the adjacent track, the same thing is done, except that the phase of the carrier is rotated in the opposite direction. On playback, the opposite phase shift to that used in recording is used in the decoding process. It should seem reasonable that this would result in normal recovery of the signal on the track we desire to read. But to see what happens to the undesired information on the adjacent track, let us look at Fig. 8.5, remembering that the arrows indicate the phase of the carrier used to record a given section of tape and not to the magnetic orientation of the tape itself.

It is seen that the undesired signal from the B track picked up by head A on playback is essentially canceled out. This might raise a couple of questions in the mind of the thoughtful reader. Let us try to anticipate them. In the first place we see that crosstalk cancellation is not obtained in line 1. This does not mean that we must "live with it" there, however. The tape is looped around the head drum so far that each head is in contact with the tape for a little more than 180° of rotation, which means that each track holds a little more than a field with overlap at the ends of the tracks. Thus the first line(s) recorded on a given track are actually an unused duplicate of the last lines of the previous field and will never appear on the screen. The second question that might be raised is even more fundamental: Why is it necessary to eliminate crosstalk at

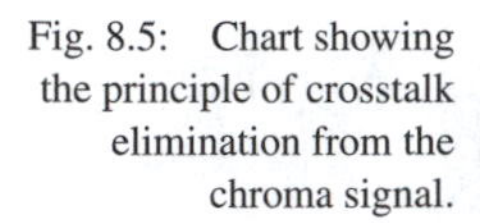

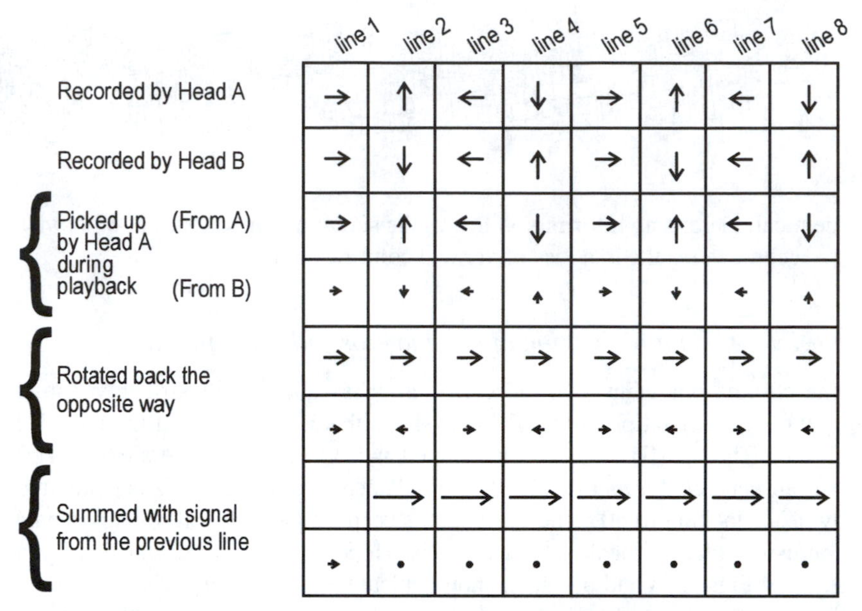

Fig. 8.5: Chart showing the principle of crosstalk elimination from the chroma signal.

all? In our effort to eliminate it from the chroma signal, we have actually mixed the signal from two adjacent lines in a field scan. Couldn't we live with the crosstalk from the adjacent track, since it will correspond to the interlaced line that is even closer to the one we're trying to read? This might well be true if the tracks were perfectly vertical on the tape, but as a consequence of the helical scanning, the information recorded on a given track for a given line is offset from the information on the adjacent track for the same point in the picture. This can be clarified by examining Fig. 8.6.

Suppose that a head picks up crosstalk from the even field track while it is supposed to be reading the odd field track. Then while it is reading line 3 of the odd field, it is getting some of line 10 of the even field. Line 3 of the odd field is the 5th line of the composite picture, and line 10 of the even field is the 20th line of the composite picture, which is a far worse situation than our "cure" produced.

Exercise 8.4 Using information already given, find the linear track length required to record one horizontal scan line.

It should be noted that Fig. 8.6 does not depict the effect of azimuth skew. If this were included, the line delineators would not be collinear as shown, but would alternate slope on either side of perpendiculars to the track edges.

From the standpoint of the circuitry needed to implement the principle illustrated in Fig. 8.5 the "signal from the previous line" is the output of a delay line of exactly 1 H (one horizontal line sweep time). The entire circuit containing both the delay line and a summer is called a *comb filter*.

To summarize what we have shown about crosstalk in the chroma, the main reason for its intolerability is that it represents information that is too far removed from the point being scanned and thus too uncorrelated. The very technique used to eliminate crosstalk works because there is usually a high degree of correlation between contents of adjacent scan lines. To the extent that they are

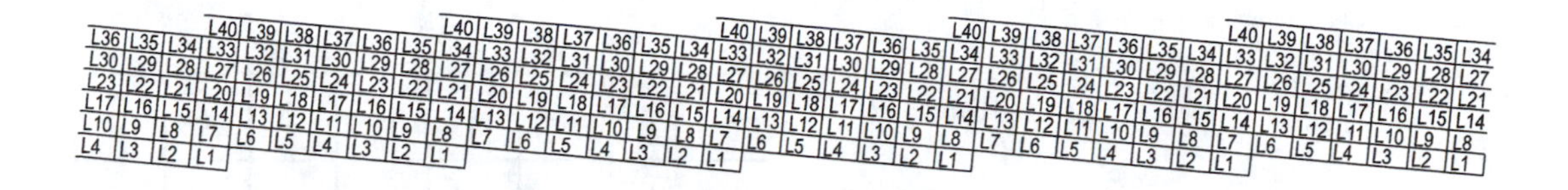

Fig. 8.6: Chart showing the relative positions of the scan lines in adjacent tracks of a VHS videotape.

identical, there is no information lost in the summing operation, but otherwise the technique results in a loss of vertical color resolution.

8.4.1.3 Circuit realization of chroma crosstalk elimination

The chroma phase inversion circuitry, as it is called, is inextricably bound up with the generation of the 629 kHz signal that is the carrier of the encoded chroma. The 4.2 MHz signal (that is mixed with 3.58 MHz to generate 629 kHz) is itself generated *from* the 629 kHz. Fundamentally, the 629 kHz is generated by phase locking to 40× the horizontal sweep frequency. If all of this seems confusing, that's probably because it is. Hopefully it can be clarified by a study of Figs. 8.7a and b, which show typical frequency-synthesis portions of the chroma circuitry in the record and playback modes, respectively. It must be emphasized that these do not represent full block diagrams of the chroma circuitry, which would include such additional circuitry as burst emphasis and de-emphasis, color killers, ACC, and other circuitry not directly related to frequency synthesis.

It may be noted that both diagrams show a signal labeled REF30 going to the address counter. This is one of the servo reference signals associated with the head cylinder. It will be discussed in more detail under that topic.

After examining Fig. 8.7b, we might well be led to ask why it must be so complex. Wouldn't it be possible, for example, to get by with just the circuitry shown in Fig. 8.8?

The answer is that although this circuitry would ideally be adequate, practically, it is of no value. This is because such effects as tape stretching, motor speed variation etc., introduce frequency and phase variations in the chroma signal that would give color errors on playback if they were not compensated for. In Fig. 8.7b, the signals shown as having a "$\pm\Delta\phi$" component are those that include these errors. Two such signals are combined in the final balanced modulator, where the error components cancel out and leave a true 3.58 MHz color signal. This is mixed with the recovered luminance before being modulated up onto a channel 3 or 4 carrier for output.

Several variants on the circuitry of Fig. 8.7b may be found, but all have the same purpose as that outlined here.

Exercise 8.5 It should be obvious from an inspection of Figs. 8.7 that there are a fairly large number of blocks that are common to both diagrams. Draw an integrated block diagram that is applicable to both the Record and Play functions, and which uses as many blocks as possible for both functions. You may of course, add as many poles of R–P switching as necessary to do the job.

The chroma frequency-synthesis circuitry we have just seen may be thought of as an electronic control system. Except for the rotating phase aspect, there is little in it to separate it conceptually from other frequency-synthesis circuitry

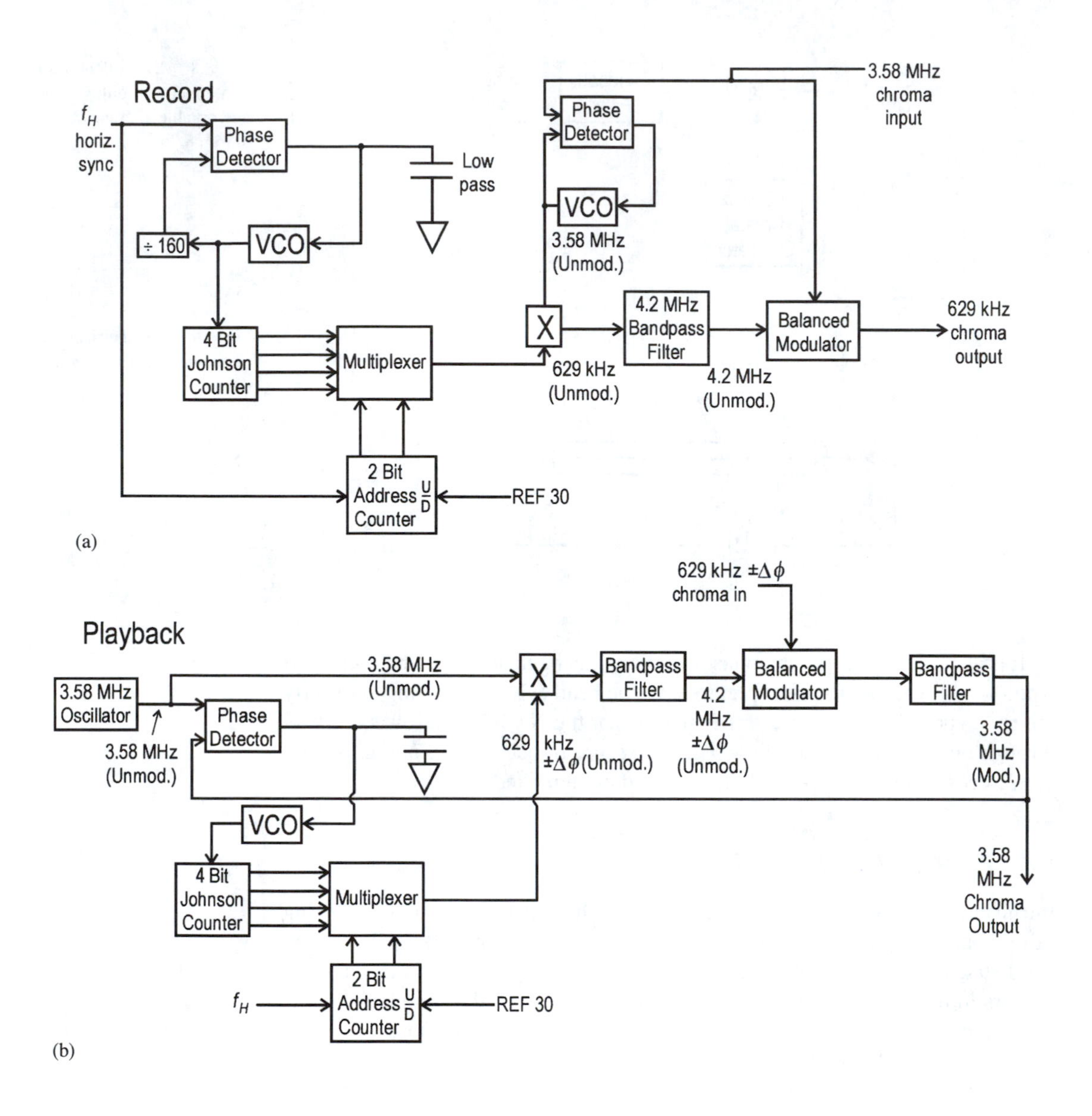

Fig. 8.7: Block diagrams of frequency-synthesis portions of the chroma circuitry: (a) record circuitry; (b) playback circuitry.

that we have already studied. We now come to our next major new concept – that of electromechanical servoing.

8.4.2 Frequency and phase lock of the capstan and head cylinders

It should seem reasonable upon reflection that, if tapes are to be transportable from one VCR to another, the extreme narrowness of the video tracks will require that some means exist to force both the capstan and head cylinder to turn at exactly the same speed as they did in the recording machine. Although it is less obvious, the same is true if they are to be played back on the *same* machine that recorded them. This is because the motor speeds may vary slightly with age, dirtiness, applied voltage, etc., not to mention such problems as a possible failure and replacement of a motor.

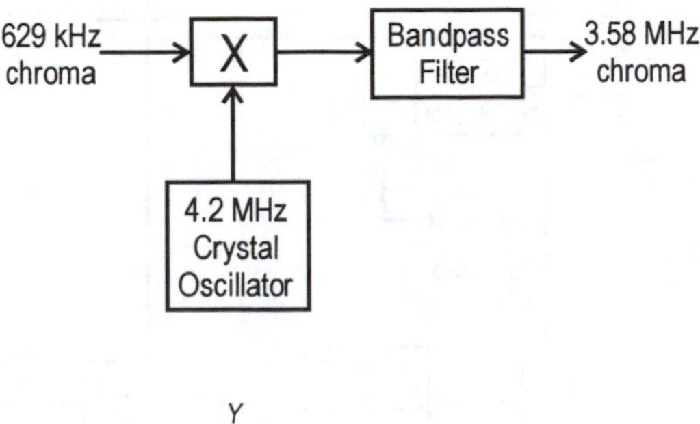

Fig. 8.8: Block diagram of a greatly simplified, but unworkable alternative to Fig. 8.7b.

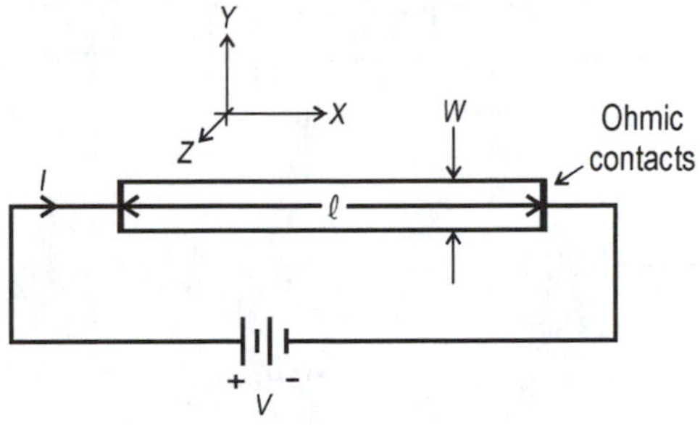

Fig. 8.9: Geometrical parameters of a Hall-effect device.

It is also necessary that the phases of these mechanical entities be correct. It would do little good, for example, to have the head cylinder rotating at exactly the right speed but starting a frame scan when the NTSC signal was halfway through one. Thus, in general, we need to secure frequency and phase lock on both Record and Playback for both the head cylinder and capstan.

8.4.2.1 *Hall-effect devices*

Intrinsic to the achievement of lock for all of these parameters is the sensing of the magnetic poles on the rotating elements. In most cases, this is done by solid-state Hall sensors. The combination of the rotating poles and the magnetic sensors in the head cylinder and capstan motors are called *frequency generators*. Since the Hall device is not covered in some undergraduate curricula, it may be worthwhile to take the time to derive the simple equation that governs its operation. We start conceptually with a reasonably extrinsic bar of semiconductor with a current flowing from end to end as shown in Fig. 8.9. If the bar is homogeneous, essentially all of the current is carried through it by majority drift current. Let us suppose that the bar is n-type, so current is carried by electrons. Then the direction shown for conventional current means that electrons will physically flow from right to left, or in the $-x$ direction.

Now let us suppose that a magnetic field pointing in the $+z$ direction is applied. By the Lorentz force law, the electrons will experience a deflection force

$$\vec{F} = q(\vec{v} \times \vec{B} + \vec{E}) = -|q|(\vec{v} \times \vec{B} + \vec{E}).$$

Since $\vec{v}$ and $\vec{B}$ are assumed orthogonal, the component of $\vec{F}$ due to $\vec{B}$ will lie along the y axis. Thus we can write

$$F_y = -|q|[v_x \cdot B_z + E_y]. \tag{8.2}$$

It might be suggested that there is no E field along the y axis, allowing us to drop the latter term. This is indeed true when no B field is applied, but it is not true in general. We can see from Eq. (8.2) that the magnetic field will force electrons toward the bottom of the bar, causing it to acquire a more negative charge than the top of the bar. There the ionized and immobile donor atoms will impart a net positive charge. The resulting charge polarization between the top and bottom will create a vertical electric field, E_y, which will oppose the magnetic deflection. Equilibrium is reached when the downward force on the electrons due to the magnetic field is exactly canceled by the upward force due to the electric field. The existence of the electric field over a distance w means that there is a voltage between the top and bottom faces. This is called the *Hall voltage* (V_H):

$$E_y = -V_H/w.$$

Inserting this into (8.2) yields

$$F_y = -|q|[v_x \cdot B_z - V_H/w].$$

The requirement that the net force on the electrons be zero means $F_y = 0$, which means

$$V_H = w \cdot v_x \cdot B_z.$$

It only remains to incorporate the relationship for electron drift velocity in terms of mobility,

$$v_x = -\mu_n \cdot E_x = \mu_n \cdot V/l,$$

so that

$$V_H = \mu_n\left(\frac{w}{l}\right)VB_z.$$

Since the Hall voltage is directly proportional to B_z, without any sort of threshold effect, it is necessary to condition the output signal in order to generate voltage pulses that correspond in time to the passage of the magnetic pole over the Hall-effect sensor. This is often accomplished by integrating a Schmitt trigger with the Hall-effect device.

Some early VCRs used a magnetic pickup coil instead of Hall-effect devices. Although this was a less-expensive method at the time, it had the drawback that its output voltage depends not only on the strength of the magnetic field to which it is subjected but also on the relative velocity of the magnet and coil, as required by Faraday's Law. This velocity dependence does not exist with the Hall-effect device, making it especially attractive where the relative velocity is low, such as in a VCR using slow playback.

8.4.2.2 Magnetic sensing within the motors

It has become common to use motors for both drum and capstan that are driven by an assembly consisting of a phased array of drive coils, an annular permanent

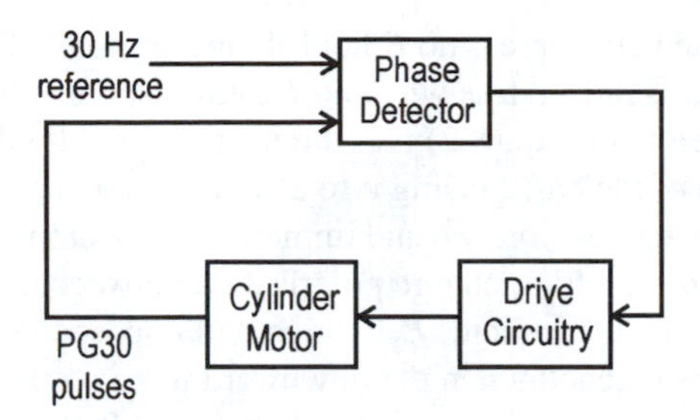

Fig. 8.10: Block diagram of the control loop used for maintaining phase lock of the head motor. For record, the 30 Hz reference is derived from the vertical sync. On playback it is derived from the crystal oscillator.

magnet, and Hall-effect devices between the coils. The latter sense the position of the magnetic poles and control the drive to the coils to give efficiency and precision of operation. These are called *printed-circuit motors*, the name reflecting the fact that all parts of one except the rotor and shaft are readily built on a PC board. Although printed-circuit motors have become the norm for VCRs, there is still no universality as to the frequency generators they drive. Because of this, we will attempt to generalize as much as possible and save the specifics for Section 8.6, where we analyze a specific VCR.

Both the drum and capstan motors have what is called a *field generator* (FG) built in. This consists of another annular permanent magnet that rotates in close proximity to a magnetic sensor. The annular magnet might typically have 60 poles, each of which will generate one pulse in the magnetic sensor as it rotates past. (The capstan motor generally has even more since it must rotate at different speeds for different tape speeds.) For a drum motor with a 60-pole FG, the magnetic sensor will generate 1,800 pulses/sec. The output of the FG is used for speed control. In record and normal playback modes the reference signal is the horizontal sync. It might seem rather awkward to compare 1,800 Hz to 15,734.26 Hz, but the comparisons are not done in a PLL but in circuitry that lends itself to scaling either or both frequencies prior to their being compared, as we shall soon see. The output of the error amplifier is used to control the motor drive circuitry.

8.4.2.3 Phase lock of the head cylinder motor

We know from our study of the integrated phase-locked loop that phase lock cannot be achieved unless frequency lock is also. Thus the phase-lock circuitry provides an output to the capstan frequency-control circuit. This circuit, as we shall see next, is not a locking type circuit. It simply provides a coarse adjustment of frequency within which the phase-lock circuitry can function to provide both phase and fine frequency control and lock.

The VCR will not contain a PLL chip as such for cylinder phase lock, but it will contain the PLL function as part of a much larger-scale IC. The functional PLL block diagram is replaced by one just a bit less direct in this application, as shown in Fig. 8.10. The phase detector is quite different in its realization than that of a general-purpose integrated PLL.

In the record mode, the incoming vertical sync pulses are, as we have seen, frequency-divided down to 30 Hz and recorded on the control track. At the same time, these 30 Hz pulses are applied to the phase detector input of Fig. 8.10 as the 30 Hz reference. The PG30 pulses come from a magnetic pickup coil that is activated by a magnet on the rotor of the cylinder motor. Since the cylinder

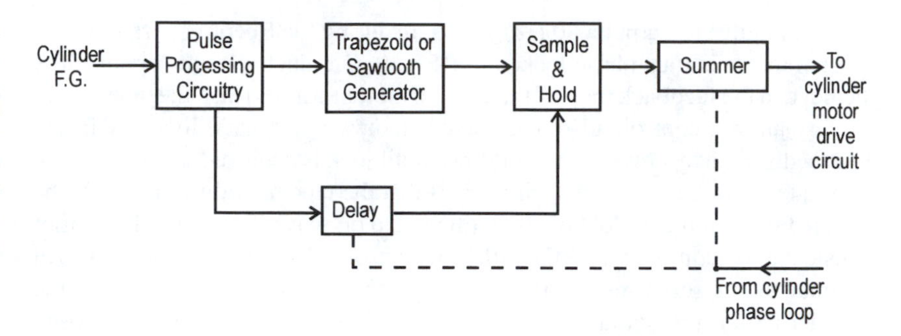

Fig. 8.11: Block diagram of a control loop commonly used for maintaining frequency lock of the head motor. The control voltage out of the phase-control loop is input to this circuit in one of the two positions shown.

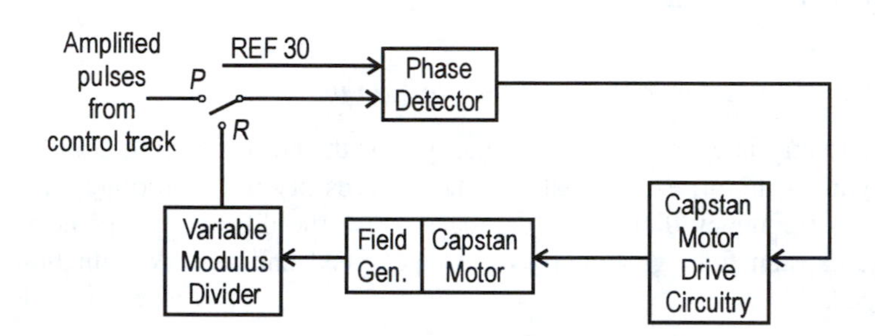

Fig. 8.12: Block diagram of the control loop used for maintaining phase lock of the capstan motor. Changing the modulus of the divider changes the capstan speed.

must rotate at 1,800 RPM (30 RPS), the pulse output frequency is 30 Hz. This is a totally separate system from the field generator. When we consider this whole picture, we see that during recording the rotation of the head cylinder is phase locked to the divided-down vertical sync. On playback, the reference signal is changed to a divided-down crystal oscillator. Usually this oscillator runs at 32,760 Hz and is divided down by 1,092 to give a 30 Hz reference for the phase detector.

8.4.2.4 Frequency control of the cylinder motor

Although considerable variety exists in the means of achievement of this circuit function, many VCRs derive the speed-control voltage from a sample and hold (S&H) circuit. This circuit receives its input from a trapezoid or sawtooth generator, which is driven by the cylinder field generator.

The S&H circuit receives a "Sample" command from the output of a delay circuit whose input is also derived from the field generator. As the cylinder motor speed changes, the "Sample" pulse will arrive at a differing point on the sawtooth and thus cause the output of the S&H to output a different control voltage to the motor drive circuitry. Somewhere in this chain, the output of the cylinder phase circuitry is input. Perhaps it is used to modify the delay for the "Sample" command, or it might be summed with the output of the S&H before application to the motor drive circuit. The former configuration is shown in Fig. 8.11 in solid lines, and the latter by a dashed line.

8.4.2.5 Phase lock of the capstan motor

The operation of this circuitry is very similar in principle to that of the cylinder motor, but with a couple of noteworthy changes as shown in Fig. 8.12. The variable-modulus divider is set by the speed-selection circuitry to have the proper modulus to output a 30 Hz pulse train during the Record operation. The capstan motor is thus phase locked to REF 30. On playback, the tape itself is included in the feedback loop. If the capstan motor turns too fast, the tape moves too fast, and the control pulses recorded thereon will not phase lock to REF30. The feedback then slows the motor down until lock is achieved. An interesting consequence of the use of this circuitry is that the tape recording speed can be changed right in the middle of a recording, and on playback, except for a brief transient to reacquire lock to the different control pulse spacing, playback will continue as the servo resets to the new tape speed. As was the case with the cylinder motor, the output of the phase control circuit is input into the speed control circuit to achieve lock.

8.4.2.6 Frequency control of the capstan motor

This circuitry is generally of the same type as the corresponding circuitry for the cylinder motor, except that it is again necessary to incorporate variable-modulus frequency dividers to compensate for the different frequencies out of the capstan field generator as the tape speed and recorder function are changed.

8.5 Remaining VCR circuitry

We have at this point either examined or alluded to all of the circuitry necessary to make our conceptual VCR function. The circuitry that we have not considered falls broadly into two categories: unelucidated circuitry and finesse circuitry.

8.5.1 Unelucidated circuitry

This category consists of circuitry that we know must be present in a VCR, but which is either so nearly identical to what is in a TV or so obviously intrinsic to VCR operation that it has not been included in any detailed discussion. Besides the purely mechanical parts of the device, we might include under this heading, the following:

8.5.1.1 Tuner or tuners and AFT

These are used in the Record mode to generate the signal to be recorded. For this reason it is not necessary to have the TV tuned to the same channel being recorded. Indeed, it is not necessary to have the TV on at all when recording.

8.5.1.2 Tuning circuitry

Since the VCR is programmable as to channel and time to start and stop recording, it is clear that some circuitry must be present to handle this function. It would be possible to provide this capability with a clock chip, a register, and a small amount of combinatorial logic. However, the job is done by a microprocessor because the microprocessor is a necessity on other grounds anyway, as will be seen when we look at the second category.

8.5.1.3 Digital display(s)

Some VCR models use one digital display as the clock and another as the tape counter. Most use one vacuum-fluorescent display for both. All have a channel indicator.

8.5.1.4 Programming capability

Some models have a remote control or a calculator-type keyboard for direct entry of channel and time information in the programming mode. Others have sequential scanners that operate when a single button is pressed until it is released. Some units have both.

8.5.1.5 Audio tape recording circuitry

This has not been studied in any detail, because it is essentially identical to audio-only tape recorders.

8.5.1.6 IF amplifier and video detector

These are shown in Fig. 8.3. As was the case for the tuner, they perform the same function as in a TV.

8.5.1.7 End switch

Since the VCR automatically stops and may even rewind the tape when it reaches the end, and stops the rewind when it reaches the other end, it obviously has some means of sensing the ends of the tape. The system used is photoelectric. The cassette housing is designed to allow the passage of light from a "well" at the midpoint to sensors on either side. When the end of either reel within the cassette is reached, light is enabled to reach the sensor on that side of the cassette. The photoelectric signal so generated triggers the stop sequence or the rewind sequence, depending on which end is reached.

8.5.1.8 Remote control

Most VCRs have IR remote controls. Except for different labels on the buttons, these are very much like that of the RCA color TV we analyzed in Chapter 7. The principles are very similar and so are not deemed worthy of separate elucidation.

8.5.1.9 RF modulator

The RF modulator takes the decoded NTSC signal from within the VCR in the playback mode and vestigial-sideband modulates it onto the proper carrier frequency for either channel 3 or channel 4. A user switch allows the selection of the weaker of the two broadcast channels as the VCR output band.

8.5.2 Finesse circuitry

The second category of circuitry we have not examined in some detail is circuitry that may not be necessary for operation but is, nevertheless, necessary for it to operate well and with some degree of longevity. It also compensates for some of what we might call the second level of nonidealities. Although some of these circuits will be examined when we look at a specific VCR, generally we do not consider them in detail because it becomes too easy to "lose sight of the forest because of all the trees." Although there is bound to be some disagreement as to which circuitry actually belongs in this category, nevertheless, here is one man's view of the matter.

8.5.2.1 Circuitry for the record operation

AGC

After separation of chroma and luminance, the luminance is passed through an AGC circuit prior to being input to the FM modulator. Excessive luminance amplitude would overdrive the modulator, causing too wide a frequency swing and complicating its separation from the chroma.

Pre-emphasis

This actually occurs in two circuits. The first is generally called nonlinear emphasis. The purpose of pre-emphasis, as discussed previously, is to emphasize the high-frequency components of a signal to be recorded in order to improve the S/N ratio of the record/playback process. The nonlinear pre-emphasis takes cognizance of the fact that the S/N ratio actually drops more rapidly than the signal level itself. Therefore, the amount of pre-emphasis added *increases* as the signal level decreases. This output passes through video amplification and clamping (its level must be held constant to insure that the FM modulator will swing over the proper frequency range). It then passes to the second pre-emphasis circuit. This one is linear as to amplitude but selectively boosts high frequencies. It is thus more nearly analogous to the types of pre-emphasis we have seen previously (Chapters 2 and 5).

White and black clipping

The second (linear) pre-emphasis circuit's boosting of the high-frequency content of the luminance signal causes relatively large spikes to be formed on the edges of the waveform. The white and black clipper provides independent adjustments of both levels to prevent overmodulation of the FM modulator, whose input is fed from this clipper.

8.5.2.2 Circuitry for the playback operation

Head switching

Since we have seen that the two heads pick up alternate video tracks from the tape, there must be selection circuitry that feeds first one head and then the other to the amplifier.

Exercise 8.6 Or must there be? What would be wrong with just summing the outputs of the two heads?

There will also be a balance control service adjustment to equalize the outputs of the two heads before passing the playback signal on to subsequent processing circuitry. The signal that performs head switching, called SW30, is derived from the PG30 signal output from the cylinder motor assembly.

Color killer

The color killer circuit is similar in intent and operation to that in a color TV. When there is no color signal, it prevents the output from the color circuitry from being applied to the TV, since it would just be noise anyway. We might then ask "Why duplicate the function? Why not just count on the TV circuitry to kill the color if there is none?" For one thing, although we attempt to minimize crosstalk between luminance and chroma, there is some, and a high noise level in the chroma can have some effect on the luminance noise. More fundamentally, if a VCR with no color killer were used with a monochrome TV (which has no color killer), the "color noise" would be intolerable for black and white playback.

Dropout compensation (DOC)

Because of imperfections on the tape, there may be short spans of time during playback when the heads pick up no signal. When we remember that a single horizontal line is recorded on a portion of the tape that is only 58 μm $\times$ 40 μm, we understand that it does not take much to cause the dropping of a line. The DOC circuit, like the comb filter, incorporates a 1 H delay line. Then if dropout is detected, the DOC circuitry not only connects the delay line output to the succeeding luminance circuitry but also connects it back to its own input. Thus the DOC circuit can keep on outputting the same line of luminance indefinitely. Practically, however, although repeating a line once is indistinguishable from the real picture in most cases, successive repetitions become increasingly obvious. Typically, if there were a long dropout, the circuit design would allow a single line to be repeated a maximum of three times. After that, the "live" material from the head is again sent on to the receiver, even if it is garbage, since five identical lines in succession are no less nonaesthetic to the viewer than the flashing white lines that would characterize dropout without DOC. In the vast majority of cases, the dropout is not long enough for this to be a problem, and the DOC circuitry performs admirably.

De-emphasis

Just as was the case for magnetic audio recording, the pre-emphasis applied to the luminance signal during the record operation goes through complementary circuitry on playback to undo what was done by the original circuitry and suppress noise in the process. Since the pre-emphasis process was dual, so must the de-emphasis process be. First the high-frequency emphasis is canceled, and then the nonlinear emphasis.

Noise cancellation

The noise reduction provided by two types of pre-emphasis is still not adequate to produce a broadcast quality picture on playback. The noise canceller works on the more or less flat segments of the video signal. The video is first passed through a high-pass filter, which will sharply attenuate the flat (longer time duration) portions of the signal. The output of the high-pass filter is then amplitude limited to remove the large spikes associated with edge transitions. The "signal" coming out of the limiting circuitry is thus almost pure noise. It is then recombined with the original video signal, but with a phase that will cause it to be subtracted from the original signal.

8.5.2.3 Circuitry for both record and playback operations

Automatic color control

The ACC circuitry is also analogous to that in a color TV. It functions during Record to stabilize the amplitude of the chroma signal fed to the down converter; on Playback it stabilizes the chroma amplitude relative to that of the luminance with which it is recombined. The gain of the ACC amplifier is governed by the output of an ACC detector. The detector may be designed to respond to the amplitude of the chroma signal itself or to that of the burst (in which case it is gated).

The dew sensor

The heart of the dew sensor circuit is a humidity-sensitive resistance that disables the VCR if the moisture level in the device is high enough that the tape might possibly stick to the rotating head.

System control

We have already alluded to the use of the microprocessor in storing the programmed record information and timekeeping. Some VCR models use a separate microprocessor for quartz-synthesized tuning. But the most important microprocessor application in the VCR is in system control. This broad function can be broken down into two parts:

- Switching the appropriate specialized circuitry in and out as the speed selector is changed.
- Properly sequencing all mechanical and electrical operations as the VCR is changed from one operational mode to another. For example, the tape must be threaded through a tortuous mechanical path when it is loaded. This is done under microprocessor control. When a tape has been playing and the stop button is pressed, a number of mechanical operations must be performed in a precise time sequence to avoid stretching the tape. The microprocessor is a natural for this job.

Speed switching

When one switches between the various possible recording speeds, an enormous number of contacts must be made and broken. The use of electronic switching greatly facilitates this, allowing simultaneous switching all over the VCR without having a myriad of wires going to a multi-gang switch that would have to be of monumental proportions. Here is just some of the circuitry that requires switching when the recording speed is varied:

(i) Burst pre-emphasis and de-emphasis are used on SP and EP only – not LP where it is not helpful.
(ii) The slower record speeds require the switching in of different equalization than SP does.
(iii) Equalization also differs between speeds for the playback amplifier.
(iv) Nonlinear pre-emphasis and de-emphasis are not used in SP – only in EP and LP.
(v) In some units, a chroma-sensitive ACC detector has been used for SP and LP and a burst-sensitive detector used for EP.
(vi) Undoubtedly the most switching is required in the servo circuit. As the record speed is changed, the capstan motor must have its speed changed. This in turn changes the frequency out of its field generator. Since the whole process of frequency and phase lock of the capstan motor is intimately tied to the frequency out of the capstan field generator, speed changing requires different division ratios in the phase-lock circuits.
(vii) All of this says nothing about special effects such as reverse action, search mode at various speeds, freeze-frame, etc. Although they are provided on the basic type of machine we have been considering, they are generally quite unsatisfactory. We will briefly discuss these effects in Section 8.7. For our purposes here, it will suffice to say that the implementation of these effects, even when they are of poor quality, requires significantly more speed-switching circuitry than we have already indicated.

8.6 A real VCR

No American electronics company has ever produced a VCR commercially. Not only was this a Japanese item from the beginning, but production and innovation of it were also almost exclusively Japanese until certain patents expired and the Koreans entered the field. All VCRs sold under an American trade name are (as of this writing) manufactured by Japanese firms. We have chosen a Sharp VC-785U VHS unit because it is well conceived and simply laid out. The main circuit board of this unit contains the great majority of the circuitry in the VCR on itself and on two smaller boards (audio and Y/C) that plug into it. The remaining circuitry is found on power supply and timer boards that are cabled to the main board and in a few very small boards that interface sensors, motors, heads, etc. to the main board.

8.6.1 Power supply

It will be beneficial to determine how much of the circuitry on the main board is for the purpose of generating, regulating, and switching of the multitudinous power supplies in this unit. To this end, we show the power supply board in Fig. 8.13 and follow its cabling onto the main circuit board.

8.6.1.1 The UR 15 V supply

This 15 V supply is very much like that of Fig. 5.22. The salient differences are that in this one:

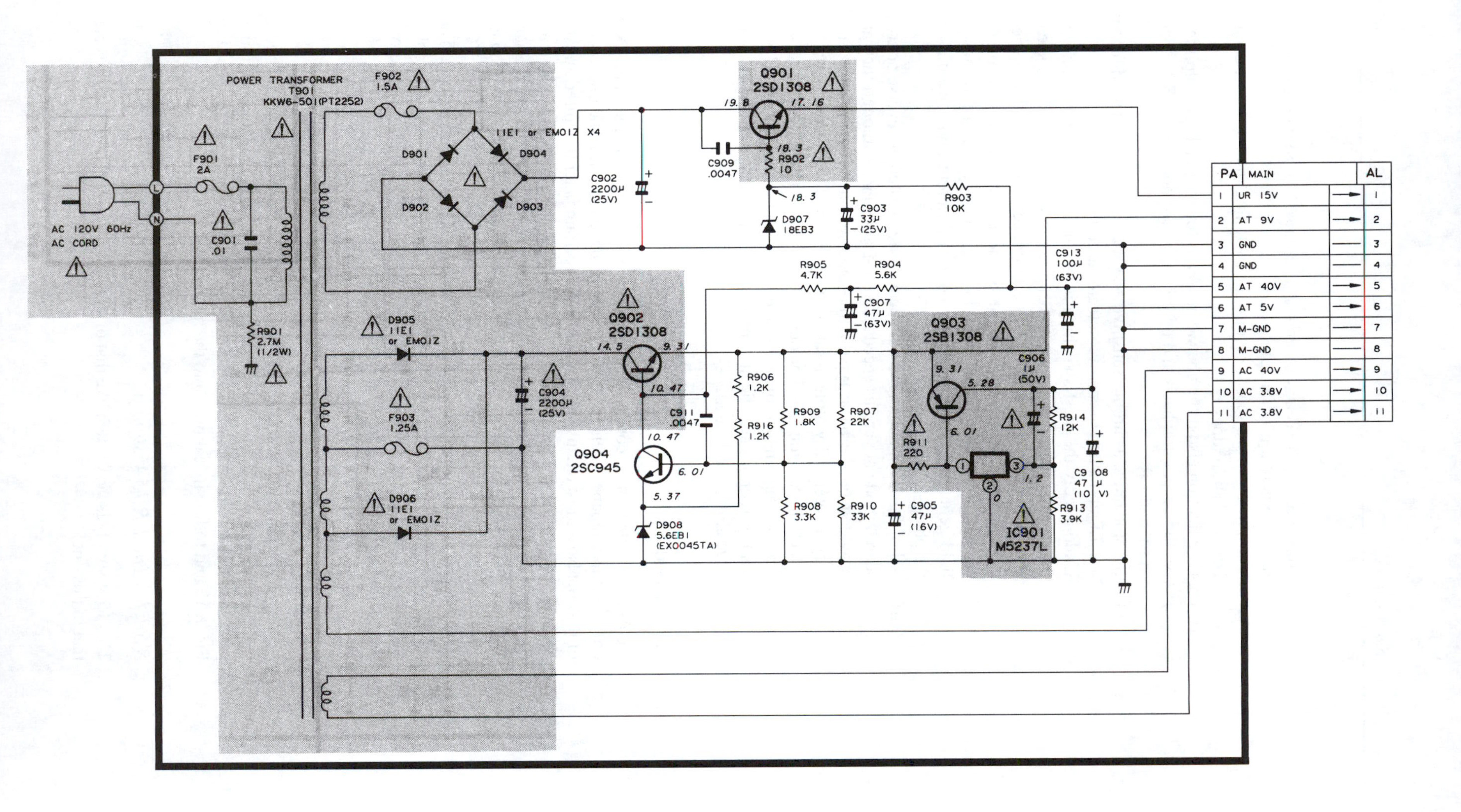

Fig. 8.13: Schematic diagram of the power-supply board in the Sharp VC-785U VCR.

- The Zener is biased by a separate supply (AT 40 V), instead of a resistor between base and emitter. The benefit of this is that the well-known factor for ripple attenuation by a Zener, r_z/R, can be made small if R is large, and the larger the supply voltage, the larger R can be and still bias the Zener correctly.
- A capacitor is added between base and collector. Its value is so small that its impedance is over 250 kΩ at 60 Hz. Thus we can safely say that it is insignificant at 60 Hz. It is obviously insignificant at DC also. It may have some effect on transient response.
- This circuit also has a small resistance added to the transistor's base lead. DC analysis shows that this resistance does not affect the output voltage. It is probably in circuit for protection of the transistor in the event of the supply being short-circuited.

The UR 15 V supply comes onto the main board at pin 1 of the power supply (AL) connector. The schematic of the main board is shown in two parts. The power supply connector is on the second part, which is shown in Fig. 8.14.

A very nice feature of the Sharp schematics is that all connectors show the direction of the signal or power flow for each pin. The UR 15 V pin on the main board leaves the schematic immediately after arriving there. This is because it is used exclusively on the first portion of that board. This will be encountered in Fig. 8.49. For now it will suffice to note that this supply is used to drive several motors in the VCR.

8.6.1.2 The AT 9 V supply

Referring back to Fig. 8.13, we see that the AT 9 V supply is derived from a full-wave rectifier. The discrete regulator looks more complex than necessary, because the Zener is biased through two series-connected 1.2 k resistors rather than a single 2.4 k one. Both legs of the output voltage divider are made up of two resistances in parallel. The upper resistance works out to be 1.67 k, and the lower is 3.0 kΩ. All of these could be replaced to good accuracy by single 5% tolerance resistors.

Exercise 8.7 Perform a DC analysis on this regulator to find V_{out}. Assume both base–emitter voltages are .7 V. Explain physically the sequence of events that will tend to raise the output voltage if it drops.

The AT 9 V supply lead enters the main board at pin 2 of the AL connector. Once on-board, it takes three paths.

(a) The first is to the PC 9 V GEN. The schematic of this stage is shown in Fig. 8.15. It shows two sets of voltages at each pin. Ordinarily this means that the voltage sets were measured under two different conditions. We saw a similar thing on the schematic of the Walkman in Chapter 3. In this case, we have no indication from the schematic of the conditions under which the voltages were measured, but it is not of great significance at this early stage of our analysis. What we *can* say with a fair degree of certainty is that since the voltages are about the same in both measurement conditions, neither of those conditions

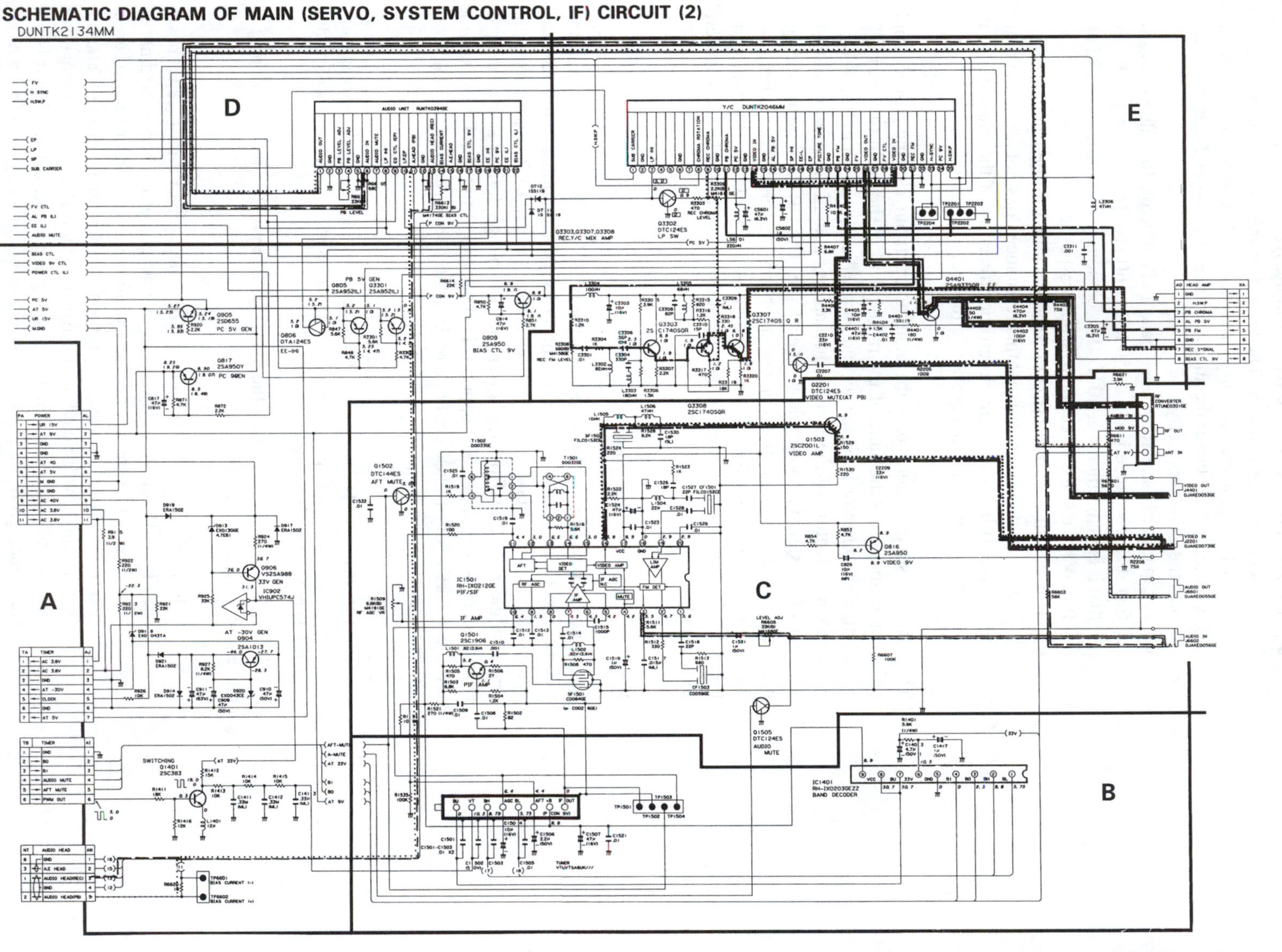

Fig. 8.14: Schematic diagram of that part of the main board that deals with power control, sound, and video (before it is split into chroma and luminance portions) in the Sharp VC-785U VCR. An electronic version of this figure is available on the world-wide web at http://www.cup.org/titles/58/0521582075.html.

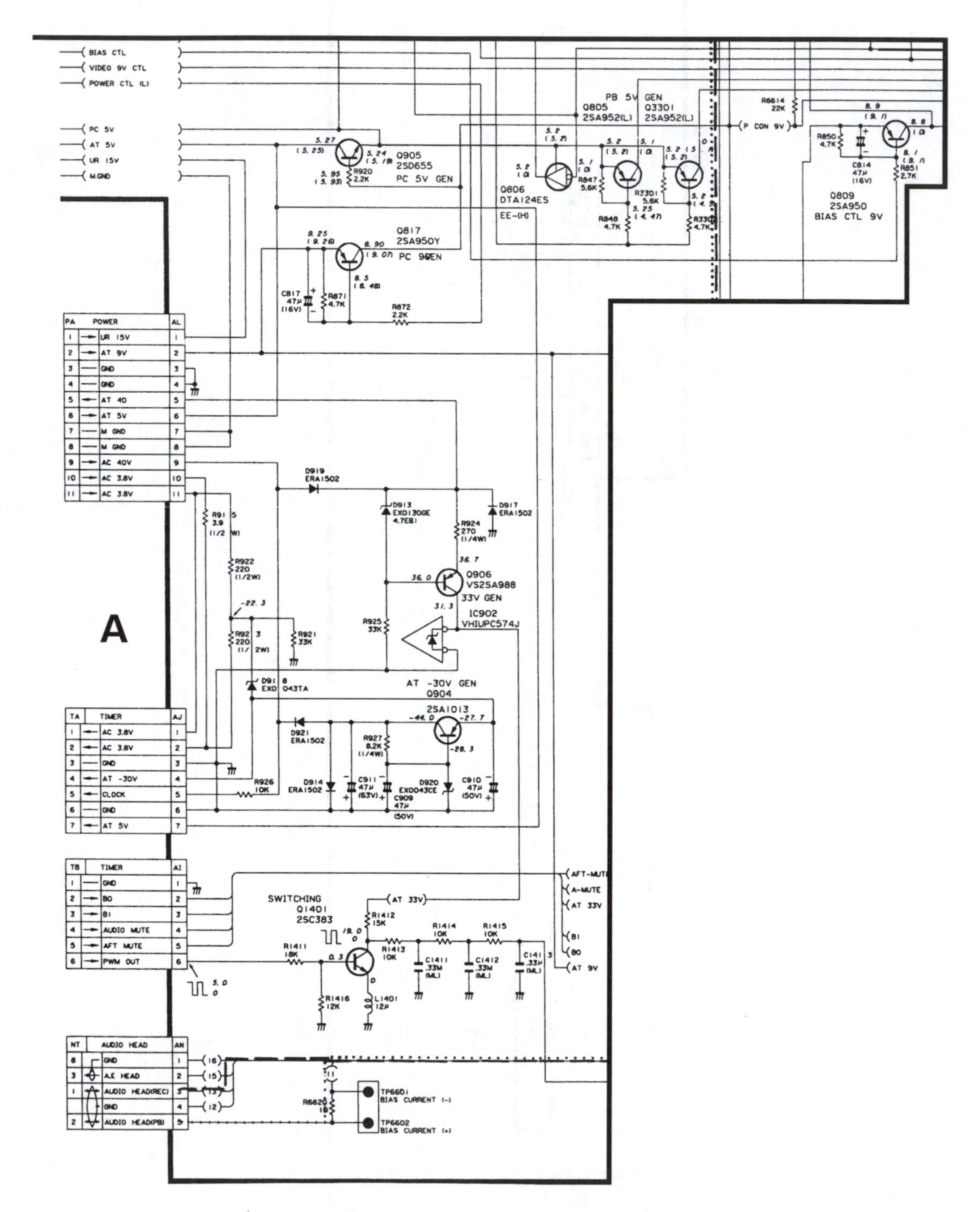

Fig. 8.14A: Power supplies.

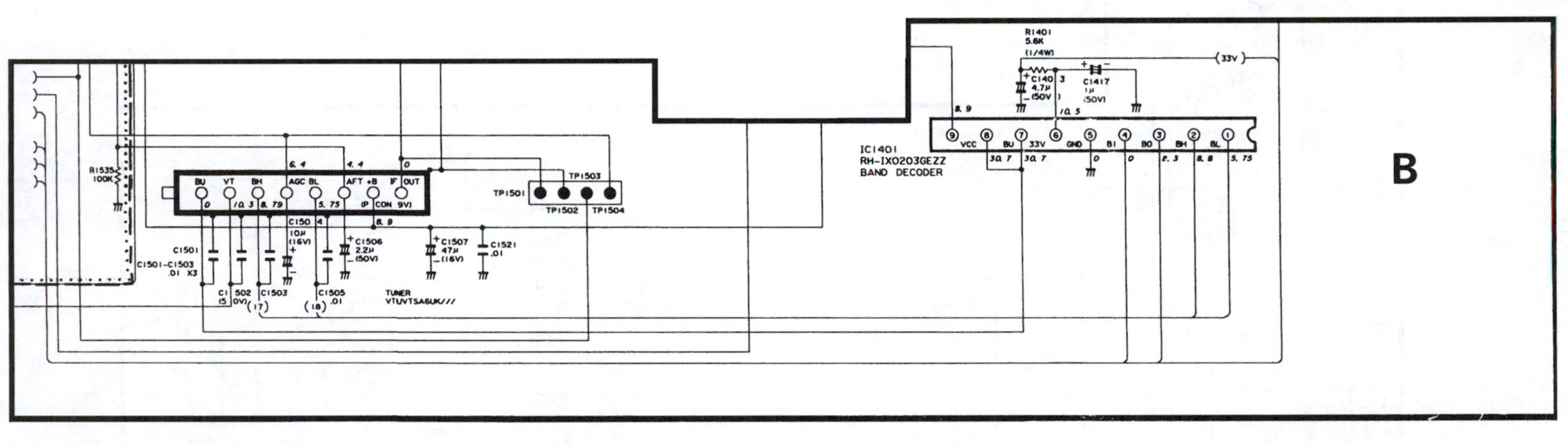

Fig. 8.14B: Tuner and band decoder.

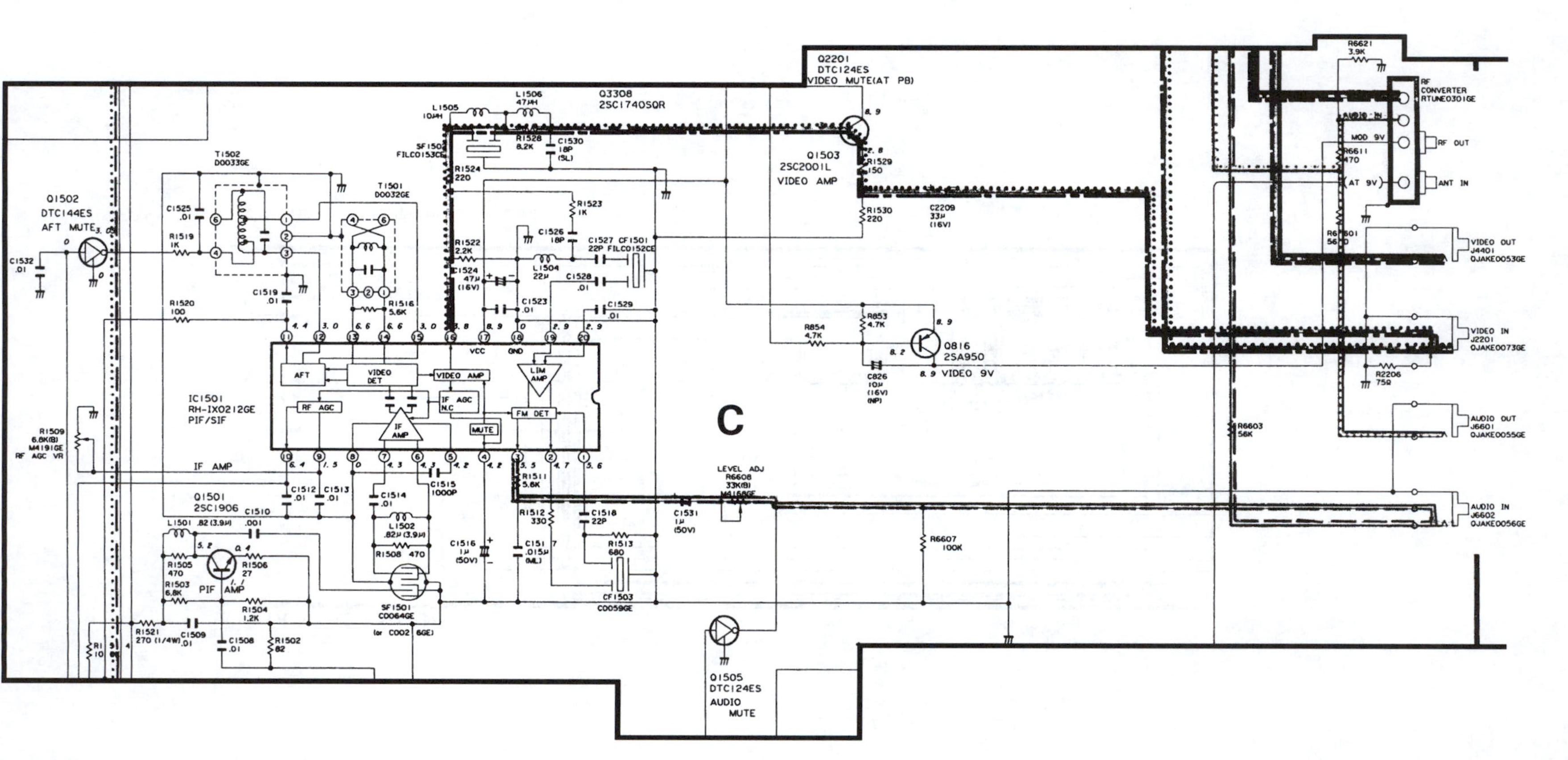

Fig. 8.14C: IF and video amplifiers plus video and audio detectors.

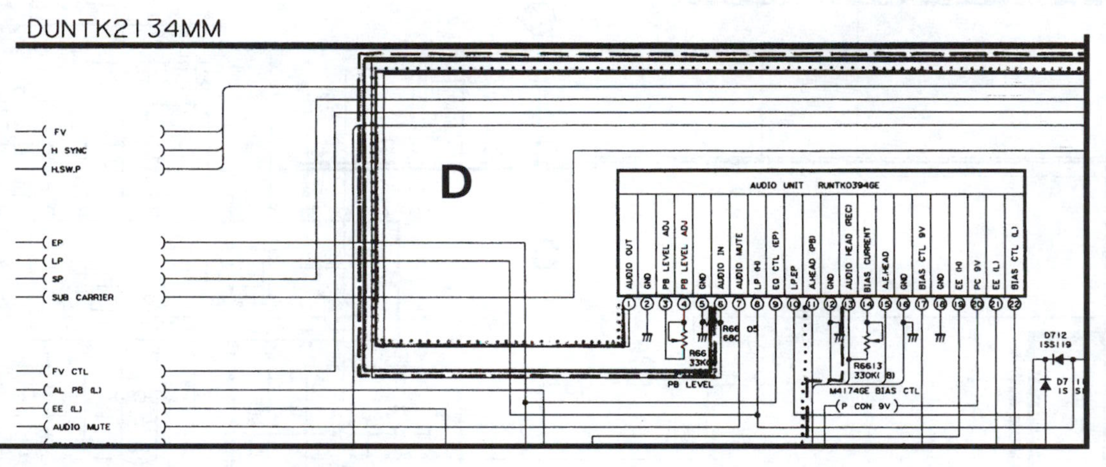

Fig. 8.14D: Connector for audio board.

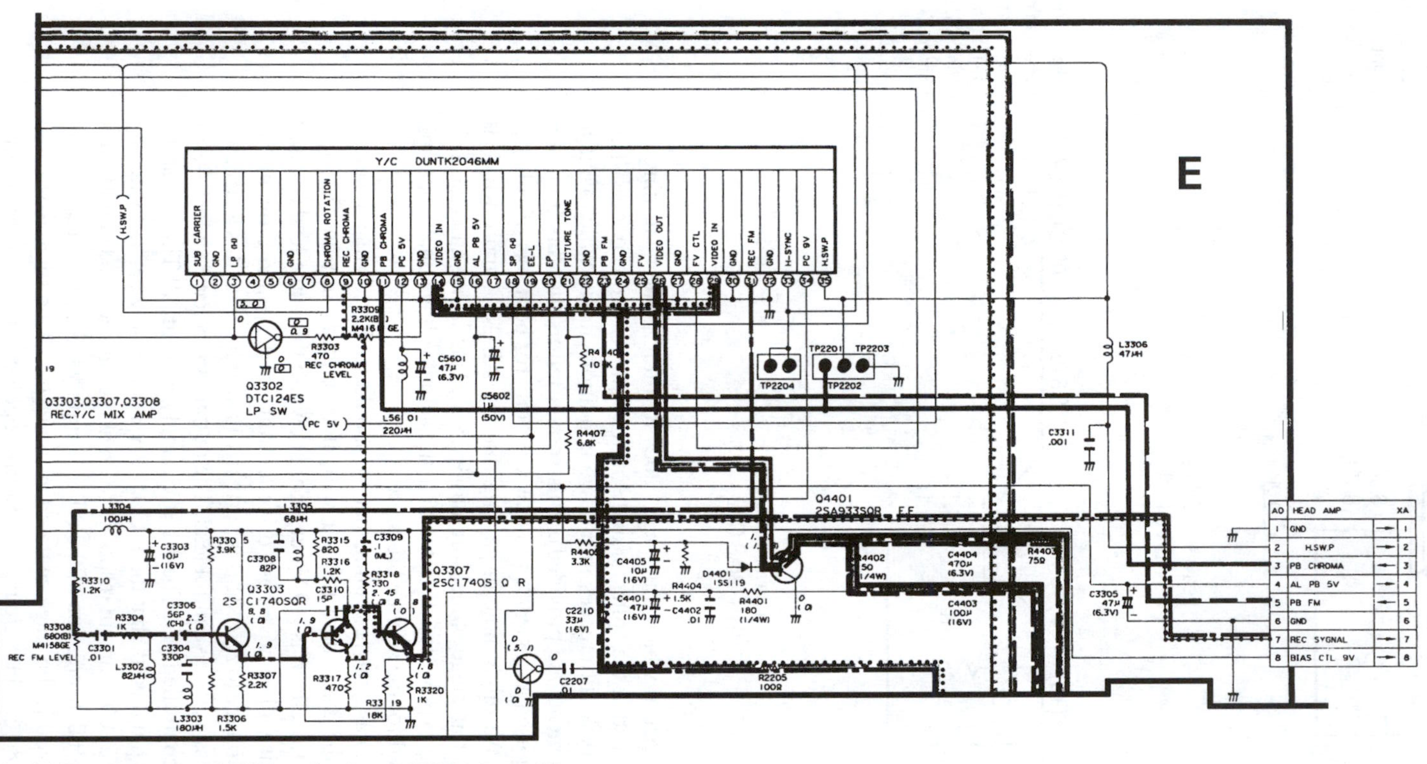

Fig. 8.14E: Connector for Y/C board and record Y/C mix amplifier.

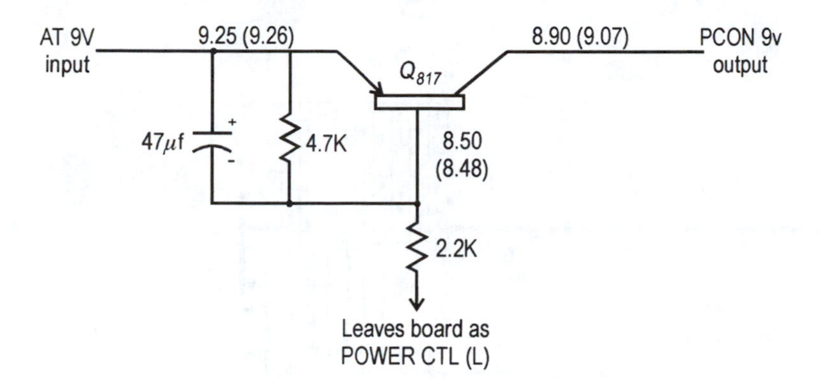

Fig. 8.15: Schematic diagram of the PC 9 V GEN stage. This is a switch that controls the PCON 9 V supply in the Sharp VC-785U VCR.

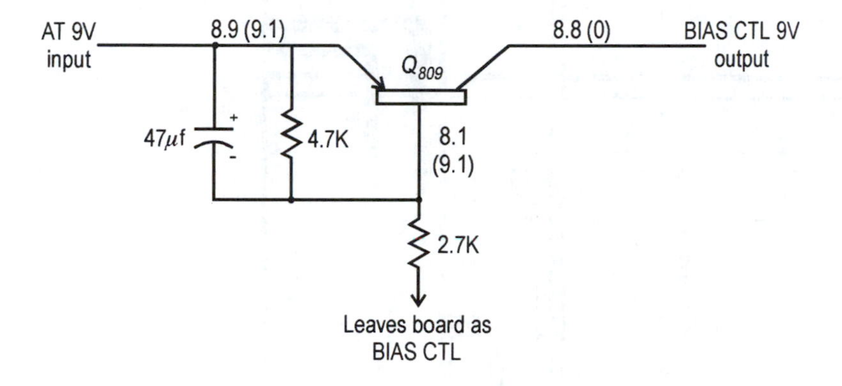

Fig. 8.16: Schematic diagram of another switching stage. This is the switch that controls the Bias Ctl. 9 V supply in the Sharp VC-785U VCR.

will be that in which this stage "does its thing." The next thing we observe is that the collector and emitter voltages are very close (.35 V and .19 V differences in the two conditions under which voltage measurements are provided). This should tell us that this transistor is being used as a switch and that it is "on" for both sets of voltages shown. We might note in passing that the AT 9 V supply which is input to this stage is measured as 9.25 V. This is not an error. It is designed to provide 9 V, but 9.25 is the actual voltage on the one unit measured for the service literature.

In any case, the next thing that we should see here is that the lower end of the 2.2 k base resistor will, if grounded, turn on Q_{817} and allow it to deliver its ~9 V output as shown. If this same lead is raised to 9 V, it will remove base drive, cutting off Q_{817} and dropping its output voltage to zero. This is perfectly in keeping with this lead being labeled as a power-control lead. The output is labeled P CON 9 V, for power-controlled 9 V. The AT 9 V supply is on at All Times that the VCR is plugged in, but the P CON 9 V is only on when activated by the Power CTL line in response to an as-yet-unspecified condition.

(b) The second destination of the AT 9 V supply load after it reaches the main board is the stage centered around Q_{809}. This stage is isolated in Fig. 8.16. This circuit is not only topologically identical to that of Fig. 8.15, but even its part values are the same except for the 2.2 k resistor being replaced by a 2.7 k one. We are therefore safe in concluding that this stage is likewise a power switch. In fact, we would not bother to duplicate the schematic if it were not for the difference in voltage shown in this case. The bias control signal applied to this circuit *does* manage to shut off this supply when it is raised high. For this circuit, we can tell from the main schematic the condition under which this power switch stage will be activated. The Bias Ctl. 9 V output is the V_{cc} supply for the three-stage amplifier composed of Q_{3303}, Q_{3307}, and Q_{3308}. This amplifier, as indicated in Fig. 8.14E, is the Y/C mix amplifier used during the record process. Thus the Bias Ctl. lead will be low during record and high otherwise.

(c) The third path for the AT 9 V line on the main board is downward to where it joins a cable that is represented by the lowest line on the schematic. Unfortunately, unlike many other schematics, this one does not indicate cables with a heavy line. In this schematic, our only indication that we have a multiconductor cable is that it makes rounded corners on the schematic, rather than the square corners shown when wires change direction. When the wire being traced is part of a cable, you must trace along the cable until it comes out again. In the present case, this happens on the far right of the schematic, where we see AT 9 V entering the RF modulator.

8.6.1.3 *The AT 40 V supply*

After the AT 9 V supply connection, the next thing we encounter while moving down the power supply connector of Fig. 8.13 is two ground connections. These are followed by the AT 40 connection. This is an input to the power supply board. We see this in Fig. 8.14A. Also in this figure we see that the AC 40 V line drives the AT 40 V supply, which is situated on the main board. Its output is then returned to the power supply board where, as we have already seen, it biases the Zeners for the UR 15 V and AT 9 V supplies. The main board circuitry that generates the AT 40 V supply is shown in Exercise 8.8.

Exercise 8.8

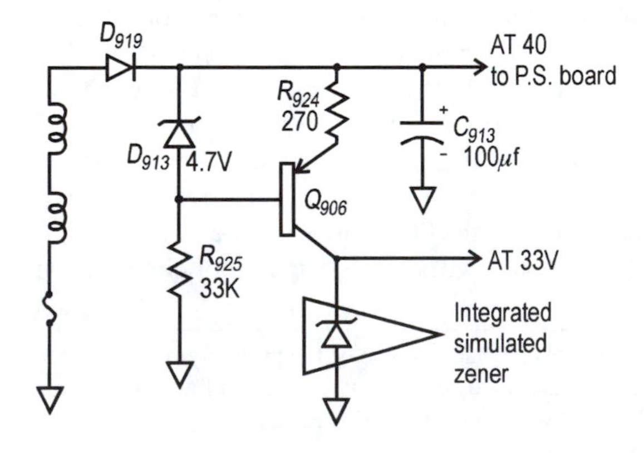

(a) What is the purpose of Q_{906}?
(b) What is the maximum current that could be drawn from the AT 33 V supply?

The AT 33V supply serves two functions. One is the V_{cc} supply for the band-decoder IC (IC_{1401}). The other is the V_{cc} supply for the PWM switching transistor, Q_{1401}, which may be found in the lower portion of Fig. 8.14A. The

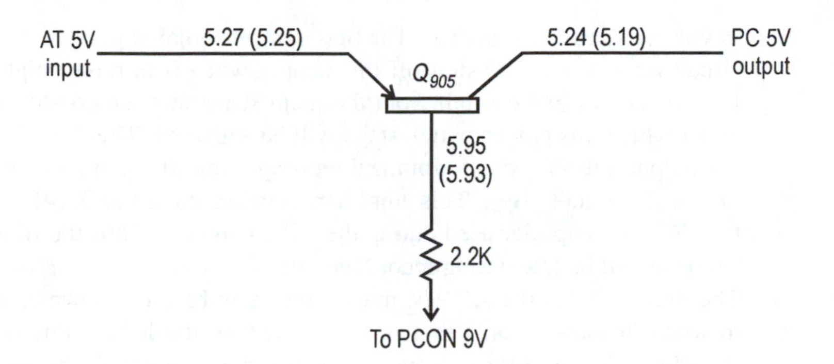

Fig. 8.17: Schematic diagram of the PC 5 V GEN stage. This is a switch that controls the PC 5 V supply in the Sharp VC-785U VCR.

PWM output of this stage is low-pass filtered to extract the DC tuning voltage for the tuner. Normally the duty cycle of this pulse train will vary, depending on the channel selected. The LPF extracts the DC component of the pulse train. If, however, V_{cc} of the switching transistor were to vary, the DC component at the LPF output would reflect not only changes in pulse duty cycle but also changes in the pulse amplitude. If, however, V_{cc} is held constant for this stage, the pulse amplitude will be also, and the tuning voltage will not be perturbed by supply variations, and the desired stations will be tuned accurately.

8.6.1.4 The AT 5 V supply

This supply is generated by another down-regulator (Q_{903} and IC_{901}) from the AT 9 V supply. This regulator servos the divided-down 5 V output to a precision 1.2 V voltage reference IC. This chip appears to be somewhat unique. If the voltage on pin 3 is less than its internal 1.2 V reference, pin 1 is pulled low. This pin is designed to be current sinking. The additional current flows from emitter to base of Q_{903}, causing it to conduct more and raise the AT 5 V output voltage.

The AT 5 V supply connects to many places in the first schematic of the main board (Fig. 8.49). It exits the second schematic (Fig. 8.14A) through pin 7 of the AJ connector (one of the timer board connectors). The only other place it connects on Fig. 8.14 is to the Q_{905} stage, which is shown in Fig. 8.17.

As was the case for the PC 9 V stage, this PC 5 V stage shows no significant voltage variation between the two conditions under which voltages are specified. Again, the function of the stage is to switch the PC 5 V supply on and off. The Power-controlled 9 V supply provides the base drive to this stage, so when the PC 9 V supply is enabled, the PC 5 V one will be also.

The output of this stage goes to a two-transistor circuit shown in Fig. 8.18.

Exercise 8.9 Aside from what seems to be the perfectly ridiculous idea of having two transistors producing the same output on two different lines, there is an obvious flaw in Fig. 8.18 – a flaw that was copied from Fig. 8.14A.

(a) What is it?

(b) Which is correct?

Of the possible explanations for the doubling of this stage, a desire to decouple the circuits driven by the two outputs seems to be most likely. The output from Q_{805} supplies power to the Y/C board, and that from Q_{3301} supplies the

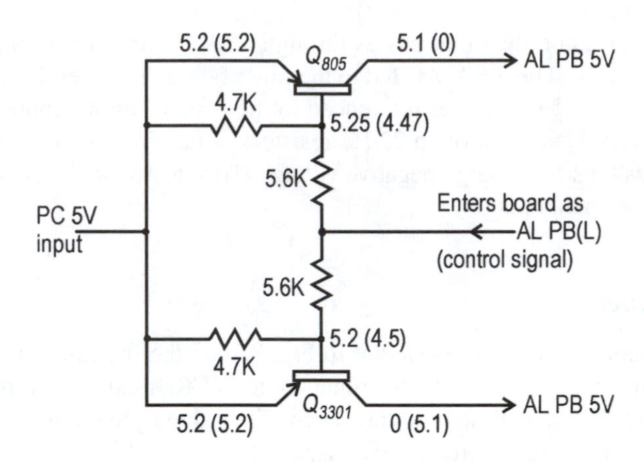

Fig. 8.18: Schematic diagram of the dual switching stages that output the AL PB 5V supplies in the Sharp VC-785U VCR.

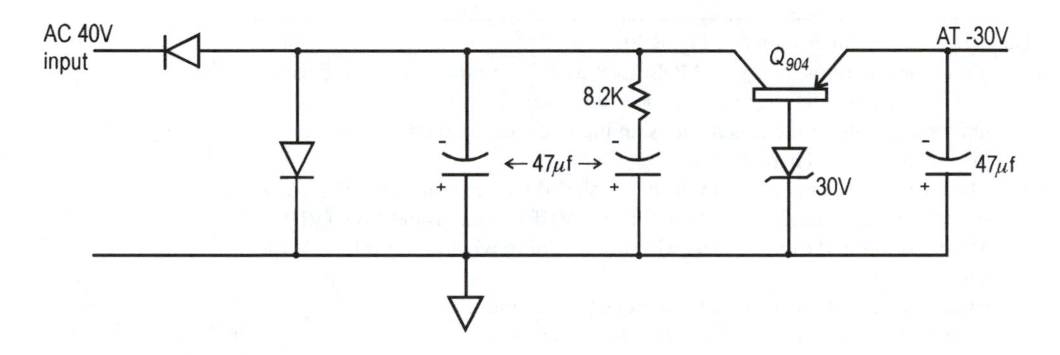

Fig. 8.19: Schematic diagram of the AT -30 V power supply in the Sharp VC-785U VCR.

head amplifier. Since the signals from the heads are small, the power for the amplifier should be as "clean" as possible. AL PB 5 V is the only supply to this amplifier.

8.6.1.5 *The AC 40 V supply*

As we proceed down the power supply connector, we pass two more ground pins, called M-GND. These go over onto the other main board schematic. They are motor ground pins. They will generally have a level of noise that should be avoided in signal-handling circuitry. Hence they have a separate ground path back to the supply.

Below the M-GND pins is the AC 40 V supply. We have already seen its use in generating the AT 40 V and AT 33 V supplies. This AC 40 V pin out of the power supply board and onto the main board is also used to do two other things. One is to provide a timing reference to the clock board through a 10 k resistor. The other is to drive the -30 V regulator by means of the circuitry shown in Fig. 8.19. This circuit does not provide constant current drive to the Zener but is of the "impedance divider" type shown Fig. 5.22.

8.6.1.6 *The AC 3.8 V supply*

There is a last pair of pins in the AL connector. They are the two leads of a transformer winding on the power supply board. One passes through directly to the timer board, and the other passes through a 3.9 Ω current-limiting resistor to the same place. The 3.8 V AC fed to the timer board, however, is changed in one other way. A 5.6 V Zener referenced to the −30 V supply applies to both of the 3.8 V AC leads (through 220 Ω resistors) a nominal −24.4 volts. Thus the AC rides on a fairly large negative voltage. This, as we shall see, is used to drive the front panel display.

8.6.2 Tuner

The schematic of the tuner is shown in Fig. 8.20. Like the tuner in the RCA color TV of Chapter 7, and like the tuners of all VCRs made today, this one is all electronic. There is little in it that is new. The salient points of the analysis, both old and new, are covered in Exercise 8.10.

Exercise 8.10 For the tuner shown in Fig. 8.20:

(a) This tuner features dual-gate MOSFETS as RF amplifiers. In both devices, one gate receives RF signal. What does the other receive?

(b) On the left side of the schematic is an input terminal called VT. What does the voltage applied here do?

(c) There are three connections to the tuner labeled BU, BH, and BL. These presumably mean "Band UHF," "Band High (VHF)," and "Band Low (VHF)." Selection of one of these bands will involve doing what to the selected connection?

(d) Exactly what will the band-select input do inside the tuner?

(e) What is the configuration of the VHF local oscillator?

(f) What is the configuration of the VHF mixer? (Be careful on this one – it's not as simple as it might seem.)

(g) What is the configuration of the UHF local oscillator?

(h) What is the configuration of the UHF mixer?

(i) How does the UHF mixer output reach the "outside world"?

(j) How is the UHF tuner tuned?

(k) What is the "B" terminal of the tuner for?

(l) Diodes D_{104}, D_{105}, and D_{106} are switching diodes. Explain what they do in the output circuitry of the VHF RF amplifier.

(m) To the right of the tuner is IC_{1401}, the "Band Decoder." Given that B0 and B1 are logic signals to this chip, what would the chip be called in digital terms?

8.6.3 The IF sections

8.6.3.1 Video IF section

We return to Fig. 8.14B to see where the signal goes after leaving the tuner, and we see it going to a single-stage amplifier first and then to an IF chip. The discrete amplifier (Q_{1501}) looks more formidable than it needs to, because there appears to be no ground in evidence. The 82 Ω resistor connected to the IF output of the tuner has its upper end grounded, as may be seen by following the lead three inches to the right of R_{1502}. The IF signal is coupled from its lower end into the base by C_{1508}. The collector supply voltage for this stage is derived from the P CON 9 V supply by way of the R_{1521}–C_{1509} decoupling network. This leaves the stage itself as shown in Fig. 8.21.

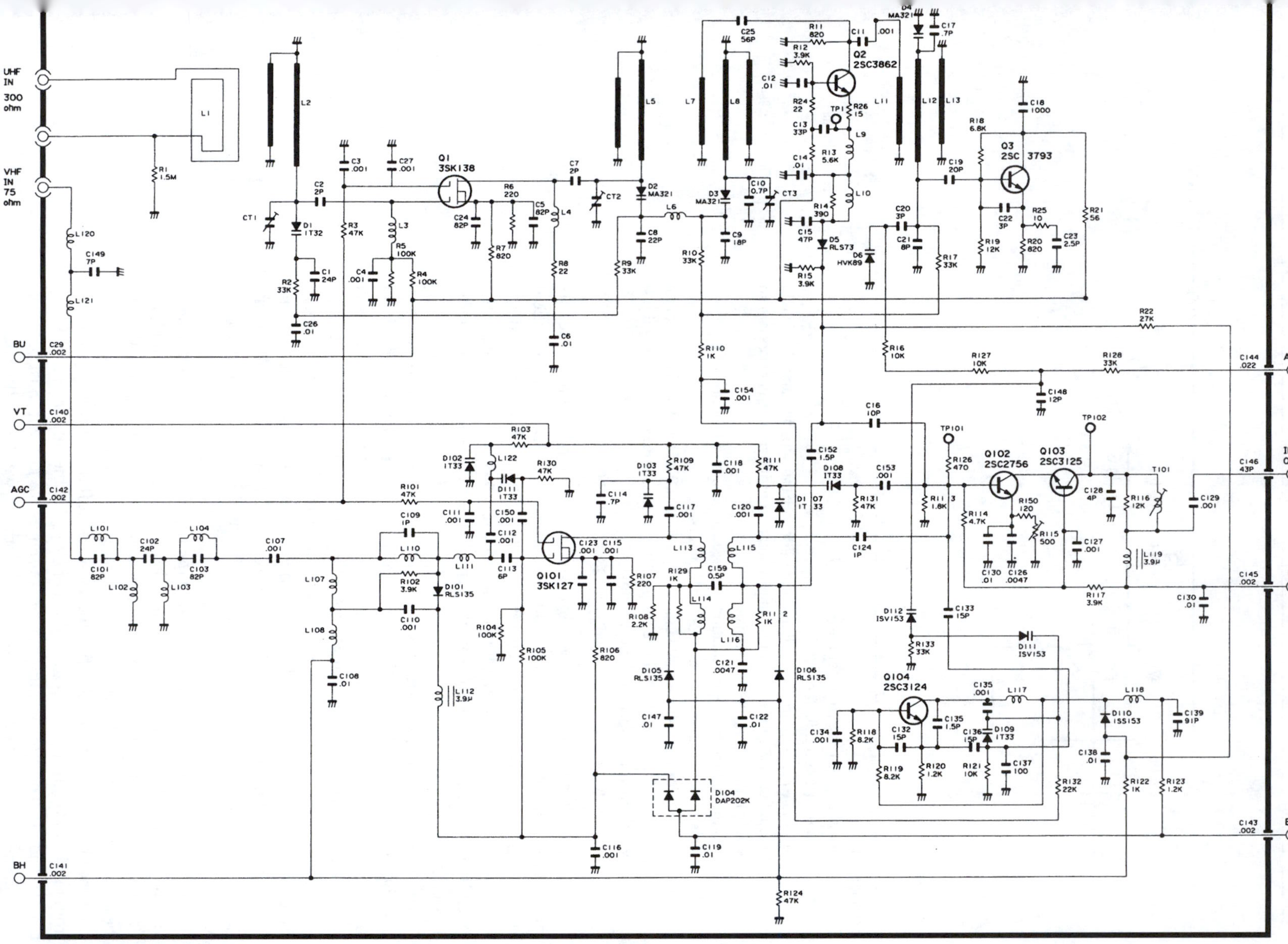

Fig. 8.20: Schematic diagram of the tuner in the Sharp VC-785U VCR.

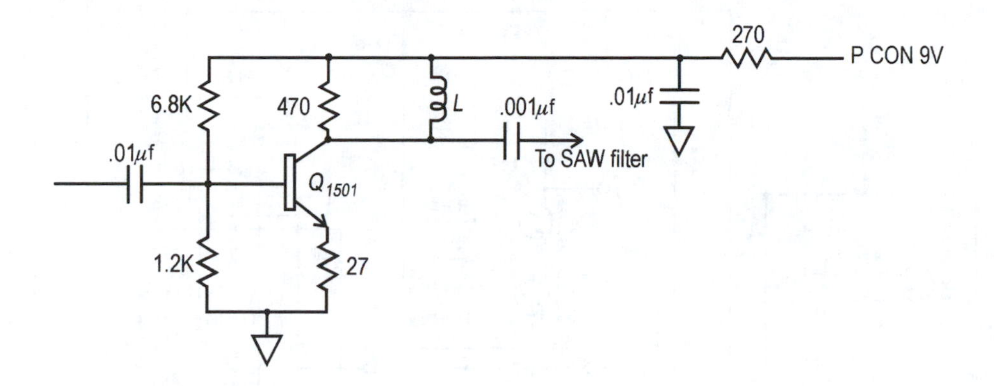

Fig. 8.21: Schematic diagram of the first stage of IF amplification in the Sharp VC-785U VCR.

Exercise 8.11

(a) Find the quiescent collector current of Q_{1501}.
(b) Find the transconductance.
(c) This stage is configured as a voltage amplifier with transconductance feedback. To validate use of the first-order approximation to its gain, it is necessary to establish that $g_{\mathrm{m}} R_{\mathrm{e}} \gg 1$. Evaluate $g_{\mathrm{m}} R_{\mathrm{e}}$.
(d) Find the first-order approximation to the stage gain as a function of complex frequency, and make a Bode amplitude plot of it.
(e) The value of the collector inductance is given as .82 μH or 3.9 μH. The parts list gives it as .82 μH. Evaluate the pole frequency for each of these L values. Which do you suspect is correct and why?
(f) What is the asymptotic gain above the pole frequency?

The SAW filter has already been discussed in Chapter 7. Here we see its input side driven by Q_{1501}. Its output side drives a parallel RL circuit identical to the one in the collector of the stage that drives it. The SAW filter will shape the video response. In particular, it will cause the response to fall off above the video IF band to eliminate noise from the part of Q_{1501}'s passband that contains no signal.

This brings us to IC_{1501}, which is reproduced in Fig. 8.22 for convenience. The signal from the SAW filter is differentially coupled to the IF amp within the chip. Differential outputs from this stage drive the video detector, which, in turn, drives the AFT and video amplifiers. The principle of AFT was discussed in Section 7.7.5.3. This one is exactly the same in principle. The resonant circuit of T_{1502} forms the heart of an $f-v$ converter, whose output becomes the AFT signal for the tuner. The "AFT mute" (otherwise known as AFT defeat) applies ~3 V to pin 12 of IC_{1501}, which kills the AFT voltage output from pin 11.

The video detector output is further amplified by the video amplifier, which, in turn, feeds it to the IF AGC circuit and to the outboard video amplifier, Q_{1503}. This should be compared to the TV circuits previously studied. In addition to these two outputs from the video section, there are also outputs to the sound and sync circuits. It is not hard to find the sound takeoff in this circuitry. It is fed from pin 16 through R_{1522}, L_{1504}, C_{1529}, and CF_{1501} to the limiting amplifier at pins 19 and 20. The sync is slightly more obscure. We know it must be there. We shall just have to be on the lookout for it. For now, we shall follow out only those video paths we can see readily.

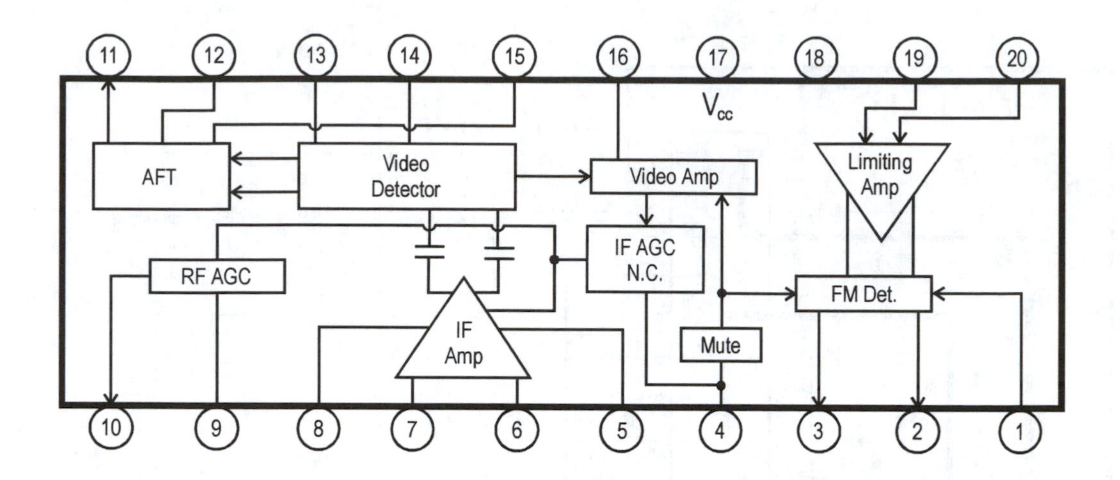

Fig. 8.22: Block diagram of IC_{1501} in the Sharp VC-785U VCR. This chip performs IF amplification, AGC generation, video detection and amplification, and sound IF limiting and detection.

The IF AGC, not surprisingly, drives the RF AGC (delayed AGC) circuitry, as well as feeding back into the IF amp, to control its gain. The input on the IF AGC block labeled N.C. is for noise canceling. When noise spikes are detected, the mute block blanks both the video and sound.

The output from pin 16 not only feeds the sound circuitry in IC_{1501}, but it also passes through SF_{1502} to the video amp Q_{1503}. SF_{1502} is mislabeled. The designation SF is for SAW filters, as SF_{1501} shows. However, the device labeled SF_{1502} is a ceramic filter as shown by its schematic and by the manufacturer's part number (compare to CF_{1503}).

The video path is through Q_{1503}, which is simply a buffer, through J_{2201} to the video input pin. If an external video source is to be recorded, it is plugged into J_{2201}. This disconnects the video signal we have been following and feeds the external video to the Y/C board instead.

8.6.3.2 Audio

The sound IF goes through a limiting amplifier (refer to Fig. 8.22) and then to an FM detector. It is unfortunate that we have no further information about this detector. It cannot be readily identified with any of the types we have already studied. The use of a ceramic filter in the circuit adds to the intrigue. It is most likely a PLL with a VCO centered on the resonator frequency. In any event, the output of the FM detector is taken from pin 3 of IC_{1501} and passed to the audio input jack, J_{6602}. This jack is analogous in form and function to the video input jack, except it lacks a counterpart to R_{2206}. This resistor provides a 75 Ω load to any external video cable plugged into the VCR. Whereas the video input jack routes internal or external video to the Y/C board, the audio input jack routes internal or external audio to the audio board. An audio mute capability is also provided (Q_{1505}).

There remain in Fig. 8.14 basically five transistors and the modulator module of circuitry we have not yet analyzed. These circuits are all fed by circuitry on the audio and Y/C boards. We must therefore look at this circuitry in proceeding along the signal path. Because it is simpler, we attack the audio first. The schematic of the audio board (which plugs into the main board) is shown in Fig. 8.23.

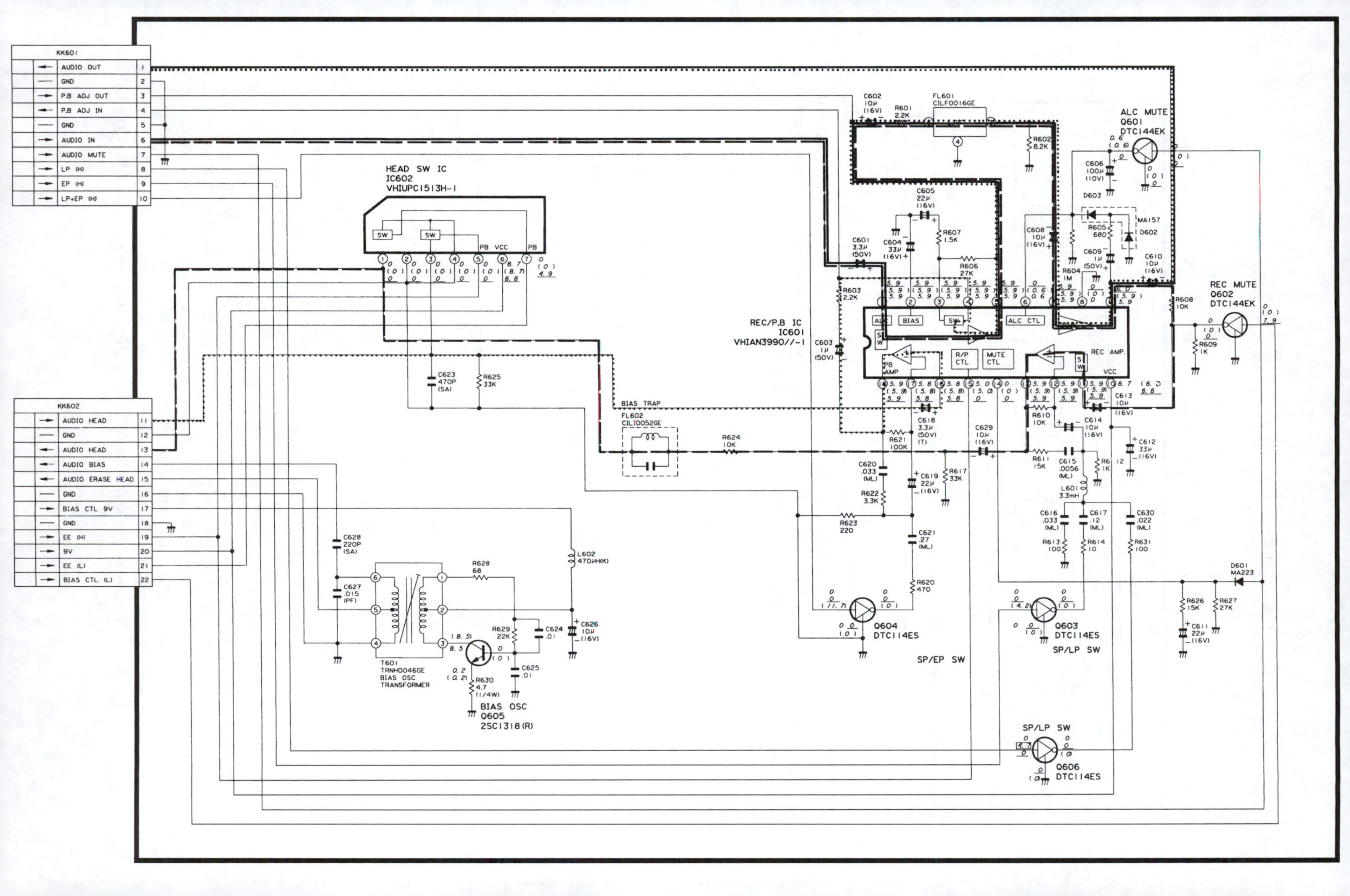
KK601
AUDIO OUT
GND
P.B ADJ OUT
P.B ADJ IN
GND
AUDIO IN
AUDIO MUTE
LP (H)
EP (H)
LP+EP (H)
KK602
AUDIO HEAD
GND
AUDIO HEAD
AUDIO BIAS
AUDIO ERASE HEAD
GND
BIAS CTL 9V
GND
EE (H)
9V
EE (L)
BIAS CTL (L)
HEAD SW IC
IC602
VHIUPC1513H-1
REC/P.B IC
IC601
VHIAN3990//-1
ALC MUTE
Q601
DTC144EK
REC MUTE
Q602
DTC144EK
BIAS TRAP
FL602
CIL10052GE
FL601
CILF0016GE
D601
MA223
D602
MA157
D603
BIAS OSC
Q605
2SC1318(R)
T601
TRNH0046GE
BIAS OSC
TRANSFORMER
Q604
DTC114ES
SP/EP SW
Q603
DTC114ES
SP/LP SW
SP/LP SW
Q606
DTC114ES
ALC CTL
BIAS
REC AMP.
PB AMP
R/P CTL
MUTE CTL
VCC
PB
L601
3.3mH
L602
470μH(K)

8.6.4 The audio board

The Audio In line (from the board just studied) enters this "sub-board" through pin 6 of the upper left connector. It enters pin 1 of the audio chip and encounters the automatic level-control stage. From here, it goes through an amplifier internal to the chip and then exits into an LPF and returns to a 20 dB amplifier.

8.6.4.1 Record mode

In the record mode, the signal out of this stage is applied through a voltage divider to the internal record amplifier of IC_{601}. The switch it passes through in so doing is enabled by the bias control 9 V supply, which we already saw in our discussion of supplies on the main board. It was found to be switched off or on by the "Bias Control" line that connects the two pages of the main board schematic. We have not yet come to the first part of the schematic, but since this line controls the "Bias Control 9 V" supply, and since Fig. 8.23 shows this supply activating the audio bias oscillator, we can be certain that it is essentially a "Record Enable" line. We shall see later that this signal originates in the system control IC (μP).

The record amplifier of IC_{601} has a feedback network consisting of nine passive elements outboard to the IC. Q_{605} and Q_{606} switch different audio equalization elements into the feedback loop of this amplifier depending on whether SP, LP, or EP has been selected as the recording speed.

Exercise 8.12 Without deriving the transfer function for the record amplifier:

(a) Reason *physically* to find the frequency for which the stage gain will be maximum if both Q_{605} and Q_{606} are off.

(b) Why is this frequency so much lower than the corresponding curve (Fig. 5.19) for the audio record equalization of the Marantz tape recorder?

After leaving the record amplifier, the signal to be recorded passes through a bias trap, past a shunt switch to ground (in IC_{602}), and back onto the main board via pin 13 of the connector (audio head). Pin 14 of the same connector carries the output of the bias oscillator back to the main board for AC bias in the record process. If we refer back to Fig. 8.14A, we will see that in the lower part of the schematic where these two signals leave the main board to go to the audio head, they are separated by only a 10 Ω resistor, R_{6620}. This would give the bias an opportunity to "back up" into the record amplifier if it were not for the bias trap. This parallel resonant circuit provides a high impedance at the bias oscillator frequency but a relatively low impedance at the audio frequencies that pass out through it.

Exercise 8.13 The bias oscillator shown in Fig. 8.23 has the transistor connected as a transconductance feedback amplifier. Assuming that pin 2 of T_{601} is at the center of the transformer primary:

(a) Find the frequency of oscillation, assuming perfect flux coupling between the windings.

(b) Find the minimum transconductance gain required for oscillation.

(c) From the schematic, estimate the quiescent collector current.

(d) Evaluate $g_m R_E$. Is it $\gg 1$?

(e) Evaluate closed-loop transconductance gain.

(f) From your result to part (a) and what you know about the bias frequency, find the value of the inductance of half the secondary winding.

The Head Switch IC, IC_{602} can, as noted, shunt the record signal to ground if activated by the control block within the chip. This block, in turn, is activated when a switch in IC_{601} connects a current source to it. We are not shown what activates the switch, but it is reasonable to suppose that it is the logical inversion of the "Record Enable" signal.

8.6.4.2 Playback mode

On playback, the signal from the audio playback head enters pin 1 of IC_{601} and enters the playback amplifier. Like the record amplifier, this one has external feedback elements, but as was the case in Chapter 5, the playback equalization network is less complex than that for record. The output of this amplifier exits the chip at pin 18 and loops back into and out of IC601 through the same path as it does on record (except for the ALC block) and finally exits the audio board at pin 1 of the connector to the main board. Once more referring to Fig. 8.14, we see that this signal goes from Fig. 8.14D through 8.14E and into 8.14C to a line-level audio output jack on the VCR and, more importantly, to an RF modulator, where it is recombined with picture information to form an NTSC signal. It is then modulated onto a channel 3 or 4 carrier to go to the antenna terminals of the TV the VCR is driving. As already noted, this allows a channel to be viewed while being recorded.

8.6.5 The Y/C board

The other sub-board plugged into the main board is the Y/C Board. This is shown in Fig. 8.24. As far as the main board is concerned, the audio and Y/C boards perform perfectly analogous functions for their respective signals. They take a signal from the tuner and prepare it for recording to tape during the record process. Then they drive the appropriate heads. On playback, they take the signals from their respective heads and condition them for delivery to the RF modulator. The audio section, as we have just seen, is pretty conventional, but as our discussion of the luminance and chroma processing in the early part of this chapter showed, the "high-tech" and unique parts of the VCR are to be found primarily on this board. When we followed the video off the main board schematic, we saw that it entered the Y/C board at two pins. One is the entry point for the luminance processing (pin 29 of the connector), and one is for chroma (pin 14).

8.6.5.1 Luminance record

We shall start this board with the luminance path. As shown in the upper left corner of Fig. 8.24A, the composite video signal goes directly to pin 4 of IC_{201}. Figure 8.25 shows a more detailed block diagram of this chip's contents. The latter figure, which was made available through the courtesy of Sharp Electronics, clears up several perplexing issues raised by omissions in Fig. 8.24.

AGC

The composite video entering this chip first goes to an AGC amplifier via pin 4 of IC_{201}. Our experience with AGC leads us to look for a rectifier whose

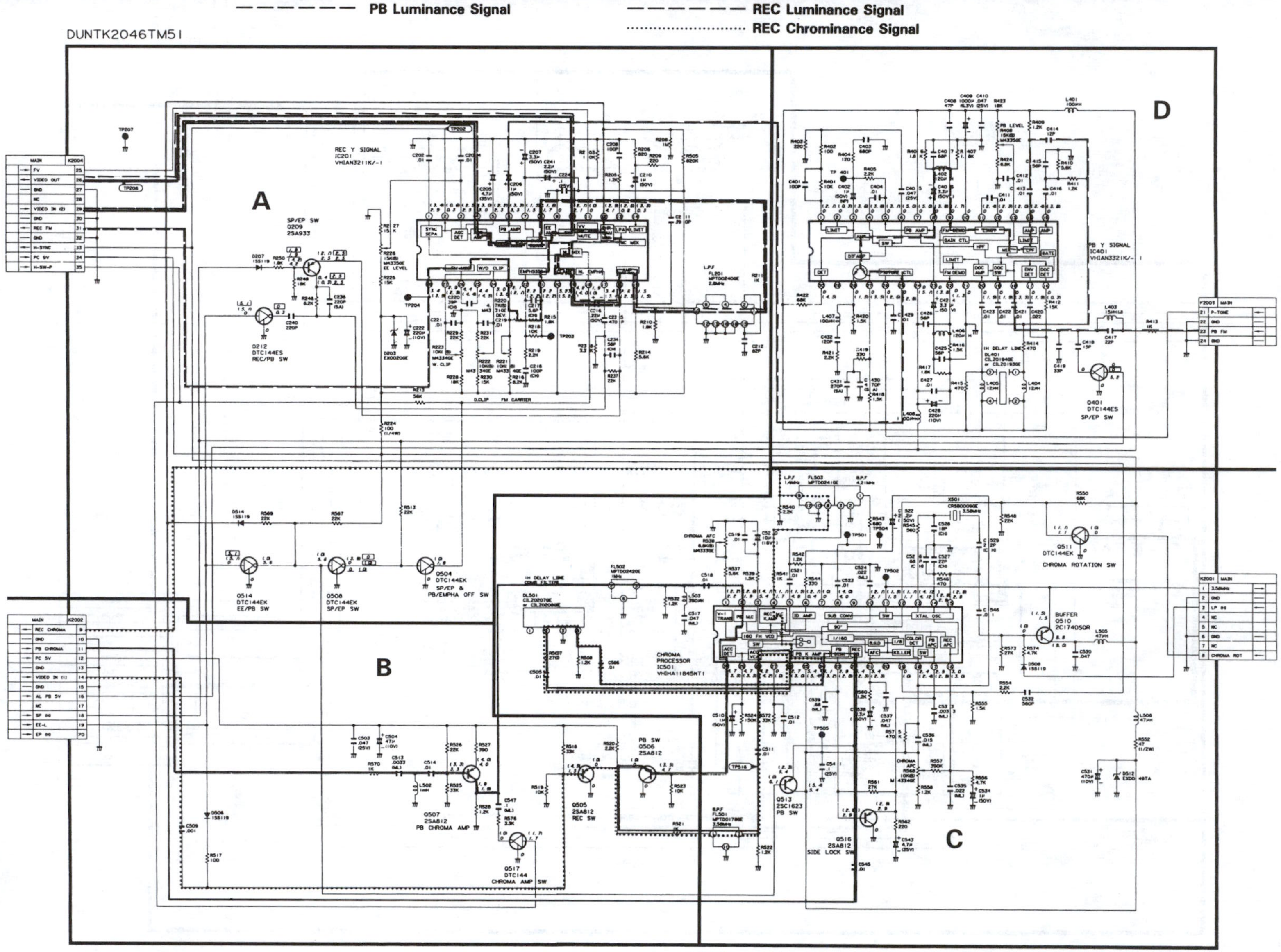

PB Luminance Signal

REC Luminance Signal

REC Chrominance Signal

DUNTK2046TM51

A

B

C

D

Fig. 8.24: Schematic diagram of the Y/C (luminance/chroma) board in the Sharp VC-785U VCR. This board plugs into the main board. An electronic version of this figure is available on the world-wide web at http://www.cup.org/titles/58/0521582075.html.

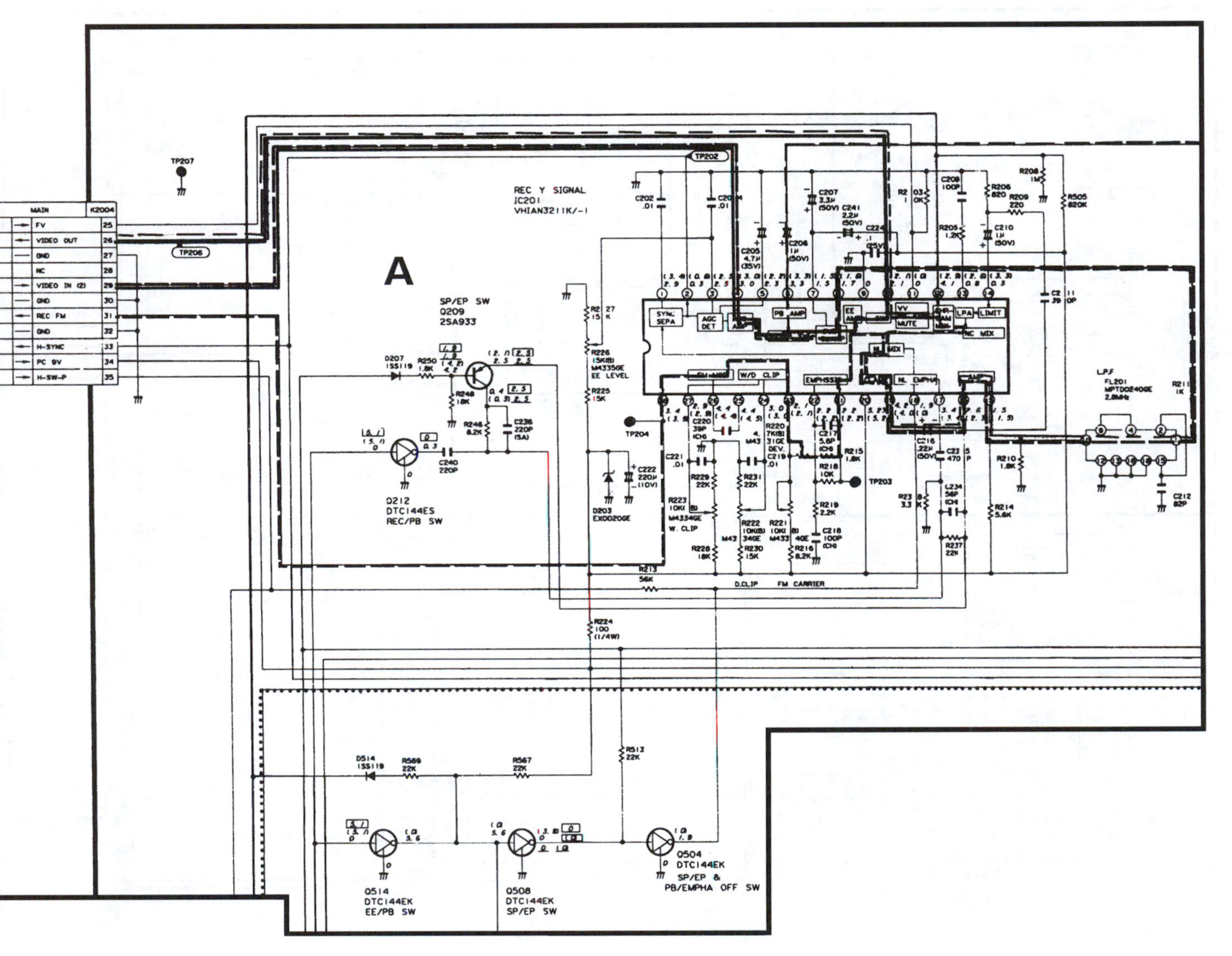

Fig. 8.24A: Luminance processing.

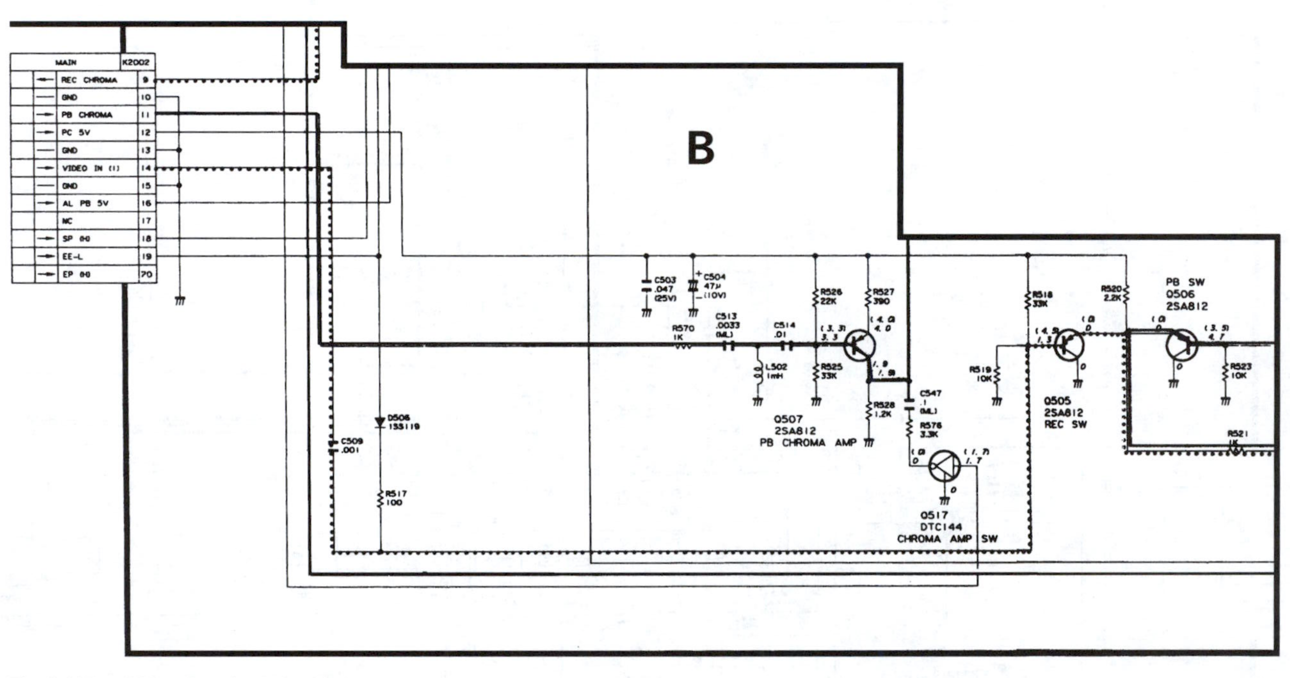

Fig. 8.24B: Video (Luminance) and chroma preamps.

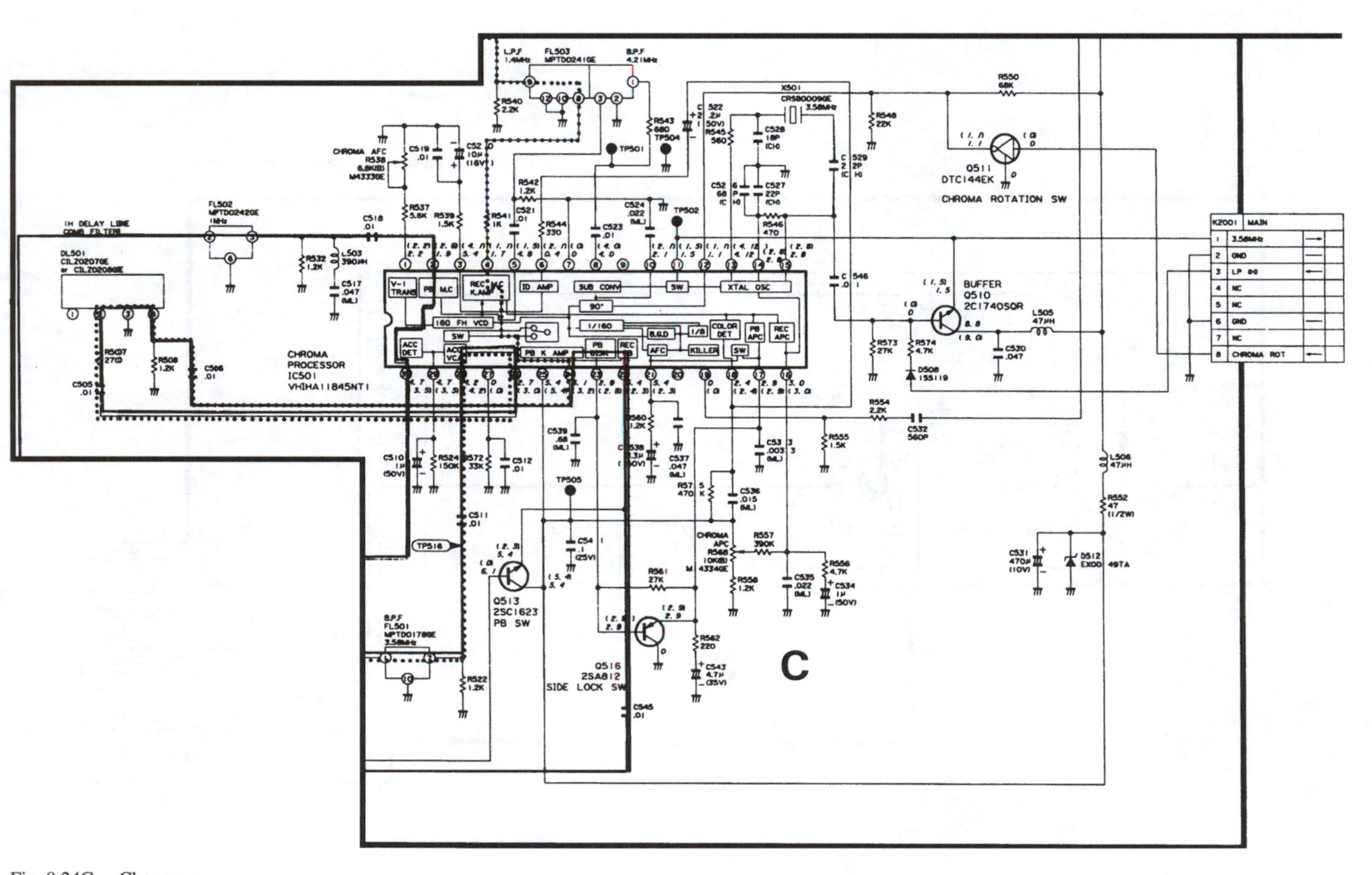

Fig. 8.24C: Chroma processor.

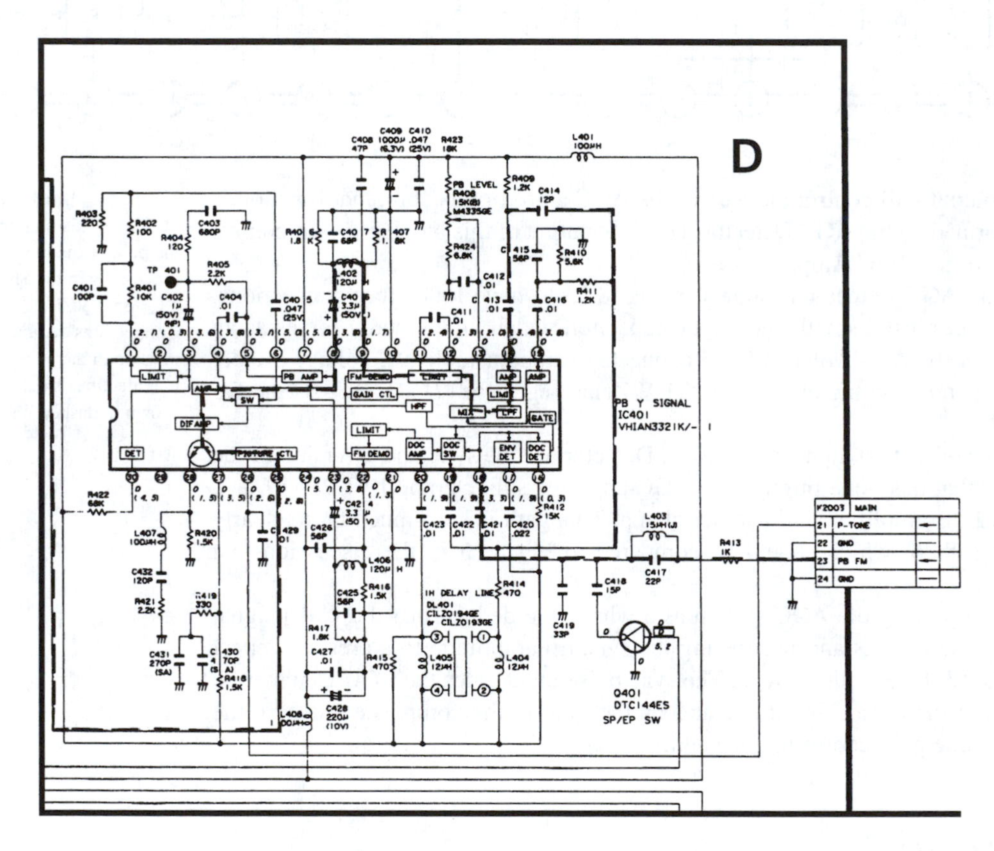

Fig. 8.24D: Playback luminance processor.

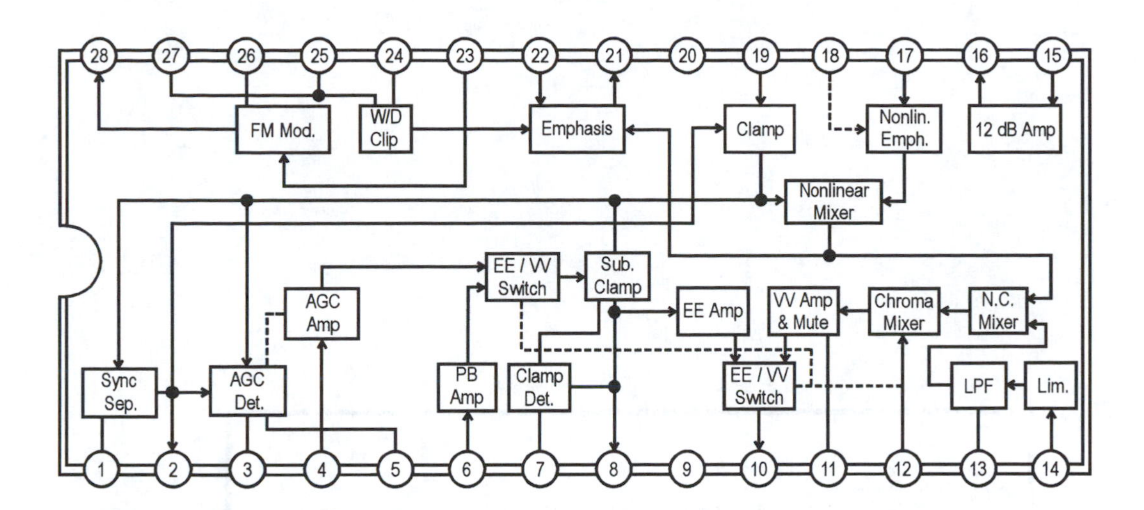

Fig. 8.25: Block diagram of IC_{201}. This chip handles the bulk of the luminance processing in the Sharp VC-785U VCR. Connections deduced to be missing in the manufacturer's data are shown in dashed lines.

DC output will control the gain of the AGC amp. In this chip, that function is performed by the AGC Detector. The only output of this block is a gain-control lead for the AGC Amp.

The AGC detector is shown in Fig. 8.25 to have four other connections. Pin 5 sees just a small electrolytic capacitor, which undoubtedly serves as an LPF for the AGC voltage. Pin 3 connects to an adjustable DC voltage divider that operates on the output of a 5.1 V Zener regulator (D_{203}) and sets the EE level.

The other two inputs to the AGC Detector are the sync and a version of the video that has gone through the AGC amp, the sub. clamp, the LPF (FL_{201}), the 12 dB amplifier, and another clamp. This path is seen much more clearly in Fig. 8.26, which should be compared to Fig. 8.25 in the discussion that follows.

Considering the AGC system as a whole, we deduce that the video signal level is held constant by servoing the sync tip amplitude to a preset level set by the EE level control, R_{226}. The sync pulse input gates the AGC Detector on during sync pulses, but it operates on the sync in the composite video, at the same time gain-controlling the entire signal.

EE/PB switch

The gain-controlled signal passes on to a switch labeled "EE/PB Switch" in Fig. 8.24A. In Fig. 8.25, which shows the block diagram of IC_{501}, it is called EE/VV. EE refers to the electric-to-electric signal path in the record mode, and PB refers to Playback mode. VV is a more general term referring to the normal signal routing inside the VCR that includes the magnetic tape. We also see a playback amplifier that feeds the switch. The next problem arises in that although we clearly have an analog switch here, in Fig. 8.24A, there is no control line shown to accomplish switching. Such a line is shown in Fig. 8.25, and it controls both EE/VV switches in that diagram. Pin 12 is shown connecting internally to the chroma mixer. As we shall see later, this is part of the playback circuitry. We recall that Q_{514} is an EE/PB switch. When the EE line is enabled low, the output of Q_{514} goes high, which delivers 4.1 V to pin 12 through R_{569}

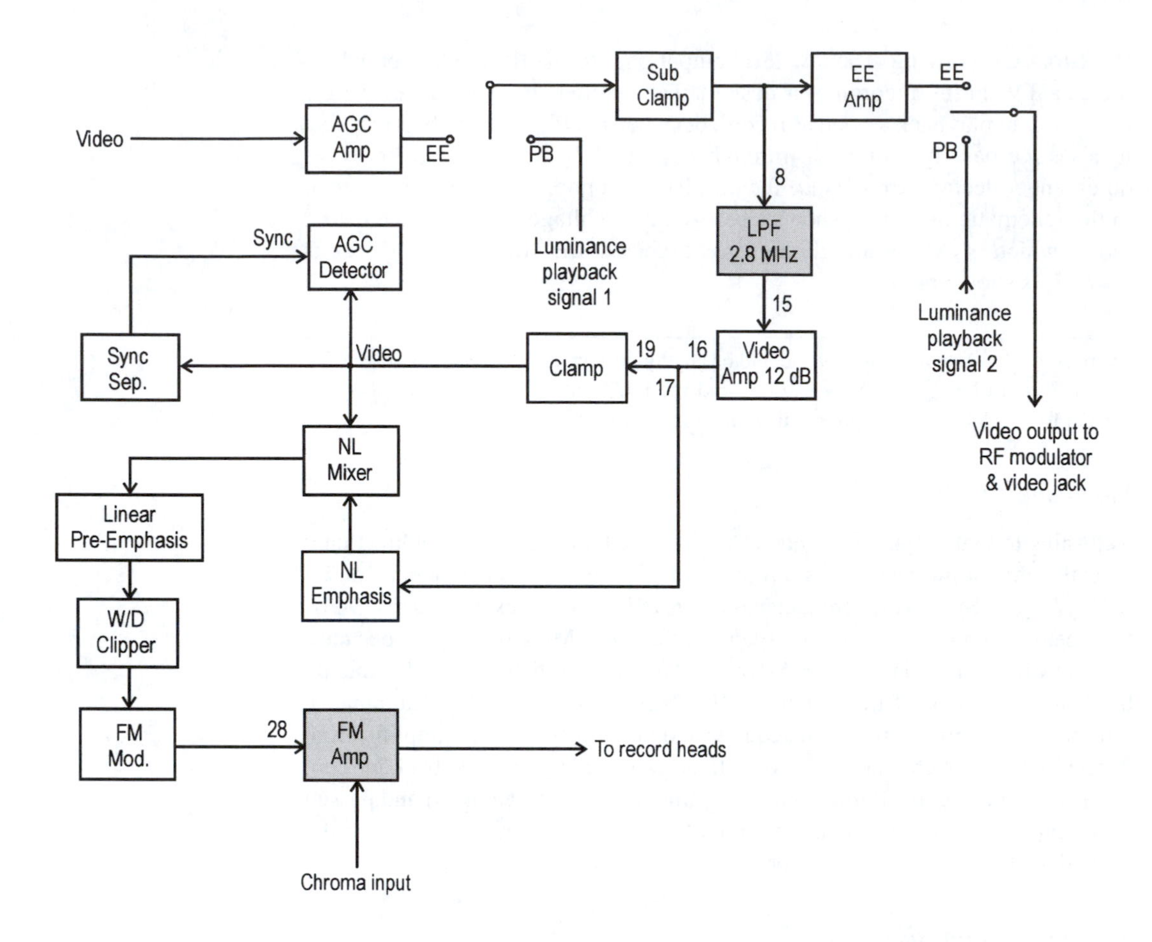

Fig. 8.26: Block diagram of the luminance record circuitry in the Sharp VC-785U VCR.

and D_{514}. This disables the chroma mixer in the EE mode, but it also causes switching between EE and PB within the chip.

Sub. clamp

The output of the first EE/VV switch passes through a sub. clamp. This stage clamps the sync tips to a level that will allow the widest possible dynamic range of the video output signal without overdriving the FM modulator. This sub. clamp stage provides separate outputs for the EE video signal and video signal which will proceed on to be recorded.

EE path

The EE output goes to an EE amplifier and then to another switch. Although it is not labeled in Fig. 8.24A, Fig. 8.25 shows us this is another EE/VV switch. In the record mode, the EE signal is selected and is routed out of pin 10 of IC_{201} and off the Y/C Board at pin 26 of the connector. Where it returns to the main board, we see that it is applied to the base of Q_{4401}, a video emitter follower. The emitter resistance of this stage is split (R_{4401} and R_{4402}) to allow a high-level output and a low-level one. The high-level output goes to the video output jack through a 75 Ω resistor to match the video cable. This output allows

VCR-to-VCR copying. The low-level output goes to the RF modulator, which drives a TV during recording, if desired. This emitter follower drives the RF modulator in playback as well as record, but when the PB 5 V supply is enabled, it raises the base bias of Q_{4401} through R_{4405} and D_{4401}. This will lower the quiescent collector current since the transistor is a pnp. There may be an error in the schematic here, as it shows a base–emitter voltage of .9 V in the record mode, which is extraordinarily high. As expected, the voltage in Playback is lower. It is shown as .7 V.

Exercise 8.14 Assuming that .9 V could be anything in the range $.85\text{ V} < V_{BE} < .95\text{ V}$ and .7 V could be $.65\text{ V} < V_{BE} < .75\text{ V}$, find the minimum factor by which the base current for ".9 V" would be greater than that for ".7 V."

Video amp

Returning to that output of the sub. clamp destined for recording, we find that it exits the chip at pin 8 and goes through an LPF with a corner frequency of 2.8 MHz (FL_{201}). Since video frequencies are usually taken as extending to 4 MHz, this seems a bit low. It is, but the span from 2.8 to 4 MHz is only .15 decades. Rolling off the response at 2.8 MHz is a "quick and dirty" way of reducing the chroma content of the signal. As Fig. 8.26 shows, the filtered luminance returns to the chip at pin 15 and acquires another 12 dB of video amplification. It again leaves the chip at pin 16, only to be capacitively coupled back into pin 19. Since it has been LP filtered and amplified, it is clamped again and passed to the inputs of the sync separator and a "Nonlinear mixer," as well as to the AGC detector, as discussed earlier.

Nonlinear emphasis

The nonlinear mixer does not derive its name from the fact that the mixer is nonlinear – all "mixers" in the communications sense are nonlinear. This mixer is not one in that sense. It is rather a summer. Its name is accounted for by the fact that it is used to sum ("mix") the main luminance signal we have been following with a side path of the same signal that has gone through nonlinear emphasis. Pin 18 receives a logic-level signal which will remove nonlinear pre-emphasis in SP mode. The path into the nonlinear pre-emphasis block is through the components between pins 16 and 17. In other than Standard Play, Q_{209} is in saturation, connecting R_{246} and C_{236} between pins 16 and 17 as well. We shall shortly resume our pursuit of the output of this "nonlinear mixer," but we shall momentarily digress.

Sync

The video output of the clamp that feeds the nonlinear mixer also feeds the sync separator, whose output leaves the chip at pin 2. From there it leaves the Y/C board at pin 33 of the connector with the main board and is applied to the servo circuitry there. The signal on pin 33 of the Y/C connector is also applied to pin 19 of IC_{501}. In getting there, it passes through a network consisting of C_{532}, R_{554}, and R_{555}. These values give a time constant of about 2 μsec. This is so much shorter than the width of even a horizontal sync pulse

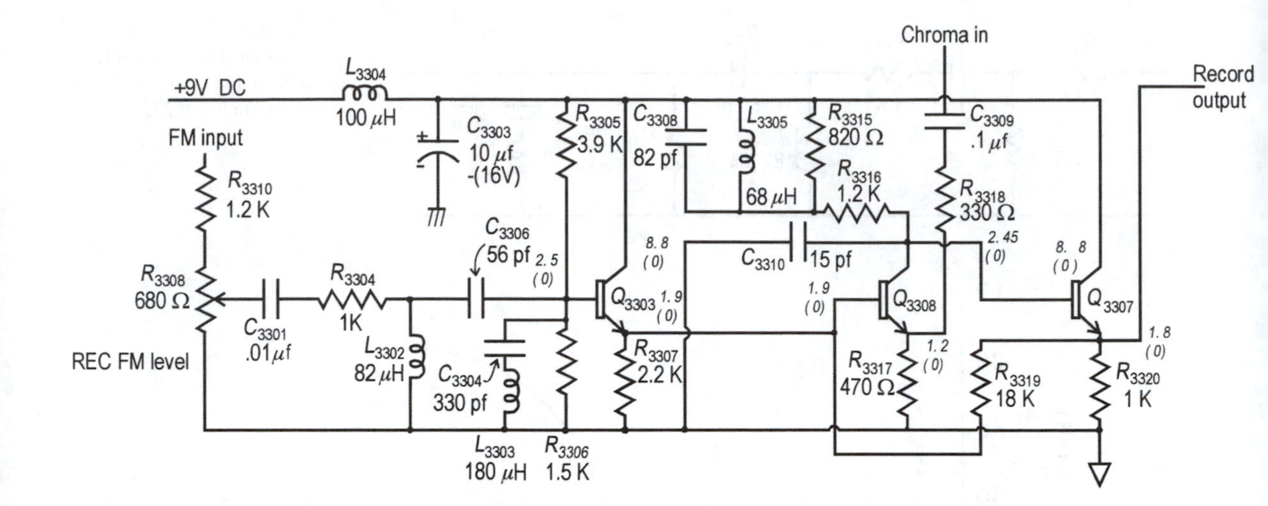

Fig. 8.27: Schematic diagram of the amplifier that combines the encoded luminance and chroma during the record process in the Sharp VC-785U VCR.

that we recognize that the network will actually function as a differentiator, delivering a sharp voltage spike to IC_{501} with each sync pulse. We shall see later what this chip uses the spikes for. The path off the Y/C board through pin 33 of the connector will be seen later to end up at the vertical sync processing stage.

Main preemphasis, W/D chip, and FM modulator

We now return to the luminance output of the "nonlinear mixer" in Figs. 8.25 and 8.26. This signal next goes through a linear pre-emphasis stage. The components connected to pins 21 and 22 of IC_{201} set the frequency response characteristics of this pre-emphasis stage. The pre-emphasis is analogous in purpose and implementation to the RIAA and NAB equalization curves for LP records and audio tape, respectively, as discussed in Sections 4.2 and 4.3. The output proceeds to the FM modulator by way of the white/dark clipper. The limiting of the white and dark shadings of the picture will correspond to a limiting of the negative and positive excursions of the luminance signal respectively. These, in turn, will determine the frequency deviation out of the FM modulator as discussed in Section 8.5.2.1.

FM amplifier and mixer

The output of the FM modulator returns to the main board via pin 31 of the connector between them. On the main board, it enters the three-stage amplifier shown in Fig. 8.27. This circuit has several interesting features. The input circuitry is a 629 kHz notch filter whose schematic is isolated in Fig. 8.28 for the case where R_{3308} is at the middle of its rotation. An analysis of this circuitry by PSPICE® is shown in Fig. 8.29. The notch rejects any chroma in the luminance signal at this point. An emitter follower (Q_{3303}) serves as an output buffer for the filter. The output of this emitter follower drives a two-stage feedback amplifier.

A first look at the circuit may raise the question, "Since the purpose of negative feedback is to produce a gain that depends only on the feedback element(s),

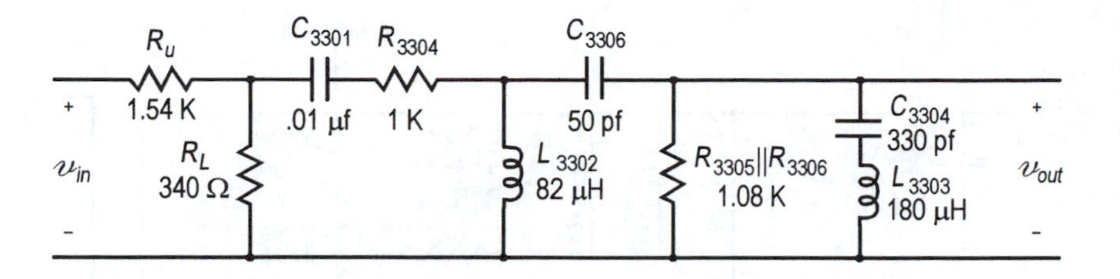

Fig. 8.28: Schematic diagram of the notch filter at the input of the amplifier shown in Fig. 8.27.

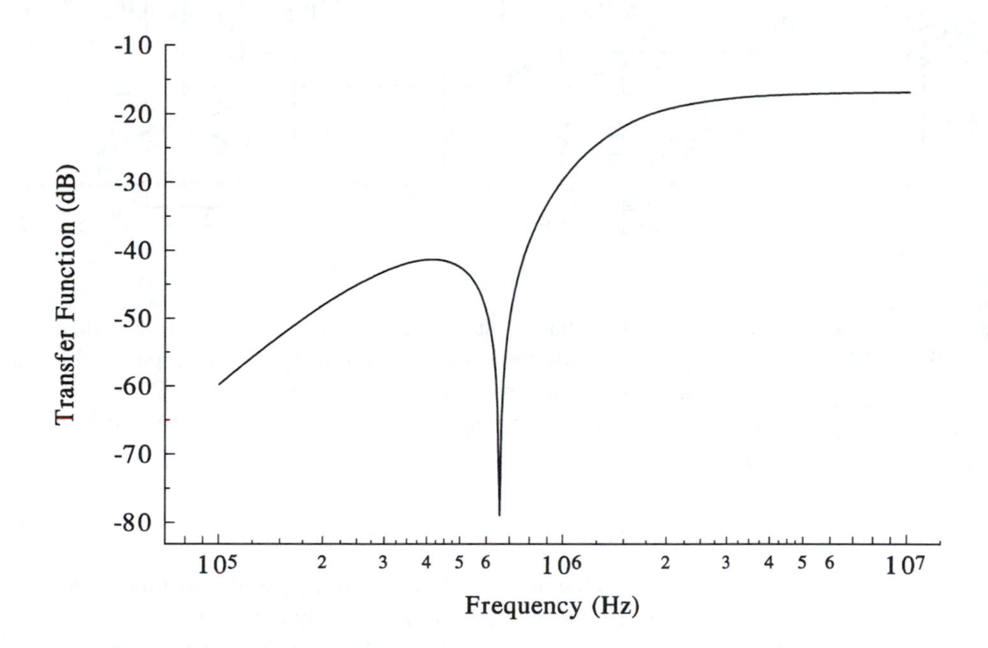

Fig. 8.29: Transfer function of the notch filter of Fig. 8.28.

and since the collector load of Q_{3308} is not part of the feedback network, why have they bothered to use frequency-selective elements in that load impedance?" This is a very good question. The answer hinges on the fact that Q_{3308} is part of two feedback loops. The transconductance feedback is most readily seen: It is provided by unbypassed emitter impedance, R_{3317}. There is also feedback due to R_{3319}, which loops both Q_{3308} and Q_{3307}. Since the first-order approximation to closed-loop gain is also calculated from the outermost loop, we might be tempted to replace our first question with a second: "Why do we consider the transconductance feedback at all, since it is an inner loop?" To answer this question, we note that the output voltage of emitter follower Q_{3307} is applied to R_{3319}, which then delivers a current to the base of Q_{3308}. This tells us that the outer loop is transimpedance feedback, which is working at cross-purposes to the transconductance feedback. Although both feedback types will lower the gain, it is not obvious which type of gain it will be! Will input impedance be low as in a transimpedance amplifier, or will it be high as in a transconductance amplifier? We will be on safer ground if we attack the problem by conventional circuit analysis, rather than by feedback concepts. The analysis will be facilitated by a couple of assumptions, to wit:

Emitter follower Q_{3307} is in circuit to buffer the Record Output from the collector of Q_{3308}. It will be assumed to have a gain of unity. Thus, to the first order, feedback from its emitter will have the same effect as feedback from the collector of Q_{3308}.

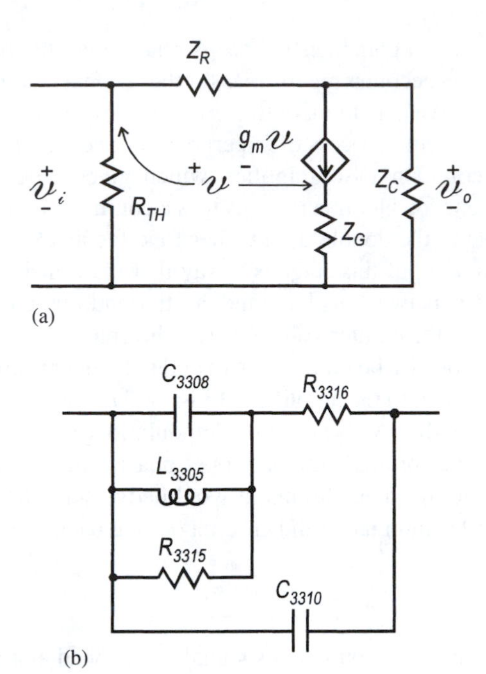

Fig. 8.30: Schematic diagram of the AC equivalent circuit used to analyze the feedback amplifier comprising the last two stages in the schematic of Fig. 8.27. (a) The amplifier itself in terms of Z_c. (b) The components comprising Z_c.

Secondly, we will assume $r_\pi \to \infty$. This is essentially equivalent to assuming that $\beta_{AC} \gg 1$.

With these two assumptions, we are able to represent the two feedback loops with their host amplifier as shown in Fig. 8.30a. Figure 8.30b shows the components which have been grouped into Z_c. In Fig. 8.30a, Z_R is the impedance producing transresistance feedback, namely $R_{3319} = 18\,\text{k}\Omega$. In like manner, Z_G is the impedance producing transconductance feedback, $R_{3317} = 470\,\Omega$.

Exercise 8.15 Calculate the voltage gain, v_o/v_i, for Fig. 8.30a, and evaluate the low-frequency asymptotic gain numerically (the actual gain will differ from this by less than 5 dB from DC to 10 MHz).

Exercise 8.16 Calculate the input impedance (exclusive of R_{TH}) for the circuit of Fig. 8.30a.

Exercise 8.17 Calculate the output impedance (exclusive of Z_c) for the circuit of Fig. 8.30a.

These analyses tell us that the gain is small and inverting (as expected). It also turns out that for the values used in this circuit, the low-frequency voltage gain is within 10% of what it would be if we had only transconductance feedback. The low-frequency input impedance will be about 5.4 kΩ. This is comparable to what it would be with no feedback, but with one significant difference: It is relatively independent of the Q point of the transistor! The low-frequency output impedance is quite close to $1/g_m$, which is lower than could reasonably be expected from transresistance feedback alone and certainly lower than what transconductance feedback gives.

The circuit of Fig. 8.27 has one more noteworthy feature. Note that the chroma record signal enters the emitter of Q_{3308}, whereas we have seen that the FM-encoded luminance is applied to the base of this same stage. This is the stage, therefore, that combines the two signals such that the FM-encoded luminance essentially becomes the AC bias for the chroma record signal, prior to magnetic recording. We may think of this process in one of two ways: First, we may think in terms of the process of superposition and say the chroma signal uses Q_{3308} as a common–base amplifier, which gives a large, noninverting gain, whereas the encoded luminance uses it as a common–emitter with a small inverting gain, due to the double dose of negative feedback on the stage. The second way of thinking of this stage is to say that the transistor amplifies the difference in voltage between its base and emitter and that the base voltage is FM (luminance) and the emitter voltage is the chroma.

Perhaps we are now in a better position to understand why an elaborate notch filter is used at the input to the amplifier of Fig. 8.27. Any chroma here would effectively appear on the FM as amplitude modulation. When we get to Q_{3308}, we want to add in the chroma with the proper phase and amplitude. If the FM enters this stage with resident chroma, it would adversely affect the decoding process that separates luminance and chroma on playback.

Head amplifier

The amplifier just discussed outputs its signal to the head amplifier board, the schematic of which is shown in Fig. 8.31. We shall discuss the various parts of this board as we encounter them. For now, we will follow the Rec FM signal from pin 7 of the connector to a class AB push-pull common–emitter amplifier composed of Q_{304} and Q_{305}. The output of this amplifier is applied to one side of each of the video heads (or more accurately, to one side of the rotating transformer that drives each head). Switches internal to IC_{301} ground the other lead from each rotary transformer through 10 Ω resistors when in Record mode.

We return for a moment to the circuitry of the Y/C board. We had left pin 28 of IC_{201} and headed back to the main board enroute to the heads with the luminance part of the record signal. We have covered all of the circuitry in this chip except that in the lower right-hand corner of Fig. 8.25. This circuitry is for luminance playback and will be covered in Section 8.6.5.3.

8.6.5.2 Chroma record.

It was previously pointed out that there were separate paths from the main board to the Y/C board for luminance and chroma. We now pick up the pursuit of the chroma record signal as it enters the Y/C board at pin 14 of its connector to the main board. As it enters, it is composite video. It is applied to the base of emitter follower Q_{505} and passes on to a 3.58 MHz bandpass filter. In spite of the fact that Q_{505} is clearly an emitter follower, it is labeled "Record Switch." This is because of pin 19 on the connector. Here we see an EE enable that is active low. When it is low, Q_{505} functions normally. When it is high, 5 V is applied to the base of Q_{505} through D_{506} and R_{517}. This pulls the base potential up high enough to cut the stage off. When enabled, the signal passes to the chroma bandpass filter and enters IC_{501}, the chroma processing chip, at pin 28. The internal contents of this IC as shown in Fig. 8.24C are so seriously in error that we show instead the block diagram supplied to the author directly by Sharp Electronics. This chip's block diagram is shown in Fig. 8.32. In reviewing the

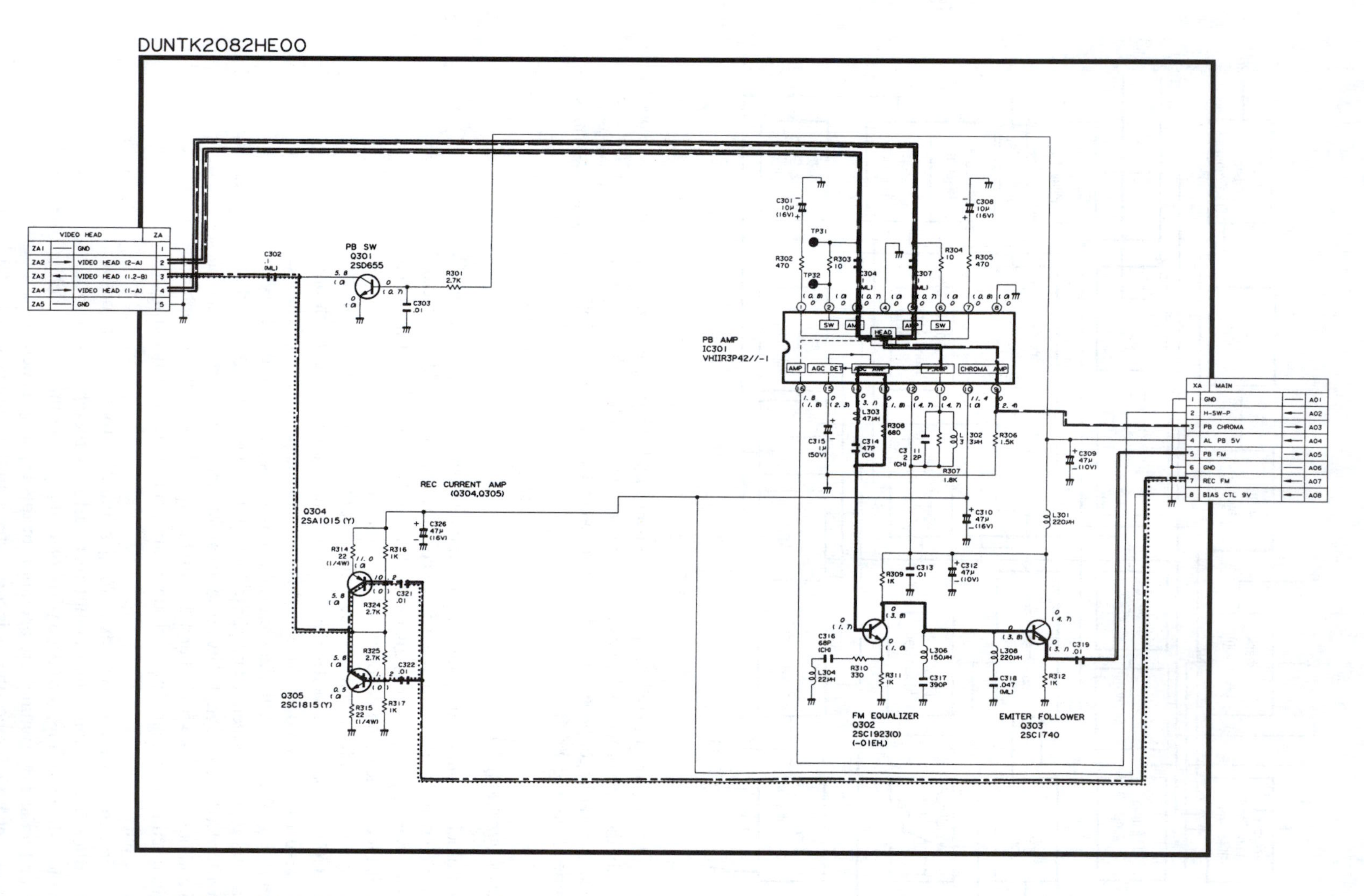

Fig. 8.31: Schematic diagram of the head amplifier board in the Sharp VC-785U VCR. ———, PB luminance signal; — — —, PB chroma signal; ··········, REC luminance signal; —— - ——, REC chroma signal.

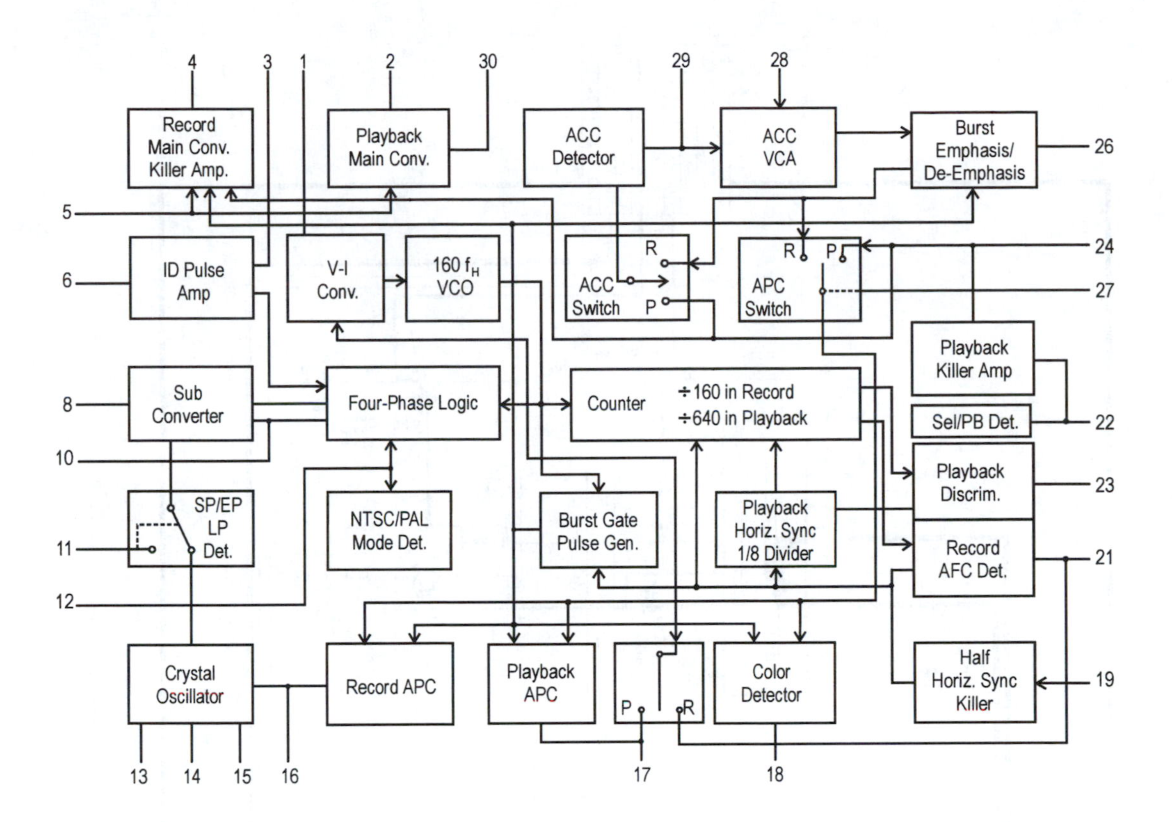

Fig. 8.32: Block diagram of IC_{501}. This chip does the bulk of the chroma processing in the Sharp VC-785U VCR.

functions of this chip, it may be helpful to consult Fig. 8.33 also, as Fig. 8.32 shows only the chip proper, whereas Fig. 8.33 is a system block diagram and gives a somewhat larger perspective. External connections to IC_{501} will still be determined from Fig. 8.24C.

ACC and burst emphasis

The first block encountered by the chroma upon entering this chip is an ACC stage, which will regulate color intensity in conjunction with the ACC detector. The loop coupling these blocks also includes the ACC switch and the burst emphasis block. The combination is isolated in Fig. 8.34.

It cannot be stated for certain, because the emphasis block has two outputs, but the ACC is probably servoed to the burst amplitude rather than to the chroma amplitude. This is commonly done, and is the better approach, since chroma amplitude will also vary with the picture. The burst is emphasized because it plays a vital role in frequency and phase locking the playback signal, and we cannot afford to have it degraded by noise. Its amplitude is therefore doubled on record and halved on playback (burst de-emphasis). Achieving burst emphasis involves nothing more than gating on some gain-control mechanism during the burst time only.

The burst-emphasized chroma exits IC_{501} at pin 26. From there it goes to a comb filter. This is the actual circuit that performs the chroma crosstalk elimination shown in Fig. 8.5. It consists of nothing more than a 1 H delay circuit and an analog summer. Its operation in this capacity is dependent on the rotating phase reference that is incorporated into the chroma signal after

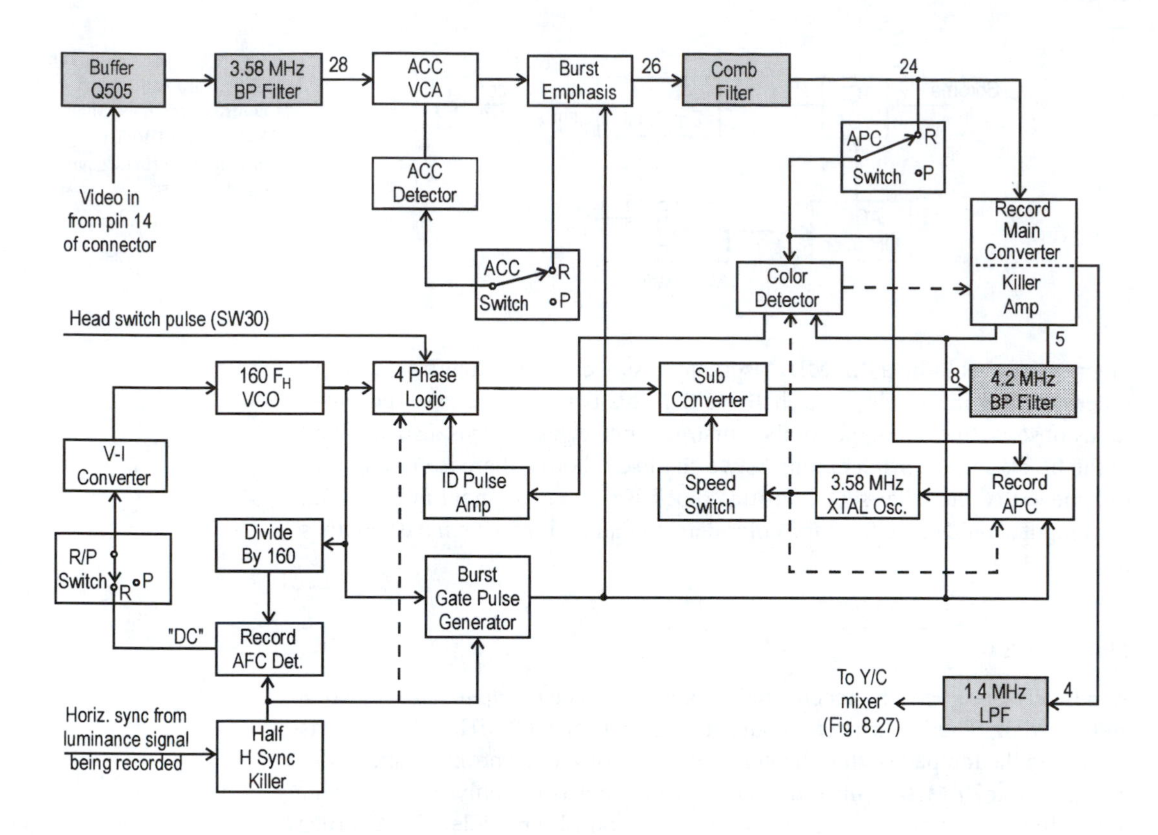

Fig. 8.33: Block diagram of the circuitry used in chroma record in the Sharp VC-785U VCR. The lines shown dashed are those deduced to be missing from the manufacturer's data. Each is discussed in the text. Connections deduced to be missing in the manufacturer's data are shown in dashed lines.

it has been translated down to 629 kHz. In the circuit of Fig. 8.24D, however, the chroma shown passing through the comb filter does not have a rotating phase reference. We might wonder why it is used here. The explanation for this again lies in one of the fundamentals of TV. In Section 7.3.2 we discussed the selection of the chroma subcarrier frequency. We saw that one of the criteria was that it be an odd multiple of half the horizontal sync frequency. The actual frequency is 227.5 f_H. The luminance signal, on the other hand, has its spectral energy concentrated at integer multiples of f_H. The comb filter delays the input signal by H and then subtracts the delayed signal from the input. The delay of H means that chroma passing through this device will be delayed by 227.5 cycles. The 227 is of no significance here, but the .5 is. It corresponds to a 180° phase shift that the chroma experiences in passing through. Thus when the chroma signal is subtracted from the signal in the non-delayed path, it reinforces it. (Subtracting an inverted signal is the same as adding a noninverted one.) The luminance signal, however, comes through the delay with no net phase shift. Thus when it is subtracted from the input, it cancels. We conclude therefore that the comb filter is used to significantly depress the luminance relative to the chroma in the chroma record circuitry.

Upon leaving this filter, the 3.58 MHz chroma returns to IC_{501} at pin 24.

Main converter/killer amp

The signal entering pin 24 is routed to the record main converter/killer amp. From comparison to Fig. 8.7a, we see that this is a balanced modulator whose

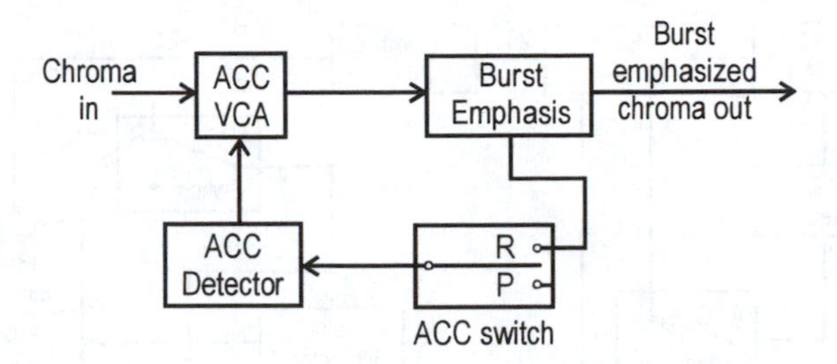

Fig. 8.34: Block diagram of the automatic color control loop in the Sharp VC-785U VCR. This circuit also provides burst emphasis.

other input should be a 4.2 MHz oscillator. As we follow along some of the other circuitry in this chip, we shall see that this is indeed the case. For now, let us observe that the output of the burst gate pulse generator is also supplied to the block under consideration. As we shall see shortly, there is good reason to believe that the killer section of this block is missing an input in Fig. 8.32. This input should come from the color detector, and it is shown in dashed lines in Fig. 8.33.

Output filter

As already observed, the record main converter receives inputs at 3.58 MHz and 4.2 MHz. Thus we expect outputs at 7.78 MHz and .62 MHz. (Actually, as we saw in the forepart of this chapter, the down-converted chroma is spectrally centered at .629 MHz.) Since the latter component is the only one we want, the higher-frequency component is removed by an LPF module with a corner frequency of 1.4 MHz. This filter module is rather unusual in that it houses two dissimilar filters: the low-pass just mentioned and a 4.2 MHz bandpass filter.

We have just followed the chroma from the input to the output of the chroma record circuitry. In analyzing the circuitry for the luminance record signal, we saw that it was joined in the emitter circuitry of Q_{3308} (Fig. 8.27) by the chroma record signal. This is the signal out of the LPF we have just discussed. Yet a great deal more must be done to get a viewable signal. We have already alluded to the 4.2 MHz that enters the record main converter. We now purpose to backtrack and see where this comes from.

Frequency synthesis PLL

We saw in Section 8.6.5.1 that the horizontal sync signal from the luminance record circuitry was passed through a differentiator and applied to the chroma record circuitry. Now that we are looking into IC_{501}, we see that this signal first reaches a block identified in Fig. 8.32 as "Half H(orizontal) Sync Killer." This might be puzzling unless we remember the nature of the time-domain video signal (Fig. 6.5). There we see that during the vertical blanking time, the horizontal sync pulses come in at twice the usual frequency (and thus with a period of .5 H) for reasons given in Section 6.1.3. Our interest in this arcane detail here is due to the fact that the horizontal sync pulse is the frequency reference for the chroma carrier generator. If this reference frequency changes with each vertical blanking pulse, it should be obvious that the consequences would be disastrous. It is, therefore, the job of the "Half Horizontal Sync Killer" to only pass every other pulse *when the frequency doubles* but to pass every pulse otherwise.

Exercise 8.18 Recognizing that the "Half H Sync Killer" is a block with just one input (differentiated horizontal sync from the sync separator) and one output, design circuitry to fit in this block and perform its proper circuit function as just discussed.

The heart of the frequency synthesis circuitry is the 160 f_H VCO, the $\div 160$ counter, and the record AFC detector. The VCO is a current-controlled oscillator, so the voltage out of the AFC block is converted to a control current by a v–i converter block. This voltage passes through one gang of an electronic Record–Playback switch on the chip just prior to entering the v–i converter. In more familiar parlance, the "Record AFC Detector" is the phase detector of this PLL frequency synthesis circuitry. A frequency 160 times f_H is 2.516 MHz. The reason for this choice of frequency will soon become clear, but its use requires a $\div 160$ counter in the VCO to phase detector feedback path to bring it down in frequency for comparison to the horizontal sync.

Burst gate pulse generator (BGPG)

The burst gate pulse is almost certainly initiated by the sync/killer block. It is less clear what ends it. It may also be terminated by what it receives from this block, but we see that the output of the VCO just discussed is also an input to the burst gate pulse block. This may be used as a time base to close off the gate pulse. Several outputs of the BGPG have been discussed earlier in this section. There is also one other, which will be discussed when we get to the color detector. For now we will conclude this subject by reiterating that the function of this block is to generate a logic-level pulse which is time-coincident with the burst. This logic pulse is used to gate other circuitry off and on – thus its name.

Exercise 8.19 How would you modify the design of Exercise 8.18 if the "Half H Sync Killer" block were also expected to output the burst that is on the back porch of the H blanking pulses? This is likely, since the output of this block is input to the BGPG.

Exercise 8.20 Since the burst has to pass through the differentiator that feeds pin 19 of IC_{501} to get to the "Half H Sync Killer," use Laplace transforms to calculate the phase and time shifts this imposes on the burst.

Four-phase logic

The output of the 160 f_H VCO also goes to a block labeled "Four-Phase Logic." This is the circuit that is used to produce the rotating phase used for recording the chroma in such a way as to cancel crosstalk (Section 8.4.1.2). What we have here is essentially a $\div 4$ counter with four outputs at phases 90° apart. These outputs are multiplexed onto a single output line.

This block is shown as having three inputs. There is probably an additional one that is not shown. As shown in Fig. 8.7, the rate of phase rotation is governed by f_H. Figure 8.32 shows no input at f_H to the four-phase logic block, though there is one at 160 f_H. It is possible that this block contains a separate $\div 160$

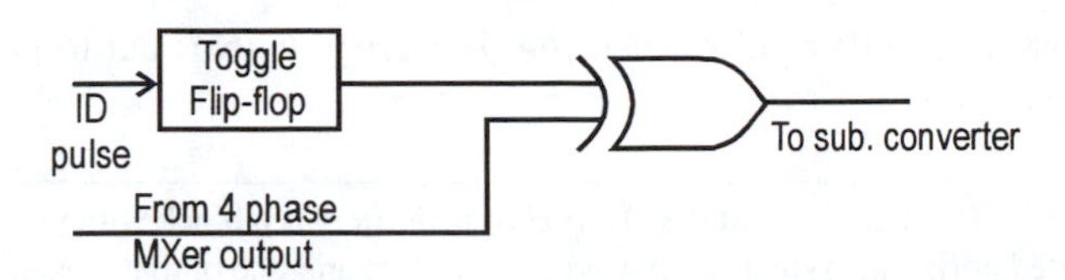

Fig. 8.35: Diagram of a logic-selectable phase inverter such as that activated by the ID pulse on playback if phase reversal is necessary.

function, but this seems less likely. It is more likely that we have an omitted connection in the block diagram. This presumed connection has been added to Fig. 8.33 as a dashed line. Another input to this block is the head-switch pulse (30 Hz), which governs the *direction* of phase rotation. This is input on pin 12. It will be recalled that the method of crosstalk elimination is based on the use of opposite phase rotations for each of the two heads when it is selected to do the recording. Thus the same pulse used to accomplish head switching is also used to switch the direction of the rotating phase generator. The process of dividing the frequency by four gives an output of 629 kHz with the previously mentioned 2.516 MHz input. This is, it will be recalled, the correct frequency for the down-converted chroma subcarrier.

The last input to this block is called the ID pulse. This comes from the output of the color detector. This input to the four-phase logic block does nothing during the record operation but is important in playback. For now it will suffice to say that it may be conceptualized as shown in Fig. 8.35. It is essentially a logic-selectable phase inverter.

Subcarrier converter and output filter

The converter block accepts as inputs the rotating-phase chroma at 629 kHz (from the four-phase logic block) and the output of a crystal oscillator at a nominal 3.58 MHz. The converter will produce outputs at $3.58 + .63 = 4.21$ MHz and $3.58 - .63 = 2.95$ MHz. The upper frequency is selected by the 4.2 MHz bandpass filter and fed into the main converter. In this block, which we have already discussed, it is combined with the burst-emphasized chroma to give the down-converted chroma at 629 kHz. On first reading, this may all seem rather nonsensical. We had 629 kHz coming out of the four-phase logic block, and this was up-converted to 4.21 MHz in the subcarrier converter and then converted back down to 629 kHz in the record main converter. It is all necessary, however. Recall that the 629 kHz out of the four-phase logic block is unmodulated, whereas that out of the record main converter contains all of the chroma information that was on the original 3.58 MHz chroma carrier.

APC circuitry

In discussing the subcarrier converter, we observed that one of its inputs was from a crystal oscillator running at a nominal 3.58 MHz. It is actually phase locked to the frequency of the chroma during record. In this section of the circuitry there is apparently another connection missing. The Record APC (phase detector) would also seem to need an input from the 3.58 MHz oscillator. This presumed connection has been added to Fig. 8.33 as another dashed line. The 3.58 MHz oscillator output should be phase-compared to the 3.58 MHz chroma to generate the phase error signal, which will vary the frequency of the 3.58 MHz oscillator slightly, if necessary to maintain phase lock. The output of the 3.58 MHz oscillator passes through a pole of electronic switching labeled 2·6/4 PAL Det. The reference to PAL reflects the fact that the chip was designed flexibly enough to be able to handle NTSC or the European PAL system. In several places this flexibility is achieved by a ternary sensing of the DC level at a chip pin, which may in general have an AC signal on it as well. A straightforward example is pin 12, where SW30 is input to the four-phase Logic section as previously discussed. This signal is also input to a block labeled NTSC/PAL Mode Det. If the VCR is an NTSC unit as is the one under study, the head switch signal applied to this pin must have a logic-low level between 0 and .6 volts and a logic-high level between 1.4 and 3.0 volts. On the other hand, if this chip is used in a PAL VCR, the logic-low level must be between 1.4 and 3.0 volts, and the logic high must be between 3.8 and 5.25 V. Detecting NTSC or PAL mode is then as simple as determining the average DC value of the head switch pulse.

The switch coupling the 3.58 MHz oscillator to the sub. converter is less straightforward in its operation, however. It connects to chip pin 11, which can be either an input or an output, but will be carrying signal either directly or indirectly from the 3.58 MHz oscillator. In both playback and record modes, it is generally toggled as shown in Fig. 8.32, and signal is fed directly from the oscillator through the switch to the sub. converter. The only exception to this is the 4-hour (LP) record mode. In this mode, a logic-high signal is placed on pin 3 of the connector between the Y/C board and the main board. This pulls up the base of Q_{510}, which is otherwise cut off. This base bias accomplishes two things. The first is that it also raises the emitter potential. The emitter potential is applied to pin 11 where (according to the dashed control line in the switch) it toggles the switch to the 4-hour position. In this position the sub. converter is looking for a 3.58 MHz input to come into pin 11. This is also provided by the elevated base bias on Q_{510}. A small (.4 V pk–pk) oscillation always appears at pin 15 as a consequence of the 3.58 MHz oscillator operation. When Q_{510} is cut off, this signal is too low to ever drive the transistor into the active region, but when it receives bias, this signal is amplified by the transistor emitter follower. The buffered oscillator output is fed to pin 11 as required. It is also fed back to the main board via pin 1 of the connector. There it passes through a two-stage transistor amplifier to the drum/capstan servo control IC. We shall have more to say on this circuitry later.

Color detector

This is the last circuit block in Figs. 8.32 and 8.33. As the name implies, its job is to determine when there is no color signal. In the record mode this is fairly straightforward. The block is shown with two inputs: the burst gate pulse and the burst-emphasized chroma. From this we would be disposed to expect that during the time when the burst should be there (as determined by the burst gate generator) the color detector examines the chroma to see if it *is* there. If it's not, the color detector should provide a logic output to the appropriate sections of the circuitry. Here we run into another little problem. The circuitry that is preeminently in need of knowing if there is no color signal is the color killer, which is shown in the same block as the record main converter. Yet no such connection is shown, although one almost certainly should be shown. This is the fourth probable missing connection in Fig. 8.32 as received by the author. It also has been added to Fig. 8.33 as a dashed line. We shall see in the color playback mode that there is one more missing connection – another color detector input. It may even be active in record, but there is no way for us to tell.

The output that *is* shown for the color detector goes to an ID amp and then to the four-phase logic. As already noted, there should never be an "ID Pulse" in record mode.

Figure 8.36 is a reproduction of Fig. 8.32, but with the blocks and connections unrelated to the chroma record function removed.

8.6.5.3 Luminance playback

Having taken a signal to be recorded all the way to the heads, we next consider how we can get it back off the tape and onto the TV. Note that in so doing, we give no thought yet to the mechanical aspects of tape and head transport and positioning. These will be considered after we finish with the electrical aspects of playback. The playback signal must, of course, begin with the heads. We therefore refer back to Fig. 8.31.

Head amplifier

The signals from the heads proceed directly to separate input pins on IC_{301}, from where they are input to a head switch (analog switch). The toggle control is a signal from the servo control portion of the main board. The output of this switch is shown feeding separate luminance and chroma amplifiers. Since we are now considering luminance playback, and since luminance and chroma are mixed coming off the tape, we must concern ourselves with how the luminance is to be removed from the amplified head signal. Certainly Fig. 8.31 shows us nothing to suggest that any such separation occurs inside the chip. The parallel RLC connected to pin 11 of IC_{301}, since it is in shunt between signal and ground, would attenuate the chroma somewhat, but it has a fairly low Q. This is necessary if it is to pass the entire luminance bandwidth. But this raises another question. The parallel resonant frequency of this network is 5.9 MHz – half again what it should be at the FM carrier frequency. Perhaps there are other impedances internal to the chip that would lower the resonant frequency into the expected range.

Exercise 8.21 How much additional shunt capacitance would be required to make the resonant frequency 3.9 MHz?

Exercise 8.22 Investigate the possibility that resonant frequency is deliberately offset from the FM carrier frequency to achieve a high-frequency boost on playback to compensate for the falling response of the head.

The luminance leaves IC_{301} primarily via pin 14 (pin 13 supplies bias to Q_{302}) and passes through a series resonant circuit tuned to 3.4 MHz, which is near the FM carrier frequency. This circuitry will strongly favor the blocking of the chroma at 629 kHz, and the passing of the luminance. The signal arriving at Q_{302} sees an amplifier every bit as complex as that of the record-head driver amplifier

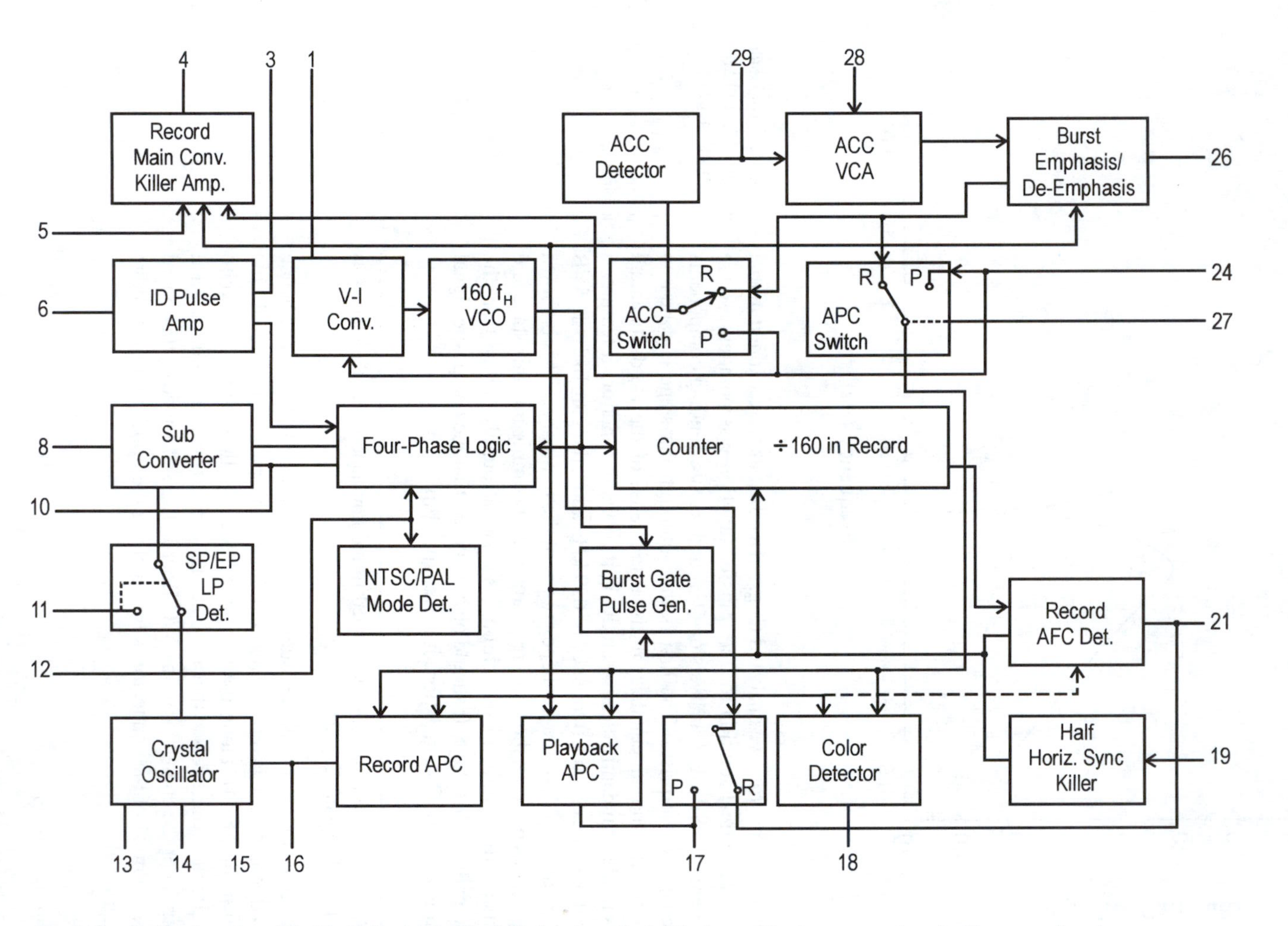

Fig. 8.36: Reduced version of the block diagram in Fig. 8.32 showing only the blocks used in chroma record mode. The connection shown in dashed lines is deduced to be missing in the manufacturer's data.

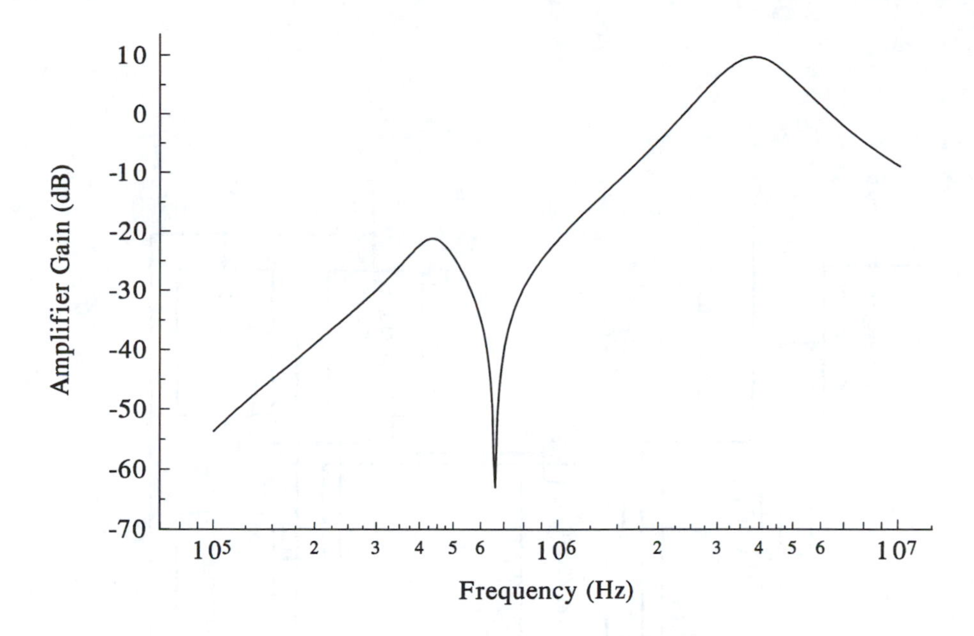

Fig. 8.37: Amplitude response of the "FM Equalizer" stage of Fig. 8.31. This stage is actually a notch filter to remove chroma from the luminance path.

in the audio tape recorder we analyzed (Fig. 5.18). Just as we reckoned in that case that an analytical solution for the transfer function was pointless, so we conclude here. A PSPICE® analysis of Q_{302} and its surrounding components yields a couple of surprises. The first is that although the stage is called an FM equalizer, it is not an equalizer in the usual sense of the word. Its actual purpose is made abundantly clear by studying its transfer function. It features a -64 dB notch at the chroma frequency of 629 kHz and a flat gain of $10 \pm .5$ dB between 3.4 and 4.4 MHz, the frequency range of the luminance. This is shown in Fig. 8.37. Thus between the series resonant circuit coming into this stage and the response of the amplifier itself, chroma is very thoroughly purged from the luminance path by this stage. It should also be noted that the composite signal passes through an AGC stage before it leaves the chip.

Exercise 8.23 Shouldn't the AGC be applied to the luminance *after* the chroma is separated? Does it matter? Explain.

The signal out of the "Equalizer" stage goes to an emitter follower, and hence off the head amp board, and on to the main board. We pick up the signal flow path again in Fig. 8.14E along the right-hand edge. Following this line onto the main circuit board, we see that it makes no contact with anything until it reaches pin 23 of the connector to the Y/C Board. The schematic of that board is Fig. 8.24. Pin 23 of the connector is found along the upper right-hand edge of this diagram also.

Phase compensation

The first circuit the playback luminance signal encounters on the Y/C board is a phase-compensation circuit composed of R_{413}, L_{403}, C_{417}, C_{418}, and C_{419}. The purpose of this circuit is to compensate for phase errors that may have been introduced in previous processing of the luminance signal. FM is not sensitive to amplitude errors, but it is to phase errors, and any circuit that does not have

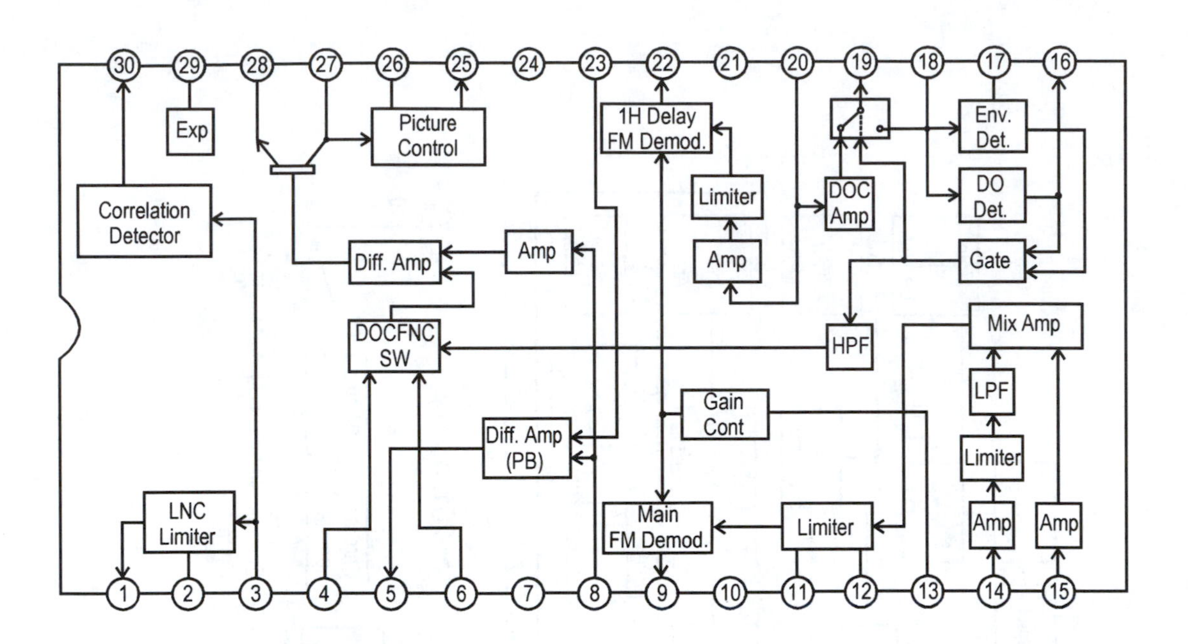

Fig. 8.38: Block diagram of IC_{401} in the Sharp VC-785U VCR. This chip provides demodulation of the FM-encoded luminance and a variety of noise-suppression circuits. It also implements dropout compensation (Section 8.5.2.2).

constant phase shift between 3.4 and 4.4 MHz will introduce a phase error. This circuit provides a falling phase response over this frequency range. Switch Q_{401} adds an additional 15 pF in parallel with the 33 pF capacitor when standard play is selected and removes it when extended play is selected. The additional capacitance changes the limits over which the phase response varies.

Dropout compensation

As was explained earlier in the chapter, circuitry is needed to preserve picture integrity when there is a loss of signal due to a localized tape defect. Dropout compensation, as well as other luminance playback circuitry, is found in IC_{401}. This IC is shown in Fig. 8.38.

Luminance is applied to a DOC detector and an envelope detector at pin 18. Since the luminance is still in FM form at this point, the amplitude should be quite constant, and the envelope detector will produce a fairly constant output. A short–time-constant network is used with the envelope detector, so the loss of a significant part of a line's worth of luminance will cause the envelope detector output to drop enough to register on the gate. Pin 16 is a pullup for the DO detector output. We may therefore conclude that this block provides a binary output that enables or disables the block labeled "Gate." The latter block is actually a gated amplifier, and the signal it gates is the output of the Env. Det.

As long as a strong signal is received from the tape, the gated amp passes the Env. Det. output to the HPF and the control line of the DOC switch. The switch will then enable the input from pin 18 to pass through and out pin 19.

If the envelope amplitude starts dropping (but not enough to cut off the gated amp) we may presume that it will hold the DOC switch in the normal position. If the signal into pin 18 drops too low, however, the DO detector will change its logic state. This will cut off the gated amplifier and will toggle the DOC switch, which will then pass the signal from pin 20 via the DOC amp and out pin 19. The signal from pin 20 is the delayed signal from the previous line. This may be seen most clearly in Fig. 8.39, which shows playback circuitry in both

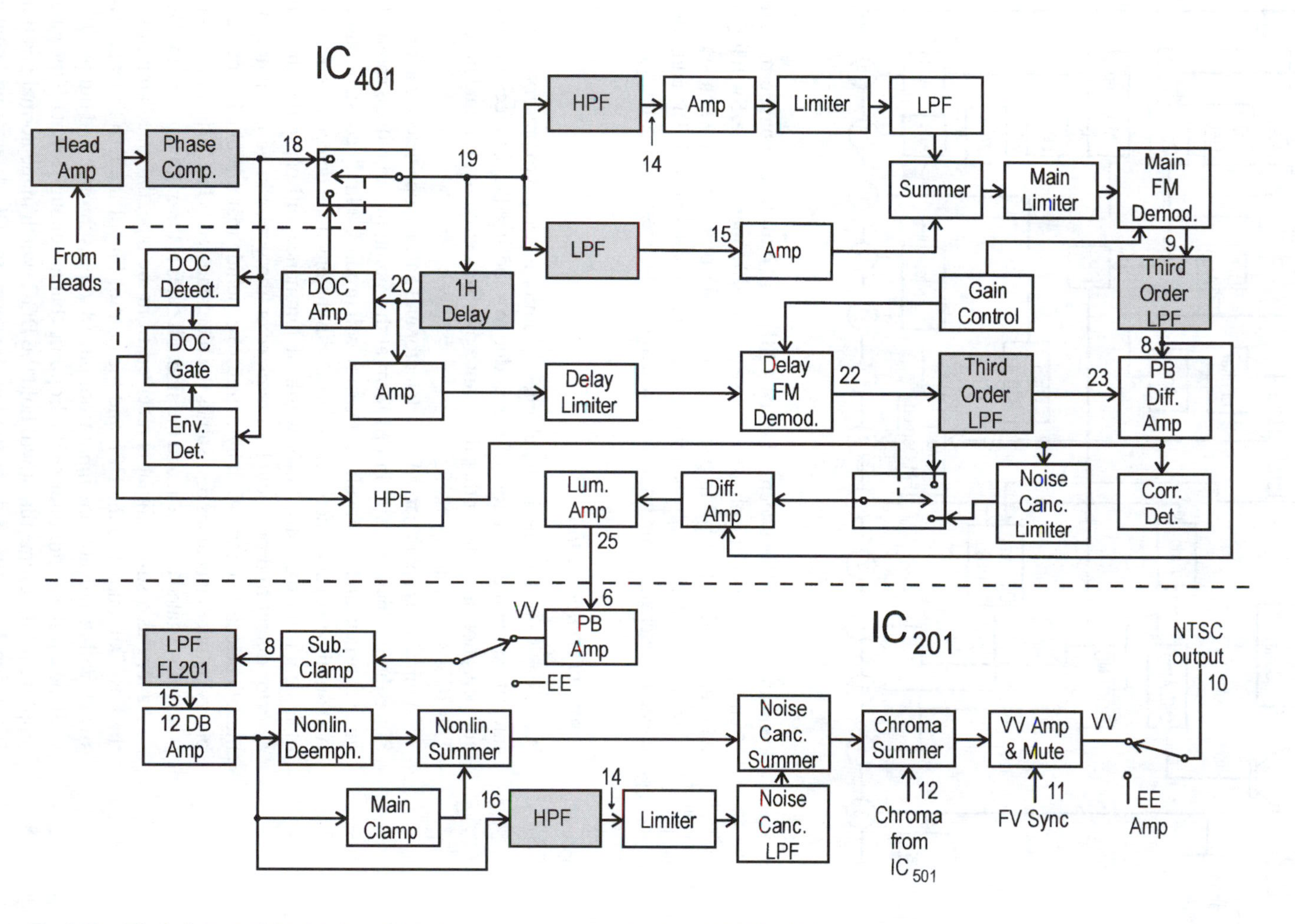

Fig. 8.39: Block diagram of the circuitry used for luminance playback in the Sharp VC-785U VCR.

IC_{401} and IC_{201}, as well as associated off-chip circuit blocks. In other words, this line is reread and passed on in place of the current one. Note that when this occurs, the output of the DOC switch at pin 19 is also connected back to the input of the delay line, so if the next line read from the tape is also defective, the previous good line can be read out again. This could go on forever, but, as noted previously, significant picture degradation results if a line is repeated more than three times. The gate at the output of the DOC detector takes care of counting repetitions and disabling the dropout switch after three recurrences.

Two-stage limiter

We should well understand the importance of FM limiting to eliminate AM noise. If we were to limit the FM signal leaving pin 19 in a conventional way, however, we would risk the loss of the low-frequency sidebands of the FM signal. Since these correspond to the high-frequency components of the demodulated signal, this would entail the loss of picture sharpness. This problem arises because of the high-frequency pre-emphasis that was applied during the recording process. The remedy is to split the FM signal into two parts by passing it through low- and high-pass filters, and initially limiting just the carrier and upper sidebands, and not the lower sidebands. This is accomplished by the blocks in the lower right-hand corner of Fig. 8.38 or the upper right-hand corner of Fig. 8.39.

The signal leaving pin 19 of this IC goes through an LPF consisting of R_{410}, R_{411}, and C_{415}. The .01 μF capacitors through which the FM-encoded luminance signal passes have an impedance of $<5\,\Omega$ at the FM frequency. This low-pass filtered signal goes into pin 15 of IC_{401}, and thence through an amplifier to another "mixer," which is actually a summer. The other input to this summer also originates at pin 19 of the IC. From there it passes through a high-pass filter consisting of C_{414} and R_{409}. Upon reentering the chip at pin 14, the carrier and upper sidebands are amplified and then pass through a limiter and an LPF before being applied to the other input of the mixer–summer. The LPF shown internal to the chip lends a bit of mystery to the circuitry, since not only is its use uncommon but because there are seemingly no good candidates for outboard filter components.

This lack would suggest that some "on-board" capacitance is used to realize the filter. This is at least possible, since the frequency here is relatively high. The best guess is that the LPF is probably in circuit to bandlimit the amplitude-limited signal, since there should be no frequency components above 4.4 MHz in it.

The output of the mixer–summer goes to the second stage of limiting, where the entire signal is limited. From there, it goes to the main FM demodulator.

Main FM demodulator

The main FM demodulator circuit shows no outboard components that we can identify readily with any of the types we have already discussed. This leads us to suspect that the FM input is converted to a PWM or PFM waveform that need only be low-pass filtered to extract the demodulated FM. This type of detector is common in VCR circuits. One such circuit is discussed in Chapter 2 under the heading of Monostable Detector. This main FM demodulator and the delay

FM modulator both have their gains set by an electronic gain control block, which is controlled in turn by a service adjustment connected to pin 13. As Fig. 8.39 indicates, the output of the FM demodulator is applied to a third-order LPF with a corner frequency of about 2.5 MHz.

The delay path

In addition to being applied to the DOC amp, as already discussed, the output of the 1 H delay line is applied to a delay path at pin 20 that subjects it to processing analogous to that given to the main signal path. Since this path is seldom used by the playback luminance and then only for three or fewer lines at a time, a single limiter is used rather than the double limiter in the main path. The separate (delay) FM demodulator is followed by an outboard third-order LPF exactly analogous to that in the main path. The outputs of the LPF in both the delay and main paths come together in a block identified as "Dif Amp (PB)" in Fig. 8.38 and "PB Amp" in Fig. 8.24D. We shall return to this block shortly, but first we have a small problem to deal with. In the luminance recording process, the last two processes before FM modulation were W/D clipping and linear pre-emphasis. The clipping merely insured that the dynamic range of the modulator would not be exceeded. The pre-emphasis, however, involved a major revision of the spectral composition of the video signal. Therefore, the first step after FM demodulation would be expected to be linear de-emphasis. No such block is seen in Fig. 8.38, either where we expect it in the circuit or in any other obvious place. We therefore provisionally conclude that it has been incorporated into some other block and is not explicitly identified.

Edge-noise canceler circuitry

As noted, both the main and delay FM modulators feed into the "Diff. Amp" (differential amplifier) block. Yet, according to the circuit trace notation that Sharp uses on their schematics, this amplifier does not lie in the main signal flow path of the demodulated luminance signal. This amplifier has as its inputs the video signals from the current line and the previous line (courtesy of the 1 H delay line). This looks suspiciously like a noise-reduction system in which two demodulated inputs are compared in the "PB Amp." Ordinarily, there should not be major differences in two adjacent scan lines of a given field, so major differences that are actually observed are likely to be noise. It would seem, then, that the "Diff. Amp" takes the difference of these two video signals and outputs mostly noise. This output passes through C_{404} and becomes one input to an analog switch. The toggle control for this switch is the DOC enable signal passed through a high-pass filter. In the absence of dropout, the switch connects this "mostly noise" signal to another differential amplifier whose second input is the main luminance signal. The differential amplifier will, of course, subtract the "mostly noise" signal from the main signal + noise. Presumably the amplitudes of the two components are such that the noise will largely be canceled out.

When there is dropout, both inputs to the "Diff. Amp" will be nearly the same, since the main and delayed signal paths will be processing the same line. Thus we might expect the output of the "Diff. Amp" to be zero. This will not be the case, however. The fact that the main path has a double limiter and the delay path has a single limiter will, as noted, mean the delay path will have less high frequency in it and thus less noise also. Thus when the two signals are combined in the PB Amp, the output will still be noise enhanced, just as it was in the absence of dropout.

Figures 8.38 and 8.39 also show a block labeled as "Correlation Detector." This block has one input from the output of the PB Amp and one output on pin 30 of the IC. Figure 8.24D shows that this pin goes to the +5 V DC through a 68 k resistor (i.e., it goes nowhere, signal-wise). Furthermore, we have the perception that any block that truly accomplishes correlation should also have *two* inputs to be correlated and an output that "reports" on correlation. Thus we are led to believe that this block is either missing an input (and very possibly an output also) or it is misnamed. It is certain that having a correlation function as part of the noise-canceling circuitry would allow the most selective removal of noise while minimizing adverse effects on the high-frequency part of the luminance signal.

The output of the "Diff. Amp" appears at pin 5 of IC_{401} and makes its way back to the two inputs of the noise-canceler switch by two paths. These are best seen by referring back to Fig. 8.24D. The one through C_{404} has already been mentioned. This is the path in the absence of dropout. The other path goes through R_{405} and C_{402}, and into the line-noise-canceler limiter (pin 3 of IC_{401}), and out pin 1. From there it passes through R_{401} and C_{401} in parallel to R_{402} and then through C_{405} to pin 6 of the IC. This leads to the other input of the switch. The additional components used in the presence of dropout will spectrally shape the noise-enhanced signal when there is dropout; these are required because the different origin of the signal in this case will dictate its having a different spectral distribution. The signal path having returned to the inside of the chip, we again refer to Figs. 8.38 and 8.39.

As we noted, the toggle control for this switch originates with the DOC signal. One or the other inputs thus appears at the switch output and is applied to another differential amplifier. Based on what has just been outlined, it should be plausible that when the "noise-rich" signal is applied to this differential amp along with the main luminance signal (which also carries some noise), the phases and amplitudes may reasonably be assumed to be such as to cancel a significant part of the noise from the main signal path.

The discrete luminance amp

Upon leaving the differential amp, the luminance signal enters a single transistor integrated into IC_{401}. We do not have to look far to see why a single transistor, rather than an amplifier block, is shown. The biasing of the base is achieved by DC coupling to the output of the differential amplifier, but all emitter and collector circuitry is external to the chip. With all of the reactive components used, we would suspect that the stage accomplishes substantial frequency response shaping and/or phase equalization.

Exercise 8.24

(a) What kind of negative feedback does this stage have?

(b) Find its low-frequency gain.
(c) Find its high-frequency gain. As a practical matter, this will never be achieved because the transistor has inadequate open loop gain at these frequencies to validate the feedback approximation. Furthermore, these frequencies are well above the luminance spectrum.

A complete frequency analysis of this circuit shows that, contrary to expectation, neither the frequency nor the phase dependence of its transfer function is strong. From DC to 10 MHz, the frequency response drops from 0 dB down to -14 dB. But this is just what a typical linear de-emphasis operation would require. All but 2 dB of this drop occurs above 100 kHz, again as de-emphasis requires. Thus the linear de-emphasis we were looking for earlier turns up at the final exit of the PB luminance signal from IC_{401}. Basically, following the FM demodulator, this set interposes an LPF and the edge-noise canceler circuit. Since both the LP filtering and de-emphasis are linear operations, their order should not matter. Performing noise canceling prior to de-emphasis means that we first suppress noise relative to the signal and then cut it further along with the higher signal frequencies. The multiplicity of components in the amplifier circuit therefore is used to produce a composite rolloff characteristic averaging just 6 dB/*decade*.

The block labeled "Picture Ctl." has a control line input on pin 26 of IC_{401}. Tracing the control line back through the Y/C connector to the main circuit board (Fig. 8.14E) shows that it is supplied by a fixed voltage divider connected to the AL PB 5 V supply. Since this supply is only activated in the Playback mode, we deduce that the luminance output here is disabled in the Record mode.

After leaving IC_{401} at pin 25, the luminance returns to IC_{201} (Fig. 8.25) at pin 6. It may help to consult both Fig. 8.25 and Fig. 8.39 to clarify the discussion that follows. The former shows the chip (IC_{201}) only, whereas the latter shows how it interfaces to off-chip components including IC_{401}. Signal entering pin 6 first encounters a playback amp and passes to the sub. clamp. Luminance signal is passed through this clamp in the record cycle, and now on playback, the recovered luminance is passed through it again to restore the black level. The luminance then leaves IC_{201} at pin 8 and is routed through an LPF with a corner frequency of 2.8 MHz (FL_{201}). From there it passes back into pin 15, through a 12 dB amplifier in the chip, and then back out again at pin 16.

Nonlinear de-emphasis

The signal we have been following was subject to nonlinear pre-emphasis when it was recorded. This boosts high frequencies just as linear pre-emphasis does, but the effect is level dependent. There is greater boost of high frequencies when the signal amplitude is low. In other words, the same principle is acting as in Dolby B noise reduction, but now it is applied to video. From pin 16 of IC_{201}, the luminance signal takes two paths to the NL mixer (summer). What we called the main chain in Dolby (main signal path) is through C_{216} and back to the main clamp at pin 19. The output of this clamp is one input to the nonlinear summer. What we called the side chain in Dolby B goes through R_{237}, C_{234}, and C_{235} to get back to pin 17 of the chip and thus into the NL emphasis block. Despite its name, it performs NL de-emphasis also. This dual use of the same block is also characteristic of Dolby B, as shown in Fig. 5.7. Logic signals are applied to the NL emphasis/de-emphasis block at pin 18 to vary the parameters based on tape speed.

This side-chain signal also enters the nonlinear summer. As Fig. 8.25 shows, at one input of the NL mixer (summer), there is a takeoff point for horizontal sync, which is separated out by the sync separator block and passes off the chip at pin 2 and off the Y/C board at pin 33 of the connector to the main board. The NL mixer–summer is part of the NL pre-emphasis/de-emphasis system – again just as it is in the Dolby system for audio.

Noise cancellation

We previously deduced that the block in IC_{401} labeled "Diff. Amp" accomplishes edge-noise reduction. Here we have another noise-reduction stage. The block is labeled "NC Mixer" in Fig. 8.25. We would be predisposed at this point to understand this to be a summer also. Its inputs are the de-emphasized luminance from the main signal flow path and another luminance signal that has not yet been de-emphasized. This signal is picked up at pin 16 of IC_{201}. It passes through a discrete HPF made up of C_{211}, R_{209}, and R_{206} and is applied to the limiter at pin 14 through a blocking capacitor, C_{210}. The HP filtering of this nonde-emphasized signal with a corner frequency of about 400 kHz will cause the signal passing to the limiter to have an enhanced level of noise. The limiter is actually a clipper. It limits the amplitude of the peaks in the noisy signal input to it. The purpose of this limiter is exactly analogous to that of the edge-noise canceler – that is, to leave sufficient amplitude at its output to cancel noise without significantly affecting legitimate transitions in the signal itself.

The limiter output passes through an LPF to the noise-canceler mixer (summer). The path we have been following supplies almost pure noise to be inverted and summed with the luminance signal from the nonlinear mixer. The output is a "cleaned-up" luminance signal.

Chroma mixer and EE/VV switching

As usual here, the mixer appellation actually refers to a summer. This is the stage where the demodulated luminance and the up-converted color are recombined. The chroma enters via pin 12. This pin is also DC sensitive. A DC level placed on it will control the EE/VV switches. When the voltage is low, VV mode (Playback) is selected. It originates with the EE (active low) signal on pin 19 of the main board connector. It is then applied to Q_{514}. This unit has a 22 K pullup resistor to the PC 9 V supply. When it is in the low state, it pulls down the drive to pin 12, switching into the VV mode.

VV amp and mute

The output of the chroma mixer is input to this stage. When, as just discussed, VV is selected, the composite NTSC signal passes through the EE/VV switch and out pin 10. We have already traced this path to the RF modulator and video output jack when we were tracing the record signal. In that case, the signal out of pin 10 came from the EE amp.

The muting essentially shorts out the video during fast forward and reverse, and briefly after Play is pressed to give the picture time to lock in.

Figure 8.39 also shows an input to pin 11 that can be traced back to pin 25 of the main board connector. It is labeled as FV, which stands for "False Vertical (Sync)." It is variously referred to as Quasi-Sync, Dummy Sync, etc. In normal operation the input is unused. It *is* used for what are sometimes called trick video effects, such as freeze frame and slow motion. In those cases, the FV is summed into the signal at this point so that there *is* clean vertical sync in this NTSC output.

8.6.5.4 Chroma playback

Referring back to Fig. 8.31, we see that this signal follows the same path as the luminance playback did until it leaves the head switch block in IC_{301}. From there it goes to the chroma amp in the same IC. Since no frequency-selective components are shown associated with this block, we may assume that we have full chroma plus luminance leaving pin 9 of this chip and leaving this board at pin 3 of the connector. We pick this signal up again at the right-hand edge of Fig. 8.14E. From there we follow it to pin 11 of the connector to the Y/C board. This takes us back to Fig. 8.24B, where we see the PB chroma signal entering at the left-hand side of the schematic.

Figure 8.40 provides an overview of the chroma playback circuitry. As usual, blocks within are white and "outboard" blocks are shaded. The actual circuitry within most blocks can be found in Fig. 8.24.

The PB chroma amp

This stage is preceded by a stage of *LC* low-pass filtering and one of *RC* low-pass filtering. The composite effect of these two filters is shown in Fig. 8.41. We see unity transmission of frequencies above 83.5 kHz. Below that frequency, response falls off at −40 dB/decade, and below 1.59 kHz, this drops to −60 dB/decade. Since the down-converted chroma being recovered from the tape has sidebands that span from about 130 kHz to 1.13 MHz, all of them will be passed unattenuated. Of course, so will the luminance, which lies between 3.4 and 4.4 MHz. This filter thus does nothing to separate the FM luminance from the chroma. The filter is followed by a single stage amplifier (Q_{507}) whose voltage gain is 3 when the chroma amp switch (Q_{517}) is off. When the switch is on, an additional 3.3 kΩ of resistance is thrown in parallel with the 1.2 kΩ collector resistance. This drops the stage gain to 2.25. The series capacitor C_{547} can be considered an AC short, since its impedance is just 14 Ω at the lowest chroma frequency and less at higher frequencies. We note that Q_{517} is driven by the head-switch pulse. This leads us to the conclusion that the outputs of the two heads are not equal, and this circuitry is used to equalize their inputs to the chroma processing circuitry.

Chroma filter

At the head amp we began looking for a means of removal of luminance from chroma. Here we finally see it. Coming out of the PB chroma amp, we encounter a 1 MHz LPF module, followed by a discrete passive notch filter. The 1 MHz bandwidth of the LPF will pass basically all of the chroma

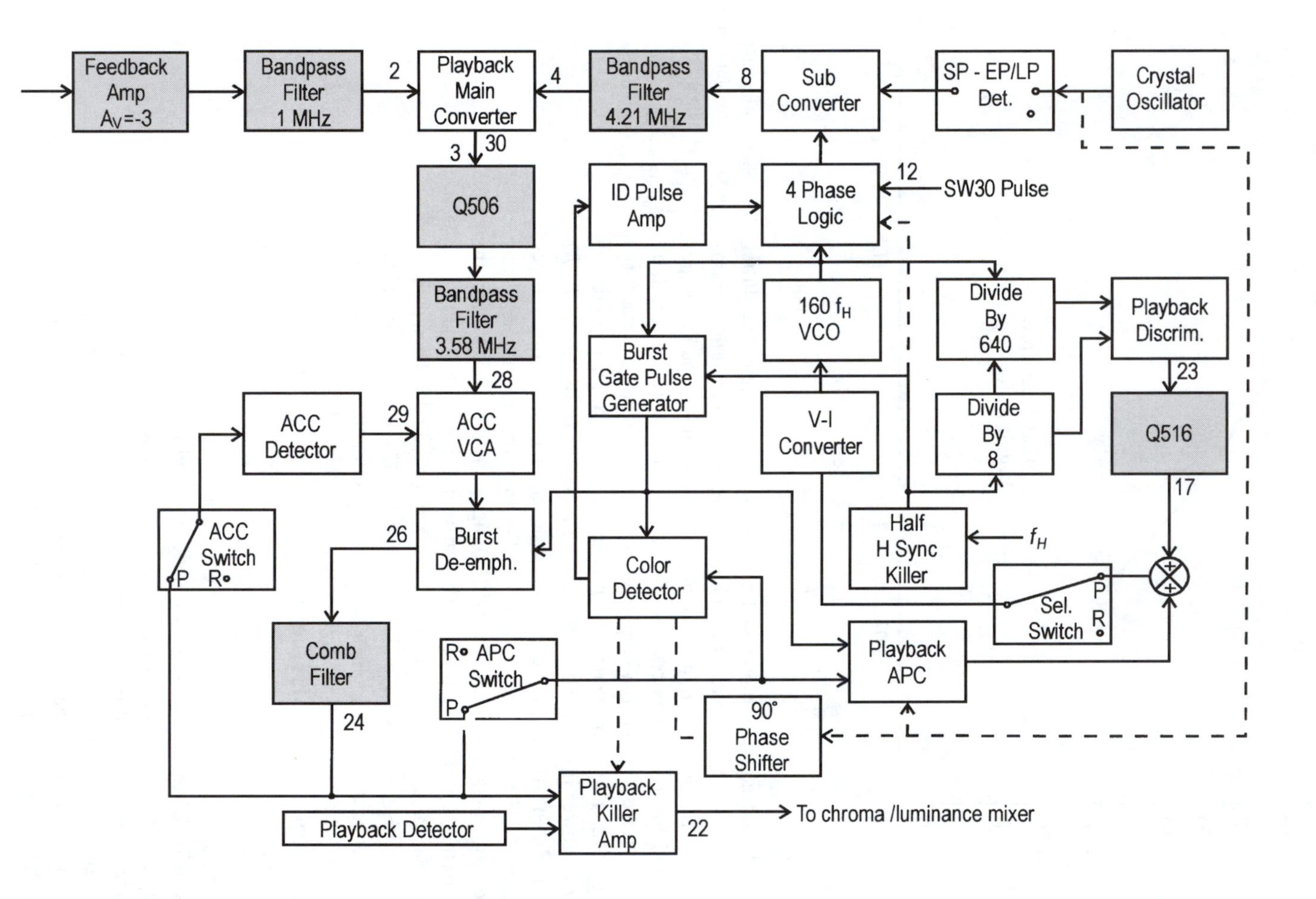

Fig. 8.40: Block diagram of the chroma playback circuitry in the Sharp VC-785U VCR. Connections deduced to be missing in the manufacturer's data are shown in dashed lines.

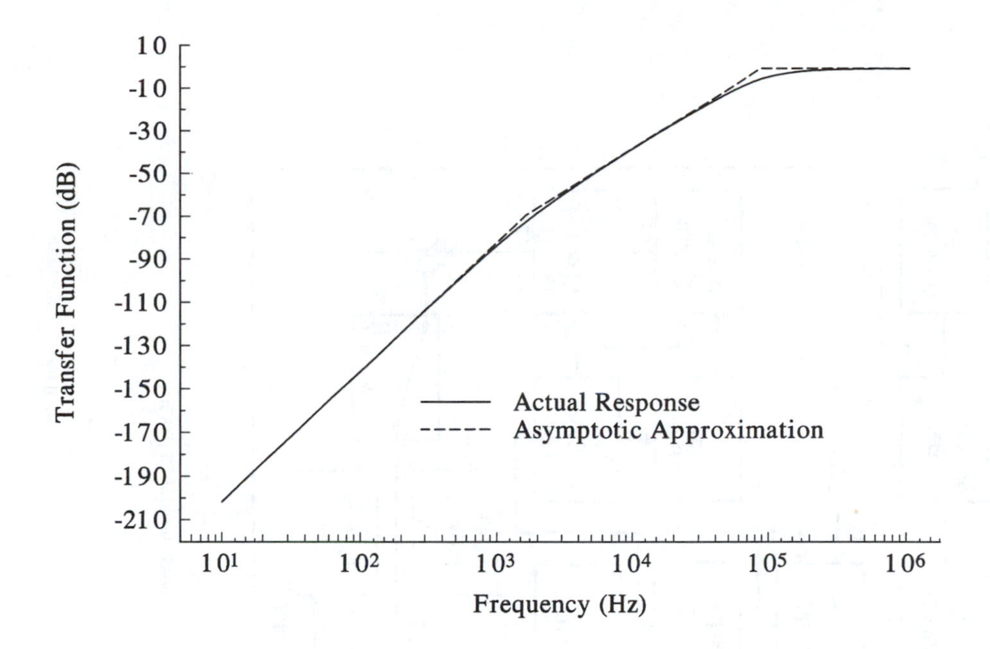

Fig. 8.41: Amplitude response of the third-order high-pass filter at the input of the playback chroma amplifier, Q_{507}, in the Sharp VC-785U VCR.

and very little of the frequency-modulated luminance. The discrete notch filter contains an inductance, two capacitors, and a resistor. It also contains two implicit resistances – namely the input resistance of pin 2 of IC_{501} and the output resistance of the LPF. The former resistance is revealed by the spec sheet to be 5 kΩ. The latter we do not have a handle on. The network thus has the form shown in Fig. 8.42. The signals coming into this network are all expected to be at 1 MHz or lower in frequency. The series LC would resonate at 37,170 Hz if it were isolated. At that frequency, both L_{503} and C_{517} have impedances of magnitude 91 Ω. Since this is so much smaller than the magnitude of the branch to the right, we suspect that C_{518} will not have any significant effect on the resonant frequency of the series LC. This suspicion is strengthened when we calculate the impedance of C_{518} at the lowest frequency in the chroma bandwidth (629 − 500 kHz = 129 kHz) and find it to be 123 Ω. This is so much less than R_i that we conclude that C_{518} is a coupling capacitor only and should be considered an AC short. Thus R_i is effectively in parallel with R_{532}, yielding a resistance just under a kilohm. This simplification is incorporated in Fig. 8.43. However, by Thevenin's Theorem, we can subsume R_o and the 1 k resistor into a single resistance, R. The resulting circuit is shown in Fig. 8.44.

This circuit has been analyzed with PSPICE® for values of R varying over two decades. It is found to consist of two superimposed responses. The first response is a first-order high-pass filter with a corner frequency that stays very close to 3 kHz no matter what the value of R is. The second response is the aforementioned notch at 37 kHz. As R is increased, the notch gets dramatically deeper and wider. The phase response spreads out greatly also. The purpose of a 37 kHz notch in the chroma path is not at all obvious.

The playback main converter and filter

The chroma playback signal finally arrives back at pin 2 of IC_{501}, where it enters the chroma playback main converter. It is the job of this block to translate VHS

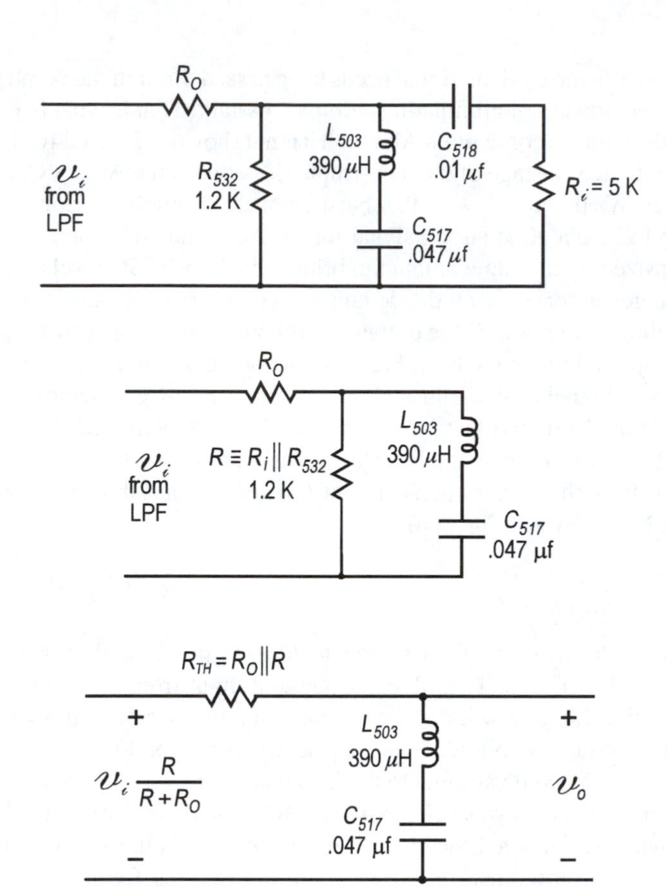

Fig. 8.42: Schematic diagram of the discrete notch filter at the output of the playback chroma amplifier in the Sharp VC-785U VCR.

Fig. 8.43: Simplification of Fig. 8.42 achieved by treating C_{518} as an AC short and thereby combining R_i and R_{532} in parallel.

Fig. 8.44: Simplification of Fig. 8.43 achieved by making a Thevenin equivalent of the resistances.

chroma at 629 kHz back up to NTSC chroma at 3.58 MHz. To do this, it mixes a 4.21 MHz signal, whose genesis we will discuss at the end of this section, with the .629 MHz chroma to yield sum and difference frequencies at 4.84 and 3.58 MHz respectively. These signals immediately leave the chip at pin 30 and are delivered to the base of Q_{506}, which is labeled as a playback switch. The label is of dubious validity. The stage is actually an emitter–follower buffer amplifier. It has the additional charming feature that it has an erroneous emitter voltage of zero listed for both the playback and record modes. Whereas an emitter voltage of zero would be consistent with the identification of the stage as a switch, it would not be consistent with the base–emitter voltage required to produce that saturation condition nor with the passage of a signal from said emitter lead to the input of BPF, FL_{501}. This filter module extracts the 3.58 MHz difference signal from the two frequencies input to it, and returns it to the ACC stage at pin 28 of IC_{501}.

The ACC loop

Although the ACC detector directly feeds the ACC voltage-controlled amplifier (VCA) in this set, the VCA does not directly feed the ACC detector. In Fig. 8.24C we can trace the VCA output as it again leaves IC_{501} at pin 26 and passes through the comb filter. As we shall see, the 4.208 MHz used to translate the chroma back to 3.58 MHz has a rotating phase reference as did the recorded chroma. Their combining in the main playback converter gives us the situation shown in

Fig. 8.5 (lines 5 and 6). This signal needs to be passed through the comb filter to complete the process of eliminating chroma crosstalk. That is what happens on this pass through the comb filter. Although it is not shown in Fig. 8.24C, Fig. 8.40 shows that there is a stage of burst de-emphasis between the ACC VCA and the comb filter. We previously saw that burst emphasis is applied prior to chroma recording to give a good strong signal for the ACC and APC loops. Here it is de-emphasized by the same amount to bring it back to NTSC levels. The burst gate pulse generator gates on the de-emphasis circuitry only during the burst.

According to Fig. 8.24C the output of the comb filter returns to IC_{501} at pin 24 and is applied to a switch. In Fig. 8.40, we see that this is identified as the ACC switch. Signal passing through it is input to the ACC detector (rectifier) and filtered by the network at pin 29 of IC_{501}. This DC voltage is then applied to the VCA to control its gain, thus closing the loop. It should be noted that the connection from the ACC switch to the ACC detector is not shown in Fig. 8.24C at all, but it is shown in Fig. 8.40.

Chroma output circuitry

The signal from the comb filter returns to IC_{501} at pin 24 and passes through the playback killer amp. This block also has an input from a record/playback detector in the chip that will disable the chroma transmission path in playback mode if the chroma amplitude is inadequate to give a good signal. If enabled, this amplifier passes the chroma to the Y/C mixer that we analyzed in Section 8.6.5.3, from whence it was delivered to the RF modulator for feeding to a TV.

We have now followed the chroma playback path from the VCR heads to the TV antenna terminals. Yet we have not covered the frequency-synthesis circuitry that makes it all work. This must be done next.

The 40 f_H source (AFC)

The heart of this circuitry is the 160 f_H VCO. This is part of a frequency-synthesis PLL. Its output goes three places, as shown in Fig. 8.45, which is extracted from Fig. 8.40. As was the case in record mode, the BGPG simply provides a timing reference for several operations that we've yet to examine. The most important path at the moment is that into the $\div$640 block. There is more to the $\div$640 than appears in the block diagram. The configuration is not uncommon in modern VCRs. This block contains an enable function that is driven by one of the outputs of the H sync 1/8 divider. Figure 8.46 shows a subset of the circuitry in Fig. 8.45, as well as giving more detail on the innards of the $\pm$640 counter block in that figure.

The 1/8 divider has two outputs. Both are at the same frequency but have different duty cycles. The output delivered to the counter has a 50% duty cycle. The period of this waveform is $8/f_H$. Thus the gate enables the VCO to the counter for half this time, or $4/f_H$. Since the input frequency is 160 f_H, the number of pulses that will pass each time the gate is enabled will be $(4/f_H) . (160\ f_H) = 640$. Obviously then, one pulse will be output from the $\pm$640 counter to the playback discriminator each time the gate is enabled – that is, for every 8 horizontal sync pulses. The other output to the discriminator comes directly from the 1/8 divider. It has a duty cycle of 12.5%. In other words, it is high for a time H out of every 8 H, which is the period of its output waveform.

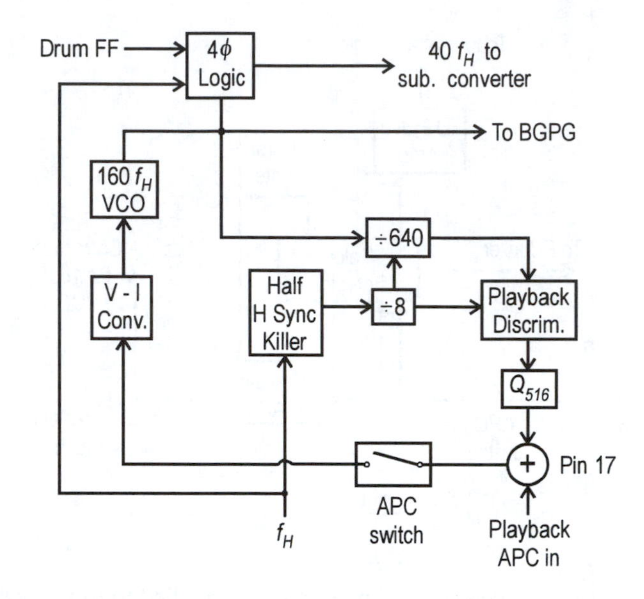

Fig. 8.45: Block diagram of the frequency-synthesis circuitry used to generate the 40 f_H signal used in both recording and playing back the chroma in the Sharp VC-785U VCR.

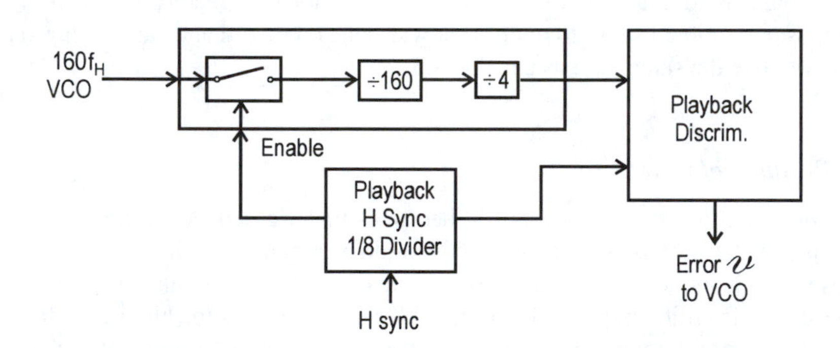

Fig. 8.46: Detail of a portion of Fig. 8.45 showing the internal structure of the ÷640 block.

These two waveforms will nominally have the same frequency, but we must not lose sight of their ultimate origin. The outputs of the 1/8 divider are derived from the f_H signal recorded on the tape along with the luminance. In general, this frequency will vary due to motor speed variations, tape stretching, temperature variation, etc. The other input to the discriminator comes from the 160 f_H VCO. In keeping with the PLL principle, we would expect to have the low-pass-filtered output of the playback discriminator supplied back to this VCO to force its frequency to be exactly 160× the instantaneous value of the horizontal sync frequency recovered from the tape. In this VCR, it is not that simple, however. The discriminator output goes through a buffer (Q_{516}). But Fig. 8.45 shows a far more profoundly fundamental variation. In it, we see that the error signal output of the playback discriminator (the AFC output) is actually summed with the error signal output of the APC circuit, which we shall cover later in this section. It is the sum that keeps the VCO on frequency to lock against the recorded sync. Basically the AFC "coarse tunes" the VCO and the APC "fine tunes" it. As was the case for chroma record, the V–I converter interfaces the error voltage to a frequency-control current for the VCO.

Finally, the 160 $f_{H'}$ output (we shall so denote the "variable" horizontal sync frequency recovered from the tape) is applied to the four-phase logic

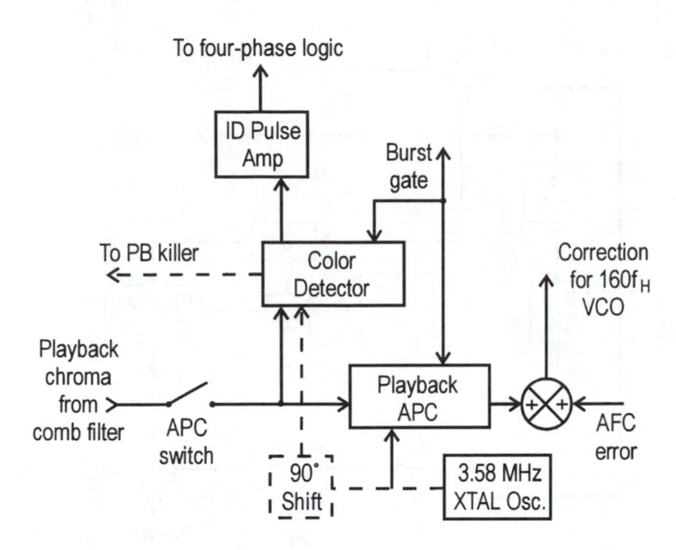

Fig. 8.47: Block diagram of the APC and color detector circuitry as connected for chroma playback. The lines shown dashed are those deduced to be missing from the manufacturer's data. Each is discussed in the text. The dashed lines are again used to indicate where we have deduced that lines are missing from Fig. 8.32.

block. Again $f_{H'}$ and the drum FF pulse are also applied to this block to step the phase rotation and its direction, just as was done for record, but with the opposite rotational sense to undo what was done on record and cancel chroma crosstalk at the same time as per Fig. 8.5.

APC and color detector

In Fig. 8.33 we showed several dashed lines that we deduced were missing from Fig. 8.32. All of these connections would be required for chroma playback also. In this mode, the crystal oscillator runs open loop. A connection should be shown from its output to both the APC and color detector blocks. In fact, the crystal oscillator output is applied to the color detector through a 90° phase shifter, which is not explicitly shown in either Fig. 8.40 or Fig. 8.32. It might be considered a part of the color detector. In fact, these blocks also share inputs of playback chroma and the burst gate pulse. The interconnection of these blocks is shown in Fig. 8.47, where we again show items missing from Fig. 8.32 in dashed lines.

This figure makes it clear that the only difference in the inputs to the APC and color detector blocks is the 90° phase shifter. The internal circuitry of the two blocks would be expected to be very similar. Both are basically phase detectors. They will give a DC output that depends on the phase difference between their inputs. The comparison takes place during the burst time. The burst gate operates on the playback chroma to extract the burst timing. Its phase is compared with that of the crystal oscillator during this time span. Since the chroma will have a burst phase reflecting the nonidealities of the playback loop, the APC output will be continuously shifting. This shifting error voltage will sum with that from the AFC circuit as previously described and modify the frequency of the 160 f_H VCO. We have not yet conceptually "closed the loop" here but shall do so shortly. For now we simply observe that if the burst phase of the playback chroma is exactly in phase with the output from the crystal oscillator, the APC phase detector output will be zero ideally.

The same type of comparison will be performed inside the color detector, but the added 90° phase shift in this path means that when the output of the APC detector is zero, the output of the color detector is maximum. The application of this output to the playback killer is perfectly reasonable. If the color detector doesn't detect color, it kills the color path on playback.

Returning to Fig. 8.40, we see that the ID output (pin 18 of IC_{501}) is amplified and applied to the four-phase logic block. This connection is necessitated by the fact that it is entirely possible to have the AFC loop locked when the burst from the playback signal and the crystal oscillator are 180° out of phase. When this happens, it is detected by the color detector, which generates an ID pulse. This pulse is amplified to logic level by the ID pulse amplifier.

Four-phase logic

This circuit, as has already been noted, performs the same function on playback as it does on record, but in the opposite sense. The block still needs an f_H frequency reference. If it does not have a separate ÷160 divider, then an f_H connection is missing from Figs. 8.40 and 8.32. In the process of producing the four separate phases, the frequency is also divided by four to give 40 f_H as the frequency of its outputs. We have shown how frequency and phase errors of the playback chroma signal generate a correction in the 160 f_H VCO. These corrections are scaled down by a factor of four also in going through the four-phase logic.

The sub. converter and BPF

The crystal oscillator that is input to the APC and color detector circuits is also input to a sub. converter where it is mixed with the 40 f_H from the four-phase logic to generate the usual sum and difference frequencies – in this case at 2.95 MHz and 4.21 MHz. The bandpass filter on its output selects the 4.21 MHz component. This component will also contain the frequency and phase corrections we have been referring to. We finally "close the loop" when we bring the "error compensating" 4.21 MHz into the playback main converter, which also receives the "error-bearing" chroma from the tape at 629 kHz. Not only do the nominal frequencies combine here to restore the chroma subcarrier to the 3.58 MHz as we have already discussed, but the frequency errors cancel out here to a large degree, thus reducing jitter and other picture instabilities to levels far below what they would be without this AFC/APC circuitry. In fact we can say that without this circuitry, the picture from the VCR would be unviewable.

Figure 8.48 reproduces Fig. 8.32, but with all of the blocks in this chip (IC_{501}) which are used only for recording chroma removed.

The recovered chroma from pin 22 of IC_{501} is then mixed with the recovered luminance in the chroma summer block of IC_{201} (see Fig. 8.39) and passed to the RF modulator for application to TV antenna terminals.

8.6.6 The servo systems

This brings us to the circuitry of Fig. 8.49. Unfortunately, it is immediately obvious that much of the most interesting circuitry is in IC_{701} and IC_{801}, for which we have not even so much as a block diagram. Nevertheless, this diagram

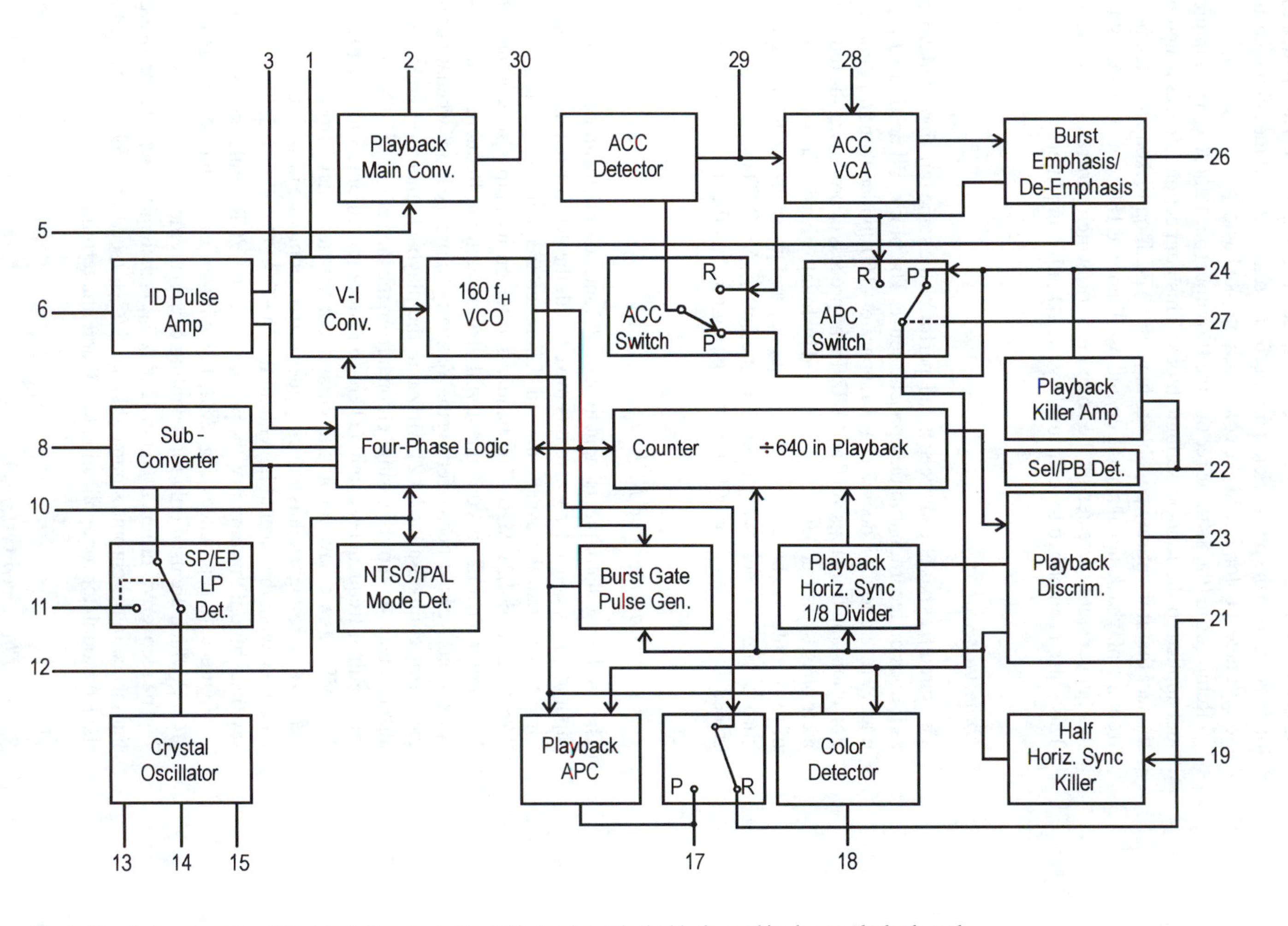

Fig. 8.48: Reduced version of the block diagram in Fig. 8.32 showing only the blocks used in chroma playback mode.

is so imposing that we hardly know where to start. Let us note that along the left edge we have the connection to the timer board, which we shall briefly examine next. Along the upper right-hand edge, we have the connections to the other half of the main board schematic (Fig. 8.14). Let us therefore start at the top here and work our way down.

In the analysis that follows, we shall have to do a lot of jumping between Fig. 8.49 (the servo circuitry), Fig. 8.14 (the rest of the main board circuitry), and Fig. 8.24 (the Y/C board).

The uppermost line in Fig. 8.49A is labeled FV, which stands for False Vertical sync. In Section 8.6.5.3 we saw the destination of this signal. Here we see that its origin can be traced back to pin 29 of IC_{701}. In normal playback, the vertical sync is reconstituted from the control track. For special effects that involve low or high-speed playback, however, the control pulses will in general not play back at the correct speed to regenerate the vertical sync. Furthermore, vertical synchronization tends to be difficult for trick effects because of noise in the vertical blanking interval.

Although tape speed varies during these special effects, the head rotational velocity remains essentially unchanged. Recall that sensors activated by the head drum generate what Sharp calls the HSWP (Head-Switch Pulse) at 30 Hz. In other machines this is called SW30. This 30 Hz is doubled in frequency to generate the FV signal. This signal provides stable vertical synchronization even during special effects. We shall look at these trick video effects in more detail in Section 8.7 of this chapter.

The next line down on Fig. 8.49A is the horizontal sync. Tracing this line back to the Y/C board shows that it originates there (not surprisingly) and is an *input* to the circuitry of Fig. 8.49. We saw that it was extracted from the combined luminance and sync in IC_{201} of the Y/C board. Tracing it into Fig. 8.49B, we see that it passes through an RC network to Q_{701}. The waveforms there show that this stage is used to bring it up to a logic-level signal. A small mystery here is why the stage is labeled V-Sync Amp, since the line feeding it is identified as H-Sync. Perhaps we get further insight by looking at the frequency response of the RC network that feeds this stage.

Exercise 8.25 Find and sketch the asymptotic amplitude response of the transfer function i_B/v_{A2} where i_B is the base current into Q_{701}.

The results of this exercise may not be conclusive in themselves. The "pass band" of this network covers both the horizontal and vertical sync *frequencies*. However the gain is about 10 dB higher at the horizontal sync frequency. On second thought, however, the network's response to the sync frequencies is not what matters. We are dealing here with *pulses*, and it is well known that the bandwidth required to transmit a pulse is at least the reciprocal of the pulse width. In Chapter 6 we found that the horizontal sync pulse has a nominal width of .08 H or 5 μsec. Thus the minimum bandwidth to transmit this effectively would be 200 kHz. The width of the vertical sync pulse is 3 H, or 190 μsec, requiring a bandwidth of just over 5 kHz. This is so low compared to the upper frequency limit of the transistor's input network that the vertical sync pulse will not only be passed but will be passed with good fidelity, producing a nice square pulse of base current (and thus output voltage).

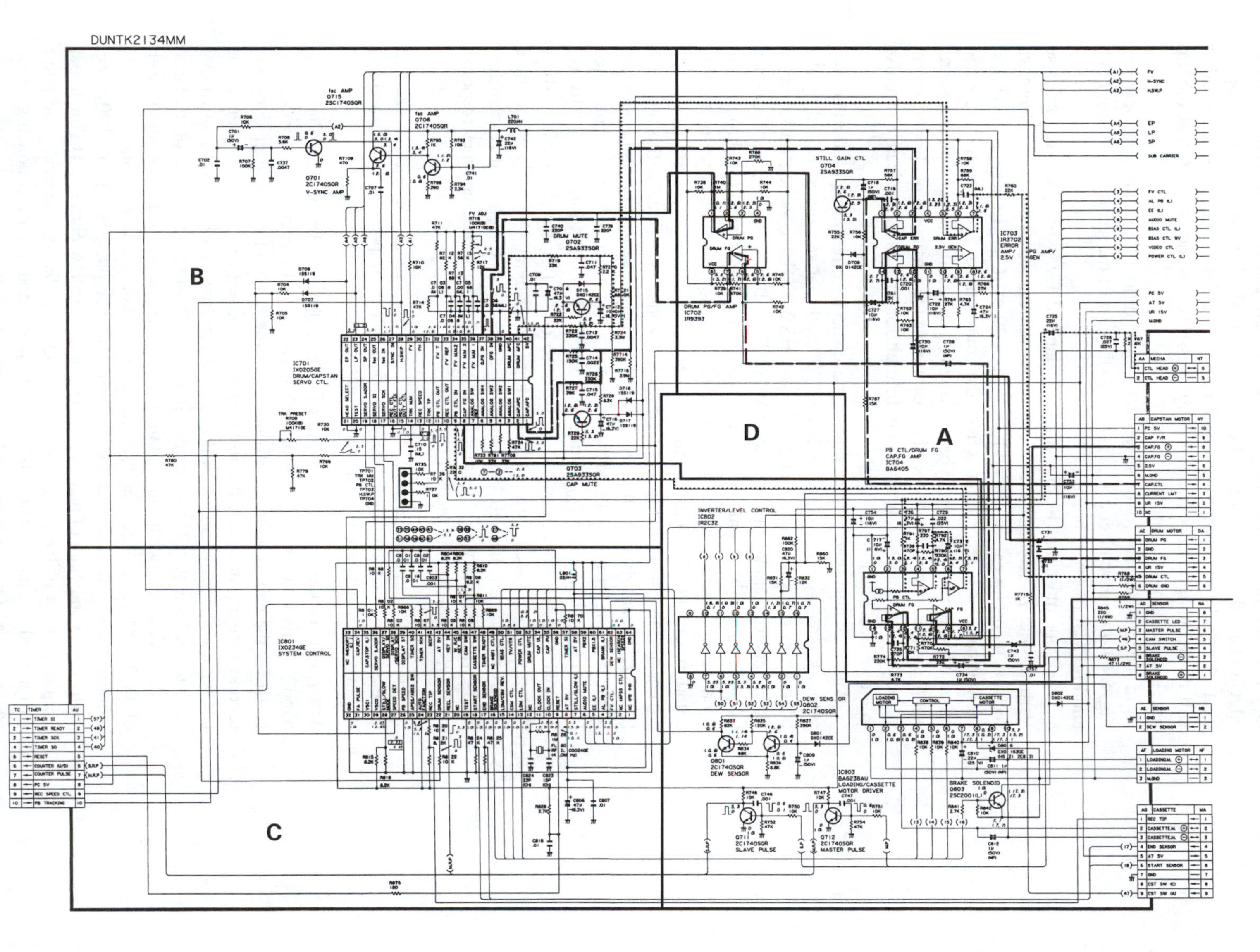

Fig. 8.49: Schematic diagram of that part of the main board that deals with the servo systems in the Sharp VC-785U VCR. An electronic version of this figure is available on the world-wide web at http://www.cup.org/titles/58/0521582075.html.

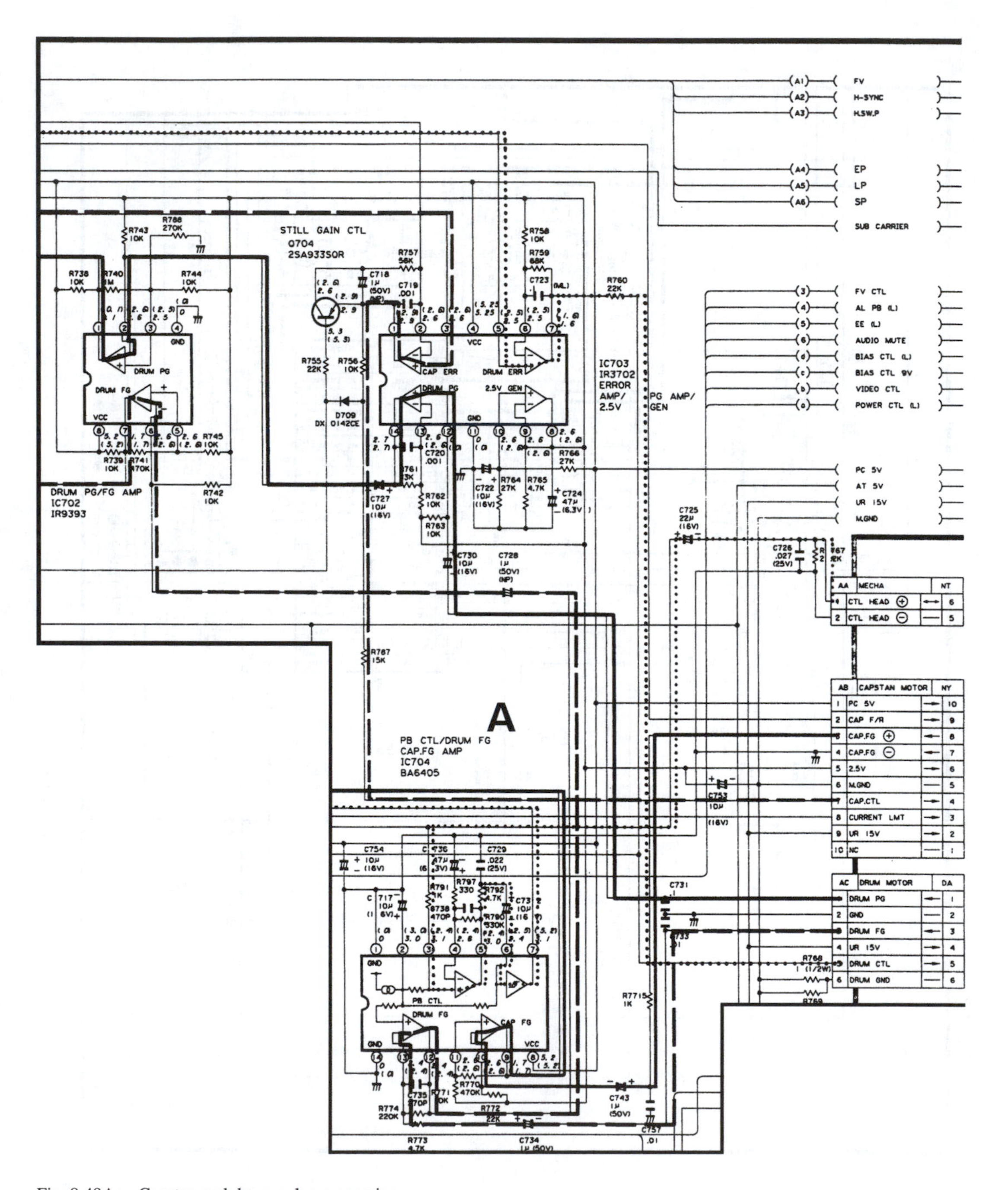

Fig. 8.49A: Capstan and drum pulse processing.

DUNTK2134MM

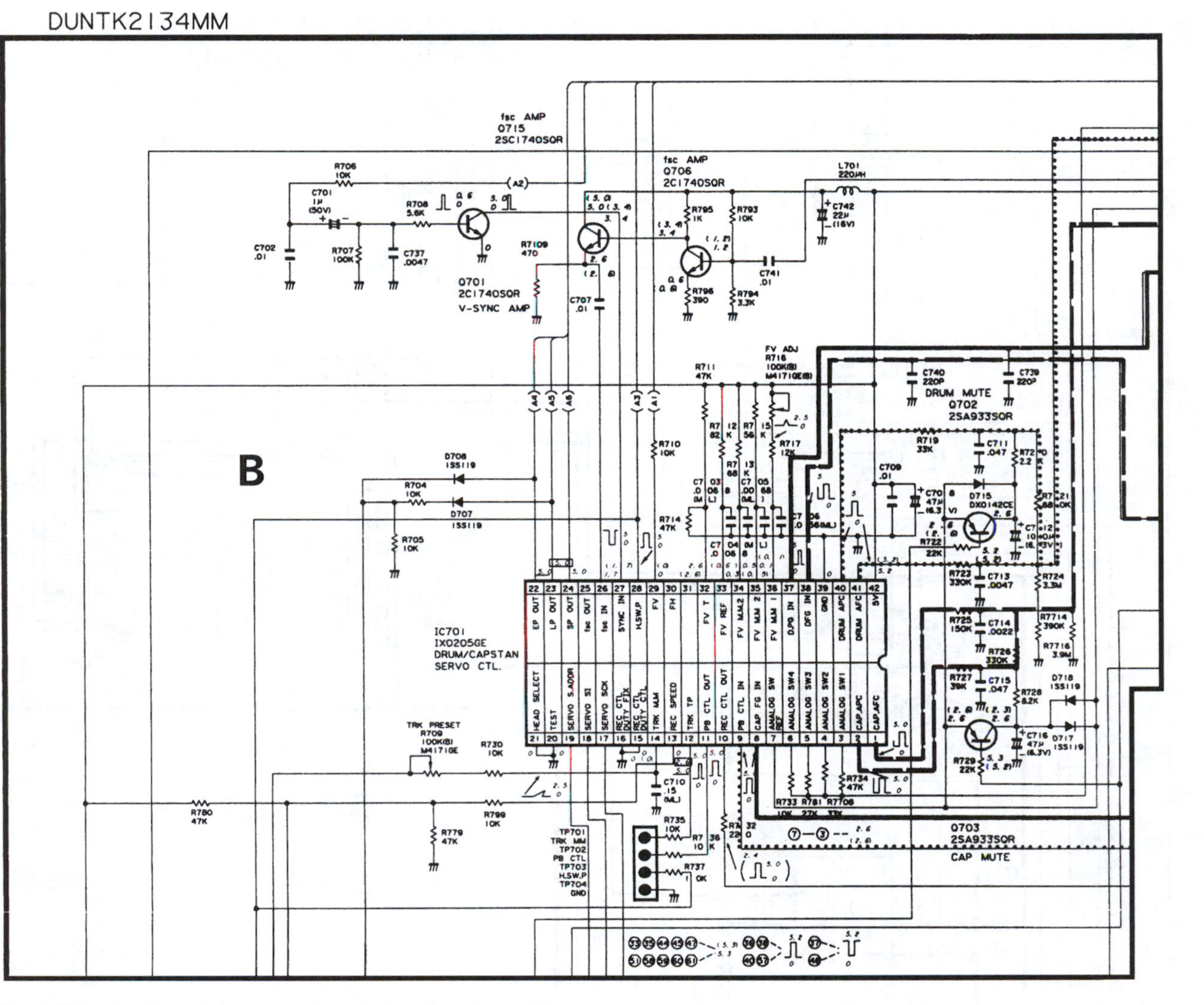

Fig. 8.49B: Capstan and drum servo control plus subcarrier amplifier.

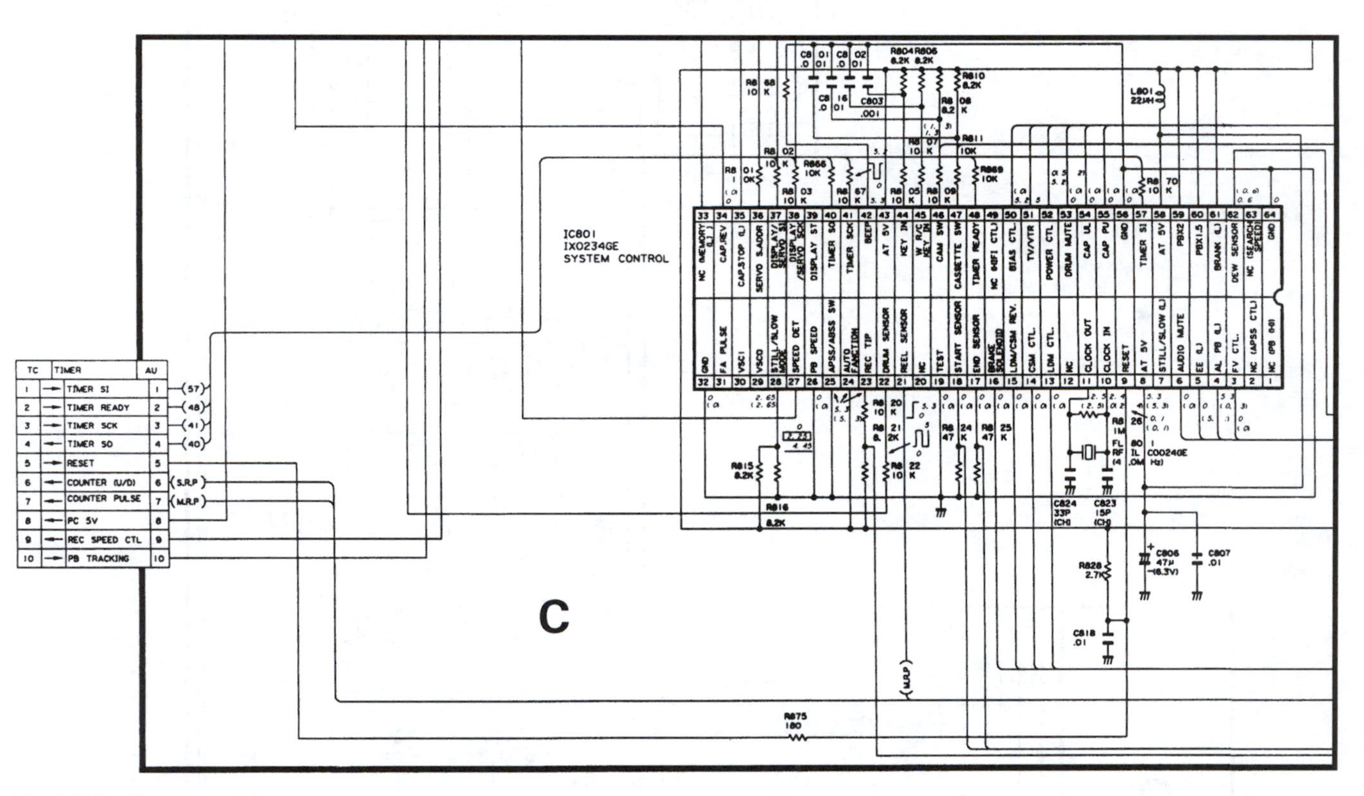

Fig. 8.49C: System control.

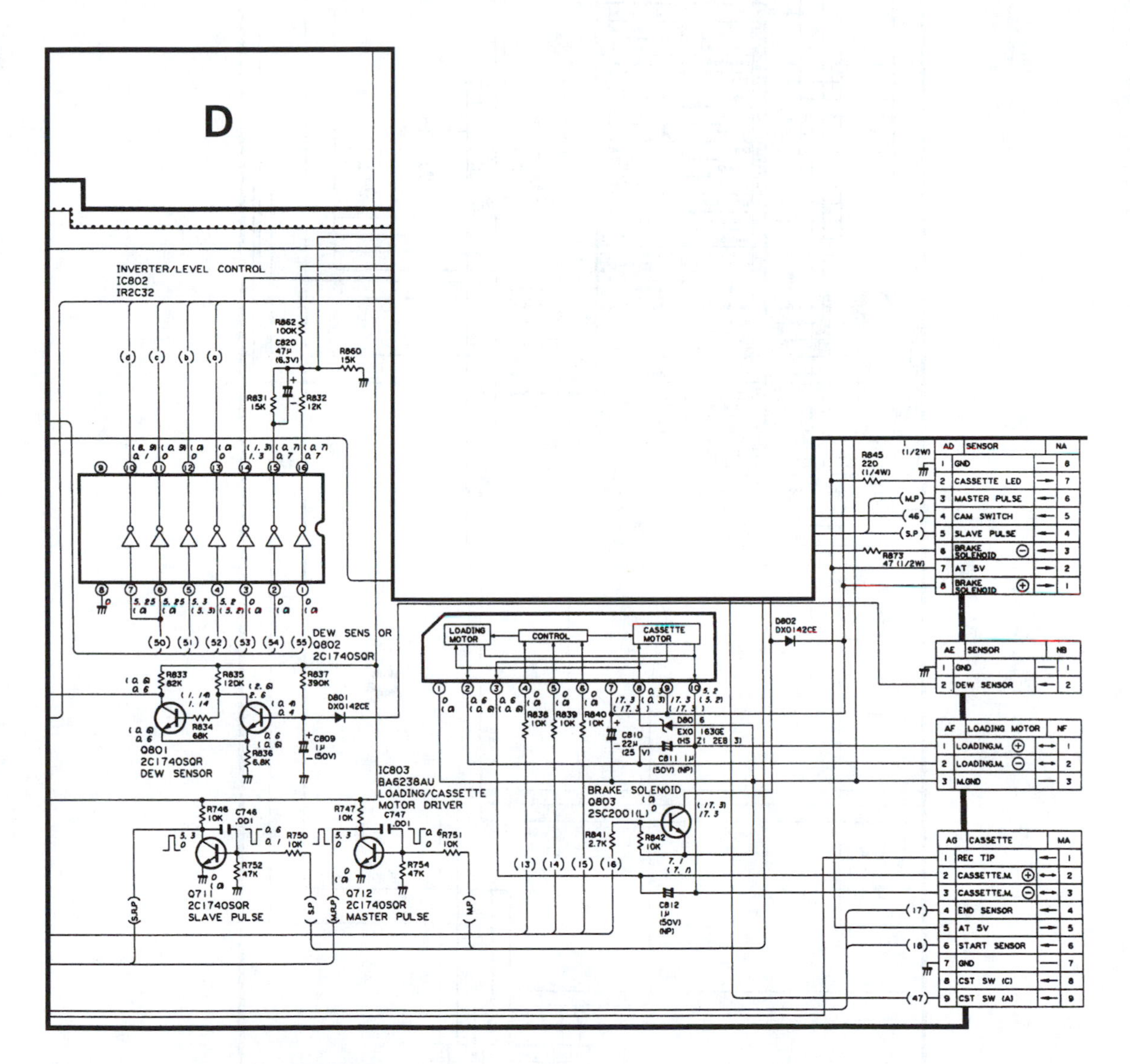

Fig. 8.49D: Miscellaneous functions.

The bandwidth to transmit the horizontal sync pulse is not available from this network, however. In fact, a time-domain analysis shows that the horizontal sync pulse, when coupled through this network, will produce a maximum amplitude response less than 9% of that which a vertical sync pulse of the same amplitude produces. This strongly suggests that it is the vertical sync pulses being processed here.

We must remember that the line labeled H-Sync actually contains both the horizontal and vertical sync. This might account for the line being labeled H-Sync and the stage it feeds being labeled as V-Sync Amplifier. Furthermore, as we shall see, the servo section needs vertical sync but not horizontal. This leads us to believe Q_{701} is properly labeled.

The next line in Fig. 8.49A is the head-switch pulse. It is again true that a comparison of this figure with Figs. 8.14D and E and 8.24 shows that this pulse originates in IC_{701} at pin 28. It is made available to the system control IC_{801} and passes to Fig. 8.14. From there, we see it coupled to the Y/C board at pins 8 and 35 and then to the head amp.

The connection to pin 8 of the Y/C board is called chroma rotation. It is used to switch the phase of the chroma on a continuous basis, as shown in Fig. 8.5.

The connection to pin 35 of the Y/C board goes two places on that board. The first is to a switch which toggles the gain of the chroma playback amplifier, Q_{507}. When Q_{517} is on, R_{576} is thrown in parallel with the resistor R_{528}. This lowers the gain of this stage. The switch is made at the end of every field, to compensate for the imbalance in the outputs of the two heads.

If we trace out the other connection to pin 35 of the Y/C board, we see that it is applied to pin 18 of IC_{201} through a 56 k resistor. The output of Q_{504} pulls this pin low on SP playback, changing the nonlinear emphasis. In so doing, it also shorts out the HSWP. The IC data sheet shows this to be another dual-purpose pin, selecting nonlinear pre-emphasis and also the HSWP signal. It gives no indication, however, of why you would ever want the HSWP to be input here. Possibly it could be used to accommodate head asymmetry by allowing different values of the NL pre-emphasis. We will see in Section 8.7 that such asymmetry is common in the design of modern machines.

8.6.6.1 Speed-switch lines

The next lines coming into Fig. 8.49A from Fig. 8.14 are the EP, LP, and SP lines. These refer, of course, to Extended Play (six hours), Long Play (four hours), and Standard Play (two hours). Tracing these lines over to IC_{701}, we see it clearly marked that these are outputs from the servo-control IC. When these lines are traced back onto Figs. 8.14D and E, they are seen to return to the audio and Y/C boards. Before we even look at those boards, we can say that because different speeds will require different record and playback equalization, we expect these logic level signals (EP, LP, and SP) will be used to switch compensation components in and out of circuit. Looking at Fig. 8.23 for the audio board, we see that our expectation is fulfilled. Diodes D_{601} and D_{602} comprise a distributed OR gate that switches record and play equalization components around IC_{601}.

Exercise 8.26 A resistor, R_{622}, is in series with D_{601}.

(a) Why is there one here and not one in series with D_{602}?

(b) What two things will the EE line (Fig. 8.23: pin 21) accomplish?

The SP line enters the Y/C board at pin 18. It is active high. When active, it tries to pull up the input of Q_{504} through R_{513}. This will cause its output to go to ground, which will in turn ground pin 18 of IC_{201}, and change the equalization. All of this is contingent upon the output of Q_{508} being in the open state. If its input is high, its output will ground out the SP drive to the input of Q_{504}. We therefore need to backtrack and see what will make the input of Q_{508} high. We see that it will be pulled up toward the PC 9 V supply by R_{567} unless pulled down by the output of Q_{514}. Thus our trace must go back to the input of this switch, which is driven directly by the EE enable. This line goes low during the record cycle. We see from the connector that it is active low. When we are recording, therefore, we have the following situation:

input of Q_{514} – low,
output of Q_{514} – pulled high, and
output of Q_{508} – low.

In this state, Q_{504} cannot respond to SP. In playback, however:

input of Q_{514} – high,
output of Q_{514} – low, and
output of Q_{508} – pulled high by SP (otherwise low).

The SP line also switches Q_{401}, which selects playback luminance equalization. A third switch controlled by SP is Q_{209}. The integrated switches we have been seeing are not used here, because this stage does not handle a logic signal, although it is controlled by one (SP). When SP is low, diode D_{207} is reverse biased, and bias current can flow from the emitter to the base of Q_{209}. This will saturate the transistor and thus connect the parallel combination of C_{236} and R_{246} in between pins 16 and 17 of IC_{201}. When SP comes high, D_{207} conducts and pulls the base of Q_{209} high enough to cut it off.

The LP line enters the Y/C board at pin 3. At pin 3 it is also applied to the input of integrated switch Q_{3302} as shown in Fig. 8.14D. When LP is high, the output of this switch goes low, connecting pin 9 of the Y/C board to ground through R_{3303}. This will attenuate the chroma output from pin 9 of the Y/C board before recombining it with the luminance. The LP signal applied to pin 3 of the Y/C connector is delivered to Q_{510} on the Y/C board. When LP is high, Q_{510} is biased on. When low, it is not. This will enable or disable the 3.58 MHz path out of pin 1 of the connector.

This subcarrier just happens to be the next line connecting the schematics of Figs. 8.14 and 8.49A. We have already seen that it originates with the crystal oscillator section of IC_{501} in Fig. 8.24C. From there it passes to Fig. 8.14D through pin 1 of the Y/C connector and then to Fig. 8.49A. It is there amplified by the CE–CC cascade of Q_{706} and Q_{715} and applied to pin 26 of IC_{701}. This signal is divided down to create REF30, which has a frequency of 30 Hz and is used as the tape speed reference on playback.

8.6.6.2 *Logic control signals*

The remaining interconnections along the upper right-hand corner of Fig. 8.49A are logic signals and power connections. We shall go through them quickly before commencing a closer look at the innards of the ICs in this figure. The first of these is FV control. It can be traced back through Fig. 8.14 onto Fig. 8.24A, but here, alas, no connection is shown – the lead dead-ends at pin 28 of the Y/C board. If we trace it back along its path, we see no side paths on any of the three schematics through which the signal passes. This signal apparently originates at pin 3 of the system control (IC_{801}) chip and goes nowhere! This is clearly in error. The FV control is almost certainly a logic signal. If so, it would control the generation or application of FV through connections that have been inadvertently omitted from the manufacturer's schematic.

The next line down is AL PB. This is an active-low logic signal (from pin 4 of IC_{801}) that switches on the PB 5V supply (discussed previously) to the luminance processing circuitry on the Y/C and head amp boards during playback mode.

The next line down is the EE logic signal. We have already discussed this signal in connection with the Y/C board. Here we see its origin at pin 5 of IC_{801}.

The next line is audio mute. This was discussed in conjunction with the audio board (Section 8.6.4.2). It is also a logic signal originating with IC_{801}.

The next four lines are logic control outputs from IC_{801} via IC_{802}. The first controls the bias control 9 V supply (which powers record mode circuitry on the Y/C and head amp boards). The next lead is redundant. A careful count of the interconnections at the left of Fig. 8.14E and at the right of Fig. 8.49A shows that the latter has one excess lead. This is the one. Either the leads in Fig. 8.49A labeled "Bias Ctl. (L)" and "Bias Ctl. 9 V" should be one and the same, or there is a lead missing from Fig. 8.14. The similarity of the labels strongly suggests that the former is the case. The next lead is the video control pin, which turns on the 9 V power to the RF modulator under the control of Q_{816}. The last part of this group of four is labeled as Power Control. This active-low logic signal turns on the PC 9 V supply, which is one of the main supplies in the VCR. The next group of four are all supply leads importing power into this section of the board.

8.6.6.3 The cylinder servo system

We now change the direction of our analysis from tracing the leads that connect the two parts of the main circuit board schematic to tracing the systems that are largely confined to this board. This preeminently includes the head and capstan servos. The cylinder (or drum) motor servo circuitry is discussed in Sections 8.4.2.3 and 8.4.2.4, where the importance of both phase and frequency lock of the motors used in the record and playback processes is discussed. The locking of the cylinder motor is achieved by having two generators on it. The PG (pulse generator) outputs pulses at a nominal 30 Hz. The FG (field generator) outputs pulses whose frequency will be 720 Hz.

Figure 8.50 delineates this system as far as we can with no information other than that given in Fig. 8.49. Both frequency and phase control for the drum are accomplished with this circuitry. The op amps are required to operate from a single +5 V supply. To keep them operating linearly requires a 2.5 V DC supply, which is ubiquitous in its appearance in Fig. 8.50. There are inputs to this circuitry from both the drum pulse generator and the drum field generator. Let us first examine the signal path from the drum PG. The first stage amplifier initially seems rather formidable but simplifies nicely when we note that the

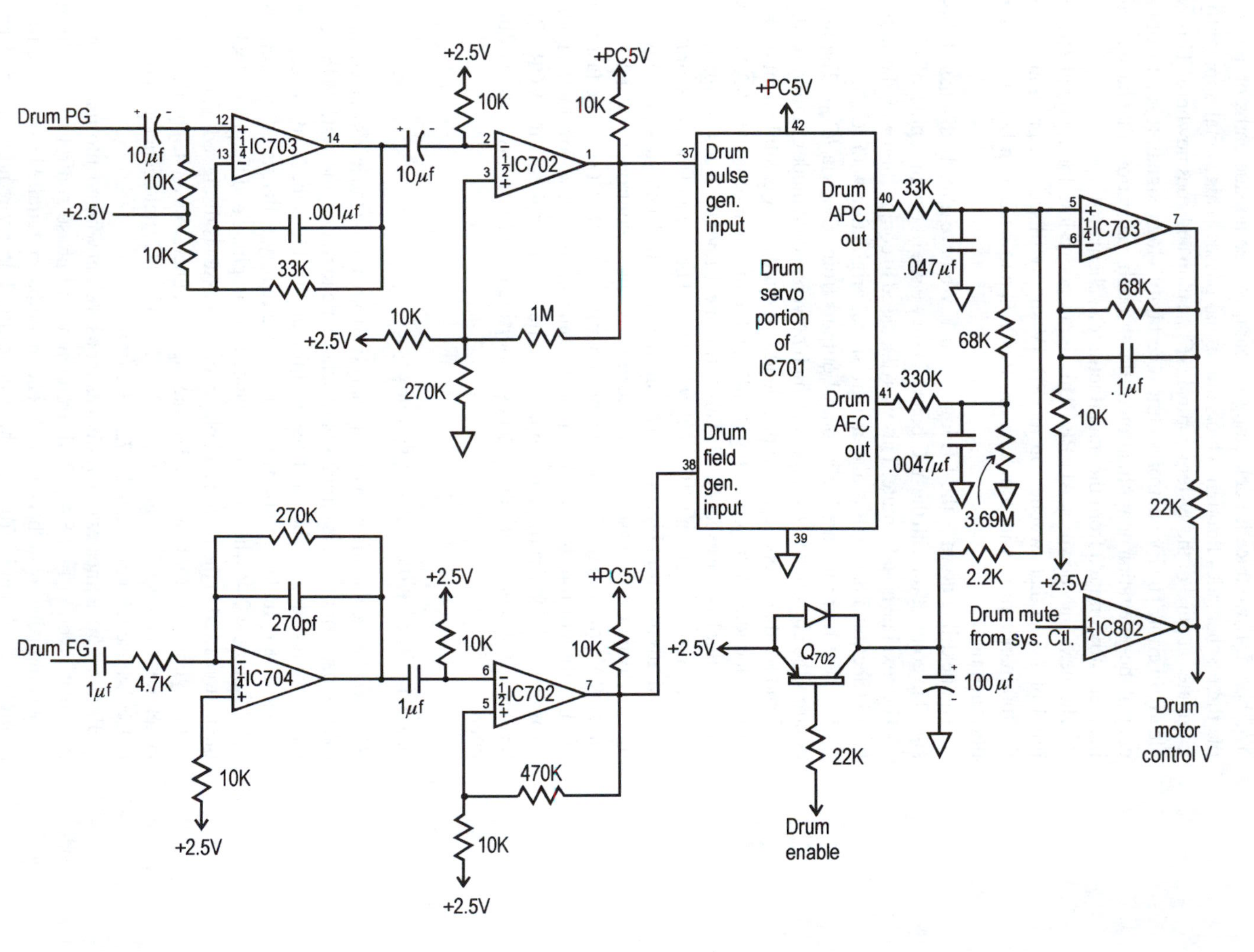

Fig. 8.50: Schematic diagram of the head motor servo circuitry in the Sharp VC-785U VCR.

2.5 V DC supply is AC ground. This "uncouples" the two 10 k resistors, and allows us to see the stage for what it is – a noninverting low-pass amplifier with AC-coupled input.

Exercise 8.27 Find the AC response of this circuitry and make a Bode amplitude plot.

This exercise shows that the input coupling network is effectively an AC short for frequencies above 1.5 Hz. The low-pass characteristic of the amplifier effectively causes the output to drop off above about 5 kHz.

The manufacturer's service information shows that the input to this stage is a slope-sided pulse of about 2 ms duration with a peak voltage of about 2 V. The first stage provides a gain of over 4, which assures that its output will be driven to the rails. This will, of course, also accomplish some squaring up of the pulse, but most of this is accomplished in the next stage. The chip used here appears to be a dual comparator with open-collector outputs. It is configured as a Schmitt trigger. The service literature shows the output to be a TTL-level pulse with a duration of <1 ms. This is applied to pin 37 of IC_{701}.

Exercise 8.28 Find the upper and lower trip voltages of the Schmitt trigger stage just discussed. Take pin 2 as the input, and ignore the 10 μF capacitor, C_{727}. The 10 k resistor from the input to $+2.5$ V just holds the input at a defined potential in the absence of other input.

Referring again to Fig. 8.50, we see that the amplifier for the drum FG signal is even more conventional than that for the drum PG. The combination of the input and feedback capacitors gives an amplifier passband from about 30 Hz to 2.5 kHz. This neatly encompasses the 720 Hz output of the drum FG. The service information shows that this signal is very close to sinusoidal and of the order of 60–70 mV pk–pk. The amplifier raises this to about 3 V pk–pk and applies this sinusoid to another Schmitt trigger nearly identical to that used for the drum PG signal. In this case, however, the facts that the input waveform is sinusoidal and the hysteresis is small combine to make the input to pin 38 of IC_{701} an almost perfect logic-level square wave.

Within IC_{701}, the two inputs we have just discussed are not compared directly. The drum PG input is first compared to another 30 Hz signal to produce a phase-error signal. This is output on pin 40. The 30 Hz reference signal is derived from the vertical sync of the signal being recorded when the VCR is in record mode. It is derived from a crystal oscillator time base in playback mode. In this unit, the color subcarrier oscillator is used. The error signal on pin 40 is low-pass filtered by the 33 k resistor and .047 μF capacitor. These give a cutoff frequency of around 100 Hz.

Exercise 8.29 Refer to Figs. 8.49A and B. Which pins of IC_{701} receive the signals that will ultimately be compared to the drum PG pulses in the drum phase-control portion of the IC?

The drum FG pulses input to pin 38 of IC_{701} are used for speed control. Back in Section 8.4.2.4, a common approach to drum frequency control was discussed in which an error signal was derived from the value of a ramp waveform when a sample pulse arrives. A simpler approach is to run the shaped drum PG output into an $f-v$ converter. This will produce an output directly proportional to the speed of the drum. The voltage thus obtained can then be used to reduce the drive to the drum motor. As was the case with the scheme discussed earlier, the drum phase error signal must be factored into the motor drive voltage also. We do not know which of these two systems (if either) is used in the Sharp VCR. In any case, we can think of the PG as providing a fine control of the drum speed, and the FG as providing coarse control.

Proceeding to the motor drive amplifier in Fig. 8.50 (the stage in the upper right corner), we see that it has four voltage inputs: the drum APC error voltage, the drum AFC error voltage, the 2.5 V source being summed into the inverting input, and a strangely configured summing-in of 2.5 V to the noninverting input. Our first reaction to this latter input might be that the 2.5 V will be summed in, regardless of whether or not Q_{702} is on. This is true, but the conduction state of this transistor will still make a large difference in what this stage does, as we shall see. We first note that the base lead of this transistor seems to go to places unknown. Actually, a close look at Figs. 8.49B and C will show that it goes to pin 33 of IC_{801}. This pin has a dual label. It says "NC MEMORY (L)." Now "NC" is universal for "no connection" and the Sharp glossary in the service manual agrees thereto. This makes no sense. We wonder if it is an NC pin or a MEMORY (L) pin. The latter makes little sense either based on the circuitry in which we found it. The best clue as to where this lead should go is given by the label on Q_{702} in Fig. 8.49B. It is identified as "Drum Mute." We take this to mean that it will stop the drum motor. We will proceed on this basis.

Exercise 8.30 Perform a DC analysis on the motor voltage amplifier shown in Fig. 8.50. Let V_s be the voltage across the 100 μF capacitor (treat it as a variable) and express the output voltage of the op amp as a linear sum of terms in V_{40}, V_{41}, V_s, and 2.5 V.

If we momentarily ignore the terms in V_o due to V_{40} and V_{41}, we have

$$V_o = 7.3\ V_s - 17.$$

This might suggest that if $V_s = 0$, $V_o = -17$, but this can never be, since the op amp is run off a single +5 V supply. What it actually means, of course, is that the op amp rails negative (to ground) and cuts off the drum motor. If $V_s = 2.5$ V, then the equation predicts $V_o = 18.3$ V. This, of course, corresponds to the positive rail. More realistic estimates of V_s are $2.5 - .6 = 1.9$ V when the transistor is off and $2.5 - .1 = 2.4$ V when it is saturated. These give V_o values of -3.1 V and .5 V respectively. Thus the roughly .5 V difference in V_s between when Q_{702} is in saturation and cutoff will move the op amp output between the negative rail and slightly into the active region. This brings us back to the effects of V_{40} and V_{41}. The complete expression for the output voltage of this stage is (ignoring the 3.69 M resistor)

$$V_o = .49\ V_{40} + .040\ V_{41} + 7.3\ V_s - 17$$

From this equation, we see that both V_{40} and V_{41} will raise V_o above the values we would predict on the basis of V_s alone. Since IC_{701} runs on a supply voltage of +5, we are safe in saying that V_{40} and V_{41} are both limited to +5 V or less. If both are +5, their contributions to V_o are then 2.45 V and .20 V respectively.

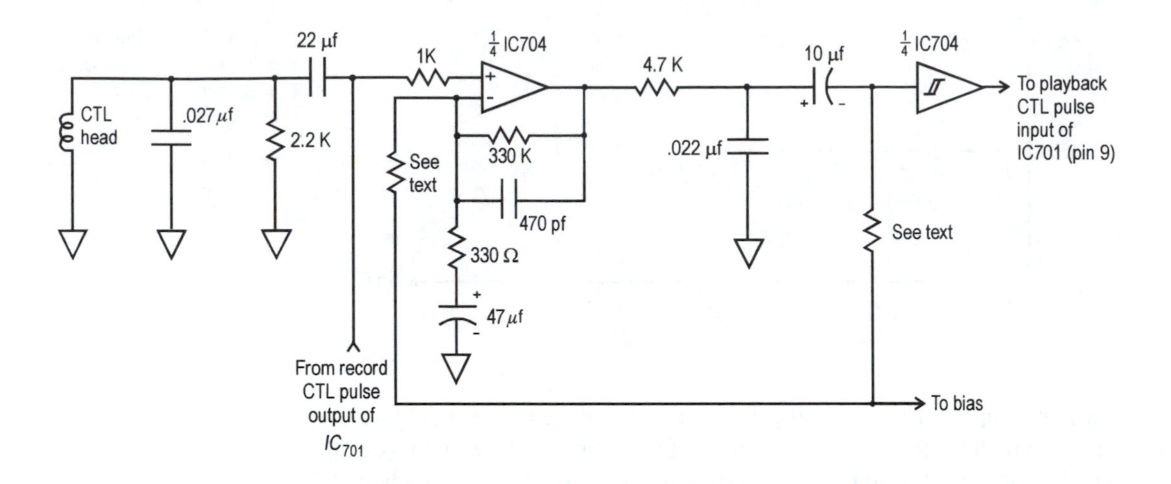

Fig. 8.51: Schematic diagram of the circuitry for processing the control track signal in the Sharp VC-785U VCR.

If these contributions are added to those of V_s, the results are $-.45$ when Q_{702} is cut off and $+3.15$ when it is saturated. Thus we see that the cutoff of Q_{702} keeps the output voltage of this stage at the ground rail regardless of what V_{40} and V_{41} are.

The output of the stage just analyzed passes through a 22 k resistor and becomes the drum speed control voltage. Figure 8.50 also shows that this voltage can be grounded out by a signal from the system control chip, (IC_{801}) via an inverter in IC_{802}.

Exercise 8.31 We have done a DC analysis of the drum control voltage amplifier. Determine the frequency effect of the .1 μF capacitor in its feedback loop and use that to deduce why it is in the circuit.

8.6.6.4 *The control signal processing circuitry*

This circuitry is shown in Fig. 8.51. It shows that the input comes from the control head on playback. (The control head writes control pulses to the tape on record.) If the tape is being played back at the same speed at which it was recorded, these pulses will be at 30 Hz. These pulses are capacitively coupled to the first amplifier stage. An analysis of this stage shows that it has the amplitude response shown in Fig. 8.52.

The 330 Ω resistor and 47 μF capacitor set the lower corner frequency, which is enough below 30 Hz so the control pulses will not be attenuated. Then why have the lower pole? The "midband" gain is large enough that the amplifier is subject to large output offsets. These will charge the 47 μF capacitor to a voltage that will hold the two op amp inputs at the same DC potential.

Exercise 8.32 Perform a DC analysis to find V_o for this stage, considering I_{BIAS} as the only op amp nonideality. Repeat the analysis if the 47 μF capacitor is replaced with a short circuit.

This analysis shows clearly one large benefit of having a capacitor in series with the path to ground. Another benefit is less obvious, because most of us are used to thinking of op amps as having dual supplies. This one, however,

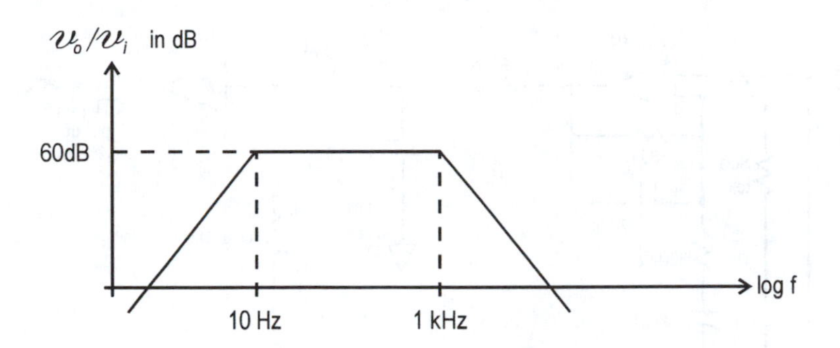

Fig. 8.52: Amplitude response of the first stage amplifier in Fig. 8.51.

has a single supply. This means that in the absence of signal, we would like the output and both the inputs to be at about half the supply voltage. In this circuit, the bias voltage applied to the noninverting input need only bias it to about 2.5 V. This will cause the op amp to initially rail positive. The 47 μF capacitor then charges through the 330 k resistor in the feedback network. When V_c reaches the same 2.5 V as is applied to the noninverting input, both input potentials are equal, and the output comes down from the positive rail.

Exercise 8.33 How long will this process take for this circuit, and why doesn't the delay have any adverse effects on the operation of the circuit?

After passing through the amplifier, the signal goes through a passive LPF with a cutoff frequency of 1.65 kHz. This effectively doubles the rolloff rate at the high end. The amplified control signal is then applied to an integrated Schmitt trigger and then proceeds to IC_{701} pin 9.

8.6.6.5 The capstan servo system

Figure 8.53 is the capstan counterpart of the drum circuitry shown in Fig. 8.50. The input circuitry is very similar. In Fig. 8.53, the capstan FG pulses go through a passive low-pass section with a cutoff frequency of about 16 kHz and then into a stage that is topologically identical to the second stage of drum FG processing in Fig. 8.50. This stage is, as we saw, a Schmitt trigger, and it delivers a square logic-level signal to pin 8 of IC_{701}.

Within the capstan servo circuitry of IC_{701}, a comparison is made between the processed control signal just studied and a 30 Hz reference signal derived from a crystal oscillator when the VCR is in playback mode. In record mode, the reference signal is compared to a divided-down signal from the capstan FG.

The capstan circuitry in IC_{701} generates APC and AFC outputs for control of the capstan. The APC output is supplemented by a DC voltage from the collector of Q_{703}. This configuration looks more than a little like that of Q_{702} in Fig. 8.50. This is hardly surprising, since in Fig. 8.49B, Q_{703} is labeled as "CAP MUTE," which strongly suggests that it performs as the capstan counterpart of the "DRUM MUTE" function attributed to Q_{702}. Current from the +2.5 V supply will flow through the 8.2 k resistor. The only question is whether it will flow through the transistor or through D_{718}. This will, of course, depend on whether or not the transistor base is allowed to remain high. If it is, Q_{703} is off and current flows through D_{718}. If the capstan stop line at the bottom of the

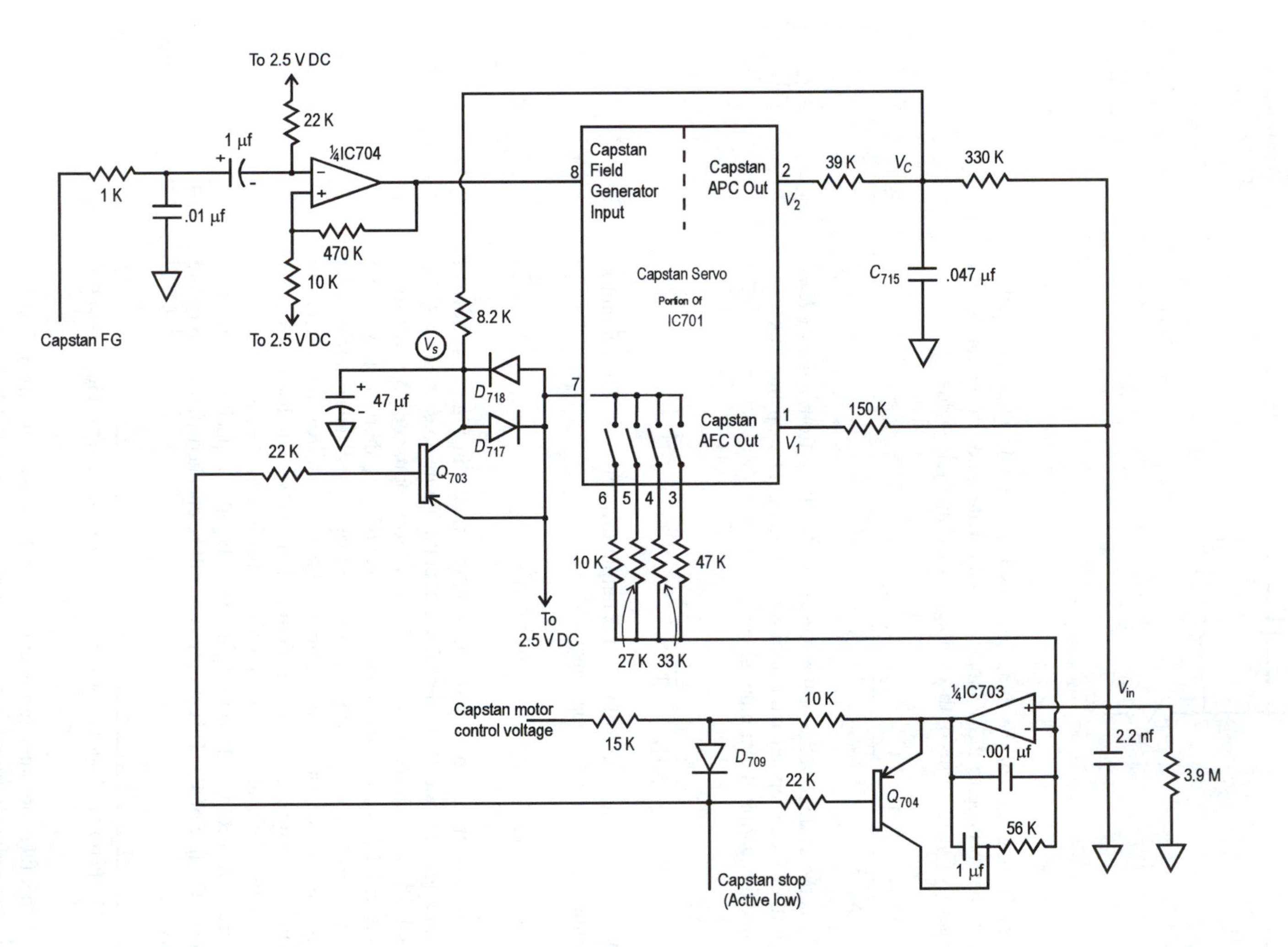

Fig. 8.53: Schematic diagram of the capstan motor servo circuitry in the Sharp VC-785U VCR.

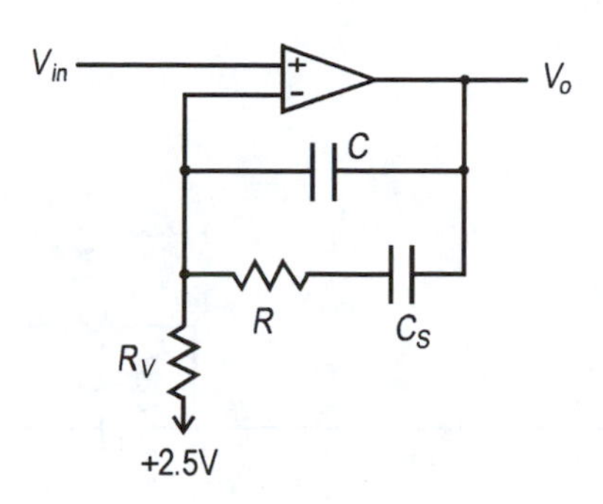

Fig. 8.54: Schematic diagram of the modified ramp generator in the capstan servo circuitry of Fig. 8.53.

page is pulled low, one of the things it does is to saturate Q_{703} and apply a slightly-higher voltage to the V_s node. We can write a node equation for the voltage across C_{715} in terms of this voltage. A DC analysis yields

$$\frac{V_s - V_c}{8.2\,\mathrm{K}} + \frac{V_2 - V_c}{39\,\mathrm{K}} = \frac{V_c - V_{in}}{330\,\mathrm{K}},$$

where V_2 is the average value of the APC output voltage from pin 2 of IC_{701} and V_{in} is the voltage applied to the noninverting input of the op amp. Likewise, the DC nodal equation at that input gives

$$\frac{V_c - V_{in}}{330\,\mathrm{K}} + \frac{V_1 - V_{in}}{150\,\mathrm{K}} = \frac{V_{in}}{3.9\,\mathrm{M}},$$

where V_1 is the average value of the AFC output from pin 1 of IC_{701}. Simultaneous solution of these two equations yields

$$V_{in} = .248\,V_s + .054\,V_2 + .674\,V_1. \tag{8.3}$$

We next look at the op amp stage, which is shown in Fig. 8.54. Several observations are necessary here. The first is that this figure does not show Q_{704}, which will short C_s on receiving the Capstan Stop command. We will solve the circuit with C_s in place and model the effect of Q_{704} shorting it by seeing what happens as $C_s \to \infty$. Secondly, we note that the resistance shown as R_v is actually one of the four resistors connected to pins 3–6. Analog switches in IC_{701}, under system control, will select one of these. We will discuss their use after completing the analysis. A third point is that in a "first-pass" analysis, we will ignore C. It is comparable in magnitude to the other capacitors we ignored to find the DC value of V_{in}. This assumption will be checked after we complete our analysis.

Exercise 8.34 Perform a Laplace transform analysis of the circuit in Fig. 8.54 to find $v_o(t)$.

We see that there are three terms in the output voltage. The first is just V_{in}, which we have already found as a function of V_s, V_1, and V_2. A second term describes a ramp of slope $(V_{in} - 2.5)/C_s R_V$. The time constant for this ramp will depend on R_V and will be between 10 and 56 ms. The last term in the answer to this exercise is $(V_{in} - 2.5)R/R_V$. Considering the possible values of

R_V, we see that the gain for this term ranges between about 1.2 and 5.6. To summarize, the circuit output is given by

$$v_o(t) = V_{in} + (V_{in} - 2.5)\left[\frac{t}{C_s R_v} + \frac{R}{R_v}\right]. \tag{8.4}$$

If we had not ignored C in this analysis, this expression would have a factor of $(1 - e^{-t/RC})$ multiplying R/R_V. The RC time constant is 56 μsec. Because this is so much faster than the time constant of the ramp, we are justified in ignoring it in a simplified analysis.

We see that when Q_{704} has its base pulled low by the Capstan Stop signal, it will saturate and short out C_s. The ramp term then drops out of the equation above, leaving a DC output of

$$V_o = V_{in} + (V_{in} - 2.5)R/R_V.$$

It is easy to show by algebraic arrangement of this equation that if $R_V =$ 10 kΩ, any $V_{in} < 2.12$ V will produce $V_o = 0$, thus shutting off the capstan drive voltage. If $R_V = 56$ kΩ, a V_{in} of 1.36 V or less will produce the same effect. The value of R_V is determined by analog switches in IC_{701}, as already mentioned. System control changes these switches to accommodate different running speeds due to different basic tape speeds (SP, LP, EP) and to allow "trick effects" such as slow and fast playback within one of the basic speed settings.

Before we leave the circuit of Fig. 8.53, we look back one more time to the voltage at the collector of Q_{703} – a voltage we have been calling V_s. In (8.3) we saw that

$$V_{in} = .248\, V_s + \cdots.$$

In (8.4) we saw how the capstan control voltage depends on V_{in}. From it we can write

$$\frac{dv_o}{dv_{in}} = 1 + \frac{t}{C_s R_V} + \frac{R}{R_V}.$$

And since $dv_{in}/dv_s = .248$,

$$\Delta V_o \approx .248[1 + t/C_s R_V + R/R_V]\Delta V_s.$$

ΔV_s would be expected to vary about .4 V as Q_{703} moves from saturation to cutoff. Hence

$$\Delta V_o \approx .1[1 + t/C_s R_V + R/R_V].$$

Since R/R_V can be >5, it should be obvious that Q_{703} is capable of changing V_o by something on the order of a volt. Thus, as was the case for the drum servo, we see that a seemingly small change in V_s can cause a more significant change in the motor control voltage.

Returning to Figs. 8.49A and C, we see two pins on the system-control IC (pins 54 and 55) that are buffered by two of the inverters in IC_{802} and are applied to a resistive summing network whose output is labeled "Current Limit." This line goes directly to the capstan motor connector. Hence the current limiting mechanism is internal to the capstan motor assembly and the outputs from IC_{801} will activate or deactivate current limiting. The mnemonics on pins 54 and 55 mean "Capstan Unload" and "Capstan Pullup" respectively.

It is more than a little disconcerting to see that the inverter between pins 2 and 15 and that between pins 1 and 16 of IC_{802} are showing output voltages that are hardly the digital inverse of their respective inputs. The probable explanation is that these are open-collector inverters in which the zero inputs put the output transistors into the open state. Then the voltages shown at the output pins would simply reflect the voltage division ratio between R_{862} (100 kΩ) and R_{860} (15 kΩ). If this is the case, the expected value of the current limit control voltage is

$$\frac{5\text{ V} \times 15\text{ K}}{100\text{ K} + 15\text{ K}} = \frac{75}{115} = .68\text{ V},$$

which is in good agreement with the value of .7 V shown on the schematic.

8.6.6.6 Tracking control

Most VCR users are aware of the tracking control, but few know what it does aside from the obvious picture improvement it can effect. If you think back over your own experience, however, you will probably recognize that adjustment is never required when playing back tape you have recorded on the same machine. Let us see why this is so. Our discussion hitherto should convince you of the necessity of having the control pulse received from the tape at the exact right time relative to the heads. Assuming that the servos are working correctly and that head and capstan speeds are correct for the tape, it is still possible to have the phasing wrong between the control pulses and the heads. Subject to the assumptions we have just made, the time delay between these two signals is a sole function of the mechanical distance along the tape path between the control and video heads. This is why tapes recorded and played back on the same machine never have tracking problems in a properly functioning machine.

If a tape is played back on a different VCR, however, the slight differences in head spacing will cause the head to scan a track parallel to the desired one, but slightly offset from it. Thus it will introduce substantial noise into the picture, which shows up as white flashes. If we could mechanically tweak the inter-head distance, we could eliminate this problem. A much more satisfactory solution, however, is to pass the control head pulses through a variable time delay circuit composed of a couple of monostable multivibrators. The tracking control circuitry ties in with the capstan servo. This accounts for pin 14 of IC_{701} being denoted as "Tracking MM." This pin has a capacitor to ground and a variable resistance to the PC 5 V supply. These are obviously timing components, and the variable resistance contains a fixed 10 k resistance, a 100 k "Tracking Preset" control (which is a service adjustment), and a 200 k tracking control, which is user-accessible.

Pin 13 of IC_{701} is a line that signals to the servos which recording speed is desired. This comes from IC_{5001} in Fig. 8.61. This chip is designated as a "Timer/Volsyn." It will be discussed in Section 8.6.8 of this chapter.

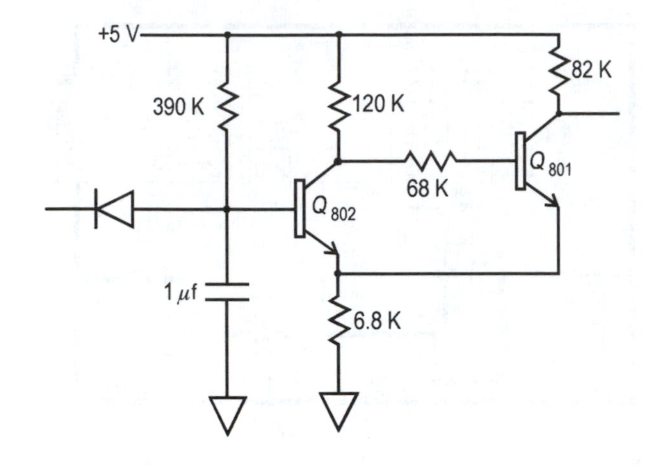

Fig. 8.55: Schematic diagram of the dew-sensor amplifier in the Sharp VC-785U VCR.

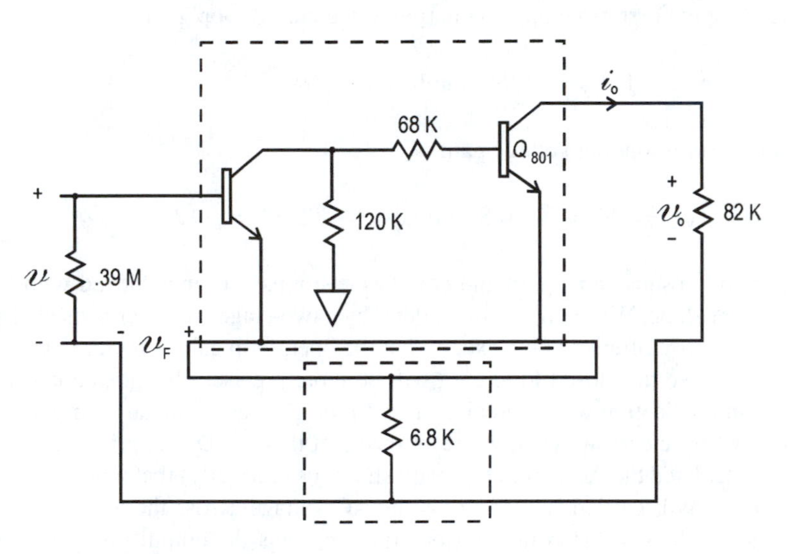

Fig. 8.56: AC equivalent circuit of the amplifier of Fig. 8.55 drawn in standard feedback form.

Exercise 8.35 Refer to pins 22–24 of IC_{701} and pin 27 of IC_{801}. What can you say about the speed-detector logic in the latter chip?

8.6.7 *Miscellaneous circuitry*

8.6.7.1 *Dew amplifier*

As previously noted, the dew sensor detects condensation in the VCR and disables it until the dew evaporates, thus eliminating the hazard of the tape clinging to the capstan and twisting around it. The dew sensor is a resistance that drops when moist. This is coupled to the amplifier shown in Fig. 8.55.

When dew is sensed, the potential at the base of Q_{802} drops. This is the input to the two-stage amplifier. If this amplifier is drawn in standard feedback form, it is as shown in Fig. 8.56.

For this amplifier the transfer function of the feedback network is

$$F = v_F/i_o = -6.8\ \text{k}\Omega.$$

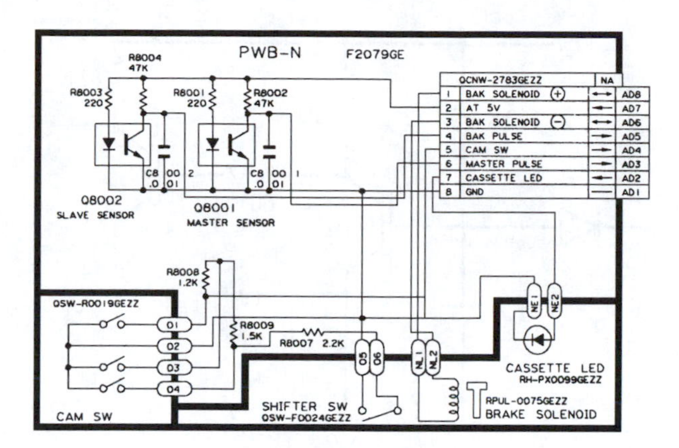

Fig. 8.57: Schematic diagram of a small printed-circuit board in the Sharp VC-785U VCR. This board contains the optocouplers that sense the direction and amount of rotation of the take-up reel of the cassette.

Therefore the first-order approximation to the closed-loop gain is

$$A_{GF} = 1/F = -1/6.8 \text{ mmhos} = i_o/v$$

and the corresponding voltage gain is

$$A_{VF} = -82\,\text{K} \cdot (1/6.8) \text{ mmhos} = -82/6.8 = -12 = v_o/v.$$

This is not a substantial gain and could as easily have been achieved with the first stage alone. We might well wonder why a two-stage amplifier is used. The analysis above offers no clue except that the open-loop gain is larger with two transistors, so the closed-loop gain will be more precise. This does not strike us as an application where precise closed-loop gain is particularly important. The mystery clears up when we recognize that this is a DC amplifier, and we have calculated the AC gain. Our result shows us (correctly) that when v goes down, v_0 will get larger. But v_0 is the AC voltage across the 82 k resistor. If this gets bigger, the collector potential, and thus the output to the system control IC (pin 62) drops. It is this dew-induced drop of DC output that justifies the use of two stages. With a single stage, a drop in the DC input causes the collector potential to rise, which is opposite to the desired effect. In fact, as the input drops, the collector voltage of the first stage ($= v_{B2}$) rises, and the emitter potential of the first stage ($= v_{E2}$) drops. Both of these effects will increase v_{BE2}, and turn on the transistor more heavily, thus pulling down its collector potential.

8.6.7.2 *The master/slave pulse circuitry*

The master/slave pulse circuitry is shown in the lower left corner of Fig. 8.49D. What we have there are essentially a pair of logic inverters. The inputs to these stages can be traced back to the circuit shown in Fig. 8.57.

The devices labeled as Slave Sensor and Master Sensor are, of course, optocouplers. What does not show in Fig. 8.57 is that both of these optocouplers are open units that can "see" outside of themselves. Both are "looking" at a disc fastened to the underside of the take-up reel of the cassette. The disc is covered with wedge-shaped segments like the cuts in a pie. The segments are alternately light and dark. The light segments reflect light back to the optocouplers, and the dark ones do not. When they "see" the reflected light, the optocouplers produce logic-level pulses. The one labeled "Master Sensor" produces a "Master Pulse," which is coupled to the input of Q_{712}, which buffers and inverts it. At the output of this stage the pulse is called the "Master Reset Pulse." It proceeds off of Fig. 8.49C via a connector pin labeled "Counter Pulse." If this pulse is followed into Fig. 8.61, we see that it is similarly denoted when it enters IC_{5001} at pin 2. It seems clear that each pulse will increment the tape counter.

Exercise 8.36 The hub in a VHS videocassette is about 1″ in diameter. When it is fully wound with tape, the diameter is about $3^{1/2}$″. If the segmented disc produces 8 pulses per revolution, estimate the change in the counter reading after running a tape all the way through the machine.

If we follow the path of the slave pulse, we see that it is in all respects analogous. When it finally reaches IC_{5001} it is labeled "Count U/D." It is manifestly obvious that the counter must know if the counter pulses it is receiving should be used to increment or decrement the counter. By analogy with counting we have seen elsewhere, we might be inclined to think that "Count U/D" down is a control signal that might be high to count up and low to count down (for example). Not only does the parallel character of the master pulse and the slave pulse militate against this, but Fig. 8.61 also shows clearly that the signals applied to pins 2 and 3 of IC_{5001} are pulse trains. From this we infer that the optocouplers are probably set up at the same radius from the center of the disc, but with one slightly ahead of the other along the direction of rotation. Thus each segment will have its leading edge sensed first by one optocoupler and then by the other. A small amount of logic in the timer/volsyn circuit could easily determine rotational direction from this and thus make a decision on whether to increment or decrement the counter.

We might note in passing that after leaving Q_{712}, the master reset pulse train is also applied to pin 21 of IC_{801}. Since this chip is the system controller, it must have inputs from and outputs to many parts of the VCR. The inputs are "supervised" by the IC. For instance, the pulse train on pin 21 tells the IC that the take-up reel is turning. Besides other pins of IC_{801} already mentioned, here are some other inputs:

Pin 23 (RecTIP – Record tab in place). This indicates that a recordable tape has been loaded into the VCR.

Pin 22 (Drum Sensor) This pin gets the head-switch pulse (HSWP) that indicates the drum is rotating.

Pin 28 (Still/Slow Mode) Since this pin is permanently tied to $V_{cc}/2$, and since the VCR doesn't run exclusively in these modes, we deduce this line is probably either a tri-state input or an "Enable," as opposed to a "Select" command. We might also note that pin 7 is not connected. Pins 17 and 18 (Start/End Sensor) these will be discussed in the next section.

We also note that the system control IC must have communication with the timer/volsyn chip and the servo control chip. These connections are made on pins 36, 37, 38, 40, 41, 48, and 57.

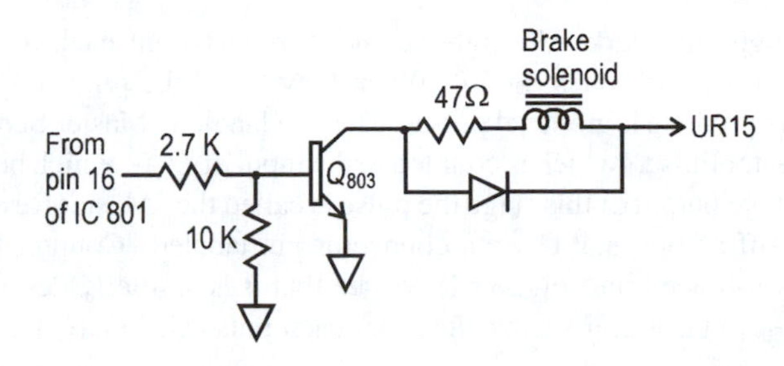

Fig. 8.58: Schematic diagram of the driver stage for the brake solenoid in the Sharp VC-785U VCR.

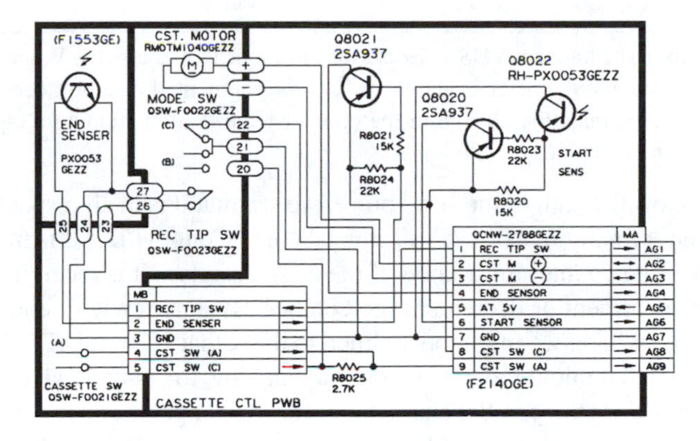

Fig. 8.59: Schematic diagram of the small printed-circuit board that contains the tape sensor circuitry in the Sharp VC-785U VCR.

8.6.7.3 Brake solenoid

Referring back to Fig. 8.57 one last time, we see a brake solenoid. This is activated on command from IC_{801} for some of the "trick effects" that will be discussed in Section 8.7. The output of pin 16 is applied to Q_{803}. The circuitry for this stage is shown in Fig. 8.58. The diode is for protecting Q_{803} from the inductive kick of the solenoid when it is cut off. It could be put across the solenoid only, but the placement shown will limit the surge current through the diode as it absorbs the inductive spike.

8.6.7.4 Tape sensors

Figure 8.59 shows the tape sensor circuitry. The schematic shows two ordinary transistors (Q_{8020} and Q_{8021}) and two phototransistors (Q_{8022} and the unidentified end sensor device). Because of the very crowded nature of the diagram, it will be useful to redraw the start and end sensor circuitry. The result is Fig. 8.60.

In spite of the fact that the end-sensing phototransistor is npn and the start sensor is pnp, the circuits are quite similar. When the videocassette is loaded, the cassette lamp or cassette LED is actually sticking up into the cassette. At the ends of the tape, there are transparent leaders that will allow light from the cassette lamp to reach the appropriate phototransistor when the tape is at one end or another. The end sensor of course enables automatic rewind at the end of playing a tape, as well as terminating fast forward at the end. The start sensor terminates rewind.

In both sensors, the reception of light by the phototransistor pulls the base of the pnp output transistor low. This causes current to flow in the emitter and out the collector to pin 17 or 18 of IC_{801}.

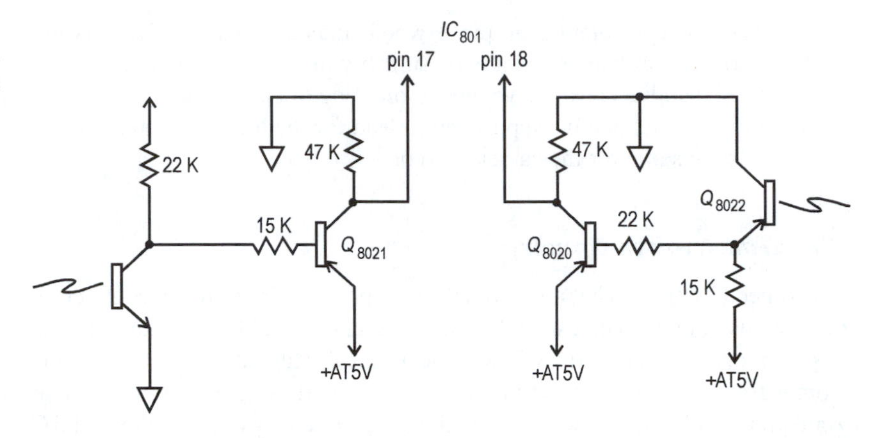

Fig. 8.60: Schematic diagram of the start and end sensing circuitry isolated from Fig. 8.59.

The remaining circuitry in Fig. 8.59 consists of just a number of switches and the cassette motor. This motor keeps slight tension on the tape after it passes through the capstan/pinch roller assembly so the tape does not "pile up." It also does fast forward and rewind. The switches shown are integral with the cassette motor and either control or are controlled by what the motor is doing at a given moment.

8.6.7.5 The motor driver IC

IC_{803} contains the control and power-driver circuitry for the cassette motor and the loading motor. The former motor was discussed in the previous section. As the name implies, the loading motor is responsible for the familiar "sucking in" of the videocassette by the VCR. What is not visible is the continued activity inside the VCR after the cassette is loaded. Through driving a mechanical structure called the *loading ring*, the motor actually pulls a substantial amount of tape from the cassette and threads it into the proper tape path. When the VCR is stopped, the cycle reverses and the tape is returned to the cassette. These loading and unloading cycles account for both the delays and the mechanical sounds from the VCR after you press Stop, Play, or Record. Again, based on your experience, you might rightly judge that in pause mode, the tape does not unthread. It is also true in pause mode that the head scan continues. This will minimize the time required to reacquire servo lock after returning to play or record modes, but it also wears out the tape quickly, since the same portion of tape keeps being scanned. Most VCRs have a timeout to end pause mode after three or four minutes.

We will make a few quick observations about IC_{803} and the surrounding circuitry:

- The control circuitry in this chip is driven by the system control IC pins 13–15.
- The loading motor is shunted by C_{811} (1 μF) and the cassette motor by C_{812} (1 μF). These do not really substitute for shunt diodes though they do reduce spikes somewhat. The main inductive spike suppression is expected to be accomplished by diodes that are part of IC_{803}.
- The UR 15 V supply is applied to pins 7 and 9. It supplies power to the motors when enabled by the chip. A 22 μF local bypass capacitor is connected from pin 7 to ground.

- The voltage notation on pin 8 would indicate that the Zener is not normally broken down. It is presumably in circuit as a clamp.
- Pin 10 applies positive voltage to one side of each motor.
- Pin 2 is the negative supply return lead for the loading motor and pin 3 is the same for the cassette motor.

8.6.8 Timer/volsyn circuitry

We first need to know what a volsyn is! In the present circuit it is a contraction of the words voltage synthesizer. We saw in Section 8.6.2 that the tuner in this VCR (as indeed in all modern VCRs) is electronically tuned. The voltage to tune it comes from the volsyn. The timer is so much a part of the user interface that we are pretty well familiar with its function. If the circuitry of Figs. 8.49B and C was so sketchy as to make analysis difficult, that in Fig. 8.61 is far more so. We will content ourselves with making a few observations and a few conjectures:

(1) It is obvious that IC_{5001} totally controls the entire VCR display and supplies all of the information on it.

(2) IC_{5001} also takes care of all interfacing to the keyboards on the VCR itself as well as on the remote. Note that pins 38–45 and 56–57 are multiplexed to handle both the display and the keyboard.

(3) The signal called "Clock" on the main board connector at the upper right-hand corner can be traced back onto Fig. 8.14A to find its origin in the power supply section. It is a 60 Hz clock derived from the power line. Q_{5001} squares it up before applying it to IC_{5001}. This signal is the time base for the clock and timer portions of the display.

(4) The reset circuitry centered around Q_{5010} and Q_{5011} presents a small challenge in that it apparently has only two connections to "the rest of the world" – one to AT 5 V through D_{5022} and one to pin 22 of IC_{5001}. Our task is made easier if we recognize that this is a power-on reset circuit.

Exercise 8.37 Describe how this circuit works, including the function of as many parts as possible.

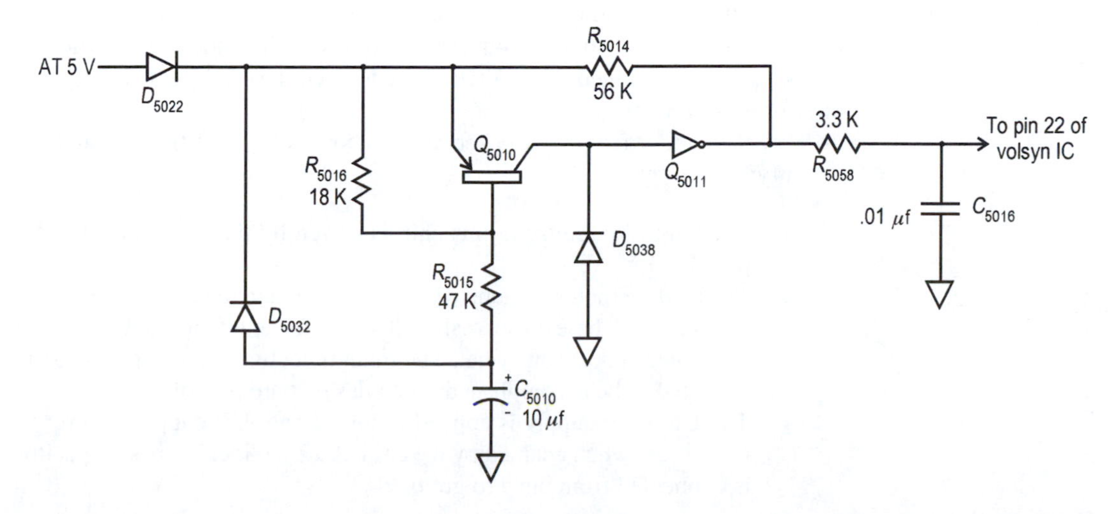

The only remaining portion of the circuitry in Fig. 8.61 that we have not touched upon is IC_{8002}, which is identified as EAROM. This chip digitally stores the tuning information for each channel. The V_{ss} supply on pin 1 supplies the power for normal operation. The V_{GG} supply on pin 2 enables user programming, if it is necessary for a service person to fine tune a channel. Besides the clock, the EAROM is connected to the volsyn circuitry by the four lines C1, C2, C3, and I/O. These suggest that the memory chip is used serially, with several cycles of address/data exchange going on to give the volsyn what it needs to synthesize the channel-tuning voltage. The pin 70 output of IC_{5001} is labeled VT, but as we follow it over to the connector on the right center of Fig. 8.61, we see that it is called "PWM Out." This tells us that the tuning voltage is PWM encoded. This makes sense, as nothing more than an LPF is required to extract the DC value of the PWM waveform. If we follow this line back onto the lower portion of Fig. 8.14A, we see that it goes through a stage of amplification and then to a three-stage LPF whose output is applied directly to the VT terminal of the tuner.

While we are in this part of the circuit, we might as well observe that the lines B0 and B1 from pins 6 and 5 of IC_{5001} also proceed from Fig. 8.14A to 8.14B and thus to the band decoder IC (IC_{1401}) in the lower right-hand corner. This chip outputs the band selector signals used by the tuner (BL, BH, and BU).

This completes our analysis of the Sharp VC-785U VCR as such. We will now proceed to look into the "trick effects," but we will do so generically rather than for this specific VCR, since the detailed information required to exegete these effects for this VCR is simply not available. This is because so many of the functions are implemented in software.

8.7 Special effects

Under this heading we would include fast, slow, and reverse playback as well as freeze frame. There are two constants in all of these effects: (1) They are all playback phenomena and (2) the head drum always spins at the same rate as in normal playback. These will be true regardless of the basic speed at which the tape was recorded. In producing these special effects, several problems must be addressed and solved. The problems of slow playback are more difficult on the whole than those for fast playback. Figure 8.62 illustrates one of the problems – that of what the head reads.

Despite the similarities, this figure should not be confused with Fig. 8.6, which was an accurately scaled representation of the layout of the tracks on the videotape and the scan lines within the tracks. Figure 8.62, on the other hand, is much more schematic. Each parallelogram in it represents an entire track. This figure does not even attempt to depict the individual scan lines. The waveform at the bottom of the page represents the head switch signal. If we consider the heavily outlined strip, we see that it represents normal play. When the head switch signal is high, head A is activated and the odd fields (tracks 1, 3, 5, etc.) are read. When the head switch signal is low, head B is activated and the even fields (tracks 2, 4, 6, etc.) are read. Even and odd field tracks are read interleaved. To illustrate, suppose that the capstan were not turning at all. The tape is still. Suppose further that at the start of track 1 head A is perfectly aligned to read it. As the drum rotates, alignment will get progressively worse, since this track was recorded with the tape moving. This is illustrated in Fig. 8.63. Now when head B comes around, it will start at the same point as head A did

SCHEMATIC DIAGRAM OF TIMER, OPERATION CIRCUIT

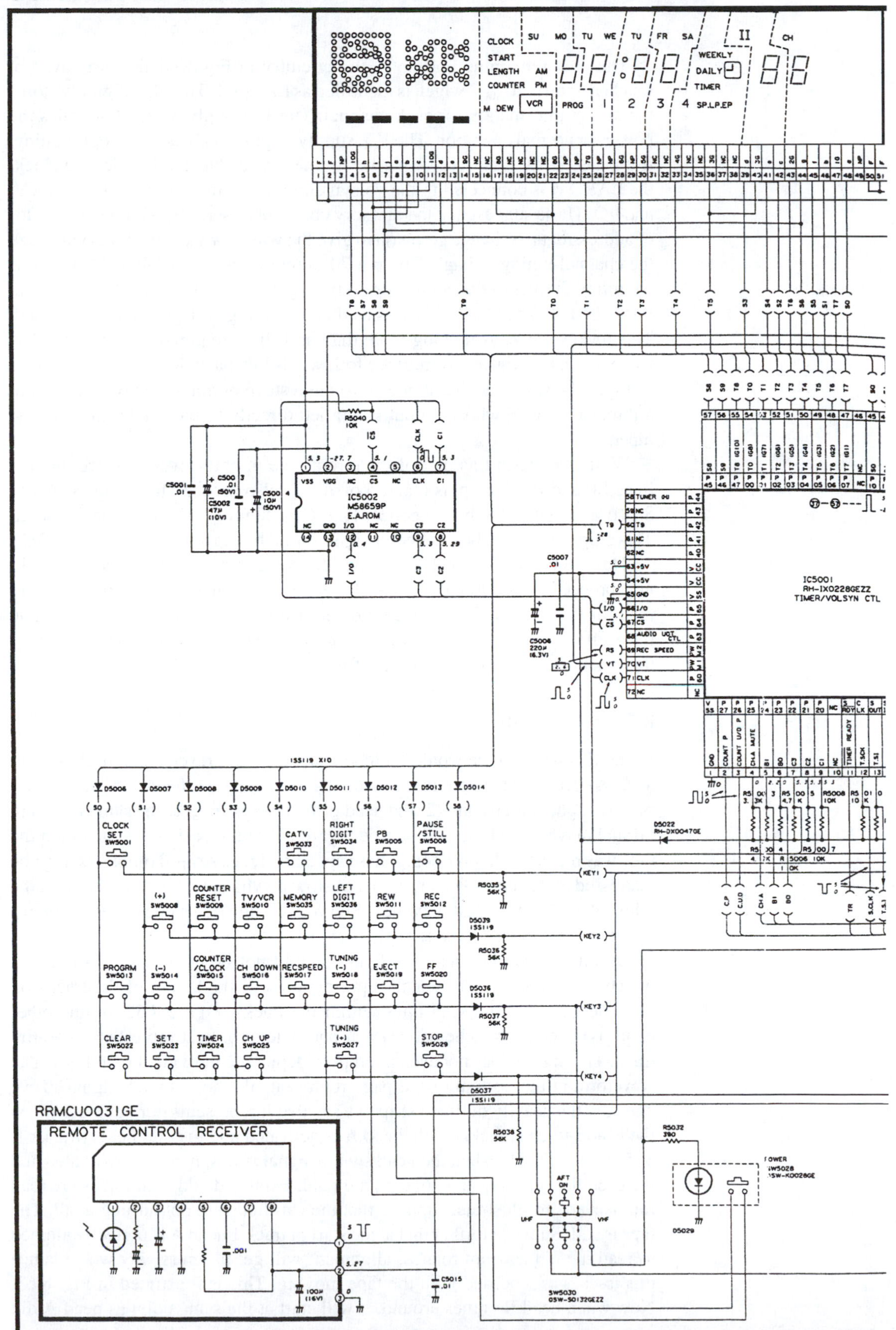

Fig. 8.61: Schematic diagram of the voltage-synthesizer circuitry in the Sharp VC-785U VCR. An electronic version of this figure is available on the world-wide web at http://www.cup.org/titles/58/0521582075.html.

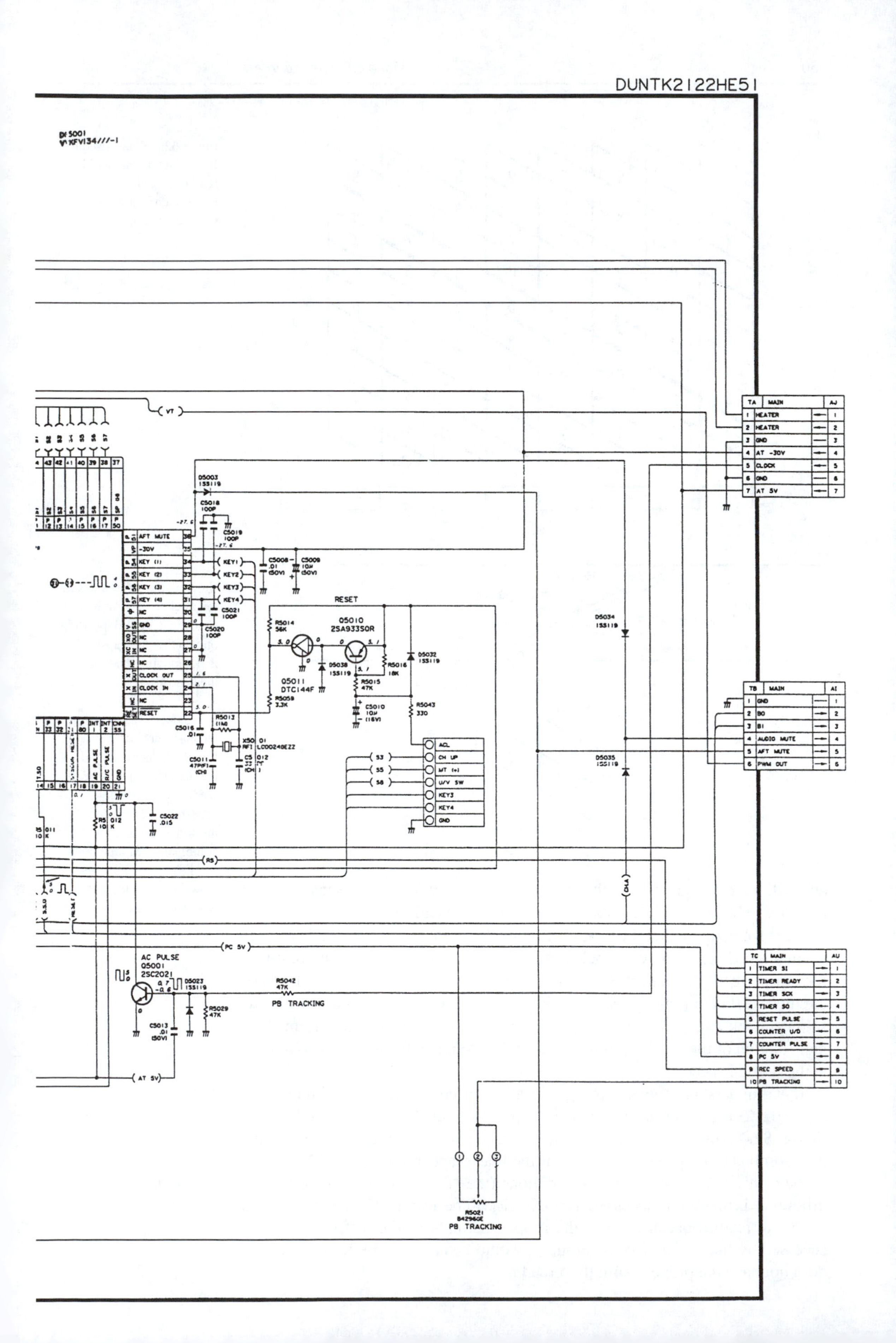
DUNTK2122HE51
RESET
Q5010
2SA933S0R
Q5011
DTC144F
AC PULSE
Q5001
2SC2021
PB TRACKING
R5021
PB TRACKING
TA MAIN AJ
HEATER
HEATER
GND
AT -30V
CLOCK
GND
AT 5V
TB MAIN AI
GND
BO
BI
AUDIO MUTE
AFT MUTE
PWM OUT
TC MAIN AU
TIMER SI
TIMER READY
TIMER SCK
TIMER SO
RESET PULSE
COUNTER U/D
COUNTER PULSE
PC 5V
REC SPEED
PB TRACKING
AFT MUTE
-30V
KEY (1)
KEY (2)
KEY (3)
KEY (4)
NC
GND
CLOCK OUT
CLOCK IN
RESET
ACL
CH UP
MT (+)
U/V SW
KEY3
KEY4
GND

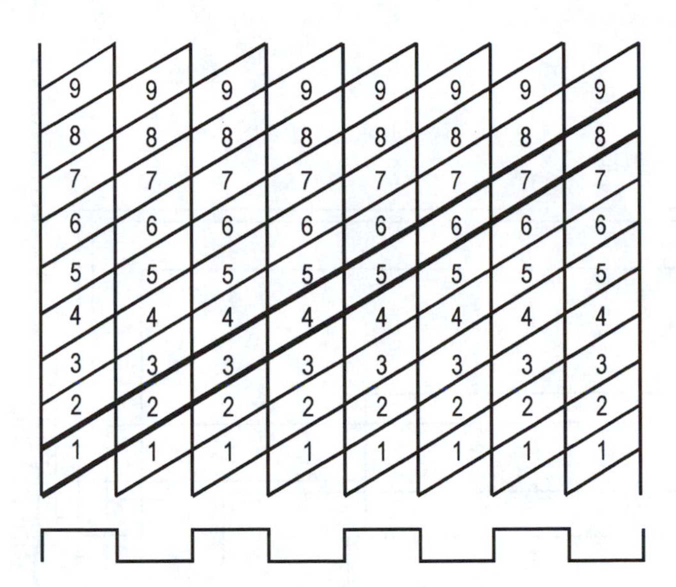

Fig. 8.62: Chart used to find which tracks will be read during slow or fast scans such as are used in "trick effects."

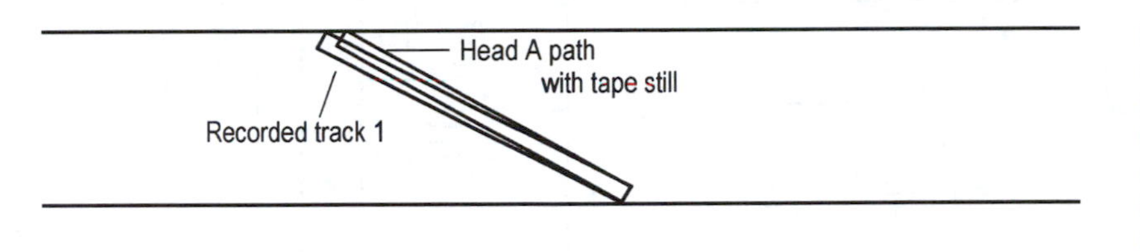

Fig. 8.63: Diagram showing the differences in the track traced out by the heads in a VCR depending on whether the capstan is or is not moving the tape as the head scans.

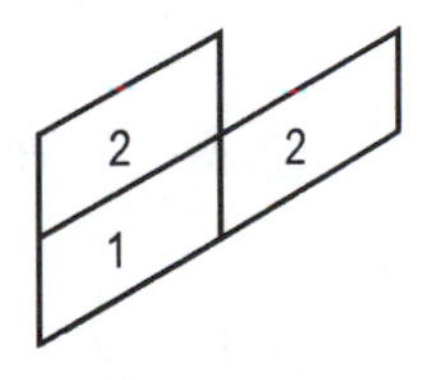

Fig. 8.64: The lower left-hand portion of Fig. 8.62. As explained in the text, if track 1 is being read, track 2 can be accessed by linear tape motion alone or by a combination of tape motion and head scanning. However, only the latter will read it properly.

and will trace out the same path. To whatever extent it lies along track 1, it will produce little or no output because of its differing head azimuth.

Next we imagine that the tape is moving at normal speed and that head A is again aligned perfectly with track 1 at the start of that track. If this is so, it will remain perfectly aligned throughout the track (assuming that the VCR servo circuits are functioning correctly). When the head B comes around, it will now start on track two. In other words, the forward motion of the tape during one half rotation of the head drum is just sufficient to move from track 1 to track 2 at the normal speed.

If we understand the foregoing, we are in a good position to understand Fig. 8.62. Consider just the three tracks in the lower left-hand corner, as depicted in Fig. 8.64. Track 1 is, of course, the starting point. Track 2 can be accessed by a combination of head scan and normal linear tape motion (the "2" segment to the right) or by linear tape motion alone (the "2" segment to the left). The difference is that if it is accessed normally, it will be accessed by head B, which is able to read it, because it has the proper azimuth, whereas if track 2 were accessed by linear tape motion alone, it would be accessed by head A, which does not have the proper azimuth to read it.

All of the special effects will actually be a combination of standard scan and nonstandard linear tape speeds and can thus be represented by straight lines of varying slope on a track chart like Fig. 8.62.

8.7.1 Accelerated play (search mode)

8.7.1.1 Tracks read

Suppose we run the tape exactly twice the normal speed. The chart that illustrates this is shown in Fig. 8.65. Here is how we interpret this figure: The two solid parallel lines which are not parallel to anything else are the path taken by the heads. Note that during time slot T1 the A head starts reading track 1, but because of the accelerated tape speed, as you proceed through T1, the head is moving away from track 1 and going over track 2. But recall that adjacent tracks are recorded with opposite azimuth angles (Section 8.4.1.1). Thus there is virtually no pickup from track 2, and the sole effect of the doubled tape speed so far is a tapering down of the output from track 1 to zero by the end of T1.

During time slot T2, the B head is active. Its azimuth is such as to allow it to read the even fields. At normal speed, it would be reading track 2. At double speed, it is passing over 3 (which has the wrong azimuth for it to read) and then into track 4. The output from track 4 tapers up as head B overlaps more of it and is at a maximum by the end of T2.

In T3, head A reads from track 5, which shows a declining output with time, just as track 1 did. Thus, though at normal speed head A reads tracks 1, 3, 5, etc., at double speed, every other track is skipped. It reads 1, 5, 9, etc. The same thing happens with head B. It reads 4, 8, 12, etc. Interleaved together, we have the sequence 1, 4, 5, 8, 9, 12, 13, etc. Half the tracks are never read. Those that are read have outputs that taper from maximum strength to zero as shown in Fig. 8.66.

A similar approach will reveal what happens for tape speeds that are higher multiples of the normal.

Exercise 8.38 Photocopy the track chart below and for a tape speed 4× normal:

(a) Use your photocopy to find the interleaved sequence of the tracks read.

(b) Draw a diagram analogous to Fig. 8.66 for this case.

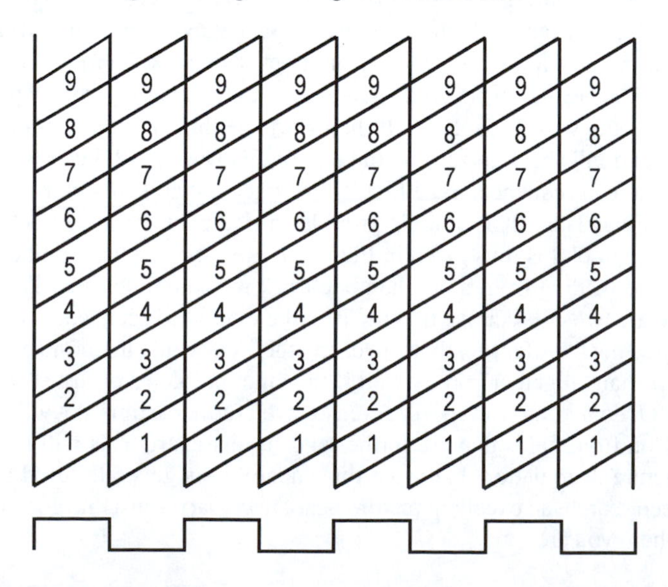

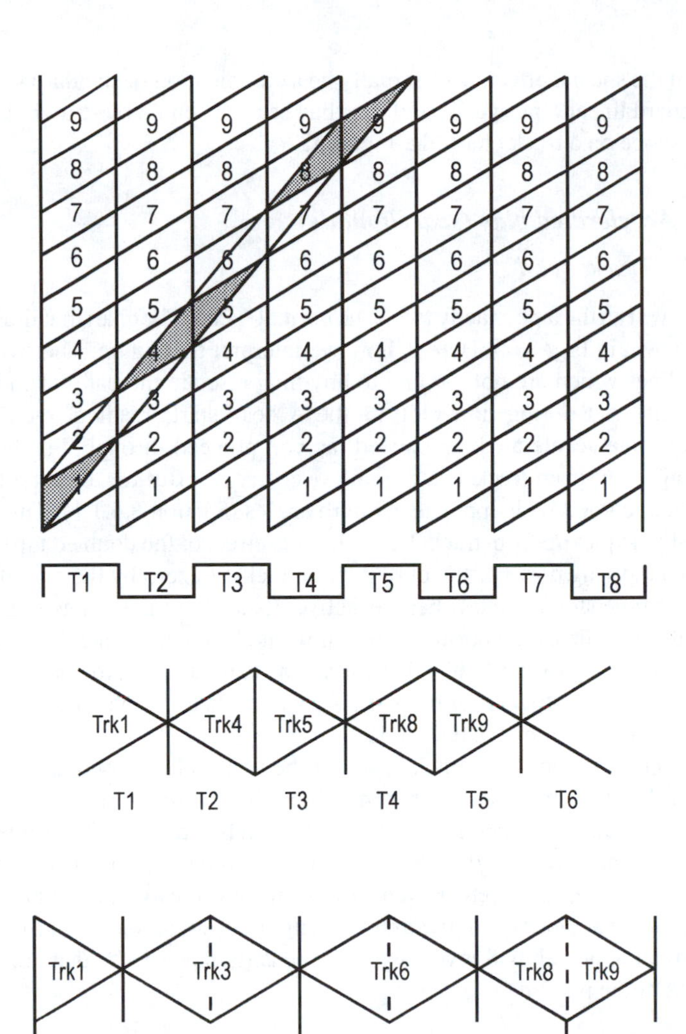

Fig. 8.65: An illustration of the use of the chart in Fig. 8.62 to analyze double-speed playback of the video tape. The parallel lines that cut across tracks show which tracks the heads will pass over. However, as explained in the text, the azimuth must be correct for a given track to actually be read by a given head.

Fig. 8.66: A signal-strength plot deduced from Fig. 8.65 and showing which tracks are actually read. The time slots indicated at the bottom correspond to those in Fig. 8.65.

Fig. 8.67: A signal-strength plot analogous to that of Fig. 8.66, but for quadruple-speed playback.

At normal tape speed, the track will be incremented by one during each time slot. We saw that during 2× speed scan it was incremented by two (from track 1 to track 3, for instance). In this exercise the tape moving at 4 times its normal speed will increment the track by 4 (from track 1 to track 5 during T1).

During T1, head A picks up a signal from track 1 that tapers down to zero one third of the way through T1. It then starts getting an output from track 3 that rises and falls back to zero by the end of T1. During T2, head B starts by executing a read sequence in the first 2/3 of track 6 that is the same as the last 2/3 of track 3. The last 1/3 of T2 provides a rising output from track 8. The signal strength plot is thus given in Fig. 8.67. Here we see that, unlike the 2× case, we are reading from more than one track in some time slots. Tracks one and three are two consecutive odd fields in the interlaced scan; and would thus be expected not to differ greatly from each other. We would therefore expect the picture to "hang together" pretty well if the other problems of fast scan can be handled. Of course, as scan speed increases even further, a time slot will contain information from fields that are further and further apart. This will inevitably cause picture degradation, but since this capability is generally used only for high-speed search, an excellent picture is not necessary – just one good enough to tell where you are.

8.7.1.2 Noise in the picture

As Figs. 8.66 and 8.67 show, there are times in search mode where the output signal gets down to zero, and even when it is not zero it is too low to give a good signal-to-noise ratio much of the time. This shows up on the screen as a horizontal noise band. This noise band can be wide or narrow and moving or stationary depending on the speedup factor. For example, the signal distribution depicted in Fig. 8.66 would lead to stationary noise bars at the top and bottom of the picture, since field scan is from top to bottom of the picture and zero output is obtained at the end (bottom) of the odd field and at the beginning (top) of the even field. This is characteristic of the performance of two-head VCRs. We shall see in Section 8.7 how this problem can be dealt with using a machine with four or more heads. We might mention at this juncture that similar problems are observed for slow playback in a two-head VCR.

8.7.1.3 Maintaining capstan servo lock

Once the tape starts moving faster (the capstan must already be moving faster to achieve this), the control pulses will also be coming faster, as will the pulses from the capstan field generator. The initial impetus for this speedup is the user selection of search mode. This produces the appropriate logic outputs from the system control microprocessor. One of these logic outputs will speed up the capstan. Another will insert programmable dividers into the control pulse and capstan FG paths so that they will both input their designed frequencies into the capstan servo circuitry as they would in normal play, thus enabling it to maintain lock.

8.7.1.4 Tweaking the drum velocity

Although we have treated the drum rotational velocity in search mode as a constant, there are good grounds for making a small adjustment to it. This is because when the tape moves faster, there is a second-order change in the tape-to-head speed, and this will cause a second-order change in the horizontal sync frequency. Because the velocity of the head relative to a fixed point is much greater than the velocity of the tape relative to the same point, the magnitude of this effect is typically only a few percent, but it is enough so that some monitors cannot handle it. Therefore, it is customary to generate a small correction to lower the drum rotational velocity slightly in forward search and to raise it in reverse search.

8.7.1.5 Horizontal skew correction

This correction is necessary because of an obscure effect encountered in search mode from LP (4-hour) playback. As the head runs across multiple tracks in search mode, it must be able to pick up the horizontal sync pulses that are "time coherent." This, in turn, requires that the pulses be lined up from track to track. This requirement is met in SP (2-hour) and EP (6-hour) modes, but not in LP, where the horizontal sync pulses are offset by .25 H from track to track. Suppose a given VCR is capable of $8\times$ search speed in the LP mode. The 262.5 lines/field would then be compressed into $\text{INT}(262.5/8) = 32$ lines, and 6.5 lines of the original field signal would be "left over." Skew correction in this case would probably consist of inserting an additional 1.5 H delay to the 6.5

H to make up another line. In other cases, there might be a smaller "leftover" that would be "gated out" in favor of an inserted sync pulse at the correct time. Of course, another alternative is just not to provide search capability at the LP speed, since LP is the least used speed (and justly so). The SP speed will clearly give a better quality picture, and EP will give the longest record time, while giving picture quality at least the equal of LP. With few exceptions, LP capability is still provided on modern machines for just two reasons: to allow playback of tapes recorded before EP was available and because the chips were designed to provide LP capability.

8.7.2 Slow playback/freeze-frame

Slow playback and freeze-frame are achieved by a play sequence consisting of periods of tape motion alternating with periods in which the tape is stationary (for slow playback) or having tape completely stationary (in the case of freeze-frame). They are grouped together because both have the tape stationary at least part of the time in playback.

The cycle actually consists of four parts: still, accelerate, run, and brake. The braking process is done electronically by reversing the voltage polarity to the capstan motor until it stops. Both the braking and acceleration are done very rapidly.

Only the video read from the tape during the time it is stopped is output from the VCR. As shown in Fig. 8.63, the stopped condition will result in less than perfect alignment between the rotating head and the recorded tracks. As was the case for the search operation, this will give less than full output from the tape at some points in the scan and result in noise bars.

The lengths of time spent in the still and run parts of the cycle will depend on the speed reduction factor and the servo response time, among other factors. Even though the capstan servo is inoperative while video is being read, it is operative between reads and must be cycled in a way that will allow it to retain lock.

8.7.3 Multiple-head machines

It has been observed that all special effects on a two-head machine are achieved only with a noisy picture, and that the noise is due to the scanning of partial tracks, which provide noise rather than signal for whatever part of the video head and/or its scan path do not lie over tape tracks recorded with the same azimuth as the head attempting to read it.

It has been observed further that VCRs with additional heads can go a long way toward eliminating this noise. The most common of the multiple-head machines are those with four heads, though there are three and five-head machines also. We shall examine briefly the two variants of the four-head machine. They have two additional things in common: The head gaps of the four heads differ in width, and circuitry in the VCR automatically selects the head that will have the largest output at a given moment.

8.7.3.1 The 90° head structure

As the name implies, this drum contains four heads spaced 90° apart. In addition to the normal channel 1 and channel 2 heads which are 180° apart, two additional

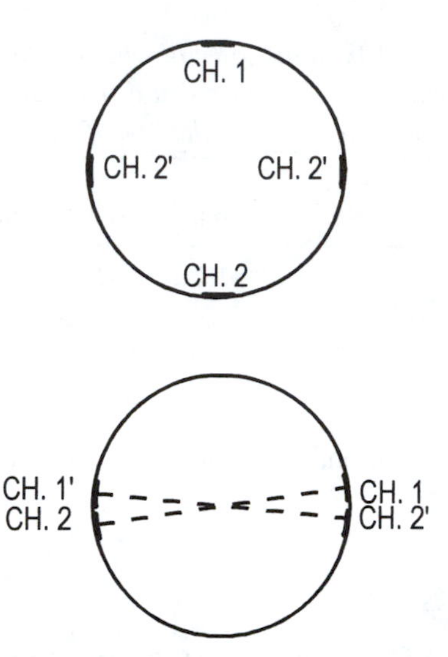

Fig. 8.68: Diagram showing the head placement in the head drum of a four-head VCR with the 90°-head structure.

Fig. 8.69: Diagram showing the head placement in the head drum of a four-head VCR with the adjacent-head structure.

heads called channel 2′ are placed at 90°. A diagram of the 90°-head structure is shown in Fig. 8.68. As the name implies, these heads both have the same azimuth as channel 2. In addition, both have a wider head gap than the channel 1 and 2 heads. The channel 1 and 2 heads do all recording and normal playback at all three tape speeds (SP, LP, and EP).

When trick play is selected, the head switch signal is phase shifted 90° to bring the channel 2′ heads into the position previously occupied by the channel 1 and 2 heads. The use of the channel 2′ heads has two consequences. The first is that since both have the same azimuth they will never pick up any signal from the odd fields. Thus the "trick play" picture will always be a noninterlaced even field picture. The second consequence is that owing to their wider head gap, the channel 2' heads will always be over some part of an even field track and will thus always pick up some of the desired signal. The fact that they will usually be over an odd field track is irrelevant, since their azimuth will not provide significant output from those tracks. This is rather a brute-force approach to the problem, but it does provide clean special-effect pictures even though their resolution suffers relative to normal play.

8.7.3.2 *The adjacent-head structure*

The adjacent-head structure is shown in Fig. 8.69. Since channels 1′ and 2′ are 180° apart (as are channels 1 and 2), each pair is capable of record and normal playback. In fact, at SP speed, both record and playback are handled by the channel 1′ and 2′ head pair, while at the LP and EP speeds, the channel 1 and 2 head pair is used. In the "trick play" modes, various combinations of heads are used based, as noted earlier, on an ongoing selection of which is giving the highest output. A breakdown of which heads are used in each trick mode is given in Table 8.1.

Note that, as was the case for the 90°-head structure, still mode reads tracks of one azimuth only. However, the read heads are not 180° apart. Furthermore, both search and slow modes will give pictures with information from both odd and even fields in the adjacent-head structure. It should be obvious that a

Table 8.1. *Breakdown of which of the heads on an adjacent-head structure are used for each of the "trick modes"*

Speed \ Mode	Search	Slow	Still
SP	2″-1-2-1	1-2-2″	2-2″
LP/EP	2-1	1-1″-2	1-1″

substantial amount of additional logic and switching circuitry is necessary to make use of this structure.

8.8 Enhancements

Since the advent of the VCR, there have been three major enhancements of its capabilities besides noise-free special effects (which we have just studied). Each is easily understood if we understand the basic operation of the unenhanced VCR.

8.8.1 The hi-fi stereo VCR

The audio track on a VCR, as we saw early in the chapter, is not read by a scanning head at all, but lies on a track just 1 mm wide at the top of the tape and is read by a separate audio head. At the low linear tape speeds used in a VCR, this narrow a track does not lend itself well to high-fidelity sound. Some VCRs actually use the 1 mm allocation to incorporate two stereo audio tracks .35 mm wide with a .30 mm guard band between them. This provides stereo sound, but fidelity is even worse. In fact, machines with this capability always have Dolby noise reduction because of the small signal obtained from such a narrow audio track.

It would be ideal if audio could be recorded and read by a scanning head. Any system that could cope with the large bandwidth requirement of a TV signal could do a magnificent job with a stereo audio signal! The problem with the concept of replacing the video signal with an audio one, however, is that the VCR circuitry uses the vertical sync pulses during the record process and the horizontal sync pulses during the playback process, and the audio signal has neither. Therefore the high-fidelity VCR must have two additional rotating (audio) heads, as well as a pseudo-sync generator. The audio heads are offset from the video heads so that the audio tracks will not conflict with the recorded sync. The audio heads are 180° apart, just as the video heads are, and the stereo audio channels are recorded with ±30° azimuth difference for the usual reasons.

Generally, machines with scanned audio capability are going to have four video heads in addition. This makes them capable of high-quality video or audio reproduction. They are thus advertised as six-head machines.

8.8.2 HQ circuitry

The HQ circuitry enhancement (or rather group of enhancements) was introduced in 1986. In its fullest sense, it consists of four enhancements designed to give sharper pictures. It should not be construed, however, that every VCR advertised as HQ (and now all do so advertise) contains all four of the enhancements. By mutual consent of VCR manufacturers, any machine that incorporates any two of them can be labeled as an HQ machine. Let us look at the enhancements individually.

8.8.2.1 *Extended white clip*

We saw in Section 8.6.5.1 that before the luminance is applied to the FM modulator, it goes through white and black-level clippers. These are used so that the output of the modulator will be within the frequency band allocated to it. The white level is translated to the upper end (4.4 MHz) of the modulated luminance spectrum. By extending the white clip level, the upper frequency limit will extend above 4.4 MHz. The benefit, of course, is that you have a wider contrast range available.

Extending the white clip level is made possible by improvements in the linearity of the FM modulator and better video tape to reproduce the higher frequencies resulting from its wider excursions. Unlike a broadcast TV signal, which *must* stay within an allocated frequency band, there is no problem with expanding the frequency range of a recorded signal provided the VCR circuitry and the tape can handle it.

8.8.2.2 *Luminance noise reduction*

Luminance noise reduction is accomplished by adding a 1 H delay line in the luminance path before the modulator and adding its output to the luminance for the present line. Since the luminance for two adjacent lines is usually nearly identical, this results in something like a doubling of the signal. Since noise is random, however, it will not be doubled. The net effect is a 3 dB improvement in S/N.

This technique does not result in as much loss of vertical resolution as might be thought, because each line still contains the information that is uniquely its own *plus* some from the previous line.

8.8.2.3 *Chroma noise reduction*

This circuit works on exactly the same principles as the luminance one just discussed.

8.8.2.4 *Detail enhancer*

We have already seen that any VCR will have at least one and usually two noise-reduction systems even without HQ. These work in part by attenuating high frequencies. But it is the high frequencies that also carry picture detail. The detail-enhancer circuit is kind of like a record-side Dolby system. Low-amplitude high-frequency signals are boosted before recording. Then when the high frequencies are cut by the noise-canceling circuitry, there is still enough high-frequency content left to give improved picture detail.

8.8.3 Super VHS

Super VHS (S-VHS) recorders appeared in the United States in 1987. They are capable of producing broadcast-quality TV images. These VCRs operate on the same principles we have already studied, but some of the numbers are changed. The chroma processing is identical to what we have studied. The luminance is still frequency modulated, but the range of modulated frequencies has been both shifted and expanded.

The luminance frequency range of VHS is 3.4–4.4 MHz, which overlaps the assigned 3.58 MHz color frequency in a broadcast TV signal. Even though the chroma is translated in the VCR, when the playback chroma and luminance are recombined in the VCR the crosstalk introduced will degrade the received picture. The S-VHS VCR provides separate outputs for luminance and chroma. Viewing this signal will require a monitor or TV equipped to handle it.

The other major change is an expansion of the modulated luminance bandwidth from 1 MHz (3.4 to 4.4 MHz) up to 1.6 MHz (5.4 to 7 MHz). This additional bandwidth allows a resolution of 430 lines, as compared to 240 for VHS. It is obvious that improved tape is needed for an S-VHS machine, and such tape is available. It is identical in appearance (though not in price!) to standard VHS tape except for a small additional hole in the bottom of the cartridge that the VCR senses to know that it is dealing with an S-VHS cartridge. If the hole is not sensed, the VCR will treat the tape as VHS and function as a normal VHS machine. We thus say that S-VHS is downward compatible with VHS but not upward compatible!

Exercise 8.39 What would you observe if you played an S-VHS tape on a VHS machine, and why?

The S-VHS system is not a "hot" consumer item at present, but it may experience a resurgence when HDTV begins to proliferate.

References

Haykin, S. 1983. *Communication Systems*, 2nd ed., p. 186. New York: Wiley.

Jorgensen, F. 1980. *The Complete Handbook of Magnetic Recording*. Blue Ridge Summit, PA: TAB Books.

Lemke, J.U. 1988. Instrumentation Recording. In *Magnetic Recording*, vol. 3, eds. C.D. Mee and E.D. Daniel, p. 223. New York: McGraw-Hill.

9

Digital audio

The strongest trend in consumer electronics at present is to encode analog signals to digital and record and process them digitally. The consumer unit must then decode the recorded digital data to output what is in most respects an excellent approximation to the original analog signal.

9.1 Video discs

Unlike magnetic tape technology, which began with audio machines (Chapter 5) and was extensively upgraded to accommodate video (Chapter 8), digital discs began with video and were then downscaled to audio.

The first digital video disc player hit the market in 1978. There were two main technologies available in the players. Each of them required its own type of disc. Each type of disc had pits on it to differentiate between logic "1" and "0." Some of the players had the pits embedded in a groove to facilitate reading by a stylus, whereas others used a dynamic tracking technique that required no grooves.

The two main technologies were the optical pickup system and the capacitance pickup system. Optical pickups use a dynamic tracking laser system to read an ungrooved disc. The "read" mechanism never touches the disc. Capacitance pickups sensed the change in capacitance between a conductive pickup element and a slightly conductive record surface. When the pickup element was over a pit, there was less capacitance. This changed the frequency of an oscillator. The frequency changes were discriminated to produce logic levels. The discs for this type of system might or might not have grooves, but the "read" mechanism did contact the disc surface.

In the U.S. market, the main contenders were the Philips (optical) and the RCA Selectavision (grooved-capacitance). Both produced excellent quality pictures. The RCA system had the advantages of lower price (about half that of optical systems) and an aggressive marketing of low-cost (<$20) video discs. The Philips system, which was produced by others as well, benefited from a public perception that discs read by a laser should last forever. Practically, this may be of little importance except to rental businesses. RCA tests showed that they could play a disc 100 times with no visible picture degradation. This is probably well over the lifetime enjoyment threshold for most people for most movies.

As it turned out, however, all of the video disc technologies were pretty well overshadowed by the VCR (Chapter 8). RCA "threw in the towel" with

Selectavision in 1984. It does not take much imagination to see why. The video disc was a "Read-Only" technology, whereas video tape is a "Read-Write" technology. It might be argued that phonograph records survived audio tape (although they did not survive compact discs!), even though they too are "Read-Only" devices. This is true enough. In fact, audiophiles perceive them as complementary – even symbiotic. The tape machine allows the recording of music to make longer programs or to make one's music of choice portable. By contrast, one survey of VCR owners showed that 80% of respondents listed their main use of the VCR as translocating TV shows to more convenient time slots. Obviously a "Read-Write" medium is required here.

9.2 Audio compact discs

History has already shown that the most important development in the early history of video discs occurred in 1979 when Philips and Sony jointly proposed what has come to be known as the digital compact disc (CD) system. The system would use the optical tracking system that had been developed by Philips for their video disc player and would incorporate the sophisticated error-correction systems developed by Sony. The proposal also spelled out detailed specifications and coding conventions for the discs. Sony produced a prototype in 1980, but it was only one of several competing digital-disc systems. By 1982, it was clear that the CD would become the world standard, and in 1983 the systems appeared in the U.S. mass market.

The CD had several advantages over the LP phonograph record. First, it is exactly 12 cm in diameter, rather than 12 inches. Secondly, it can be made capable of portable or mobile (auto) operation. Third, it has superior audio specs – particularly frequency response, noise performance, stereo separation, and dynamic range.

There are few compensating disadvantages. Originally, CDs cost almost twice what an LP cost, but that soon changed. The CD player itself, unless you get a lot of "bells and whistles" is not a great deal more expensive than a good quality turntable and cartridge and does not need replacement of an expensive stylus to protect the recordings. A few audiophiles still speak disparagingly of "CD sound." Presumably they hear something that the majority do not. Of course, the battle is now over and CDs have won. LP records are no longer made. However, there are still many LP recordings in individual collections. Not only do these represent an enormous investment in dollar terms, but in many cases they are irreplaceable in that artists they feature may never have recordings rereleased in CD format. For this reason, LP turntables are still manufactured.

9.2.1 *The disc geometry*

Figure 9.1 shows the geometric layout of the disc itself. Compared to the LP, there are a couple of things besides pits that take some getting used to. The first is that the disc is played from the inside out and rotates counterclockwise. The second is that whereas the LP is played with constant angular velocity (33 1/3 RPM), the CD is played with "constant" linear velocity. This means that since the inner tracks are less than half as long as the outer ones, the disc must change rotational velocity continuously during the course of a disc play.

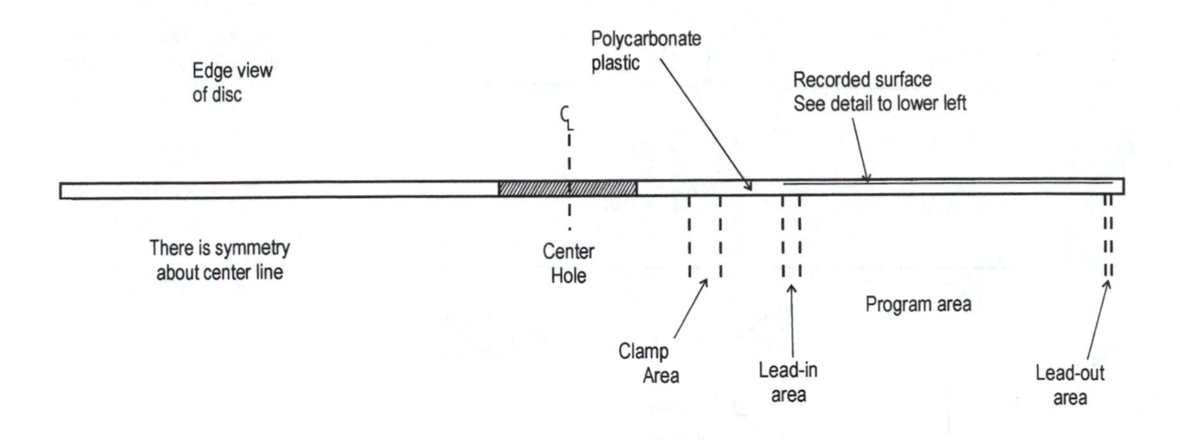

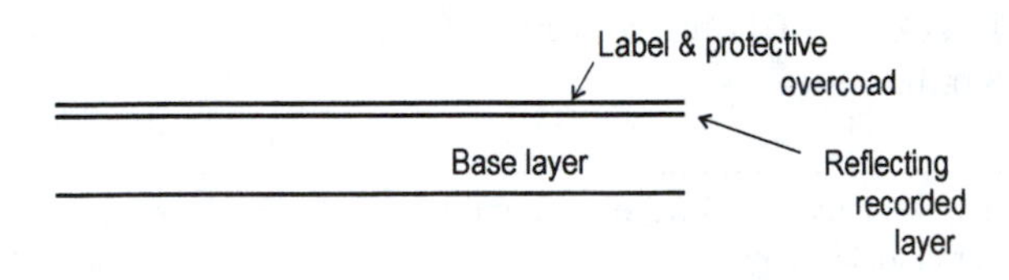

Fig. 9.1: Scale drawing of the structure and geometry of a CD. (a) Entire disc. (b) Close-up.

The rotating disc is read by a laser focused on its signal surface from the underside. The pitted surface is covered with a layer of evaporated aluminum to increase its reflectivity. The laser beam passes through the polycarbonate plastic base and hits either the bottom of a pit or the "land" (or flats) between pits. This plastic has an index of refraction $n \approx 1.5$ and the GaAlAs laser used has an IR output with $\lambda = .78\,\mu\text{m}$ in air. In a material with $n = 1.5$, $\lambda = .52\,\mu\text{m}$. Thus $\lambda/4 = .13\,\mu\text{m}$. If the pits were .13 μm deep, the light reflected from their underside would return shifted by $\lambda/2 \Rightarrow 180^\circ$ relative to reflection from the land. This reflected beam is summed with the incident beam. The reflected beam from .13 μm pits would be 180° out of phase and would destructively interfere with the incident light. That from the land would be in phase and would interfere constructively. A photodetector translates the varying light intensity to the digital pulse train. Figure 9.2 shows this setup in oversimplified form. In reality, the pits are not .13 μm deep, as suggested, but .11 μm deep. This still gives a substantial difference in photodetector signal between flats and pits. The value of .11 μm is a concession to the tracking system, which will be discussed in detail later. The very fact that an item can be mass-produced to a tolerance of about 50 Å is itself an incredible testimonial to modern technology.

It has been noted that the disc is intended to be played back with a nominally constant linear velocity (CLV). The nominal value is 1.3 m/sec, with maximum and minimum values of 1.4 and 1.2 respectively. Not only will the exact linear velocity vary from one disc to another, but there may even be significant variation between the beginning and end of a given disc. Signals encoded on the disc force its instantaneous playback velocity to equal its instantaneous record

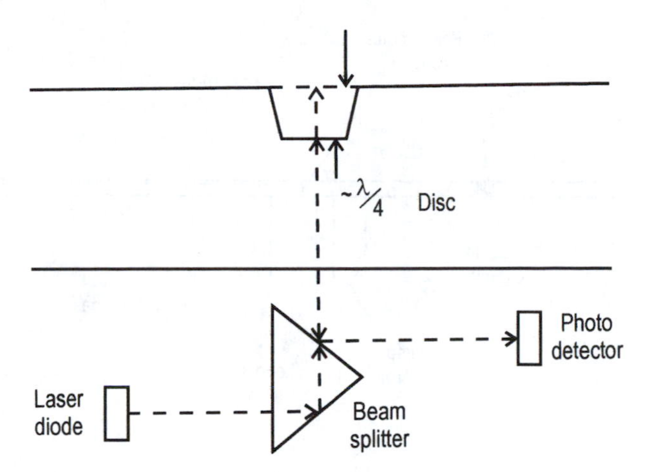

Fig. 9.2: Oversimplified drawing of the optics of a CD player.

velocity. This is analogous to the function of the control track of a VCR. As shown in Fig. 9.1, the innermost (beginning) track of a CD has a radius of 25 mm, and the ending track has a radius of 58 mm.

Exercise 9.1 Assuming a nominal linear (tangential) velocity of 1.3 m/sec and radii extrema of 25 and 58 mm, find the extrema of angular velocity in rpm.

Based on the results of Exercise 9.1, we may round off the angular velocity extrema to 500 rpm at the innermost track and 200 rpm at the outermost track.

The spacing between centers of adjacent tracks (pitch) on the disc is 1.6 μm. The pits along each track may have lengths varying between about .9 μm and 3.3 μm, as may the flats between pits. The width of the pits in the radial direction is .5 μm.

Exercise 9.2 Using information already given, determine the typical number of rotations the disc will make in playing through it.

Exercise 9.3 Using information already given, determine the typical total track length in kilometers on a CD.

Exercise 9.4 Based on the result of Exercise 9.3, find the maximum playing time of a CD turning at a constant linear velocity of 1.3 m/sec. Repeat this calculation if it were turning at the minimum spec value of 1.2 m/sec.

9.2.2 Encoding to disc

We next look at how the signal is formatted on the disc. This will unavoidably embroil us in a consideration of some of the error-correcting schemes used in the CD system. The remainder will be covered when we consider the circuitry itself.

The CD is designed to have an upper frequency response of 20 kHz. By the Nyquist criterion, digital sampling must be done at over twice this frequency. A sampling frequency of 44.1 kHz has been established, allowing a small guard band to prevent aliasing. Each digitized sample is 16 bits wide. Since there are 44,100 samples/second, we know we will need to put at least

$2 \times 16 \times 44,100 = 1.4112 \times 10^6$ bits/second down on the disc. The factor of 2 arises because we are recording stereo and each channel is digitized separately. We shall see that this is only the beginning of our need to expand the bit rate. Each sample will produce 32 bits (16/channel). The sample time is $1/44,100$ sec $= 22.7\,\mu$sec.

The first error countermeasure is to arrange these 32 bits in an alternating sequence so that 16-bit words of L data are followed by 16-bit words of R data. The reasoning here is that if part of the disc is obliterated, the error-correcting circuitry will be more likely to be able to reconstruct the program if a shorter section of each channel is affected than if a longer section of one channel is.

The sampling process is repeated six times. This will require $22.7 \times 6 = 136\,\mu$sec and will produce $32 \times 6 = 192$ bits of audio data. This is the total audio content of what is called a *frame*. These 192 bits are arranged into groups of 8 bits each. In the common CD parlance, these are called *symbols*, after the usual terminology of PCM, but could as well be called *bytes*, after the computer usage of the term. The 192 audio bits in a frame are thus constituted into frames of 24 bytes with 8 bits per byte. The pattern is

$$S_{R1A} S_{R1B} S_{L1A} S_{L1B} S_{R2A} S_{R2B} S_{L2A} S_{L2B} S_{R3A} S_{R3B} S_{L3A} S_{L3B} S_{R4A} S_{R4B}$$
$$S_{L4A} S_{L4B} S_{R5A} S_{R5B} S_{L5A} S_{L5B} S_{R6A} S_{R6B} S_{L6A} S_{L6B}.$$

In this representation, S stands for a symbol. For each S:

- The first subscript refers to which audio channel has part of its data encoded in this symbol
- The second subscript tells which of the six samples in a frame has part of its data encoded in this symbol
- The third subscript will be A or B, where A refers to the 8 most-significant bits and B refers to the 8 least-significant bits of the 16 bits that make up one sample for one audio channel.

The second error countermeasure is to delay all of the symbols whose second subscript is even by two symbols and to scramble the order of the symbols. Since the data arrive at the encoder in frames of 24 bytes, this means that the even bytes passing out of this *interleave* operation are those that arrived at the input two frames earlier. The new order of the symbols after the scrambling operation is

$$S_{R1A} S_{R1B} S_{R3A} S_{R3B} S_{R5A} S_{R5B} S_{L1A} S_{L1B} S_{L3A} S_{L3B} S_{L5A} S_{L5B} S_{R2A} S_{R2B}$$
$$S_{R4A} S_{R4B} S_{R6A} S_{R6B} S_{L2A} S_{L2B} S_{L4A} S_{L4B} S_{L6A} S_{L6B}.$$

The purpose of these operations is to make interpolation possible if two successive decoded frames prove to be uncorrectable on playback. *Interpolation* is a form of error concealment used when error correction is inadequate to recover the original signal perfectly.

Before we are done "padding" the frame with other bits, its original 192 bits will have been expanded to 588. It will still occupy the same 136 μsec in the time domain. Figure 9.3 shows the frame structure before any of the "padding."

The third error countermeasure is to implement the first stage of the Reed–Solomon error correction code (C2 encoder), which inserts a group of four

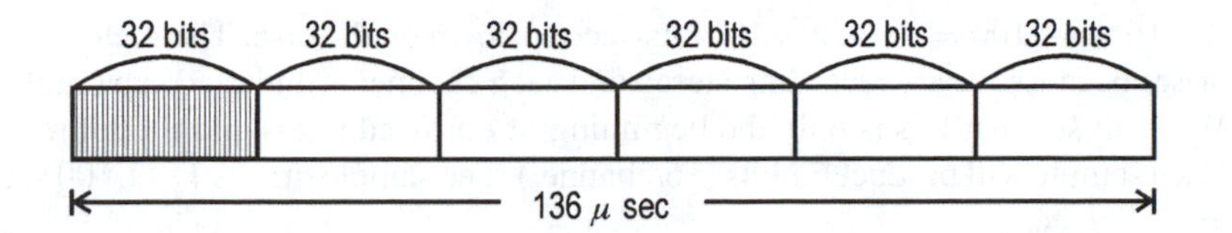

Fig. 9.3: Schematic representation of the data symbols in one frame of a CD recording.

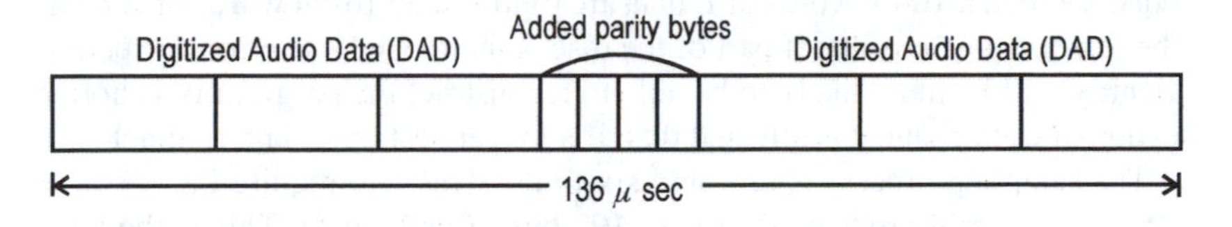

Fig. 9.4: Schematic representation of the same data as shown in Fig. 9.3, but with the addition of the four parity bytes introduced by the C2 Reed–Solomon encoder.

parity bytes into the data bit stream at the midpoint of the frame. The new data stream now has the structure shown in Fig. 9.4. The reader might wonder how the digitized audio data were shrunk to fit in less than its original 136 μsec. This is easily accomplished because all of these data and much more are stored in RAM for error-correction operations, and they can be clocked out at any desired rate. In the form shown in Fig. 9.4, we are not yet ready to clock it out, because other additions to the data stream must be made. In point of fact, the addition of four bits to be clocked out in the same time as the original 24 requires an updating of the clock frequency by a factor of (28/24). When multiplied by the original 1.4112×10^6 bits/second, the new clock rate must be 1.8816×10^6 bits/second. Successive encoding steps will require further increases in the clock rate.

The fourth error countermeasure is called *cross-interleaving*. Each of the 28 symbols leaving the C2 encoder is subjected to a further delay of $n \cdot \mathrm{D}$ where $0 \leq n \leq 27$ and D is equal to four frame times. This means that one frame's worth of data is now spread out so that parts of it can be found in every fourth one of 112 successive frames. This permits correction when many successive bits are read erroneously (*burst errors*). The C2 decoder with cross-interleaving can completely correct two erroneous symbols/frame. One completely defective frame would distribute one defective symbol into 1/4 of the 112 frames into which it is interleaved and is thus easily correctable.

The fifth error countermeasure is another stage of Reed–Solomon encoding (C1). This process adds four more parity bytes to the end of the structure shown in Fig. 9.4. It is primarily used for the correction of random errors.

The combination of two stages of Reed–Solomon encoding with an interleaving step between them form on extremely powerful error-correction technique called CIRC (Cross-Interleaved Reed–Solomon Coding). It is capable of completely correcting for the loss of over 4,000 consecutive bits due to disc damage or defect (Nakajima, Doi, Fukuda, and Iga, 1983). We shall see more of this system when we study the CD player itself.

The sixth error countermeasure is a one-symbol interleave. This spreads the encoded symbols over two frames, reducing the effect of random errors.

The next step in constructing a frame is to preface the string shown in Fig. 9.4 with a single byte of what is variously called *subcode*, *C & D*, or *Control & Display bytes*. The C & D bytes of each frame form a distributed frame structure of their own. The C & D frame is constituted as the C & D byte of 98 consecutive

Table 9.1. *Contents of the Q bit in each of the 98 blocks of a subcode frame.*

1–2	C & D frame sync
3–6	Control bits (2 or 4 channels & pre-emphasis on or off)
7–10	Address bits
11–18	Music selection #
19–26	Index # (pinpoints a location within a selection)
27–34	Minutes duration of selection playing
35–42	Seconds durations of selection playing
43–50	Frame where selection begins
51–58	Zero
59–66	Minutes duration of entire disc
67–74	Seconds durations of entire disc
75–82	Frames on entire disc
83–98	Error-correction coding

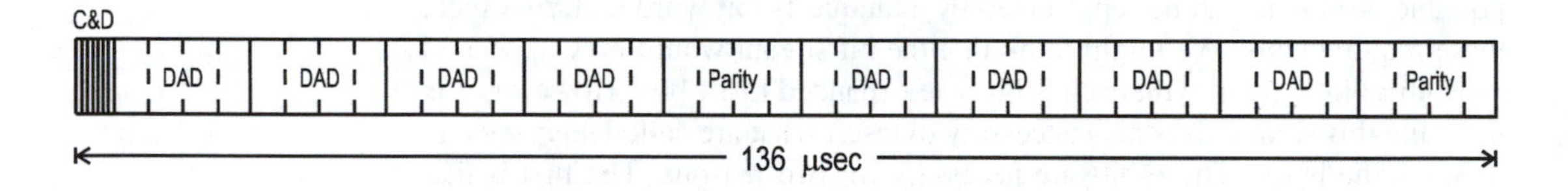

Fig. 9.5: Schematic representation of the same data as shown in Fig. 9.4, but with the addition of the four parity bytes introduced by the C1 Reed–Solomon encoder and the C & D byte.

signal frames. Of these 98 bytes, numbers 1 and 2 are sync for the C & D frame, and the remaining 96 bytes are each divided into individual bits labeled P, Q, R, S, T, U, V, and W. In current CD player designs, only the P & Q bits are used. The P bit is 0 when music is playing and 1 when it is not. The Q bit in general carries different information in each byte, as shown in Table 9.1.

One of the "cute" ideas that has been proposed is to use the remaining subcode bits to store video signals that could provide still pictures with the addition of a subcode adapter and monitor. At first this might seem ridiculous, given the high bandwidth requirements of video as compared to audio, but it is instructive to look at the numbers. A broadcast-quality TV picture may be considered to be composed of $(525)^2 \times 4/3 = 367{,}500$ pixels, since the normal raster has 525 lines/frame and an aspect ratio (horizontal/vertical) of $4/3$. Let us allocate 8 bits to encode the information for each pixel's luminance (and chroma if we chose to give up some luminance resolution). The video frame would thus require 2.94×10^6 bits. Now if the R, S, T, U, V, and W bits from each signal frame carried video information, we would need $2.94 \times 10^6/6 \approx 5 \times 10^5$ signal frames to constitute a picture. Since a signal frame is generated each 136 μsec, a broadcast-quality picture could be built up in $5 \times 10^5 \times 136 \times 10^{-6} =$ 68 sec. It is entirely possible to build up a picture with fewer pixels and fewer bits/pixel in less than a tenth of this time, especially if data compression is used, allowing up to hundreds of pictures per disc with no reduction of its music content.

In any case, the data stream with the C & D block added has the structure shown in Fig. 9.5. We now have a data stream composed of 33 bytes of 8 bits each. As we shall see, the player circuitry produces a logic 1 when and only when the laser beam moves from a pit to a flat or vice versa. The resolution

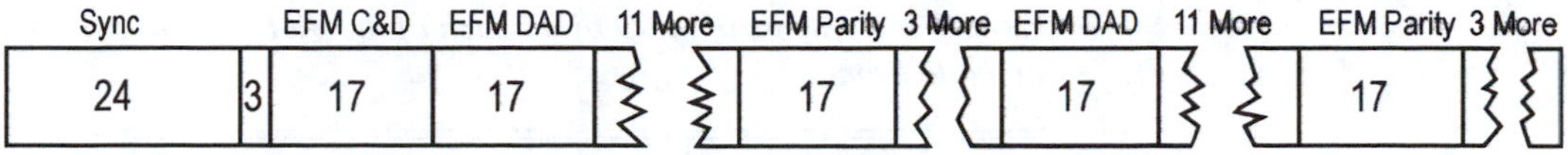

Sync EFM C&D EFM DAD 11 More EFM Parity 3 More EFM DAD 11 More EFM Parity 3 More
24 3 17 17 17 17 17

Fig. 9.6: Schematic representation of the same data as shown in Fig. 9.5, but with all of the symbols (bytes) having been EFM encoded and merged, and a 24-bit sync word appended to the beginning of the data. This then is the full 588-byte block of data that constitutes a frame in a CD recording.

of the laser beam at the signal surface imposes a requirement that a logic 0 (or 1) must always persist for at least 3 bit times. This imposes a lower limit on the length of a pit or a land. An upper limit must also be maintained. Since the CD player uses the PCM from the disc to reconstruct a clock signal, it is necessary that a logic 0 (or 1) cannot persist for longer than 11 bit times. It should be obvious that neither of these requirements can be met in general by an 8-bit byte that can range from 00000000 to 11111111. It is thus necessary to transform these bytes into another code that can meet these requirements. This code translation is called *Eight-to-Fourteen Modulation* (*EFM*). In a 14-bit word, there are $2^{14} = 16{,}384$ possible values, but only 267 of them meet the conditions just given. This number is, however, greater than 256. Thus each possible 8-bit byte can be represented by a unique 14-bit word that *does* meet these requirements. We might think that the bit stream would now appear as shown in Fig. 9.5 but with each 8-bit byte expanded to 14 bits. However, it is not quite this simple. It is also necessary to insert what are called *merging bits* between the bytes. These bits are necessary for two reasons. The first is that while our use of a 14-bit code insures that there will never be less than 3 or more than 11 identical consecutive bits in a byte that are equal, it cannot assure that these forbidden patterns will not occur in crossing from one byte to the next. The second reason for merging bits is that the transformed code should have no DC component, because this would adversely affect the accuracy of the EFM recovery. We shall have more to say about the latter issue in Section 9.6.4. The first problem can be circumvented by adding two bits and the second by adding one. Thus three merging bits are appended to each 14-bit EFM word for a total of 17. The bit stream now looks like Fig. 9.5 except that each byte of 8 bits has been replaced by one of 17 bits. Since less than half of the bits in this stream are audio data at this point, it is no longer called a data bit stream but a channel bit stream. The frame now consists of 33 bytes of 17 bits each, or 561 bits.

One step remains in the construction of a signal frame for recording on the CD. That is to preface the entire stream as it now stands with a 24-bit frame sync word and the usual 3 merging bits. The sync word is a pattern of bits that can never occur elsewhere in the frame (100000000001000000000010), so the circuitry recognizes it as frame sync. The addition of these 27 bits brings the total in the frame up to $561 + 27 = 588$ bits. The frame is now complete as shown in Fig. 9.6. After completion of one frame, the A/D converters take more samples and start on the next frame. It is the job of the player to not only unscramble this mess, but to do the best it can when errors make this impossible.

We started each frame with 192 audio bits and ended with 588 total bits. Since these 588 bits have to fit in the same 136 μsec as the original 192, the $\sim$1.4 MHz bit rate we originally deduced must be scaled up to

$$\frac{588}{192} \times 1.4112 \text{ MHz} = 4.3218 \text{ MHz}.$$

This is the standard crystal clock frequency in a CD player. Inverting the above ratio shows that only 192/588 or 32.7% of a CD's recording area is used to record music. The rest is necessary overhead.

To summarize the disc itself, we have seen that it contains six error countermeasures. Some would include EFM as a seventh because it does prevent laser read errors that would be obtained even on a hypothetical perfect disc for the laser spot size obtainable. But it is an error due more to fundamental physics than to worst-case disc problems. This distinction may seem artificial to some readers. If so, simply reckon that the disc contains seven error countermeasures. As we examine the CD player, we will see how these measures are used to reduce errors. We will see other measures that are self-contained within the player.

One last word about the disc concerns its durability. There is a prevalent myth that a CD is virtually indestructible. If you believe that, leave one lying in summer sunlight for a day or so. Because the disc is only 1.2 mm thick, warpage is a distinct possibility and it does not require much to make the disc unplayable. We shall see the reason for this in the next section. A large gouge also can obliterate enough pits to cause audible error.

9.3 The CD player

9.3.1 *Optics*

The heart of the player is the optical readout head. It is a remarkable feat of electrical, mechanical, and especially optical engineering. Figure 9.2 was called an oversimplified diagram of the optics. Figure 9.7 gives a more detailed look at a typical optical pickup.

The solid-state laser diode is designed to emit in the infrared at $\lambda = 7{,}800$ Å. When this radiation leaves the laser, the beam is diverging. Therefore it passes through a collimating lens. This is simply a convex lens with the light-emitting plane of the diode placed at its focus. As the light exits from the other side of the lens, it is moving parallel to the lens axis and is said to be collimated. The collimated light beam passes through a diffraction grating whose ruling is chosen such that the first pair of side lobes will be about 2 μm on either side of the central beam when it strikes the recorded surface. This three-beam system is the key to precision tracking, as we shall soon see. The distribution of light intensity upon exiting the grating is shown in Fig. 9.8.

The beam next enters a polarization beam splitter, which serves to split the beam into two parts. This splitter is composed of two 45° prisms with their long faces joined by a thin dielectric layer. This structure will accept the unpolarized beam from the grating and will split it into horizontally and vertically polarized components. Typically the horizontally polarized component will pass through undeflected toward the disc, and the vertically polarized one will be reflected at a right angle toward the photodiode array. Each of these "beams," of course, consists of the three-spot pattern produced by the grating. The beam destined for the disc passes through a quarter-wave plate, which will in general convert the polarization of the triplet beam from pure linear to a sum of circular and linear. If the E field of the linearly polarized beam incident on the quarter-wave plate makes an angle of 45° relative to the plate's optic axis, the output of the

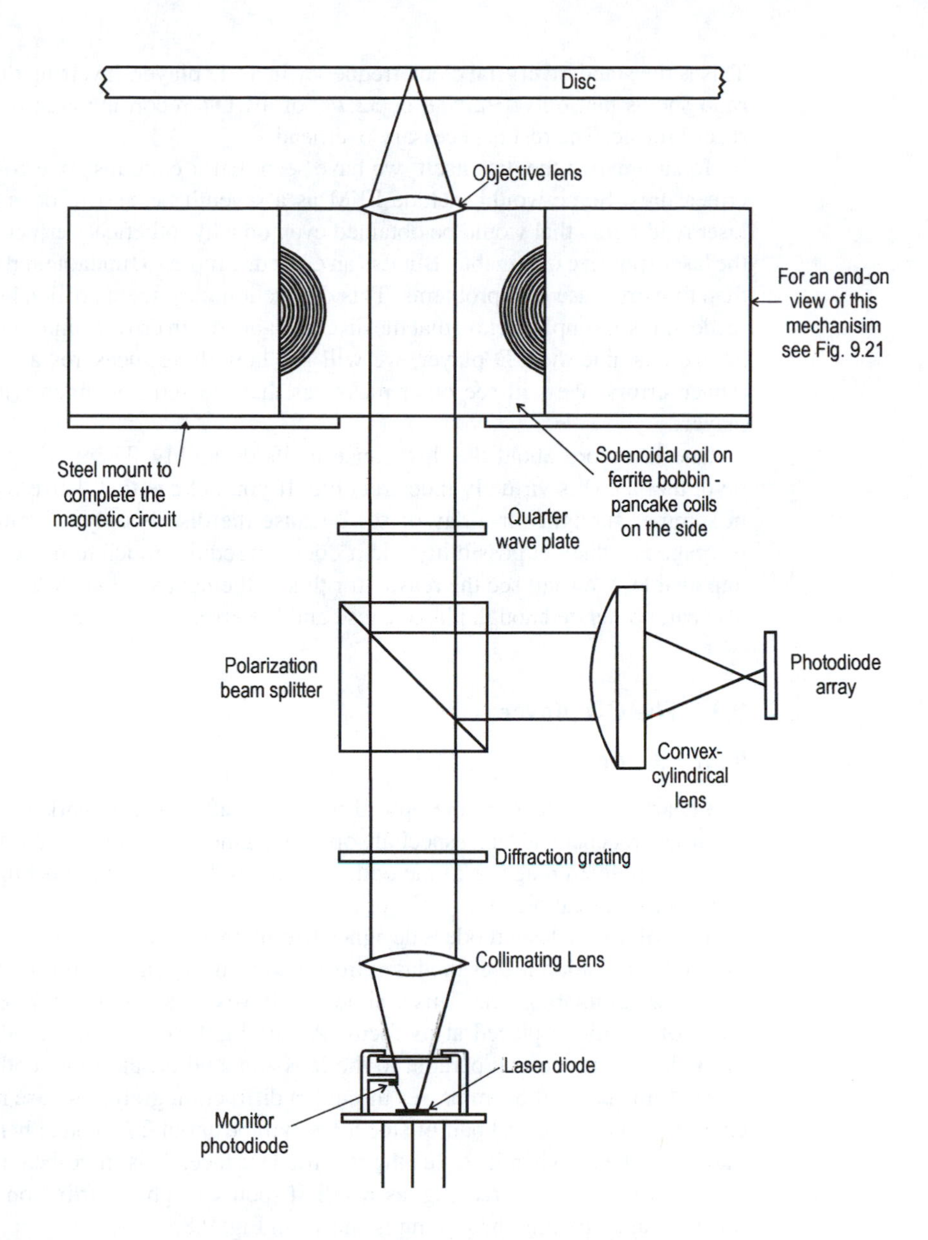

Fig. 9.7: A much more detailed and accurate version of Fig. 9.2 showing the optics of a CD player.

plate is pure circularly polarized light (Nussbaum and Phillips, 1976). This is the design objective here. The light next passes through another convex lens called the objective lens. It is the purpose of this lens to bring the parallel three-lobe beam to a focus on the recorded surface of the disc. The objective lens has a focal length of 4 mm. It is placed with its center line a little more than 4 mm from the recorded surface since the converging beam must travel through about 1.1 mm of clear plastic with $n = 1.5$ before it reaches the recorded surface, and this defocuses the beam somewhat.

Exercise 9.5 Figure (a) shows the focusing mechanics if there were no plastic between the objective lens and the recorded surface. Figure (b) shows what happens when the

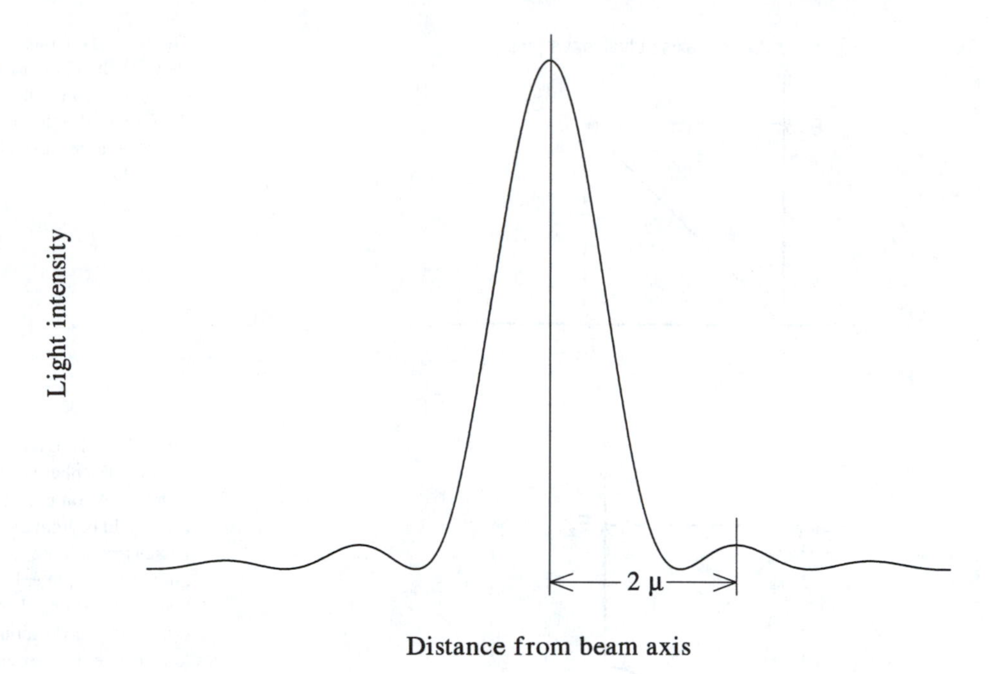

Fig. 9.8: The light intensity coming out of the diffraction grating of Fig. 9.7. The two small "blips" beside the main peak are the tracking beams.

beam must pass through plastic. Consider the lens as "thin" in the optical sense. Find x. (Hint: Remember Snell's Law?)

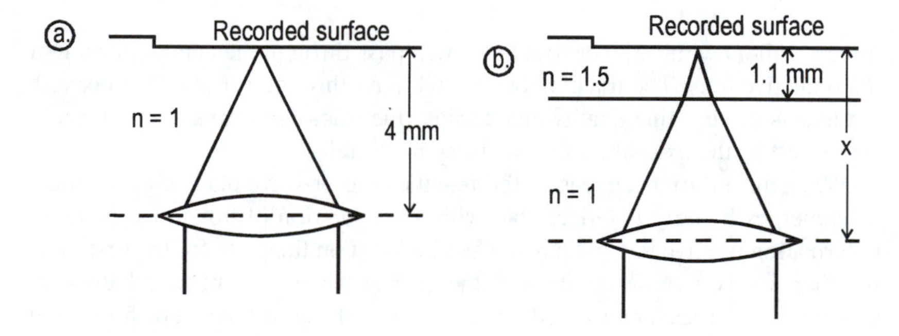

The fact that the beam is relatively defocused at the bottom of the disc means that surface flaws there will be demagnified many times (typically 1,000) at the recorded surface, greatly reducing their potential for causing errors in the sound.

The beam being focused on the recorded surface will consist, we recall, of a strong main beam and two weaker side lobes. The main beam will be focused down to a spot size between 1 μm and 2 μm when it reaches the recorded surface. These side lobes are referred to as tracking beams because their function is to facilitate keeping the main beam on a track of pits. All three beams will be reflected by the aluminized recorded surface of the disc, but since the beam dimensions are comparable to the patterns on the disc, they can be reflected with dramatically different intensities than those with which they were incident on the aluminum layer.

In any event, these three beams are reflected back through the objective lens, which recollimates them into a parallel beam or perhaps we should say into

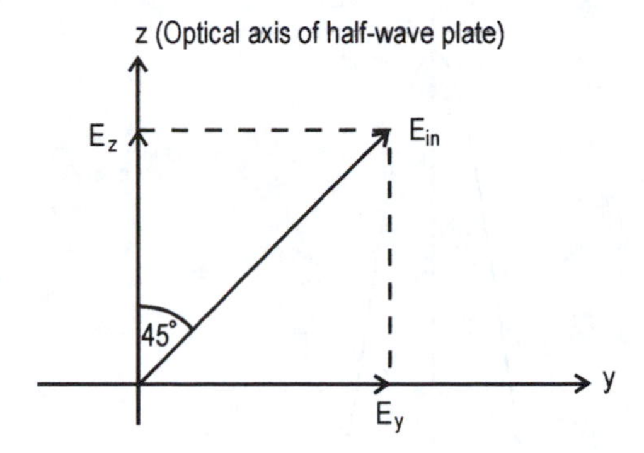

Fig. 9.9: Diagram showing the phase of the orthogonal components of the *E* field of light incident on the quarter-wave plate for the first time.

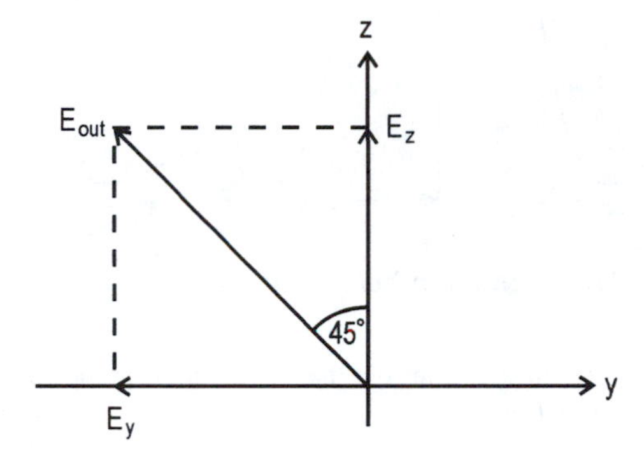

Fig. 9.10: Diagram showing the phase of the orthogonal components of the *E* field of light after two passes through the quarter-wave plate. Note that the *E* field has been rotated 90°, corresponding to a change in polarization of the light from horizontal to vertical.

three parallel beams. The central beam will pass through the central portion of the objective lens. The tracking beams will pass through off axis, but because the lens is of very fine quality and because the tracking beams are only about 20 μm off of the axis, aberrations will be negligible.

When this return beam passes through the quarter-wave plate a second time, it reemerges linearly polarized, but with a net rotation of 90° in the plane of polarization relative to what it had when incident on the plate for the first time. In other words, two passes through the quarter-wave plate make a half-wave phase shift. The reader may well ask how a 180° phase shift converts horizontal linear polarization to vertical linear polarization. Isn't this a 90° shift? Indeed it is. The answer lies in *what* is shifted by 180°, as Fig. 9.9 shows.

We will treat the two passes through the quarter-wave plate as one through a half-wave plate. As shown, the E field incident on the half-wave plate is at 45° to the optic axis, as it ideally should be. The components of E along the z and y axes will travel at different velocities through the plate because the half-wave plate is made of a material that has an anisotropic index of refraction. It is a half-wave plate because in passing through the plate, E_y will be shifted 180° relative to E_z leading to the situation shown in Fig. 9.10. As shown, E_y has been rotated 180° relative to E_z, but E_{out} is thus rotated 90° relative to E_{in}, converting its polarization in our example from horizontal to vertical.

After passing back through the quarter-wave plate, the light reflected from the disc reenters the polarization beam splitter. The fact that the light heading toward the disc is horizontally polarized and that returning is vertically polarized prevents them from interfering. When it reenters the beam splitter, its new polarization causes the triplet beam from the disc to be directed almost entirely toward the photodiode assembly. At the surface of the photodiode,

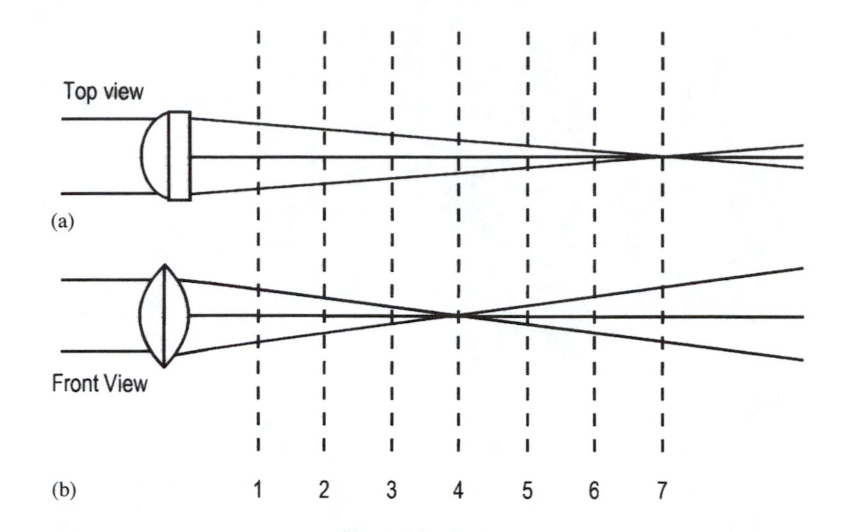

Fig. 9.11: The use of a sphero-cylindrical lens in an automatic focusing system. (a) Light polarized along the direction of the axis of the cylindrical part of the lens receives no focusing impetus from it, and so the light does not come to a focus until plane 7. (b) Light polarized normal to the axis of the cylindrical part of the lens and lying in the plane of the paper in the front view receives additional focusing impetus from the cylindrical part of the lens, and so it comes to a focus at plane 5.

it will interfere with the triplet beam from the laser, producing high and low intensities as pits and flats are read.

Before this beam reaches the photodiode assembly, it must pass through a lens assembly consisting of a convex lens designed to bring the more or less parallel beam to a focus at a point beyond the plane of the photodetector. It then passes through a cylindrical lens. This lens will have no effect on light along its axis but will give additional focusing impetus to light normal to its axis. This additional focusing power in conjunction with that from the convex lens is adequate to bring rays normal to the axis of the cylindrical lens to a focus in front of the photodetector assembly. In other words, this *sphero-cylindrical lens* assembly, as it is called, will give unequal focusing power for light in two orthogonal directions. This same type of lens is used in eyeglasses to correct regular astigmatism.

Since the light passing through this lens assembly is linearly polarized, it is necessary to place the axis of the cylindrical lens at 45° to the axis of polarization so that about half of the light will pass through the cylindrical lens along its axis and the other half will be normal to its axis. The polarized character of this light is no longer important; it has already done its work of bringing only the correct beam components into the photodetector.

The fact that the components of light along and normal to the cylindrical lens axis are focused at different distances from the photodiode assembly means, among other things, that there is no true focus for the triplet beam, but only what is called a *circle of least confusion*. The reason for this is shown in Figs. 9.11a and b. At each of the planes numbered 1 through 7, a spot could be formed on a screen inserted there. In general, the spot would have two unequal axes whose lengths would be given by the widths of the two ray diagrams in Fig. 9.11 at the plane in question. This changing shape of the spot is shown in Fig. 9.12. The three-dimensional composite of Figs. 9.11 is called *Sturm's Conoid*, and the linear distance between planes 4 and 7 is called the *interval of Sturm*. The circle of least confusion lies in plane 5.

In our application, the design goal is to place the photodiode assembly at the nominal location of plane 5. It has been mentioned, however, that the triplet beam is supposed to be sharply focused on the recorded surface of the disc by the objective lens, which then serves to collimate the beam reflected from the

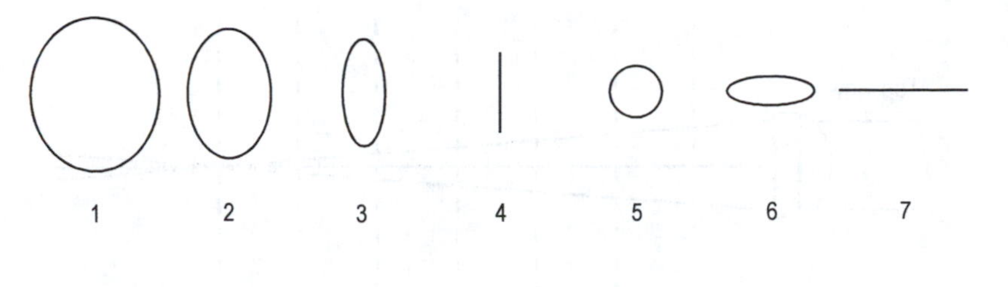

Fig. 9.12: The spot size formed at any of the planes shown in Fig. 9.11 can be determined from the two widths of the conoids shown there. At any of the planes marked in Fig. 9.11, the width of the upper conoid gives the horizontal axis and the width of the lower conoid gives the height of the spot.

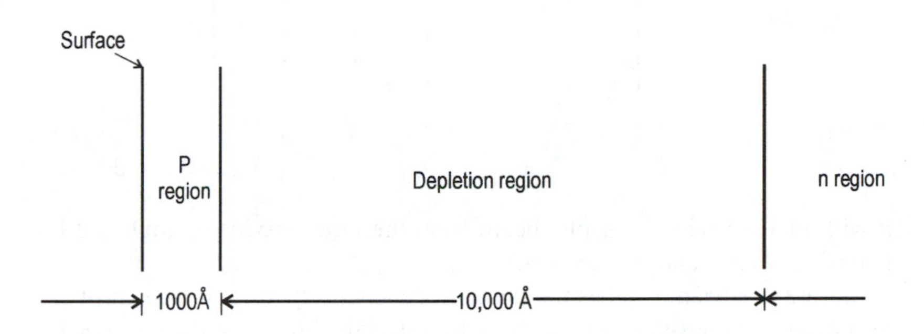

Fig. 9.13: Scale drawing showing the location and extent of the depletion region in a typical p-n junction diode.

disc. If the beam is not properly focused, the return beam will not be properly collimated but will be either slightly diverging or converging when it reaches the photodetector lens assembly. This will cause the circle of least confusion to shift along an axis normal to the photodetector, so that it no longer lies on the photodiodes. As Fig. 9.12 shows, this means that the spot shape on the detector array will thus elongate to form an ellipse or line. We shall next look at the photodetector diode array to see how this behavior is utilized.

9.3.2 Photodetectors

A normal p-n photodetector for visible light might be fabricated using a very thin surface p-layer of only 1,000 Å and with an average doping upwards of $10^{19}/\text{cm}^3$. The base layer would be much thicker and might have a doping of only $10^{16}/\text{cm}^3$. These numbers correspond to a depletion width in the surface layer on the order of 10 Å thick. Figure 9.13 shows these distances more or less to scale. The 10 Å extension of the depletion region into the p-type surface layer is too small to even show. The depletion region shown all lies in the n region.

The absorption constant in silicon of light with a wavelength of 7,800 Å is about $10^3/\text{cm}$. This governs the absorption of radiation as the light penetrates the photodiode. The intensity will vary with x (penetration) as

$$I(x) = I_0 e^{-kx} = I_0 e^{-1000x(\text{cm})},$$

where I_0 is the incident intensity. At the metallurgical junction

$$I(1{,}000\,\text{Å}) = I_0 e^{-.1} = .90 I_0.$$

Thus we conclude that the light would lose about 90% of its energy in the n-type base layer, and so 90% of the generation would take place there. If the diode

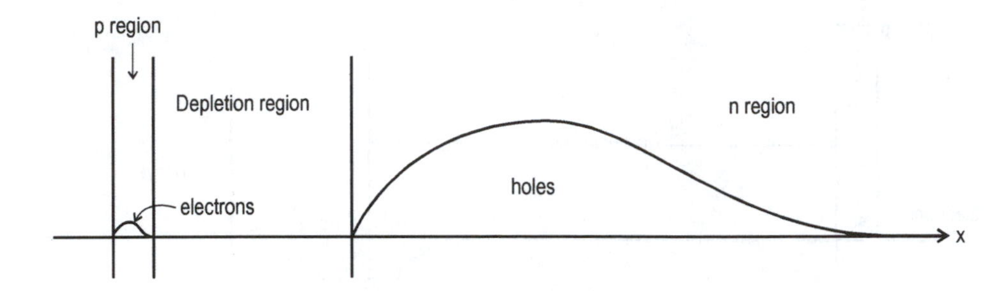

Fig. 9.14: Scale drawing of the device shown in Fig. 9.13 showing the location and relative values of excess carrier concentrations created by photoexcitation of the diode.

is operated in the short-circuit mode so that the current is directly proportional to the illumination, the minority carrier concentrations will have the general appearance shown in Fig. 9.14. Owing to the very thin p region, its high rate of recombination, and the long wavelength of the laser light, the contribution of the surface layer to photo current will be negligible. The roughly 9% of light energy lost in the depletion layer is converted to current with a quantum efficiency of nearly unity. The holes in the n region will produce current to the extent that they diffuse back to the depletion region without recombining. It should be obvious that many will, in fact, diffuse away from the depletion region. This seriously reduces the efficiency of the cell.

The distribution of holes in the n region has another effect that poses an even more serious problem in our application, however. When the light is cut off (as when it is reflected from a pit with the phase to cause destructive interference), current will continue to flow through the photodiode until all of the holes have either recombined or diffused to the depletion region. This makes the frequency response intolerably slow. Rise and fall times of the current will generally both be in excess of 1 μs. By contrast, the data are typically read off of the disc at about 700,000 translations per second. This means that the real-time duration of each reading is about 1.4 μsec. Obviously if the photodetector takes over 1 μs to rise or to fall, it will never reach the full "off" and "on" states.

Both of these problems can be alleviated by using what is called a *p-i-n structure* in the photodiodes. Here the i stands for intrinsic, although the material does not have to be purely intrinsic. It must just have a high resistivity. This region is made fairly long so that most of the generation will take place in the i material. Hence the depletion region will extend into it even further than it did into the n material – perhaps even orders of magnitude further. The fundamental governing relationship here is that the space charge must be equal on both sides of the metallurgical junction. Since the surface layer is so heavily doped, a given amount of space charge can be found within a very short distance of that side of the junction. Since the i material is very lightly doped, it is necessary to move to the right of the junction a long distance to include a space charge equal to that on the heavily doped p side.

This expanded depletion (or space charge) region does two things for us. In the first place, as shown in Fig. 9.15, most of the generation will now take place in the depletion region, and as already noted, this will be converted to current with nearly unity quantum efficiency. This will increase the cell's light sensitivity. Secondly, since the minority carrier concentration outside of the depletion region is so much smaller, there is less charge to remove when the light is cut off and so the p-i-n photodiode has a much faster response time than the p-n diode. A typical value is 1 ns.

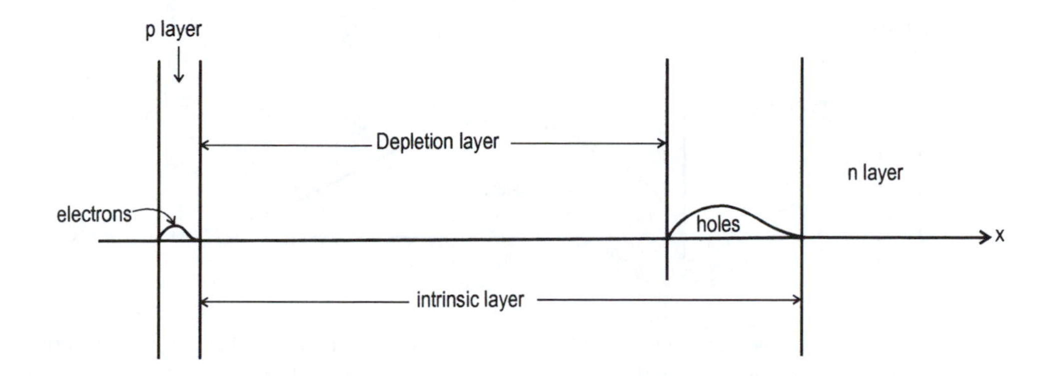

Fig. 9.15: Scale drawing showing the location and extent of the depletion region in a typical p-i-n junction diode. With this type of device, a much larger portion of the photo-generated carriers will be formed in the depletion region and will thus contribute to external photo current. Owing to the small carrier concentrations outside of the depletion region, this device will also have a much faster response time than a p-n junction.

The photodiode assembly used in a three-beam CD player contains six separate photodiodes. Four are used to respond to the main beam, and one for each of the tracking beams. Their arrangement is shown in Fig. 9.16 with the triplet beam superimposed.

To help us understand the purposes of this array, let us look at a block diagram for the remainder of the CD player's circuitry.

9.3.3 *Tracking and focus correction*

The block diagram in Fig. 9.17 shows that the data from the photodiode array comprise the signal source for the entire rest of the player. The first part of the HF preamp circuitry is an array of comparators as shown in Fig. 9.18.

By referring back to Fig. 9.16, it can be seen that the A–C and B–D pairs of photodiodes shown tied together in Fig. 9.18 are along the diagonals of the array that detects the main beam. As discussed earlier, the circle of least confusion should nominally fall on this array. But if the beam at the disc is focused in front of the recorded surface, it will already be diverging when it is reflected from the disc, and the cylindrical lens assembly will not have enough power to pull it down to the circle of least confusion at the plane of the photodiodes. In the frame of reference of Fig. 9.12, this will result in a horizontal elongation of the spot. The photodetector array is to be oriented such that this elongated spot will lie primarily on diodes A and C. Likewise, if the "read" beam comes to a focus behind the plane of the recorded disc, it will still be converging when it is reflected. This convergence will be carried through to the photodetector array and will cause the circle of least confusion to lie in front of the plane of the diodes. In the frame of Fig. 9.12, this will result in a vertical elongation of the spot on the detectors. Since we just established that a horizontal elongation will differentially illuminate diodes A and C, it follows that a vertical elongation will differentially illuminate diodes B and D. Most importantly, if the focus is correct, all four diodes will be illuminated equally, and the A–C signal will equal the B–D signal, generating no focus correction signal.

Because the depth of field of the objective lens is just $\pm 2\,\mu$m, and since disc wobble will typically be on the order of $600\,\mu$m, it should be clear that this dynamic focus control servo is an absolute must. However, we have not yet "closed the loop" on the servo.

It should also be clear from Fig. 9.18 that an excess of illumination on diodes A and C (relative to B and D) will drive the focus comparator output low. If B

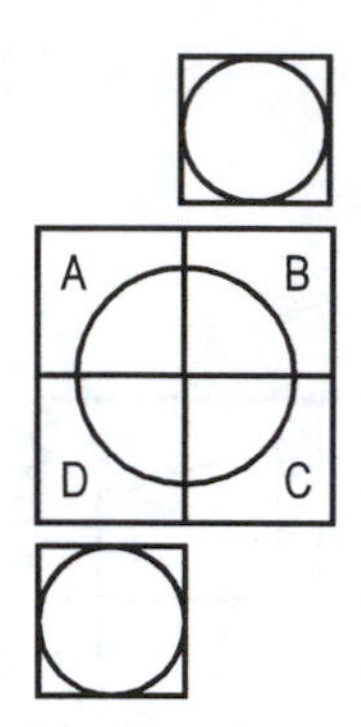

Fig. 9.16: Schematic representation of the structure of the photodiode assembly used in a three-beam CD player.

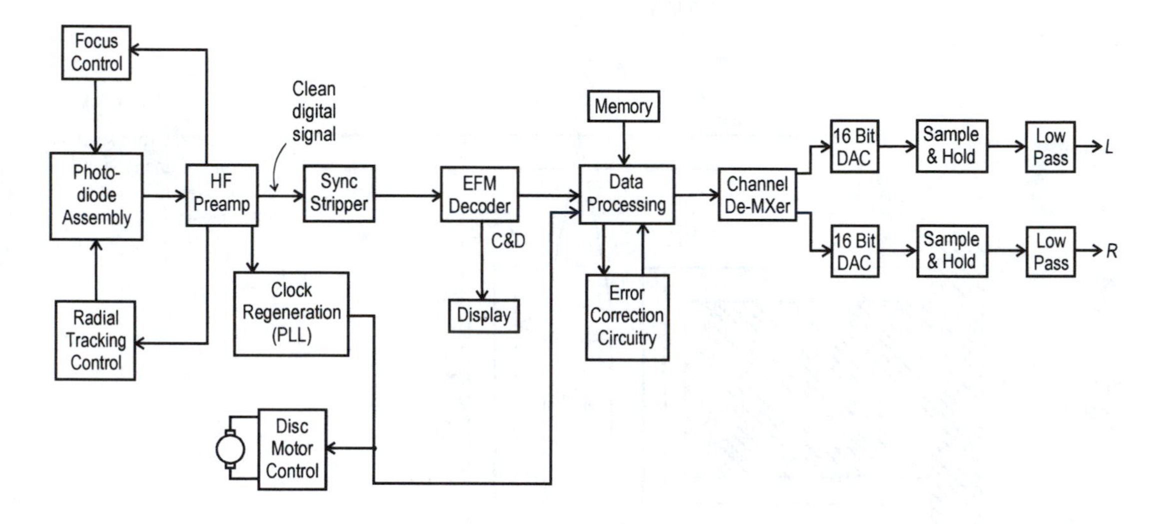

Fig. 9.17: Block diagram of a CD player.

and D are more illuminated, the comparator output will go high. This comparator's output is ultimately used to change the spacing between the objective lens and the disc's recorded surface. An example of a transducer that accomplishes this is shown in Fig. 9.19. The principle of operation of this assembly is similar to that of the voice coil in a loudspeaker. The current through the focus coil is derived from the focus comparator. As current flows through this fixed coil, the magnetic field created will interact with the field of the magnetized cylindrical shell in which the lens is mounted. Since the shell moves up and down in response to this magnetic force, the objective lens moves toward and away from the disc. This changes the focus and thus the spot shape on the four main photodiodes. When the beam is in focus on the disc, the photodiode pattern will be the circle of least confusion, and the output from diodes A and C will equal that from diodes B and D. There will thus be no error signal generated by the focus comparator. This idyllic state of affairs would seldom persist in practice for even a millisecond, owing to disc warpage and wobble. Focus correction is a dynamic ongoing process. It should be obvious that it is also possible to mount the lens in a nonmagnetic cylinder and wind the coil around it. A permanent magnet may then be fitted around the cylindrical coil form. The advantage of this is that the lens mount is lighter this way and so will have less inertia and need less drive power. The disadvantage is that the coil leads are no longer stationary.

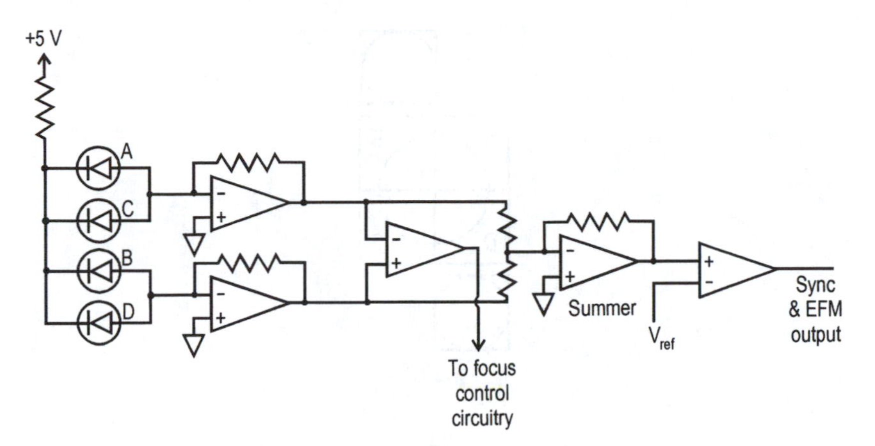

Fig. 9.18: Schematic diagram of the high-frequency preamp circuitry in a CD player.

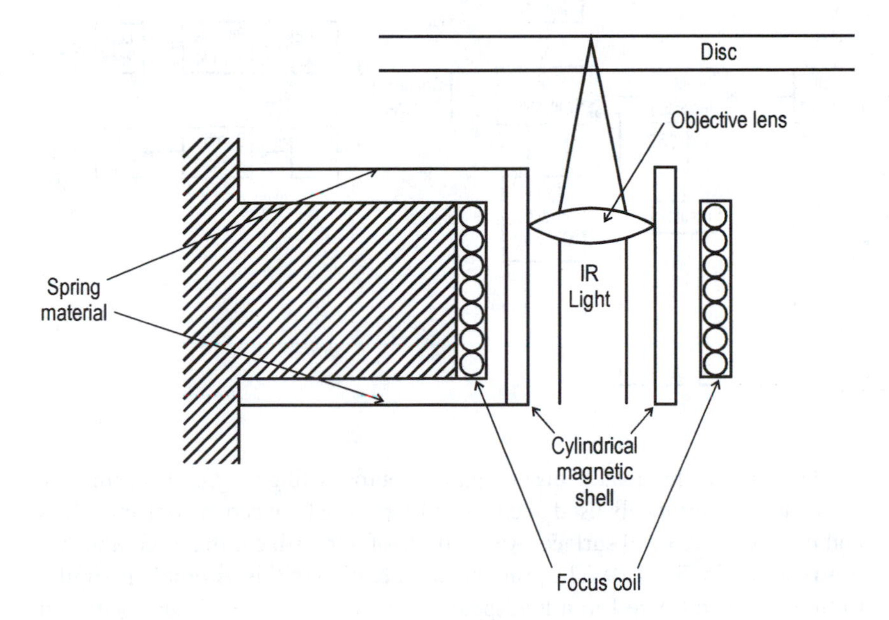

Fig. 9.19: Cross-sectional drawing of one type of focus assembly used in CD players.

It is now time for us to consider the tracking operation, which is handled by the two "satellite" photodiodes in Fig. 9.16. As already indicated, these respond to the two side beams. It remains to be seen, however, how they relate geometrically to the disc being played. This is shown in Fig. 9.20. The signal from the forward tracking photodiode is delayed by 30 μs before it is compared to that from the trailing one. If the main beam is centered on the track, the signals from both of the tracking diodes will be equal. Any difference between them is the tracking error signal, which drives the tracking transducer. This system is capable of keeping the main beam centered to within .1 μm of the center of the track. The transducer is designed to move the objective lens radially. Referring back to Fig. 9.19, we see that the springs that hold the objective lens assembly must not only accommodate up and down motion but left and right motion as well. In practice, this can be accomplished by mounting the lens between the center and the circumference of the cylindrical magnet. Rotation of the magnet will then move the lens in and out along the disc radius, providing a tracking

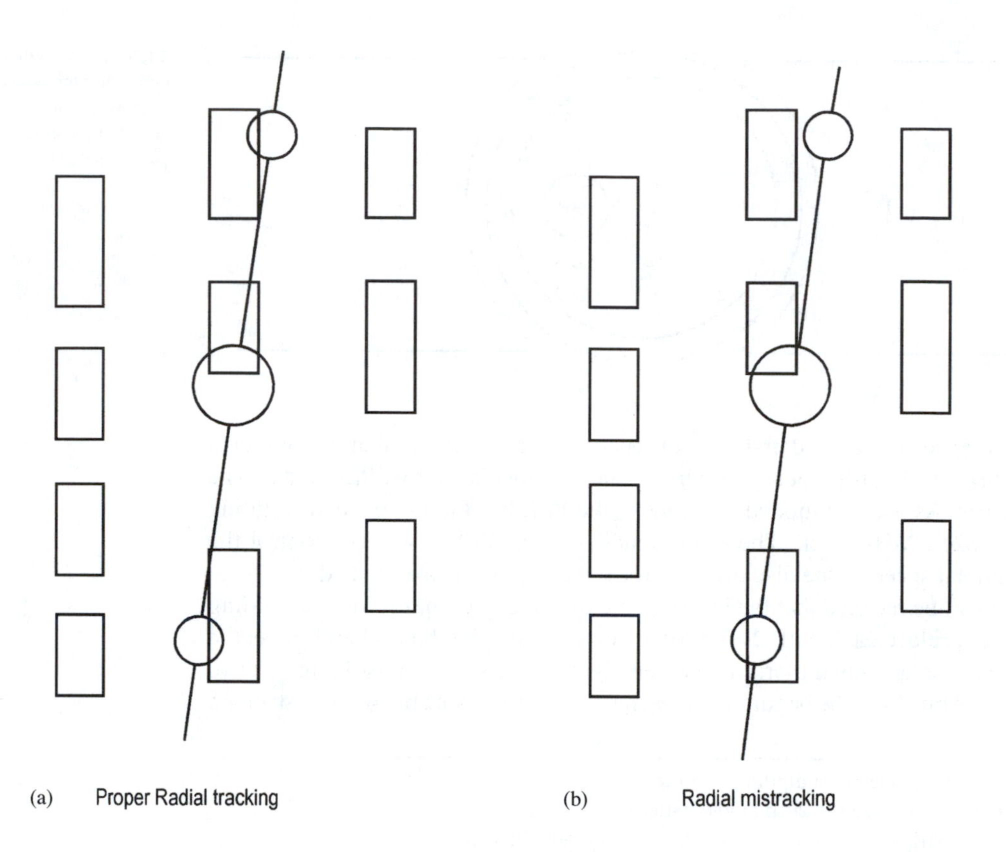

Fig. 9.20: A portion of the pit structure of a CD with the positions of the triple beam imposed upon it. (a) Proper radial tracking. Note that the tracking beams each overlap the edges of the pits by the same amount. (b) Radial mistracking. Note that the right-hand tracking beam does not "see" any part of the pit, and the left one half overlaps it. This will cause a large difference in the light received by the two tracking photocells, which is interpreted as a tracking error.

adjustment. The rotation is caused by two pairs of "pancake" coils placed at 90° intervals around the cylindrical magnet. These coils are also visible in Fig. 9.7. They are driven by the tracking error signal. This setup is shown in Fig. 9.21 for a system where the coil form and its coils move.

Coils 1 and 2 will be in series, as will 3 and 4. If 1 and 2 are excited such that their N pole is outward, and 3 and 4 have their S pole outward, the cylinder will rotate clockwise, moving the lens the same way. The opposite excitation will produce the opposite rotation. This servo system must be able to accommodate a total radial excursion of as much as 300 μm around its nominal position, primarily due to fabrication tolerances in the placement of the disc center hole. The regular and slower motion of following the tracks outward is handled by a sled drive motor whose precision and response speed fall short of being able to handle instantaneous tracking adjustments.

9.3.4 Decoding

Early CD players used only a single beam, but shock and vibration performance of these units tended to be inferior. These early models have been eliminated from the market to a large extent by the triple-beam design. It is interesting, however, that Philips has retained the single-beam design. Its performance can be made comparable to that of the three-beam design by substituting electronic sophistication for optical in the tracking circuitry.

As shown in Fig. 9.18, the summer stage combines the outputs of all four main photodiodes. Since this sum will never get down to zero, it is compared to

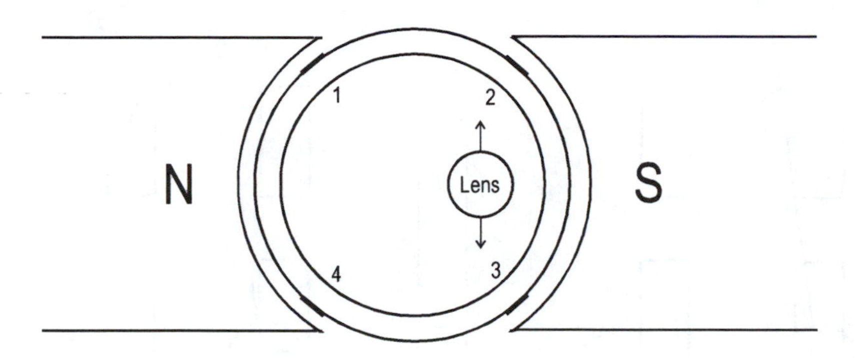

Fig. 9.21: Schematic representation of a common mechanism for changing the radial tracking of the spot on the CD.

a reference and squared up by the output comparator. The comparator output is applied to a high-frequency amplifier whose output is fed to a PLL and a sync stripper. As was mentioned previously, the PLL has the job of reconstituting the 4.3218 MHz clock. The error signal from the PLL is used to control the rotational speed of the disc drive motor such that lock is maintained.

It will be recalled that the last step in constructing frames prior to recording was to preface each with 24 bits of sync and 3 merging bits. Therefore as the player reconstructs the original stereo signal, its first step must be to remove these 27 bits from the beginning of each frame. This is done by the sync stripper.

Exercise 9.6 Design a digital circuit to:

(1) recognize the 24-bit sync pattern,
(2) strip off it and the three following merging bits, and
(3) pass on the remainder of the frame bits.

The remainder of the frame is entirely composed of 33 EFM bytes of 17 bits each. The EFM decoder (or demodulator as it is called in CD literature) will convert this to 33 binary bytes of 8 bits each. The EFM decoder need be nothing more than a serial-to -parallel shift register and a lookup table in ROM. The first of these 33 is the C & D byte. It is separated off to the display and to perform some control functions as indicated in Table 9.1. The remaining 32 bytes go to the data-processing circuitry. As discussed under the topic of encoding to disc, the CIRC encoding process is done in two steps with interleaving between them. Likewise, decoding must be done in two steps with deinterleaving between them. The contents of the error-correcting block in Fig. 9.17 are further broken down in Fig. 9.22.

The C1 decoder removes most of the random errors. All 32 of its 8-bit bytes are input to the C1 decoder, but the even ones are delayed by the duration of one frame. Thus the bytes that were not delayed in the sixth error countermeasure are delayed here and thus brought back into time alignment. The C1 decoder strips off four of the parity bytes and performs parity checks with them. It is capable of completely correcting either one or two erroneous bytes out of the 32. If three or four bytes are defective, this fact can be detected but not corrected by C1. It therefore raises error flags on the defective bytes and passes them on unchanged toward the C2 decoder. If more than four bytes are defective, C1 flags all 28 bytes that it outputs ($32 - 4 = 28$). Before getting to C2, the bytes must pass through delay circuits of *differing* lengths. This is the deinterleaving operation, and it undoes the interleaving operation performed

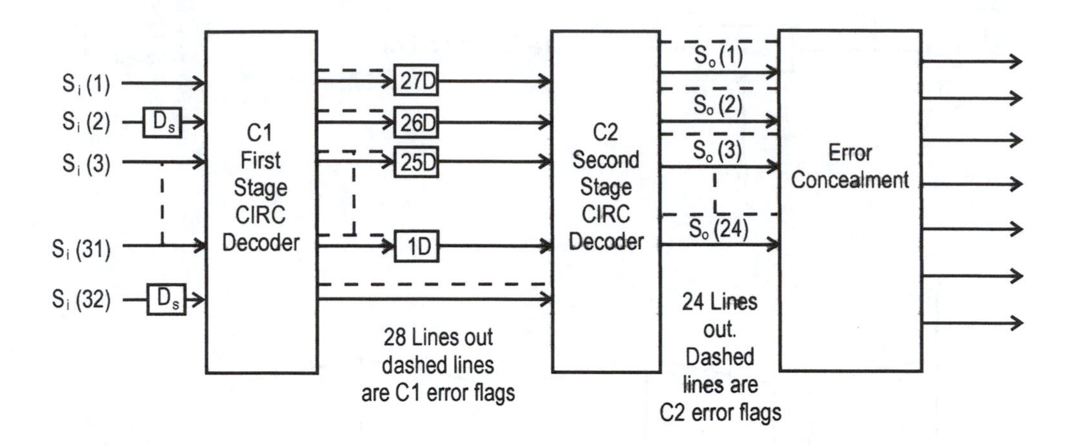

Fig. 9.22: Block diagram of the decoding system for a CD player.

during encoding. Whereas the interleaving delays introduced during encoding were $n \cdot \mathrm{D}\,(0 \leq n \leq 27)$, during decoding the delays are $(27 - n)\mathrm{D}$. Thus each symbol will come through the complementary cross-interleaving process with an additional 27 D (=108 frame times) delay. The other delays in the decoding process add a total of 3 more frames worth of delay for a total of 111. These plus the undelayed frame are the 112 over which the data are spread out.

The C2 decoder takes a second crack at decoding the errors after deinterleaving that the C1 decoder could not correct *before* deinterleaving. It will thus correct most burst errors. Because of the differing delays in the deinterleaving process, the error-flagged bytes leaving the C1 decoder at a given time will not all arrive at the C2 decoder at the same time.

The CIRC system has done all it can do. Remaining errors cannot be corrected, but they can be concealed. The seventh error countermeasure is to attempt to replace flagged words by linear interpolation between adjacent bytes that are not flagged. If there are too many consecutive flagged bytes, the previous value is held until good data are obtained again. This is one section of the player where good design particularly asserts itself. In a well-designed unit, even errors that grossly exceed the capabilities of the CIRC error correction may be made inaudible by concealment.

Next, all of the symbols corresponding to each odd-numbered sample are delayed by two frames. Thus the bytes that were not delayed in the second error countermeasure are delayed here and thus brought back into time alignment. After this the 8-bit symbols or bytes are reassembled into 16-bit ones. Referring again to Fig. 9.17, we see that the data stream is next demultiplexed to separate the L and R data.

The L and R data are delivered to 16-bit DACs, which convert back to analog. Some units may use one DAC to process one channel at a time, while storing the other in a register.

9.3.5 Sampling review

Because the output of a DAC persists only until the next conversion begins, it is necessary to run the DAC output through a Sample-and-Hold (S&H) unit. This also removes glitches in the DAC output. It may be helpful at this point to review some fundamentals of sampling theory. The S&H block will have

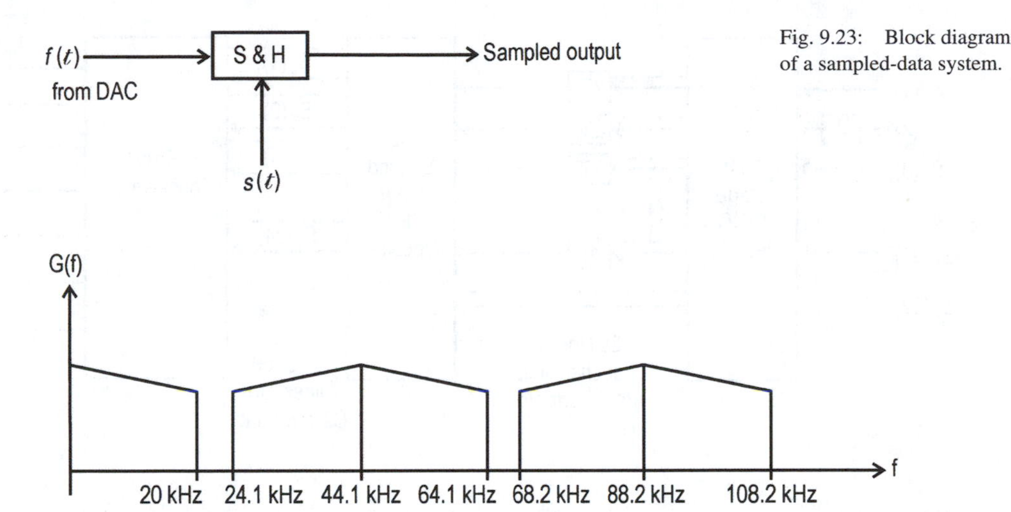

Fig. 9.23: Block diagram of a sampled-data system.

Fig. 9.24: Spectral energy diagram for the output of an S & H module after sampling at 44.1 kHz.

two inputs: the signal to be sampled (in our case, this is the DAC output signal, which we will call $f(t)$), and a sample command. This latter input is not shown in Fig. 9.17, but it is implicit. We will call this $s(t)$. This arrangement is shown in Fig. 9.23. As is usual, we will consider that $f(t)$ has been bandlimited. This is certainly the case in a CD system where the upper frequency limit is 20 kHz. The output will, of course, depend on the specific character of the sampling waveform. We shall assume that it is a periodic train of narrow pulses with width Δt and with a frequency $f_s = 44.1$ kHz. The sampled output will then be

$$\begin{aligned}\text{Output} &= f(t) \cdot \Delta t \cdot f_s \sum_{n=-\infty}^{\infty} \cos(2\pi n f_s t) \\ &= f(t) \cdot \Delta t \cdot f_s \left(1 + 2\sum_{n=1}^{\infty} \cos(2\pi n f_s t)\right) \\ &= f(t) \cdot \Delta t \cdot f_s + 2f(t) \cdot \Delta t \cdot f_s \cdot \cos(2\pi f_s t) \\ &\quad + 2f(t) \cdot \Delta t \cdot f_s \cdot \cos(4\pi f_s t) + \cdots .\end{aligned}$$

The first term on the right is just the input signal multiplied by a constant. This is the term in the output that we must recover. The second term has the form of a DSB-SC signal whose suppressed carrier frequency is f_s. The third term is the same type, but the suppressed carrier is at $2f_s$. Subsequent terms will be centered at $n \cdot f_s$. In the frequency domain, the output will have the structure shown in Fig. 9.24. No significance should be attached to the sloping tops drawn on the spectrum, as the spectrum will vary with the program material recorded. The sloping top is drawn to show that the upper frequencies in the baseband signal are translated to the lowest ones in the lower sidebands. The most important observation we make from Fig. 9.24 is that although we want to extract the baseband portion of the S&H output, we do not want any of the higher-frequency components. Conceptually this can be accomplished by passing it through a low-pass filter. But as shown, the guard band is only

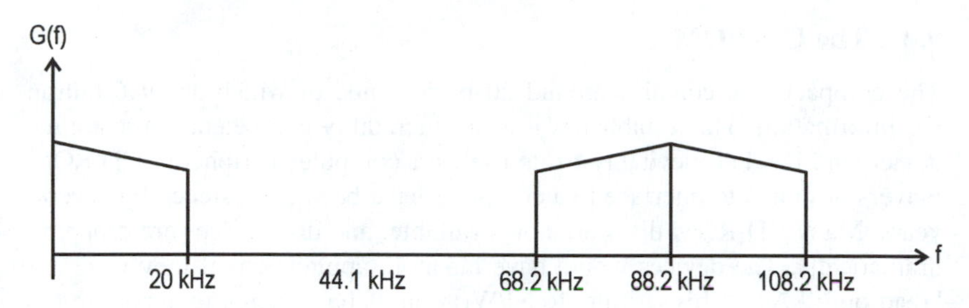

Fig. 9.25: Spectral energy diagram for the output of an S & H module with 2× oversampling.

4.1 kHz wide above a frequency of 20 kHz. A change from 20 to 24.1 kHz is a change of only about .1 decade. Since a low-pass filter is specified by its rolloff in dB/decade, it is obvious that an enormously complex filter is required to adequately suppress the unwanted sidebands. This problem is exacerbated by the fact that analog filters tend to exhibit significant phase shift near their corner frequencies. This phase error has serious adverse effects on the quality of the recovered sound. The most straightforward solution to this problem is not to build a better filter, but to sample at $2f_s$, instead of f_s. This involves interpolating one sample value between each pair of pulses recovered from the disc. Then we have

$$\begin{aligned}\text{Output} &= f(t)\cdot\Delta t\cdot 2f_s\left(1+2\sum_{n=1}^{\infty}\cos(4\pi n f_s t)\right)\\ &= 2f(t)\cdot\Delta t\cdot f_s + 4f(t)\cdot\Delta t\cdot f_s\cos 4\pi f_s t\\ &\quad + 4f(t)\cdot\Delta t\cdot f_s\cos 8\pi f_s t+\cdots,\end{aligned}$$

where f_s is still taken to be 44.1 kHz. From this expression, we see that if the sampling pulse width, Δt, is held constant, doubling the sampling frequency will double the amplitude of the recovered baseband signal. But even more importantly, the first "suppressed carrier" is now at 88.2 kHz, and thus the first unwanted sideband will never be below 68.2 kHz, as shown in Fig. 9.25.

Now the guard band between the baseband signal and the sidebands relative to 20 kHz is just over .5 decades. The price paid for this improvement is that a better and thus more costly S&H device must be used. This still does not completely solve the filter problem. A digital filter has the sharp cutoff required, but it also has spurious responses outside of the passband. The best units use a digital filter to provide a sharp rolloff above 20 kHz, followed by an analog filter to remove the spurious responses of the digital filter. An eighth-order analog filter might be typical of a good unit. Such a unit would have a response that is flat within .2 dB out to 20 kHz and a maximum phase error of 2° at 20 kHz. The technique we have just discussed is called *oversampling*. It is used in all good CD players. Some use 4× or higher oversampling, which places the lowest sidebands at 156 kHz, easing the filtering requirements even further. Each factor of 2 in oversampling also reduces quantization noise by 6 dB.

The signals out of the low-pass filters are the output signals of the CD players, and these go to a stereo amplifier for listening. Portable units that drive headphones and automobile units are also made, but for a given dollar cost, you will generally get better specs in a home-based unit.

9.4 The CD-ROM

The compact disc contains around 20 billion bits, of which about 6 billion are information. This establishes it as an incredibly competent mass storage device and has led inevitably to its use as a computer peripheral. CD-ROM players designed to interface to a computer have been in existence for several years. Many CD-ROM discs are now available, and drive prices are dropping dramatically. The devices would have far more applications if they were not "read only." As of this writing, Read/Write units have begun to appear in the mass market. They are still rather pricey, but they can probably be expected to drop to within a couple of hundred dollars in the next few years.

9.4.1 Combination machines

A natural progression has led to players capable of playback of both audio and video discs, though at present, these have not gained wide acceptance.

On another front, Philips and Sony have again teamed up to propose standards for what is called CD-I for *Compact Disc-Interactive*. This is a standard that would accommodate five different levels of sound recording, with the maximum quality being the same as the present CD audio, and lower qualities allowing much longer playing time. The standard also envisions three levels of video picture resolution, text, and sound–video synchronization. The system would be part computer (68000 series μP) and would not only be able to execute object code but would *expect* the CD-I disc to initialize it when the disc is first loaded. Finally, the standard accommodates two input devices (since it is interactive) – a mouse, and a keyboard. The video capability extends to stills and animation but not to movies. This is certainly due, at least in part, to the slow rate of data transfer (by video standards) that is locked into the standard by making it compatible with CD audio. Unlike audio CDs and CD-ROM, CD-I has not received quick acceptance. Perhaps a composite will emerge. Perhaps a completely new standard will do so. In this fastest moving area of consumer electronic technology, the only thing we can be sure of is change.

9.5 Digital audio tape

Digital audio tape recorders come in stationary-head and rotary-head versions. The primary stationary-head format is called DCC for digital compact cassette. It uses cassettes the same size and shape as analog compact cassettes. In fact, DCC players are generally able to play both analog and digital cassettes.

With digital tape recording, the problem of overdriving is much more severe than with an analog recorder. In the A/D conversion, all inputs that exceed the reference voltage will cause the output to remain at the maximum count possible. This corresponds to hard limiting in the analog domain. The DCC approaches this problem by passing the signal through an analog soft limiter before the A/D conversion. This and other compromises result in a recorder whose sound is superior to that of an analog tape recorder but inferior to that of a CD. Perhaps this accounts for the slow acceptance of such units to date.

The R-DAT (rotary digital audio tape) recorder uses a cassette about half the size of a compact cassette, and it holds a full two hours of recording. The R-DAT system draws heavily upon the technologies of both the VCR and the compact disc. It has specifications every bit as good as a CD has, but it requires some expertise to choose a record level that will never overdrive the converter. Consequently, after a short effort to market it as a consumer device, R-DAT, or just DAT as it is generally called, has reverted to being the device of choice for professional digital recording.

9.6 The Yamaha CDX510U compact disc player

Given that the CD player as it exists today was a joint development of Philips and Sony, it might be thought only reasonable for us to choose a unit by one of these manufacturers for analysis. However, a Sony consumer electronic device was already analyzed in Chapter 3. That was a Walkman personal stereo – another device they pioneered. We also analyzed part of a discrete Sony color TV in Chapter 7. We have elected not to analyze a Philips CD player because, although they make a fine unit, it is not typical of most CD players today in that it uses a single beam.

While this explains why we are not analyzing a Sony or Philips unit, it does not explain why we have chosen Yamaha. In America, Yamaha is best known for motorcycles. But their roots go further back in music than in motorcycles. In fact, even the Yamaha logo placed on motorcycles consists of three crossed tuning forks. They have made fine pianos for many years and were an early entry into the area of electronic keyboards. So their entry into the CD derby is well in keeping with their history. There is yet another reason for our choice of a Yamaha player.

The optics of a CD player described in Section 9.3.1 (including the laser and photodetector array) are always part of an integrated system that is replaced as a whole when any part of it fails. It is also the component responsible for a large majority of CD player failures. Not surprisingly, it is the most expensive component in the player, also. In the Japanese CD players, which still dominate the market, the optics modules are made, for the most part (and not surprisingly), by Japanese camera manufacturers. They typically cost \$100–\$125, installed. This is close enough to the cost of a new CD player that many owners use the occasion of a failed optics module to purchase a new player that may contain more of the features they want.

When Olympus began producing an optics module (called PICKUP assembly in the service literature) for CD players, Yamaha was the first manufacturer to incorporate the Olympus optics module into their CD players. The Olympus unit had two unique things going for it. The first was that by all reports it was the most stable (in the control system sense of the word) and easily adjusted of all pickup assemblies then available. The second was that its price was around \$50. Therefore, we have chosen the Yamaha unit because of the superior performance and value of its optical module.

The block diagram for the CDX-510U is shown in Fig. 9.26. As indicated in the lower left-hand corner, it is a remote-control unit. The remote-control response circuitry contains its own 4-bit μP. Besides this, there are 4 LSI circuits:

- the signal processor/controller IC,
- the digital filter,
- a 2 kilobyte static RAM, and
- an 8-bit system control μP.

In addition to these major chips there are numerous smaller ones as well as discrete transistors in the circuitry.

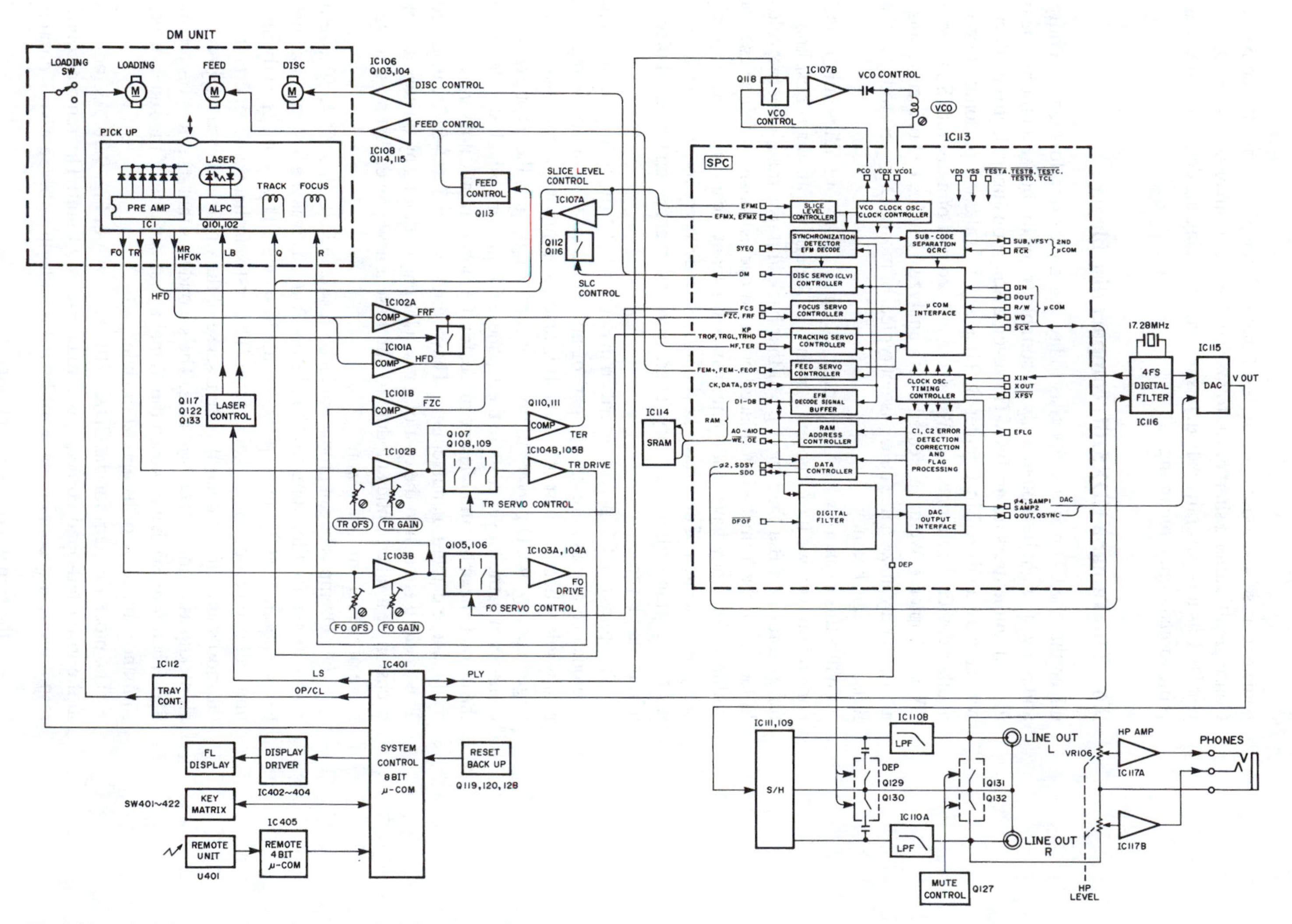

Fig. 9.26: Block diagram of the Yamaha CDX-510U CD player.

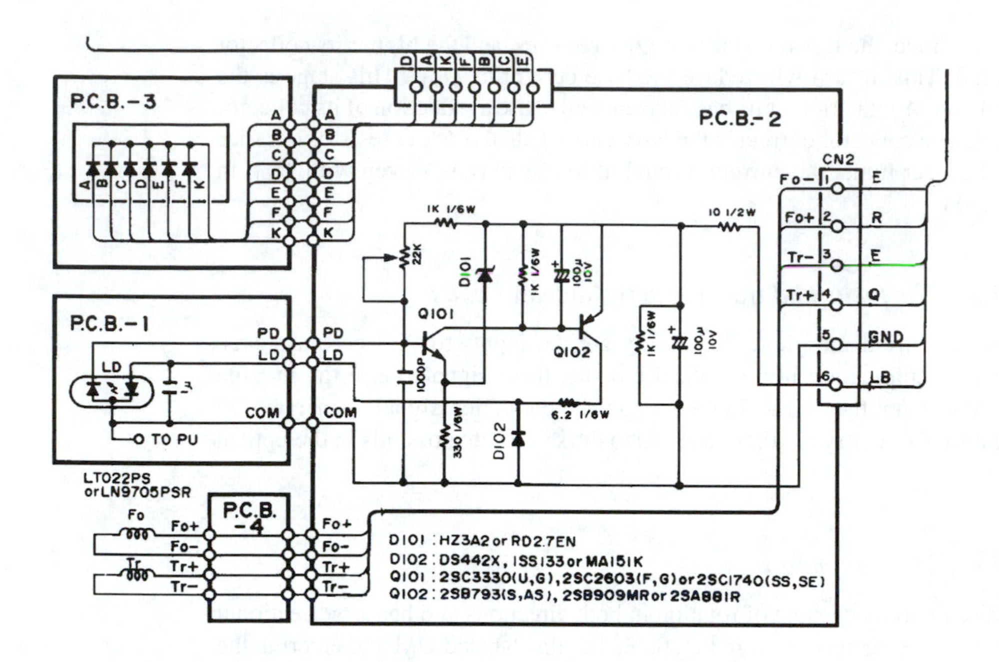

Fig. 9.27: Schematic diagram of the laser power control module of the Yamaha CDX-510U CD player.

9.6.1 The disc mechanism unit

The disc mechanism is shown in the upper left corner of the block diagram (Fig. 9.26). It includes the three motors and the optical pickup assembly. The loading motor loads the disc onto the disc motor after the user loads it into the disc tray. The feed motor provides coarse control of the radial position of the pickup assembly on the disc. We shall consider these two motors further when we come to the circuitry that drives them. The pickup assembly block shows (from left to right) the photodiode array and preamp, the laser diode and its monitor photodiode, the automatic laser power control (ALPC), and the tracking and focus servo actuator coils.

9.6.1.1 The photodiode array

This unit feeds an integrated preamp, which outputs correction signals for focus and tracking, as well as the high-frequency data signal read from the disc and some logic signals.

9.6.1.2 ALPC

Figure 9.27 shows the schematic of the ALPC circuit. The DC power is input on lead LB. This lead is a 4 V supply that will be switched off by the μP when the unit is opened to insert a disc.

As the schematic shows, the monitor photodiode receives some of the laser output. This generates a signal to the base of Q_{101}. The emitter of this transistor is held at about 2.3 V by D_{101} and the emitter resistor of Q_{101}. The base bias is jointly determined by the 22 K pot in the base circuit and the output of the monitor photodiode. The polarity of the photodiode is such as to sink current from the pot, depriving Q_{101} of base current. Thus, the greater the light output of the laser diode, the less base current Q_{101} receives, and the higher its collector potential. This in turn will reduce the base current of Q_{102}. This stage is the laser driver. A reduction of its base current will cause a reduction of its collector current. The collector current is the laser current also. A 6.2 Ω resistor in series with the laser limits the current through it to less than 1 A even with Q_{102} in saturation.

9.6.1.3 *The focus and tracking servo actuator coils*

The signals from the photodiode array are the inputs to the servo systems. After preamplification in the optical module, these signals leave the module for further amplification and processing. The correction signals generated by this external circuitry are then returned to the servo actuator coils in the optical module.

9.6.1.4 *The disc motor*

The disc motor is capable of rotation in both directions and has a bidirectional driver whose circuit is given in Fig. 9.28. The line labeled DM and entering the schematic at the lower right-hand corner is the disc motor control line from the signal processor/controller IC.

Exercise 9.7

(a) From Fig. 9.28, determine the steady-state current through the disc motor as a function of the DC voltage on line DM. This procedure will be both less painful and more accurate if feedback concepts are used.

(b) The chip that drives the DM line has only a +5 V supply with respect to ground. Find the maximum and minimum values of steady-state current through the disc drive motor.

9.6.2 *Focus circuitry*

The optical pickup head has three important signal outputs: the focus and tracking error signals and the demodulated EFM. We will consider these one at a time. The focus circuitry is shown schematically in Fig. 9.29.

The focus error signal, it will be recalled, is an analog voltage that is zero when the focus is perfect and can be positive or negative as the focal point moves above or below the reflecting plane of the disc. The first stage, IC_{103B}, provides a gain on the order of 6 or 7 for this analog voltage. The output of this stage follows two paths. The first goes through C_{110} and a resistive voltage divider to the noninverting input of IC_{101B}. The inverting input of this IC is held just below ground potential by another resistive divider. Since this comparator swings between ±9 V supplies, the network composed of R_{153}, R_{154}, and R_{155} is used to level-shift the comparator output so it will be compatible with the signal processing chip. This same circuit is used several times in this unit.

Exercise 9.8 Find the exact voltage limits of the $\overline{\text{FZC}}$ signal assuming IC_{101B} swings to +9 V and −9 V.

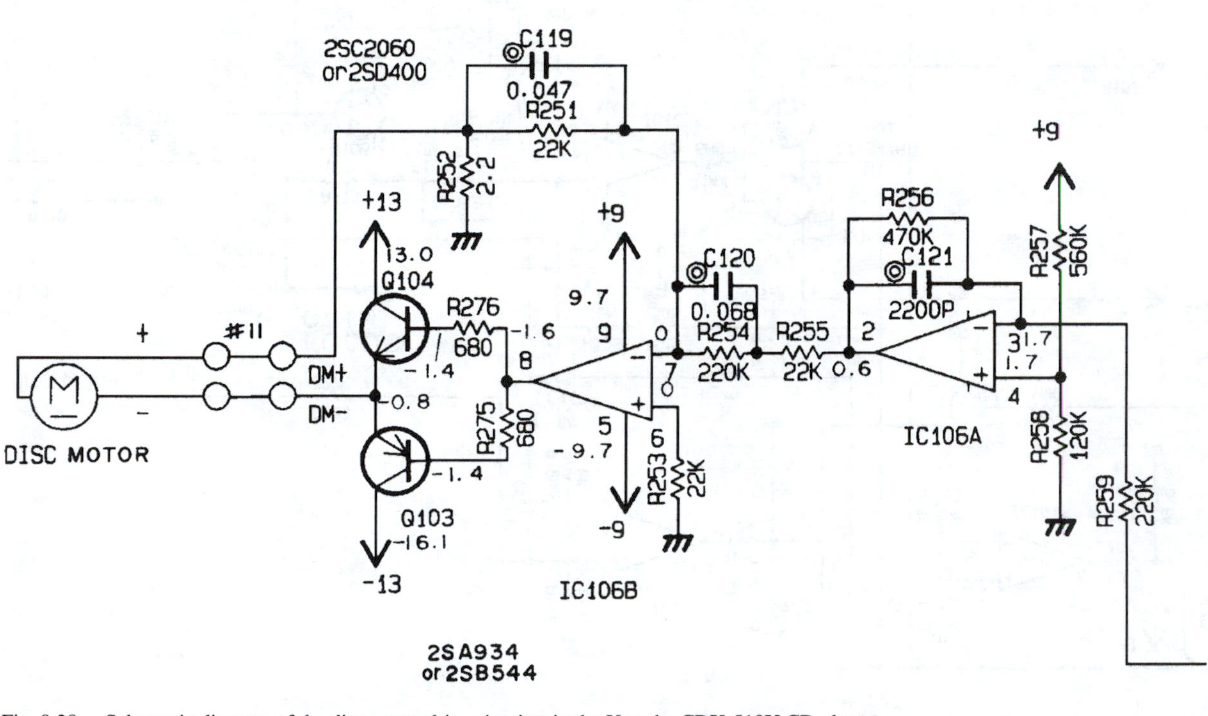

Fig. 9.28: Schematic diagram of the disc motor drive circuitry in the Yamaha CDX-510U CD player.

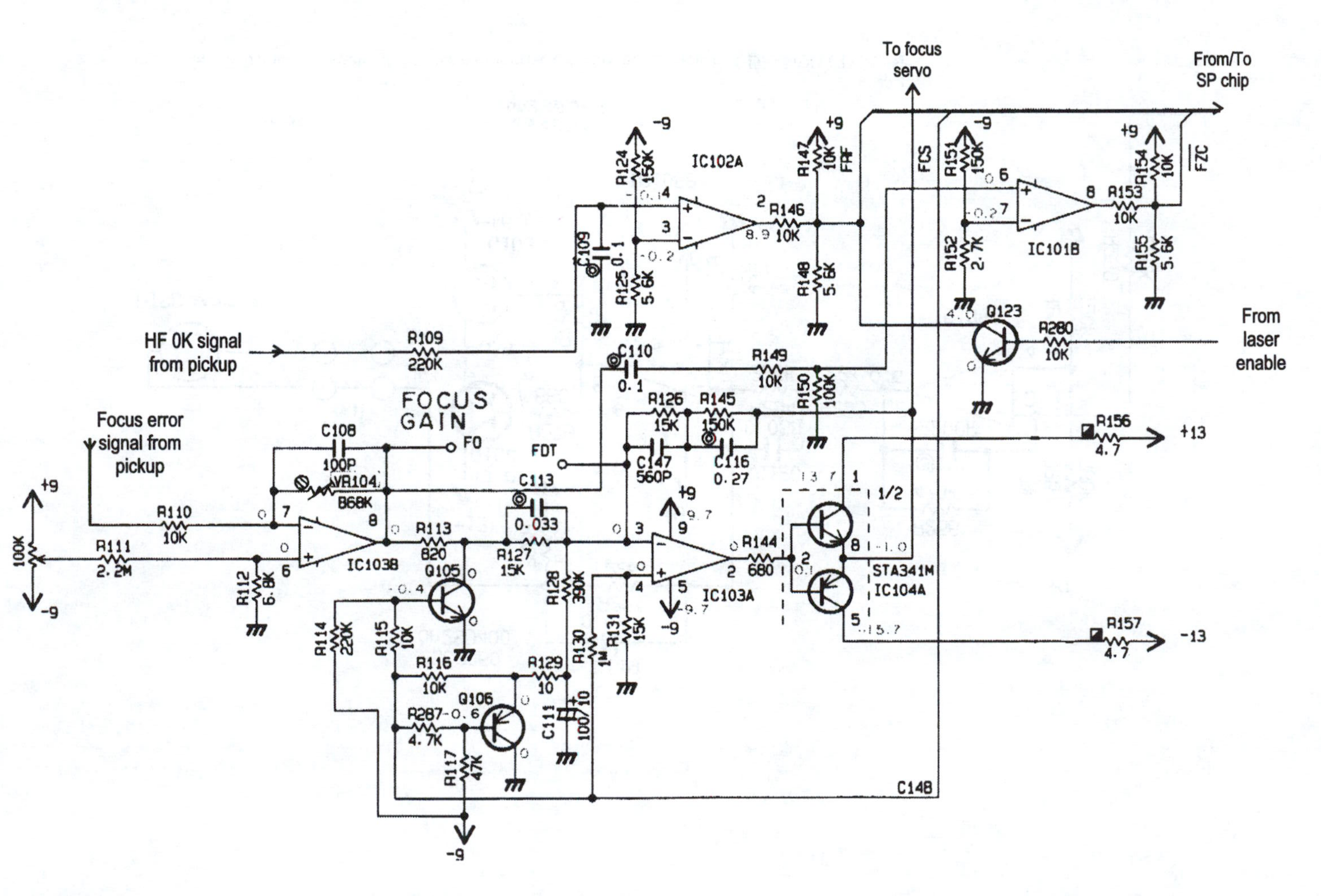

Fig. 9.29: Schematic diagram of the focus-control circuitry in the Yamaha CDX-510U CD player.

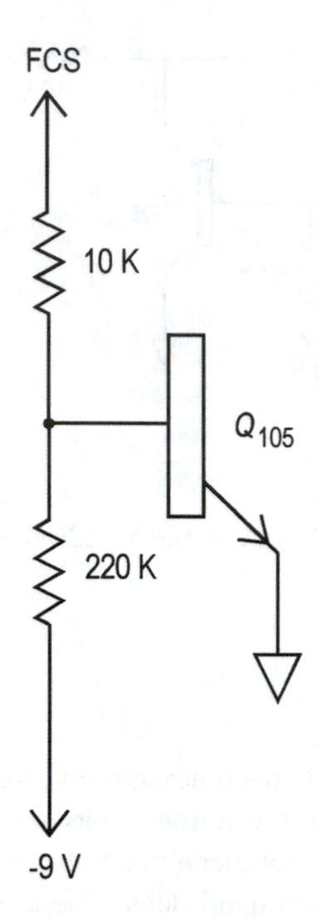

Fig. 9.30: Schematic diagram of the switching stage within the focus-control circuitry of the Yamaha CDX-510U CD player. This stage grounds out the focus-error signal during the initial search for the correct focal distance. Once this distance is found, the switch removes the short, and normal servo action commences.

When the focus servo passes through the plane of correct focus, the FO signal changes sign. By the signal path just elucidated, this will cause $\overline{\text{FZC}}$ (focus zero crossing-active low) to change logic states. This information is input to the SPC chip.

The other path taken by the output of IC_{103B} is through R_{113}. It is clear that Q_{105}, if saturated, will ground out the output of this focus amplifier. Let us see under what circumstances this could happen. The base of this transistor has a potential determined by R_{114}, R_{115}, the -9 V supply, and the FCS signal. This is a focus command put out by the SP chip when a new disc is inserted. We shall see shortly how focus is initiated. For now, we simply note that FCS is a TTL-compatible signal. The base potential of Q_{105} may thus be determined by reference to Fig. 9.30. If $I_B \approx 0$, then

$$(\text{FCS} - V_\text{B})/10\,\text{K} = (V_\text{B} + 9)/220\,\text{K}$$

$$22(\text{FCS} - V_\text{B}) = V_\text{B} + 9$$

$$(22 \cdot \text{FCS} - 9)/23 = V_\text{B}.$$

When FCS $= 0$, $V_\text{B} = -9/23 \approx -.4$ V. Then Q_{105} will be in cutoff, and it is certainly true that we can consider I_B to be zero. When FCS $= 5$ V, $V_\text{B} \approx 4.4$ V. This is obviously ridiculous. If $V_\text{B} = 4.4$ V, the B–E junction would "flame out" more or less instantaneously, since the emitter is grounded. The problem is, of course, our neglect of I_B. If I_B is nonzero, we may approximate $V_\text{BE} \approx .6$ V and the equation becomes

$$(\text{FCS} - .6)/10\,\text{K} = (.6 + 9)/220\,\text{K} + I_\text{B}.$$

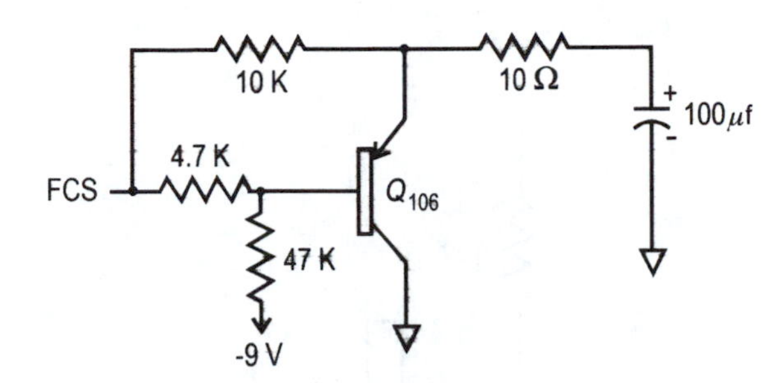

Fig. 9.31: Schematic diagram of the slow ramp generator used to find the correct plane of focus of a disc first inserted into the Yamaha CDX-510U CD player.

When FCS = 5 V,

$$(5 - .6)/10\,\text{K} = 440\,\mu\text{A} = 9.6\,\text{V}/220\,\text{K} + I_\text{B} = 44\,\mu\text{A} + I_\text{B}$$

or

$$I_\text{B} \approx 400\,\mu\text{A}.$$

Thus, when FCS goes high, the quiescent reverse bias on the B–E junction is overcome and Q_{105} goes into saturation, removing the focus error signal from the servo actuator. This does not mean that there is no input to it, however. Let us see what else the FCS command does. The circuitry around Q_{106} is a bit more involved than that around Q_{105}. The former is shown in Fig. 9.31. We will again begin the analysis by assuming $I_\text{B} \approx 0$:

$$(\text{FCS} - V_\text{B})/4.7\,\text{K} = (V_\text{B} + 9)/47\,\text{K},$$

which yields

$$V_\text{B} = (10 \cdot \text{FCS} - 9)/11.$$

When FCS = 0, $V_\text{B} = -.8$ V. The emitter voltage will be FCS (= 0 V) less any drop across the 10 k emitter resistor. Even a .02 mA emitter current would drop .2 V across this resistor, reducing V_BE to .6 V, which is essentially the threshold of conduction. Thus we may safely say that when FCS = 0, not only is our assumption of $I_\text{B} \approx 0$ a good one, but I_C and I_E are quite close to zero also. When FCS = 5 V, the equation above says $V_\text{B} = 3.7$ V. Since FCS is also applied to the emitter through the 10 k resistor, we might expect to have some emitter current. But all of this "handwaving" has been done without reference to the capacitance from emitter to ground! What actually happens when FCS jumps from 0 to 5 V? The base voltage of Q_{106} does jump to 3.7 V at once, but the emitter is held near ground potential by the capacitor. Hence $V_\text{BE} < 0$, and the transistor is cut off. This will allow all of the current flowing through the 10 k resistor to charge C – at least until V_C gets large enough to forward bias the B–E junction. In practice, this will not happen, because FCS is only brought high for 2 seconds. The capacitor charge equation is given by

$$v_\text{C}(t) = 5(1 - e^{-t/RC}) = 5(1 - e^{-t/1.\,\text{sec}}).$$

Thus, in 2 seconds, the voltage across C would reach $5(1 - e^{-2}) = 4.3$ volts. If the ramp were ever allowed to rise for this long, we would have $v_\text{C} = v_\text{EMITTER} =$

4.3V, $v_{BASE} = 3.7$ V, $V_{BE} = .6$ V, and the transistor would just reach the threshold of conduction when the SP chip would bring FCS back down to zero. This would immediately drop the base voltage back to about −.8 V. Since v_C cannot change instantly, the B–E junction would be strongly forward biased and the capacitor would discharge rapidly through the 10 Ω resistor, which is in circuit only to limit emitter current and so protect the transistor. If the circuit is operating properly, however, the charge process will be terminated in less than 2 seconds.

Basically, the circuit of Fig. 9.31 is a slow ramp generator. Its output is applied to the focus servo amplifier (IC_{103A}) through a 390 kΩ resistance. This ramp is amplified and applied to the focus servo actuator in the optical module. It makes the objective lens move smoothly from its lowest to its highest focus position. When it passes through the plane of correct focus, $\overline{FZC}$ signals this to the SP chip as previously discussed, and the ramp is terminated.

The focus servo amplifier, IC_{103A}, is followed by a complementary-symmetry class-B amplifier inside the feedback loop.

When the ramp has achieved its purpose and fallen back to a logic low, Q_{105} releases the output of IC_{103B}, and the focus servo amplifier begins to receive focus error signals from the disc and adjusts dynamically.

The only remaining circuitry in Fig. 9.29 is that which originates with the HFOK signal. This is a logic signal out of the pickup that signals that reflection has been received from the disc by the tracking photodiodes. After passing through comparator IC_{102A} and the level-shifting network following it, it emerges as the logic level signal FRF (focus reflection) and proceeds to the SP chip. The FRF line can be pulled to a logic low by Q_{123}. This will happen when the laser is disabled.

9.6.3 *Tracking circuitry*

The circuitry that keeps the pickup on track is shown in Fig. 9.32.

The tracking error signal from the pickup is applied first to a preamp consisting of IC_{102B} and its associated circuitry. In addition to the error signal, this stage receives an adjustable DC voltage from VR_{102}. Since the entire circuit is DC coupled all the way through from the input to the tracking servo actuator coil, any DC component summed in at the input will appear amplified at the coil. This will allow the output offset voltage of the op amps to be nulled out. This is a service adjustment and is not available to the user. Our initial impression is that the .012 μF capacitor in series with the input lead would hardly be expected to constitute a good short for AC tracking error signals. Let us investigate the frequency response requirements of this circuit.

Exercise 9.9 In Exercise 9.1 the maximum and minimum values of disc RPM were calculated. Convert these values to Hertz.

The component of tracking error due to the "center" hole of the disc not being really centered relative to the tracks will certainly be the lowest frequency component of the tracking error signal (besides vibration) as well as the largest. This signal will have the same periodicity as the rotation of the disc; it will therefore have a frequency between 8.3 and 3.6 Hz as the player moves from the innermost to the outermost track on the disc.

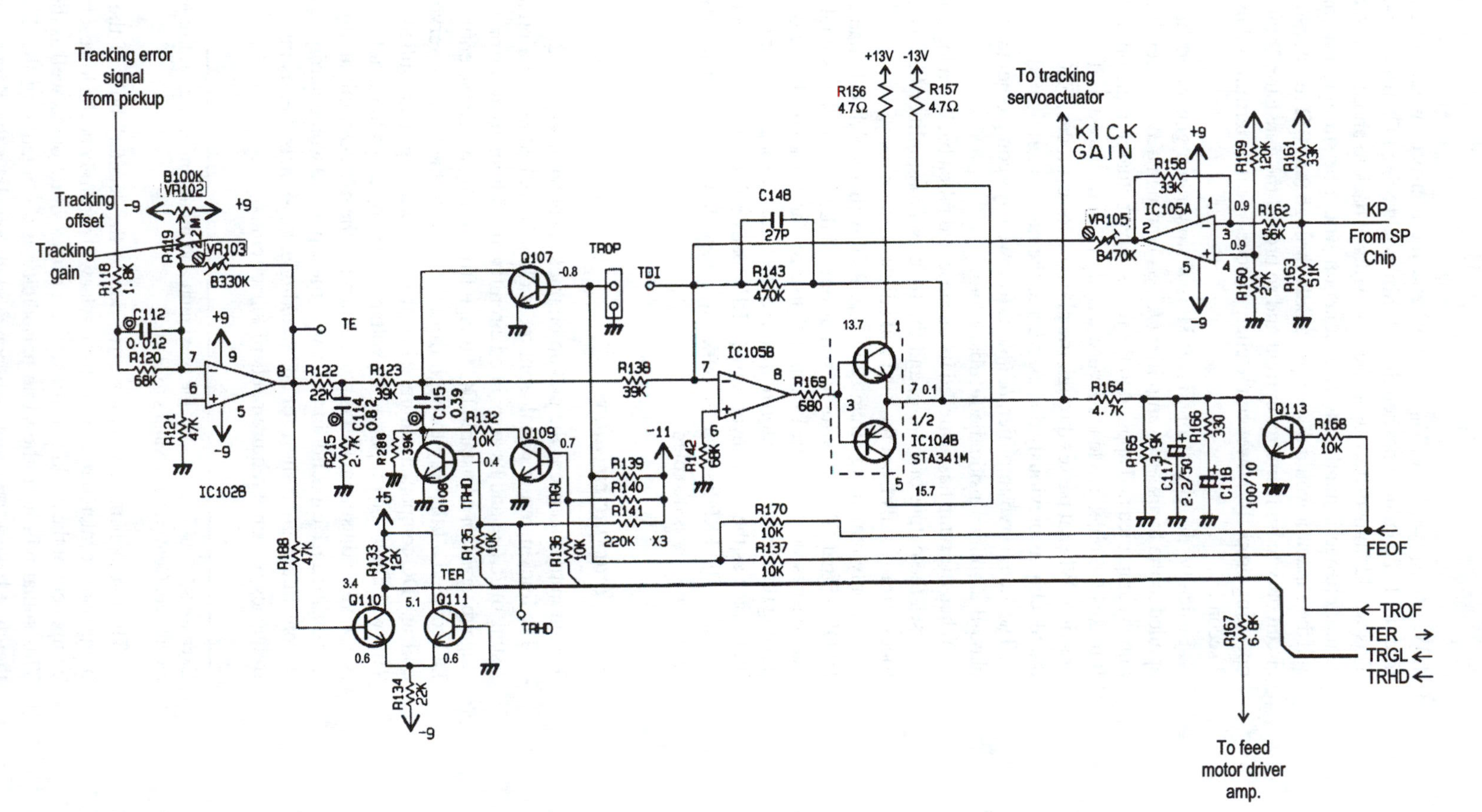

Fig. 9.32: Schematic diagram of the tracking-control circuitry in the Yamaha CDX-510U CD player.

On the other hand, according to the service information on this unit, tracking gain is to be adjusted at a frequency of 800 Hz. This is not to be construed, however, as implying that 800 Hz is the most probable frequency component in the tracking error signal. Rather, it reflects the fact that in search mode, the tracking system must function at frequencies this high and higher. Measurements taken on a typical CD player show that if a search for the last track on a disc is made while the first track is playing, the time required is about 12 seconds. This, compared to the ~60 minutes (3,600 seconds) of normal disc playing time, tells us that the search mode will produce frequencies in the tracking error signal that are about 300 times larger than those in play mode, meaning that the main frequencies in search mode will be between about 1,110 and 2,500 Hz. As we shall soon see, this is clearly reflected in the tracking circuitry for search mode. It is a simple matter to derive the AC transfer function for the first stage of Fig. 9.32.

Exercise 9.10 Find the transfer function for the circuit shown and sketch the Bode amplitude plot.

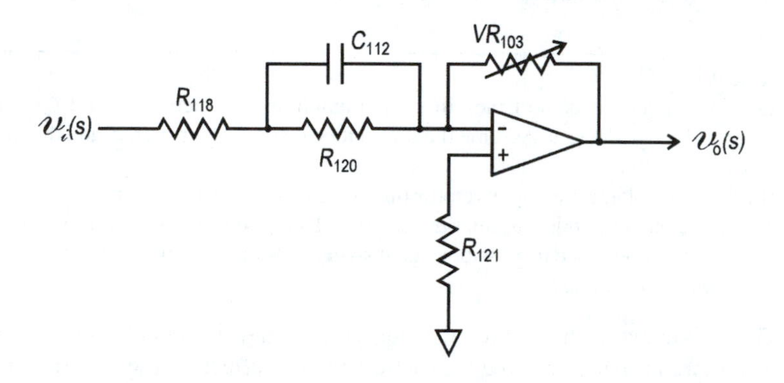

The analysis shows a zero at $1/(2\pi C_{112}R_{120}) = 196$ Hz and a pole at $1/(2\pi C_{112}R_{\rm P})$, where $R_{\rm P} = R_{118} \parallel R_{120}$. This pole has a numerical value of 7,580 Hz. The Bode plot thus shows a gain that rises by 31.7 dB between the zero and pole and constant gains below the zero and above the pole. The dB values of these gains will depend on the setting of VR_{103}, but their difference will remain at 31.7 dB. In the light of our previous discussion of the frequency spectrum of the tracking error signal, we should not be puzzled by the rising response of this stage, although we do not have the full picture yet.

A secondary function of this rising response may be to compensate for the falling mechanical response of the servo actuator. This type of problem was discussed in Chapter 4 in connection with the frequency response of a phonograph cutting head. It should be obvious that some kind of frequency compensation is necessary. A typical pickup head will have complex-conjugate poles in its mechanical response at about 25 Hz, whereas we have just seen that the head must still function out to a frequency two decades higher. If uncompensated, the servo response would have fallen 80 dB in that frequency span. Even a single zero would reduce that drop to 40 dB. The unknown factor here is that since we don't have a schematic of the "innards" of the photocell amplifier IC, we don't know if any frequency-response shaping is done there or not. If not, the zero at 196 Hz in the $IC_{102\rm B}$ stage has a double importance.

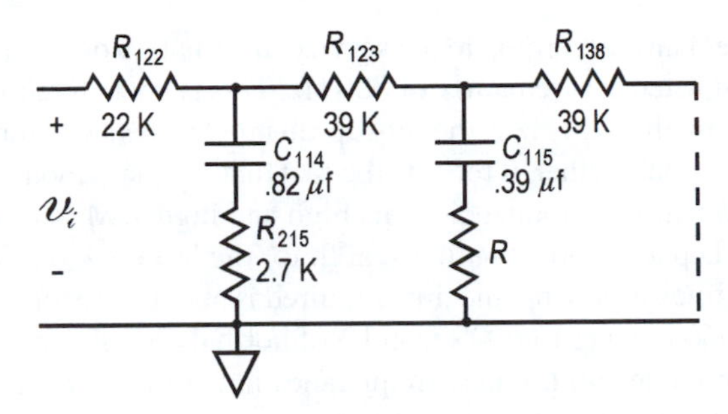

Fig. 9.33: Schematic diagram of the lower leg of a voltage divider that performs tracking gain control in the Yamaha CDX-510U CD player.

In all of this discussion, we have ignored the phase response of the tracking amplifier circuitry. This does not imply that it is unimportant. We leave it as an exercise to the student. Use $\omega_n = 150$ rad/sec and $\zeta = .045$ as the figures for the tracking response of the optical pickup head.

The output of IC_{102B} takes two paths. One is through a differential amplifier (Q_{110} and Q_{111}); the other is into a passive RC network.

Exercise 9.11

(a) Calculate the gain of the differential amplifier of which Q_{110} and Q_{111} are the active devices. Assume the transistors are perfectly matched and that their $r_\pi \ll R_{134}$.

(b) What voltage reading given for this stage is indisputably incorrect?

(c) The output of this amplifier goes to the SP chip. It is not only amplified but is conditioned as to its possible signal swing. What are the limits of signal swing on the TER line?

The RC network that shares the output of IC_{102B} is actually quite an interesting circuit, because three transistor switches affect its transfer function. These transistors are Q_{107}, Q_{108}, and Q_{109}. Note that each is held in cutoff by a 220 k base pulldown resistor to -11 volts. Signals from the SP chip can override this bias and drive each transistor into saturation. The RC network is shown in Fig. 9.33. The dashed line to the right indicates that the "downstream" side of R_{138} is connected to "virtual ground" at the input of IC_{105B}. The key element in this network is the resistor labeled R. If all of the switch transistors just mentioned are cut off, $R = R_{288} = 39\,\mathrm{k}\Omega$.

If the TRGL (tracking gain lower) line is brought high, Q_{109} saturates and brings $R_{132} = (10\,\mathrm{k}\Omega)$ into the circuit in parallel with R_{288}. Then R becomes about $8\,\mathrm{k}\Omega$. As the line name indicates, this will lower the tracking gain under control of the SP chip. With either resistor in circuit, the shape of the network transfer function is about the same. Both are low-pass transfer functions with a transmission of -8 dB at low frequencies. The response starts falling between 5 and 10 Hz and levels out again above about 100 Hz. The difference is that when $R = 39$ k, the asymptotic drop is at 20 dB/decade, whereas part of it is at 40 dB/decade when $R = 8$ k. This will make the high-frequency gain plateau out about 7 dB lower when $R = 8$ k. This gain-switching capability is used to compensate for differences in reflectivity (and thus received signal) between discs.

If the TRHD line is brought high, Q_{108} shorts out R_{288}, and $R = 0$. The low-frequency response of the network remains roughly what it was for $R = 39$ k and $R = 8$ k, but there is no longer a "plateau" at higher frequencies. The network transfer function loses a zero, and the gain approaches zero for high frequencies. The THRD line comes high when either fast forward or fast reverse is activated. It is interesting and instructive to note that when $R = 0$, the falling response of this RC network at higher frequencies almost exactly compensates for the rising response of IC_{102B}, and the resultant transmission is a constant −9.5 dB ±.5 dB between 400 Hz and 6 kHz. Beyond 6 kHz, this stage's response falls off rapidly due to the op amp itself as already discussed. It will be observed that this range of frequencies (over which the response of the circuits of Exercise 9.10 and Fig. 9.33 taken in cascade is flat in search mode) nicely flanks what we deduced to be the primary range of tracking error frequencies in search mode (i.e., 1,100 to 2,500 Hz). Thus, in search mode, the tracking signal path will have a constant gain around those frequencies that are dominant in the tracking error signal.

This signal from the RC network is not the only signal applied to the servo output amplifier in search mode, however. As shown in Fig. 9.32, there is also an input from the kick stage. The KP input line to this stage is 2.5 V in the stop mode but falls to 0 V for fast reverse and rises to 5 V for fast forward. This amplifier sees DC only.

Exercise 9.12 Find the DC output of IC_{105A} for each of the three DC input voltages that can be applied to it.

The output of the kick stage is applied to the summing node of IC_{105B}, which is also the first stage of the servo driver amp. The output stage is a class-B push-pull emitter–follower amplifier. Since the emitter follower is noninverting, the output from the emitters is in phase with the output of IC_{105B}. This allows the feedback to be taken from the emitters, thus including the transistors in the feedback loop, with the accompanying precision and gain stability available with negative feedback. The feedback around this stage consists of 470 kΩ in parallel with 27 pF. The capacitor introduces a pole at about 12.5 kHz. This is just to deliberately reduce bandwidth for frequency compensation purposes. Relative to the output of the kick stage, the gain of the servo output amplifier may be as low as 1, depending on the setting of the kick gain control, VR_{105}. The corresponding gain relative to the output of the RC network will be 470 K/ 39 K $\approx$ 12, or 21.6 dB.

A high on either the TROF (tracking offset) or FEOF (feed offset) will drive Q_{107} into saturation and completely short out the tracking-error signal path through the RC network of Fig. 9.32. This would be required when the servo mechanism was to be disabled altogether, such as when the pickup is moving to the beginning of a new disc to begin play.

This is the first indication we have of the close relationship that exists between the tracking and feed mechanisms. The feed mechanism includes a motor that drives the entire pickup assembly. The tracking mechanism involves a much lower mass system that moves within the pickup assembly. The feed mechanism may be thought of as the coarse positioning control, and the tracking mechanism as the fine positioning control. We shall soon see that the tracking error signal is one of the inputs applied to the feed motor driver.

The output of the servo driver amplifier goes to the servo actuator in the optical module and to a low-pass filter whose output goes to the feed motor driver amplifier.

Exercise 9.13

(a) Find the approximate location of the poles and zeros of the low-pass filter shown (in terms of the symbols).

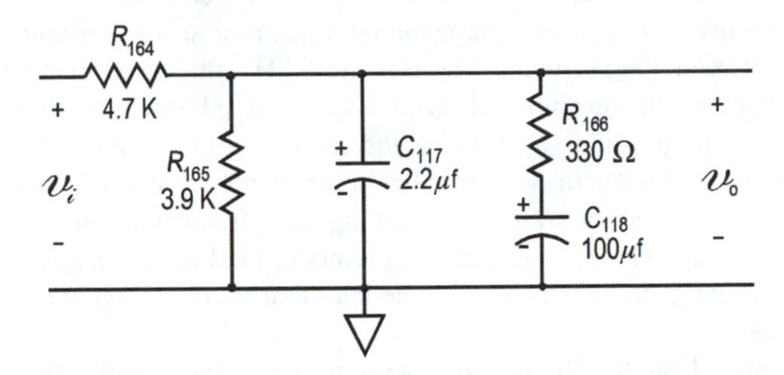

(Hint: There are two poles and one zero. The zero can be solved for exactly. The quadratic that gives the poles can be solved approximately by expanding the square root according to the binomial theorem as per the solution of Eq. (7.11a). You may assume that $2.2\,\mu\text{F} + 100\,\mu\text{F} = 100\,\mu\text{F}$ and $(1/2.2)\,\mu\text{F} + (1/100)\,\mu\text{F} = (1/2.2)\,\mu\text{F}$.)

(b) Evaluate the poles and zeros numerically.

The frequency response of this filter also reflects the need to function in normal play as well as in search mode. It has a pole at .64 Hz. In normal play, the output of this filter should be well under .64 Hz, and maximum transmission is expected. The zero at 5 Hz levels the response out again for search mode. The pole at 256 Hz starts the response falling again.

The feed motor drive amplifier is shown in Fig. 9.34. This drive circuitry is similar to that used for the tracking actuator and the disc drive motor but here the op amp stage has true differential inputs. FEM+ and FEM− are the differential outputs of the SP chip and constitute the feed motor input signals to this output amplifier.

Exercise 9.14 In Fig. 9.34, the DC voltage across the feed motor may be shown to be of the form $\text{FM} = \text{A} \cdot (\text{FEM+}) + B \cdot \hat{Q}$. Here $\hat{Q}$ is the average voltage delivered to the tracking servo actuator. This voltage comes from the LPF of Exercise 9.13.

Assuming that $(\text{FEM}-) = -(\text{FEM}+)$, find the values of A and B. Note that this assumption is equivalent to assuming that FEM+ and FEM− are the two outputs of a differential amplifier.

9.6.4 *EFM circuitry*

So much preamplification and signal processing are done in IC_1 of the pickup that there are basically three signals of interest leaving the pickup. We have just looked at the focus and tracking error signals. We now come to the so-called HF signal, which is the recovered version of the EFM signal modulated onto the disc. This HF signal exits the pickup on the HFO line. Although it is recovered by decoding digital information, it does not look very digital on a scope. The scope trace looks something like a sine wave display with no scope triggering. With proper triggering you get the "eye pattern" characteristic of PCM systems. Recall that the duration of a high or a low in the EFM signal can range between

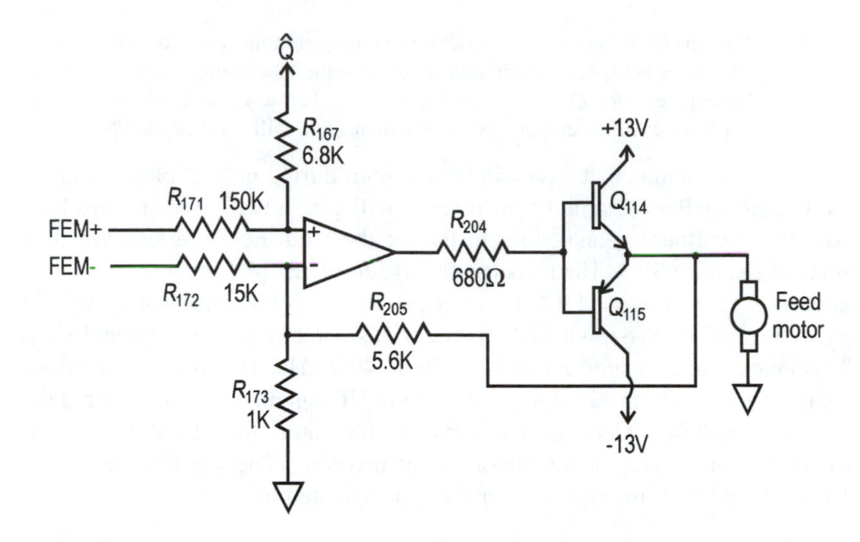

Fig. 9.34: Schematic diagram of the feed motor drive circuitry in the Yamaha CDX-510U CD player.

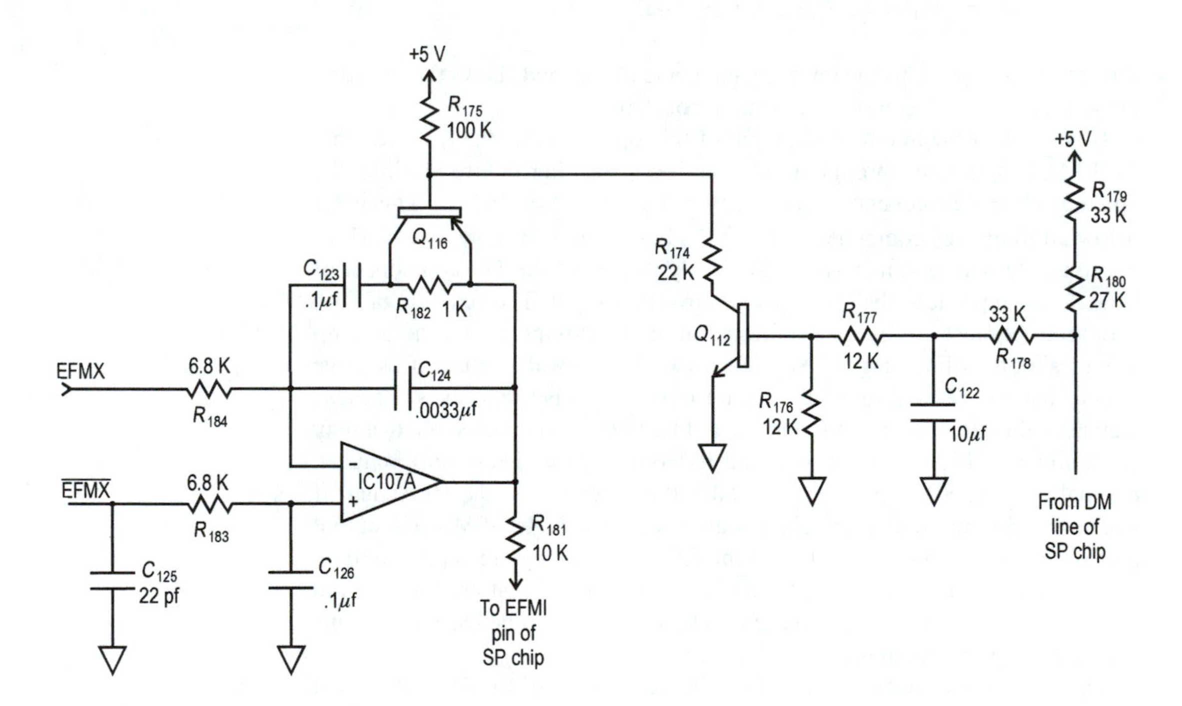

Fig. 9.35: Schematic diagram of the dynamic slicing-level circuitry used to recover the PCM data from the EFM data read off the CD in the Yamaha CDX-510U CD player.

3 and 11 channel bit times. This accounts for the multiplicity of traces. The rounding is eliminated by clipping at a dynamically selected level. The EFM circuitry is shown in Fig. 9.35. The voltage on the DM line will be a DC value between 0 and 5 V (Exercise 9.7).

Exercise 9.15

(a) Find the Thevenin voltage at the base of Q_{112} as a function of the voltage on the DM line.

The service information indicates that the DM line should be at 0 V in the forward direction, 2.5 V in the stop condition, and 5 V for reverse rotation.

(b) The service schematic gives measured voltages at most points in the circuit, but it does not specify the conditions under which the voltages were measured. Nonetheless, for Q_{116}, it gives $V_B = 4.6$ V, $V_C = V_E = 2.3$ V, and DM = 2.3 V. Deduce the conditions under which Q_{116} will short across R_{182}.

We thus conclude that R_{182} will be in circuit during normal play. This tells us that the feedback circuit of the op amp will give a two-pole one-zero low-pass transfer function analogous to that of the feed motor passive low-pass filter of Exercise 9.13. The response relative to the EFMX input shows a pole at zero, a zero at $f_z = 1/(2\pi C_{123} \cdot R_{182}) = 1{,}590$ Hz, and another pole at $f_p = 1/(2\pi C_S \cdot R_{182})$, where C_S is the series combination of C_{123} and C_{124}. This latter pole evaluates out numerically to 49.7 kHz. This is still well below even the lowest frequency component of the HF signal. The pole at zero tells us that, at least for DC, the op amp serves as integrator. Since EFMX is applied to the inverting input, the integrator will be inverting. The step response to the DC component of this signal is closely approximated by

$$v_o(t) = -V_i(R_{182}/R_{184} + t/C_{123}R_{184}). \tag{9.1}$$

The first term represents the inverting gain, and the second shows the inverting integration. $C_{123} \cdot R_{184}$ has a .68 ms time constant.

If we look at the noninverting input of this op amp (Fig. 9.39) we see that the $\overline{\text{EFMX}}$ input goes through a one-pole low-pass filter before reaching the op amp. The pole frequency is $f_p = 1/(2\pi R_{183} \cdot C_{126}) = 234$ Hz. This is far below all frequency components of $\overline{\text{EFMX}}$, besides the DC component. Thus, we conclude that this filter is provided only to extract the DC component of $\overline{\text{EFMX}}$ and apply it to the op amp's noninverting input. The function of C_{125} is obscure because we have no information on the driving circuit the SP chip provides for the $\overline{\text{EFMX}}$ signal. Not only do we not know the nature of the drive circuit, but we know little of the nature of the signals themselves. Because they are called EFMX (EFM External) and $\overline{\text{EFMX}}$, we may presume that they are nominally a logic-level signal and its complement. These are produced internally to the SP chip by a comparator that operates on the HF signal. If whatever reference voltage the comparator uses is too high, EFMX will output pulses that are too narrow relative to the EFM signal they are supposed to be reproducing. At the same time, $\overline{\text{EFMX}}$ will have pulses that are high for too long. Over the long term, these pulse trains should have equal DC components that lie halfway between logic 1 and logic 0.

Thus, when R_{183} and C_{126} extract the DC component of $\overline{\text{EFMX}}$ and apply it to the op amp's noninverting input, it is germane to ask what effect this has on the op amp output. The analysis is cumbersome, but to a good approximation it is found that the op amp's step response to this voltage is given by

$$v_o(t) = V[1 + (R_{182}/R_{184}) \cdot (1 - e^{-t/C_{124}R_{182}}) + t/C_{123}R_{184}],$$

where V is the amplitude of the voltage step. Since $R_{182}/R_{184} \approx .15$, the middle term is not the major one. Its time constant is 3.3 μsec. When it is gone, we have

$$v_o(t) = V[(R_{184} + R_{182})/R_{184} + t/C_{123}R_{184}],$$

where we have chosen this form to show that the first term represents the noninverting gain of the op amp, and the second term represents noninverting integration with a slow time constant ($C_{123}R_{184} = .68$ ms) relative to the HF signal. In fact, comparison of Eqs. (9.1) and (9.2) shows that the integration terms are identical except for the sign. In our scenario, we imagined that EFMX pulses were too narrow (due to a faulty comparator reference) and that $\overline{\text{EFMX}}$ pulses were thus too wide. This would result in the DC component of EFMX being too small and that of $\overline{\text{EFMX}}$ being too large. The lowered input to the inverting integrator will tend to make the op amp output ramp down slowly, and the excessive input to the noninverting integrator will tend to make the output ramp up rapidly. The net effect of these two tendencies will be to give an output voltage that varies as

$$v_o(t) = (V - V_i)t/.68\,\text{ms},$$

where, we recall, V is the average value of $\overline{\text{EFMX}}$ and V_i is the average value of EFMX.

This ramp is applied back to the SP chip through R_{181}. There it is summed with the incoming HF signal to shift the comparator reference level in the direction that will cause the amplitude of this ramp to approach zero. When it does, (9.1) and (9.2) show that the output of IC_{107A} will be just the average of EFMX (= the average of $\overline{\text{EFMX}}$), which is the desired comparator reference voltage.

9.6.5 Signal processing circuitry

Throughout the discussion of this unit we have made frequent reference to the signal processing (SP) chip. We now have a look at this chip and its associated circuitry. It is shown in Fig. 9.36, where labels have been provided to show the circuit functions of each chip.

We have previously observed that the CD player must reconstruct the clock from the data on the disc. This constraint, in fact, is what sets an upper limit of 11 times the channel bit time as the maximum duration of a pit or a flat. The circuit to accomplish this reconstruction is shown in the upper left-hand portion of Fig. 9.36. The amplifier portion of the oscillator is internal to the SP chip (its output is on pin VCOX and its input on pin VCOI). The op amp, IC_{107B}, does nothing more than provide the DC bias for the dual varicap, D_{103}. Its associated RC network provides the low-pass function that extracts the DC value of PCO. It may not be obvious what PCO is. This can be determined by sketching the complete PLL system that we know is used to recover the clock. The block diagram is shown in Fig. 9.37. The combination of the amplifier and its feedback network, F, is the VCO. The oscillation frequency changes as the DC voltage from the LPF changes the varicap capacitance and thus the resonant frequency. The circuitry above the dashed line is part of the SP chip. Notice that there are three connections to "off-chip" circuitry. The output of the amplifier is the VCOX pin of the SP chip. The input to the amplifier is the VCOI pin. This leaves the output of the phase comparator as the PCO pin. Thus, the mnemonic here is Phase Comparator Output.

We also observe from Fig. 9.36 that the LPF (IC_{107B} and associated circuitry) includes a transistor, Q_{118}, that operates as a switch. In the play mode, it is

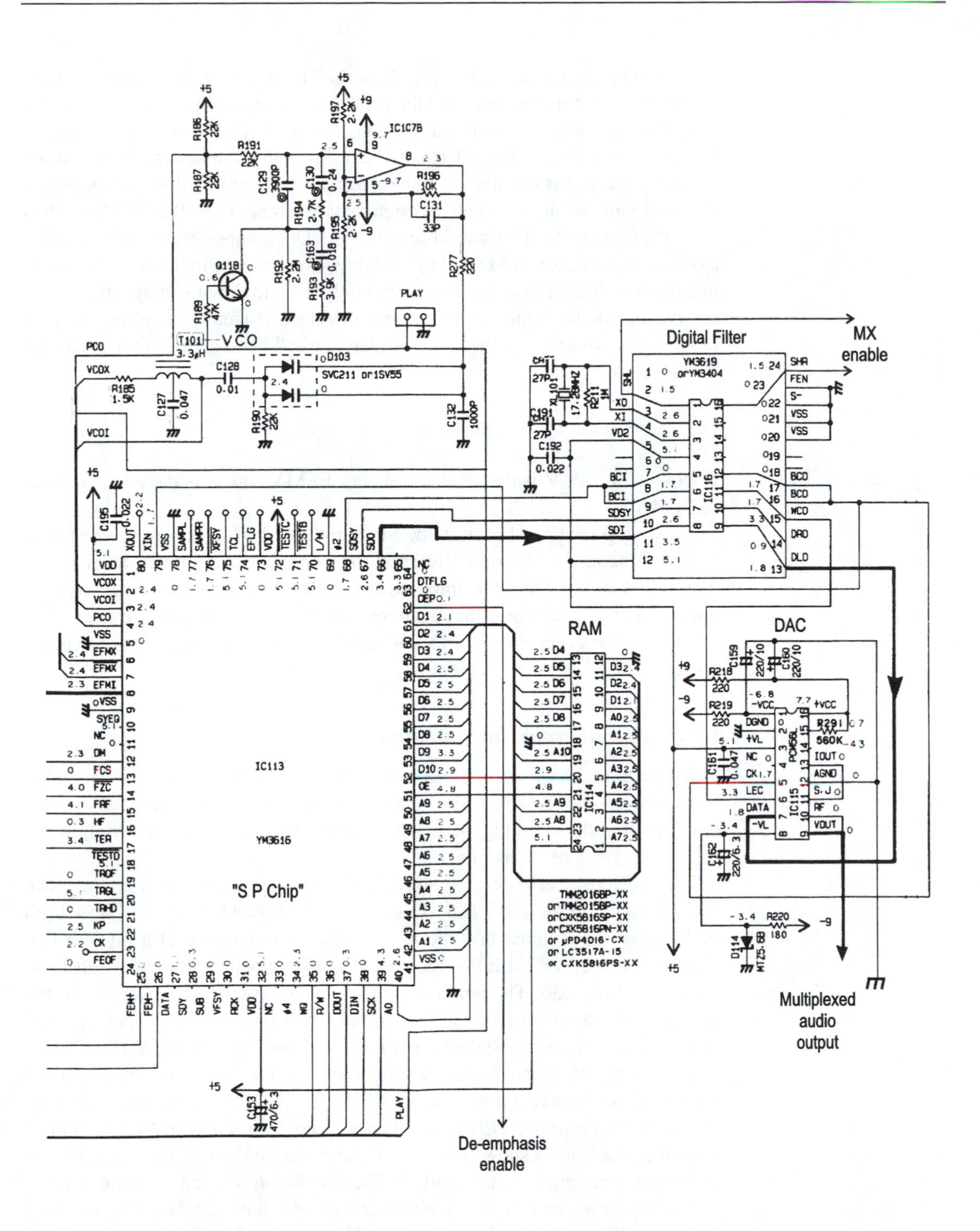

Fig. 9.36: Schematic diagram of the DSP chip in the Yamaha CDX-510U CD player and its associated circuitry.

saturated, and the LPF has a rolloff frequency of about 13 Hz. When in search mode, the bandwidth is broadened to about 200 Hz.

Moving on to the SP chip itself, we can get some appreciation of its diversity of functions by referring back to Fig. 9.26. This block diagram of the entire circuit incorporates a block diagram of the "innards" of the SP chip. In general terms, we see the following functions:

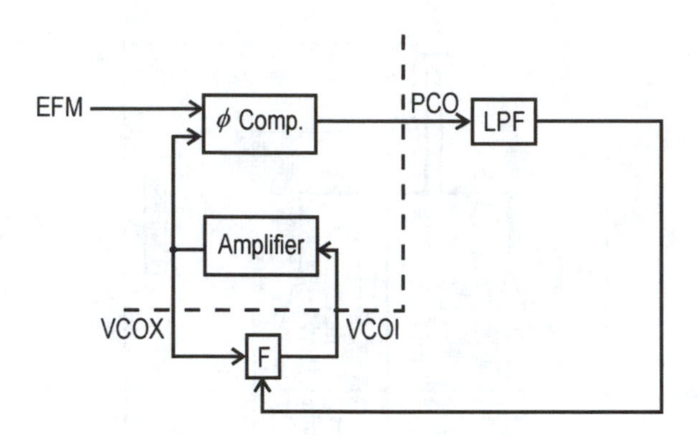

Fig. 9.37: Block diagram of the circuitry used to reconstruct the PCM clock from the EFM data in the Yamaha CDX-510U CD player.

- recovery of the EFM signal,
- recovery of the channel bit clock,
- EFM decoding,
- subcode recovery,
- error detection, correction, and concealment, and
- control sections of all servos (tracking, focus, feed rate, and disc rotation velocity).

Due to the paucity of information about this chip, we can say little more about it than what we have already said in connection with other circuitry. The RAM chip to the right of the SP chip is required for Reed–Solomon decoding and error correction. The recovered sampled audio signal appears in serial form at pin 66 of the SP chip (SDO ≡ serial data output). From there it goes to a digital filter. Besides the signal input to this filter chip, the SP chip provides a bit clock of 2.16 MHz. In this unit, sampling is done in the digital filter chip. This chip has its own crystal oscillator at 17.2872 MHz. It is presumably phase locked to 8× the input bit clock by means not shown. The chip accomplishes 4× oversampling and outputs the sampled signal at a bit rate of 8.6436 MHz, which is 4× the input bit clock rate.

Besides the sampled output, the digital filter chip also provides the bit clock output and a word clock output to the DAC, as well as deglitching signals for the S&H stage. These signals, as we shall soon see, also direct the DAC output to the *R* or *L* channel amplifiers as appropriate. It provides the logic-selectable capability to accommodate designs with either one or two DACs. This circuit uses one DAC.

The DAC is a 16-bit current-output type with an on-board op amp to convert to voltage output where required (as in this unit). It also incorporates serial/parallel conversion and a 16-bit latch. The output of this chip will be multiplexed audio alternating between the *L* and *R* channels.

9.6.6 Audio demultiplexing and amplification

Figure 9.38 shows the complete schematic of the analog audio processing block. In it we can see that the sample-and-hold block consists of two sections of a three-pole double-throw analog switch and a pair of op amps. Figure 9.39 shows one channel of this circuitry. It is clear that when the analog switch

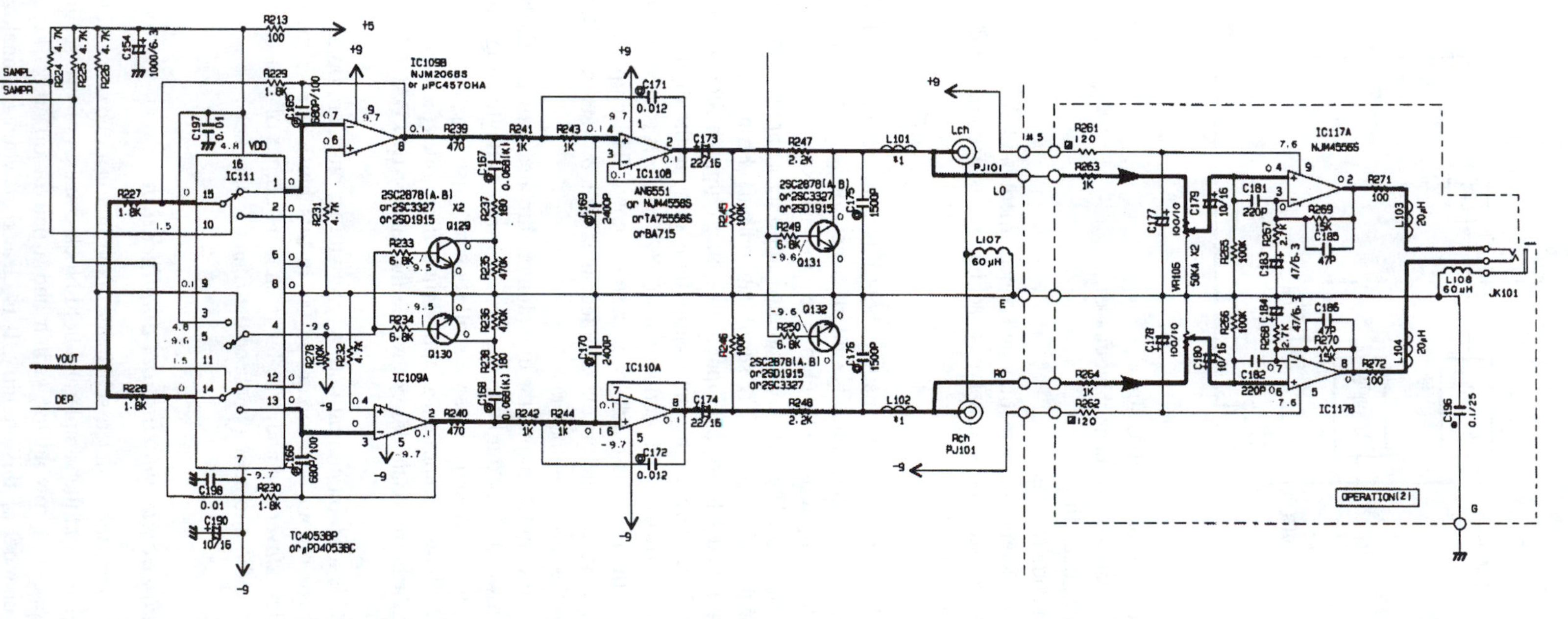

Fig. 9.38: Schematic diagram of the audio circuitry in the Yamaha CDX-510U CD player.

Fig. 9.39: Schematic diagram (extracted from Fig. 9.38) showing the use of an operational amplifier and an analog switch to provide a sample-and-hold circuit function.

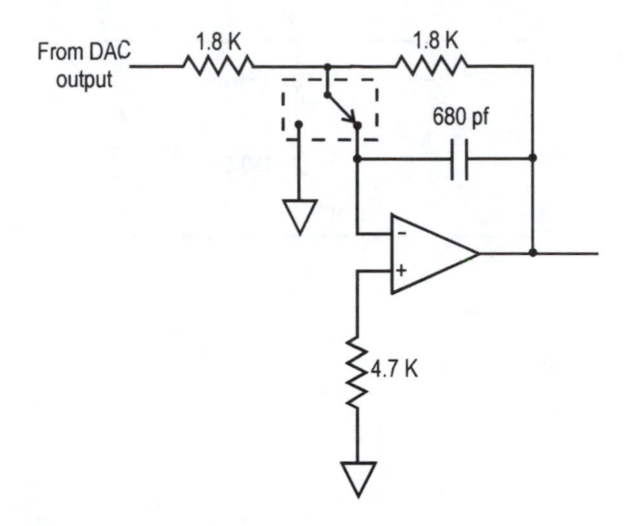

is in the position shown, the DAC output will be amplified with unity gain. The 680 pF capacitor in the feedback loop will limit the acquisition time, but insignificantly. The feedback loop time constant is 1.22 μsec. By comparison, we recall that there are 88,200 samples/second taken in the encoding process and thus reproduced in the decoding process. This requires that a D/A conversion be accomplished every $1/88{,}200 = 11.3\ \mu$sec, suggesting that there are over nine time constants available for acquisition. This is not quite the case, however, as the switching signals that the digital filter supplies to toggle the analog switches are somewhat offset in time in order to eliminate glitches. Even so, we would expect seven or eight time constants to be available for acquisition, and this should suffice for accuracy to within .1% or better.

When the D/A has completed its next conversion and it has settled, the analog switch in Fig. 9.38 will disconnect the D/A output from the summing mode and the op amp output voltage will be held by the capacitor.

Exercise 9.16

(a) The DAC operates from ± 9 V supplies. What is the maximum possible change in analog output voltage for a change in the LSB of the digital input?

(b) Suppose we require that during the "Hold" time the op amp output shift due to input bias current of the op amp be less than half the value of the LSB. Determine a maximum spec for input bias current of the op amp.

At the same time that the L channel is put into the hold mode, the R channel is put into the sample mode and the DAC output is applied to it.

Referring back to the audio schematic (Fig. 9.38), after leaving the S&H stages, the L and R signals each pass through a 470 Ω resistor to a series-connected string of three elements. The lower element is shunted by a transistor that is obviously intended to function as a switch. When the transistor is saturated, the network is as shown in Fig. 9.40. This is just a lead-lag network. It will have a pole at

$$\omega_\text{p} = 1/(470 + 180) \times .068\ \mu\text{F} ==> f_\text{p} = 3.6\ \text{kHz}$$

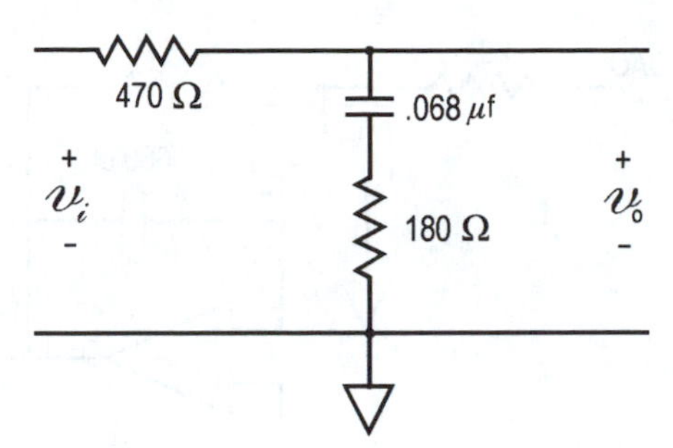

Fig. 9.40: Schematic diagram of the audio de-emphasis network used in the Yamaha CDX-510U CD player.

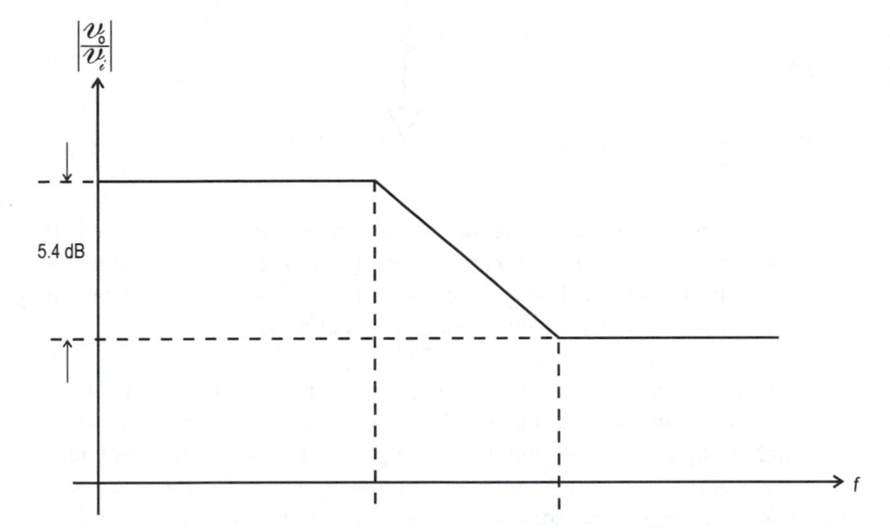

Fig. 9.41: Bode amplitude plot of the network shown in Fig. 9.40.

and a zero at

$$\omega_z = 1/180 \times .068\,\mu\text{F} ==> f_z = 13\,\text{kHz}.$$

Thus the network will have the amplitude response shown in Fig. 9.41. As shown on the block diagram of Fig. 9.26, this is a de-emphasis network. As we have previously indicated, the use of pre-emphasis on a CD is optional with the disc manufacturer. If it is used, this is reflected in the subcode. The signal-processing chip extracts the subcode and from it produces a control signal that activates the third gang of the analog switch. This section of the circuitry is shown in Fig. 9.42.

The analog switch control line is pulled up by the 4.7 k resistor unless the SPC pulls it low. When it is high, the 5 V supply is applied to the transistor base through the 6.8 k resistor. This saturates the transistor, shorting out the 470 k resistor, and thus switching in the de-emphasis circuit shown in Fig. 9.40. When the control line is low, the switch is open, and the −9 V supply is applied to the transistor bases through the 100 k resistor. This puts the transistors (one for each audio channel) into cutoff and thus leaves the 470 k resistors in circuit. This in turn moves both the pole and zero of the network down to 3 or 4 Hz, which is below the CD audio range. Thus, the network accomplishes no de-emphasis and is effectively removed from circuit.

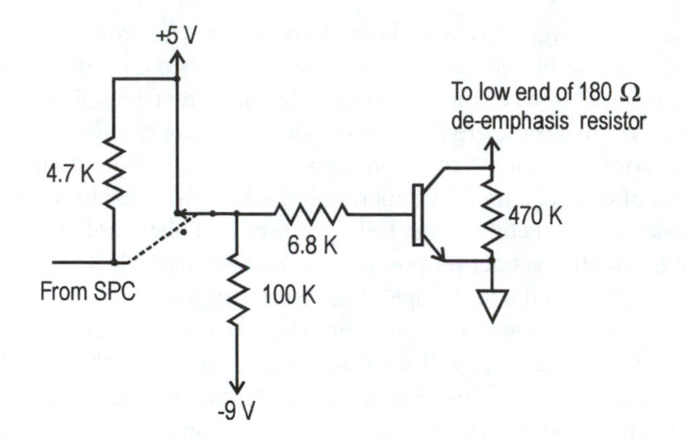

Fig. 9.42: Schematic diagram of the circuit used to switch the de-emphasis network of Fig. 9.41 in or out of the circuit on command from the SP chip.

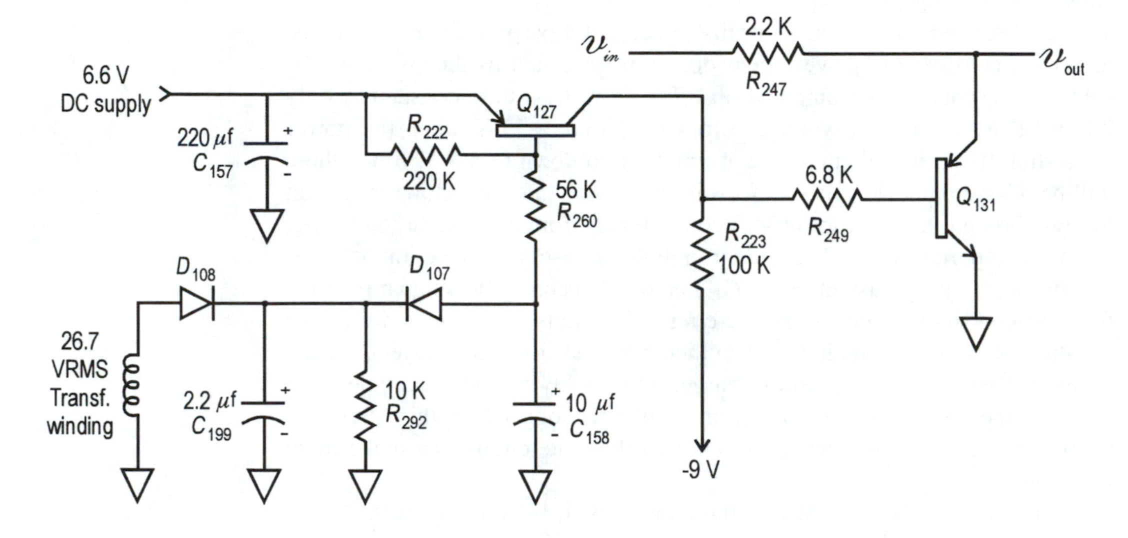

Fig. 9.43: Schematic diagram of the circuitry used to mute the audio of the Yamaha CDX-510U CD player during power-on and power-off transients.

The next circuit blocks in Fig. 9.38 are the low-pass filters. Although these are based on unity gain buffers, they are second-order filters.

Exercise 9.17

(a) Find the corner frequency of these low-pass filters.

(b) Find the phase error at 20 kHz.

The block diagram (Fig. 9.26) identifies the next circuit blocks as being for muting. This is no surprise, given the transistors' placement in the schematic of Fig. 9.38. It is clear that if the transistors turn on, the signal transmission will be zero. There is, however, something of a mystery as to how they ever could turn on. Figure 9.43 shows the relevant circuitry for one channel. Except for the 6.8 k resistor and Q_{131}, none of this circuitry is found on Fig. 9.38. The −9 V pulldown is exactly analogous in form and function to that found in Fig. 9.42. This leaves it up to Q_{127} to activate the mute if it ever will be activated. It should be obvious that in steady state, this can never happen. The voltage across R_{292} will ripple between roughly +20 and +35 volts. The voltage on the anode of D_{107} can never reach this since the only voltage applied to the transistor is

6.6 V, and even though C_{158} will charge to nearly this full voltage, D_{107} would be strongly reverse biased and thus remain open. When C_{158} is fully charged, the base current falls to zero and thus the collector current does also. This means that the switch transistors are held off by the −9 V supply.

The purpose of the mute function is to disable the audio output on power-up. This is, of course, a transient operation. Perhaps we need to reconsider the circuit under transient conditions! Before power is applied, both capacitors are uncharged. For all practical purposes, C_{199} will be charged to +35 V the first time a positive sinusoidal peak appears across the transformer, which could even be at the instant the power is turned on. The 6.6 V supply should be on-line shortly afterward, but C_{158} will need some time to charge through the base-emitter junction and R_{260}. The charging time constant will be about .5 seconds. During the charge, base current does flow in Q_{127}, and so it does conduct and activate the muting switch Q_{131}.

This discussion, however, still fails to suggest a purpose for C_{157}. Let us, therefore, consider the power-down operation. As soon as the AC power is shut off, the transformer output drops. The R_{292}–C_{199} time constant is only 22 ms. Thus, even if the voltage across C_{199} were +35 V when the power were shut off, in just about 40 ms it would drop down to 5.4 V. Since there will be 6 V across C_{158}, this will forward bias D_{107} and start allowing current to flow through it. This current will come in part from the discharge of C_{158}, but more importantly, it will come through the base–emitter junction of Q_{127}. The discharge time constant, R_{249}–C_{157} is over 1 second, allowing ample time for transients to dissipate around the circuit. This only leaves unanswered the question of why the circuit is not vulnerable to shut-down transients in the 40 ms or less it takes C_{199} to discharge. Presumably the other supplies will maintain their voltages in the operating range for longer than this, thus preventing the generation of noise pulses until the mute circuit can make them inaudible.

If the signal makes it past the mute switches, it is available at the audio output jacks. But this unit, like many others, also provides headphone output so that it may be listened to without the system amplifier. The power amplification is provided by a stereo IC chip, IC_{117}. Although this provides mostly current gain, it also provides some voltage gain.

Exercise 9.18 Referring to the schematic in Fig. 9.38, find the midband voltage gain for each channel of the headphone amplifier.

9.6.7 Control and display

Control and display functions are represented in the lower left-hand corner of Fig. 9.26. Infrared signals from the remote are detected by a photodiode whose output is fed to an integrated receiver. A block diagram of this IC is shown in Fig. 9.44. The block labeled as ABLC is automatic bias level control. It is analogous to the discrete circuitry used for the same purpose in the IR preamp of the RCA TV discussed in Section 7.7.2. The bandpass filter will screen out light signals that do not have the pulse rate produced by the remote.

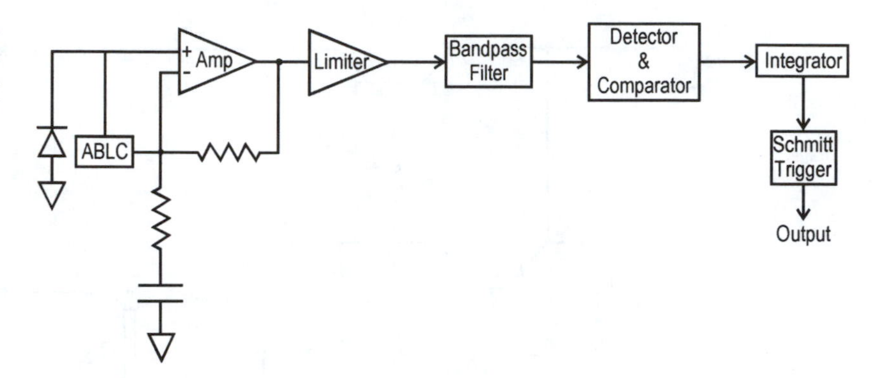

Fig. 9.44: Block diagram of the integrated IR remote-control receiver in the Yamaha CDX-510U CD player.

The output of this IC is fed to a 4-bit μP whose sole purpose is to interface the IR receiver to the 8-bit processor. It is not clear from the service information if the actual decoding of the IR pulse train takes place in the 4-bit μP or in the 8-bit one, but it is probably the former since it is more logical partitioning and reduces the demands on the 8-bit μP, which is the main system control device. Like the RCA color TV, the functions of the remote are duplicated by keys on the front panel. These keys are also input to the control μP. Other jobs of the control μP include:

- driving the chips that provide the drive for the front panel display,
- controlling the disc loading motor via a power driver IC,
- disabling the laser when the unit is open, and
- Interfacing to the SP chip, thus enabling the function selected from the front panel keyboard or the remote.

9.6.8 Power supplies

The power supplies in this unit are remarkable in two respects. The first is their quantity, and the second is that none of them uses an integrated regulator. The following DC voltage supplies are found in the player (all are derived directly from a transformer unless otherwise stated):

+32 V unregulated,
±13 V unregulated,
±9 V regulated,
+7 V regulated,
+6.2 V regulated (from +13 V),
+5 V regulated, and
±2.5 V regulated (from ±9 V).

This circuit also provides a three-transistor power-on-reset circuit, which is shown in Fig. 9.45. The first part of the reset cycle for the 4-bit μP is straightforward. As the 5 V supply rises, it is applied to the E–B junction of Q_{119} in series with $R_{198} + R_{199}$. Until $V_{\text{in}} > .5$ V, Q_{119} is cut off, and the collector floats. As V_{in} continues to rise, Q_{119} starts to conduct, and the collector potential rises

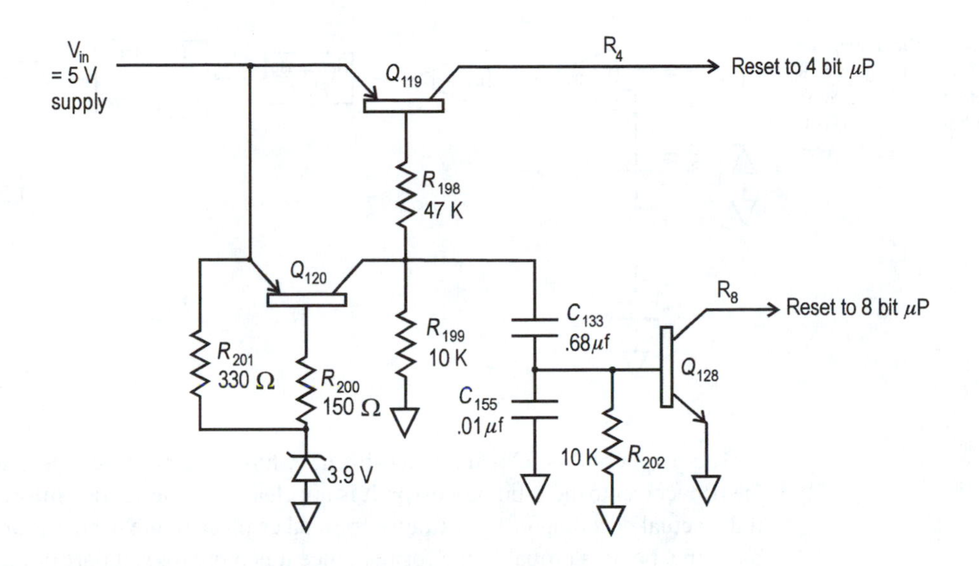

Fig. 9.45: Power-on reset circuitry for the microprocessors in the Yamaha CDX-510U CD player.

with it. When V_{in} is sufficient to fully saturate Q_{119}, the voltage on line R_4 will track the instantaneous value of V_{in}.

No base current can flow in Q_{120} until V_{in} is about 4.5 V (= 3.9 V Zener + .6 V emitter–base bias). Since the base resistor, R_{200}, is so small, Q_{120} no sooner starts to conduct than it saturates. When it does, its collector voltage rises abruptly to V_{in}. But this effectively removes base drive from Q_{119} since the potential difference across its E–B junction and R_{198} becomes zero. This situation remains unchanged even when V_{in} reaches 5 V, where it stops changing. The reset of the 4-bit μP occurs sometime before R_4 reaches its positive peak of about 4.5 V. When the R_4 line drops down to near zero, the μP is on-line and ready to work.

When Q_{120} saturates and its collector voltage jumps, this step is coupled through C_{133} to the base of Q_{128}, temporarily driving it into saturation, and pulling its collector to ground. This provides the reset for the 8-bit μP, which is active-low, as opposed to the active-high reset required by the 4-bit μP. The pulling up of the base potential of Q_{128} is short-lived, because C_{133} will charge rapidly to the new collector voltage of Q_{120} through the B–E junction of Q_{128}. When the charging is completed, Q_{128} will again be cut off and R_8 will be high, and the reset will have been accomplished.

References

Nakajima, H., Doi, T., Fukuda, A. and Iga, A. 1983. *Digital Audio Technology*, p. 134. Blue Ridge Summit, PA: TAB Books.

Nussbaum, A. and Phillips, R. 1976. *Contemporary Optics for Scientists and Engineers*, pp. 354–355. Englewood Cliffs, NJ: Prentice-Hall.

10

Telephones

Until the FCC removed the ban on connecting non-AT&T equipment to the phone lines, few people would have thought of the telephone as a consumer electronic item. Now, however, there can be little doubt that it is one. Not only did this decision unchain the creative genius of free enterprise, but it coincided in time with the emerging availability of inexpensive MSI and LSI chips. The result has been a proliferation of phones and features undreamed of just twenty years ago. The Bell System's development of Touch-Tone® dialing led to the development of the *DTMF* (dual-tone, multi-frequency) encoder and, later, decoder chips. Sharply declining prices on digital memory chips led to their incorporation in many phones for phone number storage and rapid dialing. Phones with memory may be broken down into three groups:

(1) Those that can store just a single number – the last one dialed. If that number is busy, it can be redialed with a single button.
(2) Those that store two to ten numbers. These numbers can be dialed by pushing just one of the digit buttons.
(3) Those that can store more than ten numbers. These numbers can be dialed only by pressing two buttons in sequence to form a two-decimal-digit code.

Anyone who has ever listened to a phone conversation where the speaker tried to make a comment while covering the receiver with his hand knows that this does a very poor job of masking the sound. It is therefore surprising that a concept as simple as the privacy or *mute switch* did not come into wide use until the lines were made available to all equipment that met phone company specifications.

Speakerphones have existed for some time, but they tended to be prone to acoustic feedback. Current versions, however, electronically sense where the input is coming from and open the channel only in the direction it needs to go. When the speaker changes, signal flow in the opposite direction is enabled.

Cordless phones have benefited from inexpensive transmitter and receiver ICs and have declined in price to where they are now a popular option. They are almost too popular, in fact. In some wealthy neighborhoods, there have been so many cordless phones that two related problems have become evident. The first is interference between phones, and the second is eavesdropping. Phones with built-in scrambling/descrambling circuitry have already appeared, and these are almost sure to proliferate and to become more sophisticated.

Such features as a hold button, line switching, and one-touch redialing of the last number dialed are now commonplace. Phones are also available to keep redialing a busy number automatically until they get through.

Phones that allow dialing from a large repertory of numbers by the entry of alphabetic data (person's name) are now available also. They will complete the call as soon as they get enough characters to unambiguously identify the individual being dialed. One of the most remarkable phones currently available is one that incorporates voice-recognition circuitry and can be trained to dial (on verbal command) a limited number of "callees."

Integrated telephones and phone-answering machines are now ubiquitous.

Although the picture phone was conceived of, researched, and built by AT&T decades ago, the technology was not mature enough to make it a viable consumer item. Besides this, even with the video signal encoded for bandwidth reduction, the bandwidth requirement far exceeded that required for "sound-only" transmission. Now several manufacturers have brought to the consumer market a picture phone costing less than $400. Furthermore, the signal can be transmitted in the same bandwidth as a voice channel. This is accomplished by foregoing real-time operation. The user instead strikes a pose he likes and presses a button to store it in the sending phone. The phone then converts this to an AM data train that is transmitted over a period of 5–10 seconds. When it has been completely received, it is displayed. Perhaps the most exciting development in this arena is that digital signal processing continues to raise the frequency range that can be recovered after transmission through a low-bandwidth channel. This trend may never give us real-time video over a phone line, but we will no doubt get closer than we are now.

The near future is likely to see the incorporation of dedicated microprocessors into phones with such capabilities as programmable speech synthesis to allow the phone itself to automatically initiate calls and deliver a message to lists of numbers and even more sophisticated functions.

10.1 The central office connection

Each residential customer or *subscriber* phone is connected to the central office by two leads called T and R. The T or *tip* lead is green and the R or *ring* lead is red. Remember: ring = red. These names are derived from the connector that was originally used in manual switchboards. These two lines are adequate to take care of ringing, detection of a phone being off hook, dialing response, and carrying on of a normal conversation. The central office applies a nominal 48 volts DC to each phone on a continuous basis. This voltage is of such a polarity as to make the T line more positive than the R line, and it is derived from batteries at the central office. The use of batteries assures the continued functioning of the phones in times of power outage.

It can be readily appreciated that the current demands on the batteries must be minimized in every way practical. A key aspect of this effort is that the subscriber phone draws no current when it is on hook. The means by which this is accomplished are shown in Fig. 10.1. The bell draws no current from the batteries because of the blocking capacitor between the coils. The remainder of the subscriber phone circuit is isolated by the hook switch, which is open while it is on hook.

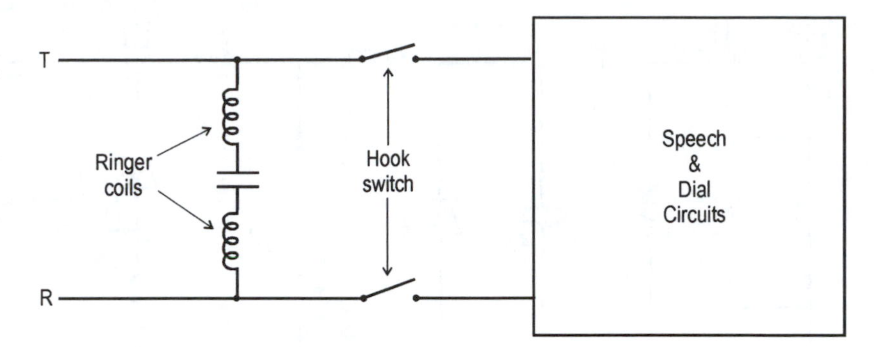

Fig. 10.1: The power-conservation scheme in a subscriber phone. The ringer draws no power from the DC line because of the blocking capacitor and the speech and dial circuits draw nothing until the phone is lifted from the hook.

The use of the capacitor in series with the ringing coils dictates the use of AC to activate them. The AC voltage is superimposed on top of the DC and is typically about 80–90 VRMS and between 16 and 60 Hz in frequency.

When the subscriber lifts the phone in response to a ring, the hook switch closes, allowing DC to flow into the phone from the batteries in the central office. This flow of DC is sensed there and used to remove the AC ringing voltage from the answered phone. The AC ringing voltage is also generated in the central office from the battery power there.

The speech and dial circuits have varied from one phone model to another. Rather than "beginning at the beginning," as we have with other consumer electronic equipment, we shall begin with the 500D telephone, some of which are still in use today. This is because we will not learn any more fundamental principles of operating by going back more than the years that the 500D has been with us.

10.2 The 500-series telephone

The original telephone in this series was the 500B, a rotary-dial desk model phone. It was superseded in 1953 by the 500D, which was a slightly modified version that not only offered improved performance but reduced cost as well.

Not only are schematics for Bell Telephones difficult to find, but when you do find them, the schematic notation is often nonstandard, particularly as regards their inclusion of mechanical structure information on the schematic. Figure 10.2 shows the 500D circuit schematically in normal notation as it is connected during a conversation.

There are two sets of switching contacts that change the schematic during other phone operations. The first set is associated with the hook switch, which senses whether or not the phone is hung up. The second is associated with the rotary dial. Not only is this dial designed for generating the dial pulses, but it has a muting switch built in which functions when dialing is in process. The notations in square boxes are terminal identifications on what the Bell System calls the 425B network. It incorporates the transformer, all resistors and capacitors, as well as varistors V_1 and V_2. It is installed in the phone as a single encased unit with screw terminals for connecting the line, the transducers, the ringer, and the switches.

The lines marked R and T are the two lines that are brought in for a normal residential telephone connection. The inductances labeled L_1, L_2, L_3, and L_4 are all actually windings on one transformer that is sometimes loosely referred

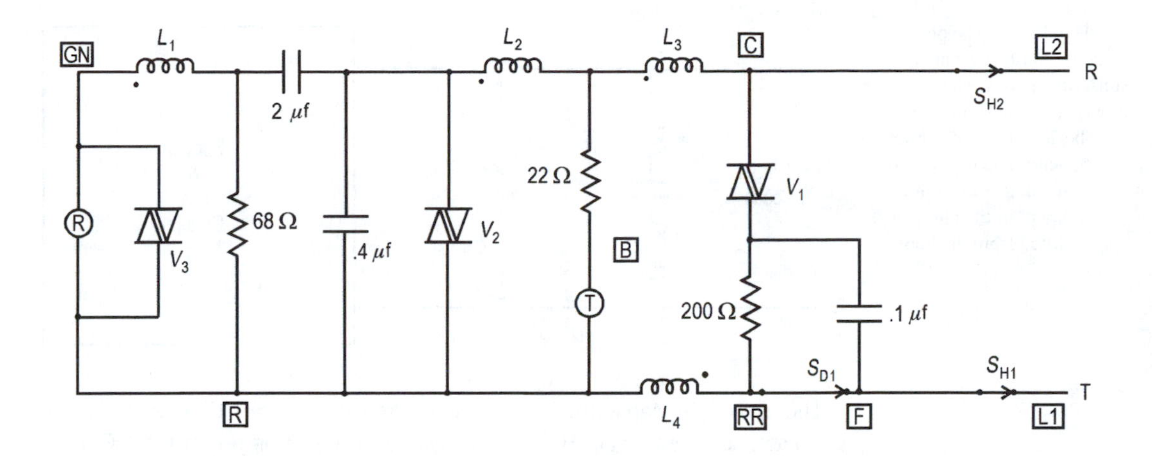

Fig. 10.2: Schematic diagram of a 500D telephone (in standard schematic notation) as it is connected during a conversation. The letters in rectangular blocks refer to terminal identifications on the 425B network.

to as "the hybrid." Strictly speaking, the hybrid is another type of transformer that converts between a 4-wire telephone connection and a 2-wire one. In our schematic, we have assigned individual inductance names to each of the windings to facilitate their identification for purposes of analysis.

Before we commence analysis we will make one change in the schematic. That will be to move L_4 from where it is found in Fig. 10.2 and place it in series with L_3. It is clear from this schematic that the same current flows through L_3 and L_4. In fact, the original Bell Labs article (Bennett, 1953) on this phone identifies both L_3 and L_4 as "1/2-primary." In addition to this change, we also show the switching contacts and part of the functional partitioning in Fig. 10.3.

The circle marked "T" is a carbon microphone transmitter, which has an effective source resistance of about 160 Ω.

The circle marked "R" is the receiver unit, which may be modeled as a 15 mH inductor and a 35 Ω resistor in series.

V_1, V_2, and V_3 are silicon carbide varistors, which are nonlinear resistors. By virtue of their nonlinearity they can provide an incremental AC resistance that is less than the DC resistance, as shown in Fig. 10.4. They are used to stabilize signal amplitudes to some degree. When amplitudes are small, the varistors have high resistance and give minimal loading, but when signals are large, they load the source and so reduce signal amplitude.

10.2.1 The dialing operation

In Fig. 10.3, S_H is the hook switch. As indicated, it has three sets of contacts. The two sets in series with T and R are open when the phone is on the hook. The third set normally shunts the receiver to assure its silence when the phone is on the hook. When the handset is lifted, this short is removed, allowing the receiver to become "live." At the same time, the other two sections of this switch close, allowing DC to flow from the batteries in the central office. Figure 10.3 shows that there are three paths through which this DC current can flow. The shortest path is through V_1 and the 200 Ω resistor. The next is through the transmitter, the 22 Ω resistance, and $L_3 + L_4$. The last is through V_2, L_2, and $L_3 + L_4$.

The flow through the subscriber set will be between 20 and 120 mA, depending primarily on the line length. The off-hook condition is thereby sensed

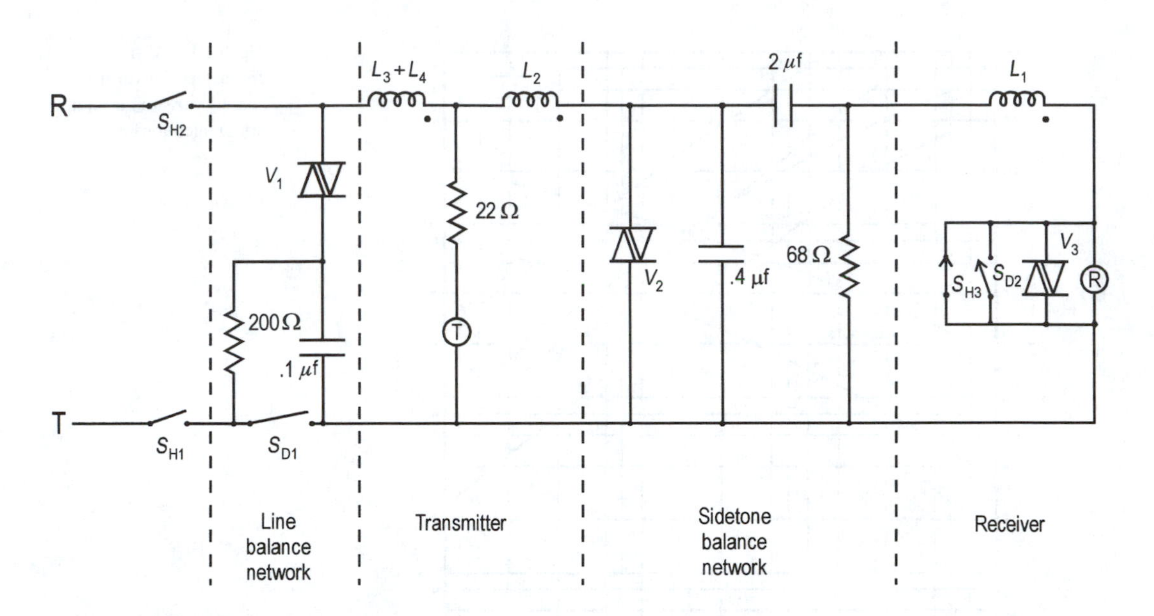

Fig. 10.3: The circuit of Fig. 10.2 showing the partitioning of the circuitry among the various phone functions. More of the switching is shown in this figure, and L_4 has been moved around to highlight the fact that it is actually in series with L_3.

at the central office, which responds by sending a dial tone. The user then dials this type of phone by manually rotating the dial by an amount corresponding to the digit desired. In so doing, a spring is wound in the dial mechanism. Neither of the dial switches (S_D in Fig. 10.3) is affected up to this point. When the dial is released, S_{D2} shorts out the receiver to mute it. It will remain shorted until the dial returns to its rest position. As it rotates back toward that position, it is under the control of a governor, which gives it a constant rotational speed. Switch S_{D1} repeatedly opens and closes during this rotation. As it does so, it interrupts the current in the loop linking the central office with the subscriber phone. Each break corresponds to an increase of one in the digit being dialed. In other words, the user who wishes to dial a "4," inserts their finger into the hole in the rotary dial that is so marked, and pulls it toward the finger stop. This action winds the spring. When the finger is released after reaching the stop, the dial rotates back towards its equilibrium position, and in so doing causes S_{D1} to open four times. These interruptions are counted at the central station to determine the digit dialed. The dialing is repeated until enough digits have been dialed to uniquely determine a phone number.

The dial is so designed that during its return rotation, S_{D1} is open 60% of the time and closed 40% of the time. This is called a 60% break ratio. The total period of a dial pulse is 100 ms.

Exercise 10.1 Referring to Fig. 10.3, we see that as soon as the handset is lifted from the switch hook, the current through $L_3 + L_4$ will begin to rise. We may no doubt safely assume that it will reach a steady-state value before dialing begins. This steady-state current will, for the most part, be flowing through the low impedance transmitter and the 22 Ω resistor, rather than through L_2 and V_2.

When S_{D1} first breaks, this relatively large current will be diverted through the 200 Ω resistor and the .1 μF capacitor that are shunted across it. This combination should hopefully produce a decay in the inductor current that will be rapid compared to the period of the dial pulse.

(a) Ignore V_1 (equivalent to assuming long lines) and find the range of values of $L_3 + L_4$ compatible with the achievement of this condition.

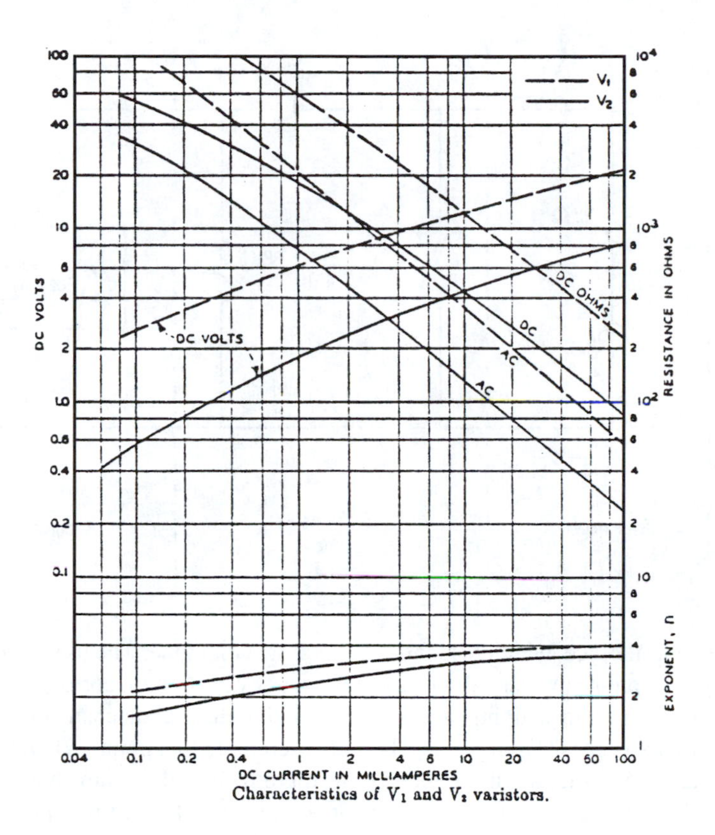

Fig. 10.4: V–I characteristics for the V_1 and V_2 varistors used in the 500D telephone.

(b) Assuming the transmitter resistance is 160 Ω find the value of $L_3 + L_4$ that would give critical damping.

The phone company equipment decodes the dial pulses to determine the number being called. It then sends a ringing signal to that phone. When it is picked up, the equipment again senses the current it draws and disconnects the ringing voltage. Finally, it connects the called phone to the calling phone.

Exercise 10.2 How could a telephone answering machine, which does not lift the phone from the hook, "answer" the phone?

10.2.2 The transmit operation

As we refer once more to Fig. 10.3, we observe first that during conversation, S_D will be as shown, but all three sections of S_H will be connected opposite to that shown. Thus the receiver is not muted and the T and R lines are connected.

We will next observe qualitatively that speaking into the transmitter will not only send a signal to the left (out onto the phone lines) but also to the right (toward the receiver). This portion of the transmitted signal fed back to the receiver is called *sidetone.* It is an important concept in telephone communication, because we gauge how loud to talk by what we hear in the receiver when we talk. If the sidetone is too loud, we quiet our voices, thus transmitting too weak a signal. If it is too soft, we tend to speak so loudly as to make the listener uncomfortable and the sound distorted. The reader's own experience with the

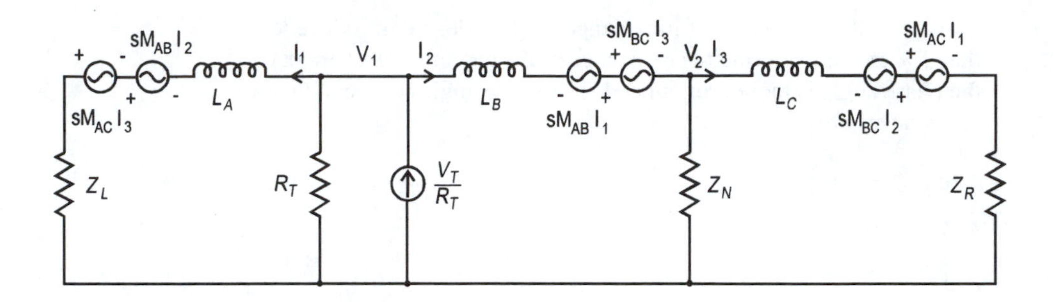

Fig. 10.5: The equivalent circuit for Fig. 10.3. This is the form of the circuit that will be used for analysis. The subscripts of the vertically oriented components correspond to the partitions in Fig. 10.3.

phone should readily demonstrate that the latter possibility is less likely to cause a degradation of intelligibility than the former. This translates into the design criterion of trying to make the sidetone quite small.

10.2.2.1 Sidetone reduction

We next proceed to show that for a given line impedance, it is at least theoretically possible to choose a sidetone balance network that will reduce the sidetone to zero. In external terms, this means that when transmitting there is no signal fed to the receiver at all. To perform this analysis, we will use the circuit of Fig. 10.5, which is derived from Fig. 10.3.

The following observations will aid in understanding this circuit.

(1) $\mathbf{Z}_{\mathrm{L}}$ is the parallel impedance of the line balance network and the phone lines.
(2) $L_{\mathrm{A}} \equiv L_3 + L_4 : L_{\mathrm{B}} \equiv L_2 : L_{\mathrm{C}} \equiv L_1$. This notation is used in many of the early phone company technical articles.
(3) The transmitter and its series 22 Ω resistance can be modeled as a larger resistance (22 Ω + 160 Ω source resistance of T = 182 Ω $\equiv R_{\mathrm{T}}$) in series with an ideal voltage source, V_{T}. In Fig. 10.5, these elements have been replaced with a Norton equivalent.
(4) $\mathbf{Z}_{\mathrm{N}}$ is the complex impedance of the sidetone balance network.
(5) $\mathbf{Z}_{\mathrm{R}}$ is the impedance of the receiver element.
(6) The inductors, L_{A}, L_{B}, and L_{C} should be represented properly as complex impedances, to take account of their winding resistances. However, measurements have shown that this is unnecessary for most purposes. To the nearest whole millihenry or ohm, the measured inductances and resistances in the 425B network have been found to be:

$$L_{\mathrm{A}} = 114 \text{ mH}, \quad R_{\mathrm{A}} = 30\ \Omega,$$
$$L_{\mathrm{B}} = 25 \text{ mH}, \quad R_{\mathrm{B}} = 15\ \Omega,$$
$$L_{\mathrm{C}} = 30 \text{ mH}, \quad R_{\mathrm{C}} = 14\ \Omega.$$

Since the voice band of frequencies extends between 300 Hz and 3 kHz, the smallest possible inductor reactances will be found by taking $2\pi \times 300$ Hz $\times$ Inductance:

$$X_{\mathrm{AMIN}} = 215\ \Omega, \quad |\mathbf{Z}_{\mathrm{AMIN}}| = \sqrt{215^2 + 30^2} = 217\ \Omega,$$
$$X_{\mathrm{BMIN}} = 47\ \Omega, \quad |\mathbf{Z}_{\mathrm{BMIN}}| = \sqrt{47^2 + 15^2} = 49\ \Omega,$$
$$X_{\mathrm{CMIN}} = 57\ \Omega, \quad |\mathbf{Z}_{\mathrm{CMIN}}| = \sqrt{57^2 + 14^2} = 58\ \Omega.$$

Thus from the standpoint of impedance magnitude, we introduce less than 5% error in ignoring the resistance of the inductors. The effect on the phase angle is more pronounced. The phase angles of the inductor impedances are

$$\phi_A = 82°,$$
$$\phi_B = 72°,$$

and

$$\phi_C = 76°,$$

whereas ignoring all of the inductor resistances will make all of these angles 90°. It must be borne in mind, however, that as we move from the bottom to the top of the voice band frequency range, all of these approximations will improve. In other words, we have calculated the worst case here. We thus conclude that we generally will be justified in ignoring the resistances of the inductors in the network.

Before leaving the subject of the network inductances, it is interesting to note that the first-order theory of inductors predicts that the ratio L/N^2 should be constant for all inductors wound on a common core, assuming perfect flux coupling. For these inductors the ratio comes out very close to $1.69 \times 10^{-4}\ \Omega \cdot \text{sec}$. In fact, using this ratio and the values for N taken from Bell specs, the inductances should be (to the nearest .1 mH)

$$L_A = 113.7\,\text{mH},$$
$$L_B = 24.4\,\text{mH},$$

and

$$L_C = 30.5\,\text{mH}.$$

These obviously compare very favorably to the measured values of 113.9 mH, 24.5 mH, and 30.1 mH.

(7) The mutual inductances M_{AB}, M_{BC}, and M_{AC} will depend on the coefficients of coupling between the respective windings. There is no reason to expect that these coefficients will all be exactly the same. However, since all inductors are wound on the same iron core there *is* reason to expect that all will be close to unity. Measurements taken with a signal generator and scope give

—	$k_{12} = .95$	$k_{13} = .95$	$k_{14} = .93$
$k_{21} = .89$	—	$k_{23} = .88$	$k_{24} = .87$
$k_{31} = 1.02$	$k_{32} = 1.03$	—	$k_{34} = .99$
$k_{41} = .99$	$k_{42} = .99$	$k_{43} = .95$	—

The error due to scope reading is ±.04. The remaining error is due to the fact that the inductors were not completely disconnected from other circuit elements in the network, allowing some loading effects. Nevertheless, these data do show that it will be quite a good approximation to regard all the coefficients of coupling as equal to unity. Thus

$$M_{AB} = \sqrt{L_A L_B} = 53\,\text{mH},$$

$$M_{BC} = \sqrt{L_B L_C} = 27\,\text{mH},$$

and

$$M_{AC} = \sqrt{L_A L_C} = 59\,\text{mH},$$

to the nearest whole millihenry.

We will be using these facts and assumptions to solve for $i_3(t)$ (actually $\mathbf{I}_3(s)$) and then to find the conditions under which it is zero. The required circuit equations are

$$\mathbf{V}_T/R_T = \mathbf{I}_1 + \mathbf{I}_2 + \mathbf{V}_1/R_T \qquad \text{(node equation at } V_1\text{)},$$
$$\mathbf{I}_2 = \mathbf{I}_3 + \mathbf{V}_2/\mathbf{Z}_N \qquad \text{(node equation at } V_2\text{)},$$
$$\mathbf{V}_1 = sL_A\mathbf{I}_1 - sM_{AB}\mathbf{I}_2 - sM_{AC}\mathbf{I}_3 + \mathbf{Z}_L\mathbf{I}_1 \qquad \text{(left loop)},$$
$$\mathbf{V}_1 = sL_B\mathbf{I}_2 - sM_{AB}\mathbf{I}_1 + sM_{BC}\mathbf{I}_3 + \mathbf{V}_2 \qquad \text{(middle loop)},$$

and

$$\mathbf{V}_2 = sL_C\mathbf{I}_3 + sM_{BC}\mathbf{I}_2 - sM_{AC}\mathbf{I}_1 + \mathbf{Z}_R\mathbf{I}_3 \qquad \text{(right loop)}.$$

The first two equations can be used to eliminate $\mathbf{V}_1$ and $\mathbf{V}_2$ from the latter three. We substitute

$$\mathbf{V}_1 = \mathbf{V}_T - R_T\mathbf{I}_1 - R_T\mathbf{I}_2 \quad \text{and} \quad \mathbf{V}_2 = \mathbf{Z}_N\mathbf{I}_2 - \mathbf{Z}_N\mathbf{I}_3$$

to get

$$\mathbf{V}_T - R_T\mathbf{I}_1 - R_T\mathbf{I}_2 = sL_A\mathbf{I}_1 - sM_{AB}\mathbf{I}_2 - sM_{AC}\mathbf{I}_3 + \mathbf{Z}_L\mathbf{I}_1,$$

$$\mathbf{V}_T - R_T\mathbf{I}_1 - R_T\mathbf{I}_2 = sL_B\mathbf{I}_2 - sM_{AB}\mathbf{I}_1 + sM_{BC}\mathbf{I}_3 + \mathbf{Z}_N\mathbf{I}_2 - \mathbf{Z}_N\mathbf{I}_3,$$

and

$$\mathbf{Z}_N\mathbf{I}_2 - \mathbf{Z}_N\mathbf{I}_3 = sL_C\mathbf{I}_3 + sM_{BC}\mathbf{I}_2 - sM_{AC}\mathbf{I}_1 + \mathbf{Z}_R\mathbf{I}_3.$$

These latter three equations can be rearranged to collect terms and thus put in matrix form:

$$\mathbf{V}_T = (R_T + sL_A + \mathbf{Z}_L)\mathbf{I}_1 + (R_T - sM_{AB})\mathbf{I}_2 - sM_{AC}\mathbf{I}_3, \tag{10.1a}$$

$$\mathbf{V}_T = (R_T - sM_{AB})\mathbf{I}_1 + (R_T + sL_B + \mathbf{Z}_N)\mathbf{I}_2 + (sM_{BC} - \mathbf{Z}_N)\mathbf{I}_3, \tag{10.1b}$$

$$0 = -sM_{AC}\mathbf{I}_1 + (-\mathbf{Z}_N + sM_{BC})\mathbf{I}_2 + (\mathbf{Z}_N + sL_C + \mathbf{Z}_R)\mathbf{I}_3. \tag{10.1c}$$

From Cramer's Rule, $\mathbf{I}_3$ will be zero if the cofactor matrix in its numerator is zero:

$$\begin{vmatrix} \mathbf{V}_\mathrm{T} & (R_\mathrm{T} + sL_\mathrm{A} + \mathbf{Z}_\mathrm{L}) & (R_\mathrm{T} - sM_\mathrm{AB}) \\ \mathbf{V}_\mathrm{T} & (R_\mathrm{T} - sM_\mathrm{AB}) & (R_\mathrm{T} + sL_\mathrm{B} + \mathbf{Z}_\mathrm{N}) \\ 0 & -sM_\mathrm{AC} & (-\mathbf{Z}_\mathrm{N} + sM_\mathrm{BC}) \end{vmatrix} = 0.$$

Expanding, we get

$$\begin{aligned} &\mathbf{V}_\mathrm{T}(R_\mathrm{T} - sM_\mathrm{AB})(-\mathbf{Z}_\mathrm{N} + sM_\mathrm{BC}) + \mathbf{V}_\mathrm{T}(R_\mathrm{T} - sM_\mathrm{AB})(-sM_\mathrm{AC}) \\ &\quad - \mathbf{V}_\mathrm{T}(R_\mathrm{T} + sL_\mathrm{B} + \mathbf{Z}_\mathrm{N})(-sM_\mathrm{AC}) - \mathbf{V}_\mathrm{T}(R_\mathrm{T} + sL_\mathrm{A} + \mathbf{Z}_\mathrm{L}) \\ &\quad \times (-\mathbf{Z}_\mathrm{N} + sM_\mathrm{BC}) = 0 \end{aligned}$$

or

$$\begin{aligned} &-R_\mathrm{T}\mathbf{Z}_\mathrm{N} + sM_\mathrm{BC}R_\mathrm{T} + sM_\mathrm{AB}\mathbf{Z}_\mathrm{N} - s^2M_\mathrm{AB}M_\mathrm{BC} - sM_\mathrm{AC}R_\mathrm{T} \\ &\quad + s^2M_\mathrm{AB}M_\mathrm{AC} + sM_\mathrm{AC}R_\mathrm{T} + s^2M_\mathrm{AC}L_\mathrm{B} + sM_\mathrm{AC}\mathbf{Z}_\mathrm{N} + R_\mathrm{T}\mathbf{Z}_\mathrm{N} \\ &\quad + sL_\mathrm{A}\mathbf{Z}_\mathrm{N} + \mathbf{Z}_\mathrm{L}\mathbf{Z}_\mathrm{N} - sM_\mathrm{BC}R_\mathrm{T} - s^2M_\mathrm{BC}L_\mathrm{A} - sM_\mathrm{BC}\mathbf{Z}_\mathrm{L} = 0, \end{aligned}$$

which can be solved for $\mathbf{Z}_\mathrm{N}$ to give

$$\begin{aligned} \mathbf{Z}_\mathrm{N} &= \frac{sM_\mathrm{BC}(sL_\mathrm{A} + \mathbf{Z}_\mathrm{L} + sM_\mathrm{AB}) - sM_\mathrm{AC}(sL_\mathrm{B} + sM_\mathrm{AB})}{s(M_\mathrm{AB} + M_\mathrm{AC}) + sL_\mathrm{A} + \mathbf{Z}_\mathrm{L}} \\ &= \frac{s^2(M_\mathrm{BC}L_\mathrm{A} - M_\mathrm{AC}M_\mathrm{AB}) + s^2(M_\mathrm{BC}M_\mathrm{AB} - M_\mathrm{AC}L_\mathrm{B}) + sM_\mathrm{BC}\mathbf{Z}_\mathrm{L}}{s(M_\mathrm{AB} + M_\mathrm{AC}) + sL_\mathrm{A} + \mathbf{Z}_\mathrm{L}}. \end{aligned}$$

Since the coefficients of coupling are being taken as unity, the terms in both parentheses in the numerator are zero and

$$\begin{aligned} \mathbf{Z}_\mathrm{N} &= \frac{sM_\mathrm{BC}\mathbf{Z}_\mathrm{L}}{s(M_\mathrm{AB} + M_\mathrm{AC} + L_\mathrm{A}) + \mathbf{Z}_\mathrm{L}} \\ &= \frac{.027s\mathbf{Z}_\mathrm{L}}{.226s + \mathbf{Z}_\mathrm{L}}. \end{aligned} \tag{10.2}$$

All of the parameters on the right-hand side of (10.2) are fixed parameters of the "hybrid," except $\mathbf{Z}_\mathrm{L}$. This tells us that for a given $\mathbf{Z}_\mathrm{L}$ we can in theory find a network impedance, $\mathbf{Z}_\mathrm{N}$, that will reduce sidetone to zero. In practice, $\mathbf{Z}_\mathrm{L}$ can vary over a wide range both as to magnitude and phase, and the network used is simply a compromise. Nevertheless, the use of the network typically drops sidetone 10–12 dB below what it would be without sidetone suppression.

Physically of course, the key to sidetone balance is that the voltages induced in L_C by L_A and L_B essentially cancel the voltage developed across $\mathbf{Z}_\mathrm{N}$. Not only does this reduce sidetone, but it also means that little of the transmitter signal power is wasted in the receiver, allowing more for transmission.

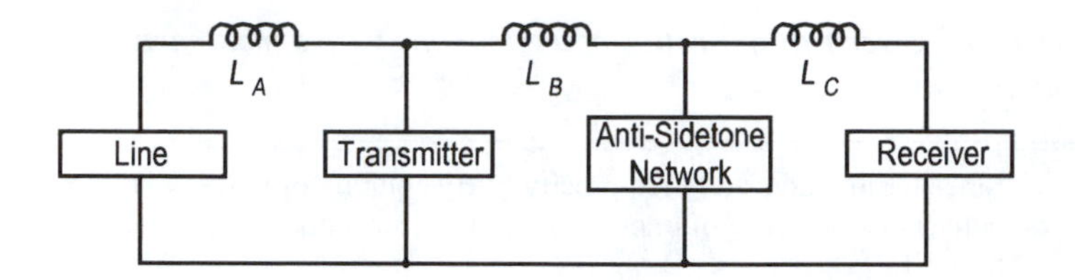

Fig. 10.6: Simplified diagram of the 500D telephone, which will be used for analysis of the power interrelationships among the various signals in the phone.

10.2.3 Power interrelationships

The ability to achieve sidetone rejection is just one consequence of the fact that the phone uses what is called a *maximum output network*. Let us redraw Fig. 10.5, showing only the components that are explicitly wired into the circuit. This schematic not only exhibits a certain pleasing symmetry but allows us to make two very important observations:

(1) We may think of the transformer (L_A, L_B, and L_C) as a four-port network, with one of the boxes shown in Fig. 10.6 connected to each port.

(2) The lossless character of the four-port network is a necessary (though not sufficient) condition for it to be a maximum output network. The implications of the use of such a network in the telephone have been exhaustively considered (Campbell and Foster, 1920) and nicely summarized in the forepart of a more recent paper (Means, 1980).

For our purposes, the most important property of such a network is that the power transfer between any two ports is the same as that between the other two ports. The lengthy calculation just completed said that if we know the line impedance, we can choose an anti-sidetone network impedance that will allow zero power transfer from the transmitter to the receiver. But by the principle just stated, this means there will also be zero power transfer from the line to the anti-sidetone network under those circumstances. This might not be expected on superficial physical grounds. It is quite desirable, however, because it says that since none of the signal power from the line is delivered to the anti-sidetone network and none is lost in the transformer (it being considered lossless), it must all go to the receiver and transmitter.

We propose to use this property of a maximum output network (as well as its other properties) to show that we need concerns ourselves with only two parameters. We will use $|S_{AB}|^2$ to denote the power transfer from termination A to termination B. A and B will be L (line), T (transmitter), N (network), or R (receiver). There are only sixteen mathematical possibilities, which we enumerate in Table 10.1, which should be read from left to right.

This result shows that if the anti-sidetone network is designed to give zero power transmission from the phone transmitter to its receiver, the only two unique power transmission ratios may be taken to be from the transmitter to the line and from the line to the receiver. We have already noted in effect that $|S_{LT}|^2 + |S_{LR}|^2 = 1$. When cast in this form, we might think we would like to make $S_{LT} = 0$ so that no line signal is wasted driving the transmitter. However, by reciprocity, $S_{LT} = S_{TL}$, and we could not tolerate a zero value of S_{TL} or else no signal from the transmitter would ever get out onto the line.

Table 10.1. *Power-transfer ratios of a telephone with a maximum output network*

	Since termination delivers no power to itself	By reciprocity of transformer	By equality of power transfer ratios	If there is no sidetones
S_{LL}	0	0	0	0
S_{LT}	S_{LT}	S_{TL}	S_{TL}	S_{TL}
S_{LN}	S_{LN}	S_{LN}	S_{TR}	0
S_{LR}	S_{LR}	S_{LR}	S_{LR}	S_{LR}
S_{TL}	S_{TL}	S_{TL}	S_{TL}	S_{TL}
S_{TT}	0	0	0	0
S_{TN}	S_{TN}	S_{TN}	S_{LR}	S_{LR}
S_{TR}	S_{TR}	S_{TR}	S_{TR}	0
S_{NL}	S_{NL}	S_{LN}	S_{TR}	0
S_{NT}	S_{NT}	S_{TN}	S_{LR}	S_{LR}
S_{NN}	0	0	0	0
S_{NR}	S_{NR}	S_{NR}	S_{TL}	S_{TL}
S_{RL}	S_{RL}	S_{LR}	S_{LR}	S_{LR}
S_{RT}	S_{RT}	S_{TR}	S_{TR}	0
S_{RN}	S_{RN}	S_{NR}	S_{TL}	S_{TL}
S_{RR}	0	0	0	0

We also observe from the foregoing chart that $S_{LR} = S_{TN}$, or $|S_{LR}|^2 = |S_{TN}|^2$. This relationship shows that we must be willing to deliver some transmitter power to the anti-sidetone network or we will not be able to receive any signal from the line.

Calculations show that over reasonable ranges of frequency (300–3,300 Hz) and line impedance (200–1,200 Ω), $|S_{TL}|^2$ ranges between .09 and .32, with the higher transmission ratios being obtained at the higher values of f and $\mathbf{Z}_L$. These ratios were calculated with zero sidetone, which required modifying the anti-sidetone network each time $\mathbf{Z}_L$ was changed. The results must therefore be considered as an approximation to the actual situation where there is a single network that must approximate sidetone null under all frequency and line conditions.

10.3 The Touch-Tone® telephone

This innovation was introduced in 1963, just ten years after the 500D. It was a significant development for many reasons, some of which are:

(1) It was the first time that an active device was used in a subscriber phone. This device was a transistor used in the Touch-Tone® oscillator.

(2) Touch-Tone® dialing affords an enormous speed improvement as compared to rotary dialing. The actual improvement will depend on the number being dialed, but a ten fold increase in dialing speed would be typical.

(3) It required a level of electronic sophistication at the central office that was well beyond anything that had been in use previously. As a consequence, it spread relatively slowly throughout the country, coming

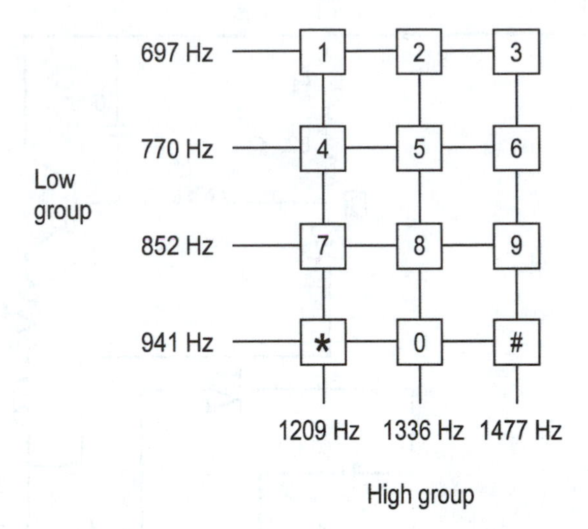

Fig. 10.7: Schematic representation of the DTMF keyboard showing which tones will be generated for a given keypress.

into new service areas only as the central office was equipped to handle it.

(4) It opened the door to a host of new uses for the telephone such as electronic banking. This might have been possible in theory with rotary dialing, but the time required to enter a transaction by rotary dialing would simply have precluded its acceptance by much of the public. This and other new uses of Touch-Tone® are based on the capability of using the tones for signaling to equipment at the receiver end after the call has been answered.

10.3.1 Dialing

As applied to the subscriber home phone, Touch-Tone® is simply digit dialing by the transmission of tones, rather than by a sequential making and breaking of the line circuit. For reasons of noise immunity, the Touch-Tone® dialing system generates two different tones as each digit button is pressed. This explains the first part of the name Dual-Tone Multi-Frequency dialing. The Touch-Tone® keypad is arranged as a 4×3 matrix. Each row and each column is associated with a distinct frequency. Therefore pressing a button on the keyboard selects one row (low group) and one column (high group) frequency. This arrangement is shown in Fig. 10.7.

The seemingly strange frequencies in each group are each larger than the next lower one by a factor of about 1.105. Hence there is a margin of about $\pm 5\%$ in each frequency. This is adequate to allow for the tolerances in the oscillator frequencies ($\pm 1.5\%$ maximum) and the tolerances in the center frequencies of the tone detector circuitry in the central office ($\pm 2\%$ maximum) while still delivering enough signal into the detection bandwidth of the tone detectors to be properly decoded. The basic configuration of the oscillator is common-collector with transformer-coupled feedback from the emitter. Ever mindful of power drain considerations, the Bell System designed the dialer so that it draws no power until the handset is lifted. A current is then allowed to flow through the tuned circuits to store up magnetic energy. When a number key is

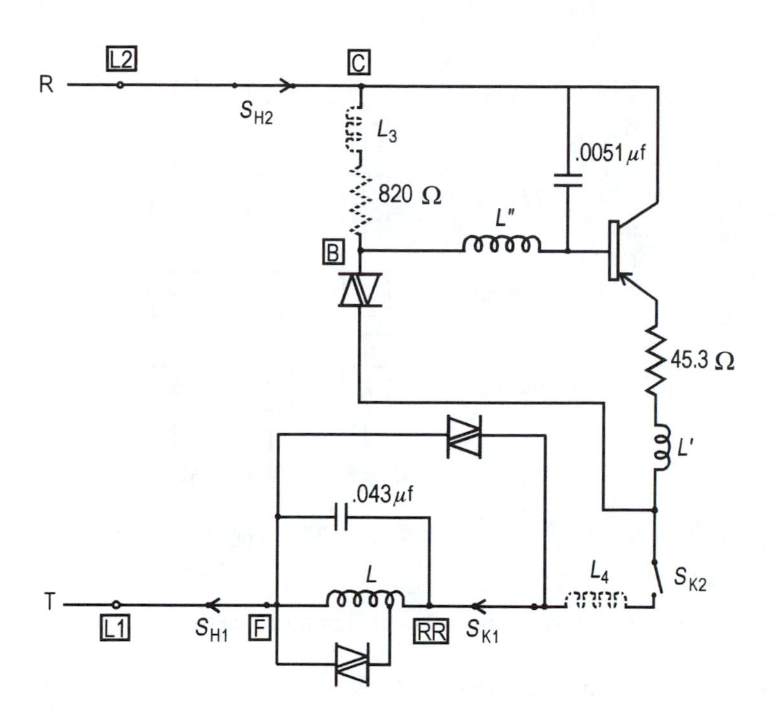

Fig. 10.8: Schematic diagram of the oscillator portion of a 2500 series Touch-Tone® telephone.

pressed, the first switching action performed is to select a tap on the inductor for the high and low group oscillators. Then another set of contacts (which are activated as the latter part of the mechanical sequence for any keystroke) performs four functions. The two that concern us here are that the transistor is biased on and that the current through the tuned circuit is interrupted to shock-excite the oscillator. This assures that the tone will start instantly when the key is fully depressed. Figure 10.8 shows the oscillator portion of a phone in the 2500 series, one of the modern Touch-Tone® phones. Elements shown in dashed lines are parts of the remainder of the phone that serve a dual purpose. Terminals of the 425 K network are shown in squares. The 425 K network is the same as the 425B, except that: (1) The .1 μF capacitor previously brought out to terminal F is omitted. Terminal F is still found on the exterior of the 425 K network where it is used as a tie point. (2) The 22 Ω resistor connected to terminal F is replaced with an 820 Ω one. The connections of the switches are shown at an instant when the handset has been lifted and a number keystroke has begun. Thus a tap on L has already been selected. As the stroke continues, S_{K2} will close, allowing emitter current to flow, and S_{K1} will open, shock-exciting the oscillator. Of course, the tuned circuit will still be magnetically coupled to the rest of the circuit even after the switch opens; it will thus continue to hold the oscillator on frequency. The emitter current then flows through the varistor connected between terminal F and the left end of L_4. The transistor receives base bias through L_3, and emitter current flows through L_4. The inductances L, L', and L'' are three windings on a coupled inductor.

In most oscillators, the amplitude of oscillation is stabilized by the transistor being driven into nonlinearity. In this design, measures are taken to insure that the transistor itself does *not* significantly enter the nonlinear region of operation. This is achieved by the application of a varistor across part of the tuned circuit. As the amplitude of oscillation increases, the varistor resistance decreases, killing the Q, and thus lowering the amplitude again. This linear operation of the transistor is very important because, in practice, the two tones required for DTMF operation do not come from two separate oscillators but from one oscillator that oscillates simultaneously at two frequencies! Since superposition applies in a linear circuit, this is perfectly feasible. Two separate coupled inductors are used, with corresponding windings connected in series. Thus in Fig. 10.8, L, L', and L'' are each actually the corresponding windings of two transformers connected in series.

It might be suspected that these series-connected windings would cause shifts in the oscillation frequencies calculated for single-frequency oscillation. This is, in fact, the case, but the magnitude of the effect is only about a 3 Hz shift for each frequency in the worst case.

The oscillator output is coupled out onto the phone lines by virtue of the voltage induced across L_4 as the emitter current flows through it.

When no keys are being pressed, the active phone circuitry is basically the same as for the 500-series phone.

Exercise 10.3 Using Figs. 10.2 and 10.8, draw a complete schematic of the circuitry in a 2500-series phone.

10.4 Electronic telephones

Many of the nonlinear telephone operations that the Bell System has always accomplished with varistors are better accomplished with active electronics. Likewise, the elaborate sequence of switching operations performed by both rotary and Touch-Tone® dialers can be eliminated with a properly designed electronic DTMF oscillator.

A wide variety of telephone-related chips is now available including some single-chip telephones. In any single-chip approach, there must be a trade-off of flexibility versus price. By far the largest number of single-chip telephones find their way into inexpensive "throw-away" phones, and thus the designs tend to optimize cost rather than flexibility. Motorola, however, has an Electronic Telephone Circuit (ETC) that incorporates all stand-alone phone functions. They also have a version capable of interfacing to a microcomputer. Figure 10.9 shows a block diagram of the MC34011, which is the nonmicroprocessor version. Typical "outboard" parts values to make a working phone are also shown. The chip is built using I^2L technology, which allows it to operate correctly down to very low line voltage of 1.4 V. The diagram shows three switches (besides the DTMF keyboard). All are part of the hook switch and are shown in the "on-hook" condition.

10.4.1 Ringer circuitry

The connections to the phone lines are shown in the upper right-hand corner of Fig. 10.9. As shown in the "on-hook" condition, the DC voltage on the phone line is blocked by C_{17} on the input side of the bridge. This serves the same purpose as the capacitor between the two ringer coils in the Bell System phones we have studied. When a ringing voltage appears on the line, however, there is a very significant difference. The 6.8 k resistor in series with the 1 μF capacitor

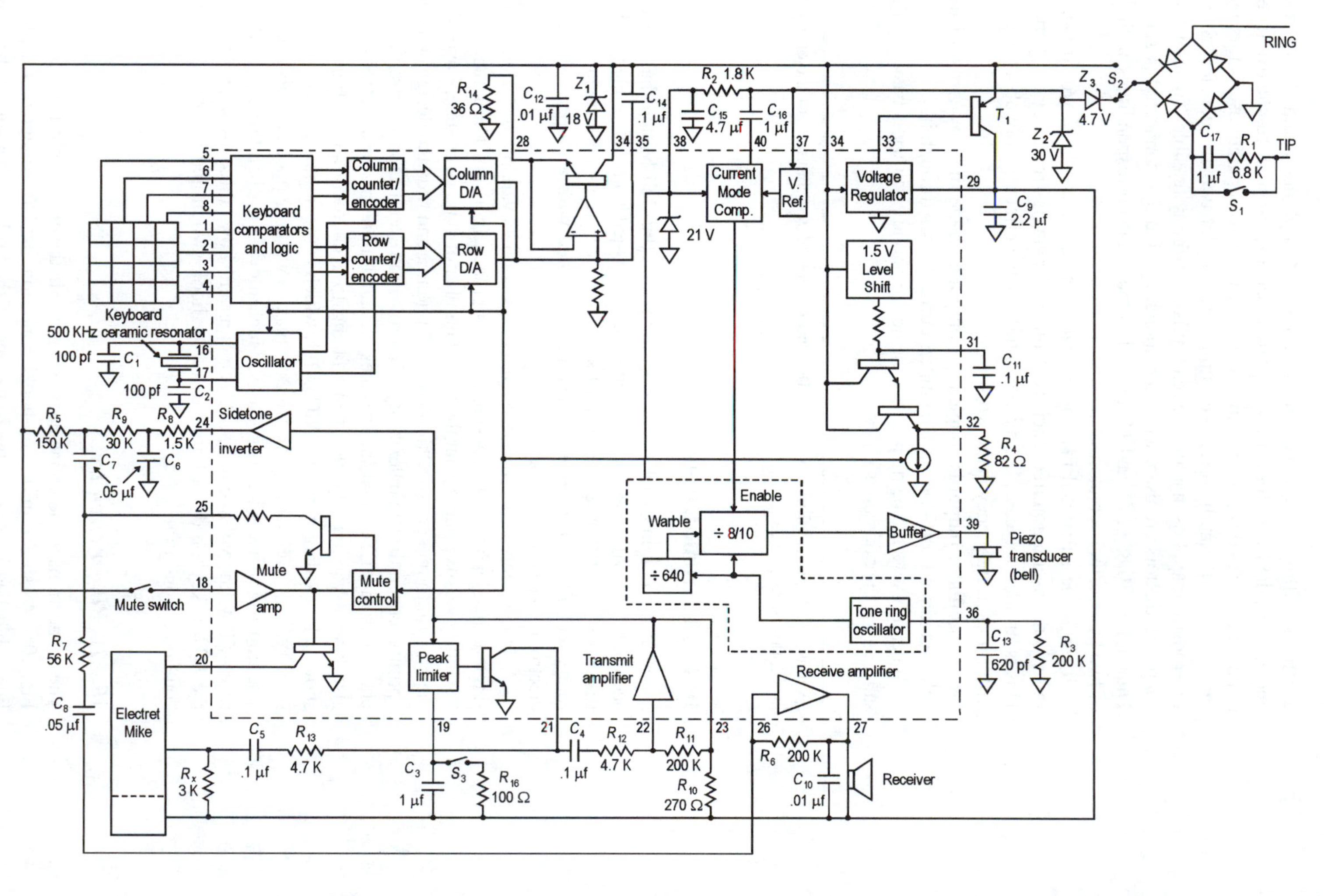

Fig. 10.9: Block diagram of the Motorola MC34011 electronic telephone circuit and the associated outboard components required to make a fully functioning telephone.

puts a very low current drain on the ringing voltage generator compared to a bell. The ringing voltage is rectified by the full-wave diode bridge. This not only assures the proper polarity of voltage for subsequent circuitry but also loads the ringing-voltage generator symmetrically on both half-cycles (as does a bell). The rectified ringing voltage is applied to a Zener diode voltage divider, which limits the instantaneous voltage sent to the ringing circuit to 30 V. When the rectified voltage is less than 35 V, the 4.7 V Zener subtracts from it, and the 30 V Zener remains in the open condition. This means that the instantaneous value of the pulsating voltage applied to pin 37 can be anything up to 30 V. From this voltage, a reference of 1.6 V is derived with its most positive terminal at pin 37 and its most negative terminal as one input to the current mode comparator. Thus the right-hand input to the comparator will be 1.6 V below the potential of pin 37. Resistor R_2 is connected between pins 37 and 38 and is called a current-sense resistor. Note that its right end is connected to a 30 V Zener external to the chip, and its left end to a 21 V Zener internal to the chip. Thus in the presence of a ringing voltage, this resistor may have up to 9 V across it. The left end also connects to the other input of the current-mode comparator. Thus, when the ringing voltage is sent, the left-hand input of the comparator is 9 V below the potential of pin 37. This more than suffices to trip the comparator and enable the 8/10 divider. The tone-ring oscillator will run at a frequency set by the parallel RC on pin 36. A typical frequency is about 7,500 Hz. This is divided by 640 to make a warble frequency of about 12 Hz. Thus twelve times a second, the modulus of the main divider is switched between 8 and 10. Hence output of this divider block is being switched between about 940 and 750 Hz respectively. This warbling oscillation is the ringer tone that is enabled when the potential at the left input of the comparator is below that at the right input. As indicated on the block diagram, the DC current to run all of the oscillator and divider circuitry flows through R_2 and is smoothed by the 21 V Zener and C_{15}. Once the oscillator starts, this additional current flow will increase the voltage drop across R_2 and thus assure that the comparator stays on, even though the voltage at pin 37 drops because of the increased current flow through R_1 and C_{17}, which are in series with the Tip line. This is why a current-mode comparator is used. Since the voltages at pins 37 and 38 are pulsating to some degree, the 1 μF capacitor connected to pin 40 provides some low-pass filtering of the difference voltage between the comparator inputs. It also filters out transient spikes on the lines that might otherwise trigger the oscillator briefly. These transients are not a problem with a mechanical bell both because their amplitude is not generally large enough to activate it and because their duration is generally too short to allow the mechanical inertia of the bell to be overcome.

10.4.2 Power supplies

Once the phone has rung and the subscriber has responded by lifting the handset, the hook switch is toggled and the rest of the chip's circuitry is enabled. The lifting of the handset removes the ringing voltage being sent by the central office. This leaves only a DC voltage between the Tip and Ring lines. (However, we shall see that both Transmit and Receive signals will also appear on these lines after speech starts.) The hook-switch shorts out R_1 and C_{17}, applying the full DC voltage on the lines to the diode bridge. The S_2 section of the hook-switch also toggles, connecting the output of the bridge to Z_1, which clamps the V^+ line to 18 V or less. The capacitor in parallel with Z_1 (C_{12}) is one of several components that are present to meet Bell System specs for impedance. It also inhibits oscillation in the voltage regulator. Said regulator appears to be conventional in all ways except for how low its output is designed to be. It receives an input $\leqq 18$ V from the V^+ line (pin 34), and it uses T_1 as a linear series regulating pnp pass transistor to increase current capability. This transistor's base is driven by the regulator output (pin 33).

The output voltage of the pass transistor is taken from the collector and is sampled by the regulator (pin 29) for comparison to its internal reference. The only unusual thing about this is that the design value of the output voltage (pin 29) is only 1.1 V. We have noted that the chip will function down to line voltages of 1.4 V. When the line voltage drops low enough so that the 1.1 V output cannot be maintained with T_1 operating in normal active mode, it is driven into saturation to keep the output as high as possible. Capacitor C_9 also inhibits oscillation in the voltage regulator.

To understand the function of the 1.5 V level shift block, we need to recall that in general the V^+ line (pin 34) will contain both DC and AC (speech) signals. Thus if we can express

$$V^+(t) = V_{\text{DC}} + v_{\text{AC}}(t),$$

then after the level shift we will have

$$v_{\text{shift}}(t) = V_{\text{DC}} + v_{\text{AC}}(t) - 1.5.$$

The resistor in series with the base of the Darlington acts in conjunction with C_{11} to make an LPF that attenuates the AC reaching the base. Thus

$$V_{\text{base}} = V_{\text{DC}} - 1.5.$$

Obviously no significant current will flow through the Darlington until V_{base} is large enough to forward bias both base–emitter junctions – call it 1.5 V. Thus conduction begins when

$$1.5 \leq V_{\text{DC}} - 1.5$$

or

$$V_{\text{DC}} \geqq 3 \text{ V}$$

as per spec.

The removal of the AC signal from the base of the Darlington effectively assures that there will be no AC signal at its emitter, and thus the AC on the V^+ line is not loaded by the Darlington even when it conducts.

When the DC component of V^+ reaches ~3 V, we have seen that the Darlington will conduct. This will load the V^+ line more heavily, causing the DC voltage on it to drop, and thus easing the job of the voltage regulator. The degree of loading is determined by R_4. Hence the Darlington essentially forms a sort of shunt regulator that works in conjunction with the series regulator to maintain a stable 1.1 V output. The current sink in parallel with R_4 is a "dummy load." When the Darlington conducts, this sink draws current until the keypad is pressed. Then it is shut off. The idea here is that because the DTMF circuitry draws its power from the regulator (thus holding the V^+ line down) when it is off, the dummy load draws a comparable amount of current instead, thus maintaining a more nearly constant load and promoting a better-regulated output voltage. In fact, these measures produce a typical load regulation of 6 mV.

10.4.3 Speech circuits

10.4.3.1 Receive operation

Received speech signals travel from the phone lines through the diode bridge to the V^+ line. It might be supposed that passing through the nonlinear diodes would distort the speech signals, but the diodes are already forward biased by the DC current from the line, so bi-directional signal travel through them is only a case of small-signal (linear) signal swing about a much larger DC quiescent point. The signal on V^+ passes through R_5, C_7, R_7, and C_8 to reach the input of the receiver amplifier (pin 26). A feedback resistor (R_6) goes from there to the amplifier's output (pin 27). A simple calculation will show that for frequencies in the voice band the capacitors may be considered as AC shorts (they are included to block DC and low-frequency noise on the line), and the voltage gain of the amplifier is about .4.

Exercise 10.4 Prove that if C_6, C_7, and C_8 can be considered as shorts for AC, the receiver amplifier will have a voltage gain of about .4.

This might seem so small a gain as to be pointless, but let us show that it is not so unreasonable. In the first place, the input impedance of this amplifier is on the order of 170 kΩ. The receiver device that it drives has an impedance of 300 Ω. Thus a fairly large power gain of 20 dB is delivered by the amplifier. Secondly, the receiver amplifier is fabricated in I^2L. It runs on a V_{cc} of 1.1 V. Thus it cannot have much output swing without running into saturation or cutoff. The output is, in fact, biased at about .6 V, near the midpoint of the 1.1 V supply. Thirdly, we shall show later that the transmit signal reaches the phone lines through the collector-to-emitter path of the receiver. It is inevitable that there is some receiver signal coming from the line through T_1 and down to the low end of T_1. Since the receiver amplifier is apparently an inverting type, the signals at the two terminals of the receiver element are 180° out of phase and will increase the signal voltage seen by the receiver without a need to increase the supply voltage. This is the principle of the bridge amplifier. This may not be an effect of major significance, however, because any signal coming through T_1 would have an effective source impedance equal to the line impedance. This is typically greater than the 300 Ω impedance of the receiver, and so the receiver might represent a load heavy enough to cause attenuation of the signal reaching the load by this path.

Frequency compensation of the receiver amplifier is accomplished by C_{10}. Between C_7 and R_7 on the input path of the amplifier is a connection to pin 25 of the chip, which is used for partially muting the receiver during DTMF dialing.

10.4.3.2 Transmit operation

Speaking into the transmitter will produce a signal on the phone lines. This will, of course, be picked right back up by the receive circuitry just discussed. Thus the ETC also incorporates sidetone–reduction circuitry. But unlike phones using a passive network, the ETC is not bound by inflexible interrelationships between the various power transfer ratios.

The speech enters the *electret* microphone where it is converted to an electrical signal. These microphones are based on a dielectric that has a permanent charge locked into it. The microphone may have the dielectric in the form of a film that forms the diaphragm of the mike, or the dielectric may be fixed and a separate diaphragm used. In either case the electret will generally have one surface metallized. In every case the speech varies the inter electrode spacing and thus the capacitance. Since Q is fixed, the voltage across the microphone varies as

$$v = Q/C.$$

The capacitance is proportional to the reciprocal of the plate spacing (d) so

$$v \propto d.$$

The plate spacing should vary nearly linearly with the sound pressure, so the output voltage, though small, transduces the sound into voltage with low distortion. The low-amplitude microphone output requires amplification, but that is no problem on the chip. Electret microphones come in two varieties – those with two and those with three terminals. The third terminal, if present, is a line for DC bias only. The two-terminal version derives its bias through R_X. The ETC can accommodate both types. In Fig. 10.9, the lower terminal of the mike can be seen to connect back to pin 29 of the chip, which is the output of the 1.1 V regulator. This is the connection that is omitted with two-terminal mikes. The upper connection to the mike (pin 20 of the ETC) is the ground return path. This path passes through a transistor that completes the path when saturated and opens it when cut off. When cut off, no mike signal is delivered to the phone. This transistor is, in turn, controlled by the mute circuitry, which will disable the mike when dialing or when manually selected by the mute switch.

The middle lead of the microphone is the signal output lead. It proceeds to the input of a transmit amplifier that is configured almost exactly like the receive amplifier, even to the point of having a divided input impedance (C_5–R_{13}/C_4–R_{12}) with provision for the midpoint being pulled down by a transistor. In this case, the transistor is driven by the output of a peak limiter whose input is the output of the transmit amplifier. If the speech level is too high, the peak limiter turns on the transistor and pulls down the amplifier input line, reducing the input to the transmit amplifier. The components in this input line also perform some bandlimiting. They cause the low-frequency response of the transmit amplifier to start falling off at almost exactly 300 Hz. The peak limiter's attack and release time constants (see Section 5.4) can be varied by C_3. When the phone is replaced on hook, S_3 discharges C_3 through R_{16}. When the phone is removed from the hook, the charging of C_3 suppresses a click.

The output of the transmit amplifier takes two paths. The first is through R_{10} and thus into the phone lines through T_1 and the diode bridge. This resistor governs the amplitude of the transmit signal sent directly onto the phone lines. The signal from the transmit amplifier also goes to the sidetone inverter. The signal at pin 24 is thus roughly 180° out of phase with the transmit signal that the receiver amplifier gets from the phone line through R_5. The reactance of the phone lines, however, does cause some phase shift in the signal sent out through pin 23 before it returns to the receive amplifier input circuitry. The phase shift introduced by R_8 and C_6 compensates for this and ideally causes the currents through R_5 and R_9 to be exactly 180° apart. The values of these two resistors will thus govern the relative amplitudes of the sidetone and anti-sidetone signals respectively. The uncanceled sidetone component delivered to the receiver also flows through it and back out onto the phone lines just as does the direct component through R_{10}, thus reinforcing it.

10.4.4 DTMF dialer circuitry

The ETC is capable of accommodating a keypad with four rows and four columns, although for most telephones, only a three-column keypad would be used. This is a seven-contact device. The keyboard logic detects the connection of a row and column wire and generates an enable signal, which starts a master oscillator whose frequency is set by an inexpensive 500 kHz ceramic resonator. This frequency is divided down to given approximations that are within .2% of the standard DTMF frequencies.

The 500 kHz from the oscillator is fed to the row and column counters. Keyboard logic has set the modulus of each counter. The frequency-divided pulse trains go to encoders, which are basically ROMs that have a sine wave function encoded into them. The counter outputs address the ROM sequentially and cause a digital approximation to a sine wave to appear at the ROM output pins. The D/A converters convert the ROM output to a stepwise approximation to a sine wave. The circuitry in the ETC produces a 16-level approximation. The lower the modulus the counters are set for, the higher will be their output frequency, and the higher will be the frequency of the output sinusoids. All of the sinusoids are integer submultiples of the 500 kHz oscillator frequency with the modulus ranging from 718 down to 338.

The D/A converters are of the current-output variety and are fed to the summing node of an amplifier in the inverting mode. This amplifier consists of an op amp and a separate transistor. Because the op amp feeds the transistor in the common–emitter configuration, the output taken from the collector is phase inverted from that at the op amp output. This necessitates using the noninverting input of the op amp as the summing node to get overall negative feedback. The feedback is between the collector and the summing node of the op amp and consists of C_{14} and an internal resistor. The unbypassed resistance in the emitter lead (R_{14}) controls the amplitude of the DTMF signal through the transistor. The collector is connected directly to the V^+ line and thus, through the diode bridge, to the phone line.

10.5 Comparisons

Having completed the analysis of the ETC, we may look at the whole picture and wonder why. Even with the LSI chip, many more components are present than in a standard Bell System Touch-Tone® phone. How can we therefore justify making the change? There are several reasons, and we shall enumerate some of them. But before we do, let us note that the matter of how many components a circuit contains is seldom a criteria in the selection of a circuit. Much more important considerations are generally

(1) cost,
(2) reliability, and
(3) performance.

Our first instinct might be to equate cost with parts count. This is not completely valid. Printed circuit techniques now allow machine loading and soldering of an entire board without operator intervention. Labor is a very important part of cost. Besides this, there is the obvious truth that many inexpensive parts may well cost less than a few more expensive ones.

Let us look at a few specific advantages of the electronic phone.

(1) The Touch-Tone® keypad is at the same time a marvel of engineering and a mechanical monstrosity. The concept of a strobe or enable is used in both the mechanical and electronic phones, but it is so much easier to generate electronically. Mechanically it requires one more addition to the multiplicity of switch contacts that are sequentially activated with each keypress. The electronic keypad is very simple and inexpensive by comparison.
(2) The tone generator itself does away with the large, heavy and expensive ferrite-core inductors used in the Touch-Tone® design.
(3) The large, heavy, power-hungry bell has been replaced by a small inexpensive piezo transducer that is just as "attention-getting," but much less jarring.
(4) The use of electronic amplification allows weak signals to be boosted to a usable level. With old Bell-System equipment, the hearing-impaired subscriber had to get an electronic amplifier. With the ETC, R_6 need only be increased to increase the signal at the receiver.
(5) The electronic phone adds the mute function. This not only makes user-selected mute available, but it replaces a multiplicity of switch sequences that were designed to mute the Touch-Tone® phone during dialing.
(6) With the ETC, sidetone null can be adjusted to the level desired, without affecting the other power-transmission ratios in the phone.
(7) The electret microphone is far smaller and lighter than the carbon mike it replaces. Perhaps more importantly, it has less distortion and requires virtually no power from the phone lines. The carbon mike has low cost, ruggedness, and high output to its credit. The electret mike has low cost and ruggedness. Its low output is readily amplified electronically to create a comparable level of speech signal, but it is not power-hungry like the carbon mike.
(8) Some energy savings are also realized in the receiver. The usual Bell-System receiver has an impedance of about 100 Ω at 1 kHz. The ETC is optimized for use with a 300 Ω transducer.
(9) Although we have not analyzed the companion chip, the MC34010, it incorporates a microprocessor interface to enable computer storage and dialing of phone numbers.

These are some of the more prominent advantages of the ETC over the traditional discrete component phone design. The present rapid advances in telecommunications will undoubtedly "spill over" into subscriber telephones as time goes on.

10.6 Cordless and cellular phones

One distinction between these types is that the cordless phone is basically a home phone while the cellular phone is designed for travel.

Cordless phones may or may not allow dialing from the remote unit or handset. Those that do not will have a keypad on the *base unit*. The functional circuitry is identical in all respects to the phones we have studied, but both the remote and base units each contain an FM transmitter and receiver. The base unit usually transmits at about 1.7 MHz and the remote transmits between 44.8 and 49.9 MHz. The use of the two separate frequencies allows *duplex* operation (the user of the remote can both talk and hear at the same time). If the remote has a keypad, the DTMF tones are FM-encoded just like the speech and sent to the base unit, which demodulates them and sends them out onto the lines. Because the receiver/transmitter combination in the base unit draws significant power, it is not operated from the phone lines but draws power from the AC line, making it vulnerable to power outages.

10.6.1 Cellular phone – concept overview

The basic concept of the *Advanced Mobile Phone Service* (*AMPS*®) or cellular phone network as it is more commonly called, is that a geographical region can be divided up into smaller regions called *cells*. Each cell contains equipment to transmit to and receive from mobile phones in its general area.

The spectral allocation for this service lies in two bands. There are a total of 666 channels, each of which is an FM channel 30 kHz wide. The signal bandwidth is 12 kHz. Cellular phones broadcast roughly between 825 and 845 MHz and receive between 870 and 890 MHz, thus allowing for full-duplex conversation on each channel. The spacing between the two bands (exactly 45 MHz for all channels) is often chosen as the frequency of the first IF amplifier. This choice allows one frequency-synthesis system to be used to generate both the transmission frequency and the local oscillator signal on receive.

The geographical cells are hexagonal in shape. This shape has been chosen on two counts. The first is that it is a close-packed structure that completely covers an area, leaving nothing that is not within a cell. The second is that it is a fair approximation to a circle. This is important because when a cell is first established, an omnidirectional transmitting antenna is placed as close to its center as possible. Thus the signal will reach out equally to all points on a circle centered at the transmitter. The hexagonal perimeter actually used will have nearly the same signal strength along its perimeter.

Exercise 10.5 For a transmitter feeding an omnidirectional antenna at the center of a hexagon, by how many dB will the signal be down at the vertex (furthest from the center) versus at the midpoint of a side (closest to the center). State your assumptions.

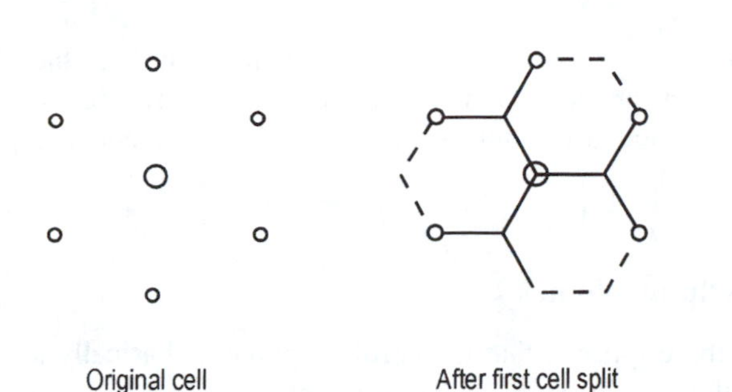

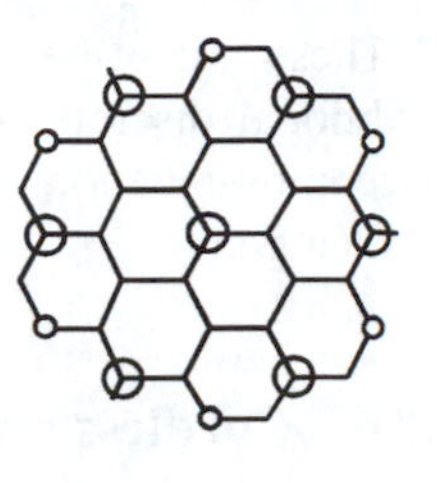

Fig. 10.10: The progression of cell structure used in a cellular phone system as more users are added and the cell size decreases.

In general, of course, it will not be possible to situate the antenna exactly at the center of a cell, but the standard is that it must be sited within 25% of the cell radius from the center.

The transmitted power is limited to about 10 W. This in turn limits the maximum cell radius to 8 miles. Each cell's transmitter is allocated a number of channels out of the 666 available. This number will be between 16 and 144 and will determine the maximum number of simultaneous phone conversations the cell can handle. Transmitters in nearby cells will each have a different set of channels allocated. If an omnidirectional antenna is used, a channel assigned to one cell can be assigned to another that is at least six cell radii away without significant interference.

As the average number of calls handled by a cell approaches the number of channels it has been assigned, either more channels will be assigned to it or the cell will be split. The fixed transceiver, which was originally at the cell's center, is now at the vertex of three cells as shown in Fig. 10.10. The omnidirectional antenna is now replaced by three directional ones with a 120° radiation pattern. Equally importantly, the power delivered to each antenna is reduced. This reduces the allowable spacing between two cells using the same channel to 4.6 cell radii and thus accommodates many more users. When a mobile unit is served by one of these split cells, a setup signal is sent to the mobile to reduce its transmitted power to match the range of the cell's transmitter. The smaller cell radius and lower power mean that the channels assigned to each cell can be assigned to other cells closer to it than before cell splitting.

In theory, this cell splitting could be continued indefinitely as long as each split was accompanied by a reduction in radiated power. In practice, there is a lower limit of about one mile on cell radius that is set by the paucity of suitable fixed sites – especially in metropolitan areas where demand is the greatest. Another reason for this limitation will be seen in the next section.

10.6.2 Cellular phone – system implementation

Since the purpose of the network is to provide mobile communications, and since the cells may be as small as a mile in radius, it is manifestly obvious that some users will be crossing over cell boundaries while engaged in a phone conversation. Let us examine how this is handled. In addition to multiple voice channels, each cell site has one or more setup channels. If a user wants to place a phone call, the entry of a number causes the mobile circuitry to seize the strongest available setup channel. Over the same channel, the cell site will transmit a voice channel assignment to the mobile phone. The internal frequency synthesizer will thus set the phone to transmit and receive on the assigned channel. All of this happens essentially instantaneously. The ringing signal and subsequent conversation are then transmitted over the voice channel, freeing the setup channel almost immediately. Each cell site is connected to a *Mobile Traffic Switching Office* (*MTSO*), which is in turn connected to a central telephone office switching system. But the MTSO does much more than switching. As the conversation proceeds, the serving cell site examines the signal level from the mobile user every few seconds. If it drops too low to assure clean communication, this information is conveyed to the computer at the MTSO, which will then allocate an available voice channel in the new cell into which the user has traveled. The new channel assignment is transmitted from the new servicing cell over a setup channel. The phone's frequencies are automatically switched so rapidly that the user is not even aware of it.

In receiving, the incoming call is received by the MTSO from the central phone office and a signal sent out over a setup channel. When the mobile unit recognizes its own phone number, it acknowledges on the strongest setup channel available. When the cell site receives the acknowledgment over its setup channel, it transmits a voice channel assignment over that channel. When the phone's frequencies have been programmed, the call is "patched through" from the MTSO through the cell. *Hand-off* proceeds as when the mobile originates the call whenever the signal received in the servicing cell drops too low. As the cell size shrinks, hand-off must occur more and more frequently. This puts an increasing burden on the computer in the MTSO. This is the other factor that limits the minimum size of a cell.

The MTSO also keeps call records and handles billing.

A mobile unit that leaves its home area and moves into the area served by a different cellular system altogether is called a *roamer*. Although long-distance lines could be used to transmit to an MTSO and the user could be contacted there, this is not generally done without the approval of the caller (for whom it will be a toll call) and the receiver (who may not want his whereabouts known).

10.6.3 Cellular phone receive circuitry

Like the VCR and compact disc players, the cellular phone is a product of space-age technology. Both the MTSO and the mobile unit would be unthinkably complex without VLSI devices. Obviously then, the modern devices have no discrete precursors.

The cellular phone will probably never attain the sales of VCRs. This is because, even with the cellular system, there is not, at present, adequate channel capacity for as many people to own one. Furthermore, since the cellular phone must necessarily have a monthly service charge on top of the charge for calls made, it will remain somewhat of an elitist item. The present monthly charge is typically in the \$30 to \$40 range. This pays for the purchase and maintenance of equipment and sites for the cells and the MTSO.

For these reasons, we will not give a detailed analysis of the cellular phone but will focus on the hardware system overview, while providing a little deeper coverage of some of the discrete RF circuitry. We will analyze a unit from the 1400 series manufactured by Oki Telecom. This unit was chosen because a number of surveys have chosen it as the one with the best sound quality. All of the 1400 series have the same hardware. The differing features of the various units in the series are realized with software. The service manual for this series of phones partitions the circuitry into four parts: the power supply, the audio, the logic, and the RF (or radio portion as Oki calls it). The RF is the only portion of the circuitry we will be examining. A block diagram is given in Fig. 10.11.

Our analysis begins and ends with the antenna, which is at the top center of the diagram. Some cellular phones have two antennas. One is used for receiving only, and the other for both transmitting and receiving. The obvious question of why two receiving antennas are used can be answered by considering the fading problem. In the middle of the transmit band, the wavelength is 36 cm. When the two antennas are spaced apart by at least several quarter wavelengths, the signal strengths received by them may differ enormously. The received signal strength (*RSSI*) signal is fed to the logic unit. Whenever it drops 10 dB below its mean value, the other antenna is selected by the logic unit via a *diversity switch*. If neither antenna has adequate signal, it keeps toggling until one brings the received signal strength back up above threshold.

The signal from the antenna is passed through a duplexer, which keeps the transmitter signal from swamping the sensitive receiver circuitry. It can be well appreciated that this circuitry must have strong frequency-discriminating capability. The schematic of this unit is not given in the service manual. It is sold as a single module.

The received signal passes through the discrete RF amplifier and a SAW filter and thence into the first mixer. The SAW filter provides further rejection of any signal from the transmitter that gets past the duplexer, as well as any other out-of-band signals that the duplexer is not designed to remove. Figure 10.12 shows the schematic diagram of the discrete RF amplifier. Other than the small size of the capacitors required for frequencies as high as those handled by this stage, the circuitry is unique in only three respects:

(1) Unfortunately, the Oki service literature is not of the same high quality as their phones. Note that in the emitter circuit of the amplifier, there are two capacitors, and one of them is marked "RESERVED," which is another way of saying "We are not going to tell you the value of this one." It would be bad enough to have this on one component, but a cursory scan of the schematic of the RF section reveals no less than seventeen components with a similar notation. Furthermore, there are a number of other components whose values are not specified, but which do not bear even this notation. The service manual contains parts lists so detailed as to call out even two pieces of double-stick tape, but it does not give the values of these "RESERVE" components either.

(2) The collector circuit contains two ferrite beads. We saw these used previously in a discrete monochrome TV, and in Section 6.3.10.5 we showed that they can be treated as a small inductance. Beads used in a circuit dealing with frequencies near 1 GHz would have to have much lower inductance than those used in the TV, but we are given no clue as to what their typical value might be. In fact, in spite of the very detailed parts lists already alluded to, Oki has not seen fit to even call out the beads as a component at all, much less to give a value for them. An educated guess would be that they are of the order of .1 nH. We shall return to the matter of the beads in the next exercise.

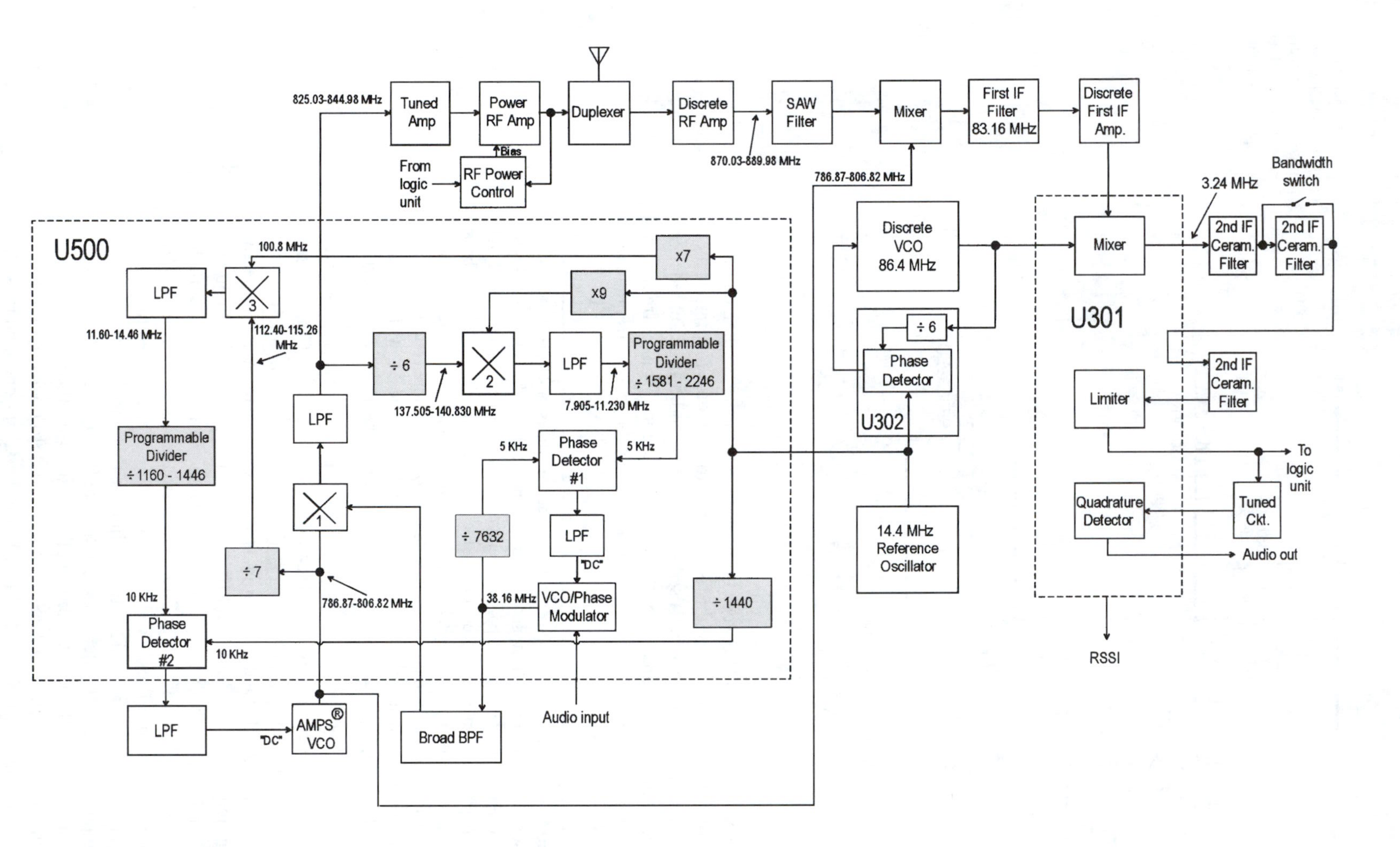

Fig. 10.11: Block diagram of the RF portion of the circuitry in a 1400 series cellular phone from Oki Telecom. The shaded blocks are all frequency multipliers and dividers. Since the moduli of these blocks are not specified, these must be understood as nothing more than one set of values that would generate the internal frequencies deduced for the circuit.

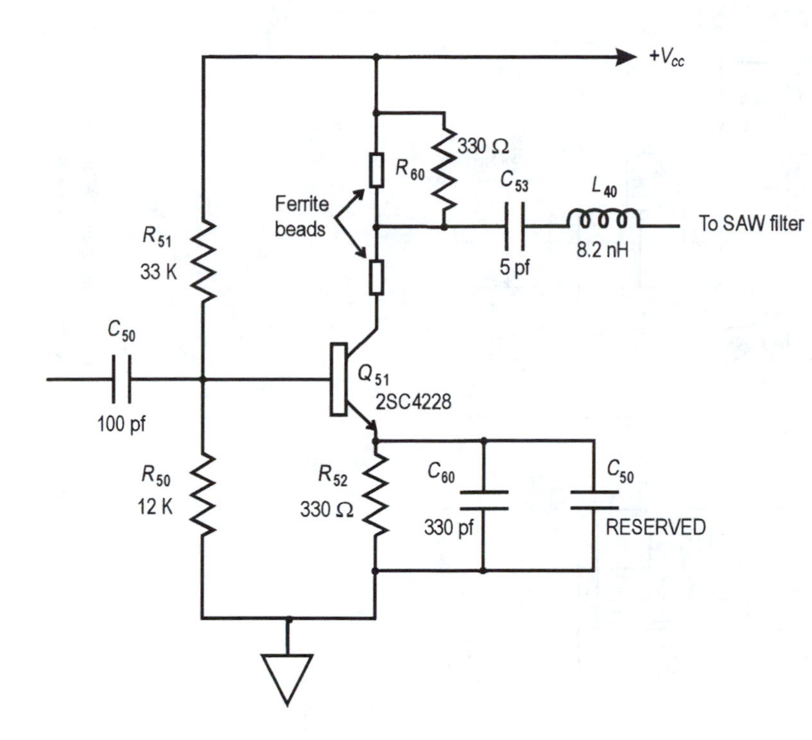

Fig. 10.12: Schematic diagram of the first RF amplifier in the "Receive" circuitry of the 1400 series cellular phone from Oki Telecom.

(3) There is a series LC between the collector and the SAW filter. Our first conjecture might be that this is designed to provide a low impedance at the receive frequencies used in the system and a high impedance elsewhere. Unfortunately, when we calculate the resonant frequency of the LC, we find it is about 100 MHz below the receive frequencies.

Exercise 10.6 Starting with Fig. 10.12, assume that the transistor can be modeled as a current source going from collector to emitter (ground) and that the SAW filter input impedance can be considered to be pure capacitance. Designate the inductance of the beads as L_B. Find a symbolic expression for the frequency at which V/I will be a maximum, where V is the voltage delivered to the input of the SAW filter and I is the current out of the transistor. The effect of considering the whole coupling circuit, including the input impedance of the SAW filter, rather than just the LC will be to raise the resonant frequency – presumably into the band for which the amplifier is designed.

From this stage, the signal proceeds to an integrated mixer where it is mixed with a signal which originates in a block called the AMPS® VCO. The frequency range of this VCO is never indicated in the service literature, but it can be inferred by the fact that when its output is mixed with the received signal, the difference frequency must be 83.16 MHz, since that is given as the frequency of the first IF filter. On that basis we conclude that the AMPS® VCO oscillates between 786.87 and 806.82 MHz. Strictly speaking, the same difference frequency could be obtained if the VCO output were above the received frequency, but we shall see that other considerations eliminate this possibility.

The output of the first IF filter goes to a discrete IF amplifier, which appears no different from a small-signal audio amplifier, except for the capacitor sizes.

From this amplifier, the signal enters an IC (U301) from which it ultimately emerges as demodulated audio. The first block the signal encounters within

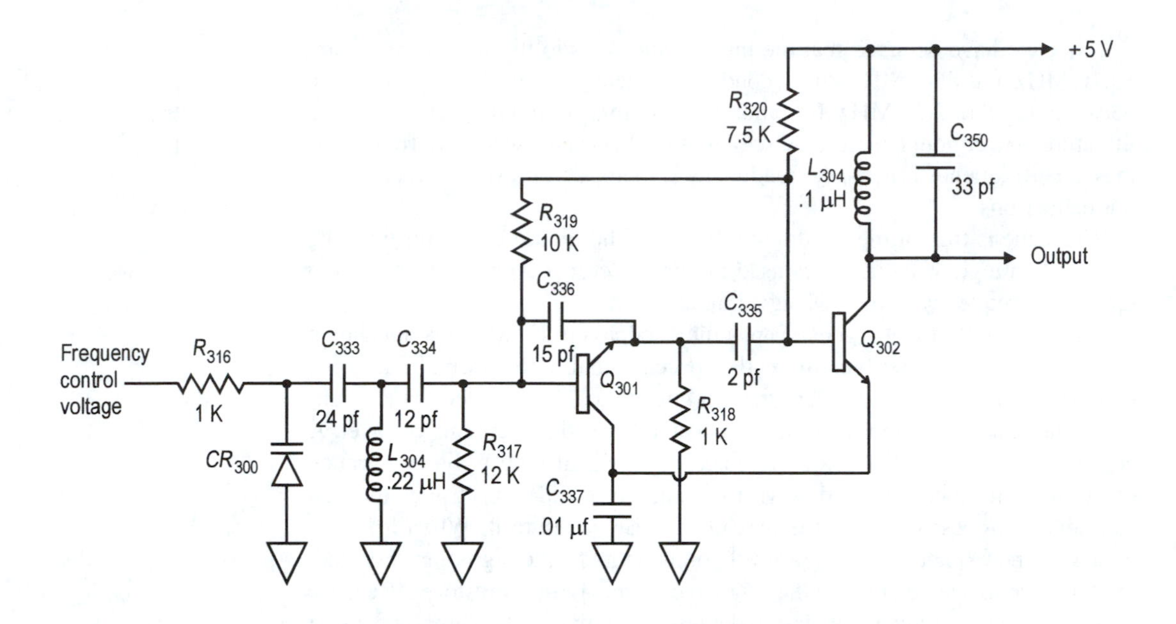

Fig. 10.13: Schematic diagram of the discrete VCO (86.4 MHz) used for the first IF translation in the 1400 series cellular phone from Oki Telecom.

U301 is a second mixer, whose other input frequency is not specified in the service literature. This unspecified frequency comes from a discrete VCO. This circuit is shown in Fig. 10.13, where it has been redrawn to show that it is basically another example of an oscillator whose base amplifier is an emitter follower.

The oscillator is followed by a common-emitter tuned amplifier. It might appear at first glance that there is feedback between the emitter of Q_{302} and the collector of Q_{301}. This is not the case, however. C_{337} has a reactance at the frequency of this oscillator of well under an ohm, and hence it must be considered an AC short. This single capacitor is used to provide the AC ground return for both stages. Direct current flows from the 5 V supply, through L_{304}, and into the collector of Q_{302}. It leaves Q_{302} as emitter current and enters Q_{301} as collector current. Upon exiting the emitter of that transistor, it returns to ground through R_{318}. Thus DC considerations dictate that the collector of Q_{301} and the emitter of Q_{302} be connected together, so a capacitor which bypasses one to ground must bypass the other as well.

The main frequency-determining elements are C_{333}, C_{334}, L_{304}, and the varicap, CR_{300}. However, the varicap capacitance is unknown, so it will not give us a "fix" on the VCO frequency. But the load of Q_{302} does. It is a parallel LC circuit that resonates at 87.6 MHz. U302 receives the output of this oscillator stage. The pinout on U302 shows that one pin is labeled "DIV," which indicates that there is a frequency divider in the chip. Since the 14.4 MHz oscillator is also input to this chip, it seems certain that the VCO is divided by some integer and phase locked to 14.4 MHz. Six times 14.4 MHz is 86.4 MHz, which is very close to the resonant frequency of the VCO load. We thus postulate that the VCO runs at 86.4 MHz.

Exercise 10.7 In the circuit of Fig. 10.13, isolate the components that are part of the feedback network of the oscillator and show their interconnection. Group like components that are in series or parallel. Justify any omissions. Your network should contain eight components and four nodes, besides ground. Draw it in the simplest form possible.

Since we have deduced that the input frequencies to the second mixer are 83.16 MHz and 86.4 MHz, the second IF frequency should be the difference between these, or 3.24 MHz. It is another annoying shortcoming of the service literature that, although there are three ceramic filters at the second IF frequency, the center frequency is not specified for any of them. This prevents our checking our deductions.

The same is true of the quadrature detector. The failure of the service data to give the range of inductances used in this detector prevents our bracketing the center frequency of the FM signal input into it.

The audio out of U301 goes to an audio bandpass filter, which is not shown in Fig. 10.11. This filter bandlimits the speech signal to the normal telephone bandwidth: 300–3,300 Hz. From there it goes to a 2:1 expander (which "undoes" the compression in the transmitter) and drives the telephone receiver. A compression of 2:1 means, for example, that a signal with a dynamic range of 80 dB will be compressed down to a range of 40 dB. As it goes through subsequent processing it has the inevitable noise added to it. When it finally comes to the expander, the high-level components of the signal are boosted, the low-level ones are cut, and the original dynamic range is restored. Because noise will be a low-level signal, it is depressed relative to the signal, resulting in what is typically a very substantial improvement in the signal-to-noise ratio. This is also the basis of the dbx audio noise reduction system discussed in Section 5.5.1.

The U301 chip also generates the RSSI signal that is fed to the logic unit. FM signal coming out of the limiter is also fed to the logic unit, which extracts setup data from it. On the basis of the setup signal received from the cell and extracted from the receive channel just discussed, the logic unit programs the frequency synthesizer through the data bus, and on the basis of the RSSI it sets the output power level.

10.6.4 Cellular phone transmit circuitry

The great bulk of the transmit circuitry is the frequency synthesizer. In analyzing this circuitry, we are stymied by the fact that the greater part of the relevant circuitry is within U500 IC. This is apparently a proprietary chip. The only information as to the internal structure available in the service literature is that it contains two phase-locked loops. Even this turns out to be not exactly true. Every indication is that it contains one and a half PLLs – that is, one VCO and two phase detectors. An external VCO (called the AMPS® VCO in Fig. 10.11) completes the other loop. The internal VCO also serves as a phase modulator. The U500 IC also contains numerous frequency dividers, multipliers, and low-pass filters (LPFs). It should be mentioned that each of the LPFs will have at least one capacitor external to the chip, but these are not shown in the block diagram.

In spite of the paucity of information about the chip, we can get some insight from the small portion of the circuitry that is external to the IC, from the identifying labels on some of the pins, and from a knowledge of the general approach that is used for frequency synthesis in cellular telephony. For comparison purposes, Fig. 10.14 shows a frequency synthesis circuit from a classic paper published in the *Bell System Technical Journal* (Fisher, 1979). As noted

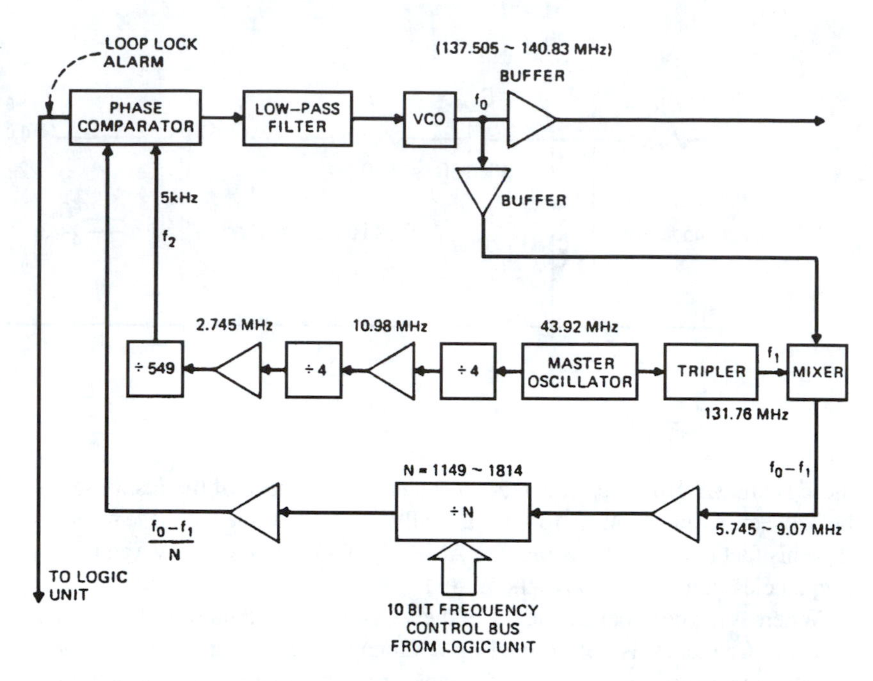

Fig. 10.14: Block diagram of the frequency-synthesis circuitry used by the Bell System in their tests of the cellular phone system.

in the caption to Fig. 10.11, the blocks within U500 that are white are those we can be pretty certain about. Those shaded in gray are those which are less certain. We may be reasonably sure about the fact that a given block is a frequency divider or multiplier but not sure about its modulus.

We begin with the AMPS® VCO. We have already seen the output of this module used in the first IF conversion of the receive process. For the transmit process, it goes to a mixer, whose output is low-pass filtered to extract the difference frequency. The sum frequency becomes the transmit signal. We previously deduced that the AMPS® VCO operates from 786.87 to 806.82 MHz, because this is one of the two frequency ranges that would give the first IF frequency specified for the IF filter. Since we project that this is one of the two inputs to mixer 1, and we also know the output frequency of this mixer, this tells us what the frequency of the other input must be. This turns out to be 38.16 MHz, which is found by subtracting the lowest frequency into the mixer (786.87 MHz) from the lowest frequency out of it (825.03 MHz).

Now let us suppose for the sake of argument that the AMPS® VCO is outputting a frequency range of 953.19 to 973.14 MHz. This would still provide the specified first IF frequency of 83.16 MHz, but this frequency range input to the mixer we have just analyzed in the transmit frequency synthesizer would require that the other input to the mixer (from the phase modulator/VCO) be at a frequency of 128.16 MHz (953.19 − 825.03) in order to get the proper frequency range for transmitting. The issue then becomes one of whether we can tell if the phase modulator/VCO in U500 runs closer to 38.16 MHz or 128.16 MHz. This is easy to answer. Figure 10.15 shows the schematic of the broad bandpass filter that passes the output of the VCO/phase modulator to the mixer. DC supplies have been treated as ground in creating this drawing. Although analysis shows that the Q of the filter, the gain in the passband, and the center frequency will all depend on the exact values of R_o and R_i, neither of

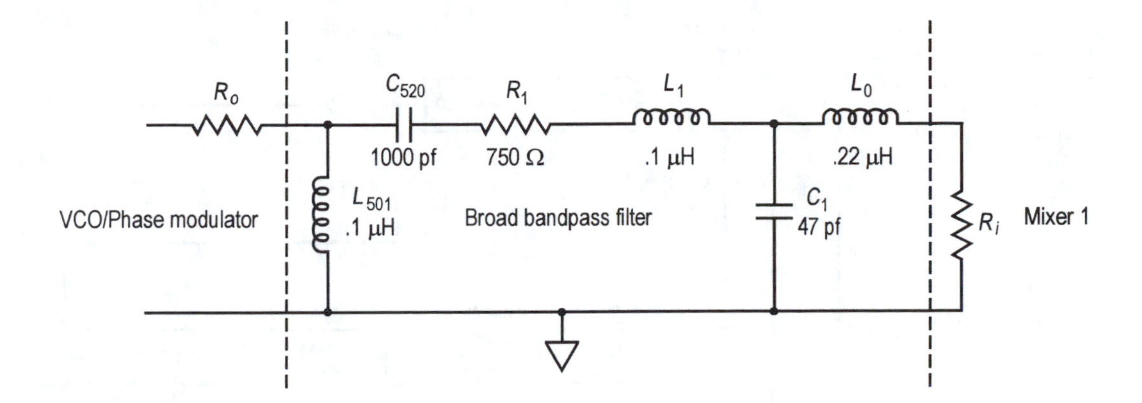

Fig. 10.15: Schematic diagram of the broad bandpass filter used in the 1400 series cellular phone from Oki Telecom. The passband of this filter pins down the frequency of the VCO/phase modulator fairly conclusively. R_o represents the output impedance of the VCO and R_i represents the input impedance of the mixer.

these is known. However, it is readily shown that the peak of the response will lie between 30 and 40 MHz for virtually all reasonable values of these resistors. It is this fact that assures us that the AMPS® VCO operates below the receive frequencies rather than above them.

Whereas the nominal frequency of the VCO/phase modulator is 38.16 MHz, its exact frequency is determined by a varicap. The varicap, in turn, has its capacitance determined by the audio input from the microphone (which phase modulates the 38.16 MHz carrier) and by the voltage from phase detector 1. We know that, in equilibrium, the two inputs to this phase detector must be locked in frequency, but we don't know *what* frequency. In drawing Fig. 10.11, it has been conjectured that these inputs will be 5 kHz, because that is the value used in the standard circuit of Fig. 10.14. If this is the case, then the 38.16 MHz out of the VCO/phase modulator would have to be divided by 7,632 as shown to deliver 5 kHz to one input of phase detector 1. The other input comes from a programmable divider whose frequency is set by the logic circuitry module. The range of moduli of this divider will depend not only on the input frequencies to the phase detector but also on the frequencies input to this divider. It is here where our proposed circuitry is most speculative.

Following the pattern shown in Fig. 10.14, we expect the transmit frequencies to be divided by 6 and applied to mixer 2, yielding a frequency range of 137.505–140.830 MHz. But we have no handle on the other frequency to this mixer. Figure 10.11 leads us to believe that it comes from the 14.4 MHz reference oscillator multiplied up by a factor of 9. This would give us a second input to mixer 2 at a frequency of 129.6 MHz. These frequency inputs would provide sum and difference frequencies on the mixer output that will be far enough apart to be separated easily. The difference frequency is selected by the LPF and input to the programmable divider.

A similar situation pertains in the circuitry at the far left of the block diagram. Again, by analogy from Fig. 10.14, mixer 3 receives inputs from the reference oscillator (multiplied by 7) and from the AMPS® VCO (divided by 7). Again, the actual numerical values of these ratios are speculative. Therefore, so are the output frequency of the mixer and the moduli of the programmable divider. We could have again postulated that the inputs to phase detector 2 are also 5 kHz, but that would have required that both frequency dividers feeding it have moduli twice as large as those shown. The output of this phase detector feeds the AMPS® VCO and holds it on frequency.

The net effect of all of this circuitry is to produce frequencies for receive and transmit that are as accurate as the frequency of the 14.4 MHz reference oscillator.

10.6.5 The logic unit

The hardware of the logic unit is straightforward. However, what it accomplishes in software is nothing short of amazing. It performs the functions of:

(1) Processing the setup data sent from the MTSO and setting the transmit and receive frequencies of the phone accordingly.
(2) Muting transmitter or receiver as needed.
(3) Varying the transmitted power from the phone in response to the received signal strength.
(4) Cleaning up received setup signals and distinguishing them from voice signals.
(5) Providing diversity switch logic (where used).
(6) Performing A/D conversion of the RSSI signal.

In addition, the logic unit performs many tasks we have not even considered, such as the conversion of setup data generated in the CPU in the NRZ format to Manchester encoding, clock regeneration, detection of supervisory audio tones (SAT), generation of signaling tones (ST), and recognition of its own number when called.

The reader interested in learning much more about the entire cellular system is referred to the January 1979 volume of the *Bell System Technical Journal.* which is devoted in its entirety to the AMPS® system.

References

Bennett, A.F. 1953. An Improved Circuit for the Telephone Set. *Bell System Technical Journal* 32 (3): 611–626.

Campbell, G.A. and Foster, R.M. 1920. Maximum Output Networks for Telephone Substation and Repeater Circuits. *AIEE Transactions* 39: 231–280.

Fisher, R.E. 1979. AMPS: A Subscriber Set for the Equipment Test. *Bell System Technical Journal* 58 (1): 123–143.

Means, D.R. 1980. A Class of Lossless, Reciprocal Anti-Sidetone Networks for Telephone Sets. *Bell System Technical Journal* 59 (8): 1483–1492.

Index

Roman type indicates isolated occurrences of the word or phrase.
Italic type indicates sections dealing with the word or phrase.
Bold type indicates chapters dealing with the word or phrase.

FRANKLIN PIERCE COLLEGE LIBRARY
00113464